Phyllostomid Bats

Phyllostomid Bats

A Unique Mammalian Radiation

Edited by

Theodore H. Fleming,
Liliana M. Dávalos, and
Marco A. R. Mello

The University of Chicago Press :: Chicago and London

The University of Chicago Press, Chicago 60637
The University of Chicago Press, Ltd., London
© 2020 by The University of Chicago
Published 2020
Printed in the United States of America

29 28 27 26 25 24 23 22 21 20   1 2 3 4 5

ISBN-13: 978-0-226-69612-6 (cloth)
ISBN-13: 978-0-226-69626-3 (e-book)
DOI: https://doi.org/10.7208/chicago/9780226696263.001.0001

Library of Congress Cataloging-in-Publication Data

Names: Fleming, Theodore H., editor. | Davalos, Liliana M., editor. | Mello, Marco
    A. R. (Marco Aurelio Ribeiro), editor.
Title: Phyllostomid bats : a unique mammalian radiation / edited by Theodore H.
    Fleming, Liliana M. Davalos and Marco A. R. Mello.
Description: Chicago ; London : The University of Chicago Press, 2020. | Includes
    bibliographical references and index.
Identifiers: LCCN 2019058303 | ISBN 9780226696126 (cloth) |
    ISBN 9780226696263 (ebook)
Subjects: LCSH: Phyllostomidae.
Classification: LCC QL737.C57 P49 2020 | DDC 599.4/5—dc23
LC record available at https://lccn.loc.gov/2019058303

♾ This paper meets the requirements of ANSI/NISO Z39.48-1992 (Permanence of
Paper).

We dedicate this book to past, present,
and future students of this fascinating group
of mammals

CONTENTS

Section 1 **Introduction**

1   Overview of This Book   *3*

Theodore H. Fleming, Liliana M. Dávalos, and Marco A. R. Mello

2   Setting the Stage: Climate, Geology, and Biota   *7*

Theodore H. Fleming

Section 2 **Phylogeny and Evolution**

3   Phylogeny, Fossils, and Biogeography:
The Evolutionary History of Superfamily Noctilionoidea
(Chiroptera: Yangochiroptera)   *25*

Norberto P. Giannini and Paúl M. Velazco

4   Diversity and Discovery: A Golden Age   *43*

Andrea L. Cirranello and Nancy B. Simmons

5   Fragments and Gaps: The Fossil Record   *63*

Nancy B. Simmons, Gregg F. Gunnell, and Nicolas J. Czaplewski

6   Phylogenetics and Historical Biogeography   *87*

Liliana M. Dávalos, Paúl M. Velazco, and Danny Rojas

7   Adapt or Live: Adaptation, Convergent Evolution,
and Plesiomorphy   *105*

*Liliana M. Dávalos, Andrea L. Cirranello, Elizabeth R. Dumont,
Stephen J. Rossiter, and Danny Rojas*

8   The Evolution of Body Size in Noctilionoid Bats   *123*

Norberto P. Giannini, Lucila I. Amador, and R. Leticia Moyers Arévalo

Section 3   Contemporary Biology

9   Structure and Function of Bat Wings: A View from
the Phyllostomidae   151
Sharon M. Swartz and Justine J. Allen

10   The Relationship between Physiology and Diet   169
Ariovaldo P. Cruz-Neto and L. Gerardo Herrera M.

11   Sensory and Cognitive Ecology   187
Jeneni Thiagavel, Signe Brinkløv, Inga Geipel, and John M. Ratcliffe

12   Reproduction and Life Histories   205
Robert M. R. Barclay and Theodore H. Fleming

13   Patterns of Sexual Dimorphism and Mating Systems   221
Danielle M. Adams, Christopher Nicolay, and Gerald S. Wilkinson

Section 4   Trophic Ecology

14   The Omnivore's Dilemma: The Paradox of
the Generalist Predators   239
Claire T. Hemingway, M. May Dixon, and Rachel A. Page

15   Vampire Bats   257
John W. Hermanson and Gerald G. Carter

16   The Ecology and Evolution of Nectar Feeders   273
Nathan Muchhala and Marco Tschapka

17   The Frugivores: Evolution, Functional Traits, and Their Role
in Seed Dispersal   295
Romeo A. Saldaña-Vázquez and Theodore H. Fleming

Section 5   Population and Community Ecology

18   Roosting Ecology: The Importance of Detailed Description 311
Armando Rodríguez-Durán

19   Population Biology   325
Theodore H. Fleming and Angela M. G. Martino

20   Community Ecology   347
Richard D. Stevens and Sergio Estrada-Villegas

21   Network Science as a Framework for Bat Studies   373
Marco A. R. Mello and Renata L. Muylaert

22   Contemporary Biogeography   391
Richard Stevens, Marcelo M. Weber, and Fabricio Villalobos

Section 6   Conservation

23   Challenges and Opportunities for the Conservation of
       Brazilian Phyllostomids   413

       Enrico Bernard, Mariana Delgado-Jaramillo, Ricardo B. Machado,
       and Ludmilla M. S. Aguiar

24   Threats, Status, and Conservation Perspectives for
       Leaf-Nosed Bats   435

       Jafet M. Nassar, Luis F. Aguirre, Bernal Rodríguez-Herrera,
       and Rodrigo A. Medellín

       Contributors   457
       Subject Index   459
       Taxonomic Index   467

Introduction

1

Theodore H. Fleming,
Liliana M. Dávalos,
and Marco A. R. Mello

Overview of This Book

Introduction

Adaptive radiations have always fascinated biologists because they provide dramatic evidence of the power of natural selection and opportunism in the evolution of life on Earth. Although the concept of adaptive radiation has itself evolved over its century-long history, contemporary definitions agree with Givnish (2015, 301) who emphasized "the rise of a diversity of ecological roles and associated adaptations within a lineage, accompanied by an unusually high level or rate of accumulation of morphological/physiological/behavioral disparity and ecological divergence compared with sister taxa or groups with similar body plans and life histories." These features have made Darwin's finches, Hawaiian honeycreepers, African Rift Valley cichlids, and Australian marsupials into textbook examples of radiations whose species have evolved to fill a variety of ecological niches via changes in their behavior, dietary choices, morphology, and other aspects of their life histories. Certainly, the evolution of New World leaf-nosed bats (Phyllostomidae) qualifies as an adaptive radiation. While their closest relatives in the chiropteran superfamily Noctilionoidea—Mormoopidae—exhibit modest taxonomic and ecological diversity (2 genera, 18 species), the phyllostomids (60 genera, 216 species) have radiated into an unprecedented variety of ecological and behavioral types in the past 36 Ma. Thus, several decades after the first monographs of New World leaf-nosed bats (Baker et al. 1976, 1977, 1979), a wealth of new scholarship has accumulated that merits review.

This book is about the adaptive radiation of these abundant and ecologically highly diverse bats. Reflecting this abundance, anyone aiming to capture bats using Japanese mist nets set at ground level throughout the tropical and subtropical Americas can hardly avoid catching one or more species of phyllostomids in very short order. For example, in his survey of tropical rainforest bats at La Selva, Costa Rica, in 1986, Fleming and his field crew caught 94 individuals of 11 phyllostomid species in 10 ground-level nets in 3.5 h (THF, unpubl. data). These species represented 6 of the 11 phyllostomid subfamilies (clades); 3 species of frugivorous *Carollia* (Carolliinae) dominated the captures. Food habits of these bats included insects, blood, vertebrates, nectar and pollen, and fruit. In their much more extensive survey of bats in a lowland rainforest in northern French Guiana, Nancy Simmons and husband Rob Voss recorded a similar abundance but much higher species richness of phyllostomids. Forty-eight of the 59 species they captured in ground-level nets were phyllostomids, and *C. perspicillata*, *Phyllostomus elongatus* (an omnivore), and *Artibeus obscurus* (a frugivore) were caught most frequently (Simmons and Voss 1998). Recent syntheses on phyllostomid diversity show that single localities in Brazil's Atlantic Forest can surpass the mark of 60 bat species (Bergallo et al. 2003; Muylaert et al. 2017; Stevens 2013). Phyllostomids also dominate bat assemblages throughout the Caribbean islands, where they are found in massive cave-roosting colonies alongside several species of mormoopids. At hot caves such as La Chepa in the Dominican Republic, 6 out of the 10 species are phyllostomids, and captures of *Macrotus waterhousii* and *Monophyllus redmani* outnumber those of mormoopids (Núñez-Novas et al. 2016). From these and many other studies, there can be no doubt that phyllostomid bats are among the most common and species-rich mammals in lowland Neotropical forests.

Because of their abundance, ecological diversity, and ease of capture and maintenance in captivity, phyllostomids have been intensively studied by many research groups in recent decades. Much of the early research on these bats was summarized in a landmark, three-volume monograph edited by Robert Baker, J. Knox Jones Jr., and Dilford Carter in the late 1970s (Baker et al. 1976, 1977, and 1979). Since then, our knowledge about the evolution, general biology, and ecology of these bats has increased tremendously, and a book summarizing these findings has been long overdue. The overall aim of this book, therefore, is to review our current knowledge about these bats, particularly from a phylogenetic perspective. Compared to the 1970s, our understanding of phylogenetic relationships within this family and the timing of the appearance of its evolutionary clades is much deeper. We can now use this information to better understand the evolution of all aspects of the biology of these bats. We can ask, for example, what were the ancestral states of morphology, physiology, and behavior in this family, and how long did it take for the

specialized adaptations associated with blood, nectar, and fruit feeding to evolve? What is the historical biogeography of the family and how long have contemporary assemblages of species been interacting with each other and, in the case of nectar and fruit eaters, with their food plants?

While this book clearly focuses on only one of the 20 families of bats, it should have a much broader appeal than simply to bat biologists. Because of the unique biological and ecological diversity found in this family, which easily comprises the most diverse diet of any mammalian family, it can serve as a model system for understanding many aspects of the adaptive radiation of mammals during the latter half of the Cenozoic era. As chapters in this book reveal, a plethora of recent field, laboratory, and bioinformatics studies have provided us with an unprecedented view of how this group of mammals has responded to geological, climatic, and biological changes and challenges that have occurred in the Americas over the past 30 Ma. Moreover, because these bats provide many important ecosystem services such as insect control, pollination, seed dispersal, and nutrient input to caves, as well as negative economic effects resulting from vampire attacks on livestock, these animals can have enormous ecological and economic impacts on tropical and subtropical ecosystems of the Americas (Kunz et al. 2011). As a result, their conservation and protection is critical for the health of Neotropical ecosystems. In sum, phyllostomid bats have played an extremely important role in the evolutionary and ecological dynamics of these systems for tens of millions of years.

The 24 chapters of this book are arranged in six major sections, beginning with a review of the physical and biological environment into which these bats evolved and ending with a review of their conservation status and needs. In the sections in between, chapters discuss the evolutionary history of these bats, their major physiological and behavioral adaptations, and their ecological roles in the web of life. Wherever appropriate, adaptations are discussed from a phylogenetic perspective to emphasize the evolutionary pathways that this adaptive radiation has taken. Major takeaway messages from these sections and their chapters include

1. Knowledge of the biology of bats of the New World family Phyllostomidae has advanced tremendously since Baker et al.'s seminal

monographs. The range of disciplines that now use phyllostomid bats as a model system is huge and encompasses not only many traditional biological fields such as systematics, behavior, and ecology but also transdisciplinary fields such as complexity and robotics.

2. Phyllostomid bats have undergone a much broader evolutionary radiation in terms of species richness, body size, ecology, physiology, and behavior than their noctilionoid relatives and other bat taxa and most other mammalian taxa. Ancestrally insectivorous, the basic foraging mode of this family involves gleaning prey from vegetation instead of aerial hawking. This foraging mode is probably the key adaptation that allowed these bats to evolve into a diverse array of feeding niches, including sanguinivory, carnivory, nectarivory, and frugivory. Biogeographical history and ecological interactions among trophically similar species and their food resources seem to be key for understanding this radiation.

3. The roles played by phyllostomid bats in the web of life go far beyond those traditionally considered. For instance, these bats now receive much attention from public health authorities for being reservoirs of emerging diseases such as coronaviruses. Their relative contribution to different ecosystem services together with those of other animals such as birds and reptiles is now much better understood. They are also considered outstanding models for cutting-edge studies on biomimetics, ranging from drug development to aerodynamics and communication.

4. Deforestation, habitat fragmentation, and roost disturbance are major threats to phyllostomid bats, and an important focus on their conservation has emerged in the last few decades. Major bat conservation networks have developed within Latin America and the Caribbean with an emphasis on basic research, public education, and local conservation actions. Despite this activity, much remains to be done to increase public awareness and appreciation for the ecological and evolutionary importance of phyllostomid bats.

Acknowledgments

All chapters in this book were reviewed by at least two people in addition to at least one editor. We thank all of our reviewers for their constructive comments and suggestions. We also thank two outside readers of the entire manuscript for their suggestions for improving this book. A number of institutions and agencies contributed funds that helped subsidize the book's publication. These include the home institutions of at least eight of our senior authors. Finally, we thank Yvonne Zipter for her careful copyediting of this book.

References

Baker, R. J., J. K. Jones, Jr., and D. C. Carter (eds.). 1976. Biology of bats of the New World family Phyllostomatidae, Part 1. Texas Tech University Press, Lubbock.

Baker, R. J., J. K. Jones, Jr., and D. C. Carter (eds.). 1977. Biology of bats of the New World family Phyllostomatidae, Part 2. Texas Tech University Press, Lubbock.

Baker, R. J., J. K. Jones, Jr., and D. C. Carter (eds.). 1979. Biology of bats of the New World family Phyllostomatidae, Part 3. Texas Tech University Press, Lubbock.

Bergallo, H. G., C. E. L. Esbérard, M. A. R. Mello, V. Lins, R. Mangolin, G. G. S. Melo, M. Baptista et al. 2003. Bat species richness in Atlantic Forest: what is the minimum sampling effort? Biotropica, 35:278–288.

Givnish, T. J. 2015. Adaptive radiation versus "radiation" and "explosive diversification": why conceptual distinctions are fundamental to understanding evolution. New Phytologist, 207:297–303.

Kunz, T. H., E. Braun de Torrez, D. Bauer, T. A. Lobova, and T. H. Fleming. 2011. Ecosystem services provided by bats. Annals of the New York Academy of Sciences 1223:1–38.

Muylaert, R., R. Stevens, C. E. L. Esbérard, M. A. R. Mello, G. Garbino, L. Varzinczak, D. Faria et al. 2017. Atlantic bats: a dataset of bat communities from the Atlantic Forests of South America. Ecology, 98:3227–3227.

Núñez-Novas, M. S., Y. M. León, J. Mateo, and L. M. Dávalos. 2016. Records of the cave-dwelling bats (Mammalia: Chiroptera) of Hispaniola with an examination of seasonal variation in diversity. Acta Chiropterologica, 18:269–278.

Simmons, N. B., and R. S. Voss. 1998. The mammals of Paracou, French Guiana: a Neotropical lowland rainforest fauna. Part I: Bats. Bulletin of the American Museum of Natural History 237.

Stevens, R. D. 2013. Gradients of bat diversity in Atlantic Forest of South America: environmental seasonality, sampling effort and spatial autocorrelation. Biotropica, 45:764–770.

2

Theodore H. Fleming

Setting the Stage
Climate, Geology, and Biota

The antecedents of the Phyllostomidae—other members of the chiropteran super-family Noctilionoidea—date from the Middle Eocene (about 42 million years ago [Ma] or earlier, in the case of Myzopodidae; chap. 3, this vol.), and the phyllostomids date from the Late Eocene (about 36 Ma) (Rojas et al. 2016; Teeling et al. 2005). Phyllostomids and two of its close relatives, Noctilionidae and Mormoopidae, evolved in tropical and subtropical habitats in the New World, probably in tropical North America (Arita et al. 2014; Morgan et al. 2012; chap. 6, this vol.). Their current distributions include the southwestern United States, Mexico, Central and South America south to Argentina, and the West Indies. What was this vast area like when phyllostomids began to evolve? How were New World landmasses—continents and islands—configured and how much has this changed in the intervening millennia? Similarly, how have New World climates and biotas changed during the past 30–40 Ma, and how have the major Neotropical biomes and their biotas responded to climate change? And, more specifically, what were the consequences of all of these changes for the evolution of the phyllostomids? What physical and biological opportunities and constraints did these bats face during their adaptive radiation? I aim to address these questions in this chapter, which is meant to set the stage for detailed discussions of the evolution of these fascinating bats. As a final point, I use the term "noctilionoid" here to refer just to three of the seven extant families of superfamily Noctilionoidea—Noctilionidae, Mormoopidae, and Phyllostomidae—discussed in chapter 3.

Climate, Geology, and Biomes

The world during the Cenozoic era is generally described as having two major climatic phases: a "greenhouse" phase during the Paleocene and Eocene, when global temperatures were high and polar icecaps were absent, and an "icehouse" phase from

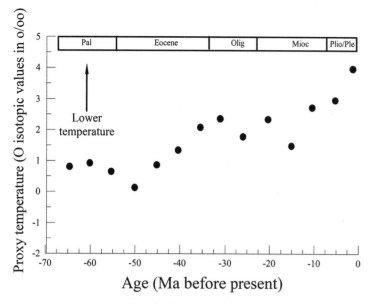

Figure 2.1. Generalized estimates of global temperatures during the Cenozoic era based on benthic δ^{18} O (in ‰). Lower values of δ^{18} O indicate warmer temperatures. Phyllostomids first evolved in the Oligocene and radiated extensively in the Miocene. Adapted from Blois and Hadly (2009). Abbreviations: Pal = Paleocene; Olig = Oligocene; Mioc = Miocene; Plio/Ple = Pliocene/Pleistocene.

the Oligocene on, when global temperatures generally decreased and one (the Antarctic) and then both poles became ice covered (fig. 2.1). Noctilionoid bats began to evolve at the end of the greenhouse phase and diversified during the icehouse phase. At the peak of the greenhouse phase in the Early Eocene, atmospheric CO_2 levels exceeded 1,000 ppmv and midlatitude global air temperatures are estimated to have been 4–5°C higher than before or after that time (Gingerich 2006; Zeebe and Zachos 2013). After the Eocene climatic optimum (52–50 Ma), global air and sea temperatures began to decline, and latitudinal temperature gradients and seasonality became more pronounced. Smaller temperature increases occurred again at the end of the Oligocene (26–24 Ma) and in mid-Miocene (17–14 Ma) before steadily decreasing during the rest of the Miocene and into the major glacial periods of the Pliocene and Pleistocene (Bowen and Zachos 2010; Laurento et al. 2015; Zachos et al. 2001, 2008). At the peak of the icehouse phase during the Pliocene and Pleistocene, average air temperatures in lowland Brazil and the Andes, major centers of phyllostomid diversity, were 5–8°C cooler than today (Graham 2011).

Global climatic changes in temperature and precipitation have had a profound effect on the distributions of plants and animals, and as a result, the extent and distributions of major biomes such as tropical forests have fluctuated greatly during the Cenozoic. In the Late Cretaceous and Early Cenozoic, for example, tropical forests of increasingly modern composition covered most of South, Central, and North America as far north as Greenland (Morley 2000, 2007). At 79 Ma on Ellesmere Island, Arctic Canada, the tropical flora included members of Anacardiaceae, Bombacaceae (now part of Malvaceae), Fabaceae, Lauraceae, Rubiaceae, and Rutaceae; the fauna included alligators, an early tapir, and "prosimian" primates (Eberle and Greenwood 2012; McKenna 1980).

As reviewed in detail by Graham (2011), decreasing atmospheric CO_2 levels, along with decreasing global air and sea temperatures and increasing climatic seasonality beginning in the Late Eocene, are hypothesized to have led to the evolution of a greater diversity of vegetation types, including early versions of tropical dry forests, savannas, and deserts. Reinforcing the effects of temperature changes was the uplift of the major mountains in western North America, Central America, and western South America. This orogeny changed patterns of wind and water flow and created rain shadows favoring the development of arid vegetation types. By the end of the Eocene, the Rocky Mountains and Sierra Madre Mountains were half their current height, but the Andes did not begin to rise substantially until the Miocene. The northern Andes attained its current heights within the last 5 Ma (Garzione et al. 2008; Graham 2009; Hoorn et al. 2010). Prior to the rise of the Andes, swamps, lagoons, and riverine habitats were widespread in the Amazon Basin. Major rivers in northern South America such as the Amazon and Orinoco flowed west to the Pacific Ocean until mid-Miocene; they began to flow east to the Atlantic Ocean with the rise of the Andes (Hoorn et al. 2010).

Finally, during the mid-Miocene and the Pliocene and beyond, air temperatures at the poles were low enough to support the formation of extensive ice sheets. Ice masses first formed in Antarctica in the Oligocene and, by mid-Miocene, in the Arctic. Cooler air temperatures and strong climatic seasonality from the mid-Miocene on led to the spread of tropical dry forests and grasslands and the formation of modern deserts (Graham 2011).

In addition to changes in climate, tectonic processes have profoundly affected the diversity and distribution of Earth's biota. Seafloor spreading and increased volcanism in the North Atlantic in the Late Paleocene and Early Eocene, for instance, resulted in the massive release of methane and CO_2 into the atmosphere which caused Earth's average temperature and the atmosphere's CO_2 concentration to increase to unprecedented heights (Zachos et al. 2008) (fig. 2.1). In the mid- to Late Eocene when superfamily Noctilionoidea and the phyllostomids began to evolve, rates of seafloor spreading were decreasing, and continents of the Western Hemisphere were close to their current positions. South America was well-separated from Africa by about 80 Ma; South America and Antarctica were beginning to separate (at about 36 Ma); and South and North America were not yet connected via the Panamanian land bridge, which dates from at least 3.5 Ma (and possibly much older; Hoorn and Flantua 2015; Leigh et al. 2014; Ramírez et al. 2016). However, there is evidence that short-distance overwater dispersal for bats and other animals between Central and South America was possible at least three times prior to the Panamanian closure: (1) at 50–40 Ma when island blocks of the eastward drifting Caribbean Plate emerged between these two land masses; (2) at 35–33 Ma via the GAARlandia land bridge between northern South America and the Greater Antilles along the Aves Ridge; and (3) at 25–20 Ma when the Central American peninsula approached northwest South America (Almendra and Rogers 2012; Olmstead 2013 and included references).

Modern versions of the Greater Antilles did not yet exist at 40 Ma, so the early evolution and diversification of members of superfamily Noctilionoidea must have taken place on the mainland of North, Central, and South America. The West Indies includes two geological units: (1) the Greater Antilles, which lie on the Caribbean Plate and which attained their present positions at about 25 Ma; and (2) the Lesser Antilles, which is a double arc of volcanic islands along the eastern margin of the Caribbean Plate; these islands date from 40–45 Ma (the northeast outer arc) and 25–20 Ma (the northwest inner arc) (Fleming et al. 2009, and included references; Graham 2011). At an age of 22–30 Ma, Puerto Rico is the oldest member of the Greater Antilles, and Jamaica (10 Ma) is the youngest.

Colonization of the West Indies by noctilionoid bats was probably episodic and likely occurred during periods of low sea levels beginning in the mid-Miocene (Dávalos 2009). Estimates of the age of molecular divergences of island bats from mainland taxa are about

15–14 Ma for West Indian *Mormoops* and *Pteronotus* (Mormoopidae); about 15 Ma for *Erophylla* and its relatives (Glossophaginae); about 14 Ma for *Monophyllus* (Glossophaginae); and about 10 Ma for *Ardops* and its relatives (Stenodermatinae) (Dávalos and Turvey 2012; Rojas et al. 2016). These colonizations produced 9 endemic genera and 22 endemic species in the West Indies. Members of the *Ardops* clade were able to successfully colonize mainland habitats in Central and South America.

Although early phyllostomids (and mormoopids) probably evolved in North America, the bulk of phyllostomid diversification likely occurred in the lowlands of South America (Arita et al. 2014; Morgan et al. 2012; chap. 6, this vol.). Highest species and generic richness currently resides in the lowlands of northwestern South America, an area that has been covered by tropical forest throughout the Cenozoic. We can use an extensive survey of Venezuelan mammals reported by Handley (1976) to highlight the association between different tropical habitats and the diversity of these bats. Except for the Mormoopidae, whose species richness in Venezuela and elsewhere peaks in tropical dry forests, peak species richness in noctilionoid lineages in that country (which I assume is representative of northern South America) currently occurs in moist or wet evergreen forests followed by tropical dry forests; relatively few species live in drier, open habitats such as deserts, thorn forests, and savannas (fig. 2.2). Furthermore, except for subfamilies Glossophaginae and Stenodermatinae,

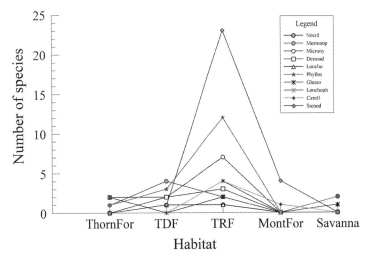

Figure 2.2. Habitat distributions of noctilionoid clades in Venezuela based on data in Handley (1976). Abbreviations: ThornFor = thorn forest; TDF = tropical dry forest; TRF = tropical rain forest; MontFor = montane forest.

Table 2.1. Elevational distributions of noctilionoid bats in Venezuela based on data in Handley (1976)

Clade	Number of Species In	
	Lowlands	Uplands
Noctilionidae	2	0
Mormoopidae	5	0
Micronycterinae	7	0
Desmodontinae	3	0
Lonchorhininae	2	0
Phyllostominae	13	0
Glossophaginae	4	3
Lonchophyllinae	5	0
Carolliinae/Glyphonycterinae*	4	1
Stenodermatinae	18	10

Note: Here species were classified either as "lowland," when most of their distribution occurs below 1,500 m, or "upland," when most of their distribution occurs above 1,500 m. Clades come from Baker et al. (2003).

* Includes *Rhinophylla*.

which contain a number of upland or montane species (especially in the genera *Anoura* and *Sturnira*, respectively), noctilionoid lineages are mostly restricted to lowland habitats at or below 1,500 m (table 2.1). Unlike nectar-feeding hummingbirds and fruit-eating tanagers that radiated extensively in the northern Andes as it began to rise in the Late Miocene (Burns and Naoki 2004; McGuire et al. 2014), the radiation of plant-visiting phyllostomid bats in the Andes has been rather modest. In summary, most of the extant diversity of New World noctilionoids is found on the Neotropical mainland in lowland forested habitats, which have existed in one form or another throughout the Americas since the Late Cretaceous. A minority of species has evolved in more recent dry or arid habitats, in the mountains, or in the West Indies (fig. 2.3).

Finally, what effects, if any, did Cenozoic geological and climatic changes, especially those from the Miocene on, have on diversification rates of phyllostomids? Were rates of speciation appreciably higher, for example, during the Quaternary (the last 2.6 Ma) when periods of major glaciations caused continuous lowland forests to become fragmented than in the Neogene (23.0–2.6 Ma)? More generally, did major landscape reconfigurations in the latter half of the Cenozoic lead to concordant higher rates of diversification in Neo-

tropical birds and bats? Recent phylogenetically based analyses of many lineages of Neotropical birds and the phyllostomids do not support the hypothesis that rates of speciation in these animals are concordant and correlated with major changes in lowland South American landscapes, including those associated with Quaternary glaciation. Instead, they support the hypothesis that the major driver of speciation in these animals has been dispersal from large, long-held geographic ranges into new areas based on lineage-specific traits such as innate dispersal ability (Rojas et al. 2016; Rull 2015; Smith et al.. 2014). Whereas the accumulation of new species in Neotropical birds and bats has been relatively constant, rather than episodic, in the Neogene and Quaternary, different lineages have had different speciation rates. In birds, for instance, understory species with relatively limited dispersal abilities have had higher diversification rates than more mobile canopy species. In contrast, within phyllostomids, sternodermatine bats, many of which are canopy feeders, have had a notably higher speciation rate than other lineages (Dumont et al. 2012). Two of the three most species-rich genera in this subfamily, *Artibeus* (22 species) and *Platyrrhinus* (21 species), feed mostly in the canopy whereas *Sturnira* (23 species) feeds primarily on understory shrubs. Taxonomy comes from chapter 4, this volume, and dietary information comes from Lobova et al. (2009).

The Biotic Setting

Important biotic elements of the world in which the phyllostomids have evolved include food resources (e.g., insects, nectar, and fruit), potential competitors for those resources (e.g., other insectivores, nectarivores, and frugivores), and predators. What biotic elements were already present in the Late Eocene at the beginning of the phyllostomid radiation, how did those elements change through time, and what were the effects of these changes on the evolving phyllostomids?

Let's first consider insects and their chiropteran predators. Eighteen of the 20 extant families of bats are wholly insectivorous, or nearly so, and this includes other members of superfamily Noctilionoidea (see chap. 3, this vol.). Insects have clearly been the main food resources for bats throughout their evolution. Phylum Arthropoda and its several subphyla and many orders and families are an ancient group, and many of the orders typically consumed by insectivorous bats

Figure 2.3. Photos of several habitats in which phyllostomids are common: (A) lowland tropical rain forest, (B) lowland tropical dry forest, (C) montane wet forest, and (D) Sonoran Desert. Common phyllostomids in these habitats include: A, *Carollia perspicillata, Phyllostomus elongatus, Rhinophylla pumilio,* and *Artibeus obscurus*; B, *Carollia brevicauda, Phyllostomus elongatus, Platyrrhinus vittatus,* and *Uroderma magnirostrum*; C, *Anoura geoffroyi, Anoura caudifer, Micronycteris megalotis,* and *Sturnira ludovici*; D, *Leptonycteris yerbabuenae* and *Macrotus californicus*. Photo credits: (A) R. Voss; (B) J. Nassar; (C) T. Fleming; (D) T. Fleming.

(e.g., Orthoptera, Diptera, Lepidoptera, Coleoptera) evolved long before the Cenozoic era (Misof et al. 2014). Thus, the earliest extant families of insectivorous bats (e.g., Rhinolophidae, Hipposideridae, Emballonuridae), which date from the Late Paleocene or Early Eocene (Teeling et al. 2005), likely had a full spectrum of nocturnal insects on which to feed (Connor and Cocoran 2012).

Lowland Neotropical rainforests currently harbor up to nine families of bats, eight of which are nearly totally insectivorous (table 2.2). Five of those families evolved before the appearance of the Noctilionoidea, whose families first appeared at about 36 Ma. Although the earliest phyllostomids likely did not evolve in lowland rainforests, these habitats currently house the greatest phyllostomid diversity, and they were undoubtedly important in the early evolution of this family. Thus it is instructive to review the species richness, general morphological characteristics, and feeding niches of

these families to see where insectivorous phyllostomids fit into the general guild of nocturnal insect predators. The three oldest and geographically most widely distributed families (Emballonuridae, Vespertilionidae, and Molossidae) are represented by up to 10 species in some Neotropical lowland rainforests, but this is less than half the number of insectivorous or partially insectivorous species of phyllostomids found in these habitats (table 2.2). Insectivorous phyllostomids are also far more species-rich than their noctilionoid relatives (chap. 3, this vol.). Thus, in terms of species diversity, insectivorous phyllostomids are certainly not marginal members of the chiropteran insect-eating guild. Instead, they are likely to be major consumers of insects in Neotropical forests and have been since the Early Miocene (Giannini and Kalko 2005; Kalka and Kalko 2006).

Bat foraging behavior and diets are strongly influenced by body size and wing shape (as well as by echolocation abilities; Geipel et al. 2013; Schnitzler

Table 2.2. Families of insectivorous bats that co-occur with phyllostomids in Neotropical rainforest habitats

Superfamily	Family/Subfamily	No. of Species/Habitat	Distribution	Age (Ma)	Principal Foraging Mode(s)
Emballonuroidea	Emballonuridae	3–10	Pantropical	52	Aerial insectivores (low or high, slow to moderately fast flying)
Noctilionoidea	Noctilionidae	0–2	Neotropics	36	Trawlers/aerial insectivores (low, slow flying)
	Mormoopidae	0–3	Neotropics	36	Aerial insectivores (low, slow flying)
	Phyllostomidae/Phyllostominae[a]	9–25	Neotropics	36	Gleaning insectivores (low, slow flying)
	Thyropteridae	0–2	Neotropics	40	Aerial insectivores/gleaners (low, slow flying)
	Furipteridae	0–1	Neotropics	36	Aerial insectivores (low, slow flying)
Vespertilionoidea	Natalidae	0–1	Neotropics	50	Aerial insectivores (low, slow flying)
	Vespertilionidae	2–8	Cosmopolitan	47	Diverse methods but aerial insectivores: *Myotis* (low, slow); *Eptesicus* (moderately high, slow); *Lasiurus* (high, fast)
	Molossidae	1–9	Pantropical	47	Aerial insectivores (high, fast flying)

Note: The foraging designations "low" and "high" refer to location above ground where foraging takes place. Low foragers usually feed within or near closed forest and along watercourses; high foragers usually forage above the forest canopy. Data come from Norberg and Rayner (1987), Simmons and Voss (1998), and Teeling et al. (2005).

[a] Includes Micronycterinae, Lonchorhininae, Phyllostominae (*sensu stricto*), and Glyphonycterinae of Baker et al. (2003).

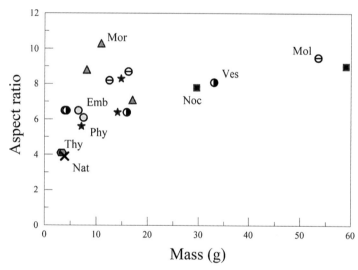

Figure 2.4. Ordination of body size and wing aspect ratio in noctilionoid clades and other insectivorous Neotropical bats based on data in Norberg and Rayner (1987). Each point represents a species. Abbreviations: Emb = Emballonuridae; Mol = Molossidae; Mor = Mormoopidae; Nat = Natalidae; Noc = Noctilionidae; Phy = Phyllostomidae; Thy = Thyropteridae; Ves = Vespertilionidae. In order of increasing size, the phyllostomids are *Micronycteris megalotis*, *Macrotus californicus*, and *Gardnernycteris crenulatum*.

are significantly larger—and they tend to have similar aspect ratio wings to those of similar-sized species. Aspect ratio denotes general wing shape. High aspect ratio wings (e.g., in Mormoopidae and Molossidae) are generally long and narrow and are associated with fast, agile flight; wings of lower aspect ratio are shorter and broader for slower, more maneuverable flight (Norberg and Rayner 1987).

Rather than body size or wing anatomy, however, what sets insectivorous phyllostomids apart from co-occurring insectivores is their foraging mode (table 2.2). Whereas nearly all other New World insectivorous bats capture their prey via aerial pursuit (hawking), most insectivorous phyllostomids capture their prey by gleaning them from vegetation; aerial hawking is rare in this family (Norberg 1998; Norberg and Rayner 1987). The aerial hawkers differ among themselves in terms of flight speeds, maneuverability, and foraging locations (e.g., relatively close to the ground, near water surfaces, or above forest canopies) and thus likely differ in their insect prey, but phyllostomids use a different tactic to capture insects and probably have done so from the beginning of their evolution. Basal members of this family, bats of the genus *Macrotus* from arid and dry tropical habitats of Mexico, the southwestern United States, and Greater Antilles, glean insects from vegetation and the ground. Their prey includes orthopterans,

and Kalko 2001), and in this regard, insectivorous phyllostomids are not statistical outliers (fig. 2.4). Some of them tend to be relatively small compared with a few co-occurring insectivores—noctilionids (*sensu stricto*) and certain vespertilionids and molossids beetles, moths, butterflies, and cicadas. In the Bahamas, a common prey item of *M. "waterhousii"* (a complex of currently undescribed species; Fleming et al. 2009) is the large black witch moth (*Ascalapha odorata*, Erebidae), with a wingspan of up to 16 cm and whose body weighs about 1 g (THF, pers. obs.). Common insects in the diets of gleaning tropical phyllostomids include katydids, crickets, lepidopteran larvae, beetles, and phasmids (Giannini and Kalko 2005; Kalka and Kalko 2006; Kalko et al. 1999). These are typically taken in the forest understory, but *Lampronycteris brachyotis* also gleans insects in and above the forest canopy (Weinbeer and Kalko 2004). *Macrophyllum macrophyllum* is apparently unique among phyllostomines because it is an aerial hawker and also trawls for insects on the surface of water (Weinbeer and Kalko 2007; Weinbeer et al. 2013). It should be noted that fruit has also been reported in the diets of a number of species of phyllostomines (*sensu lato*), and pollen is common on the fur of *Phyllostomus discolor* (Giannini and Kalko 2005; Heithaus et al. 1975; chap. 14, this vol.). The fact that gleaning insectivores sometimes visit flowers or eat fruit gives us a hint about how the morphologically more specialized phyllostomid nectarivores and frugivores could have evolved.

Schnitzler and Kalko (2001) describe the challenges that bats face when foraging for insects in different situations and the echolocation strategies they use to capture prey (also see chap. 11, this vol.). For gleaners, these challenges include hunting for stationary prey, sometimes in the midst of dense vegetation. These kinds of "cluttered" environments make prey especially hard to locate via echolocation. Phyllostomids are the primary gleaners in the New World, whereas bats in the families Nycteridae (slit-faced bats) and Megadermatidae (false vampire bats), both of which include vertebrates as well as insects and other arthropods in their diets, are their Old World ecological counterparts in terms of foraging mode (see chap. 14: fig. 14.2, this vol.). Nycteridae contains about 16 species classified in a single genus and occurs mostly in open savannas in Africa; one species occurs in Southeast Asia. Megadermatidae contains six genera and six species and occurs in Africa, southern Asia, the East Indies, and Australia. Like phyllostomids, members of these families possess noseleaves, and their nasally emitted echolocation call characteristics include production of low-intensity, frequency-modulated calls of very short duration (see

chap. 11, this vol., for a full discussion of echolocation in the phyllostomids). But only in the New World have gleaning bats moved beyond insectivory or carnivory into a much wider variety of feeding niches and much greater species richness. Why is this?

In summary, insectivorous phyllostomids differ from their non-phyllostomid relatives and other co-occurring families of insectivorous bats in foraging mode and diet. It appears that they have been successful in exploiting a food niche—insects resting and living on vegetation—that was largely untapped by other kinds of bats. And their gleaning mode of foraging exposed them to other kinds of prey (e.g., sleeping lizards and birds, frogs, and other vertebrates) that eventually led to the evolution of mostly carnivorous species such as *Trachops cirrhosus*, *Chrotopterus auritus*, and *Vampyrum spectrum* (chap. 14, this vol.). Furthermore, searching vegetation closely for insect prey also exposed gleaning insectivores to additional sources of food, including flowers and their nectar and fruits. As suggested by Gillette (1976), this novel foraging style thus opened up additional resource niche space for exploitation. A gleaning mode of foraging therefore appears to be a "key evolutionary innovation" (*sensu* Simpson 1953) that was critical for the evolutionary success of phyllostomid bats. This foraging mode reduced competition for food with other insectivorous bats and exposed them to entirely new kinds of food resources not used by other New World bats. But, as exemplified by the low ecological and species diversity of Old World nycterid and megadermatid bats, a gleaning foraging mode does not necessarily promote an extensive adaptive radiation. Other factors in addition to foraging mode must lie behind the enormous evolutionary success of the phyllostomids. We will explore these other factors throughout this book.

Vampire bats evolved early in the history of Phyllostomidae, by about 26 Ma (Rojas et al. 2016). Blood-feeding bats are utterly unique among mammals, and their evolution involved a host of morphological, physiological, and behavioral adaptations (chap. 15, this vol.). According to Baker et al. (2012), these changes occurred rather rapidly, over a period of about 4 Ma. It is not yet clear what selective pressures were behind these changes. But again, it is not hard to imagine a gleaning insectivore occasionally removing large insects or ectoparasites from sleeping mammals, much like oxpeckers (Buphagidae) that remove ticks from African ungulates today (e.g., Nunn et al. 2011; Plantan et al.

2013). According to Fenton (1992), however, a more plausible scenario is that Oligocene protovampires fed on insects and their larvae in the wounds of large mammals and birds. At that time South America had a rich fauna of large archaic mammals, including ground sloths, liptopterns (similar to horses, antelopes, and camelids), notoungulates (very diverse in size and morphology), pyrotheres (elephant-like), and astrapotheres (rhinoceros-like), as well as large birds such as ratites, phorusrachids (terrestrial carnivores), vultures, and so forth (Simpson 1980; Vizcaino et al. 2012). Modern mammalian herbivores and carnivores did not arrive in South America until closure of the Panamanian land bridge. According to this scenario, the evolutionary progression from protovampires to true vampires would have involved a switch from insect feeding to feeding on fluids from wounds attracting those insects and their larvae to actually creating wounds with modified, robust upper canines and incisors (Fenton 1992). Present-day vampires are uncommon in primary rainforests (Fleming et al. 1972; Simmons and Voss 1998), undoubtedly reflecting the low density of their potential prey. They are much more common in open habitats and around human habitations containing high densities of domestic mammals such as cows, horses, goats, and pigs (a preferred prey type; Bobrowiec et al. 2015) as well as chickens. This implies that vampires, unlike many forest-dwelling phyllostomids, probably evolved in forest-savanna ecotones where herds of herbivorous mammals and other large animals were common.

More easily understood than the evolution of vampires is the switch from eating insects to eating nectar (and sometimes pollen) and fruit. As mentioned above, many current gleaning insectivorous phyllostomids eat fruit occasionally, and less morphologically specialized glossophagines (*sensu* Koopman 1981) such as species of *Glossophaga* and *Anoura* (with the exception of *A. fistulata*; Muchhala 2006) consume many insects (Clare et al. 2014; Gardner 1977; chap. 16, this vol.). So a mixed diet of insects and plant resources is not uncommon in this family. By about 18.5 Ma, dedicated nectar feeders were evolving, perhaps initially in South America but also in the West Indies relatively early in their history (Dávalos 2009). As seen in table 2.3, most of the plant families providing nectar and fruit resources for phyllostomids had evolved long before the Early Miocene when plant-visiting bats first appeared. So a rich potential food base existed for gleaning (and

eventually hovering) bats to exploit. Most of these families were insect-pollinated ancestrally, and vertebrate pollination in them, either by birds or bats, is a derived pollination system (Fleming and Kress 2013). Vertebrate pollination as a basal pollination system occurs in very few families (e.g., only Caryocaraceae [with 2 genera and 25 species] in the New World) (Fleming and Kress 2013).

Phylogenetically based studies of three Neotropical families that contain many bat-pollinated species— Cactaceae, Agavaceae, and Bromeliaceae—can give us insights into when glossophagine bats became important pollinators in these families. The Cactaceae (about 130 genera and 1,500 species) evolved in the south-central Andes of South America about 32 Ma, and its crown groups date from about 27 Ma. Insect pollination is ancestral in its three major subfamilies, and bat pollination likely evolved from insect pollination (and not from hummingbird pollination) in several advanced tribes of columnar cacti in subfamily Cactoideae (Hernández-Hernández et al. 2014). Advanced cactoids date from 13 Ma, and bat-pollinated columnar species date from 10–5 Ma (Hernández-Hernández et al. 2014). Thus, bats have been important pollinators in this family for only about the last one-third of its history.

Members of the genus *Agave* of the century plant (sub)family (Agavoideae, which is now placed in the Asparagaceae) contains about 208 species, many of which are bat pollinated. The subfamily probably evolved in Mexico 26–22 Ma; the genus *Agave* is about 10 million years old; and agaves underwent two pulses of diversification at 8–6 Ma and 3.5–2 Ma (Good-Avila et al. 2006). As in the Cactaceae, insect pollination is basal in Agaves, and bat pollination is derived. Bats have thus been interacting with this family as pollinators for less than 10 Ma, that is, for about half the (sub)family's history.

The pineapple family Bromeliaceae (58 genera and about 3,140 species) dates from about 100 Ma and first evolved in the Guyana Shield of northeastern South America. The vast majority of its species, however, evolved 10–5 Ma as the family spread into the Andes, Amazonia, Central America, the Caribbean, and the Brazilian Shield (Givnish et al. 2014). Like Cactaceae and *Agave*, the ancestral pollination mode of bromeliads is insect pollination; hummingbird pollination of epiphytic species (where most bromeliad diversity

Table 2.3. The ages of major families of plants and animals that actually (plants) or potentially (animals) interact with plant-visiting phyllostomids

Lineage[a]	Family (No. of genera)[b,c]	Distribution	Age (Ma)[d]	Remarks
A. Nectar plants:				
Monocots	Agavaceae (3)	Pantropical	23	Bat-pollinated species from ca. 5 Ma
	Arecaceae (3)	Pantropical	119	
	Bromeliaceae (6)	Neotropics	111	Bat-pollinated species from ca. 5 Ma
Basal eudicots	Cactaceae (26)	Neotropics	30	Bat-pollinated species from ca. 10 Ma
Rosids	Fabaceae (22)	Cosmopolitan	79	
	Euphorbiaceae (3)	Cosmopolitan	69	
	Malvaceae (18)	Cosmopolitan	54	Includes Bombacaceae
	Campanulaceae (3)	Cosmopolitan	41	
Asterids	Acanthaceae (4)	Pantropical	67	
	Asteraceae (3)	Cosmopolitan	51	
	Bignoniaceae (6)	Pantropical	67	
	Gentianaceae (4)	Cosmopolitan	64	
	Gesneriaceae (6)	Pantropical	78	
	Marcgraviaceae (3)	Neotropics	64	
	Rubiaceae (3)	Cosmopolitan	78	
	Solanaceae (7)	Cosmopolitan	88	
B. Fruit plants:				
Basal angiosperms	Piperaceae (2)	Pantropical	96	
Monocots	Arecaceae (3)	Pantropical	119	
	Bromeliaceae (3)	Neotropics	111	
Basal eudicots	Cactaceae (25)	Neotropics	30	Bat fruits from ca. 15 Ma
Asterids	Gesneriaceae (6)	Pantropical	78	
	Sapotaceae (9)	Pantropical	102	
	Solanaceae (7)	Pantropical	88	
Rosids	Cucurbitaceae (2)	Pantropical	65	
	Fabaceae (6)	Pantropical	79	
	Moraceae (4)	Pantropical	87 (for *Ficus*)	
	Myrtaceae (11)	Pantropical	68	
C. Nectar feeders:				
Aves, Apodiformes	Trochilidae (104)	Neotropics	42/22	
Mammalia, Chiroptera	Phyllostomidae (17)	Neotropics	Glossophaginae: 21.4/17.4 Lonchophyllinae: 18.5/11.0	
D. Fruit eaters:				
Aves, Trogoniformes	Trogonidae (6)	Pantropical	55	Only frugivorous in the Neotropics; canopy feeders
Aves, Piciformes	Capitonidae (2)	Neotropics	13	Canopy feeders; fig specialists
	Ramphastidae (5)	Neotropics	13	Canopy feeders of large fruit

(continued)

Table 2.3. Continued

Lineage[a]	Family (No. of genera)[b,c]	Distribution	Age (Ma)[d]	Remarks
Aves, Passeriformes	Cotingidae (33)	Neotropics	39	Mostly canopy feeders
	Pipridae (13)	Neotropics	64	Understory feeders of small fruit
	Turdidae (24)	Cosmopolitan	27	Mostly canopy feeders
	Thraupidae (50)	Neotropics	15	Mostly canopy feeders
Mammalia, Chiroptera	Phyllostomidae (22)	Neotropics	Carolliinae: 18.5/17.2 Stenodermatinae: 18.5/ 16.2	*Carollia* and *Sturnira* Understory feeders; *Artibeus* and relatives canopy feeders
Mammalia, Primates	Cebidae (6)	Neotropics	25	Canopy feeders
	Atelidae (5)	Neotropics	25	Canopy feeders
Mammalia, Carnivora	Procyonidae (6)	Nearctic, Neotropics	28	Canopy feeders

Note: Ages for plant families generally represent stem clade ages; in many cases, crown clade ages are considerably younger. Data come from Baker et al. (2012), Fleming and Kress (2013), Fleming et al. (2009), Lobova et al. (2009), McGuire et al. (2014), and Muscarella and Fleming (2007).

[a] Based on APG III (2009) in which five major angiosperm lineages are recognized: basal angiosperms, monocots, basal eudicots, and advanced eudicots (Asterids and Rosids).

[b] Genera pollinated by bats.

[c] Genera dispersed by bats.

[d] The two dates indicate stem and crown ages.

currently resides) is extensive and dates from about 15 Ma; and bat pollination has evolved from hummingbird pollination independently at least three times in the last 5 Ma. Thus, as in the previous two families, bats are relatively recent pollinators in this family and have been doing so for only the last 5% of its history.

It is likely that the evolution of bat pollination in these three families is typical of how it evolved in other families: many bat flowers have evolved from insect flowers (both diurnal and nocturnal insects), while others have evolved from hummingbird flowers. Examples of the former scenario can be found in the diverse tropical genus *Ruellia* (Acanthaceae) in which bat pollination has evolved rather infrequently from either bee or hummingbird pollination (Tripp and Manos 2008). Examples of the latter scenario can be found in the Gesneriaceae of the West Indies and Central America in which hummingbird pollination is ancestral and from which bat pollination evolved (Martén-Rodríguez et al. 2009, 2010). In most cases, the evolution of bat pollination has probably occurred relatively recently— certainly from the mid-Miocene on—in a family's history.

Although two phyllostomid subfamilies (Glossophaginae and Lonchophyllinae) contain most spe-

cies of nectar bats, one genus in the Phyllostominae (*Phyllostomus*) also contains frequent flower visitors. *Phyllostomus* bats and other more opportunistic flower-visiting bats (e.g., members of the mostly frugivorous genera *Carollia* and *Artibeus*) are generally larger than glossophagines and lonchophyllines, do not have highly extensible tongues, and do not hover at flowers; instead, they land on flowers before beginning to feed. As discussed in detail by von Helversen and Winter (2003; also see chap. 16, this vol.), these size and foraging differences strongly affect the kinds of flowers visited by these bats. The larger, non-hovering species tend to visit large, easily accessible cup-shaped or shaving brush flowers (e.g., those in the Bombacoideae [e.g., *Ochroma*, *Bombax*, *Ceiba*] as well as *Hymenaea* [Fabaceae], *Marcgravia*, *Mabea* [Euphorbiaceae], etc.), whereas hovering nectar bats visit smaller, more tubular flowers that can be accessed only by hovering nectar feeders (Fleming and Muchhala 2008; Simmons and Wetterer 2002; von Helversen and Winter 2003). Large nocturnal flowers are currently also visited by marsupials and certain primates, and it has been suggested (e.g., Sussman and Raven 1978; von Helversen and Winter 2003) that these kinds of flowers were originally pollinated by non-volant mammals. Marsupials certainly evolved long before phyllostomid bats in the Neotropics, but primates might have arrived in South America at about the same time or just after phyllostomids were beginning to evolve. Thus, early nectar-feeding phyllostomids probably had large, mammal-pollinated flowers to feed at, but it wasn't until the younger genera of nectar feeders had evolved specialized wing morphology that permitted hovering flight and long, extensible tongues for probing deeply into flowers that hummingbird- or hawkmoth-pollinated flowers were readily accessible to them.

Frugivory is much more common than nectarivory in phyllostomid bats and other kinds of vertebrates (Fleming and Kress 2013). Major reasons for this include: (1) the energy density of fruit in most tropical habitats is orders of magnitude greater than that of nectar; (2) fruit eating doesn't require as specialized trophic morphology as nectar-feeding; and (3) nectar feeders are generally restricted to small body sizes whereas fruit eaters exhibit a much greater range of body sizes (and cover nearly the entire size spectrum of contemporary terrestrial mammals and birds) (Fleming and Kress 2013). Highly frugivorous phyllostomids occur in subfamilies Carolliinae and Stenodermatinae that began to evolve about 18 Ma (Rojas et al. 2016; table 2.3). Members of both subfamilies are very abundant in contemporary habitats, and stenodermatids have diversified more extensively than any other phyllostomid subfamily in the last 15 Ma (Dumont et al. 2012). As mentioned earlier, the most morphologically specialized members of the latter subfamily (*Ardops* and its relatives) evolved in the Greater Antilles and then colonized the Neotropical mainland about 7–6 Ma (Dávalos 2009; Dumont et al. 2014). *Carollia* bats feed primarily in forest understories and eat soft fruit, whereas many stenodermatines are canopy feeders that eat both soft and hard fruits (chap. 17, this vol.). Virtually all of the plant families that are important fruit sources for these bats are much older than the two frugivorous clades and include basal angiosperms as well as advanced eudicots (table 2.3). *Carollia* bats feed extensively on species in the very old pantropical family Piperaceae, most of which are shrubs or treelets in the New World. Based on the mid-Miocene evolution of carolliines, it is likely that much of the high species diversity of Neotropical *Pipers*, which is much greater than that of Paleotropical *Pipers* (Jaramillo and Callejas 2004), dates from this time period. In contrast,

canopy-feeding stenodermatines feed extensively on advanced eudicot families (e.g., the rosid Moraceae and its relatives).

Other vertebrate frugivores present in Neotropical forests at the time of the evolution of fruit-eating phyllostomids included several families of birds and mammals. Most of the frugivorous birds present in the Miocene were relatively large (up to 900 g in toucans) canopy feeders that harvested fruit either by aerial hawking (e.g., trogons, cotingids) or while perched on a branch (toucans) or on the ground (e.g., cracids [Cracidae]) (Moermond and Denslow 1985; table 2.3). Aerial hawkers generally harvest single fruits at a time, whereas perchers eat either single or multiple fruits per visit to a fruiting tree. Many perch- or ground-feeding birds also eat fruit piecemeal and hence are not limited in the size of fruit they eat by their gape width or body size (Fleming and Kress 2013). The only old family of Neotropical birds that contains understory feeders are manakins (Pipridae), which are all small birds (7–26 g). Tanagers (Thraupidae; 8–114 g) are the only speciose group of relatively small canopy-feeding birds whose evolutionary age is similar to that of frugivorous phyllostomids (table 2.3). As discussed below, the fruit diets of manakins and tanagers are very different from those of understory- and canopy-feeding phyllostomids. Arboreal frugivorous mammals present in the Miocene included a variety of primates as well as an early procyonid (*Cyonasua*) (Eizirik 2012; Goin et al. 2012). Most of these are large species (up to about 12 kg in Atelidae) that eat fruit whole or piecemeal in forest canopies. Figs (Moraceae) are often important components of their diets (Terborgh 1983).

Dietary overlap between frugivorous phyllostomid bats, birds, and other mammals currently tends to be low and probably has been low throughout the evolution of these bats. For example, Dinerstein (1986) reported that only 13 of 169 species (8%) of fruit eaten by birds at Monteverde, Costa Rica, were also eaten by bats. Similarly, Gorchov et al. (1995) found that birds and bats shared only six fruit species (i.e., only about 7% of all of the fruit eaten by these taxa) in a Peruvian rainforest. In contrast, in their extensive review of the fruit eaten by bats and other vertebrates in central French Guiana, Lobova et al. (2009) reported that of the 111 species of fruit eaten by bats, 6 species (5%) were shared with birds, 27 (24%) were shared with arboreal mammals (marsupials, primates, and procyonids), and 12 (11%) were shared by all three groups; only 66 species (59%) were eaten exclusively by bats. Plant families shared with primates included Araceae, Chrysobalanaceae, and Moraceae; families shared with birds included Arecaceae, Cecropiaceae (now included in Urticaceae), Marcgraviaceae, and Melastomataceae. Families with fruit eaten exclusively by bats in this area included Clusiaceae, Cyclanthaceae, Piperaceae, and Solanaceae, although birds also eat some of these fruits in other areas (e.g., Gorchov et al. 1995; Parolin et al. 2013; Ripperger et al. 2014; Sarmento et al. 2014). These data seem to suggest that the evolution of frugivory in phyllostomid bats has, to a large extent, occurred independently of frugivory in other vertebrate groups (Fleming and Kress 2013). Even in families shared by birds and bats (e.g., Cecropiaceae, Moraceae), bird and bat fruits are distinct in terms of morphology, nutrition, and scent, which suggests that they have had relatively long, independent evolutionary histories (Charles-Dominique 1986; Hodgkinson et al. 2013; Kalko et al. 1996; Lomáscolo et al. 2008, 2010). Nonetheless, given the great evolutionary ages of many of the plant families providing fleshy fruit for phyllostomids, which have been major fruit consumers for only about 20 Ma, and other frugivorous vertebrates, one wonders who were the ancestral vertebrate consumers of these fleshy-fruited families? Because many of these families have pantropical distributions (exceptions include Bromeliaceae and Cactaceae; table 2.3), they could have evolved their fruit characteristics by interacting with vertebrate frugivores somewhere other than the Neotropics (e.g., in Africa or Southeast Asia). Detailed timed phylogenies, coupled with historical biogeographic analyses, are needed to answer this question. An example of this approach is the fig family (Moraceae), one of the most important fruit sources for vertebrates worldwide (Harrison 2005; Shanahan et al. 2001; Xu et al. 2011). The genus *Ficus* dates from about 87 Ma, and its oldest sections occur in the Old World. Members of New World section Americana date from about 34–20 Ma, during the period when New World primates and frugivorous phyllostomid bats were first evolving (Cruad et al. 2012; Xu et al. 2011). But its ancestral fruit characteristics probably evolved with Paleotropical frugivores in the Late Cretaceous. The morphological and nutritional characteristics of Neotropical figs have evolved since the Early Miocene in response to selection pressures from birds (e.g., toucans, barbets, tanagers), primates,

and stenodermatine bats (Lomáscolo et al. 2008, 2010).

Conclusions

Much has changed physically and biotically in the New World in the last 30 Ma when phyllostomid bats were evolving. Dramatic changes in Earth's climate, including decreasing temperatures and increasing seasonality and the rise of the Andes in South America, the Sierra Madre mountains in Mexico, and so forth, during this period have led to a greater diversity of habitats such as montane forests, tropical dry forests, grasslands and savannas, and deserts than existed earlier in the warmer, less seasonal Cenozoic when much of the world was covered in tropical evergreen forests. Modern versions of these habitats and their angiosperm components date from the Miocene, by which time most modern families of birds and mammals had also evolved. The adaptive radiation of phyllostomid bats, therefore, took place in an increasingly modern and diverse world. Perhaps to avoid competition with other kinds of insectivorous bats, they evolved a gleaning foraging mode that exposed them to a much wider array of potential food types than just insects and small vertebrates. Over a period of about 10 Ma, they evolved novel feeding modes (at least for New World bats) such as vertebrate carnivory, sanguinivory, nectarivory, and frugivory (Monteiro and Nogueira 2011). In the case of the plant visitors, which represent over 70% of the species diversity in this family today, they initially took advantage of food resources (nectar and fruit) that had evolved with other kinds of animals (primarily insects and hummingbirds in the case of flowers; birds and arboreal mammals in the case of fruit). But via the process of generalized coevolution (Fleming and Kress 2013), phyllostomid bat-plant interactions ultimately produced flowers and fruit bearing distinctive suites of traits (i.e., flower and fruit syndromes) that reflect the morphological, sensory, physiological, and behavioral characteristics of these bats. No other family of bats has undergone such a dramatic adaptive radiation. As a result, phyllostomid bats today provide a broader variety of ecosystem services, such as insect control, pollination, and seed dispersal, than other kinds of Neotropical bats. Ironically, however, the success of one of its most unique clades—vampire bats—threatens the existence of many other phyllostomids as well as other families of New World bats. Unfortunately, vampire bats are important reservoirs for the rabies virus and sometimes transmit this virus to domestic animals and humans (Messenger et al. 2003). As a result, bat eradication via misguided vampire "control" programs in Latin America, coupled with deforestation and general habitat destruction, threatens to undo tens of millions of years of chiropteran evolution in a few human generations. It is our moral obligation not to let this happen.

Acknowledgments

My long-term association with phyllostomid bats has been supported by research funds provided by the Smithsonian Institution, the U.S. National Science Foundation, National Geographic Society, Earthwatch Institute, U.S. Fish and Wildlife Foundation, Ted Turner Foundation, and the Universities of Missouri–Saint Louis and Miami. I thank three anonymous reviewers for their comments and suggestions about this chapter.

References

Almendra, A. L., and D. S. Rogers. 2012. Biogeography of Central American mammals, patterns and processes. Pp. 203–229 *in*: Bones, Clones, and Biomes (B. D. Patterson and L. P. Costa, eds.). University of Chicago Press, Chicago.

APG III. 2009. An update of the Angiosperm Phylogeny Group classification for the orders and families of flowering plants: APG III. Botanical Journal of the Linnean Society, 161:105–121.

Arita, H. T., J. Vargas-Baron, and F. Villalobos. 2014. Latitudinal gradients of genus richness and endemism and the diversification of New World bats. Ecography, 37:1024–1033.

Baker, R. J., O. R. P. Bininda-Emonds, H. Mantilla-Meluk, C. A. Porter, and R. A. van den Bussche. 2012. Molecular time scale of diversification of feeding strategy and morphology in New World leaf-nosed bats (Phyllostomidae): a phylogenetic perspective. Pp. 385–409 *in*: Evolutionary History of Bats: Fossils, Molecules and Morphology (G. F. Gunnell and N. B. Simmons, eds.). Cambridge University Press, Cambridge.

Baker, R. J., S. R. Hoofer, C. A. Porter, and R. A. Van den Bussche. 2003. Diversification among New World leaf-nosed bats: an evolutionary hypothesis and classification inferred from digenomic congruence of DNA sequences. Occasional Papers, Museum of Texas Tech University, 230:1–32.

Blois, J. L., and E. A. Hadly. 2009. Mammalian response to Cenozoic climatic change. Annual Review of Earth and Planetary Sciences, 37:181–208.

Bobrowiec, P. E. D., M. R. Lemes, and R. Gribel. 2015. Prey preference of the common vampire bat (*Desmodus rotundus*, Chiroptera) using molecular analysis. Journal of Mammalogy, 96:54–63.

Bowen, G. J., and J. C. Zachos. 2010. Rapid carbon sequestration at the termination of the Palaeocene-Eocene Thermal Maximum. Nature Geoscience, 3:866–869.

Burns, K. J., and K. Naoki. 2004. Molecular phylogenetics and biogeography of Neotropical tanagers in the genus *Tangara*. Molecular Phylogenetics and Evolution, 32:838–854.

Charles-Dominique, P. 1986. Inter-relations between frugivorous vertebrates and pioneer plants: *Cecropia*, birds, and bats in French Guyana. Pp. 119–135 *in*: Frugivores and Seed Dispersal (A. Estrada and T. H. Fleming, eds.). Dr. W. Junk, Dordrecht, Netherlands.

Clare, E. L., H. R. Goerlitz, V. A. Drapeau, M. W. Holderied, A. M. Adams, J. Nagel, E. R. Dumont, P. D. N. Hebert, and M. Brock Fenton. 2014. Trophic niche flexibility in *Glossophaga soricina*: how a nectar seeker sneaks an insect snack. Functional Ecology, 28:632–641.

Connor, W. E., and A. J. Corcoran. 2012. Sound strategies: the 65-million-year-old battle between bats and insects. Annual Review of Entomology, 57:21–39.

Cruaud, A., N. Ronsted, B. Chantarasuwan, L. S. Chou, W. L. Clement, A. Couloux, B. Cousins et al. 2012. An extreme case of plant-insect codiversification: figs and fig-pollinating wasps. Systematic Biology, 61:1029–1047.

Dávalos, L. M. 2009. Earth history and the evolution of Caribbean bats. Pp. 96–115 *in*: Island Bats: Evolution, Ecology, and Conservation (T. H. Fleming and P. A. Racey, eds.). University of Chicago Press, Chicago.

Dávalos, L. M., and S. T. Turvey. 2012. West Indian mammals, the old, the new, and the recently extinct. Pp. 157–202 *in*: Bones, Clones, and Biomes (B. D. Patterson and L. M. Costa, eds.). University of Chicago Press, Chicago.

Dinerstein, E. 1986. Reproductive ecology of fruit bats and the seasonality of fruit production in a Costa Rican cloud forest. Biotropica, 18:307–318.

Dumont, E. R., L. M. Dávalos, A. Goldberg, S. E. Santana, K. Rex, and C. C. Voigt. 2012. Morphological innovation, diversification and invasion of a new adaptive zone. Proceedings of the Royal Society B–Biological Sciences, 297:1797–1805.

Dumont, E. R., K. Samadevam, I. Grosse, O. M. Warsi, B. Baird, and L. M. Dávalos. 2014. Selection for mechanical advantage underlies multiple cranial optima in New World leaf-nosed bats. Evolution, 68:1436–1449.

Eberle, J. J., and D. R. Greenwood. 2012. Life at the top of the greenhouse Eocene world—a review of the Eocene flora and vertebrate fauna from Canada's High Arctic. Geological Society of America Bulletin, 124:3–23.

Eizirik, E. 2012. A molecular view on the evolutionary history and biogeography of Neotropical carnivores (Mammalia, Carnivora). Pp. 123–142 *in*: Bones, Clones, and Biomes (B. D. Patterson and L. M. Costa, eds.). University of Chicago Press, Chicago.

Fenton, M. B. 1992. Wounds and the origin of blood-feeding in bats. Biological Journal of the Linnean Society, 47:161–171.

Fleming, T. H., C. K. Geiselman, and W. J. Kress. 2009. The

evolution of bat pollination: a phylogenetic perspective. Annals of Botany, 104:1017–1043.

Fleming, T. H., E. T. Hooper, and D. E. Wilson. 1972. Three Central American bat communities: structure, reproductive cycles, and movement patterns. Ecology, 53:555–569.

Fleming, T. H., and W. J. Kress. 2013. The Ornaments of Life: Coevolution and Conservation in the Tropics. University of Chicago Press, Chicago.

Fleming, T. H., and N. Muchhala. 2008. Nectar-feeding bird and bat niches in two worlds: pantropical comparisons of vertebrate pollination systems. Journal of Biogeography, 35:764–780.

Gardner, A. L. 1977. Feeding habits. Special Publications of the Museum, Texas Tech University 13:293–350.

Garzione, C. N., G. D. Hoke, J. C. Libarkin, S. Withers, B. MacFadden, J. Eiler, P. Ghosh, and A. Mulch. 2008. Rise of the Andes. Science, 320:1304–1307.

Geipel, I., K. Jung, and E. K. V. Kalko. 2013. Perception of silent and motionless prey on vegetation by echolocation in the gleaning bat *Micronycteris microtis*. Proceedings of the Royal Society B–Biological Sciences, 280: 20122830.

Giannini, N. P., and E. K. V. Kalko. 2005. The guild structure of animalivorous leaf-nosed bats of Barro Colorado Island, Panama, revisited. Acta Chiropterologica, 7:131–146.

Gillette, D. D. 1976. Evolution of feeding strategies in bats. Tebiwa, 18:39–48.

Gingerich, P. D. 2006. Environment and evolution through the Paleocene-Eocene thermal maximum. Trends in Ecology and Evolution, 21:246–253.

Givnish, T. J., M. H. J. Barfuss, B. Van Ee, R. Riina, K. Schulte, R. Horres, P. A. Gonsiska et al. 2014. Adaptive radiation, correlated and contingent evolution, and net species diversification in Bromeliaceae. Molecular Phylogenetics and Evolution, 71:55–78.

Goin, F. J., J. N. Gelfo, L. Chornogubsky, M. O. Woodburne, and T. Martin. 2012. Origins, radiations, and distribution of South American mammals. Pp. 20–50 *in*: Bones, Clones, and Biomes (B. D. Patterson and L. M. Costa, eds.). University of Chicago Press, Chicago.

Good-Avila, S. V., V. Souza, B. S. Gaut, and L. E. Eguiarte. 2006. Timing and rate of speciation in *Agave* (Agavaceae). Proceedings of the National Academy of Sciences, USA, 103:9124–9129.

Gorchov, D. L., F. Cornejo, C. F. Ascorra, and M. Jaramillo. 1995. Dietary overlap between frugivorous birds and bats in the Peruvian Amazon. Oikos, 74:235–250.

Graham, A. 2009. The Andes: A geological overview from a biological perspective. Annals of the Missouri Botanical Garden, 96:371–385.

Graham, A. 2011. A Natural History of the New World: The Ecology and Evolution of Plants in the Americas. University of Chicago Press, Chicago.

Handley, C. O., Jr. 1976. Mammals of the Smithsonian Venezuelan Project. Brigham Young University Science Bulletin Biological Series, 20:1–91.

Harrison, R. D. 2005. Figs and the diversity of tropical rainforests. Bioscience, 55:1053–1064.

Heithaus, E. R., T. H. Fleming, and P. A. Opler. 1975. Patterns of foraging and resource utilization in seven species of bats in a seasonal tropical forest. Ecology, 56:841–854.

Hernández-Hernández, T., J. W. Brown, B. Schlumpberger, L. E. Eguiarte, and S. Magallon. 2014. Beyond aridification: multiple explanations for the elevated diversification of cacti in the New World Succulent Biome. New Phytologist, 202:1382–1397.

Hodgkison, R., M. Ayasse, C. Haberlein, S. Schulz, A. Zubaid, W. A. W. Mustapha, T. H. Kunz, and E. K. V. Kalko. 2013. Fruit bats and bat fruits: the evolution of fruit scent in relation to the foraging behaviour of bats in the New and Old World tropics. Functional Ecology, 27:1075–1084.

Hoorn, C., and S. Flantua. 2015. An early start for the Panama land bridge. Science, 348:186–187.

Hoorn, C., F. P. Wesselingh, H. ter Steege, M. A. Bermudez, A. Mora, J. Sevink, I. Sanmartin et al. 2010. Amazonia through time: Andean uplift, climate change, landscape evolution, and biodiversity. Science, 330:927–931.

Jaramillo, M. A., and R. Callejas. 2004. Current perspectives on the classification and phylogenetics of the genus *Piper* L. Pp. 179–198 *in*: *Piper*: A Model Genus for Studies of Phytochemistry, Ecology, and Evolution (L. A. Dyer and A. D. N. Palmer, eds.). Kluwer Academic/Plenum Publishers, New York.

Kalka, M., and E. K. V. Kalko. 2006. Gleaning bats as underestimated predators of herbivorous insects: diet of *Micronycteris microtis* (Phyllostomidae) in Panama. Journal of Tropical Ecology, 22:1–10.

Kalko, E. K. V., D. Friemel, C. O. Handley, and H. U. Schnitzler. 1999. Roosting and foraging behavior of two Neotropical gleaning bats, *Tonatia silvicola* and *Trachops cirrhosus* (Phyllostomidae). Biotropica, 31:344–353.

Kalko, E. K. V., E. A. Herre, and C. O. Handley, Jr. 1996. The relation of fig fruit syndromes to fruit-eating bats in the New and Old World tropics. Journal of Biogeography, 23:565–576.

Koopman, K. F. 1981. The distributional patterns of New World nectar-feeding bats. Annals of the Missouri Botanical Gardens, 68:352–369.

Lauretano, V., K. Littler, M. Polling, J. C. Zachos, and L. J. Lourens. 2015. Frequency, magnitude and character of hyperthermal events at the onset of the Early Eocene Climatic Optimum. Climate of the Past, 11:1313–1324.

Leigh, E. G., A. O'Dea, and G. J. Vermeij. 2014. Historical biogeography of the Isthmus of Panama. Biological Reviews, 89:148–172.

Lobova, T. A., C. K. Geiselman, and S. Mori. 2009. Seed Dispersal by Bats in the Neotropics. New York Botanical Garden, Bronx, New York.

Lomáscolo, S. B., D. J. Levey, R. T. Kimball, B. M. Bolker, and H. T. Alborn. 2010. Dispersers shape fruit diversity in *Ficus* (Moraceae). Proceedings of the National Academy of Sciences, USA, 107:14668–14672.

Lomáscolo, S. B., P. Speranza, and R. T. Kimball. 2008. Correlated evolution of fig size and color supports the dispersal syndromes hypothesis. Oecologia, 156:783–796.

Martén-Rodríguez, S., A. Almarales-Castro, and C. B. Fenster. 2009. Evaluation of pollination syndromes in Antillean

Gesneriaceae: evidence for bat, hummingbird and generalized flowers. Journal of Ecology, 97:348–359.

Martén-Rodríguez, S., C. B. Fenster, I. Agnarsson, L. E. Skog, and E. A. Zimmer. 2010. Evolutionary breakdown of pollination specialization in a Caribbean plant radiation. New Phytologist, 188:403–417.

McGuire, J. A., C. C. Witt, J. V. Remsen Jr., A. Corl, D. L. Rabosky, D. L. Altshuler, and R. Dudley. 2014. Molecular phylogenetics and the diversification of hummingbirds. Current Biology, 24:910–916.

McKenna, M. C. 1980. Eocene paleolatitude, climate, and mammals of Ellesmere Island. Palaeogeography Palaeoclimatology Palaeoecology, 30:349–362.

Messenger, S. L., C. E. Rupprecht, and J. S. Smith. 2003. Bats, emerging virus infections, and the rabies paradigm. Pages 622–679 in: Bat Ecology (T. H. Kunz and M. B. Fenton, eds.). University of Chicago Press, Chicago.

Misof, B., S. Liu, K. Meusemann et al. 2014. Phylogenomics resolves the timing and pattern of insect evolution. Science, 346:763–767.

Moermond, T. C., and J. S. Denslow. 1985. Neotropical avian frugivores: patterns of behavior, morphology, and nutrition. Ornithological Monographs, 36:865–897.

Monteiro, L. R., and M. R. Nogueira. 2011. Evolutionary patterns and processes in the radiation of phyllostomid bats. BMC Evolutionary Biology, 11:137.

Morgan, G. S., N. J. Czaplewski, G. F. Gunnel, and N. B. Simmons. 2012. Evolutionary history of the Neotropical Chiroptera: the fossil record. Pp. 105–161 in: Evolutionary History of Bats: Fossils, Molecules, and Morphology (G. F. Gunnell and N. B. Simmons, eds.). Cambridge University Press, Cambridge.

Morley, R. J. 2000. Origin and Evolution of Tropical Forests. Wiley, New York.

Morley, R. J. 2007. Cretaceous and Tertiary climate change and the past distribution of megathermal rainforests. Pp. 1–32 in: Tropical Rainforest Responses to Climate Change (M. B. Bush and J. R. Flenley, eds.). Springer, New York.

Muchhala, N. 2006. Nectar bats stows huge tongue in its rib cage. Nature, 444:701–702.

Muscarella, R., and T. H. Fleming. 2007. The role of frugivorous bats in tropical forest succession. Biological Reviews, 82: 573–590.

Norberg, U. M. 1998. Morphological adaptations for flight in bats. Pp. 93–108 in: Bat Biology and Conservation (T. H. Kunz and P. A. Racey, eds.). Smithsonian Institution Press, Washington, DC.

Norberg, U. M., and J. M. V. Rayner. 1987. Ecological morphology and flight in bats (Mammalia; Chiroptera): wing adaptations, flight performance, foraging strategy, and echolocation. Philosophical Transactions of the Royal Society of London Series B–Biological Sciences, 316:335–427.

Nunn, C. L., V. O. Ezenwa, C. Arnold, and W. D. Koenig. 2011. Mutualism or parasitism? Using a phylogenetic approach to characterize the oxpecker-ungulate relationship. Evolution, 65:1297–1304.

Olmstead, R. G. 2013. Phylogeny and biogegraphy in Solanaceae, Verbenaceae, and Bignoniaceae: a comparison of continental and intercontinental diversification patterns. Botanical Journal of the Linnean Society, 171:80–102.

Parolin, P., F. Wittmann, and L. V. Ferreira. 2013. Fruit and seed dispersal in Amazonian floodplain trees—a review. Ecotropica, 19:15–32.

Plantan, T., M. Howitt, A. Kotze, and M. Gaines. 2013. Feeding preferences of the red-billed oxpecker, *Buphagus erythrorhynchus*: a parasitic mutualist? African Journal of Ecology, 51:325–336.

Ramírez, D. A., D. A. Foster, K. Min, C. Montes, A. Cardona, and G. Sadove. 2016. Exhumation of the Panama basement complex and basins: implications for the closure of the Central American seaway. Geochemistry Geophysics Geosystems, 17:1758–1777.

Ripperger, S. P., E. W. Heymann, M. Tschapka, and E. K. V. Kalko. 2014. Fruit characteristics associated with fruit preferences in frugivorous bats and saddle-back tamarins in Peru. Ecotropica, 20:53–63.

Rojas, D., O. M. Warsi, and L. M. Dávalos. 2016. Bats (Chiroptera: Noctilionoidea) challenge a recent origin of extant Neotropical diversity. Systematic Biology, 65:432–448.

Rull, V. 2015. Pleistocene speciation is not refuge speciation. Journal of Biogeography, 42:602–609.

Sarmento, R., C. P. Alves-Costa, A. Ayub, and M. A. R. Mello. 2014. Partitioning of seed dispersal services between birds and bats in a fragment of the Brazilian Atlantic Forest. Zoologia, 31:245–255.

Schnitzler, H. U., and E. K. V. Kalko. 2001. Echolocation by insect-eating bats. Bioscience, 51:557–569.

Shanahan, M., S. So, S. G. Compton, and R. Corlett. 2001. Fig-eating by vertebrate frugivores: a global review. Biological Reviews, 76:529–572.

Simmons, N. B., and R. S. Voss. 1998. The mammals of Paracou, French Guiana: a Neotropical lowland rainforest fauna. Part I: Bats. Bulletin of the American Museum of Natural History, 237:1–219.

Simmons, N. B., and A. L. Wetterer. 2002. Phylogeny and convergence in cactophilic bats. Pp. 87–121 in: Columnar Cacti and Their Mutualists: Evolution, Ecology, and Conservation (T. H. Fleming and A. Valiente-Banuet, eds.). University of Arizona Press, Tucson.

Simpson, G. G. 1953. The Major Features of Evolution. Columbia University Press, New York.

Simpson, G. G. 1980. Splendid Isolation: The Curious History of South American Mammals. Yale University Press, New Haven, CT.

Smith B.T., J. E. McCormack, M. Cuervo, M. J. Hickerson, A. Aleixo, C. W. Burney, X. Xie et al. 2014. The drivers of tropical speciation. Nature, 515:406–409.

Sussman, R. W., and P. H. Raven. 1978. Pollination by lemurs and marsupials: an archaic coevolutionary system. Science, 200:731–736.

Teeling, E. C., M. S. Springer, O. Madsen, P. Bates, S. J. O'Brien, and W. J. Murphy. 2005. A molecular phylogeny for bats illuminates biogeography and the fossil record. Science, 307:580–584.

Terborgh, J. 1983. Five New World Primates: A Study in Comparative Ecology. Princeton University Press, Princeton, NJ.

Tripp, E. A., and P. S. Manos. 2008. Is floral specialization an evolutionary dead-end? Pollination system transitions in *Ruellia* (Acanthaceae). Evolution, 62:1712–1736.

Vizcaino, S. F., G. H. Cassini, N. Toledo, and M. S. Bargo. 2012. On the evolution of large size in mammalian herbivores of Cenozoic faunas of southern South America. Pp. 76–101 *in*: Bones, Clones, and Biomes (B. D. Patterson and L. M. Costa, eds.). University of Chicago Press, Chicago.

von Helversen, O., and Y. Winter. 2003. Glossophagine bats and their flowers: cost and benefit for flower and pollinator. Pp. 346–397 *in*: Bat Ecology (T. H. Kunz and M. B. Fenton, eds.). University of Chicago Press, Chicago.

Weinbeer, M., and E. K. V. Kalko. 2004. Morphological characteristics predict alternate foraging strategy and microhabitat selection in the orange-bellied bat, *Lampronycteris brachyotis*. Journal of Mammalogy, 85:1116–1123.

Weinbeer, M., and E. K. V. Kalko 2007. Ecological niche and phylogeny: the highly complex echolocation behavior of the trawling long-legged bat, *Macrophyllum macrophyllum*. Behavioral Ecology and Sociobiology, 61:1337–1348.

Weinbeer, M., E. K. V. Kalko, and K. Jung. 2013. Behavioral flexibility of the trawling long-legged bat, *Macrophyllum macrophyllum* (Phyllostomidae). Frontiers in Physiology, 4.

Xu, L., R. D. Harrison, P. Yang, and D. R. Yang. 2011. New insight into the phylogenetic and biogeographic history of genus *Ficus*: vicariance played a relatively minor role compared with ecological opportunity and dispersal. Journal of Systematics and Evolution, 49:546–557.

Zachos, J. C., G. R. Dickens, and R. E. Zeebe. 2008. An early Cenozoic perspective on greenhouse warming and carbon-cycle dynamics. Nature, 451:279–283.

Zachos, J., M. Pagini, L. Sloan, E. Thomas, and K. Billups. 2001. Trends, rhythms, and aberrations in global climate 65 Ma to present. Science, 292:686–693.

Zeebe, R. E., and J. C. Zachos. 2013. Long-term legacy of massive carbon input to the Earth system: Anthropocene versus Eocene. Philosophical Transactions of the Royal Society A–Mathematical Physical and Engineering Sciences, 371: 20120006.

SECTION 2

Phylogeny and Evolution

Phylogeny, Fossils, and Biogeography

The Evolutionary History of Superfamily Noctilionoidea (Chiroptera: Yangochiroptera)

Norberto P. Giannini
and Paúl M. Velazco

Introduction

Noctilionoids provide the natural frame to understand the evolutionary origins of phyllostomids, the focus of the present volume. Five Neotropical families, the Thyropteridae, Furipteridae, Noctilionidae, Mormoopidae, and Phyllostomidae, evolved in the New World from one common ancestor (Hoofer et al. 2003; Teeling et al. 2005) and sparked one of the greatest mammalian diversifications on record (Dumont et al. 2012). There is only one noctilionoid fossil family, Speonycteridae†, which is known from the Oligocene of Florida and has been recovered as sister to a clade including Noctilionidae, Mormoopidae, and Mystacinidae (Czaplewski and Morgan 2012). One Australasian family, the New Zealand endemic Mystacinidae, has been consistently recovered as sister of the Neotropical noctilionoid diversification (Amador et al. 2016; Hoofer et al. 2003; Meredith et al. 2011; Rojas et al. 2016; Teeling et al. 2005; Van Den Bussche and Hoofer 2004). Another Old World family, the Malagasy endemic Myzopodidae, occupies controversial locations in the chiropteran tree and may join the group (e.g., Teeling et al. 2005) or be positioned as a highly relevant, but more distant, outgroup (e.g., Amador et al. 2016; Meredith et al. 2011). Each of these families exhibits unique ecological, behavioral, and morphological adaptations to highly diverse lifestyles. In this chapter, we revisit the systematic and biogeographic history of the group in the light of new phylogenies that depict the evolution of these bats with ever-increasing accuracy and support, as well as the expanding fossil record that now includes spectacular specimens unearthed in continents where living relatives no longer occur. We also provide a brief account of the biology of noctilionoid families as a useful comparative framework for understanding the evolution and ecology of phyllostomids and for interpreting the broad evolutionary significance of the noctilionoid diversification.

Phylogenetic Relationships

The current composition and internal relationships of Noctilionoidea have been controversial in several aspects. A close relationship between Noctilionidae, Mormoopidae, and Phyllostomidae has been recognized since Miller (1907), and this constitutes the invariable core of the superfamily. As currently understood, Noctilionoidea includes a more complex array of families, some of which were previously allocated to other superfamilies, particularly Nataloidea, and also the large Vespertilionoidea (Mystacinidae; Simmons 1998). In essence, most recent molecular phylogenetic work resulted in previously nataloid families being transferred to Vespertilionoidea (Natalidae) and to Noctilionoidea (Myzopodidae, Thyropteridae, and Furipteridae), and one previously vespertilionoid family (Mystacinidae) being transferred to Noctilionoidea. Interestingly, in one of the latest morphological hypotheses of bat interfamilial relationships (Gunnell and Simmons 2005), Myzopodidae was sister to Thyropteridae, Furipteridae, and Natalidae; this phylogeny placed Nataloidea sister to a noctilionoid clade in which Mystacinidae was sister to Noctilionidae, Mormoopidae, and Phyllostomidae

(see Gunnell and Simmons 2005, fig. 1). That is, these two clades together represented one of the current compositions of Noctilionoidea with seven families (as in Teeling et al. 2005) to the exclusion of just one family, Natalidae. Thus, either Natalidae represents a morphotype convergent with basal noctilionoids or the nataloid morphotype is basal to Yangochiroptera, with nataloid lineages basal to Vespertilionoidea (Natalidae), Noctilionoidea (Thyropteridae), and even Emballonuroidea (if *Myzopoda* is confirmed as sister to the latter, as in Amador et al. [2016]).

Most recent, comprehensive molecular studies (e.g., Amador et al. 2016; Meredith et al. 2011; Teeling et al. 2005) converge in a number of phylogenetic results (fig. 3.1):

1. The existence of a core noctilionoid clade composed of six families. These include the New Zealand endemic Mystacinidae (Australasian considering all fossils), and the Neotropical Thyropteridae, Furipteridae, Noctilionidae, Mormoopidae, and Phyllostomidae (Agnarsson et al. 2011; Amador et al. 2016; Meredith et al. 2011; Teeling et al. 2005).

2. The position of Mystacinidae. Except one study (Agnarsson et al. 2011, in which *Mystacina* was sister to Thyropteridae), most recent phylogenies recovered *Mystacina* as sister to the five Neotropical families with high support (Amador et al. 2016; Meredith et al. 2011; Teeling et al. 2005).

3. The sister relationship of Noctilionidae and Furipteridae. Previously considered a nataloid family, the diminutive Furipteridae, with just two species in two genera (*Amorphochilus* and *Furipterus*) has been consistently recovered as sister to *Noctilio*. These two families last shared a common ancestor in the latest Eocene (ca. 36 Ma; Amador et al. 2016).

4. The sister relationship of Mormoopidae and Phyllostomidae. The close connection of these two families was reflected in the traditional taxonomy in the relatively late recognition of Mormoopidae as a separate family from Phyllostomidae (not until the revision of Smith [1972]; see historical background in Simmons and Conway [2001]). Most molecular (e.g., Amador et al. 2016; Eick et al. 2005; Rojas et al.

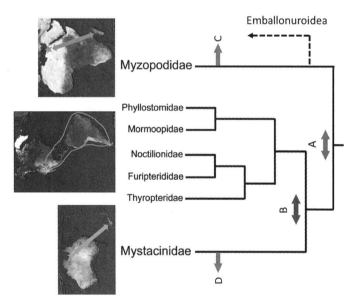

Figure 3.1. Phylogenetic and broad biogeographic relationships of noctilionoid bats. Bold branches indicate relationships according to Teeling et al. (2005). Dotted branch indicates alternative emballonuroid affinities of Myzopodidae (Amador et al. 2016). Biogeographic events discussed in the text include dispersal and dispersal followed by vicariance: (A) transatlantic dispersal, (B) trans-Antarctic dispersal followed by vicariance, (C) dispersal to Madagascar, (D) dispersal to New Zealand. Modified from Gunnell et al. (2014).

2016; Shi and Rabosky 2015; Teeling et al. 2005; cf. Agnarsson et al. 2011) and morphological (e.g., Gunnell and Simmons 2005) studies have confirmed either or both the familial status of these clades, and their sister relationship.

In turn, phylogenies differ in two major topics:

1. The inclusion of *Myzopoda* as a member of Noctilionoidea. Probably, this is the most important phylogenetic issue in the superfamily given its profound biogeographic implications (see below). This family has been considered a member of the clade Nataloidea (see above). Myzopodidae has been recovered as the oldest branch of an expanded noctilionoid clade in Teeling et al. (2005; fig. 3.1), and as sister to Vespertilionoidea in Eick et al. (2005), Meredith et al. (2011), Miller-Butterworth et al. (2007), and Van Den Bussche and Hoofer (2004). In the most recent and comprehensive molecular phylogeny of bats (Amador et al. 2016), *Myzopoda* was linked to Emballonuroidea (Nycteridae + Emballonuridae), which in turn was sister to a Noctilionoidea composed of the core six families with Mystacinidae as the most basal branch. Both Amador et al. (2016) and Meredith et al. (2011) recovered a sister relationship of Emballonuroidea and Noctilionoidea, with identical composition, which suggests that only the position of *Myzopoda* is controversial at this level. Amador et al. (2016) discussed additional evidence that may support a myzopodid connection with emballonuroids, including character complexes (Carter et al. 2008; Volleth 2013) and the recent finding of old (up to 37 my) myzopodid fossils from continental Africa (see "Biogeography" below; Gunnell et al. 2014).

2. The position of Thyropteridae. Among the Neotropical noctilionoids, thyropterids have fluctuated among studies around three main nodes: (1) sister to Mystacinidae (Agnarsson et al. 2011; Eick et al. 2005); (2) sister to all other Neotropical noctilionoids (e.g., Meredith et al. 2011; Rojas et al. 2016); and (3) sister to the furipterid—noctilionid clade (fig. 3.1; Amador et al. 2016; Hoofer et al. 2003; Teeling et al. 2005). At present, the first hypothesis of

thyropterid affinities can be safely disregarded as it was not recovered in any more recent comprehensive phylogeny (e.g., Amador et al. 2016; Shi and Rabosky 2015), but it remains difficult to decide between the other two hypotheses.

Biogeography

Noctilionoid bats are characteristic of the southern continents. At an extremely broad paleogeographic scale (Gondwana vs. Laurasia), the first reconstruction of the biogeographic history of the superfamily in the frame of its current phylogenetic composition recognized a wide Gondwanan distribution with a single origin from Laurasia dated ca. 52 Ma (Teeling et al. 2005). However, a major ambiguity emerged when only a slightly finer biogeographic scale was used in Teeling et al. (2005), specifically by mapping continents (instead of supercontinents) as biogeographic units, even when this was more sensible given the advanced stage of major landmass breakup by the time of bat origins, shortly after the K-Pg boundary (see Teeling et al. 2005). This equivocal assignment of biogeographic pattern at the continental level involved myzopodids and mystacinids but, in fact, comprised the backbone of the entire yangochiropteran subtree (see Teeling et al. 2005, fig. 3). At present, no definitive hypothesis about the biogeographic history of noctilionoids has arisen. Here we examine alternative biogeographic scenarios in the light of current phylogenies and the fossil record.

On the basis of a phylogenetic position nested in Laurasiatheria (e.g., Meredith et al. 2011; O'Leary et al. 2013), and the age and location of oldest fossils (ca. 55 Ma in Great Britain and France), bats likely originated in Western Europe (Hand, Sigé et al. 2015; Smith et al. 2012). Bats rapidly spread across the globe: fossil evidence indicates that by the Early to Middle Eocene (Ypresian–Early Lutetian ages, 56–46 Ma) bats were present in all the major landmasses they occupy today (Smith et al. 2012), thus becoming the first truly cosmopolitan placental group. In this process, bats must have crossed considerable oceanic spans to reach distant localities where they were found as fossils. For instance, Vastan bats (Early Eocene of India) belong to four typically European bat families (Archaeonycteridae†, Hassianycteridae†, Icaronycteridae†, and Palaeochiropterygidae†), plus one family indeterminate (indet.); assuming a Laurasian origin (see above), these bats must have crossed the (at that time) wide open Tethys Sea to reach the vagrant, northward-moving Indo-Pakistan subcontinent (see Smith et al. 2012). Likewise, bats reached other major isolated landmasses between the Early to Middle Eocene: South America (taxon indet.; Tejedor et al. 2005), Africa (*Tanzanycteris mannardi†*, included in a family of its own by Gunnell et al. [2003] but now considered a hipposiderid; G. Gunnell pers. comm.) and Australia (*Australonycteris clarkae†*, family indet.; Hand et al. 1994), and some hypotheses predict their presence in Antarctica by that time (see below). Thus, transoceanic dispersal was a common biogeographic process in bats, and the available evidence indicates that this process continues to occur today (e.g., megabats storm blown to New Zealand from Australia in historical times; Daniel 1975). Therefore, it is theoretically realistic to propose biogeographic scenarios for bats as a group on the basis of intercontinental dispersal. The alternative biogeographic process, vicariance, has also been invoked to explain the fascinating biogeography of noctilionoid bats (see below).

Several lines of evidence indicate that all southern major landmasses represent the geographic framework for the noctilionoid diversification (Gunnell et al. 2014). Two different kinds of biogeographic hypotheses can be derived from the known distribution of noctilionoid bats, both fossil and extant. In one hypothesis, myzopodids are the basal noctilionoid bats and historically occupied one major southern landmass, Africa (Gunnell et al. 2014). With mystacinids as the next branch splitting from the noctilionoid backbone, present in New Zealand and in Australia as fossils (see below), the role of eastern Gondwanan landmasses in the origin of the noctilionoid diversification becomes evident, followed by a westward dispersal to South America via Antarctica (Gunnell et al. 2014). This hypothesis is depicted in figure 3.1; in support of this hypothesis is the temperate wet-forest habitat continuity across southern continents during the Paleogene (Hand, Sigé et al. 2015; see below). The most likely source of this dispersal (and dispersal to New Zealand) probably is the Australian continent, where old mystacinids are found (Hand et al. 1998, 2005). This trans-Antarctic route would appear exceedingly rare, but certainly it would not be unique among mammals and other vertebrates. For instance, Australasian marsupials have been proposed to originate from dispersal in the oppo-

site direction, from South America to Australia (Beck 2008), but recent phylogenetic findings suggest that one part of the Australidelphian clade, the microbiotheriid lineage that includes the living *Dromiciops*, may represent a back dispersal to South America, the only continent where the group has been recorded (Beck et al. 2008). Interestingly, palaeognath birds show a similar biogeographic pattern with tinamous (Tinamidae, Neotropical) consistently recovered as sister to moas (Dinornithidae†, Zealandia; see Mitchell et al. 2014; Phillips et al. 2010). Therefore, microbiotheres and palaeognaths constitute additional cases that make the westward noctilionoid dispersal route very plausible (but see below).

An alternative hypothesis is an African origin of noctilionoids, as suggested by Lim (2009) and based on *Myzopoda* being sister to the remaining families. Lim (2009) reconstructed Africa as the area of origin of Noctilionoidea and suggested a first transoceanic dispersal from Africa to South America. Next, this lineage dispersed to Australia via Antarctica following the eastward route more commonly reconstructed for australidelphian marsupials (Lim 2009). Hand et al. (2001, 2007) suggested that the vicariant event that separated Australia from Antarctica/South America ca. 35 million years ago caused the isolation of the mystacinid lineage in Australia, and from Australia mystacinids dispersed to Zealandia during the Miocene, where they survive today.

The second biogeographic hypothesis is one that excludes myzopodids because they are related to vespertilionoids (e.g., Meredith et al. 2011), or to emballonuroids (Amador et al. 2016), or to Yangochiroptera (Hoofer et al. 2003). This is in line with the suggestion of Hoofer et al. (2003) of an Old World origin of myzopodids and a New World origin of Noctilionoidea, and it reduces the speculation about the biogeography of basal noctilionoids to one event: the split between Neotropical and Australasian noctilionoids from a wide southern distribution. Independent diversification followed after the separation of Australia from Antarctica/South America (Hand et al. 2001, 2007). Interestingly, Amador et al. (2016) recovered *Myzopoda* as sister to emballonuroids, and this whole clade sister to noctilionoids. This suggests an Atlantic-based split, this time between emballonuroids (inclusive of *Myzopoda*) and noctilionoids (fig. 3.1).

Overview of Noctilionoid Families

The six, or seven, extant noctilionoid families (plate 1) are characterized by high diversity in terms of taxonomy, morphology, behavior, feeding and roosting ecology, and mating system (table 3.1). As a group, noctilionoids exhibit some of the most striking morphological adaptations among mammals (e.g., nose-leaves, suction disks, additional ear structures, greatly elongated hindlimbs, greatly reduced fingers, extensive facial integumentary folds), and some extraordinary habits (e.g., convergently evolved quadrupedal locomotion, roosting by adhesion, trawling, leking behavior). Some of these families are ancient, with molecular divergence time estimates that go back to the Eocene for stem groups; however, some crown-group estimates are considerably younger, even recent (table 3.1), and constitute evidence of an ongoing diversification process. The fossil record is poor, with few or no specimens of Eocene or Oligocene age, and a scant representation

Table 3.1. Aspects of evolutionary and ecological diversity in noctilionoid families, from diverse sources

Family	Exclusively fossil genera	Recent genera	Recent species	Body mass range (g)	Geographic distribution	Oldest fossil record	Stem age (Ma)	Crown age (Ma)	Dietary items	Foraging behavior	Natural roosting sites	Mating system
Myzopodidae	1	1	2	8–9	Madagascar (fossils in continental Africa)	Latest Eocene	53	1	Insects	Aerial hawking	Unrolled leaves	Harem polygyny
Mystacinidae	2	1	2	10–40	New Zealand (fossils in Australia)	Oligocene-Miocene boundary	49	?	Terrestrial invertebrates, insects, pollen, nectar, fruit, wood	Ground search, aerial hawking, gleaning	Tree cavities, burrows, caves (in extinct populations)	Resource-based lek
Thyropteridae	1	1	5	3–6	Tropical C and S America	Miocene	45	14	Insects	Gleaning, aerial hawking	Unrolled leaves, dead foliage	Harem polygyny
Furipteridae	0	2	2	3–5	Tropical C and S America	Pliocene	36	?	Insects	Aerial hawking, gleaning	Cavities, hollow logs, caves	?
Noctilionidae	0	1	2	30–60	Tropical C and S America, Antilles	Miocene	36	4	Insects, fish, shrimp	Trawling, aerial hawking	Cavities, tree holes, caves	Harem polygyny
Mormoopidae	1	2	19	8–23	Subtropical and Tropical N, C and S America, Antilles	Oligocene	40	35	Insects	Aerial hawking	Caves, hot caves	Harem polygyny
Phyllostomidae	2	60	216	4–235	Subtropical and Tropical N, C and S America, Antilles	Miocene	40	35	Insects, blood, vertebrates, nectar, pollen, fruit, leaves	Gleaning, aerial hawking, trawling	Foliage, foliage tents, roots, tree cavities, caves, Hot caves	Harem polygyny
Speonycteridae†	1	0	0	?	Florida	Oligocene	?	30	Insects	?	Caves	?

Note: Point estimates of stem age and crown age from Amador et al. (2016). Stem age is the age at which the family split from its sister group; crown age is the age at which the extant species of a family last shared a common ancestor. Abbreviations: g = grams, Ma = million years before present, C = Central, N = North, S = South.

in younger geological periods (table 3.1). This is most striking in the most diverse family, Phyllostomidae, with 60 extant genera and just two exclusively fossil genera recorded to date (table 3.1): the 12–13 Ma old, Middle Miocene (Laventan) Phyllostominae *Notonycteris*† and *Palynephyllum*†, originally placed in Glossophaginae (Morgan and Czaplewski 2012) but recovered in Lonchophyllinae by Dávalos et al. (2014). However, the fossil record has contributed key pieces to understand the evolutionary biogeography of noctilionoids, particularly the finding of family members in continents they do not inhabit today, but may have been their land of origin, as in the cases of Myzopodidae, Mystacinidae, and Mormoopidae (see above).

As a group, these families exploit an extraordinary variety of food resources (table 3.1). Among known mammalian diets, only sap feeding (present in some primates and megabats) and mycophagy (consumption of fungi typical of some marsupials) appears to be absent, but fungal spores have been detected in abundance in stomach contents of *Mystacina* (Carter and Riskin 2006). While some feeding habits that are common among other mammalian groups are rare among noctilionoids (herbivory, granivory), at least one feeding habit is unique among mammals (hematophagy), and several habits are more highly specialized in noctilionoids than in most other mammalian groups (frugivory, nectarivory).

Echolocation is one major source of evolutionary variation in noctilionoids; many echolocation call characteristics vary considerably across noctilionoid families, including duty cycle, call type and intensity, pulse duration and repetition rate, and harmonic characteristics. However, call variation may have originated from basic low duty-cycle, broadband, frequency-modulated, downward-sweep, multiharmonic, short-duration, high-intensity calls emitted through the mouth, with a distinct terminal phase, because the majority of, as well as the most basal, families observe some or all of these characteristics. Departures from this echolocation scheme are quite remarkable in the group. For instance, phyllostomids are characterized by low-intensity calls emitted directionally through the noseleaf and used to detect and glean stationary food items, so these calls generally lack a terminal buzz (but see Weinbeer and Kalko 2007). Species of the *Pteronotus parnellii* complex emit calls with a long constant-frequency segment (Macías et al. 2006). The latter is a high-duty

cycle, Doppler-compensated echolocation system used to detect fluttering targets (e.g., flying large-winged insects such as lepidopterans) and its architecture and functionality represents a remarkable case of convergence with calls emitted by Old World rhinolophoid bats (Macías et al. 2006 and citations therein). Gleaning noctilionoids, and phyllostomids in particular, use echolocation calls integrated with visual, non-ultrasonic acoustic, and olfactory cues to handle food items (e.g., Kalko and Condon 1998; Korine and Kalko 2005). Flexibility is frequent, and diverse noctilionoid species are able to greatly modify their typical call parameters (e.g., phyllostomids are known as "whispering bats" because of their low-intensity calls, but loud calls have been detected in different species; Brinkløv et al. 2009). Below is an account of noctilionoid families in which we deal with various aspects of their history, ecology, and behavior in more detail.

Myzopodidae

Myzopodids are African bats presently restricted to Madagascar but also recorded from the Early Pleistocene of Tanzania (*Myzopoda africana*† from the Olduvai Gorge; Gunnell et al. 2015) and the Eocene-Oligocene transition of Egypt (*Phasmatonycteris phiomensis*† and *P. butleri*† from the Fayum Depression; Gunnell et al. 2014). Myzopodids represent an ancient lineage that likely originated in the Early Eocene (dated ca. 53 Ma by Amador et al. 2016). Extant species are similar and include the eastern and western suckerfooted bats, *M. aurita* and *M. schliemanni*, respectively, which are allopatric Malagasy endemics (Goodman et al. 2007). However, genetic analyses detected four well-supported (but unresolved) clades among populations of *Myzopoda* (Russell et al. 2008).

Sucker-footed bats are small echolocating bats that average 8.6 g in *M. aurita* and 9.9 g in *M. schliemanni* (Goodman et al. 2007) and exhibit an extraordinary array of uniquely derived morphological characters including a mushroom-shaped process in the ear base, long funnel-shaped pinnae, and large, sessile pads in the thumb and foot that allow adhesion to large leaves with smooth surfaces that are chosen as roosting sites (Goodman et al. 2007). Partially uncoiled leaves of a lowland wet forest pioneer plant, the traveler's tree, *Ravenala madagascariensis* (Strelitziaceae), are used frequently to roost in a head-up position; the bats stick to the leaf with the thumb and sole adhesive disks (Ralisata et al. 2015) and move up the leaf using the stiff tail as a prop (observed in captivity; Göpfert and Wasserthal 1995). This represents a remarkable convergence with the New World disk-winged bat *Thyroptera tricolor* (see below). However, *Myzopoda* species secrete a substance enabling a wet-adhesion mechanism (Riskin and Racey 2010), in contrast to the suction mechanism demonstrated in thyropterids (see below). Individuals are able to use unfurling *Ravenala* leaves for just a few (1–12) days and then need to switch roosts; up to 36 individuals were found in a single leaf roost (Ralisata et al. 2015). The elevational ranges (below 1,000 masl) of *Myzopoda* and *Ravenala* overlap (see Goodman et al. 2007). The association of *Myzopoda* and *Ravenala* and potentially other smooth-leaved plants, such as the palm *Raphia* (Arecaceae) and palmlike *Pandanus* (Pandanaceae), indicates the use of second-growth understory in forested, even degraded habitats (Goodman et al. 2007). *Myzopoda* may have benefited from deforestation and regrowth of extensive secondary forests. Both species are more common than previously thought (Goodman et al. 2007), so in terms of conservation, species of *Myzopoda* have been recategorized as "least concern" (Jenkins et al. 2008a, 2008b).

Echolocation in *Myzopoda aurita* consists of frequency-modulated (FM), long duration (up to 23 ms) calls with repetitions spaced up to 108 ms; two to four harmonics were present; the second harmonic was the strongest and decreased from 42 to 24 KHz in a rather shallow FM sweep likely used to detect fluttering targets, ending by a steep FM sweep (Göpfert and Wasserthal 1995). Echolocation call structure suggested that *Myzopoda* species are aerial hawking bats that capture soft-bodied insects. This was also suggested by fecal samples of a single captured individual of *M. aurita* that contained remains of microlepidopterans (Göpfert and Wasserthal 1995) and was confirmed with more substantial dietary data from *M. schliemanni* showing a predominance of Blattaria and Lepidoptera and few other insect remains (Rajemison and Goodman 2007).

Mystacinidae

Mystacinid bats have been restricted to New Zealand since the Miocene (Hand et al. 2013; Hand, Lee et al. 2015), but as a group they have their roots in ancient Australia (Hand et al. 1998, 2005). Four taxa, including species of *Icarops*† and one mystacinid indet., are known from Riversleigh, Queensland, and other localities widespread in Australia, spanning an age range from 26 to 12 Ma (latest Oligocene–Middle Miocene; Hand et al. 1998, 2005). In New Zealand, together with two small mystacinids indet., the monotypic *Vulcanops*†, and the first records of *Mystacina* (*M. miocenalis*†) are from Middle Miocene age (19–16 Ma; Hand et al. 2013, 2018; Hand, Lee et al. 2015). Pleistocene records are from cave deposits (Worthy and Holdaway 2002), a type of roost unknown in extant *Mystacina* (see Carter and Riskin 2006).

A single species of *Mystacina*, *M. tuberculata*, survives today; another recent, larger species, *M. robusta*†, presumably went extinct in the decade of the 1960s (Lloyd 2001). *Mystacina tuberculata* inhabits four islands of New Zealand, including both large north and south islands; its local distribution is highly correlated with the distribution of large, continuous, native forest stands, where roosting, foraging, and breeding occurs (Carter and Riskin 2006). Six genetically distinct lineages have been detected, only roughly corresponding with the three named subspecies based on morphology; genetic groups support range-expansion population hypotheses following forest recovery from catastrophic volcanic events and fast recovery from refugia following glacial periods (Lloyd 2003). In spite of its isolated, endemic status, and its declining population trend (classified as "vulnerable" by the IUCN; O'Donnell 2008), *M. tuberculata* is one of the best-known bats generally (see recent reviews in Carter and Riskin 2006; Hand et al. 2009; Hand, Lee et al. 2015; Lloyd 2001; Lloyd and Tucker 2003).

Mystacina is a specialized quadrupedal bat (Schutt and Simmons 2006). Progression on land by *M. tuberculata* has been kinematically described as a symmetrical lateral-sequence walk (Riskin et al. 2006). *Mystacina* uses the same gait at all speeds (up to 0.95 m s^{-1}; Carter and Riskin 2006). These bats echolocate while walking but there is no energy-saving biomechanical link between call emission and gait (Parsons et al. 2010). Many unique aspects of morphology correlate with a terrestrial lifestyle, including a reduced propatagium, a wing-folding socket in the plagiopatagium, a basal talon in each claw, and robust and/or highly mobile postcranial elements (Carter and Riskin 2006; Hand et al. 2009). Adaptations to terrestrial locomotion are ancient and already appear in postcranial remains of the

Miocene Australian mystacinid *Icarops aenae*† (Hand et al. 2009). Equipped this way, *Mystacina* is the only extant bat that searches for food predominantly on the ground, where most dietary items are obtained. Still, intermediate aerodynamic features allow *Mystacina* to forage aerially, glean arthropods off vegetation, and commute between forested foraging sites across open land (Carter and Riskin 2006).

While most bats are remarkably specialized in one broad food type (e.g., insects), *Mystacina tuberculata* is a true omnivore: its diet includes ground-dwelling invertebrates, aerial insects, fruits, nectar, and pollen, and also carrion in *M. robusta* (Lloyd 2001). Species of *Mystacina* also ingest wood; ferns and fungi may be also part of their diet (Carter and Riskin 2006). *Mystacina* is a flexible forager and uses olfaction, passive listening, and echolocation to capture prey (Jones et al. 2003). These bats dive into leaf litter and burrow in search of terrestrial invertebrates; aerial insect capture and gleaning off vegetation and the ground are also part of *Mystacina*'s hunting capabilities (Carter and Riskin 2006). The dentition of *Mystacina* is typical of insectivorous bats, but a greater allocation of tooth area at the anterior end of the tooth row is similar to that of frugivorous bats (Hand et al. 2009). *Mystacina* forages for fleshy diaspores, including succulent bracts and true fruits from plants of at least three families: Pandanaceae, Asteliaceae, and Elaeocarpaceae (Daniel 1979; Lloyd 2001). Flower products, including nectar, pollen, and flower parts, are taken from many plants, principally those in the Asteliaceae, Balanophoraceae, Myrtaceae, and Proteaceae (Daniel 1979; Lloyd 2001). Nectar feeding is facilitated by reduction of the crowded lower incisors so that the tongue, extensible and with a brush of long papillae on the tip, protrudes between the long lower canines (Daniel 1976). *Mystacina* is an important pollinator of the wood rose *Dactylanthus taylorii* (Balanophoraceae), an endangered terrestrial parasitic plant (see Wood et al. 2012).

As is typical in temperate bats, *Mystacina* uses torpor for up to 120 h in response to low ambient temperatures; however, this depends on roosting conditions: individuals in communal roosts only rarely enter torpor, while all individuals roosting solitarily enter torpor during winter (Czenze et al. 2017). Roosting sites are in the forest interior, a temperate rainforest habitat characterized by large native trees covered with epiphytes and with deep leaf litter (Lloyd 2001). Communal roosts

are located in large tree cavities and can harbor up to 6,000 bats of both sexes (Lloyd 2001). These cavities may be extensively modified by chewing the wood (Carter and Riskin 2006). Roosting behavior is highly variable seasonally, and solitary roosting is frequent; in the case of solitary roosting, small tree cavities are preferred, but large tree holes and even burrows can be used (Carter and Riskin 2006). Males of *M. tuberculata* attract females during the breeding season from "singing roosts" in which aggregations of males constitute true leks that correspond to the "resource-based" lek model, in which the resource is the communal roosting site used by gregarious females (Toth et al. 2015). The vocalizations are audible to humans but they also have ultrasonic components (Toth and Parsons 2013).

Mystacina is a low-duty-cycle echolocating bat. In *M. tuberculata*, calls are multiharmonic, with one to three harmonics and occasionally a fourth, faint harmonic, a fundamental harmonic with the most energy, which characteristically sweeps, depending on the harmonic (Parsons 2001). Short-duration FM calls are repeated at 12.4–3.8 Hz, with frequencies averaging at 27, 48, and 78 kHz for first, second, and third harmonics, respectively (see Carter and Riskin 2006 and citations therein). Bats of this species use echolocation calls to approach prey, but they also use passive listening and olfaction with ground prey (Carter and Riskin 2006; Jones et al. 2003). Echolocation is also used for navigation between and within forested habitat, usually when bats are flying less than 2 m above the ground (Lloyd 2001).

Thyropteridae

Disked-winged bats are sister to a clade that includes all Neotropical noctilionoids, all sharing a most recent common ancestor ca. 42.1 Ma (Rojas et al. 2016), or sister to the furipterid-noctilionid clade from which it may have split ca. 45.3 Ma (Amador et al. 2016). Five species are currently recognized in the only extant genus of the family: *Thyroptera devivoi, T. discifera, T. lavali, T. tricolor,* and *T. wynneae* (Velazco et al. 2014). *Thyroptera discifera* and *T. tricolor* are relatively common and widely distributed in tropical Central and South America, whereas the other three species are rare and occur only in South America (Velazco et al. 2014). Fossil remains of *T. lavali* and *T. tricolor* are known from the Miocene La Venta deposits in

Colombia (Czaplewski 1996, 1997; Czaplewski and Campbell 2017; Czaplewski et al. 2003). Additionally, one fossil genus and species (*Amazonycteris divisus†*) was recently described from an isolated upper first molar found in upper Miocene deposits in western Amazonia, Brazil (Czaplewski and Campbell 2017). Crown clade Thyropteridae was dated ca. 14 Ma (based on three species), with a stem clade dated ca. 45 Ma (Amador et al. 2016). Like myzopodids, thyropterid species are characterized by the presence of adhesive disks on the sole of the foot and base of the thumb, but anatomical and evolutionary studies concur that the adhesive disks in these two groups evolved convergently (Rojas et al. 2016; Schliemann 1970, 1971). Thyropterid species can be grouped into two categories based on the roost they use: roosting on half-unrolled new leaves of large understory monocots (*T. tricolor*) or roosting under dead leaves or palm fronds (remaining species; Velazco et al. 2014). *Thyroptera tricolor* roosts are suitable for suction-based attachment in which the bat attaches head-up plant species known to be used as roosts include *Heliconia* (Heliconiaceae), *Musa* (Musaceae), *Calathea* (Marantaceae), and *Phenakospermum* (Strelitziaceae; Reyes-Amaya et al. 2016; Velazco et al. 2014). The remaining species roost on the rough surfaces of dried leaves of *Musa*, *Cecropia* (Cecropiaceae), or palm fronds (Arecaceae; Velazco et al. 2014), and probably have to use their claws to attach. Since these are downward-opening roosts they probably hang head down (Riskin and Fenton 2001; Velazco et al. 2014).

Thyropterids are small (3–6 g) gleaning insectivores that, in the case of *T. tricolor*, have to switch roosts every night because the unfurling new leaves rapidly lose the small-diameter conical form these bats prefer (Vonhof and Fenton 2004). Because of the need to switch roosts daily, *T. tricolor* choses areas where there is a high density of leaves (Vonhof et al. 2004). Although using ephemeral roosts, *T. tricolor* form cohesive multiyear groups of approximately five to six individuals (Chaverri 2010; Chaverri and Gillam 2015; Findley and Wilson 1974). In order to maintain the cohesiveness of these groups, *T. tricolor* has developed a complex communication system that allows them to locate each other during flight or while searching for roosts, including the use of leaves as acoustic horns to increase sound amplification of the incoming and outgoing social calls to locate roosts and group members (Chaverri and Gillam 2010, 2013, 2015; Chaverri et al. 2010). In the fam-

ily, the call structure is known only from *T. discifera* and *T. tricolor*. In *T. tricolor*, two kinds of calls are known, "inquiry" calls, emitted mostly with a frequency within the 20–25 kHz range by flying bats looking for roosts or mates, and "response" calls, emitted mostly with a frequency within the 50–60 kHz range by individuals that already have encountered a roost in response to an inquiry call (Chaverri and Gillam 2013). In *T. discifera*, there are broadband FM calls with single calls consisting of the fundamental frequency (50 kHz) plus two upper harmonics (at 100 and 150 kHz; Tschapka et al. 2000).

Furipteridae

Bats of the Neotropical family Furipteridae are sister to the clade containing bats of the family Noctilionidae, both sharing a most recent common ancestor ca. 36.1 Ma (Amador et al. 2016; Rojas et al. 2016). Two species are currently recognized in the two monotypic genera: *Amorphochilus schnablii* and *Furipterus horrens* (Gardner 2008). These two species do not occur in sympatry. *Amorphochilus schnablii* is relatively common and is distributed west of the Andes from southern Ecuador through Peru to northern Chile (Gardner 2008). *Furipterus horrens* is rare but widely distributed in tropical Central and South America (Gardner 2008). Both species are represented in the fossil record; *Amorphochilus schnablii* is known from Quaternary deposits in northwestern Peru (Morgan and Czaplewski 1999) and *F. horrens* is known from Quaternary deposits in Brazil (Czaplewski and Cartelle 1998; Salles et al. 2014). Furipterids are small (3–5 g) bats characterized by having a blunt muzzle with anteriorly directed nostrils, short, triangular tragus, functional abdominal mammae, and a reduced thumb, which is enclosed in the wing membrane and bears a minute claw (Gardner 2008; Simmons and Voss 1998). Despite their similar diet, with both species feeding specifically on moths, these two species exhibit contrasting natural history characteristics. *Amorphochilus* inhabits arid regions throughout its distribution and prefers to roost in abandoned human constructions, culverts, and caves where colony sizes range from a few individuals to up to 300 (Gardner 2008; Ibáñez 1985). *Furipterus*, in contrast, occurs in lowland tropical rainforests in Amazonia and Central America, where it roosts in or under fallen trees (LaVal 1977; Voss et al. 2016), but it can be found also roosting in caves, tunnels, or human constructions (Fabián 2008; Uieda et al. 1980). The colony sizes range from a single individual up to 250 individuals (Uieda et al. 1980).

Furipterids are gleaning insectivores and their call structure is quite similar. Both species are characterized by having calls of short duration of extremely high frequency (70–210 kHz) and downward broadband FM calls (Falcão et al. 2015). The echolocation call of *F. horrens* is one of the highest in frequency among Neotropical bats (Falcão et al. 2015).

Noctilionidae

This family last shared a common ancestor with its sister Furipteridae ca. 36 Ma (Amador et al. 2016). Noctilionidae is composed of just two currently recognized species, the lesser (*Noctilio albiventris*) and greater bulldog, or fisherman (*N. leporinus*) bats (Simmons 2005), which diverged ca. 4 Ma (Amador et al. 2016). Substantial genetic variation exists in populations of *N. albiventris* across the Neotropics (Lewis-Oritt et al. 2001). The two species are abundant and widely distributed in tropical South America and most of Central America, with the greater bulldog bat also reaching Mexico and extending over most of the Caribbean (Hood and Jones 1984; Hood and Pitocchelli 1983). These species are specialized in aquatic environments where they predominantly feed over streams and other water bodies (Hood and Jones 1984; Hood and Pitocchelli 1983). These bats roost in mixed-sex groups chiefly in hollow trees but also in caves (Hood and Jones 1984; Hood and Pitocchelli 1983).

Both *Noctilio* species feed on insects (Hood and Jones 1984; Hood and Pitocchelli 1983) but *N. leporinus* adds crustaceans and a great proportion of both pelagic and freshwater fish on a seasonal basis (Brooke 1994). The fishing habit is derived with respect to insectivory (Lewis-Oritt et al. 2001) and confirms that carnivory in bats (the habit of species eating vertebrates) can be seen as an extreme of a gradient of animalivory (Giannini and Kalko 2005). *Noctilio* species are flexible foragers able to use aerial hawking to capture insects (Brooke 1994), but morphologically and behaviorally they are among the most highly specialized trawling bats; that is, they collect prey from the water surface using their hind feet (Brooke 1994). These adaptations are more pronounced in *N. leporinus*, which

exhibits greatly elongated hindlimbs, feet, digits, claws, and calcar, which all function in trawling behavior: the calcar retracts the uropatagium and the hindlimbs are deployed so that the feet trawl the water surface and are elevated when contacting prey (Schnitzler et al. 1994). Large prey such as fish are transferred to cheek pouches and consumed in a night roost (Hood and Jones 1984).

Prey capture strategies are highly flexible in bulldog bats. Both *Noctilio* species use the same repertoire of prey hunting techniques but in different frequency. *Noctilio albiventris* uses predominantly aerial hawking over water and land and initiates dips from high search flights above water, whereas *N. leporinus* initiates dips from both high and low search flights, and uses raking phases (see below), much more frequently (Kalko et al. 1998). Echolocation calls are emitted through the mouth and are used in prey detection and capture in all hunting modes, with typical search, approach, and terminal phases and with Doppler compensation in the terminal phase (Kalko et al. 1998). In both species, call structure is similar, with mixed FM and quasi-constant-frequency components variously combined depending on the specific task. As expected (see Jones 1999), the smaller *N. albiventris* produces higher frequency calls (66–72 kHz of the quasi-constant-frequency component; Kalko et al. 1998) than the larger *N. leporinus* (at 52.8–56.2 kHz; Kalko et al. 1998). Prey is detected by echo glints produced by reflection from the fish body breaking the surface and the associated disturbance (Schnitzler et al. 1994). Raking is an exploratory or memory-guided foraging bout with no specific target, in which the bat trawls the water surface continuously using search-phase echolocation calls. Foraging efficiency in the greater bulldog bat is low, with individuals catching fish in one out of 50–200 passes over a feeding spot (Schnitzler et al. 1994), but the nutritional reward of fish is likely greater than that of insects.

Mormoopidae

Mormoopids are Neotropical bats that are sister to phyllostomids, sharing a most recent common ancestor ca. 39 Ma (Amador et al. 2016; Rojas et al. 2016). Mormoopids as a group exhibit several peculiar characters such as expanded, ornate lips with narial structures fused to the upper lip, and wings joined in the body midline giving the appearance of bare-backed bats in the subgenus *Pteronotus*. Two genera are currently recognized within the family, *Mormoops* and *Pteronotus*, but additionally, one fossil genus and species from the Oligocene of Florida (USA) that is more closely related to *Mormoops* is in the process of being described (Czaplewski and Morgan 2012; Morgan and Czaplewski 2012; Pavan and Marroig 2016, 2017). Three species are recognized in *Mormoops*, two extant (*M. blainvillei* and *M. megalophylla*) and one fossil (*M. magna†*) (Pavan and Marroig 2016). All three species of *Mormoops* have a fossil record that dates back to the Pleistocene of North America, South America, and the West Indies (Czaplewski and Cartelle 1998; Fracasso and Salles 2005; Morgan 2001; Morgan and Czaplewski 2012; Salles et al. 2014; Silva Taboada 1974; Velazco et al. 2013). *Mormoops magna†* is known from Pleistocene humeral remains from Cuba and the Dominican Republic (Silva Taboada 1974; Velazco et al. 2013). Extant populations of *M. blainvillei* are distributed in the Greater Antilles and adjacent small islands (Simmons 2005), whereas *M. megalophylla* is known to occur from southern USA southward into northern South America (Rezsutek and Cameron 1993). Seventeen species are currently recognized in *Pteronotus*, 17 extant and 2 fossil (Pavan and Marroig 2016; Pavan et al. 2018; Van Den Hoek Ostende et al. 2018). *Pteronotus trevorjacksoni†* and *P. pristinus†* are the only two known fossil species in the genus; they have been recovered in Pleistocene deposits in Jamaica (Van Den Hoek Ostende et al. 2018) and in Cuba and Florida (USA; Morgan 1991; Morgan and Emslie 2010; Silva Taboada 1974), respectively. The combined distributional range of the extant species of *Pteronotus* extends from Mexico southward into South America to Bolivia and Central Brazil (Patton and Gardner 2008; Pavan and Marroig 2017; Reid 2009; Simmons 2005). The systematics and taxonomy of the different species of *Pteronotus* have improved greatly over the past decade, which reflects higher species diversity than previously thought (see Clare et al. 2013; Guevara-Chumacero et al. 2010; Gutiérrez and Molinari 2008; Pavan and Marroig 2016, 2017; Thoisy et al. 2014). Pavan and Marroig (2017) estimated that *Pteronotus* originated approximately 16 Ma and that most of the extant diversity is the result of cladogenetic events likely caused by Pleistocene sea-level variations and climatic oscillations. Mormoopid bats roost exclusively in large hot caves or abandoned mine shafts, with colonies ranging from a few individuals (solitary individuals are also recorded) up to 800,000 (Bateman and Vaughan 1974; Lancaster and Kalko 1996; Mancina 2005; Rezsutek and Cameron 1993).

Mormoopids are insectivorous species that use aerial hawking to capture aerial insects (Patton and Gardner 2008; Reid 2009). Echolocation calls among mormoopids exhibit a great range of variation (Macías et al. 2006). All mormoopid bats, except species of the *P. parnellii* complex, use low-duty cycle echolocation calls. *Mormoops* echolocation calls are characterized by a short (<4 ms), steep downward FM sweep, while most species of *Pteronotus* (e.g., *P. macleayii, P. quadridens*) have calls characterized by a short constant-frequency segment followed by a downward FM sweep (see Macías et al. 2006, Mancina et al. 2012). Maximum frequency of the main harmonic varies from 64 to 82 kHz in sympatric Cuban mormoopids (Mancina et al. 2012). Species of the *P. parnellii* complex are the only ones in the New World that use high-duty cycle echolocation calls (Smotherman and Guillén-Servent 2008). These high-duty cycle echolocation calls employ a Doppler-shift compensation to separate emitted pulse and received echo in frequency (Herd 1983). Call structure in the *P. parnelli* complex is radically different from other mormoopids and noctilionoids in general, with an initial short upsweep, followed by a long (ca. 15 ms) constant-frequency component at ca. 60 kHz, and finished with a long terminal FM sweep (Mancina et al. 2012). These calls are used for detecting the wingbeat of insects and are remarkably similar to Old World high-duty cycle bats.

Speonycteridae†

Speonycterid bats belong to the only currently recognized noctilionoid fossil family (Czaplewski and Morgan 2012). The two species in this family are known from three localities in Florida. *Speonycteris aurantiadens†*, known from upper and lower dentition, dentaries, and a proximal fragment of a left femur, has been found in localities in the Late (Arikareean) and Early (Whitneyan) Oligocene, whereas *S. naturalis†* is only known from a single upper molar from the Crystal River Formation (Whitneyan), Early Oligocene (Czaplewski and Morgan 2012). Phylogenetic analysis found *S. aurantiadens†* to be sister to the clade including Noctilionidae, Mormoopidae, and Mystacinidae (Czaplewski and Morgan 2012).

Evolutionary Significance of the Noctilionoid Diversification

Noctilionoid bats represent one of the greatest mammalian radiations. These bats are characteristic of the southern continents, and several controversial aspects remain to be resolved in order to fully appreciate their evolutionary history. While phylogenetic patterns tend to stabilize in several key nodes, biogeographic patterns are open to debate around the controversial relationships that await resolution. However, nothing prevents us from fully appreciating the magnificent potential for evolutionary change and adaptation in noctilionoid bats.

Of particular interest is the case of *Mystacina*, a singular taxon that emerges as a key evolutionary and biogeographic natural experiment. Many rare and/or convergent features shared with unrelated bat species are found in *Mystacina*. Harem polygyny is the mating system exhibited by the vast majority of bats, including noctilionoids (Toth and Parsons 2013 and references therein); *Mystacina* is one of only two bat species (the other being the African pteropodid *Hypsignathus monstrosus*) confirmed to exhibit lek breeding behavior (Toth et al. 2015).

Mystacina is the only fully temperate bat in the world capable of pollinating flowering plants (Cummings et al. 2014). As an ecological interactor, *Mystacina* has strong and unusual links with both plants and animals. The sequential flowering of the plants most frequently visited by *Mystacina* suggests that their phenology has been shaped by the interaction with this bat (Cummings et al. 2014). Also as a pollinator, *Mystacina* shares the flower products of the terrestrial wood rose *Dactylanthus* (see above) with an unusual potential competitor, the rare and critically endangered flightless parrot, the kakapo (*Strigops habroptilus*, Strigopidae; see Wood et al. 2012). Other ecological interactions may extend into foraging habits unexplored by other bats, as suggested by the presence of abundant fern and fungal spores in stomach contents (Carter and Riskin 2006). Perhaps only the diet of *Phyllostomus hastatus* (Phyllostomidae) among all bats, which includes insects, vertebrates, nectar, pollen, and fruit (reviewed in Giannini and Kalko 2005) is nearly as broad as that of *Mystacina tuberculata*.

Extraordinary habits may require extraordinary morpho-functional adaptations. Apart from numerous apomorphic character states, a suite of traits evident in *Mystacina* make it one of just three extant bat genera, together with *Desmodus* (Phyllostomidae) and *Cheiromeles* (Molossidae), that independently evolved the capability of specialized quadrupedal locomotion (Schutt and Simmons 2006). Much of the feeding behavior and resource exploitation in this genus depends on its terrestrial habits. Still, *Mystacina* and the other two quadrupedal bat genera retained seemingly unaltered flight capabilities, in spite of two of them being island-dwelling forms (*Mystacina* and *Cheiromeles*). By contrast, birds have frequently, almost routinely evolved flightlessness, particularly on islands (e.g., Slikas et al. 2002) including New Zealand, most notably the moas (Dinornithidae) and the kakapo (Strigopidae).

Interestingly, *Mystacina*'s roosting habits in tree cavities are typical of volant species, and its mating system depends on roost choice (see above). *Mystacina*, however, is inextricably linked to its temperate rainforest environment from ancient times. As compellingly stated by Hand, Lee et al. (2015, 1), "The majority of the plants inhabited, pollinated, dispersed or eaten by modern *Mystacina* were well-established in southern New Zealand in the early Miocene." This environment corresponds to the Austral temperate forest dominated by *Nothofagus* (Nothofagaceae), *Podocarpus* (Podocarpaceae), and other characteristically southern plants that are distributed today around subantarctic landmasses, including Australia, New Zealand, South America and Antarctica, and are remarkably absent from Africa (Hill 2001). This environmental continuity across continents supports biogeographic hypotheses of Gondwanan origins and transcontinental vicariance and/or dispersals of noctilionoids (see above). However, it does not explain the nontrivial lack of noctilionoid bats in Patagonia, which shares the same environment with Zealandia, and where other Gondwanan taxa are present; for example, australidelphians, represented by microbiotheres (most notably by the extant *Dromiciops*) and tinamous (see above).

Basal noctilionoids went extinct in their core Gondwanan areas (myzopodids in continental Africa, mystacinids in Australia) and survived only as island relics in Madagascar and Zealandia. Traditional hypotheses that explain these biogeographic patterns, primarily referring to mystacinids, involved evolution in competition-free environments; the recent finding of one fossil non-volant mammal of uncertain affinities in the Miocene of Zealandia (Worthy et al. 2006) slightly weakens this perspective. In addition, the presence of postcranial adaptations to quadrupedal locomotion in Miocene mystacinids from Australia, where terrestrial mammals were abundant, led Hand et al. (2009) to discard the terrestrial mammal competition hypothesis and to suggest that loss of suitable habitat (temperate rainforest) to severe climate deterioration may have caused extinction in Australia. Likewise, climate change causing aridification in eastern Africa probably had a comparable effect on continental myzopodids (Gunnell et al. 2014).

A different history took place in the Neotropics. Here the noctilionoid lineage flourished, principally in the largest tropical rainforest area in the world, both lowland and montane, and in the complex tropical archipelago setting of the Caribbean. Five families exhibit remarkable adaptations, including astonishing variation in body mass spanning the whole range of echolocating bats (3–235 g; see chap. 8, this vol.), and the greatest dietary spectrum observed in any comparable mammalian group and contain some of the most highly specialized bats, morphologically, ecologically, and behaviorally. Neotropical noctilionoids primitively are specialized animalivores (*sensu* Giannini and Kalko 2005). One such specialization is gleaning, which was the basic foraging style of basal animalivorous phyllostomids (Baker et al. 2012; Schnitzler and Kalko 1998). And nested within animalivorous gleaners, large groups emerged that specialized in the exploitation of the formidable plant resources available in the world's most diverse rainforest. These groups included highly derived nectarivorous and frugivorous forms that paralleled the remarkably similar diversification of pteropodid bats in the Old World tropics (see Amador et al. 2016; Fleming 2005). Specialized gleaning may be the initial, key behavioral adaptation in noctilionoids that may have unleashed the tremendous diversification of phyllostomids. It is interesting to note that bats from non-noctilionoid families that invaded the New World tropics at different times (e.g., diclidurine emballonurids at ca. 32 Ma, molossids in several dispersal waves since at least 30 Ma, vespertilionids since 12 Ma; data from Amador et al. 2016) are all basically aerial hawkers (although some vespertilionids evolved trawling [e.g., *Myotis albescens*] and gleaning [e.g., *Histiotus* spp.] habits). This may have limited the diversification of some noctilionoid families, most significantly mormoopids, and may have left the niche space reserved for gleaners,

principally to derived phyllostomids. This is a highly speculative hypothesis, but it finds preliminary support when noting that non-noctilionoid gleaners such as the vespertilionid *Histiotus* (currently a subgenus of *Eptesicus*; Amador et al. 2016; see Hoofer et al. 2003) diversified mostly in para- or extratropical areas of South America where noctilionoid diversity is lowest.

Conclusions

Noctilionoid bats represent one of the largest mammalian diversifications that took place in the southern continents. Remarkable transcontinental biogeographic events are required to explain the known current and fossil distribution of noctilionoid taxa. The noctilionoid history is ancient, around 50 Ma old, and is highly heterogeneous, with immensely long ghost lineages, as well as tight, superdiverse clades that speciated in just a few million years. The most remarkable diversification in feeding habits occurred during this history; morphological correlates of dietary changes have been identified (e.g., Dumont et al. 2012) and represent some of the most significant evolutionary adaptations on record. Much remains to be investigated in this fascinating clade of southern bats.

Acknowledgments

We thank the editors Ted Fleming, Liliana Dávalos, and Marco Mello for their kind invitation to contribute this chapter. We thank one anonymous reviewer and the late Gregg Gunnell for their careful review, which greatly improved the manuscript. We are indebted to Marco Tschapka, Stuart Parsons, and Merlin Tuttle (through Merlin Tuttle's Bat Conservation) who kindly contributed their pictures for our portrait gallery in plate 1.

References

Agnarsson, I., C. M. Zambrana-Torrelio, N. P. Flores-Saldana, and L. J. May-Collado. 2011. A time-calibrated species-level phylogeny of bats (Chiroptera, Mammalia). PLOS Currents, 3:RRN1212. DOI: 10.1371/currents.RRN1212.

Amador, L. I., R. L. Moyers Arévalo, F. C. Almeida, S. A. Catalano, and N. P. Giannini. 2016. Bat systematics in the light of unconstrained analyses of a comprehensive molecular supermatrix. Journal of Mammalian Evolution, 25:1–34.

Baker, R. J., O. R. P. Bininda-Emonds, H. Mantilla-Meluk, C. A.

Porter, and R. A. Van Den Bussche. 2012. Molecular timescale of diversification of feeding strategy and morphology in New World leaf-nosed bats (Phyllostomidae): a phylogenetic perspective. Pp. 385–409 *in:* Evolutionary History of Bats: Fossils, Molecules and Morphology (G. F. Gunnell and N. B. Simmons, eds.). Cambridge University Press, Cambridge.

Bateman, G. C., and T. A. Vaughan. 1974. Nightly activities of mormoopid bats. Journal of Mammalogy, 55:45–65.

Beck, R. M. D. 2008. A dated phylogeny of marsupials using a molecular supermatrix and multiple fossil constraints. Journal of Mammalogy, 89:175–189.

Beck, R. M. D., H. Godthelp, V. Weisbecker, M. Archer, and S. J. Hand. 2008. Australia's oldest marsupial fossils and their biogeographical implications. PLOS ONE, 3:e1858.

Brinkløv, S., E. K. V. Kalko, and A. Surlykke. 2009. Intense echolocation calls from two "whispering" bats, *Artibeus jamaicensis* and *Macrophyllum macrophyllum* (Phyllostomidae). Journal of Eperimental Biology 212:11–20.

Brooke, A. P. 1994. Diet of the fishing bat, *Noctilio leporinus* (Chiroptera: Noctilionidae). Journal of Mammalogy, 75:212–218.

Carter, A. M., S. M. Goodman, and A. C. Enders. 2008. Female reproductive tract and placentation in sucker-footed bats (Chiroptera: Myzopodidae) endemic to Madagascar. Placenta, 29:484–491.

Carter, G. G., and D. K. Riskin. 2006. *Mystacina tuberculata*. Mammalian Species, 790:1–8.

Chaverri, G. 2010. Comparative social network analysis in a leaf-roosting bat. Behavioral Ecology and Sociobiology, 64:1619–1630.

Chaverri, G., and E. H. Gillam. 2010. Cooperative signaling behavior of roost location in a leaf-roosting bat. Communicative and Integrative Biology, 3:599–601.

Chaverri, G., and E. H. Gillam. 2013. Sound amplification by means of a horn-like roosting structure in Spix's disc-winged bat. Proceedings of the Royal Society B–Biological Sciences, 280.

Chaverri, G., and E. H. Gillam. 2015. Repeatability in the contact calling system of Spix's disc-winged bat (*Thyroptera tricolor*). Royal Society Open Science, 2.

Chaverri, G., E. H. Gillam, and M. J. Vonhof. 2010. Social calls used by a leaf-roosting bat to signal location. Biology Letters, 6:441.

Clare, E. L., A. M. Adams, A. Z. Maya-Simões, J. L. Eger, P. D. N. Hebert, and M. B. Fenton. 2013. Diversification and reproductive isolation: cryptic species in the only New World high-duty cycle bat, *Pteronotus parnellii*. BMC Evolutionay Biology, 13:26.

Cummings, G., S. Anderson, T. Dennis, C. Toth, and S. Parsons. 2014. Competition for pollination by the lesser short-tailed bat and its influence on the flowering phenology of some New Zealand endemics. Journal of Zoology, 293:281–288.

Czaplewski, N. J. 1996. *Thyroptera robusta* Czaplewski, 1966, is a junior synonym of *Thyroptera lavali* Pine, 1993 (Mammalia: Chiroptera). Mammalia, 60:153–156.

Czaplewski, N. J. 1997. Chiroptera. Pp. 410–430 *in:* Vertebrate Paleontology in the Neotropics: The Miocene Fauna of La Venta, Colombia (R. F. Kay, R. H. Madden, R. L. Cifelli, and J. J. Flynn, eds.). Smithsonian Institution Press, Washington, DC.

Czaplewski, N. J., and C. Cartelle. 1998. Pleistocene bats from cave deposits in Bahia, Brazil. Journal of Mammalogy, 79:784–803.

Czaplewski, N. J., and K. E. Campbell, Jr. 2017. Late Miocene bats from the Juruá river, state of Acre, Brazil, with a description of a new genus of Thyropteridae (Chiroptera, Mammalia). Contributions in Science, 525:55–60.

Czaplewski, N. J., and G. S. Morgan. 2012. New basal noctilionoid bats (Mammalia: Chiroptera) from the Oligocene of subtropical North America. Pp. 162–209 *in:* Evolutionary History of Bats: Fossils, Molecules, and Morphology (G. F. Gunnell and N. B. Simmons, eds.). Cambridge University Press, Cambridge.

Czaplewski, N. J., M. Takai, T. Naeher, N. Shigehara, and T. Setoguchi. 2003. Additional bats from the middle Miocene La Venta fauna of Colombia. Revista de la Academia Colombiana de Ciencias, 27:263–282.

Czenze, Z. J., R. M. Brigham, A. J. R. Hickey, and S. Parsons. 2017. Cold and alone? Roost choice and season affect torpor patterns in lesser short-tailed bats. Oecologia, 183:1–8.

Daniel, M. J. 1975. First record of an Australian fruit bat (Megachiroptera: Pteropodidae) reaching New Zealand. New Zealand Journal of Zoology, 2:227–231.

Daniel, M. J. 1976. Feeding by the short-tailed bat (*Mystacina tuberculata*) on fruit and possibly nectar. New Zealand Journal of Zoology, 3:391–398.

Daniel, M. J. 1979. The New Zealand short-tailed bat *Mystacina tuberculata*: a review of present knowledge. New Zealand Journal of Zoology 6:357–370.

Dávalos, L. M., P. M. Velazco, O. M. Warsi, P. D. Smits, and N. B. Simmons. 2014. Integrating incomplete fossils by isolating conflictive signal in saturated and non-independent morphological characters. Systematic Biology, 63:582–600.

Dumont, E. R., L. M. Dávalos, A. Goldberg, S. E. Santana, K. Rex, and C. C. Voigt. 2012. Morphological innovation, diversification and invasion of a new adaptive zone. Proceedings of the Royal Society B–Biological Sciences, 279:1797–1805.

Eick, G. N., D. S. Jacobs, and C. A. Matthee. 2005. A nuclear DNA phylogenetic perspective on the evolution of echolocation and historical biogeography of extant bats (Chiroptera). Molecular Biology and Evolution, 22:1869–1886.

Fabián, M. 2008. Quirópteros do bioma Caatinga, no Ceará, Brasil, depositados no Museu de Ciências Naturais da Fundação Zoobotânica do Rio Grande do Sul. Chiroptera Neotropical, 14:354–359.

Falcão, F., J. A. Ugarte-Núñez, D. Faria, and C. B. Caselli. 2015. Unravelling the calls of discrete hunters: acoustic structure of echolocation calls of furipterid bats (Chiroptera, Furipteridae). Bioacoustics, 24:175–183.

Findley, J. S., and D. E. Wilson. 1974. Observations on the Neotropical disk-winged bat, *Thyroptera tricolor* Spix. Journal of Mammalogy, 55:562–571.

Fleming, T. H. 2005. The relationship between species richness of vertebrate mutualists and their food plants in tropical and subtropical communities differs among hemispheres. Oikos, 111:556–562.

Fracasso, M. P. A., and L. Salles. 2005. Diversity of Quaternary bats from Serra da Mesa (State of Goiás, Brazil). Zootaxa, 817:1–19.

Gardner, A. L. 2008. Family Furipteridae Gray, 1866a. Pp. 389–392 in: Mammals of South America, Volume 1: Marsupials, Xenarthrans, Shrews, and Bats (A. L. Gardner, ed.). University of Chicago Press, Chicago.

Giannini, N. P., and E. K. V. Kalko. 2005. The guild structure of animalivorous leaf-nosed bats of Barro Colorado Island, Panama, revisited. Acta Chiropterologica, 7:131–146.

Goodman, S. M., F. Rakotondraparany, and A. Kofoky. 2007. The description of a new species of Myzopoda (Myzopodidae: Chiroptera) from western Madagascar. Mammalian Biology–Zeitschrift für Säugetierkunde, 72:65–81.

Göpfert, M. C., and L. T. Wasserthal. 1995. Notes on echolocation calls, food and roosting behaviour of the Old World sucker-footed bat Myzopoda aurita (Chiroptera, Myzopodidae). Mammalian Biology–Zeitschrift für Säugetierkunde, 60:1–8.

Guevara-Chumacero, L. M., R. López-Wilchis, F. F. Pedroche, J. Juste, C. Ibáñez, and I. D. L. A. Barriga-Sosa. 2010. Molecular phylogeography of Pteronotus davyi (Chiroptera: Mormoopidae) in Mexico. Journal of Mammalogy, 91:220–232.

Gunnell, G. F., P. M. Butler, M. Greenwood, and N. B. Simmons. 2015. Bats (Chiroptera) from Olduvai Gorge, Early Pleistocene, Bed I (Tanzania). American Museum Novitates, 3846:1–36.

Gunnell, G. F., B. F. Jacobs, P. S. Herendeen, J. J. Head, E. Kowalski, C. P. Msuya, F. A. Mizambwa, T. Harrison, J. Habersetzer, and G. Storch. 2003. Oldest placental mammal from sub-Saharan Africa: Eocene microbat from Tanzania—evidence for early evolution of sophisticated echolocation. Palaeontologia Electronica, 5 (2): 1–10.

Gunnell, G. F., and N. B. Simmons. 2005. Fossil evidence and the origin of bats. Journal of Mammalian Evolution, 12:209–246.

Gunnell, G. F., N. B. Simmons, and E. R. Seiffert. 2014. New Myzopodidae (Chiroptera) from the Late Paleogene of Egypt: emended family diagnosis and biogeographic origins of Noctilionoidea. PLOS ONE, 9:e86712.

Gutiérrez, E. E., and J. Molinari. 2008. Morphometrics and taxonomy of bats of the genus Pteronotus (subgenus Phyllodia) in Venezuela. Journal of Mammalogy, 89:292–305.

Hand, S. J., M. Archer, and H. Godthelp. 2001. New Miocene Icarops material (Microchiroptera: Mystacinidae) from Australia, with a revised diagnosis of the genus. Memoir-Association of Australasian Palaeontologists, 25:139–146.

Hand, S. J., M. Archer, and H. Godthelp. 2005. Australian Oligo-Miocene mystacinids (Microchiroptera): upper dentition, new taxa and divergence of New Zealand species. Geobios, 38:339–352.

Hand, S. J., R. Beck, M. Archer, N. B. Simmons, G. F. Gunnell, R. P. Scofield, A. J. D. Tennyson, V. L. De Pietri, S. W. Salisbury, and T. H. Worthy. 2018. A new, large-bodied omnivorous bat (Noctilionoidea: Mystacinidae) reveals lost morphological and ecological diversity since the Miocene in New Zealand. Scientific Reports, 8:235.

Hand, S. J., R. Beck, T. Worthy, M. Archer, and B. Sigé. 2007. Australian and New Zealand bats: the origin, evolution, and extinction of bat lineages in Australia. Journal of Vertebrate Paleontology, 27:86A.

Hand, S. J., D. E. Lee, T. H. Worthy, M. Archer, J. P. Worthy, A. J. D. Tennyson, S. W. Salisbury et al. 2015. Miocene fossils reveal ancient roots for New Zealand's endemic Mystacina (Chiroptera) and Its rainforest habitat. PLOS ONE, 10: e0128871.

Hand, S. J., P. Murray, D. Megirian, M. Archer, and H. Godthelp. 1998. Mystacinid bats (Microchiroptera) from the Australian Tertiary. Journal of Paleontology, 72:538–545.

Hand, S. J., M. Novacek, H. Godthelp, and M. Archer. 1994. First Eocene bat from Australia. Journal of Vertebrate Paleontology, 14:375–381.

Hand, S. J., B. Sigé, M. Archer, G. F. Gunnell, and N. B. Simmons. 2015. A new Early Eocene (Ypresian) bat from Pourcy, Paris Basin, France, with comments on patterns of diversity in the earliest chiropterans. Journal of Mammalian Evolution, 22:343–354.

Hand, S. J., V. Weisbecker, R. Beck, M. Archer, H. Godthelp, A. Tennyson, and T. Worthy. 2009. Bats that walk: a new evolutionary hypothesis for the terrestrial behaviour of New Zealan's endemic mystacinids. BMC Evolutionary Biology, 9:169.

Hand, S. J., T. H. Worthy, M. Archer, J. P. Worthy, A. J. D. Tennyson, and R. P. Scofield. 2013. Miocene mystacinids (Chiroptera, Noctilionoidea) indicate a long history for endemic bats in New Zealand. Journal of Vertebrate Paleontology, 33:1442–1448.

Herd, R. M. 1983. Pteronotus parnellii. Mammalian Species, 209:1–5

Hill, R. S. 2001. Biogeography, evolution and palaeoecology of Nothofagus (Nothofagaceae): the contribution of the fossil record. Australian Journal of Botany, 49:321–332.

Hood, C. S., and J. K. Jones, Jr. 1984. Noctilio leporinus. Mammalian Species, 216:1–7.

Hood, C. S., and J. Pitocchelli. 1983. Noctilio albiventris. Mammalian Species, 197:1–5.

Hoofer, S. R., S. A. Reeder, E. W. Hansen, and R. A. Van Den Bussche. 2003. Molecular phylogenetics and taxonomic review of noctilionoid and vespertilionoid bats (Chiroptera: Yangochiroptera). Journal of Mammalogy, 84:809–821.

Ibáñez, C. 1985. Notes on Amorphochilus schnablii Peters (Chiroptera, Furipteridae). Mammalia, 49:584–587.

Jenkins, R. K. B., A. R. Rakotoarivelo, F. H. Ratrimomanarivo, and S. G. Cardiff. 2008a. Myzopoda aurita, The IUCN Red List of Threatened Species 2008: e.T14288A4429060.

Jenkins, R. K. B., A. R. Rakotoarivelo, F. H. Ratrimomanarivo, and S. G. Cardiff. 2008b. Myzopoda schliemanni, The IUCN Red List of Threatened Species 2008: e.T136465A4295196.

Jones, G. 1999. Scaling of echolocation call parameters in bats. Journal of Experimental Biology, 202:3359–3367.

Jones, G., P. I. Webb, J. A. Sedgeley, and C. F. J. O'Donnell. 2003. Mysterious Mystacina: how the New Zealand short-tailed bat (Mystacina tuberculata) locates insect prey. Journal of Experimental Biology, 206:4209–4216.

Kalko, E. K. V., and M. A. Condon. 1998. Echolocation, olfaction and fruit display: how bats find fruit of flagellichorous cucurbits. Functional Ecology, 12:364–372.

Kalko, E. K. V., H.-U. Schnitzler, I. Kaipf, and A. D. Grinnell. 1998. Echolocation and foraging behavior of the lesser bulldog bat, *Noctilio albiventris*: preadaptations for piscivory? Behavioral Ecology and Sociobiology, 42:305–319.

Korine, C., and E. K. V. Kalko. 2005. Fruit detection and discrimination by small fruit-eating bats (Phyllostomidae): echolocation call design and olfaction. Behavioral Ecology and Sociobiology, 59:12–23.

Lancaster, W. C., and E. K. V. Kalko. 1996. *Mormoops blainvillii*. Mammalian Species, 544:1–5.

LaVal, R. 1977. Notes on some Costa Rican bats. Brenesia, 10:77–83.

Lewis-Oritt, N., R. A. Van Den Bussche, and R. J. Baker. 2001. Molecular evidence for evolution of piscivory in *Noctilio* (Chiroptera: Noctilionidae). Journal of Mammalogy, 82: 748–759.

Lim, B. K. 2009. Review of the origins and biogeography of bats in South America. Chiroptera Neotropical, 15:391–410.

Lloyd, B. D. 2001. Advances in New Zealand mammalogy 1990–2000: short-tailed bats. Journal of the Royal Society of New Zealand, 31:59–81.

Lloyd, B. D. 2003. The demographic history of the New Zealand short-tailed bat *Mystacina tuberculata* inferred from modified control region sequences. Molecular Ecology, 12:1895–1911.

Lloyd, B. D., and P. Tucker. 2003. Intraspecific phylogeny of the New Zealand short-tailed bat *Mystacina tuberculata* inferred from multiple mitochondrial gene sequences. Systematic Biology, 52:460–476.

Macías, S., E. C. Mora, and A. García. 2006. Acoustic identification of mormoopid bats: a survey during the evening exodus. Journal of Mammalogy, 87:324–330.

Mancina, C. A. 2005. *Pteronotus macleayii*. Mammalian Species, 778:1–3.

Mancina, C. A., L. García-Rivera, and B. W. Miller. 2012. Wing morphology, echolocation, and resource partitioning in syntopic Cuban mormoopid bats. Journal of Mammalogy, 93:1308–1317.

Meredith, R., J. Janečka, J. Gatesy, O. Ryder, C. Fisher, E. Teeling, A. Goodbla et al. 2011. Impacts of the Cretaceous terrestrial revolution and KPg extinction on mammal diversification. Science, 334:521–524.

Miller, G. S., Jr. 1907. The families and genera of bats. Bulletin of the United States National Museum, 57: xvii+1–282.

Miller-Butterworth, C. M., W. J. Murphy, S. J. O'Brien, D. S. Jacobs, M. S. Springer, and E. C. Teeling. 2007. A family matter: conclusive resolution of the taxonomic position of the long-fingered bats, *Miniopterus*. Molecular Biology and Evolution, 24:1553–1561.

Mitchell, K. J., B. Llamas, J. Soubrier, N. J. Rawlence, T. H. Worthy, J. Wood, M. S. Lee, and A. Cooper. 2014. Ancient DNA reveals elephant birds and kiwi are sister taxa and clarifies ratite bird evolution. Science, 344:898–900.

Morgan, G. S. 1991. Neotropical Chiroptera from the Pliocene and Pleistocene of Florida. Bulletin of the American Museum of Natural History, 206:176–213.

Morgan, G. S. 2001. Patterns of extinction in West Indian bats. Pp. 369–406 *in*: Biogeography of the West Indies: patterns and perspectives (C. A. Woods and F. E. Sergile, eds.). CRC Press, Boca Raton.

Morgan, G. S., and N. J. Czaplewski. 1999. First fossil record of *Amorphochilus schnablii* (Chiroptera: Furipteridae), from the late Quaternary of Peru. Acta Chiropterologica, 1:75–79.

Morgan, G. S., and N. J. Czaplewski. 2012. Evolutionary history of the Neotropical Chiroptera: the fossil record. Pp. 105–161 *in*: Evolutionary History of Bats: Fossils, Molecules, and Morphology (G. F. Gunnell and N. B. Simmons, eds.). Cambridge University Press, Cambridge.

Morgan, G. S., and S. D. Emslie. 2010. Tropical and western influences in vertebrate faunas from the Pliocene and Pleistocene of Florida. Quaternary International, 217:143–158.

O'Donnell, C. 2008. *Mystacina tuberculata*. The IUCN Red List of Threatened Species 2008: e.T14261A4427784. Downloaded on 28 February 2017.

O'Leary, M. A., C. P. Bloch, J. J. Flynn, T. Gaudin, A. Giallombardo, N. P. Giannini, S. L. Goldberg et al. 2013. The placental mammal ancestor and the post–K-Pg radiation of placentals. Science, 339:662–668.

Parsons, S. 2001. Identification of New Zealand bats (*Chalinolobus tuberculatus* and *Mystacina tuberculata*) in flight from analysis of echolocation calls by artificial neural networks. Journal of Zoology, 253:447–456.

Parsons, S., D. K. Riskin, and J. W. Hermanson. 2010. Echolocation call production during aerial and terrestrial locomotion by New Zealand's enigmatic lesser short-tailed bat, *Mystacina tuberculata*. Journal of Experimental Biology, 213:551–557.

Patton, J. L., and A. L. Gardner. 2008. Family Mormoopidae Saussure, 1860. Pp. 376–384 *in*: Mammals of South America, Volume 1: Marsupials, Xenarthrans, Shrews, and Bats (A. L. Gardner, ed.). University of Chicago Press, Chicago.

Pavan, A. C., P. E. D. Bobrowiec, and A. R. Percequillo. 2018. Geographic variation in a South American clade of mormoopid bats, *Pteronotus* (Phyllodia), with description of a new species. Journal of Mammalogy, 99:624–645.

Pavan, A. C., and G. Marroig. 2016. Integrating multiple evidences in taxonomy: species diversity and phylogeny of mustached bats (Mormoopidae: *Pteronotus*). Molecular Phylogenetics and Evolution, 103:184–198.

Pavan, A. C., and G. Marroig. 2017. Timing and patterns of diversification in the Neotropical bat genus *Pteronotus* (Mormoopidae). Molecular Phylogenetics and Evolution, 108:61–69.

Phillips, M. J., G. C. Gibb, E. A. Crimp, and D. Penny. 2010. Tinamous and moa flock together: mitochondrial genome sequence analysis reveals independent losses of flight among ratites. Systematic Biology, 59:90–107.

Rajemison, B., and S. M. Goodman. 2007. The diet of *Myzopoda schliemanni*, a recently described Malagasy endemic, based on scat analysis. Acta Chiropterologica, 9:311–313.

Ralisata, M., D. Rakotondravony, and P. A. Racey. 2015. The relationship between male sucker-footed bats *Myzopoda aurita* and the traveller's tree *Ravenala madagascariensis* in south-eastern Madagascar. Acta Chiropterologica, 17:95–103.

Reid, F. A. 2009. A Field Guide to the Mammals of Central America and Southeast Mexico. 2nd ed. Oxford University Press, New York.

Reyes-Amaya, N., J. Lozáno-Flórez, D. Flores, and S. Solari. 2016. Distribution of the Spix's disk-winged bat, *Thyroptera tricolor* Spix, 1823 (Chiroptera: Thyropteridae) in Colombia, with first records for the middle Magdalena Valley. Mastozoologia Neotropical, 23:127–137.

Rezsutek, M., and G. N. Cameron. 1993. *Mormoops megalophylla*. Mammalian Species, 448:1–5.

Riskin, D. K., and M. B. Fenton. 2001. Sticking ability in Spix's disk-winged bat, *Thyroptera tricolor* (Microchiroptera: Thyropteridae). Canadian Journal of Zoology, 79:2261–2267.

Riskin, D. K., S. Parsons, W. A. Schutt, G. G. Carter, and J. W. Hermanson. 2006. Terrestrial locomotion of the New Zealand short-tailed bat *Mystacina tuberculata* and the common vampire bat *Desmodus rotundus*. Journal of Experimental Biology, 209:1725–1736.

Riskin, D. K., and P. A. Racey. 2010. How do sucker-footed bats hold on, and why do they roost head-up? Biological Journal of the Linnean Society, 99:233–240.

Rojas, D., O. M. Warsi, and L. M. Dávalos. 2016. Bats (Chiroptera: Noctilionoidea) challenge a recent origin of extant Neotropical diversity. Systematic Biology, 65:432–448.

Russell, A. L., S. M. Goodman, I. Fiorentino, and A. D. Yoder. 2008. Population genetic analysis of *Myzopoda* (Chiroptera: Myzopodidae) in Madagascar. Journal of Mammalogy, 89:209–221.

Salles, L. O., J. Arroyo-Cabrales, A. C. M. Lima, W. Lanzelotti, F. A. Perini, P. M. Velazco, and N. B. Simmons. 2014. Quaternary bats from the Impossível-Ioiô cave cystem (Chapada Diamantina, Brazil): humeral remains and the first fossil record of *Noctilio leporinus* (Chiroptera, Noctilionidae) from South America. American Museum Novitates, 3798:1–31.

Schliemann, H. 1970. Bau und funktion der haftorgane von *Thyroptera* und *Myzopoda* (Vespertilionoidea, Microchiroptera, Mammalia). Zeitschrift für wissenschaftliche Zoologie, 181:353–400.

Schliemann, H. 1971. Die Haftorgane von *Thyroptera* und *Myzopoda* (Microchiroptera, Mammalia) — Gedanken zu ihrer Entstehung als Parallelbildungen. Journal of Zoological Systematics and Evolutionary Research, 9:61–80.

Schnitzler, H.-U., and E. K. V. Kalko. 1998. How echolocating bats search and find food. Pp. 183–196 *in*: Bat Biology and Conservation (T. H. Kunz and P. A. Racey, eds.). Smithsonian Institution Press, Washington, DC.

Schnitzler, H.-U., E. K. V. Kalko, I. Kaipf, and A. D. Grinnell. 1994. Fishing and echolocation behavior of the greater bulldog bat, *Noctilio leporinus*, in the field. Behavioral Ecology and Sociobiology, 35:327–345.

Schutt, W. A., Jr., and N. B. Simmons. 2006. Quadrupedal bats: form, function and evolution. Pp. 145–159 *in*: Functional and Evolutionary Ecology of Bats (A. Zubaid, G. F. McCracken, and T. H. Kunz, eds.). Oxford University Press, New York.

Shi, J. J., and D. L. Rabosky. 2015. Speciation dynamics during the global radiation of extant bats. Evolution, 69:1528–1545.

Silva Taboada, G. 1974. Fossil Chiroptera from cave deposits in central Cuba, with descriptions of two new species and the first West Indian record of *Mormoops megalophylla*. Acta Zoologica Cracoviensia, 19:33–73.

Simmons, N. B. 1998. A reappraisal of interfamilial relationships of bats. Pp. 3–26 *in*: Bat Biology and Conservation (T. H. Kunz and P. A. Racey, eds.). Smithsonian Institution Press, Washington, D.C.

Simmons, N. B. 2005. Order Chiroptera. Pp. 312–529 *in*: Mammal Species of the World: A Taxonomic and Geographic Reference (D. E. Wilson and D. M. Reeder, eds.). Johns Hopkins University Press, Baltimore.

Simmons, N. B., and T. M. Conway. 2001. Phylogenetic relationships of mormoopid bats (Chiroptera: Mormoopidae) based on morphological data. Bulletin of the American Museum of Natural History, 258:1–97.

Simmons, N. B., and R. S. Voss. 1998. The mammals of Paracou, French Guiana: a Neotropical lowland rainforest fauna. Part 1: Bats. Bulletin of the American Museum of Natural History, 237:1–219.

Slikas, B., S. L. Olson, and R. C. Fleischer. 2002. Rapid, independent evolution of flightlessness in four species of Pacific Island rails (Rallidae): an analysis based on mitochondrial sequence data. Journal of Avian Biology, 33:5–14.

Smith, J. D. 1972. Systematics of the chiropteran family Mormoopidae. University of Kansas Museum of Natural History Miscellaneous Publications, 56:1–132.

Smith, T., J. Habersetzer, N. B. Simmons, and G. F. Gunnell. 2012. Systematics and paleobiogeography of early bats. Pp. 23–66 *in*: Evolutionary History of Bats: Fossils, Molecules and Morphology (G. F. Gunnell and N. B. Simmons, eds.). Cambridge University Press, Cambridge.

Smotherman, M., and A. Guillén-Servent. 2008. Doppler-shift compensation behavior by Wagner's mustached bat, *Pteronotus personatus*. Journal of the Acoustical Society of America, 123:4331–4339.

Teeling, E. C., M. S. Springer, O. Madsen, P. Bates, S. J. O'Brien, and W. J. Murphy. 2005. A molecular phylogeny for bats illuminates biogeography and the fossil record. Science, 307:580–584.

Tejedor, M. F., N. J. Czaplewski, F. J. Goin, and E. Aragón. 2005. The oldest record of South American bats. Journal of Vertebrate Paleontology, 25:990–993.

Thoisy, B., A. C. Pavan, M. Delaval, A. Lavergne, T. Luglia, K. Pineau, M. Ruedi, V. Rufray, and F. Catzeflis. 2014. Cryptic diversity in common mustached bats *Pteronotus* cf. *parnellii* (Mormoopidae) in French Guiana and Brazilian Amapa. Acta Chiropterologica, 16:1–13.

Toth, C. A., T. E. Dennis, D. E. Pattemore, and S. Parsons. 2015. Females as mobile resources: communal roosts promote the adoption of lek breeding in a temperate bat. Behavioral Ecology, 26:1156–1163.

Toth, C. A., and S. Parsons. 2013. Is lek breeding rare in bats? Journal of Zoology, 291:3–11.

Tschapka, M., A. P. Brooke, and L. T. Wasserthal. 2000. *Thyroptera discifera* Chiroptera: Thyropteridae: a new record for Costa Rica and observations on echolocation. Mammalian Biology–Zeitschrift für Säugetierkunde, 65:193–198.

Uieda, W., I. Sazima, and A. Storti-Filho. 1980. Aspectos da biologia do morcego *Furipterus horrens* (Mammalia, Chiroptera, Furipteridae). Revista Brasileira de Biologia, 40:59–66.

Van Den Bussche, R., and S. Hoofer. 2004. Phylogenetic relationships among recent chiropteran families and the importance of choosing appropriate out-group taxa. Journal of Mammalogy, 85:321–330.

Van Den Hoek Ostende, L. W., D. Van Oijen, and S. K. Donovan. 2018. A new bat record for the late Pleistocene of Jamaica: *Pteronotus trevorjacksoni* from the Red Hills Road Cave. Caribbean Journal of Earth Science, 50:31–35.

Velazco, P. M., R. Gregorin, R. S. Voss, and N. B. Simmons. 2014. Extraordinary local diversity of disk-winged bats (Thyropteridae: *Thyroptera*) in northeastern Peru, with the description of a new species and comments on roosting behavior. American Museum Novitates, 3795:1–28.

Velazco, P. M., H. O'Neill, G. F. Gunnell, S. B. Cooke, R. Rimoli, A. L. Rosenberger, and N. B. Simmons. 2013. Quaternary bat diversity in the Dominican Republic. American Museum Novitates, 3779:1–20.

Volleth, M. 2013. Of bats and molecules: chromosomal characters for judging phylogenetic relationships. Pp. 129–146 *in*: Bat Evolution, Ecology, and Conservation (R. A. Adams and S. C. Pedersen, eds.). Springer, New York.

Vonhof, M., and M. Fenton. 2004. Roost availability and population size of *Thyroptera tricolor*, a leaf-roosting bat, in north-eastern Costa Rica. Journal of Tropical Ecology, 20:291–305.

Vonhof, M., H. Whitehead, and M. Fenton. 2004. Analysis of Spix's disc-winged bat association patterns and roosting home ranges reveal a novel social structure among bats. Animal Behaviour, 68:507–521.

Voss, R. S., D. W. Fleck, R. E. Strauss, P. M. Velazco, and N. B. Simmons. 2016. Roosting ecology of Amazonian bats: evidence for guild structure in hyperdiverse mammalian communities. American Museum Novitates, 3870:1–43.

Weinbeer, M., and E. K. V. Kalko. 2007. Ecological niche and phylogeny: the highly complex echolocation behavior of the trawling long-legged bat, *Macrophyllum macrophyllum*. Behavioral Ecology and Sociobiology, 61:1337–1348.

Wood, J. R., J. M. Wilmshurst, T. H. Worthy, A. S. Holzapfel, and A. Cooper. 2012. A lost link between a flightless parrot and a parasitic plant and the potential role of coprolites in conservation paleobiology. Conservation Biology, 26:1091–1099.

Worthy, T. H., and R. N. Holdaway. 2002. The Lost World of the Moa: Prehistoric Life of New Zealand. Indiana University Press, Bloomington.

Worthy, T. H., A. J. D. Tennyson, M. Archer, A. M. Musser, S. J. Hand, C. Jones, B. J. Douglas, J. A. McNamara, and R. M. D. Beck. 2006. Miocene mammal reveals a Mesozoic ghost lineage on insular New Zealand, southwest Pacific. Proceedings of the National Academy of Sciences, USA, 103:19419–19423.

4

Diversity and Discovery

A Golden Age

Andrea L. Cirranello
and Nancy B. Simmons

Introduction

E. O. Wilson (2004) called taxonomy "the pioneering exploration of life on a little known planet." Central to all aspects of biological research—from evolutionary biology to ecology and conservation—the discovery, description, and naming of species is critical to the work of nearly every biologist. From providing a common language for scientific communication to helping formulate testable predictions concerning such biological phenomena as behavior, gene flow, and adaptive radiation, taxonomic hypotheses are at the core of modern biology (Agnarsson and Kuntner 2007). With ongoing threats to global biodiversity caused by climate change and human activity, including land-use changes, introduction of invasive species, and overexploitation of resources, timely discovery and description of extant species has been recognized as central to protecting remaining biodiversity. This need has recently put taxonomy in the spotlight, with new efforts to stem the decline in the number of taxonomists and increase support and funding for taxonomic research (e.g., Cotterill and Foissner 2010; Pearson et al. 2011). Modern efforts to delineate and describe species often include the application of phylogenetic methods, seeking to place new species within a context of evolutionary relationships using morphological, morphometric, and molecular data. These studies generally conceive of species and higher-level taxa (e.g., genera) as equivalent to clades, bringing taxonomy and systematics—the study of diversification and evolutionary relationships—into closer alignment. Importantly, the focus on phylogeny allows us to make predictions about species that are poorly known to science, increasing our ability to protect rare species known from few specimens and sparse ecological data.

Bats are the second most speciose order of mammals, second only to rodents in extant species diversity (Simmons 2005). Miller (1907, 2) speculated that there may be as many as 2,000 species of bats, a claim that may have sounded outrageous in the early twentieth century but now seems prescient. More recent estimates suggest that, overall, bat diversity may be underestimated by roughly 40–50% (Clare et al. 2011; Francis et al. 2010). With nearly 1,375 currently recognized species (Simmons and Cirranello unpublished), taxonomists have a challenging road ahead.

Phyllostomid bats are widely regarded as the most ecologically diverse family of mammals and presently are second only to vespertilionid bats in total number of species (Dávalos et al. 2014; Dumont et al. 2012; Rojas et al. 2016; Simmons 2005). Although the golden age of species discovery is often imagined as long past, since 2005 (when the last edition of *Mammal Species of the World* was published) more than 50 new phyllostomid species have been described or raised from synonymy—a phenomenal number for a family as well-studied as Phyllostomidae. Four new genera have also been recognized—two previously unknown to science. The pace of species discovery in this new golden age exceeds the pace of discovery at any previous time, even when new species were being actively cataloged from European expeditions and exploration in the Americas. As of this writing, we recognize 216 valid species in 60 genera and 11 subfamilies within Phyllostomidae (see app. 4.1). This is a significant increase from the 159 species in 54 genera and 7 subfamilies recognized by Simmons (2005) just a little over a decade ago. Elevation of species from synonymy and the discovery of new species has raised the perceived diversity of the family by more than 25%—an increase unmatched within the larger Chiropteran families, making Central and South America a hotbed of species discovery for bats. The increase in species diversity is not uniformly spread throughout the family but is, instead, concentrated within several recently revised genera (e.g., *Sturnira*, *Platyrrhinus*) evaluated in the context of phylogenetic analyses. In addition to the improvements in phylogenetic research, faunal-based studies focusing on poorly known areas and regional conservation priorities have also increased our understanding of diversity within Phyllostomidae. Below, we provide a brief overview of the history and pace of discovery and classification of phyllostomid bats and provide a discussion of some of the factors responsible for this golden age. We also provide an updated summary classification for Phyllostomidae that provides an easy reference for readers of other chapters in the book.

Overview of Species Discovery and Classification

The European age of exploration began in the 1400s and concluded in the 1700s, but surprisingly few phyllostomid species were described by naturalists prior to 1800 despite numerous expeditions in the Americas. The first descriptions of phyllostomids appeared in Willem Piso's *Historia Naturalis Brasiliae*, published in 1648. Piso, a Dutch physician, traveled in Brazil studying tropical diseases and collecting animals and plants. He described fat, gray bats with soft, long hair under the name *Vespertilio cornutus*, noting that these bats roosted in coconut palms (Piso 1648). Hans Sloane, an Irish physician, in his travels to Jamaica in 1687, collected what he believed to be one of the smaller bats originally described by Piso. He wrote, "A Batt [*sic*] with an Ear like Processe [*sic*] over its Snout . . . They are both very frequent in the Caves among the Woods, in old Houses and Walls, etc. . . . They are said to feed on prickly Pears" (Sloane 1725). Sloane's collection later formed the nucleus of the mammal collection of the British Museum (Thomas 1906). Another Dutch zoologist, Albertus Seba, figured two phyllostomids in his beautifully illustrated *Thesaurus*. Although Seba did not travel to the Americas, he asked sailors and ships' physicians to bring him specimens; he described the common American bat as tailless, with ears that were larger and longer than the European bat, and a noseleaf (Seba 1734, table LV: fig. 2, and p. 90). Seba described in detail the rare "Chienne qui vole" or flying dog, noting tendons in the wing, as well as features of the noseleaf and ears (Seba 1734, table LVIII: fig. 1, and p. 93). These two species were the first phyllostomids to which Linnaeus applied binomial names in his *Systema Naturae*, published in 1758. Linnaeus (1758) gave the name *Vespertilio perspicillata* (= *Carollia perspicillata*) to Seba's common American bat (now commonly known as "Seba's short-tailed bat"), and the name *Vespertilio spectrum* (= *Vampyrum spectrum*) to Seba's flying dog.

Notable to all early taxonomists was the noseleaf of phyllostomid bats, clearly a novelty. A description by Edwards (1751, 201) of a small bat from Jamaica is of interest as Edwards speculated on the function of the noseleaf:

> The upper Figure is principally remarkable for the Flap on its Nose, which when newly taken out of

Spirits was pliable, (by which Means I think I have discovered its Use) and would cover the Nostrils and fix its Point into a Notch in the under Lip, by which it locked up the Nose and Mouth. This I take to be a Contrivance to prevent the Transpiration of its Juices in its torpid or sleeping State . . . for this Genus are, I believe, all Sleepers at certain Times.

Edwards mistakenly believed this bat to be the same described by Sloane and Piso; but a subsequent exceptionally detailed description by Pallas (1766) gave a new binomial name to this species as *Vespertilio soricinus* (= *Glossophaga soricina*). Pallas (1767) also applied the name *Vespertilio hastatus* (= *Phyllostomus hastatus*) to Buffon's chauve-souris fer-de-lance, bringing to four the number of still valid phyllostomid species described prior to 1800. And in 1799, Lacépède recognized the first phyllostomid genus, *Phyllostoma* (= *Phyllostomus*).

In the early 1800s, the number of phyllostomid species known to naturalists rapidly increased (fig. 4.1). Between 1801 and 1850, a total of 35 species that we still recognize as valid were described by various European naturalists. Geoffroy (1803) described *Phyllostoma crenulatum* (= *Gardnerycteris crenulatum*) from a specimen from the collections of the Dutch Stadholder William V, which the French had seized as spoils of war. This practice, previously frowned upon and thought inappropriate, resulted in substantial expansion of the Muséum National in Paris where Geoffroy worked.

One estimate is that 21% of birds and mammals that entered the Muséum between 1793 and 1809 did so as spoils of war (Lipkowicz 2013). Geoffroy (1810) continued to add to the number of known species of phyllostomids. Geoffroy formally introduced binomial names for bats that had been described by Felix Azara (1809) in his book on the quadrupeds of Paraguay, which was based on Azara's travels and collections in the last two decades of the 1700s. Geoffroy listed all the then-recognized species of phyllostomids under the genus name *Phyllostoma*. The four new species that Geoffroy recognized from Azara's work were *Phyllostoma lineatus* (= *Platyrrhinus lineatus*), *Phyllostoma lilium* (= *Sturnira lilium*), *Phyllostoma rotundum* (= *Desmodus rotundus*), and *Phyllostoma elongatus* (= *Phyllostomus elongatus*).

By 1850, representatives of 10 of the 11 subfamilies that we recognize in the current classification (app. 4.1) had been described, with the sole exception being Lonchophyllinae. Many of the new species were named by collectors and naturalists who had traveled to the Americas, such as Spix (1823), who traveled to Brazil in 1817 and described four new species (current name combinations: *Diphylla ecaudata*, *Artibeus planirostris*, *Trachops cirrhosus*, and *Tonatia bidens*) based on his collections (as well as the synonym *amplexicaudata* for *Glossophaga soricina*), and Tschudi, who in his *Fauna Peruana* (1844) described four species that remain valid (current name combinations: *Anoura peruana*,

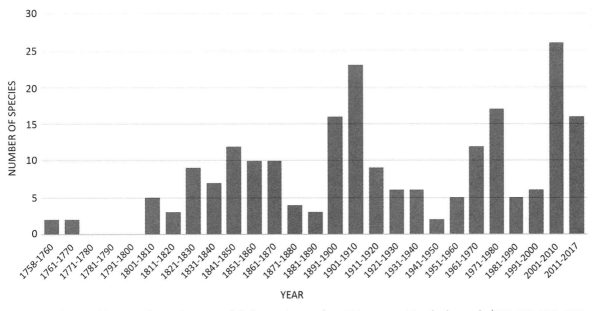

Figure 4.1. The pace of discovery of currently recognized phyllostomid species from 1758 to present. Note the three peaks (1891–1910, 1961–1980, and 2001–present) as well as the increasing pace of discovery since 2001.

Choeronycteris mexicana, *Sturnira erythromos*, and *S. oporaphilum*, as well as the synonym *innominatum* for *Phyllostomus discolor*). Other major contributors included museum-based researchers such as John Edward Gray (British Museum of Natural History) and Wilhelm Peters (Berlin Zoological Museum), who added substantially to the tally of phyllostomids based on studies of the collections that were under their care. Gray added 11 phyllostomid species during this time period, and Peters named 10 species (see app. 4.1).

The number of genera recognized began to increase even more in the early 1800s (see fig 4.2). Rafinesque (1815) coined the genus name *Vampyrum*, while Geoffroy (1818a; 1818b) first applied the generic names *Glossophaga* and *Stenoderma* in binomial classification, bringing to four the number of phyllostomid genera recognized by 1820. Over the next three decades, that number rapidly increased to 22 genera (fig. 4.2), necessitating a change in how species and genera were grouped and resulting in increasingly complicated classifications. Often, early nineteenth-century classifications had multiple genera appearing in a single family of bats, with these genera serving a function similar to the subfamilies of today (e.g., Oken [1816], with *Phyllostoma* and *Stenoderma*), or classifications included a single genus of bats that was divided into

multiple subgroups (e.g., Cuvier [1817], with the genus *Vespertilio*, but divided into 12 groups including Phyllostomes and Sténodermes). Goldfuss (1820) was the first to classify bats into multiple families, a practice that has been continued, in varying forms, in all subsequent classifications. Goldfuss (1820) recognized the family Phyllostomata for most phyllostomids (and most other leaf-nosed bats), save *Stenoderma*, which he grouped with the family Noctiliones. However, other types of divisions were also common in the first half of the nineteenth century. For example, Gray (1821), focusing on dietary habits in his classification, placed phyllostomids in the family Noctilionidae in the order Insectivorae. In his work on the animals of Brazil, Spix (1823) focused on noseleaves instead of diet and used the name Istiophori for the phyllostomids that he identified and described. Gray (1825) first used the name Phyllostomina for a subfamily including phyllostomids and *Rhinopoma*, and authorship of this name and all coordinate names date from this work. Desmodontinae was named in 1840 by Wagner, and Glossophaginae in 1845 by Bonaparte.

Between 1851 and 1900, an additional 43 phyllostomid species were described that are still valid today (fig. 4.1, app. 4.1), bringing to 83 the number of species recognized. An additional 17 species were de-

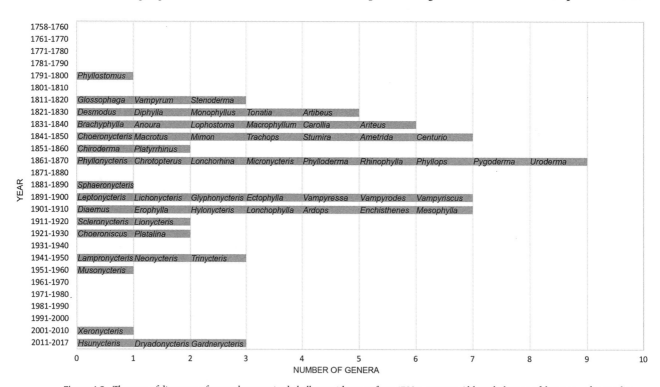

Figure 4.2. The pace of discovery of currently recognized phyllostomid genera from 1758 to present. Although the rate of discovery of genera has declined over time, new genera continue to be discovered.

scribed that have since been recognized as synonyms of older names. For example, four synonyms of *Artibeus jamaicensis* were published during the second half of the nineteenth century: *carpolegus* (Gosse 1851), *coryi* (Allen 1890), *eva* (Cope 1889) and the nomen nudum *macleayii* (Dobson 1878). Peters described an astonishing 14 new species between 1851 and 1900, which is roughly 30% of all of the valid phyllostomid species described during those years. His descriptions included several species now placed in *Platyrrhinus* (*helleri*, *infuscus*, and *vittatus*), *Uroderma bilobatum*, and *Sphaeronycteris toxophyllum* (Peters 1860, 1866, 1880, 1882). As Peters's publications demonstrate, museumbased work on existing collections contributed the bulk of the new species described by European scientists in the nineteenth century. As Dobson's *Catalogue of the Chiroptera in the Collection of the British Museum* makes clear, specimen holdings of phyllostomids had increased dramatically throughout the 1800s. In the preface to the first edition, Gunther noted that 418 specimens of all bat species were held by the British Museum in 1843 (Dobson 1878). This number had risen to 2,666 by 1878 (Dobson 1878). Many of these specimens were purchased, and some were collected by individuals who donated them to the museum. Not all notable work however, was taking place in European museums. Juan Gundlach, a trained taxonomist and German-born naturalist, moved to Cuba in 1839 and advanced the natural sciences in both Cuba and Puerto Rico (Vilaró 1897). Gundlach (1861) described *Erophylla sezekorni* and *Phyllonycteris poeyi* from his Cuban collections, which later formed the basis of the collections of the Cuban National Museum (Vilaró 1897).

The period between 1851 and 1900 brought a burst of description, with 19 genera—32% of all currently valid genera—named during this time period. Roughly half of these new genera were stenodermatines (fig. 4.2). The ground work for the higher-level classification that we still use today was first articulated by Gervais (1854, 1856). Gervais (1854) restricted the family to taxa that we now include in Phyllostomidae and divided the members into four groups corresponding to the genera *Phyllostoma*, *Glossophaga*, *Desmodus*, and *Stenoderma*. Two years later, Gervais (1856) formalized this arrangement by grouping species and genera of phyllostomids into four subfamilies: Desmodina, Vampyrina, Glossophagina, and Stenodermina, the subfamily taxonomy used today. The name Stenoder-

matinae, and all coordinate names, dates from this work. Gray (1866) introduced the name Lonchorhinina as yet another tribe, and Dobson's arrangement of bats "according to their natural affinities" offered yet another higher-level arrangement. Dobson (1875, 1878) divided the family Phyllostomidae into Lobostominae (= Mormoopidae) and Phyllostominae. Placement of mormoopids within Phyllostomidae was not uncommon in the nineteenth century. Dobson followed Peters (1865) in dividing Phyllostominae into Desmodontes, Vampyri (phyllostomines plus *Carollia* and *Rhinophylla*), Glossophagae, and Stenodermata (including *Brachyphylla*).

A peak of phyllostomid species description occurred in the years 1901–1910, with more than 50 species described during this period. Of these, 23 species remain valid, while the remainder have been synonymized. The number of species descriptions dropped sharply over the next several decades (fig. 4.1), but by 1950, a total of 129 of the 216 currently recognized species had been described. Oldfield Thomas at the British Museum described nine new species of phyllostomids between 1901 and 1910 alone. Thomas himself was involved in increasing the pace of species discovery, funding collecting trips to South America and presenting the collections to the British Museum (Thomas 1906). Beginning in 1894, the number of accessions to the British Museum increased to nearly 2,000 new specimens across all mammalian groups per year, up from an average of roughly 350 specimens per year prior to this date (Thomas 1906). During this period, Thomas described the first lonchophylline, *Lonchophylla mordax*, from collections made by the French Alphonse Roberts in Brazil. Knud Anderson, also at the British Museum, described four new species of *Artibeus*, and the Americans Gerrit S. Miller, at the U.S. National Museum, and J. A. Allen, curator of mammals and birds at the American Museum of Natural History, added five phyllostomid species descriptions (see app. 4.1).

After 1901, the number of new genera of phyllostomids described each decade began to decline (see fig. 4.2), with the highest number—seven new genera—described during 1901–1910. Three of the new genera (*Ardops*, *Erophylla*, and *Diaemus*) were described by Miller (1906). Over the course of two decades, Thomas (1903, 1913, 1928) described all the genera of lonchophyllines recognized until the twenty-first century. In addition to describing five species and three genera,

Miller made important contributions to our overall understanding of bats, publishing a comprehensive classification, including comments on each genus (Miller 1907). In this work, Miller refined the classification of phyllostomids, retaining Chilonycterinae (= Mormoopidae) within Phyllostomidae and recognizing an additional six subfamilies: Phyllostominae, Phyllonycterinae, Glossophaginae, Hemiderminae (= Carolliinae; see Miller 1924), Sturnirinae, and Stenodermatinae. Simpson's (1945) influential classification mainly synonymized genera within Miller's (1907) classification without substantially changing other aspects of the classification scheme. For example, Simpson (1945) synonymized *Hylonycteris* with *Choeronycteris*, *Mesophylla* with *Ectophylla*, and *Phyllops* with *Stenoderma*.

Forty-five phyllostomid species were described between 1951 and 2000; just two species described during these years have been synonymized: *Leptonycteris longala* Stains 1957 with *L. nivalis* (Saussure 1860) and *Leptonycteris sanborni* Hoffmeister 1957 with *L. yerbabuenae* Martínez and Villa-R. 1940. Few of the papers describing new taxa during this time were the result of comprehensive taxonomic reviews; instead, new species were described as they were found in the course of other research projects. Hernández-Camacho and Cadena-G. (1978) revised *Lonchorhina* and introduced the new species *L. marinkellei*. Davis and Carter (1978) reviewed the *Tonatia silvicola* (= *Lophostoma silvicolum*) species complex and named the species (current name combinations) *Lophostoma occidentalis* and *Lophostoma evotis*. Nearly all of these new species descriptions were the result of faunal studies (e.g., Genoways and Williams 1980; Handley 1960). The majority of the contributions in species description during these years were made by researchers in North America, a trend appearing in the last half century as American institutions such as the Field Museum (founded 1893), U.S. National Museum (founded 1846), and American Museum of Natural History (founded 1869) had assembled valuable collections and expertise. Twenty new species were described by researchers affiliated with these three major American museums; only two were described by researchers at the British Museum. Ten new species descriptions were authored or coauthored by researchers at Central or South American institutions.

From 1951 to 2000, only one new genus of phyllostomid bat was described (fig. 4.2)—*Musonycteris*, discovered in Colima, Mexico (Schaldach and McLaughlin 1960). However, during the late twentieth century, higher-level phyllostomid classification underwent dramatic changes as classifications explicitly based on phylogeny were produced by numerous authors. Analyses of different data sets often produced widely different phylogenetic trees and, therefore, very different classifications. For example, Griffiths (1982) named the subfamily Lonchophyllinae for the genera *Lionycteris*, *Lonchophylla*, and *Platalina* on the basis of a morphological investigation, but other morphological studies did not support this result (Wetterer et al. 2000). Similarly, Van Den Bussche (1992) named Macrotinae and Miconycterinae on the basis of a restriction-site analysis, but this result was contradicted by morphological data, which continued to support a classification very similar to that proposed by Miller (1907).

A remarkable 26 species of phyllostomids were described between 2001 and 2010. This number is higher than in any previous decade. From 2011 to our writing of this chapter (late 2017), 16 additional species have been described, suggesting that this rapid pace of discovery will continue. Some new species have been described in the context of taxonomic revisions. For example, eight new species of *Platyrrhinus*—roughly 40% of the diversity of the genus—were described, and three more species were raised from synonymy, by Velazco and colleagues in the context of a revision published as a series of papers (Velazco 2005; Velazco and Gardner 2009a, 2009b; Velazco and Lim 2014; Velazco et al. 2010). Similar revisionary studies have also increased perceived diversity in *Sturnira* (two new species descriptions and four species raised from synonymy; Velazco and Patterson 2013, 2014; app. 4.1) and in *Vampyressa* (two new species; Tavares et al. 2014; app. 4.1). Other genera (e.g., *Carollia*, *Anoura*, *Micronycteris*, and *Lonchophylla*) have seen diversity climb by more than 35% (app. 4.1). More than 75% of new species descriptions since 2001 involved researchers based not in U.S. or European institutions but, rather, in Central and South American countries where phyllostomid diversity is high—for example, Mexico, Colombia, Brazil, Ecuador, and Peru.

Given the rapid drop in generic descriptions after 1910 (fig. 4.2), it is also notable that four new genera have been named since 2000. Two of the genera were described based on specimens collected in the field (*Xeronycteris* and *Dryadonycteris*; Gregorin and Ditchfield 2005; Nogueira et al. 2012), and two descriptions were the result of taxonomic revisions that used molecular techniques (*Hsunycteris* and *Gardnerycteris*; Hurtado and Pacheco 2014; Parlos et al 2014). Higher-level phylogenies using both morphological and molecular data (e.g., Dávalos et al. 2012) have led to increasing clarification regarding the relationships of phyllostomids, more robust support for some groups, and recognition that the traditional subfamily classification was insufficient. The number of subfamilies has subsequently risen to 11 (Carolliinae, Desmodontinae, Glossophaginae, Glyphonycterinae, Lonchophyllinae, Lonchorhininae, Macrotinae, Micronycterinae, Phyllostominae, Rhinophyllinae, Stenodermatinae), with two formally named in 2016: Glyphonycterinae and Rhinophyllinae (Baker et al. 2016). Given the increase in the number of recognized phyllostomid species, genera, and subfamilies in the past two decades (fig. 4.1), we are clearly in a golden age of discovery and description.

Factors Contributing to the Golden Age

Introduction

Below, we address several factors that we feel have perhaps made the largest contribution to the increasing pace of phyllostomid discovery. Our choice reflects our own areas of expertise and is not intended to be exhaustive. Numerous other factors—from the global economy to improved scientific literacy—have doubtless greatly influenced the new golden age.

Phylogeny and Geography

While our system of classification dates from 1758, the attention paid to the evolutionary relationships of the organisms being classified is new. It is possible to practice taxonomy (description, naming, and classification) without systematics (the study of evolutionary relationships). However, the current surge in species descriptions seen in phyllostomids is due in no small part to phylogenetic studies, some of which use newer techniques that can only now be widely applied (see below). In these studies, a named species is equated with a specific clade or monophyletic lineage. As an example, the genus *Platyrrhinus* Saussure 1860 currently comprises 21 species. Five species were described before 1900, three between 1900 and 1950, five between 1951 and 2000, and eight have been described between

2001 and the present. Prior to 1987, the evolutionary relationships of these species were unknown, and ten species were recognized. Owen (1987) conducted the first species-level phylogenetic analysis but was unable to fully resolve relationships within the genus. In 2005, Velazco conducted the second phylogenetic analysis of species within the genus. Using morphological characters and multiple specimens from across the species' ranges, Velazco (2005) recovered a tree in which multiple species of *Platyrrhinus* either did not appear to be monophyletic (e.g., *Platyrrhinus dorsalis*) or were sister taxa with distinct ranges (e.g., *Platyrrhinus vittatus*). Velazco (2005) described four new species based on this phylogenetic analysis. Additional analyses of molecular data uncovered paraphyly in *Platyrrhinus helleri*, which Velzaco and Patterson (2008) then split into *P. helleri* and *P. incarum* based on the species ranges. Velzaco and Gardner (2009a) raised *P. aquilus* and *P. umbratus* to specific level and identified yet another new species, *P. nitelina*. Finally, Velazco et al. (2010) used molecular, morphometric, and morphological data to tease apart the *Platyrrhinus helleri* species complex, ultimately recognizing two additional species (*P. angustirostris* and *P. fusciventris*). All of the later studies were framed in the context of phylogenetic analyses informed by molecular data.

However, the role played by phylogeny is vastly underestimated if we simply look at species diversity. In recent years, one of the prime goals of the systematic community has been to reconstruct the tree of life, the unique branching pattern or "framework phylogeny" of all life. If we think of the increase in phyllostomid diversity in terms of our understanding of higher-level branching pattern of this tree, we begin to appreciate the scale of our current discoveries. As an example, several genera have been split in recent decades (e.g., *Tonatia/Lophostoma*; *Mimon/Gardnerycteris*; Hurtado and Pacheco 2014; Lee et al. 2002), and several new branches have been added to the tree (*Dryadonycteris*, *Xeronycteris*; Gregorin and Ditchfield 2005; Nogueira et al. 2012). Each of these discoveries has increased our understanding of trait evolution, improved the predictive ability of our trees, and aided in biodiversity assessments. In addition, researchers continue to recover clades in phylogenetic trees to which we do not apply names. The number of clades of this type in Phyllostomidae is very large. In a fully resolved tree of all 216 phyllostomid species (tips) currently recognized, there would be 215 nodes and hence a total of 431 potential taxa that could have formal names. Tips correspond with species, all of which are formally recognized in our classification; however, of the 215 nodes, we recognize only 43% with a formal name at the generic, subtribal, tribal, or subfamilial level (app. 4.1). When relationships that are not formally designated with names—but which are nonetheless recognized as "real" based on phylogenies—are taken into account, the extent of current golden age in terms of recognition of diversity patterns can be seen to vastly exceed any previous age of discovery (Donoghue and Alverson 2000).

New attention to phylogeny has overturned traditional views of phyllostomid ecological diversity and evolution. Traditional views were based on morphological data, wherein taxa that shared feeding specializations formed either subfamilies or clades. For example, Wetterer et al. (2000) recognized six subfamilies (Desmodontinae—sanguinivores, Phyllostominae—insectivores and omnivores, Brachyphyllinae—nectarivores, Glossophaginae—nectarivores, Carolliinae—frugivores, and Stenodermatinae—frugivores) in a phylogenetic analysis based on morphological data. However, more recent trees based on analyses of both morphological and molecular data (e.g., Dávalos et al. 2014) provide a much more complex picture of phyllostomid evolution, upending traditional notions of relationship. In the new trees, no fewer than five clades (recognized as subfamilies below; app. 4.1) include former "phyllostomines" (Macrotinae, Micronycterinae, Lonchorhininae, Phyllostominae, and Glyphonycterinae), while two clades (also subfamilies) are composed of former "carolliines" (Carolliinae and Rhinophyllinae). The nectar feeders are split into two clades in these trees (Lonchophyllinae and Glossophaginae), supporting Griffiths's (1982) hypothesis that lonchophyllines are not closely related to the remaining nectar feeders. In fact, they appear to be the sister taxon of a clade that includes Glyphonycterinae, Carolliinae, Rhinophyllinae, and Stenodermatinae (see Rojas et al. 2016). This much more "complicated" tree has led researchers to new questions regarding ecological and anatomical convergence, morphological character saturation, and gene expression (e.g., Dávalos et al. 2014) that would not have been part of a phyllostomid research agenda decades ago.

Geography has also become of greater importance in understanding phyllostomid diversity. Early scientific specimens of phyllostomids collected from the Americas had poor geographic data. For example, the type locality for both *Vampyrum spectrum* and *Carollia perspicillata* is "Surinam." The species descriptions (Linnaeus, 1758) simply state "Habitat in America." Descriptions into the mid-1800s often continued to give poor information on the type locality. Type localities for *Sturnira erythromos* and *S. oporaphilum* are recorded as "Peru." After about 1850, more detail began to be recorded by collectors, and states, cities, or rivers were reported as collecting localities. Today, accurate measurements of meters above sea level along with latitude and longitude appear with new species descriptions. In part, the addition of more detailed collecting information, especially elevation, latitude, and longitude, comes as the result of increasing technological sophistication (see below) but also in response to the biodiversity crisis. Gaps in our knowledge of species ranges or of the fauna of a region need to be ameliorated before we can begin to effectively employ conservation measures. In a number of cases, subspecies recognized along geographic lines have subsequently been shown to be distinct species; high-quality geographic data on type localities may be critical for determining proper application of existing names.

Countries throughout the Americas have begun to work to catalog their biodiversity to set conservation priorities, advance economic potential, and create national pride in local resources. An official taxonomic list of bats has been published by a working group of experts in Brazil (Nogueira et al. 2014), a checklist for the phyllostomids of Colombia has been published (Mantilla-Meluk et al. 2009), and numerous smaller-scale regional or park-based studies provide inventories across the Americas (e.g., Aguirre et al. [2010] in Bolivia; Hice et al. [2004] in Peru; Lee et al. [2008] in Ecuador; Lim and Engstrom [2001] in Guyana; Paglia et al. [2012] in Brazil; Timm and McClearn [2007] in Costa Rica; Wynne and Pleytez [2005] in Belize). Nevertheless, few sites have been adequately surveyed, and local lists remain incomplete (Voss and Emmons 1996). Such gaps in knowledge of biodiversity on local scales hamper conservation efforts, management initiatives, and regional biodiversity analyses. However, even fairly well-surveyed areas can yield surprises. The phyllostomid bat fauna of the Atlantic Forest of Brazil, a biodiversity hotspot and one of the most threatened ecosystems worldwide (Brito 2004), has been intensely studied for more than two decades (e.g., Aguiar and Marinho-Filho 2004; Faria 2006; Gorreson and Willig 2004; Marinho-Filho 1996; Sazima et al. 1999). However, this scrutiny failed to reveal the existence of *Dryadonycteris*, which is apparently rare, until 2012 when it was described (Nogueira et al. 2012).

New Tools

For much of the more than 250 years that we have been applying binomial names to organisms, we have been practicing "analog taxonomy" in which data and descriptions appeared only on paper in books and journals while specimens and associated data were stored in carefully curated collections. Today, researchers operate in a digital world where many publications are available instantly via the internet, rapid publication platforms exist (e.g., *Zootaxa*), data associated with specimens can be downloaded from online databases, and there are suggestions currently to allow species description from electronic images, completely eliminating the type specimen (although see Ceríaco et al. [2016] for counterarguments). Phylogenetic analyses always involve use of computer algorithms. We now live in the age of digital taxonomy, with many new tools that aid in discovery of species and other taxa and allow rapid dissemination of new information globally.

Beginning with the use of the personal computer and the development of software for use on desktop machines in the early 1980s, there has been a revolution in what is possible in systematics research. Publication of phylogenetic trees grew by 15–20% per year between 1989 and 1991 across the tree of life (Sanderson et al. 1993). Early phylogenetics programs such as PHYLIP (distributed since 1980; Felsenstein 2005) could handle few taxa and ran slowly on older machines with little processing power. Trees and data matrices were only available in print form and had to be hand copied or reentered into your personal computer to build on to studies or repeat analyses. Computing power significantly limited the scope of phylogenetic analyses. Today, both trees and data sets are available online via websites such as TreeBASE (Piel et al. 2002), Dryad (Vision 2010), and MorphoBank (O'Leary and Kaufman 2012). Clusters of machines, accessible via the internet (e.g., CIPRES; Miller et al. 2010), can run analyses with hundreds of taxa in days. Similarly, the more powerful computers available today and widespread availability of programs such as R (R Core Team 2008) have enabled the analysis of shape—and discrimination among species—using morphometrics (e.g., Velazco and Gardner 2009b). These powerful tools have made possible the burst of species discovery seen since 2000. However, there remain few archived data sets on sites such as Dryad (18 results for "Phyllostomidae"), TreeBASE (four results for "Phyllostomidae"), and MorphoBank (eight project results for "Phyllostomidae"). Limited data archiving and sharing hampers basic research into phyllostomid systematics and taxonomy, especially given the large data sets now in use. To replicate or add to such a study would require many hours of investigator time to hand enter data.

In addition to archiving data used in publications, efforts are underway to digitize specimen collections, including specimen data, images of specimens, and geo-references of collection localities. These range from initiatives at museums, such as the American Museum of Natural History (where 52 type specimens of phyllostomids have been digitized), to national efforts such as the iDigBio website (www.idigbio.org), to global enterprises such as the Global Biodiversity Information Facility (GBIF). Currently, iDigBio, part of a 10-year effort to digitize biological collections in the United States (https://www.idigbio.org), archives records of more than 73 million specimens and 16 million images. However, the mandate for this database is limited to the United States, so few phyllostomids are included in this effort. The GBIF, however, has a global scope, storing geo-reference data, images, and information on specimens from museums and currently includes 611 generic and species-level search results for "Phyllostomidae." These repositories allow widespread access to researchers and the public alike; however, challenges related to infrastructure, methods and workflows, and standards remain to be resolved (see Beaman and Cellinese 2012).

The advent of inexpensive digital cameras, software for image processing, wireless data transfer, powerful personal computers, and the internet have revolutionized the use of images for documentation in science. Rather than being limited in the number of images a journal will publish, the limiting factor is now often the amount of time a scientist is willing to spend on collecting and processing images. In addition to standard photography, more widespread use of computed tomography (CT) scanning equipment has led to online resources, including Morphosource (www .morphosource.org) and the Digital Morphology website (www.digimorph.org), where CT scans can be deposited and archived. Digimorph includes several CT scans of bats, including *Carollia perspicillata*. Even excellent written descriptions of morphology can be confusing in the absence of images. Archiving large numbers of images documenting morphology is now possible either through websites such as MorphoBank (O'Leary and Kaufman 2012), where they can be submitted and associated with a taxonomic description of a new species (e.g., Project 761: Nogueira et al. 2012, with 51 associated media), or as part of an online phylogenetic data matrix, where homology statements are visually documented (e.g., Project 891: Dávalos et al. 2014, with 3,150 associated media).

Continuing improvements to molecular techniques, the standardization of tissue collection practices, and the continued archiving of sequences (e.g., GenBank) have all made the collection and sharing of molecular data easier than the sharing of morphological or morphometric data. Although it is tempting to attribute the surge in new phyllostomid species descriptions to the increasing use of molecular data, this does not appear to be the case. Out of the 30 papers that described new species since 2005, only 9, or 30%, included some type of analysis of molecular data (i.e., Fonseca et al. 2007; Jarrín-V and Kunz 2011; Larsen et al. 2011; Molinari et al. 2017; Siles et al. 2013; Solari and Baker 2006; Velazco and Patterson 2014; Velazco et al. 2010; Woodman 2007). Not surprisingly, papers that used molecular data more often raised species out of synonymy. From a total of 10 papers that suggested such taxonomic changes within Phyllostomidae, 90% used molecular techniques (Hoffman and Baker 2003; Hoofer et al. 2008; Larsen et al. 2010; Mantilla-Meluk 2014; Solari et al. 2009; Velazco and Cadenillas 2011; Velazco and Patterson 2013; Velazco and Patterson 2014; Velazco and Simmons 2011). The two genera new to science, *Xeronycteris* and *Dryadonycteris*, were identified on the basis of morphology (Gregorin and Ditchfield 2005; Nogueira et al. 2012), while the two new genera named in revisionary studies, *Hsunycteris* and *Gardnerycteris*, were identified as distinct lineages on the basis of phylogenetic analyses of molecular data (Hurtado and Pacheco 2014; Parlos et al. 2014). Similarly, molecular data has been crucial in identifying the branching pattern of the family resulting in the identification of new subfamilies (Baker et al. 2003).

Conclusions

We are in a golden age of discovery for phyllostomids, one that is not limited to species or genera but extends to the discovery of clades at all levels within the family. Fueled by an active and well-trained group of experts, powerful computers, new molecular techniques, and networks that have made data sharing fast and global, we expect continued species discovery as well as ongoing improvements in our understanding of the phyllostomid evolutionary tree. From slow beginnings in the late 1700s, the pace of phyllostomid discovery increased as specimen collections were amassed in Europe in the 1800s. By 1850, representatives of 10 of the 11 subfamilies that we recognize (app. 4.1) had been named, and 22 genera described. However, it was not until the early 1900s that a peak of discovery occurred, with more than 20 still valid species descriptions between 1901 and 1910. After this point, the descriptions of new genera declined. Institutions in the Americas that were founded in the late 1800s began to make important contributions in the twentieth century, and the center of research activity and species descriptions for phyllostomids shifted to North American institutions by 1950. Species discovery continued, with at least four new species described per year until 2001. Since 2001, the majority of published species descriptions have been authored or coauthored by researchers from Central and South American institutions, areas with the highest phyllostomid diversity.

How many species of phyllostomid bats might there be? If Miller (1907) was correct and there are roughly 2,000 extant species of bats, we might expect the "real" species diversity of phyllostomids to be approximately 320 species. This suggests that roughly 33% of phyllostomid species are currently undiscovered. Clare et al.'s (2011) barcoding survey of phyllostomids suggested a similar figure, that is, that roughly 34% of phyllostomid species remain to be described (see their tables 3–7). Intriguingly, the Clare et al. (2011) figure is based on cryptic diversity within species, not on the collection of species new to science; these numbers should therefore be regarded as a conservative estimate of phyllostomid species diversity.

Appendix 4.1: Classification

Our preferred classification of phyllostomid bats is provided below. Author attribution is given for each name. Species that have been described or raised from synonymy since the last publication of *Mammal Species of the World* (Simmons 2005) are shown in bold. Comments are given when appropriate, and subfamilies are simply listed in alphabetical order. This species list is also available online at http:www.batnames.org, where it is regularly updated. Detailed discussion of molecular and morphological characters that support the higher-level taxa may be found in Baker et al. (2016) and Cirranello et al. (2016).

Family Phyllostomidae Gray 1825
Subfamily Carolliinae Miller 1924
Comment: Restricted to the genus *Carollia* by
    Baker et al. 2016
    Genus *Carollia* Gray 1838
        **Carollia benkeithi Solari and Baker
        2006**
        Carollia brevicauda (Schinz 1821)
        Comment: Includes *C. colombiana*;
        see Zurc and Velazco 2010.
        Carollia castanea H. Allen 1890
        **Carollia manu Pacheco, Solari, and
        Velazco 2004**
        **Carollia monohernandezi Cuartas,
        Muñoz, and González 2004**
        Carollia perspicillata (Linnaeus 1758)
        Carollia sowelli Baker, Solari, and
        Hoffmann 2002
        Carollia subrufa (Hahn 1905)
Subfamily Desmodontinae J. A. Wagner 1840
    Tribe Desmodontini J.A. Wagner 1840
        Genus *Desmodus* Wied-Neuwied 1826
        Desmodus rotundus (É. Geoffroy 1810)
        Genus *Diaemus* Miller 1906
        Diaemus youngi (Jentink 1893)
    Tribe Diphyllini Baker, Solari, Cirranello, and
        Simmons 2016
        Genus *Diphylla* Spix 1823
        Diphylla ecaudata Spix 1823
Subfamily Glossophaginae Bonaparte 1845
    Tribe Brachyphyllini Gray 1866
        Subtribe Brachyphyllina Gray 1833

Genus *Brachyphylla* Gray 1834
    Brachyphylla cavernarum Gray 1834
    Brachyphylla nana Miller 1902
Subtribe Phyllonycterina Miller 1907
    Genus *Erophylla* Miller 1906
        Erophylla bombifrons (Miller 1899)
        Erophylla sezekorni (Gundlach 1861)
    Genus *Phyllonycteris* Gundlach 1861
    Comment: Although *Phyllonycteris*
        major was included in the last
        volume of *Mammal Species of the*
        World, this species is known only
        from subfossil material and is not
        included here.
        Phyllonycteris aphylla (Miller 1898)
        Phyllonycteris poeyi Gundlach 1861
Tribe Choeronycterini Solmsen 1998
    Subtribe Anourina Baker, Solari, Cirranello,
        and Simmons 2016
    Genus *Anoura* Gray 1838
        Anoura aequatoris (Lönnberg 1921)
        **Anoura cadenai Mantilla-Meluk and
        Baker 2006**
        **Anoura carishina Mantilla-Meluk
        and Baker 2010**
        Anoura caudifer (É. Geoffroy 1818)
        Anoura cultrata Handley 1960
        **Anoura fistulata Muchhala, Mena,
        and Albuja 2005**
        Anoura geoffroyi Gray 1838
        Anoura latidens Handley 1984
        Anoura luismanueli Molinari 1994
        Anoura peruana (Tschudi 1844)
    Subtribe Choeronycterina Solmsen 1998
    Genus *Choeroniscus* Thomas 1928
        Choeroniscus godmani (Thomas 1903)
        Choeroniscus minor (Peters 1868)
        Choeroniscus periosus Handley 1966
    Genus *Choeronycteris* Tschudi 1844
        Choeronycteris mexicana Tschudi 1844
    **Genus Dryadonycteris Nogueira, Lima,
        Peracchi, and Simmons 2012**
        **Dryadonycteris capixaba Nogueira,
        Lima, Peracchi, and Simmons
        2012**
    Genus *Hylonycteris* Thomas 1903
        Hylonycteris underwoodi Thomas 1903

Genus *Lichonycteris* Thomas 1895
    ***Lichonycteris degener* Miller 1931**
    Lichonycteris obscura Thomas 1895
Genus *Musonycteris* Schaldach and
    McLaughlin 1960
    Musonycteris harrisoni Schaldach and
    McLaughlin 1960
Genus *Scleronycteris* Thomas 1912
    Scleronycteris ega Thomas 1912
Tribe Glossophagini Bonaparte 1845
    Genus *Glossophaga* É. Geoffroy 1818
        Glossophaga commissarisi Gardner
        1962
        Glossophaga leachii Gray 1844
        Glossophaga longirostris Miller 1898
        Glossophaga morenoi Martínez and
        Villa-R. 1938
        Glossophaga soricina (Pallas, 1766)
    Genus *Leptonycteris* Lydekker 1891
        Leptonycteris curasoae Miller 1900
        Leptonycteris nivalis (Saussure 1860)
        Leptonycteris yerbabuenae Martínez
        and Villa-R. 1940
    Genus *Monophyllus* Leach 1821
        Monophyllus plethodon Miller 1900
        Monophyllus redmani Leach 1821
Subfamily Glyphonycterinae Baker, Solari,
    Cirranello, and Simmons 2016
    Genus *Glyphonycteris* Thomas 1896
        Glyphonycteris behnii (Peters 1865)
        Glyphonycteris daviesi (Hill 1965)
        Glyphonycteris sylvestris Thomas 1896
    Genus *Neonycteris* Sanborn 1949
        Neonycteris pusilla (Sanborn 1949)
    Genus *Trinycteris* Sanborn 1949
        Trinycteris nicefori (Sanborn, 1949)
Subfamily Lonchophyllinae Griffiths 1982
    Tribe Hsunycterini Parlos, Timm, Swier,
    Zeballos, and Baker 2014
    **Genus *Hsunycteris* Parlos, Timm, Swier,
    Zeballos, and Baker 2014**
        ***Hsunycteris cadenai* (Woodman and
        Timm 2006)**
        ***Hsunycteris pattoni* (Woodman and
        Timm 2006)**
        Hsunycteris thomasi J. A. Allen 1904
    Tribe Lonchophyllini Griffiths 1982
    Genus *Lionycteris* Thomas 1913

        Lionycteris spurrelli Thomas 1913
    Genus *Lonchophylla* Thomas 1903
        Lonchophylla bokermanni Sazima,
        Vizotto, and Taddei 1978
        ***Lonchophylla chocoana* Dávalos
        2004**
        ***Lonchophylla concava* Goldman
        1914**
        Lonchophylla dekeyseri Taddei, Vizotto,
        and Sazima 1983
        ***Lonchophylla fornicata* Woodman
        2007**
        Lonchophylla handleyi Hill, 1980
        Lonchophylla hesperia G. M. Allen
        1908
        ***Lonchophylla inexpectata* Moratelli
        and Dias 2015**
        Lonchophylla mordax Thomas 1903
        ***Lonchophylla orcesi* Albuja and
        Gardner 2005**
        ***Lonchophylla orienticollina* Dávalos
        and Corthals 2009**
        ***Lonchophylla peracchii* Dias, Esbé-
        rard and Moratelli 2013**
        Lonchophylla robusta Miller 1912
    Genus *Platalina* Thomas 1928
        Platalina genovensium Thomas 1928
    **Genus *Xeronycteris* Gregorin and
    Ditchfield 2005**
        ***Xeronycteris vieirai* Gregorin and
        Ditchfield 2005**
Subfamily Lonchorhininae Gray 1866
    Genus *Lonchorhina* Tomes 1863
        Lonchorhina aurita Tomes 1863
        Lonchorhina fernanadezi Ochoa and
        Ibáñez 1982
        Lonchorhina inusitata Handley and
        Ochoa 1997
        ***Lonchorhina mankomara* Mantilla-
        Meluk and Montenegro 2016**
        Lonchorhina marinkellei Hernández-
        Camacho and Cadena-G. 1978
        Lonchorhina orinocensis Linares and
        Ojasti 1971
Subfamily Macrotinae Van Den Bussche 1992
    Genus *Macrotus* Gray 1843
        Macrotus californicus Baird 1858
        Macrotus waterhousii Gray 1843

Subfamily Micronycterinae Van Den Bussche 1992

    Genus *Lampronycteris* Sanborn 1949

        Lampronycteris brachyotis (Dobson 1879)

    Genus *Micronycteris* Gray 1866

        Micronycteris brosseti Simmons and Voss 1998

        Micronycteris buriri Larsen, Siles, Pederson, and Kwiecinski 2011

        Micronycteris giovanniae Baker and Fonseca 2007

        Micronycteris hirsuta (Peters 1869)

        Micronycteris matses Simmons, Voss, and Fleck 2002

        Micronycteris megalotis (Gray 1842)

        Micronycteris microtis Miller 1898

        Micronycteris minuta (Gervais 1856)

        Comment: Includes *M. homezorum* Pirlot 1967; see Ochoa and Sanchez 2005, Porter et al. 2007, and Solari 2008.

        Micronycteris sanborni Simmons, 1996

        Micronycteris schmidtorum Sanborn 1935

        Micronycteris yatesi Siles and Brooks 2013

Subfamily Phyllostominae Gray 1825

    Tribe Macrophyllini Gray 1866

        Genus *Macrophyllum* Gray 1838

            Macrophyllum macrophyllum (Schinz 1821)

        Genus *Trachops* Gray 1847

            Trachops cirrhosus (Spix 1823)

    Tribe Phyllostomini Gray 1825

        Genus *Gardnerycteris* Hurtado and Pacheco 2014

        Gardnerycteris crenulatum (É. Geoffroy 1803)

        Gardnerycteris keopckeae (Gardner and Patton 1972)

        Genus *Lophostoma* d'Orbigny 1836

        Lophostoma brasiliense Peters 1866

        Lophostoma carrikeri (J. A. Allen 1910)

        Comment: Includes *L. yasuni* Fonseca and Pinto 2004; see Camacho et al 2016.

        Lophostoma evotis (Davis and Carter 1978)

        Lophostoma kalkoae Velazco and Gardner 2012

        Lophostoma occidentalis (Davis and Carter 1978)

        Lophostoma schulzi (Genoways and Williams 1980)

        Lophostoma silvicolum d'Orbigny 1836

    Genus *Phylloderma* Peters 1865

        Phylloderma stenops Peters 1865

    Genus *Phyllostomus* Lacépède 1799

        Phyllostomus discolor Wagner 1843

        Phyllostomus elongatus (É. Geoffroy 1810)

        Phyllostomus hastatus (Pallas 1767)

        Phyllostomus latifolius (Thomas 1901)

    Genus *Tonatia* Gray 1827

        Tonatia bidens (Spix 1823)

        Tonatia saurophila Koopman and Williams 1951

    Tribe Vampyrini Bonaparte 1837

        Genus *Chrotopterus* Peters 1865

        Chrotopterus auritus (Peters 1856)

        Genus *Mimon* Gray 1847

        Mimon bennettii (Gray 1838)

        Mimon cozumelae Goldman 1914

        Genus *Vampyrum* Rafinesque 1815

        Vampyrum spectrum (Linnaeus 1758)

Subfamily Rhinophyllinae Baker, Solari, Cirranello, and Simmons 2016

    Genus *Rhinophylla* Peters 1865

        Rhinophylla alethina Handley 1966

        Rhinophylla fischereae Carter 1966

        Rhinophylla pumilio Peters 1865

Subfamily Stenodermatinae Gervais 1856

    Tribe Stenodermatini Gervais 1856

        Subtribe Artibeina H. Allen 1898

        Genus *Artibeus* Gray 1838

        Artibeus aequatorialis Anderson 1906

        Artibeus amplus Handley 1987

        Artibeus anderseni Osgood 1916

        Artibeus aztectus K. Andersen 1906

        Artibeus bogotensis K. Andersen 1906

        Artibeus cinereus (Gervais, 1856)

        Artibeus concolor Peters 1865

        Artibeus fimbriatus Gray 1838

        Artibeus fraterculus Anthony 1924

        Artibeus glaucus Thomas 1893

Artibeus gnomus Handley 1987

Artibeus hirsutus K. Andersen 1906

Artibeus inopinatus Davis and Carter 1964

Artibeus jamaicensis Leach 1821

Artibeus lituratus (Olfers 1818)

Artibeus obscurus (Schinz 1821)

Artibeus phaeotis (Miller 1902)

Artibeus planirostris (Spix 1823)

Artibeus ravus (Miller 1902)

Artibeus rosenbergi Thomas 1897

Artibeus schwartzi Jones 1978

Artibeus toltecus (Saussure 1860)

Artibeus watsoni Thomas 1901

Comment: Includes *A. incomitatus* Kalko and Handley 1994; see Solari et al. 2009.

Subtribe Ectophyllina Baker, Solari, Cirranello, and Simmons 2016

Genus *Ectophylla* H. Allen 1892

Ectophylla alba H. Allen 1892

Subtribe Enchisthenina Baker, Solari, Cirranello, and Simmons 2016

Genus *Enchisthenes* K. Andersen 1906

Enchisthenes hartii (Thomas 1892)

Subtribe Stenodermatina Gervais 1856

Genus *Ametrida* Gray 1847

Ametrida centurio Gray 1847

Genus *Ardops* Miller 1906

Ardops nichollsi (Thomas 1891)

Genus *Ariteus* Gray 1838

Ariteus flavescens (Gray 1831)

Genus *Centurio* Gray 1842

Centurio senex Gray 1842

Genus *Phyllops* Peters 1865

Phyllops falcatus (Gray 1839)

Genus *Pygoderma* Peters 1863

Pygoderma bilabiatum (Wagner 1843)

Genus *Sphaeronycteris* Peters 1882

Sphaeronycteris toxophyllum Peters 1882

Genus *Stenoderma* É. Geoffroy 1818

Stenoderma rufum Desmarest 1820

Subtribe Vampyressina Baker, Solari, Cirranello, and Simmons 2016

Genus *Chiroderma* Peters 1860

Chiroderma doriae Thomas 1891

Chiroderma improvisum Baker and Genoways 1976

Chiroderma salvini Dobson 1878

Chiroderma trinitatum Goodwin 1958

Chiroderma villosum Peters 1860

Chiroderma vizottoi Taddei and Lim 2010

Genus *Mesophylla* Thomas 1901

Mesophylla macconnelli Thomas 1901

Genus *Platyrrhinus* Saussure 1860

Platyrrhinus albericoi Velazco 2005

Platyrrhinus angustirostris Velazco, Gardner, and Patterson 2010

Platyrrhinus aquilus (Handley and Ferris 1972)

Platyrrhinus aurarius (Handley and Ferris 1972)

Platyrrhinus brachycephalus (Rouk and Carter 1972)

Platyrrhinus chocoensis Alberico and Velasco 1991

Platyrrhinus dorsalis (Thomas 1900)

Platyrrhinus fusciventris Velazco, Gardner, and Patterson 2010

Platyrrhinus guianensis Velazco and Lim 2014

Platyrrhinus helleri (Peters 1866)

Platyrrhinus incarum (Thomas 1912)

Platyrrhinus infuscus (Peters 1880)

Platyrrhinus ismaeli Velazco 2005

Platyrrhinus lineatus (Geoffroy 1810)

Platyrrhinus masu Velazco 2005

Platyrrhinus matapalensis Velazco 2005

Platyrrhinus nigellus (Gardner and Carter 1972)

Platyrrhinus nitelinea Velazco and Gardner 2009

Platyrrhinus recifinus (Thomas 1901)

Platyrrhinus umbratus (Lyon 1902)

Platyrrhinus vittatus (Peters 1860)

Genus *Uroderma* Peters 1866

Uroderma bakeri Mantilla-Meluk 2014

Uroderma bilobatum Peters 1866

Uroderma convexum Lyon 1902

Uroderma davisi Baker and Mc-
    Daniel 1972
Uroderma magnirostrum Davis 1968
Genus *Vampyressa* Thomas 1900
  **Vampyressa elisabethae Tavares,
    Gardner, Ramírez-Chaves, and
    Velazco 2014**
Vampyressa melissa Thomas 1926
Vampyressa pusilla (Wagner 1843)
  **Vampyressa sinchi Tavares, Gardner,
    Ramírez-Chaves, and Velazco
    2014**
Vampyressa thyone Thomas 1909
Genus *Vampyriscus* Thomas 1900
  Comment: Previously considered a
    subgenus of *Vampyressa*, but raised
    to generic level based on a phyloge-
    netic analysis by Hoofer et al. 2008.
Vampyriscus bidens (Dobson 1878)
Vampyriscus brocki (Peterson 1968)
Vampyriscus nymphaea (Thomas
    1909)
Genus *Vampyrodes* Thomas 1900
Vampyrodes caraccioli (Thomas 1889)
  Vampyrodes major G.M. Allen 1908
Tribe Sturnini Miller 1907
Genus *Sturnira* Gray 1842
  **Sturnira adrianae Molinari, Bustos,
    Burneo, Camacho, Moreno, and
    Férmin 2017**
  Sturnira angeli de la Torre 1966
  Comment: Includes *S. thomasi* de
    la Torre and Schwartz 1966; see
    Velazco and Patterson, 2013, who,
    as first revisers, selected *angeli* as
    the correct name for this species
    following ICZN (1999) guidelines.
Sturnira arathomasi Peterson and
    Tamsitt 1968
  **Sturnira bakeri Velazco and Patter-
    son 2014**
Sturnira bidens Thomas 1915
Sturnira bogotensis Shamel 1947
  **Sturnira burtonlimi Velazco and
    Patterson 2014**
Sturnira erythromos (Tschudi 1844)
  Sturnira hondurensis Goodwin 1940

**Sturnira koopmanhilli McCarthy,
  Albuja, and Alberico 2006**
Sturnira lilium (Geoffroy 1810)
Sturnira ludovici Anthony 1924
Sturnira luisi Davis 1980
Sturnira magna de la Torre 1966
Sturnira mistratensis Vega and Cadena
  2000
Sturnira mordax (Goodwin 1938)
Sturnira nana (Gardner and O'Neill
  1971)
Sturnira oporaphilum (Tschudi 1844)
Sturnira parvidens Goldman 1917
Sturnira paulsoni de la Torre 1966
**Sturnira perla Jarrín-V and Kunz
  2011**
**Sturnira sorianoi Sánchez-
  Hernández, Romero-Almaraz,
  and Schnell 2005**
Sturnira tildae de la Torre 1959

Acknowledgments

This contribution has grown out of a research collaboration that began decades ago. We wish to thank our collaborators over the years, most especially Liliana Dávalos, Norberto Giannini, Sergio Solari, and Paúl Velazco. To the editors, who invited our participation in this volume, we offer our thanks. We are grateful to Paúl Velazco and our two anonymous reviewers whose comments and critiques improved this paper.

References

Agnarsson, I., and M. Kuntner. 2007. Taxonomy in a changing world: seeking solutions for a science in crisis. Systematic Biology, 56:531–539.

Aguiar, L. M. de Souza, and J. Marinho-Filho. 2004. Activity patterns of nine phyllostomid bat species in a fragment of the Atlantic Forest in southeastern Brazil. Revista Brasileira de Zoologia, 21:385–390.

Aguirre, L. F., C. J. Mamani, K. Barbosa-Márquez, and H. Mantilla-Meluk. 2010. Lista actualizada de los murciélagos de Bolivia. Revista Boliviana de Ecología y Conservación Ambiental, 27:1–7.

Allen, J. A. 1890. Description of a new species of bat of the genus *Carollia*, and remarks on *Carollia brevicauda*. Proceedings of the United States National Museum, 13:201–298.

Azara, F. 1809. Voyage dans l'Amérique Méridionale. Dentu. Paris.

Baker, R. J., S. R. Hoofer, C. A. Porter, and R. A. Van Den Bussche. 2003. Diversification among New World leafnosed bats: an evolutionary hypothesis and classification inferred from digenomic congruence of DNA sequence. Occasional Papers, Museum of Texas Tech University, 230:1–32.

Baker, R., S. Solari, A. Cirranello, and N.B. Simmons. 2016. Higher level classification of phyllostomid bats with a summary of DNA synapomorphies. Acta Chiropterologica, 18:1–38.

Beaman, R., and N. Cellinese. 2012. Mass digitization of scientific collections: new opportunities to transform the use of biological specimens and underwrite biodiversity science. ZooKeys, 209:7–17.

Bonaparte, C. L. 1845. Catalogo metodico dei mammiferi Europei. Milan, L. di Giacomo Pirola.

Brito, D. 2004. Lack of adequate taxonomic knowledge may hinder endemic mammal conservation in the Brazilian Atlantic Forest. Biodiversity and Conservation, 13:2135–2144.

Camacho, M. A., D. Chavez, and S. F. Burneo. 2016. A taxonomic revision of the Yasuni round-eared bat, *Lophostoma yasuni* (Chiroptera: Phyllostomidae). Zootaxa, 4114:246–260.

Ceríaco, L. M. P., E. E. Gutiérrez, and A. Dubois. 2016. Photography-based taxonomy is inadequate, unnecessary, and potentially harmful for biological sciences. Zootaxa, 4196:435–445.

Cirranello, A., N. B. Simmons, S. Solari, and R. J. Baker. 2016. Morphological diagnoses of higher-level phyllostomid taxa (Chiroptera: Phyllostomidae). Acta Chiropterologica, 18:39–71.

Clare, E. L., B. K. Lim, M. B. Fenton, and P. D. N. Hebert. 2011. Neotropical bats: estimating species diversity with DNA barcodes. PLOS ONE 6:e22648.

Cope, E. D. 1889. On the Mammalia obtained by the Naturalist Exploring Expedition to Southern Brazil. American Naturalist, 23 (266): 128–150.

Cotterill, F. P. D, and W. Foissner. 2010. A pervasive denigration of natural history misconstrues how biodiversity inventories and taxonomy underpin scientific knowledge. Biodiversity and Conservation, 19:291–303.

Cuvier, G. 1817. Le règne animal distribue d'après son organisation, pour servir de base à l'histoire naturelle des animaux et d'introduction à l'anatomie comparee, tomus 1. Deterville, Paris.

Dávalos, L. M., A. L. Cirranello, J. H. Geisler, and N. B. Simmons. 2012. Understanding phylogenetic incongruence: lessons from phyllostomid bats. Biological Reviews, 87:991–1024.

Dávalos L. M., P. M. Velazco, O. M. Warsi, P. D. Smits, and N. B. Simmons. 2014. Integrating incomplete fossils by isolating conflicting signal in saturated and non-independent morphological characters. Systematic Biology, 63:582–600.

Davis, W. B., and D. C. Carter. 1978. A review of the round-eared bats of the *Tonatia silvicola* complex, with descriptions of three new taxa. Occasional Papers, Museum of Texas Tech University, 53:1–12.

Dobson, G. E. 1875. Conspectus of the suborders, families, and genera of Chiroptera, arranged according to their natural affinities. Annals and Magazine of Natural History ser. 4, 16:345–357.

Dobson, G. E. 1878. Catalogue of the Chiroptera in the collection of the British Museum. British Museum, London.

Donoghue, M. J., and W. S. Alverson. 2000. A new age of discovery. Annals of the Missouri Botanical Garden, 87:110–126.

Dumont, E. R., L. M. Dávalos, A. Goldberg, C. C. Voigt, K. Rex, and S. E. Santana. 2012. Morphological innovation, diversification and the invasion of a new adaptive zone. Proceedings of the Royal Society Series B–Biological Sciences, 279:1797–1805.

Edwards, G. 1751. A natural history of birds, part IV. College of Physicians, London.

Faria, D. 2006. Phyllostomid bats of a fragmented landscape in the north-eastern Atlantic Forest, Brazil. Journal of Tropical Ecology, 22:531–542.

Felsenstein J. 2005. PHYLIP (phylogeny inference package). Seattle: Department of Genome Sciences, University of Washington.

Fonseca, R. M., S. R. Hoofer, C. A. Porter, C. A. Cline, D. A. Parish, F. G. Hoffmann, and R. J. Baker. 2007. Morphological and molecular variation within little big-eared bats of the genus *Micronycteris* (Phyllostomidae: Micronycterinae) from San Lorenzo, Ecuador. Pp. 721–724 *in:* The quintessential naturalist: honoring the life and legacy of Oliver P. Pearson (D. A. Kelt, E. P. Lessa, J. Salazar-Bravo, and J. L. Patton, eds.). University of California Publications in Zoology 134.

Francis, C. M., A. V. Borisenko, N. V. Ivanova, J. L. Eger, B. K. Lim, A. Guillén-Servent, S. V. Kruskop, I. Mackie, and P. D. N. Hebert. 2010. The role of DNA barcodes in understanding and conservation of mammal diversity in Southeast Asia. PLOS ONE, e12575.

Genoways, H. H., and S. L. Williams. 1980. Results of the Alcoa Foundation–Suriname Expeditions. 1. A new species of bat of the genus *Tonatia* (Mammalia: Phyllostomatidae). Annals of the Carnegie Museum, 49:203–211.

Geoffroy, É. 1803. Catalogue des Mammifères du Muséum National d'Histoire Naturelle. Paris.

Geoffroy, É. 1810. Sur les Phyllostomes et les Mégadermes, deux genres de la famille des chauve-souris. Annales du Muséum National d'Histoire Naturelle, 15:157–198.

Geoffroy, É. 1818a. Description des mammifères qui se trouvent en Egypte. Pp. 99–144 *in:* Description de l'Égypte, ou, Recueil des observations et des recherches qui ont été faites en Égypte pendant l'expédition de l'armée française, tome 2. Imprimerie Imperiale, Paris.

Geoffroy, É. 1818b. Sur de nouvelles chauve-souris sous le nom de Glossophages. Mémoires du Muséum d'Histoire Naturelle, 4:411–418.

Gervais, P. 1854. Histoire naturelle des mammifères, partie I. L. Curmer, Paris.

Gervais, P. 1856. Deuxième mémoire. Documents zoologiques pour servir à la monographie des chéiroptères Sud–Américains. Pp. 25–88 *in:* Mammifères (P. Gervais , ed.). Pp. 1–116 *in* Animaux nouveaux ou rares recueillis pendant l'expédition dans les parties centrales de l'Amérique du Sud, de Rio de Janeiro à Lima, et de Lima au Para; executée par ordre du gouvernement français pendant les années 1843 à 1847, sous la direction du Comte Francis Castelnau, volume 1, tome 2 (F. Castelnau, ed.). P. Bertran, Paris. [dated 1855, but published 1856; see Annals and Magazine of Natural History, ser 7, 8:164].

Goldfuss, G. A. 1820. Handbuch der zoologie, theil. 2. Johann Leonhard Schrag, Nurnberg.

Gorresen, P. M., and M. R. Willig. 2004. Landscape responses of bats to habitat fragmentation in Atlantic Forest of Paraguay. Journal of Mammalogy, 85:688–697.

Gosse, P. H. 1851. A Naturalist's Sojourn in Jamaica. Longman, Brown, Green, and Longmans, London.

Gray, J. E. 1821 On the natural arrangement of vertebrose animals. London Medical Repository, 15:296–310.

Gray, J. E. 1825. An attempt at a division of the family Vespertilionidae into groups. Zoological Journal, 2:242–243.

Gray, J. E. 1866. Revision of the genera of Phyllostomidae, or leaf-nosed bats. Proceedings of the Zoological Society of London, 1866:111–118.

Gregorin, R., and A. D. Ditchfield. 2005. New genus and species of nectar-feeding bat in the tribe Lonchophyllini (Phyllostomidae: Glossophaginae) from northeastern Brazil. Journal of Mammalogy, 86:403–414.

Griffiths, T. A. 1982. Systematics of the New World nectar-feeding bats (Mammalia, Phyllostomidae), based on the morphology of the hyoid and lingual regions. American Museum Novitates, 2742:1–45.

Gundlach, J. 1861. Monatsberichte der Königlichen Preussische Akademie des Wissenschaften zu Berlin, 1860:817–819. [Dated 1860, but published 1861].

Handley, C. O. 1960. Descriptions of new bats from Panama. Proceedings of the United States National Museum, 112:459–479.

Hernández-Camacho, J., and A. Cadena-G. 1978. Notas para la revision del genero Lonchorhina (Chiroptera, Phyllostomidae). Caldasia, 12:199–251.

Hice, C. L., P. M. Velazco, and M. R. Willig. 2004. Bats of the Reserva Nacional Allpahuayo-Mishana, northeastern Peru, with notes on community structure. Acta Chiropterologica, 6:319–334.

Hoffmann, F. G., and R. J. Baker. 2003. Comparative phylogeography of short-tailed bats (Carollia: Phyllostomidae). Molecular Ecology, 12:3403–3414.

Hoofer, S. R., S. Solari, P. A. Larsen, R. D. Bradley, and R. J. Baker. 2008. Phylogenetics of the fruit-eating bats (Phyllostomidae: Artibeina) inferred from mitochondrial DNA sequences. Occasional Papers, Museum of Texas Tech University, 277:1–15.

Hurtado, N., and V. Pacheco. 2014. Análisis filogenético del género Mimon Gray, 1847 (Mammalia, Chiroptera, Phyllostomidae) con la descripción de un nuevo género. Therya 5:751–791.

Jarrín-V, P., and T. H. Kunz. 2011. A new species of Sturnira (Chiroptera: Phyllostomidae) from the Choco forest of Ecuador. Zootaxa, 2755:1–35.

Lacépède, B. G. 1799. Tableau des divisions, sous-divisions, ordres, et genres des mammifères. Chez Plasson, Paris.

Larsen, P. A., M. R. Marchán-Rivedeneira, and R. J. Baker. 2010. Taxonomic status of Andersen's fruit-eating bat (Artibeus jamaicensis aequatorialis) and revised classification of Artibeus (Chiroptera: Phyllostomidae). Zootaxa, 2648:45–60.

Larsen, P. A., L. Siles, S. C. Pedersen, and G. G. Kwiecinski. 2011. A new species of Micronycteris (Chiroptera: Phyllostomidae) from Saint Vincent, Lesser Antilles. Mammalian Biology–Zeitschrift für Säugetierkunde, 76:687–700.

Lee, T. E., S. F. Burneo, M. R. Marchán, S. A. Roussos, and R. S. Vizcarra-Vásconez. 2008. The mammals of the temperate forests of Volcán Sumaco, Ecuador. Occasional Papers, Museum of Texas Tech University, 276:1–10.

Lee, T. E., Jr., S. R. Hoofer, and R. A. Van Den Bussche. 2002. Molecular phylogenetics and taxonomic revision of the genus Tonatia (Chiroptera: Phyllostomidae). Journal of Mammalogy, 83:49–57.

Lim, B. K., and M. D. Engstrom. 2001. Species diversity of bats (Mammalia: Chiroptera) in Iwokrama Forest, Guyana, and the Guianan subregion: implications for conservation. Biodiversity and Conservation, 10:613–657.

Linnaeus, C. 1758. Systema naturae per regna tria naturae, secundum classis, ordines, genera, species cum characteribus, differentiis,synonymis, locis, vol. 1, 10th ed. Laurentii Salvii, Stockholm.

Lipkowicz, E. 2013. The "elephant in the room": the impact of the French seizure of the Dutch Stadholder's collection on relations between Dutch and French naturalists. Pp. 101–109 in: Of Elephants and Roses (S. Prince ed.). American Philosophical Society Memoirs, vol. 267.

Mantilla-Meluk, H. 2014. Defining species and species boundaries in Uroderma (Chiroptera: Phyllostomidae) with a description of a new species. Occasional Papers of the Museum of Texas Tech University, 325:1–25.

Mantilla-Meluk, H., A. M. Jiménez-Ortega, and R. J. Baker. 2009. Phyllostomid bats of Colombia: annotated checklist, distribution, and biogeography. Occasional Papers, Museum of Texas Tech University, 261:1–18.

Marinho-Filho, J. 1996. Distribution of bat diversity in the southern and southeastern Brazilian Atlantic Forest. Chiroptera Neotropical, 2:51–54.

Miller, G. S. 1906. Twelve new genera of bats. Proceedings of the Biological Society of Washington, 19:83–86.

Miller, G. S. 1907. The families and genera of bats. Bulletin of the U.S. National Museum, 57:1–282.

Miller, G. S. 1924. List of North American recent mammals. Bulletin of the U.S. National Museum, 128:1–673.

Miller, M. A., W. Pfeiffer, and T. Schwartz. 2010. Creating the CIPRES Science Gateway for inference of large phylogenetic trees. Pp. 1–8 in: Gateway Computing Environments Workshop (GCE), 2010. IEEE, [Piscataway, NJ].

Molinari, J., X. E. Bustos, S. F. Burneo, M. A. Camacho, S. A. Moreno, and G. Fermín. 2017. A new polytypic species of yellow-shouldered bats, genus Sturnira (Mammalia: Chiroptera: Phyllostomidae), from the Andean and coastal mountain systems of Venezuela and Colombia. Zootaxa, 4243:75–96.

Nogueira, M. R., I. P. Lima, R. Moratelli, V. Tavares, R. Gregorin, and A. L. Peracchi. 2014. Checklist of Brazilian bats, with comments on original records. Checklist, 10 (4): 808–821.

Nogueira, M. R., I. P. Lima, A. L. Peracchi, and N. B. Simmons. 2012. New genus and species of nectar-feeding bat from the Atlantic Forest of southeastern Brazil (Chiroptera: Phyllostomidae: Glossophaginae). American Museum Novitates, 3747:1–30.

Ochoa, J. G., and J. H. Sánchez. 2005. Taxonomic status of *Micronycteris homezi* (Chiroptera, Phyllostomidae). Mammalia, 69:323–335.

O'Leary, M. A., and S. Kaufman. 2012. MorphoBank 3.0: web application for morphological phylogenetics and taxonomy. http://www.morphobank.org.

Oken, L. 1816. Lehrbuch der Naturgeschichte, theil 3, abt., 2. August Schmidt, Jena.

Owen, R. D. 1987. Phylogenetic analyses of the bat subfamily Stenodermatinae (Mammalia: Chiroptera). Special Publications, Museum of Texas Tech University, 26:1–65.

Paglia, A. P., G. A. B. da Fonseca, A. B. Rylands, G. Herrmann, L. M. S. Aguiar, A. G. Chiarello, Y. L. R. Leite et al. 2012. Lista anotada dos Mamíferos do Brasil, 2nd edição. Occasional Papers in Conservation Biology 6:1–76.

Pallas, P. S. 1766. P. S. Pallas medicinae doctoris Miscellanea zoologica: quibus novae imprimis atque obscurae animalium species describuntur et observationibus iconibusque illustrantur. Petrum van Cleef, Hagae Comitum.

Pallas, P. S. 1767. Spicilegia zoologica, continens quadrupedium, avium, amphibiorum, piscium, insectorum, molluscorum aliorumque marinorum, fasciculos decem. Gottl. August. Lange, Berolini.

Parlos, J. A., R. M. Timm, V. J. Swier, H. Zeballos, and R. J. Baker. 2014. Evaluation of the paraphyletic assemblages within Lonchophyllinae, with description of a new tribe and genus. Occasional Papers of the Museum of Texas Tech University, 320:1–23.

Pearson, D. L., A. L. Hamilton, and T. L. Erwin. 2011. Recovery plan for the endangered taxonomy profession. BioScience, 61:58–63.

Peters, W. 1860. Neue Beiträge zur Kenntnifs der Chiropteren. Monatsberichte der Königlichen Preussische Akademie des Wissenschaften zu Berlin, 1859:222–225. [Dated 1859, but published 1860].

Peters, W. 1865. Vorlage von Abbildungen zu einer Monographie der Chiropteren, und Ubersicht der von ihm befolgten systematischen Ordnung der hieher gehorigen Gattungen. Monatsberichte der Königlichen Preussische Akademie des Wissenschaften zu Berlin, 1865:265–258.

Peters, W. 1866. Fernere Mittheilungen zur Kenntnifs der Fleiderthiere, namentlich über Arten des Leidener und Britischen Museums. Monatsberichte der Königlichen Preussische Akademie des Wissenschaften zu Berlin, 1866:672–681.

Peters, W. 1880. Mittheilung über neue Flederthiere. Monatsberichte der Königlichen Preussische Akademie des Wissenschaften zu Berlin, 1880:258–259. [Dated 1881, but published 1880].

Peters, W. 1882. Über *Sphaeronycteris toxophyllum*, eine neue Gattung und Art der frugivoren blattnasigen Fliederthiere, aus dem tropischen America. Sitzungsberichte der Königlich Preussischen Akademie der Wissenschaften zu Berlin, 1882:987–990.

Piel, W. H., M. J. Donoghue, and M. J. Sanderson. 2002. TreeBASE: a database of phylogenetic knowledge. Pp. 41–47, *in*: To the interoperable "Catalog of Life" with partners Species 2000 Asia Oceanea (J. Shimura, K. L. Wilson, and D. Gordon, eds.). Research Report from the National Institute for Environmental Studies No. 171, Tsukuba, Japan.

Piso, W. 1648. Historia naturalis Brasiliae, auspicio et beneficio illustriss. I. Mauritii Com. Nassau . . . adornata: in qua non tantum plantæ et animalia, sed et indigenarum morbi, ingenia et mores describuntur et iconibus supra quingentas illustrantur. Ionnes de Laet, Antwerpianus.

Porter, C. A., S. R. Hoofer, C. A. Cline, F. G. Hoffmann, and R. J. Baker. 2007. Molecular phylogenetics of the phyllostomid bat genus *Micronycteris* with descriptions of two new subgenera. Journal of Mammalogy, 88:1205–1215.

Rafinesque, C. S. 1815. Analyse de la nature ou tableau de l'univers et des corps organisés. Jean Barravecchia, Palermo.

R Core Team. 2016. R: a language and environment for statistical computing. R Foundation for Statistical Computing, Vienna, Austria. http://www.R-project.org/.

Rojas, D., O. M. Warsi, and L. M. Dávalos. 2016. Bats (Chiroptera: Noctilionoidea) challenge a recent origin of extant Neotropical diversity. Systematic Biology, 65:432–448.

Sanderson, M. J., B. G. Baldwin, G. Bharathan, C. S. Campbell, C. Von Dohlen, D. Ferguson, J. M. Porter, M. F. Wojciechowski, and M. J. Donoghue. 1993. The growth of phylogenetic information and the need for a phylogenetic data base. Systematic Biology, 42:562–568.

Sazima, M., S. Buzato, and I. Sazima. 1999. Bat-pollinated flower assemblages and bat visitors at two Atlantic Forest sites in Brazil. Annals of Botany, 83:705–712.

Schaldach, W. J., and C. A. Mc Laughlin. 1960. A new genus and species of glossophagine bat from Colima Mexico. Los Angeles County Museum Contributions in Science, 37:1–8.

Seba, A. 1734. Locupletissimi rerum naturalium thesauri accurata descriptio, et iconibus artificiosissimis expressio, per universam physices historiam: opus, cui, in hoc rerum genere, nullum par exstitit. Book 1 (Tomus I) Amstelaedami: Apud J. Wetstenium, & Gul. Smith, & Janssonio-Waesbergios, 1734–1765.

Siles, L., D. M. Brooks, H. Aranibar, T. Tarifa, V. M. R. Julieta, J. M. Rojas and R. J. Baker. 2013. A new species of *Micronycteris* (Chiroptera: Phyllostomidae) from Bolivia. Journal of Mammalogy 94:881–896.

Simmons, N. B. 2005. Chiroptera. Pp. 312–529 *in*: Mammal Species of the World: A Taxonomic and Geographic Reference (D. E. Wilson and D. M. Reeder, eds.). Johns Hopkins University Press, Baltimore.

Simpson, G. G. 1945. The principles of classification and a classification of mammals. Bulletin of the American Museum of Natural History, 85:1–350.

Sloane, H. 1725. A voyage to the islands Madera, Barbadoes, Nieves, St. Christophers and Jamaica, vol. 2. London.

Solari, S. 2008. Mistakes in the formation of species-group names for Neotropical bats: *Micronycteris* and *Sturnira* (Phyllostomidae). Acta Chiropterologica 10:380–382.

Solari, S., and R. J. Baker. 2006. Mitochondrial DNA sequence, karyotypic, and morphological variation in the *Carollia castena* species complex (Chiroptera: Phyllostomidae) with description of a new species. Occasional Papers, Museum of Texas Tech University, 254:1–16.

Solari, S., S. R. Hoofer, P. A. Larsen, A. D. Brown, R. J. Bull, J. A. Guerrero, J. Ortega, J. P. Carrera, R. D. Bradley, and R. J. Baker. 2009. Operational criteria for genetically defined species: analysis of the diversification of the small fruit-eating bats, *Dermanura* (Phyllostomidae: Stenodermatinae). Acta Chiropterologica, 11:279–288.

Spix, J. 1823. Simiarum et vespertilionum brasiliensum: species novae ou histoire naturelle des espèces nouvelles des singes et de chauve-souris, observées et recueille pendant le voyage dans l'intérieur du Brésil. F. S. Hubschmann, Munich.

Tavares, V. Da C., A. L. Gardner, H. E. Ramírez-Chaves, and P. M. Velazco. 2014. Systematics of *Vampyressa melissa* Thomas, 1926 (Chiroptera: Phyllostomidae), with descriptions of two new species of *Vampyressa*. American Museum Novitates, 3813:1–27.

Thomas, O. 1903. Two new glossophagine bats from Central America. Annals and Magazine of Natural History ser. 7, 11:286–289.

Thomas, O. 1906. Mammals. Pp. 3–66 *in*: The history of the collections contained in the natural history departments of the British Museum. Vol. 2. Longmans and Co., London.

Thomas, O. 1913. On a new glossophagine bat from Colombia. Annals and Magazine of Natural History ser. 8, 12:270–271.

Thomas, O. 1928. The Godman-Thomas expedition to Peru. VII. The mammals of the Rio Ucayali. Annals and Magazine of Natural History ser. 10, 2:249–265.

Timm, R. M., and D. K. McClearn. 2007. The bat fauna of Costa Rica's Reserva Natural Absoluta Cabo Blanco and its implications for bat conservation. Pp. 303–352 *in*: The Quintessential Naturalist: Honoring the Life and Legacy of Oliver P. Pearson. University of California Publications in Zoology, 134.

Tschudi, J. 1844. Untersuchungen über die Fauna Peruana. St. Gallen.

Van Den Bussche, R. A. 1992 Restriction-site variation and molecular systematics of New World leaf-nosed bats. Journal of Mammalogy, 73:29–42.

Velazco, P. M. 2005. Morphological phylogeny of the bat genus *Platyrrhinus* Saussure, 1860 (Chiroptera: Phyllostomidae) with the description of four new species. Fieldiana Zoology, 105:1–53.

Velazco, P. M., and R. Cadenillas. 2011. On the identity of *Lophostoma silvicolum occidentalis* (Davis and Carter, 1978) (Chiroptera: Phyllostomidae). Zootaxa, 2962:1–20.

Velazco, P. M., and A. L. Gardner. 2009a. A new species of *Platyrrhinus* (Chiroptera: Phyllostomidae) from western Colombia and Ecuador, with emended diagnoses of *P. aquilus*, *P. dorsalis*, and *P. umbratus*. Proceedings of the Biological Society of Washington, 122:249–281.

Velazco, P. M., and A. L. Gardner. 2009b. Systematics of the *Platyrrhinus helleri* species complex (Chiroptera: Phyllostomi-

dae) with descriptions of two new species. Zoological Journal of the Linnean Society, 159:785–812.

Velazco, P. M., A. L. Gardner, and B. D. Patterson. 2010. Systematic of the *Platyrrhinus helleri* complex (Chiroptera: Phyllostomidae), with description of two new species. Zoological Journal of the Linnean Society, 159:785–812.

Velazco, P. M., and B. K. Lim. 2014. A new species of broad-nosed bat *Platyrrhinus* Saussure, 1860 (Chiroptera: Phyllostomidae) from the Guianan Shield. Zootaxa, 3796:175–193.

Velazco, P. M., and B. D. Patterson. 2008. Phylogenetics and biogeography of the broad-nosed bats, genus *Platyrrhinus* (Chiroptera: Phyllostomidae). Molecular Phylogenetics and Evolution, 49:749–759.

Velazco, P. M., and B. D. Patterson. 2013. Diversification of the yellow-shouldered bats, genus *Sturnira* (Chiroptera, Phyllostomidae) in the New World tropics. Molecular Phylogenetics and Evolution, 68:683–698.

Velazco, P. M., and B. D. Patterson. 2014. Two new species of yellow-shouldered bats, genus *Sturnira* Gray, 1842 (Chiroptera, Phyllostomidae) from Costa Rica, Panama and western Ecuador. ZooKeys, 402:43–66.

Velazco, P. M., and N. B. Simmons. 2011. Systematics and taxonomy of great striped-faced bats of the genus *Vampyrodes* Thomas, 1900 (Chiroptera: Phyllostomidae). American Museum Novitates, 3710:1–35.

Vilaró, J. 1897. Sketch of John Gundlach. Appleton's Popular Science Monthly, March, 691–697.

Vision, T. 2010. The Dryad Digital Repository: published evolutionary data as part of the greater data ecosystem. Available from Nature Precedings: http://hdl.handle.net/10101/npre.2010.4595.1.

Voss, R. S., and L. H. Emmons. 1996. Mammalian diversity in Neotropical lowland rainforests: a preliminary assessment. Bulletin of the American Museum of Natural History 230:1–115.

Wagner, J. A. 1840. Die Säugethiere in Abbildungen nachder Natur mit Beschriebungen von Dr. Johann Christian Daniel von Schreber, Supplementband, Erst Abtheilung. [Part 1 of Wagner's supplement to Schreber's multivolume work]. Erlangen: in der Expedition des Schreber'schen Säughthier[e] und des Esper'schen Schmetterlingwerkes, etc.

Wetterer, A. L., M. V. Rockman, and N. B. Simmons. 2000. Phylogeny of phyllostomid bats (Mammalia: Chiroptera): data from diverse morphological systems, sex chromosomes, and restriction sites. Bulletin of the American Museum of Natural History 248:1–200.

Wilson, E. O. 2004. Taxonomy as a fundamental discipline. Philosophical Transactions of the Royal Society B, 359:739.

Woodman, N. 2007. A new species of nectar-feeding bat, genus *Lonchophylla*, from western Colombia and western Ecuador (Mammalia: Chiroptera: Phyllostomidae). Proceedings of the Biological Society of Washington, 120:340–358.

Wynne, J. J., and W. Pleytez. 2005. Sensitive ecological areas and species inventory of Actun Chapat cave, Vaca Plateau, Belize. Journal of Cave and Karst Studies, 67:148–157.

Zurc, D., and P. M. Velazco. 2010. Análisis morfológico y morfométrico de *Carollia colombiana* Cuartas et al. 2001 y *C. monohernandezi* Muñoz et al. 2004 (Phyllostomidae: Carollinae) en Colombia. Chiroptera Neotropical, 16:549–567.

5

Fragments and Gaps

The Fossil Record

Nancy B. Simmons,
Gregg F. Gunnell,
and Nicolas J. Czaplewski

Introduction

Recent calibrated analyses of relationships and diversification of noctilionoid bats place the origin of the phyllostomid lineage (separation from its sister taxon, Mormoopidae) in the Late Eocene at about 36 Ma, and date the beginning of speciation of extant lineages within the family to the Early Oligocene at about 30 Ma (e.g., Rojas et al. 2016; Teeling et al. 2005). However, the Paleogene fossil record of phyllostomids is nearly entirely lacking, and the earliest known fossils that can be confidently assigned at the generic or specific level are Miocene in age. Because all living phyllostomids as well as all members of the families most closely related to them (Mormoopidae, Noctilionidae, Furipteridae, and Thyropteridae; Rojas et al. 2016; Teeling et al. 2005) occur only in the Neotropics, it can be assumed that Phyllostomidae originated somewhere in this region. Fossils of the earliest phyllostomids, however, remain frustratingly elusive.

The bat fossil record in general is relatively poor even though quite good records exist for some places and time periods (Habersetzer and Storch 1987; Maitre 2014; Simmons and Geisler 1998; Smith et al. 2007, 2012). Eiting and Gunnell (2009) estimated that as much as 88% of the bat record was missing at a global level. Similarly, Teeling et al. (2005) found that 73% of originations of the bat lineages utilized in their analyses were underestimated, indicating a poor fossil record. Noctilionoids were found to have an even worse record, with 86% of lineage originations underestimated (Teeling et al. 2005).

Table 5.1 shows the known Tertiary fossil record of Phyllostomidae. Only three of these records have thus far been found to be complete and/or diagnostic enough to form the basis for naming fossil species. Eight of the other nine records are so fragmentary that they cannot even be assigned to a subfamily, while the ninth one can be recognized as a phyllostomine but the generic assignment is ambiguous. Given that Phyllostomidae diverged more than 30 million years ago, this is very poor fossil

Table 5.1. Paleogene-Neogene Phyllostomidae Records

Subfamily	Genus	Species	Localities	Age	Countries	Authors
Phyllostomid, indet.	Phyllostomid, indet.	sp. indet.	CTA-27	Late Middle Eocene, Early Barrancan, 40.94–41.6 Ma	Peru	Antoine et al. 2016, suppl.
	Gen. nov.	sp. nov.		Early Miocene, 20.9 Ma	Panama	Morgan et al. 2013
	Gen. indet.	sp. indet.	Gran Barranca	Early Miocene, 20 Ma	Argentina	Czaplewski 2010
	†Notonycteris	magdalenensis	Duke Locality 22, San Nicolás; Chepe Site	Middle Miocene, 13–11.5 Ma	Colombia	Savage 1951; Czaplewski 1997, Czaplewski et al. 2003b
	†Notonycteris	sucharadeus	Duke Locality 22, San Nicolás; Chepe Site	Middle Miocene, 13–11.5 Ma	Colombia	Czaplewski 1997; Czaplewski et al. 2003b
	Tonatia or Lophostoma	sp. indet.	Duke Locality CVP 2-4	Middle Miocene, 13–11.5 Ma	Colombia	Czaplewski 1997; Czaplewski et al. 2003b
Lonchophyllinae	†Palynephyllum	antimaster	Duke Locality 22, San Nicolás; Monkey Beds	Middle Miocene, 13–11.5 Ma	Colombia	Czaplewski 1997; Czaplewski et al. 2003b; Yohe et al. 2015

Note: There are poorly preserved chiropteran fossils described from the Early Oligocene of Oregon (Brown 1959), the Early Miocene Vedder Locality of California (Hutchison and Lindsay 1974), and the Late Miocene Big Cat Quarry of California (James 1963) that have been suggested to be phyllostomids. None of these records is definitive, and we here recognize them as Chiroptera indet., based on personal observations and Czaplewski et al. (2008).

† = extinct taxon.

representation of what was likely a widespread and diverse radiation across the Americas and the Caribbean.

During much of the Cenozoic, the New World comprised three geographically separated areas: North America (including Mexico and Central America), the West Indies, and the island continent of South America. Prior to the Great American Biotic Interchange (GABI)—which apparently began for terrestrial mammals with occasional overwater dispersals as early as the Early Miocene (~23–20 Ma)—accelerated in the Late Miocene to Pliocene (~9–4 Ma) and peaked with the final closure of the Isthmus of Panama after 2.8 Ma (O'Dea et al. 2016). Thus, the mammalian faunas of North and South America were quite different throughout most of the Cenozoic (Bloch et al. 2016; Carrillo et al. 2015; Gunnell and Woodburne, in press; Leigh et al. 2014; Montes et al. 2015; Woodburne 2010).

Although the Neotropics today support a fauna that includes many common bat species that range across Central America and northern South America (and, in some cases, Caribbean Islands as well; Simmons 2005), this was probably not always the case. Clearly the terrestrial mammalian faunas of these continents

were quite different prior to the GABI (Carrillo et al. 2015; Gunnell and Woodburne, in press; Leigh et al. 2014), and this is probably likely for the bat faunas as well.

Chronologies of paleontological sites in the Americas are frequently complicated by the absence of radiometric dates for many fossil-bearing units, and correlations are often accomplished by reference to formally defined provincial land-mammal ages. Land-mammal ages are essentially biostratigraphic zones defined by the first appearances of particular species in the fossil record. Because North and South America had such different faunas throughout much of the Cenozoic, different biostratigraphic zonations have been developed for the two regions. Figure 5.1 provides a summary of zones and correlations between North American Land Mammal Ages (NALMA; used in North America and often in the Caribbean as well) and South American Land Mammal Ages (SALMA). The distribution of phyllostomid fossils across these temporal zones is indicated in figure 5.1 by stars.

Many bat fossils are Quaternary in age—apparently originating from the Pleistocene (2.6 Ma–10,000 years)

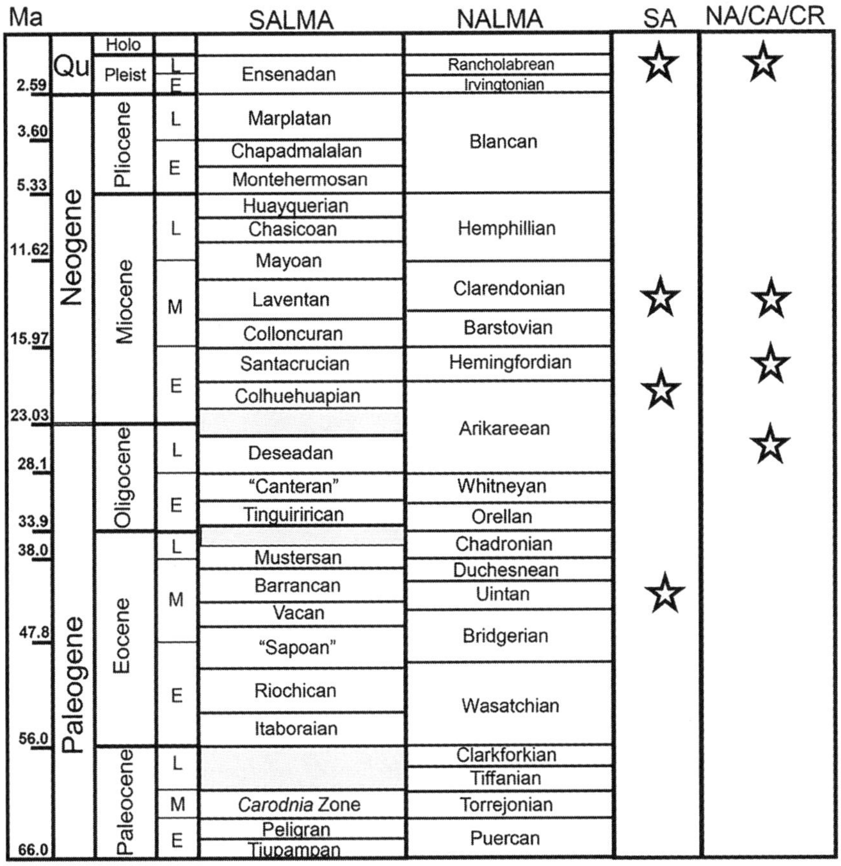

Ma				SALMA	NALMA	SA	NA/CA/CR
	Qu	Holo				☆	☆
2.59		Pleist	L / E	Ensenadan	Rancholabrean / Irvingtonian		
3.60	Neogene	Pliocene	L	Marplatan	Blancan		
			E	Chapadmalalan			
5.33				Montehermosan			
		Miocene	L	Huayquerian / Chasicoan	Hemphillian		
11.62				Mayoan			
			M	Laventan	Clarendonian	☆	☆
15.97				Colloncuran	Barstovian		☆
			E	Santacrucian	Hemingfordian	☆	☆
				Colhuehuapian		☆	
23.03					Arikareean		☆
28.1	Paleogene	Oligocene	L	Deseadan			
			E	"Canteran"	Whitneyan		
33.9				Tinguirirican	Orellan		
38.0			L	Mustersan	Chadronian		
		Eocene	M	Barrancan	Duchesnean		
					Uintan	☆	
				Vacan			
47.8				"Sapoan"	Bridgerian		
			E	Riochican	Wasatchian		
56.0				Itaboraian			
		Paleocene	L		Clarkforkian / Tiffanian		
			M	Carodnia Zone	Torrejonian		
66.0			E	Peligran / Tiupampan	Puercan		

Figure 5.1. Distribution of phyllostomid fossils in relation to South and North American biostratigraphy. Stars indicate occurrences of phyllostomid fossils (one locality or a group of localities from similar time periods). Abbreviations: Ma—million years ago; SALMA—South American Land Mammal Age; NALMA—North American Land Mammal Age (applicable to North America, Central America and the Caribbean); SA—South American records; NA/CA/CR—North American, Central American, and Caribbean records; Qu—Quaternary. Dates from Gradstein et al. 2012.

or even more recent times. Dating of these specimens, particularly those from caves, is especially difficult. Many specimens from caves are often termed "subfossils" because it is not clear if they are hundreds, or hundreds of thousands, of years old. In the absence of firm radiometric dates, such specimens are treated here as Pleistocene/Holocene in age to highlight the ambiguity involved in dates for this material.

Anatomical Terminology, Abbreviations, and Dental Comparisons

Most diagnostic chiropteran fossils, including those of phyllostomids, consist of fragmentary dental remains and humeri. In the following discussion, lower teeth are indicated with a lower case letter and a number; for example, p2 designates the lower second premolar. Upper teeth are indicated with an upper case letter and number; for example, M1 designates the upper first molar. We use the traditional premolar numbering system of p2, p3, and p4 for taxa that retain three lower premolars following most recent authors (e.g., Gunnell et al. 2015; Hand et al. 2015; Ravel et al. 2014; Smith et al. 2012)

rather than the p1, p4, and p5 system advocated by O'Leary et al. (2013). Humerus terminology can be found in Czaplewski et al. (2008).

Comparisons of fossils discussed in this chapter with specimens of many species of noctilionoid bats were made using the illustrated data matrix of dental characters published by Dávalos et al. (2014) and made available online in MorphoBank (www.morphobank .org) as project P891. MorphoBank is a public online database for assembling and managing morphological matrices (O'Leary and Kaufman 2012). Each cell in the matrix is documented with a labeled image that can be zoomed into or downloaded to better observe the structure in question. Morphological observations based on specimens published in MorphoBank project P891 are attributed to Dávalos et al. (2014, P891) in the discussion below.

The Oldest Phyllostomid Fossils

Most of the oldest known fossils that can be clearly identified as phyllostomids are of Early Miocene age (table 5.1). Although an isolated phyllostomid tooth has been reported from the late Middle Eocene of Peru (Antoine et al. 2016, supplementary data), the specimen has yet to be described in detail. If the identification is confirmed, this specimen would extend the phyllostomid fossil record to a much earlier period, the Early Barrancan SALMA (ca. 41 Ma). The oldest fossil unambiguously attributable to Phyllostomidae was recently discovered in Panama and is as yet undescribed; it is a phyllostomine and occurs in two localities of Arikareean (fig. 5.1) NALMA age, dating to about 21 Ma and 19 Ma, respectively (Morgan et al. 2013). A slightly younger phyllostomine of an indeterminate genus from the Gran Barranca of Patagonian Argentina is nearly contemporaneous. This record consists of an isolated tooth and is associated with a mammalian fauna of the Colhuehuapian SALMA with dates of approximately 20 Ma (Czaplewski 2010).

The phyllostomid fossil record becomes a bit more diverse by the Middle to Late Miocene. The famously rich Middle Miocene vertebrate fauna from La Venta, Colombia, preserves the best Neogene fossil record of phyllostomids among a fauna of over 150 vertebrate taxa (Czaplewski 1997; Czaplewski et al. 2003b; Kay et al. 1997). The La Venta fauna as currently understood includes 14 bat taxa representing 6 families; among these are 4 phyllostomids, all of which are known only from isolated teeth and/or dentary fragments (Czaplewski et al. 2003b). The phyllostomine fauna from La Venta includes a species of *Lophostoma* or *Tonatia* and two species belonging to the extinct genus *Notonycteris*, *N. magdalenensis* and *N. sucharadeus* (Czaplewski 1997; Czaplewski et al. 2003b; Savage 1951). In addition, the fauna includes *Palynephyllum antimaster*, a taxon that was originally described as a member of Glossophaginae but is now referred to Lonchophyllinae (Czaplewski et al. 2003b; Dávalos et al. 2014). The La Venta fauna is from the Laventan SALMA, and all of the phyllostomid fossils originate from levels of the Villavieja Formation dating between ca. 12.5 and 12.2 Ma (Flynn et al. 1997; Guerrero 1997).

The fruit-eating subfamilies of Phyllostomidae are traditionally thought to have no pre-Pleistocene fossil record. However, a tooth of early Late Miocene age (Mayoan SALMA) from Contamana, Peru, attributed to the primate *Cebuella* sp. by Marivaux et al. (2016, figs. 4K–O, 6B), appears to us be the p3 of a member of Stenodermatini. If confirmed, this tooth would consti-

tute the earliest record of a stenodermatine, at approximately 11 Ma.

Tooth Patterns and Diets of Early Phyllostomids

Bats have a heterodont dentition characterized by precise occlusion and complex shearing and crushing/grinding functions, features that have led to evolution of dental morphologies that vary consistently with diet (Evans et al. 2007; Freeman 1984, 1988, 1998, 2000; Santana, Strait et al. 2011; Ungar 2010). Bat researchers typically distinguished "animal-eating" or "animalivorous" taxa from those that feed partly or entirely on plant products (Freeman 1984, 1998, 2000; Norberg and Fenton 1988; Rex et al. 2010, Santana, Geipel et al. 2011). In this context, animalivory is an umbrella term that covers both insectivorous and carnivorous species, as well as those that fall somewhere in between (Freeman 1984, 1998, 2000; Norberg and Fenton 1988; Santana, Geipel et al. 2011; Santana, Strait et al. 2011; Simmons et al. 2016).

It has long been assumed that the primitive dietary habit for phyllostomids is animalivory, and in recent years this has been confirmed by more formal optimization exercises using phylogenies (e.g., Dávalos et al. 2014; Ferrarezzi and Gimenez 1996; Wetterer et al. 2000). Dilambdodont cheek teeth that have high, pointed cusps and sharp cristae are generally indicative of an animalivorous diet (Freeman 1988, 1998). In small animalivorous bats, the diet usually consists mostly of insects and other arthropods (Kalka and Kalko 2006; Norberg and Fenton 1988, although also see Rex et al. 2010). As body size increases, some species with dilambdodont molar teeth often include small vertebrates in their diet. For example, large phyllostomines (e.g., *Vampyrum spectrum*, *Chrotopterus auritus*, *Trachops cirrhosus*) are known to regularly consume small vertebrates including birds, lizards, frogs, mice, and sometimes even other bats (Freeman 1984, 1998; Norberg and Fenton 1988; Santana, Geipel et al. 2011, Santana, Strait et al. 2011).

Evolution of carnivory in bats is typically associated with elongation of the metastylar shelf and relative elongation of the postmetacrista on M1 and M2 (Freeman 1984, 1998; Hand 1985). Large-bodied species that regularly include terrestrial vertebrate prey in the diet have proportionally longer crests (especially the postmetacrista) on the upper molars than their smaller insectivorous relatives (Dávalos et al. 2014, P891; Freeman 1984, 1998; Hand 1985; Simmons et al. 2016). In concert with elongation of the postmetacrista, the paracone and metacone are often located closer together (Freeman 1984, 1998), thus reducing the relative length of the postparacrista and premetacrista (Dávalos et al. 2014, P891; Simmons et al. 2016). The W-shaped ectoloph in these species is highly asymmetrical on M1 and M2, which are usually the only teeth to preserve all four branches of the W (preparacrista, postparacrista, premetacrista, postmetacrista). In contrast, insectivorous bats (and those that are more omnivorous) typically have a more symmetrical W-shaped ectoloph in which the preparacrista and postmetacrista are subequal in length and the postparacrista and the premetacrista are subequal in length (Freeman 1984, 1998; Simmons et al. 2016).

Most of the fossils of Miocene phyllostomids discovered to date exhibit dental morphology and body size that today characterize primarily insectivorous animalivores. These fossils include specimens referred to as *"Tonatia or Lophostoma species indeterminate"* (apparently about the size of the larger extant species of *Tonatia* and *Lophostoma*) and several isolated phyllostomid (or potentially phyllostomid) teeth of unknown affinities (Czaplewski 1997, 2010; Czaplewski et al. 2003b; Hutchinson and Lindsay 1974). For example, the sole available m3 from Gran Barranca, Argentina, is similar in morphology to that seen in extant insectivorous animalivores in the genera *Lophostoma* and *Tonatia*, but the extinct bat was much larger than living members of those genera, about the size of modern *Phyllostomus hastatus*.

In contrast, members of the Miocene genus *Notonycteris* were different. These bats were quite large for phyllostomids—*N. magdalenensis* was somewhat larger than extant *Chrotopterus* while *N. sucharadeus* was slightly smaller—and both species show dental specializations associated with carnivory (Czaplewski 1997; Czaplewski et al. 2003b; Savage 1951). The undescribed Panamanian bat was about the size of *Phyllostomus hastatus* and *Chrotopterus auritus*, with teeth resembling those of *Notonycteris* (Morgan et al. 2013). A series of phylogenetic analyses including *Notonycteris* found strong support for grouping of this taxon in a clade with *Vampyrum* and *Chrotopterus* within the phyllostomid tree (Czaplewski et al. 2003b; Dávalos

et al. 2014). These results indicate that carnivory—or at least a mixed diet including small vertebrates as well as insects—had evolved in the phyllostomid lineage (Phyllostominae *sensu stricto*) by the Middle Miocene.

Phyllostomid nectarivores have molars that are dilambdodont but characterized by a buccolingually narrow stylar shelf with the paracone and metacone located near the labial edge of the tooth resulting in a flat, wide W-shaped ectoloph (Dávalos et al. 2014, P891; Freeman 1995, 1998; Simmons et al. 2016; Ungar 2010). Premolars and molars in both the upper and lower dentitions generally appear buccolingually compressed so that each tooth is longer mesiodistally than it is wide (Dávalos et al. 2014, P891; Freeman 1998; Ungar 2010). The Miocene species *Palynephyllum antimaster*, which exhibits these morphological traits, was originally described as a member of Glossophaginae *sensu lato* (Czaplewski et al. 2003b) but later was shown to most likely represent an early member of crown Lonchophyllinae (Dávalos et al. 2014). Regardless of phylogenetic placement, it clearly groups among phyllostomids that are primarily nectarivorous and that appear to compose a phenotypic optimum incompatible with eating hard foods such as beetles or figs (Dumont et al. 2012, 2014; Monteiro and Nogueira 2011; Yohe et al. 2015). However, although typically considered nectarivorous, extant bats in this clade regularly consume varying quantities of insects (Clare et al. 2014; Rex et al. 2010; Yohe et al. 2015).

Yohe et al. (2015) examined correlations between skull length, body size, and diet across nectarivorous phyllostomids (including both glossophagines and lonchophyllines) and developed models using Bayesian methods to infer the trophic level of the fossil. They concluded that, similar to robust extant lonchophyllines known for their strong bites, the diet of *Palynephyllum* probably included insects and fruit in addition to nectar (Yohe et al. 2015). This is consistent with dental similarities between *Palynephyllum* and the known extant omnivore *Hylonycteris* (Tschapka 2004; Yohe et al. 2015). Yohe et al. (2015) additionally concluded that the ecological niche occupied by *Palynephyllum*—that of a large, robust omnivore feeding on both nectar and insects—reflects an intermediate stage between strict insectivory and more intense obligate nectarivory.

Frugivorous phyllostomids typically have molars with flat, bulbous crowns and lower, blunter cusps and crests than seen in animalivorous species (Dávalos et al.

2014, P891; Ungar 2010). In the premolar dentition, the teeth are similarly blunt and tend to lack well-defined shearing crests (Dávalos et al. 2014, P891). None of the fossils from La Venta show this morphology. However, the isolated p3 from the early Late Miocene of Contamana, Peru, reported by Marivaux et al. (2016) appears to resemble that of a fruit-feeding phyllostomid instead of a primate. It is particularly similar to the labially flattened, shelfless, single-cusped, highly trenchant lower premolars of the tribe Stenodermatini. Together with the rest of the toothrow, these teeth form an important part of the rounded "melon baller" dental arcade common among members of Stenodermatini. This suggests that frugivory may have also been an ecological niche filled by phyllostomids in the Miocene.

Quaternary Phyllostomids

By far the best-represented fossil phyllostomids are those found in Pleistocene and Holocene deposits. For ease of discussion, we have divided these into three biogeographic regions: North and Central America; the Greater and Lesser Antilles; and South America. Tables 5.2–5.4 summarize the record of these younger fossils.

North and Central America

The Quaternary record from North and Central America (NA/CA; table 5.2) comes from at least 35 different localities in four different countries (United States, Mexico, Belize and Nicaragua). The total number of fossil and subfossil phyllostomid taxa known from this region is 27, with 23 taxa having been assigned to the specific level. Of these 23 taxa, only 35% are restricted to the NA/CA region, while the other 65% are also found either in South America (12 out of 23) or the Antilles (6 out of 23). The most common subfamilies represented in NA/CA during the Quaternary are Stenodermatinae (7 species), Desmodontinae (5 species), and Glossophaginae (5 species). Notably absent from the NA/CA record are Lonchophyllinae, Lonchorhininae, and Micronycterinae.

Greater and Lesser Antilles

The Quaternary record from the Greater and Lesser Antilles (GLA—table 5.3) comes from numerous localities on more than ten different islands (Bahamas,

Cuba, Puerto Rico, Jamaica, Antigua, West Indies, Hispaniola, Barbados, Lesser Antilles, Haiti, and Trinidad and Tobago; see table 5.3). The total number of fossil and subfossil phyllostomid taxa known from this region is 26, with 25 taxa having been assigned to the specific level. Of these 25 taxa, 72% are restricted to this region, while the other 28% are also found either in South America (4 out 25) or NA/CA (6 out of 25). Given the relative isolation of the Caribbean islands, the endemic nature of the bat fauna is not surprising. The most common subfamilies represented in the GLA during the Quaternary are Glossophaginae (11 species) and Stenodermatinae (9 species), while the other four subfamilies present are represented by two species at most (Desmodontinae) or only a single species (Lonchorhininae, Macrotinae, and Phyllostominae). Notably absent from the GLA record are Carolliinae, Lonchophyllinae, and Micronycterinae. The distribution patterns and biogeographic history of phyllostomines (and other mammals) across the Antilles is very complex and has yet to be completely understood (Rojas et al. 2016 and references therein).

South America

The Quaternary record from South America (SA—table 5.4) comes from at least 25 different localities in seven different countries (Brazil, Surinam, Argentina, Venezuela, Peru, Colombia, and Paraguay; table 5.4). The total number of fossil and subfossil phyllostomid taxa known from this region is 47, with 33 taxa having been assigned to the specific level. Of these 33 taxa, 58% are restricted to South America while the other 42% are also found either in NA/CA (12 out of 33) or the Antilles (4 out of 33). The most common subfamilies represented in SA during the Quaternary are Stenodermatinae (10 species), Phyllostominae (9 species), and Glossophaginae (4 species), while the other five SA subfamilies are less diverse. The only subfamily absent from SA that is known elsewhere is Macrotinae.

Paleobiogeography

In the twentieth century, much of our understanding of the Great American Biotic Interchange (GABI) was reconstructed from the fossil records of temperate parts of North and South America, far from the tropics and the crux of the actual site of the interchange in Central

Table 5.2. Quaternary Phyllostomidae from North and Central America

Subfamily	Genus	Species	Localities	Age	Countries	Authors
Carolliinae	*Carollia*	*brevicauda*	Loltún Cave	Late Pleistocene–Early Holocene	Mexico	Arroyo-Cabrales and Alvarez 2003
	Carollia	sp.	Cebada Cave	Late Pleistocene	Belize	Czaplewski et al. 2003a
Desmodontinae	*Desmodus*	†*archaeodaptes*	Inglis 1A; Haile 16A, 21A	Early to Middle Pleistocene	USA (Florida)	Morgan 1991; Morgan and Emslie 2010; Morgan et al. 1988
	Desmodus	†cf. *draculae*	Loltún Cave	Late Pleistocene	Mexico	Arroyo-Cabrales and Alvarez 1990 2003; Arroyo-Cabrales and Polaco 2003; Hatt et al. 1953;
	Desmodus	†*draculae*	Cebada Cave	Late Pleistocene	Belize	Czaplewski et al. 2003a
	Desmodus	*rotundus*	Terlingua Cave; Loltún Cave	Late Pleistocene–Early Holocene	Mexico, USA (Texas)	Arroyo-Cabrales and Alvarez 2003; Arroyo-Cabrales and Polaco 2003; Cockerell 1930; Martin 1972; Ray and Wilson 1979
	Desmodus	†*stocki*	San Josecito Cave, Nuevo León; Haile 1A, 11B; Arredondo 2A; Reddick 1; Potter Creek Cave; Cueva San Josecito; Cueva La Presita; Tlapacoya; Arkenstone Cave; New Trout Cave; SMI Loc. 261 (Bay Point Cliff); U-Bar Cave; San Miguel Island; Sierra Diablo Cave	Early Pleistocene–Early Holocene	Mexico, USA (Arizona, California, Florida, New Mexico, West Virginia)	Alvarez 1972; Arroyo-Cabrales and Polaco 2003; Brodkorb 1959; Cushing 1945; Czaplewski and Peachey 2003; Grady et al. 2002; Gut 1959; Guthrie 1980; Harris 1993, 2016; Hutchison 1967; Jones 1958; Martin 1972; Morgan 1991, 2002; Morgan and Emslie 2010; Olsen 1960; Polaco and Butrón 1997; Ray et al. 1988; Webb 1974; Webb and Wilkins 1984
	Diphylla	*ecaudata*	Loltún Cave	Late Pleistocene–Early Holocene	Mexico	Arroyo-Cabrales and Alvarez 2003; Arroyo-Cabrales and Polaco 2003
Glossophaginae	*Choeronycteris*	*mexicana*	Cueva de San Josecito	Late Pleistocene	Mexico	Tschudi 1844–45; Arroyo-Cabrales and Polaco 2003
	Glossophaga	*soricina*	Loltún Cave	Late Pleistocene	Mexico	Alvarez 1982; Arroyo-Cabrales and Alvarez 2003; Arroyo-Cabrales and Polaco 2003; Hatt et al. 1953
	Leptonycteris	*nivalis*	Mt. Orizaba, Veracruz; Cueva de El Abra; Cueva San Josecito	Late Pleistocene	Mexico	Arroyo-Cabrales and Polaco 2003; Dalquest and Roth 1970; Martin 1972; (Saussure) 1860
	Leptonycteris	*yerbabuenae*	Yerbabuena; Cueva de La Presita	Late Pleistocene–Early Holocene	Mexico	Arroyo-Cabrales 1992; Arroyo-Cabrales and Polaco 2003; Martínez and Villa-R. 1940; Polaco and Butrón 1997

(continued)

Table 5.2. Continued

Subfamily	Genus	Species	Localities	Age	Countries	Authors
Macrotinae	*Macrotus*	*californicus*	Old Fort Yuma; Terlingua Cave; Cueva La Presita	Late Pleistocene–Early Holocene	USA (California, Texas), Mexico	Arroyo-Cabrales and Polaco 2003; Baird 1858; Polaco and Butrón 1997; Ray and Wilson 1979
	Macrotus	**waterhousii**	Terlingua Cave	Late Pleistocene–Early Holocene	USA (Texas)	Cockerell 1930; Martin 1972
Phyllostominae	**Chrotopterus**	**auritus**	Lara Cave; Spukil Cave; Loltún Cave	Late Pleistocene–Early Holocene	Mexico	Arroyo-Cabrales and Alvarez 1990, 2003; Arroyo-Cabrales and Polaco 2003; Hatt et al. 1953
	Mimon	**bennettii**	Spukil Cave	Late Pleistocene	Mexico	Arroyo-Cabrales and Polaco 2003; Hatt et al. 1953
	Tonatia	*saurophila*	Cebada Cave	Late Pleistocene–Early Holocene	Belize	Czaplewski et al. 2003a
Stenodermatinae	*Artibeus*	*cinereus*	Rancho del Cielo	Late Pleistocene–Early Holocene	Mexico	Koopman and Martin 1959
	Artibeus	**jamaicensis**	Cueva de El Abra; Loltún Cave; St. Michel Cave; Spukil Cave; Lara Cave; Has Cave; Coyok Cave; Chacaljas Cave; Chicken-Itza Natural Water Well	Late Pleistocene–Early Holocene	Mexico	Alvarez 1976 1982; Arroyo-Cabrales 1992; Arroyo-Cabrales and Alvarez 2003; Arroyo-Cabrales and Polaco 2003; Dalquest and Roth 1970; Hatt et al. 1953; Leach 1821; Martin 1972
	Artibeus	**lituratus**	Cebada Cave	Late Pleistocene	Belize	Czaplewski et al. 2003a
	Centurio	*senex*	Chinandega, Realejo; Cebada Cave; Inferno; Loltún Cave	Late Pleistocene–Early Holocene	Nicaragua, Belize, Mexico	Arroyo-Cabrales and Alvarez 2003; Czaplewski et al. 2003a; Gray 1842; Koopman and Martin 1959;
	Chiroderma	*villosum*	Loltún Cave	Late Pleistocene–Early Holocene	Mexico	Arroyo-Cabrales and Alvarez 2003; Arroyo-Cabrales and Polaco 2003
	Dermanura	sp.	Cebada Cave	Late Pleistocene	Belize	Czaplewski et al. 2003a
	Enchisthenes	**hartii**	Inferno	Late Pleistocene–Early Holocene	Mexico	Koopman and Martin 1959
	Sturnira	**lilium**	Loltún Cave	Late Pleistocene–Early Holocene	Mexico	Arroyo-Cabrales and Alvarez 1990 2003; Arroyo-Cabrales and Polaco 2003
	Sturnira	sp.	Cebada Cave	Late Pleistocene	Belize	Czaplewski et al. 2003a

Note: Taxa in bold are present in more than one biogeographic region.

† = extinct taxon.

Table 5.3. Quaternary Phyllostomidae from the Antilles

Subfamily	Genus	Species	Localities	Age	Countries	Author
Desmodontinae	*Desmodus*	*†puntajudensis*	Cueva del Centenario de Lenin; Cueva de Paredones; Cueva Lamas; Cuevas Blancas	Late Pleistocene–Early Holocene	Cuba	Orihuela 2011; Suárez 2005; Wołoszyn and Mayo 1974; Wołoszyn and Silva Taboada 1977
	Desmodus	*rotundus*	Cueva Lamas	Late Pleistocene–Early Holocene	Cuba	Koopman 1958; Martin 1972
Glossophaginae	*Brachyphylla*	*cavernarum*	St. Vincent; Cueva Catedral; Cueva Monte Grande; Cueva de Clara; Cueva del Perro; Cueva de Silva; Antigua; Banana Hole; Burma Quarry	Late Pleistocene–Late Holocene	Lesser Antilles Puerto Rico, West Indies, Bahamas, Antigua	Anthony 1918; Choate and Birney 1968; Gray 1833; Olson and Pregill 1982; Pregill et al. 1988; Reynolds et al. 1953; Wing et al. 1968
	Brachyphylla	*nana*	Pinar del Rio; King Cave; Ashton Cave; Coleby Bay Cave; Banana Hole; Conch Bar Cave; Jacksons Bay Caves; Cueva de los Masones; Cueva del Jagüey; Daiquiri Cave; Camaguey; Cueva de Centenario de Lenin; Cueva Grande de Judas; Cueva de Paredones; St. Michel; Isle of Pines; Dairy Cave; Portland Cave; Oleg's Bat Cave; Bodden Cave; Crab Cave; Patton's Fissure; Pollard Bay Cave	Late Pleistocene–Early Holocene	Cuba, Bahamas, Hispaniola, Jamaica, Grand Cayman, Cayman Brac	Anthony 1919; Arredondo 1970; Koopman and Mayo 1970; Martin 1972; McFarlane et al. 2002; Miller 1902, 1929; Morgan 1989 1994; Peterson 1917; Ruibal 1955; Silva Taboada 1974; Simmons 2005; Velazco et al. 2013; Williams 1952; Wołoszyn and Silva Taboada 1977
	Erophylla	*bombifrons*	Bayamon; El Dudu; Oleg's Bat Cave	Late Pleistocene–Early Holocene	Puerto Rico, Hispaniola	(Miller) 1899; Simmons 2005; Velazco et al. 2013
	Erophylla	*sezekorni*	King Cave; Ashton Cave; Banana Hole; East Cave; Hunts Cave; Sir Harry Oakes Cave; Cueva de los Masones; Cueva del Jagüey; Dairy Cave; Jacksons Bay Caves; Portland Cave; Great Exuma Island; Camaguey; Bodden Cave; Miller's Cave; Patton's Fissure; Pollard Bay Cave	Late Pleistocene–Early Holocene	Bahamas, Cuba, Jamaica, Grand Cayman, Cayman Brac	Arredondo 1970; (Gundlach in Peters) 1861; Koopman and Ruibal 1955; Koopman et al. 1957; Martin 1972; McFarlane et al. 2002; Morgan 1989, 1994; Silva Taboada 1974; Williams 1952
	Leptonycteris	*curasoae*	Willemstad, Curaçao	Late Pleistocene–Early Holocene	West Indies	Miller 1900b
	Monophyllus	*†frater*	Cueva Catedral	Late Pleistocene–Early Holocene	Puerto Rico	Anthony 1917, 1918, 1925; Choate and Birney 1968; Martin 1972
	Monophyllus	*plethodon*	St. Michael Parish; Cueva de Clara; Cueva del Perro; Cueva Catedral	Late Pleistocene–Early Holocene	Barbados, Lesser Antilles, Puerto Rico	Martin 1972; Miller 1900a; Simmons 2005

(continued)

Table 5.3. Continued

Subfamily	Genus	Species	Localities	Age	Countries	Author
Glossophaginae (cont.)	*Monophyllus*	*redmani*	King Cave; Ashton Cave; Banana Hole; East Cave; Conch Bar Cave; Jacksons Bay Caves; Cueva de los Masones; Cueva del Jagüey; Cueva de Clara; Cueva Catedral; Cueva Monte Grande; Cueva Grande de Judas; Cueva de Paredones; Portland Cave; Camaguey; El Dudu; Oleg's Bat Cave	Late Pleistocene–Early Holocene	Jamaica, Bahamas, Cuba, Puerto Rico, Hispaniola	Anthony 1918 1925; Choate and Birney 1968; Koopman 1955; Koopman and Ruibal 1955; Leach 1821; Martin 1972; McFarlane et al. 2002; Miller 1929; Morgan 1989; Reynolds et al. 1953; Silva Taboada 1974; Velazco et al. 2013; Williams 1952; Wołoszyn and Silva Taboada 1977
	Monophyllus	sp.	Burma Quarry	Late Holocene	Antigua	Pregill et al. 1988
	Phyllonycteris	*aphylla*	Wallingford Cave; Dairy Cave; Portland Cave	Late Pleistocene–Early Holocene	Jamaica	Koopman and Williams 1951; Martin 1972; Williams 1952
	Phyllonycteris	*†major*	Cueva Catedral; Cueva del Perro; Burma Quarry	Late Pleistocene–Late Holocene	Puerto Rico, Antigua	Anthony 1917, 1918; Choate and Birney 1968; Martin 1972; Pregill et al. 1988; Simmons 2005
	Phyllonycteris	*poeyi*	Matanzas, Canimar; Banana Hole; Hunts Cave; Sir Harry Oakes Cave; Cueva de los Masones; Cueva del Jagüey; Cueva Grande de Judas; Cueva Tenebrosa; Daiquiri Cave; Camaguey; El Dudu; Oleg's Bat Cave; Patton's Fissure	Late Pleistocene–Early Holocene	Cuba, Bahamas, Hispaniola, Cayman Brac	Koopman and Williams 1951; Martin 1972; (Miller) 1898; Morgan 1994; Williams 1952
Lonchorhininae	*Lonchorhina*	*aurita*	Trinidad	Late Pleistocene	West Indies	Tomes 1863
Macrotinae	*Macrotus*	*waterhousii*	Ashton Cave; Coleby Bay Cave; King Cave; Banana Hole; East Cave; Hunts Cave; Sir Harry Oakes Cave; Conch Bar Cave; Jacksons Bay Caves; Cueva de los Masones; Cueva del Jagüey; Cueva del Centenario de Lenin; Cueva Grande de Judas; Cueva de Paredones; Big and Little Exuma Islands; Cueva de Clara; Daiquiri Cave; Dairy Cave; Portland Cave; Camaguey; Agouti Cave; Bodden Cave; Dolphin Cave; Miller's Cave; Patton's Fissure; Peter Cave; Pollard Bay Cave; Shearwater Cave 1 and 2; Spot Bay Cave	Late Pleistocene–Early Holocene	Haiti, Bahamas, Cuba, Puerto Rico, Grand Cayman, Cayman Brac	Anthony 1919; Choate and Birney 1968; Gray 1843; Koopman 1951; Koopman and Ruibal 1955; Koopman et al. 1957; Martin 1972; McFarlane et al. 2002; Morgan 1989, 1994; Silva Taboada 1974; Williams 1952; Wołoszyn and Silva Taboada 1977
Phyllostominae	*Tonatia*	*saurophila*	Wallingford Cave; Dairy Cave	Late Pleistocene–Early Holocene	Jamaica	Koopman and Williams 1951; Martin 1972

Table 5.3. Continued

Subfamily	Genus	Species	Localities	Age	Countries	Author
Stenodermatinae	*Ariteus*	*flavescens*	Orange Valley, St. Ann Parish; Dairy Cave; Jacksons Bay Cave	Late Pleistocene–Early Holocene	Jamaica	Anthony 1919; (Gundlach in Peters) 1861; Koopman and Ruibal 1955; Martin 1972; Morgan 1989; Silva Taboada 1974; Velazco et al. 2013; Wołoszyn and Silva Taboada 1977
	Artibeus	*anthonyi*	Cueva del Centenario de Lenin; Cueva de los Indios; Cueva Grande de Judas; Moza; Cueva Tenebrosa; Cueva de Paredones	Late Pleistocene–Early Holocene	Cuba	Wołoszyn and Silva Taboada 1977
	Artibeus	*jamaicensis*	Cueva de Clara; Cueva del Perro; Culebra Island; Cueva Monte Grande; Cueva Grande de Judas; Daiquiri Cave; Camaguey; Jacksons Bay Caves, Cueva del Centenario de Lenin; Diquini Cave	Late Pleistocene–Early Holocene	Jamaica, Puerto Rico, Cuba, Haiti	Anthony 1918 1919; Choate and Birney 1968; Koopman and Ruibal 1955; Leach 1821; Martin 1972; Reynolds et al. 1953; Wołoszyn and Silva Taboada 1977
	†Cubanycteris	*silvai*	Cueva GEDA	Late Pleistocene	Cuba	Mancina and García-Rivera 2005
	Enchisthenes	**hartii**	Port of Spain	Late Pleistocene–Early Holocene	Trinidad and Tobago	(Thomas) 1892
	Phyllops	*falcatus*	Habana; Daiquiri Cave; Camaguey; Isla de la Juventud; Oleg's Bat Cave	Late Pleistocene–Early Holocene	Cuba, Hispaniola	Anthony 1919; (Gray) 1839; Koopman and Ruibal 1955; Martin 1972; Silva Taboada and Wołoszyn 1975; Simmons 2005; Velazco et al. 2013; Wołoszyn and Silva Taboada 1977
	Phyllops	*†silvai*	Cueva El Abron	Late Pleistocene	Cuba	Suárez and Díaz-Franco 2003
	Phyllops	*†vetus*	Daiquiri Cave	Late Pleistocene–Early Holocene	Cuba	Anthony 1917, 1919; Martin 1972; Wołoszyn and Silva Taboada 1977
	Stenoderma	*rufum*	Cueva de Clara; Cueva del Perro; Cueva de Catedral	Late Pleistocene–Early Holocene	Puerto Rico	Anthony 1918; 1925; Choate and Birney 1968; Desmarest 1820; Martin 1972; Simmons 2005

Note: Taxa in bold are present in more than one biogeographic region.

† = extinct taxon.

Table 5.4. Quaternary Phyllostomidae from South America

Subfamily	Genus	Species	Localities	Age	Countries	Authors
Carolliinae	*Carollia*	*brevicauda*	Espirito Santo	Late Pleistocene–Early Holocene	Brasil	Schinz 1821
	Carollia	*perspicillata*	Surinam, Minas Gerais, Bahia	Late Pleistocene	Surinam, Brasil	Cartelle and Abuhid 1994; Czaplewski and Cartelle 1998; Fracasso and Salles 2005; (Linnaeus) 1758
	Carollia	sp.	Serra da Mesa	Late Pleistocene	Brasil	Fracasso and Salles 2005
Desmodontinae	*Desmodus*	†cf. *draculae*	Centinela del Mar	Late Holocene	Argentina	Pardiñas and Tonni 2000
	Desmodus	†*draculae*	Cueva del Guácharo; São Paulo; Bahia	Late Pleistocene	Venezuela, Brasil	Cartelle and Abuhid 1994; Czaplewski and Cartelle 1998; Fracasso and Salles 2005; Morgan et al. 1988; Trajano and di Vivo 1991
	Desmodus	*rotundus*	Asunción; Minas Gerais; Bahia; Serra da Mesa; Cueva de la Brújula	Late Pleistocene–Early Holocene	Paraguay, Brasil, Venezuela	Cartelle and Abuhid 1994; Czaplewski and Cartelle 1998; Fracasso and Salles 2005; (É. Geoffroy) 1810; Linares 1970
	Desmodus	sp.	Cueva del Guacharo; Serra da Mesa; Jatun Uchco	Late Pleistocene–Early Holocene	Venezuela, Brasil, Peru	Fracasso and Salles 2005; Linares 1968; Shockey et al. 2009
	Diphylla	*ecaudata*	Sao Francisco River; Bahia	Late Pleistocene–Early Holocene	Brasil	Cartelle and Abuhid 1994; Czaplewski and Cartelle 1998; Fracasso and Salles 2005; Spix 1823
Glossophaginae	*Anoura*	*caudifer*	Campos do Goitas Cazas; Minas Gerais	Late Pleistocene	Brasil	Cartelle and Abuhid 1994; Czaplewski and Cartelle 1998; Fracasso and Salles 2005; (É. Geoffroy) 1818
	Anoura	*geoffroyi*	Rio de Janeiro, Serra da Mesa	Late Pleistocene	Brasil	Fracasso and Salles 2005; Gray 1838
	Anoura	sp.	Rio de Janeiro; Jatun Uchco	Late Pleistocene–Early Holocene	Peru	Shockey et al. 2009
	Glossophaga	*soricina*	Minas Gerais; Cueva de Quebrada Honda	Late Pleistocene	Surinam, Brasil, Venezuela	Cartelle and Abuhid 1994; Czaplewski and Cartelle 1998; Fracasso and Salles 2005; Linares 1968; (Pallas), 1766
	Glossophaga	sp.	Serra da Mesa	Late Pleistocene	Brasil	Fracasso and Salles 2005
	Leptonycteris	*curasoae*	Cueva de Los Murciélagos	Late Pleistocene–Early Holocene	Venezuela	Rincón 2001
Lonchophyllinae	*Lionycteris*	*spurrelli*	Condoto, Serra da Mesa	Late Pleistocene	Colombia, Brasil	Fracasso and Salles 2005; Thomas 1913
	Lonchophylla	*mordax*	Lamarao, Bahia	Late Pleistocene	Brasil	Cartelle and Abuhid 1994; Czaplewski and Cartelle 1998; Thomas 1903
	Platalina	*genovensium*	Lima; Jatun Uchco	Late Pleistocene–Early Holocene	Peru	Shockey et al. 2009; Thomas 1928
Lonchorhininae	*Lonchorhina*	*aurita*	Serra da Mesa	Late Pleistocene	Brasil	Fracasso and Salles 2005
Micronycterinae	*Micronycteris*	*megalotis*	São Paulo; Minas Gerais; Serra da Mesa	Late Pleistocene	Brasil	Cartelle and Abuhid 1994; Czaplewski and Cartelle 1998; Fracasso and Salles 2005; (Gray) 1842

Table 5.4. Continued

Subfamily	Genus	Species	Localities	Age	Countries	Authors
Micronycterinae (*cont.*)	*Micronycteris*	sp.	Bahia; Serra da Mesa	Late Pleistocene	Brasil	Cartelle and Abuhid 1994; Czaplewski and Cartelle 1998; Fracasso and Salles 2005
	Cf. *Micronycteris*	sp.	Inciarte Tar Pit	Late Pleistocene–Early Holocene (25–46 Ka)	Venezuela	Czaplewski et al. 2005
Phyllostominae	**Chrotopterus**	**auritus**	Minas Gerais; Bahia; Serra da Mesa; São Paulo; Iporanga; Sangão Archaeological Site; Garivaldino Archaeological Site; Impossível-Ioiô Cave	Late Pleistocene–Early Holocene	Brasil	Ameghino 1907; Cartelle and Abuhid 1994; Czaplewski and Cartelle 1998; Fracasso and Salles 2005; Hadler et al. 2010; Paula Couto 1953; (Peters) 1856; Salles et al. 2014
	Lophostoma	cf. *silvicolum*	Inciarte Tar Pit; Serra da Mesa	Late Pleistocene–Early Holocene (25–46 Ka)	Venezuela, Brasil	Czaplewski et al. 2005; Fracasso and Salles 2005
	Lophostoma	*silvicolum (occidentalis)*	Talara Tar Seeps	Late Pleistocene (14 Ka)	Peru	Czaplewski 1990; Davis and Carter 1978; Velazco and Cadenillas 2011
	Mimon	**bennettii**	São Paulo; Bahia; Serra da Mesa	Late Pleistocene	Brasil	Cartelle and Abuhid 1994; Czaplewski and Cartelle 1998; Fracasso and Salles 2005; (Gray) 1838
	Mimon	*crenulatum*	Bahia; Minas Gerais; Serra da Mesa	Late Pleistocene	Brasil	Cartelle and Abuhid 1994; Czaplewski and Cartelle 1998; Fracasso and Salles 2005; É. Geoffroy 1803
	Phylloderma	†sp. n.	Serra da Mesa	Late Pleistocene	Brasil	Fracasso and Salles 2005
	Phyllostomus	*discolor*	Mato Grasso, Cuiaba; Cueva de Quebrada Honda; Bahia; Serra da Mesa	Late Pleistocene–Early Holocene	Brasil, Venezuela	Wagner 1843; Cartelle and Abuhid 1994; Czaplewski and Cartelle 1998; Fracasso and Salles 2005; Linares 1968
	Phyllostomus	*hastatus*	Surinam; Minas Gerais, Bahia; Serra da Mesa; Cueva de Quebrada Honda	Late Pleistocene–Early Holocene	Surinam, Brasil, Venezuela	Cartelle and Abuhid 1994; Czaplewski and Cartelle 1998; Fracasso and Salles 2005; Linares 1968; (Pallas) 1767
	Tonatia	*bidens*	Sao Francisco River; Minas Gerais, Bahia, Gruta do Ioiô	Late Pleistocene	Brasil	Cartelle and Abuhid 1994; Castro et al. 2014; Czaplewski and Cartelle 1998; Fracasso and Salles 2005; (Spix) 1823
	Tonatia	sp. indet.	Bahia; Serra da Mesa	Late Pleistocene	Brasil	Cartelle and Abuhid 1994; Czaplewski and Cartelle 1998; Fracasso and Salles 2005
	Trachops	*cirrhosus*	Inciarte Tar Pit; Serra da Mesa, Gruta do Ioiô	Late Pleistocene–Early Holocene (25–46 Ka)	Venezuela, Brasil	Castro et al. 2014; Czaplewski et al. 2005; Fracasso and Salles 2005; (Spix) 1823
Stenodermatinae	**Artibeus**	**cinereus**	Pará, Belem	Late Pleistocene–Early Holocene	Brasil	(Gervais) 1856
	Artibeus	**jamaicensis**	Bahia; Cueva de Quebrada Honda; Laguna Santa	Late Pleistocene–Early Holocene	Brasil, Venezuela	Cartelle and Abuhid 1994; Czaplewski and Cartelle 1998; Fracasso and Salles 2005; Linares 1968

(continued)

Table 5.4. Continued

Subfamily	Genus	Species	Localities	Age	Countries	Authors
Stenodermatinae (*cont.*)	*Artibeus*	*lituratus*	Asunción; Minas Gerais	Late Pleistocene	Paraguay, Brasil	Cartelle and Abuhid 1994; Czaplewski and Cartelle 1998; Fracasso and Salles 2005; (Olfers) 1818
	Artibeus	sp.	Serra da Mesa	Late Pleistocene	Brasil	Fracasso and Salles 2005
	Chiroderma	*doriae*	Minas Gerais	Late Pleistocene–Early Holocene	Brasil	Cartelle and Abuhid 1994; Czaplewski and Cartelle 1998; Fracasso and Salles 2005; Oprea and Wilson 2008; Thomas 1891
	Chiroderma	sp.	Bahia	Late Pleistocene	Brasil	Cartelle and Abuhid 1994; Czaplewski and Cartelle 1998; Fracasso and Salles 2005
	Chiroderma	*villosum*	Brazil	Late Pleistocene–Early Holocene	Brasil	Peters 1860
	Platyrrhinus	*lineatus*	Asunción; Minas Gerais	Late Pleistocene	Paraguay, Brasil	Cartelle and Abuhid 1994; Czaplewski and Cartelle 1998; Fracasso and Salles 2005; (É. Geoffroy) 1810
	Platyrrhinus	sp.	Bahia; Serra da Mesa	Late Pleistocene	Brasil	Cartelle and Abuhid 1994; Czaplewski and Cartelle 1998; Fracasso and Salles 2005
	Pygoderma	*bilabiatum*	Sao Paulo; Garivaldino Archaeological Site	Early Holocene	Brasil	Hadler et al. 2010; Wagner 1843
	Sphaeronycteris	*toxophyllum*	Mérida	Late Pleistocene	Venezuela	Linares 1998; Peters 1882
	Sturnira	*lilium*	Asunción; Minas Gerais; Lapa da Escrivania	Late Pleistocene–Early Holocene	Paraguay, Brasil	Cartelle and Abuhid 1994; Czaplewski and Cartelle 1998; Fracasso and Salles 2005; (É. Geoffroy) 1810; Winge 1893
	Sturnira	sp.	Serra da Mesa	Late Pleistocene	Brasil	Fracasso and Salles 2005
	Uroderma	*bilobatum*	Sao Paulo; Cueva de Quebrada Honda	Late Pleistocene–Early Holocene	Venezuela	Linares 1968; Peters 1866
Phyllostomid, indet.	Gen. et sp. indet.		Inciarte Tar Pit	Late Pleistocene–Early Holocene (25–46 Ka)	Venezuela	Czaplewski et al. 2005

Note: Taxa in bold are present in more than one biogeographic region.

† = extinct taxon.

America. More recent paleontological findings from northern South America (Carlini et al. 2008; Jaramillo et al. 2015) and Mexico–Central America (Bloch et al. 2016; Flynn et al. 2005; Woodburne et al. 2006) are emending the GABI story. Unsurprisingly, the tropical record engenders modifications to the timeline and extent of the GABI, especially its earlier phases. Concurrently, reconstructions of the critical geologic-tectonic-paleogeographic history of Central America, the Central American Seaway, and the isthmus region are being revised with new data. (Bloch et al. 2016 and references therein; Montes et al. 2015; but see also O'Dea et al. 2016). This is an active research area, hence any summary or conclusions made at present must be regarded with caution. As presently understood, the GABI is largely a phenomenon of vertebrate—particularly mammalian—paleobiogeography. Plants probably moved between the continents relatively independently of the presence of a dry land bridge (Cody et al. 2010). Even the migrations of vertebrates other than mammals are less investigated (e.g., frogs, birds; Bacon et al. 2015; Pinto-Sánchez et al. 2012; Weir et al. 2009). The improving fossil record in northern South America and southern North America, and molecular phylogenetic reconstructions, will ultimately enhance our understanding.

Guided by a strict interpretation of the fossil record and firmly dated sedimentary deposits, the Great American Biotic Interchange began as early as about the Oligocene-Miocene boundary (with the monkey *Panamacebus*, the snake *Boa* [see below], and an undescribed Panama phyllostomid). The GABI became more obvious in the Late Miocene between about 9 and 4 Ma with occasional species swimming or rafting the narrow Central American Seaway(s) (from south to north: the megalonychid and mylodontid ground sloths *Thinobadistes*, *Pliometanastes*, *Glossotherium*, and *Zacatzontli*, the pampathere *Plaina*, the glyptodont *Glyptotherium*, the capybara *Neochoerus*, and the terror bird *Titanis*; and from north to south: the megalonychid sloth *Megalonyx*, the procyonid *Cyonasua*, and the sigmodontine rodent *Auliscomys*), and peaked in the Pliocene after 2.8 Ma with the completion of the Isthmus of Panama (Bloch et al. 2016; Carillo et al. 2015; Cione et al. 2015; Leigh et al. 2014; McDonald and Carranza-Castañeda 2019; Montes et al. 2015; O'Dea et al. 2016; Webb 2006; Woodburne 2010). To date, the first mammal known to cross from South America to Panama is *Panamacebus transitus*, a recently described monkey found in Panama in beds of Early Miocene age, about 23–20 Ma, together with an otherwise totally North American vertebrate fauna (Bloch et al. 2016). This earliest South American invader apparently crossed a narrow strait or straits whose changing paleogeographic configuration is becoming less obscure but is still far from clear (Bloch et al. 2016). Snakes of the genus *Boa* made the same crossing from South America no later than about 19 Ma (Head et al. 2012). Deeper penetrations of mammals across the isthmus into Central America and northern South America are yet unknown until the Late Miocene.

Migration between Central and South America by bats is largely unknown because of a mostly blank fossil record. A new Early Miocene phyllostomine, as yet undescribed, was also a potential early participant in the interchange, occurring in Panamanian localities aged about 21 Ma and 19 Ma, together with typical late Arikareean and early Hemingfordian (Early Miocene) North American mammals (Morgan et al. 2013). The Panamanian bat fossils constitute the earliest confirmed occurrence of Phyllostomidae in North America, barely north of the Central American Seaway. Whether these bats were North American or South American autoch-

thons is unclear; the oldest confirmed fossil record of Phyllostomidae in South America, of another phyllostomine (genus indeterminate) at Gran Barranca in Patagonia at about 20 Ma, is nearly as old (Czaplewski 2010). A tropical, transisthmian center of diversification seems likely.

Taken literally, the earlier record of a phyllostomine in Panama suggests that the family Phyllostomidae appeared first in North America (i.e., Central America) and only later expanded its range into South America. Given the sparse and scattered record, however, this conclusion may be premature and arbitrary. The Panama fossil phyllostomid is the only bat amid a host of otherwise undoubtedly North American autochthonous mammals in the late Arikareean–early Hemingfordian NALMA of Panama (ca. 19 Ma; MacFadden et al. 2014). The Cenozoic fossil assemblages from near Contamana in Amazonian Peru (Antoine et al. 2016) have not yet been intensively studied taxon by taxon, but no North American immigrants have yet been recognized in the Early Miocene at this latitude in South America. The presence of a possible phyllostomid at one of the Contamana localities of late Middle Eocene age (Early Barrancan SALMA, ca. 41 Ma; Antoine et al. 2016, supplement), once studied in detail, could potentially establish the family far earlier in South America than the Panamanian Early Miocene record and would suggest the family's arrival in Panama from the south rather than an autochthonous origin in the north. Such a scenario would suggest that the family Phyllostomidae originated in South America and only later dispersed into North America and the Caribbean.

After an apparent lull of at least 10 million years following the monkey *Panamacebus* making the crossing from south to north and the new phyllostomine occurring in Panama, a minor pulse of mammalian immigration occurred in both directions in the Late Miocene and Early Pliocene (see above). So far as known, these immigration events did not include phyllostomids, but almost certainly these bats were making additional overwater and overland dispersals. Although a few fossils of phyllostomids occur in South America in later Miocene localities as mentioned above (see fig. 5.1), phyllostomids are unconfirmed anywhere in North and Central America until well into the late phase of the GABI in the latest Pliocene–Early Pleistocene.

One notable North American record is that of the extinct desmodontine *Desmodus archaeodaptes*, which has been recorded in a few localities in northern Florida, USA, of latest Blancan and early Irvingtonian NALMAs (Morgan and Czaplewski 2012). These records set a minimum age limit on establishment of desmodontines in North America. This vampire bat is a highly derived taxon, hinting at a long history as a separate lineage (also supported by molecular data that suggest that desmondontines diverged from other phyllostomids in the Oligocene; Rojas et al. 2016). Unfortunately, there is no earlier fossil record of Desmodontinae to inform an assessment of its area of origin or potential routes of interchange. There is also no record yet of phyllostomids (or any other bats) from the Paleogene or Neogene of the West Indies that might help elucidate trans-Caribbean routes of interchange.

Extinctions

Table 5.5 lists all phyllostomid species that are known to be extinct. Given the relatively poor nature of the phyllostomid fossil record, it is difficult to conclude much about extinction patterns in the family. However, it does seem that there are remarkably few known extinctions—only 15 species—given the extant diversity of the family, which is now over 200 species (Cirranello and Simmons, chap. 4, this vol.). It is not surprising that species that existed in the Miocene (e.g., *Notonycteris magdalenensis, N. sucharadeus, Palynephyllum antimaster*) should now be extinct, especially since these taxa are early representatives of radiations that include numerous extant species (Phyllostominae and Lonchophyllinae, respectively) that subsequently replaced them. More recent extinctions require further explanation. Among taxa known to have become extinct since the Pleistocene, two interesting extinction patterns emerge: a proportionally large number of these extinctions (60%) took place in the Antilles, and vampires of the genus *Desmodus* suffered very high levels of extinction (40% of known recent extinctions affected species in this genus) compared with other lineages.

It has long been recognized that postglacial extinctions affected bats in the Caribbean (Dávalos and Russell 2012; Dávalos and Turvey 2012). Explanations for these extinctions have ranged from habitat changes (Pregill and Olson 1981) and competition (Koopman and Williams 1951; Williams 1952) to sea-level rise

Table 5.5. Extinct phyllostomid taxa

Subfamily	Genus	Species	Age	Biogeographic Region
Phyllostomid indet.	†Gen. nov.	sp. nov.	Early Miocene, 20.9 Ma	Central America
	†Notonycteris	magdalenensis	Middle Miocene, 13–11.5 Ma	South America
	†Notonycteris	sucharadeus	Middle Miocene, 13–11.5 Ma	South America
Desmodontinae	Desmodus	†archaeodaptes	Early to Middle Pleistocene	North America
	Desmodus	†draculae	Late Pleistocene–Late Holocene	South America, North America, Central America
	Desmodus	†puntajudensis	Late Pleistocene–Early Holocene	Cuba
	Desmodus	†stocki	Early Pleistocene–Early Holocene	North America
Glossophaginae	Monophyllus	†frater	Late Pleistocene–Early Holocene	Antilles
	Phyllonycteris	†major	Late Pleistocene–Late Holocene	Antilles
Lonchophyllinae	†Palynephyllum	antimaster	Middle Miocene, 13–11.5 Ma	South America
Phyllostominae	Phylloderma	†sp. nov.	Late Pleistocene	South America
Stenodermatinae	†Cubanycteris	silvai	Late Pleistocene	Antilles
	Phyllops	†silvai	Late Pleistocene	Antilles
	Phyllops	†vetus	Late Pleistocene–Early Holocene	Antilles

† = extinct taxon.

caused by non-anthropogenic climate change (Dávalos and Turvey 2012; Morgan 2001). Dávalos and Russell (2012) found that island area was a significant predictor of bat species richness in the Bahamas and Greater Antilles during all time periods, but not in the Lesser Antilles during the Last Glacial Maximum (LGM). Their analysis of possible causes suggested that the relatively small number of documented fossil species in the Lesser Antilles may explain the independence of richness from island area in LGM estimates for the Lesser Antilles, leading them to suggest that more fossil species remain to be discovered from Quaternary deposits on these islands. Overall, Dávalos and Russell (2012) concluded that deglaciation, and resulting sea-level changes, explains bat extinction in the Caribbean. Many of these extinctions were local in nature (i.e., extinctions of populations on islands, not entire species), but some, notably those affecting island endemics, were true species extinctions.

Dávalos and Russell (2012) did not separately evaluate extinction patterns in different families, but phyllostomids clearly loom large in their study: 30 (42%) of the 71 species in their data set were phyllostomids. Seven (23%) of these 30 phyllostomid species are extinct, with only one still extant on the mainland. Reanalyzing the Dávalos and Russell (2012) data for phyllostomids alone is beyond the scope of this chapter, but our observation that a relatively large proportion of the overall extinctions of fossil/subfossil phyllostomid species took place in the Antilles (table 5.5) suggests that the factors highlighted by those authors—climate change and concomitant sea-level changes—might have played a significant role in producing this pattern.

The subfamily Desmodontinae (vampires) today comprises three species, each of which is placed in a separate genus (*Desmodus*, *Diphylla*, and *Diaemus*). However, vampires were once more diverse—there were five species of *Desmodus* alive in the Pleistocene but seldom found in the same places, with only *D. rotundus* occasionally overlapping with other species—only one of which survives today (table 5.5). Some of the extinct *Desmodus* species have been found in association with fossils of extinct NA Pleistocene megafauna, and it has been suggested that these bats may have depended on large Pleistocene mammals as a food source (Czaplewski et al. 2003a; Guthrie 1980). If so, it is possible that the Pleistocene extinctions eliminated some vampire bat species via the mechanism of eliminating

some of their key food sources. Other vampires—those that lived in the Antilles—may have been pushed to extinction by other factors including climate change and sea-level rise as noted above.

Conclusions

The fossil record of Phyllostomidae is relatively poor and consists of fragmentary fossils separated by broad geographic and temporal gaps. The record of the family definitely extends to the Early Miocene (~21–19 Ma) of Panama and may extend as far back as the late Middle Eocene (~41 Ma) of southern South America, but overall it is surprisingly incomplete before the Pleistocene. Crown group members of the extant subfamilies Phyllostominae and Lonchophyllinae are known from dental remains from Miocene localities in South and Central America. The teeth of most Miocene phyllostomids exhibit morphology and are of a size suggesting that these bats were insectivores, but some may have been at least partially carnivorous or omnivorous, indicating that ecological diversification in the family was already underway in the Miocene. Large numbers of phyllostomid fossils are known from Pleistocene and Holocene deposits in Central America, South America, and the Caribbean, but overall the fossil record of Phyllostomidae is poor. No Late Miocene or Pliocene fossil phyllostomids are known, hence there is a ~10 million–year gap in the fossil record of the family. There are relatively few documented extinctions in Phyllostomidae, but those that are known disproportionally seem to have affected Antillean taxa and vampire bats (Desmodontinae). It is not yet clear if these represent real patterns or are a result of paucity of the fossil record.

References

Alvarez, T. 1972. Nuevo registro para el vampiro del Pleistoceno *Desmodus stocki* de Tlapacoya, Mexico. Anales de la Escuela Nacional de Ciencias Biológicas, 19:163–165.

Alvarez, T. 1976. Apuntes para la Arqueología: restos óseos rescatados del cenote sagrado de Chichen Itza, Yucatán. Cuadernos de Trabajo, Departamento de Prehistoria, Instituto Nacional de Antropología e Historia, 15:19–39.

Alvarez, T. 1982. Restos de mamíferos recientes y pleistocénicos procedentes de las Grutas de Loltún, Yucatán, México. Cuaderno de Trabajo, Departamento de Prehistoria, Instituto Nacional de Antropología e Historia, 26:7–35.

Ameghino, F. 1907. Notas sobre una pequeña colección de mamíferos procedentes de lasgrutas calcáreas de Iporanga, en el estado de São Paulo, Brazil. Revista do Museu Paulista, 7:59–124.

Anthony, H. E. 1917. Two new fossil bats from Porto Rico. Bulletin of the American Museum of Natural History, 37:565–568.

Anthony, H. E. 1918. The indigenous land mammals of Porto Rico, living and extinct. Memoirs of the American Museum of Natural History, n.s., 2:331–435.

Anthony, H. E. 1919. Mammals collected in eastern Cuba in 1917: with descriptions of two new species. Bulletin of the American Museum of Natural History, 41:625–643.

Anthony, H. E. 1925. Scientific survey of Porto Rico and the Virgin Islands. Part 1: Mammals of Porto Rico, living and extinct—Chiroptera and Insectivora. New York Academy of Science, 9:1–96.

Antoine, P.-O., M. A. Abello, S. Adnet, A. J. Altamirano-Sierra, P. Baby, G. Billet, M. Boivin et al. 2016. A 60-million-year Cenozoic history of western Amazonian ecosystems in Contamana, eastern Peru. Gondwana Research, 31:30–59.

Arredondo, O. 1970. Dos nuevas especies subfosiles de mamíferos (Insectivora: Nesophontidae) del Holoceno Precolombino de Cuba. Memoria de la Sociedad de Ciencias Naturales La Salle, 30:122–152.

Arroyo-Cabrales, J. 1992. Sinópsis de los murciélagos fósiles de México. Revista de la Sociedad Mexicana de Paleontología, 5:1–14.

Arroyo-Cabrales, J., and T. Alvarez. 1990. Restos óseos de murciélagos procedentes de las excavaciones en las grutas de Loltún. Colección Científica, Instituto Nacional de Antropología e Historia, México, 194:1–103.

Arroyo-Cabrales, J., and T. Alvarez. 2003. A preliminary report of the late Quaternary mammal fauna from Loltún Cave, Yucatán, Mexico. Pp. 262–272 in: Ice Age Cave Faunas of North America (B. W. Schubert, J. I. Mead, and R. W. Graham, eds.). Bloomington: Indiana University Press.

Arroyo-Cabrales, J., and O. J. Polaco. 2003. Caves and the Pleistocene vertebrate paleontology of Mexico. Pp. 273–291 in: Ice Age Cave Faunas of North America (B. W. Schubert, J. I. Mead, and R. W. Graham, eds.). Bloomington: Indiana University Press.

Bacon, C. D., D. Silvestro, C. Jaramillo, B. T. Smith, P. Chakrabarty, and A. Antonelli. 2015. Biological evidence supports an early and complex emergence of the Isthmus of Panama. Proceedings of the National Academy of Sciences, 112:6110–6115.

Baird, S. F. 1858. Description of a phyllostome bat from California, in the museum of the Smithsonian Institution. Proceedings of the Academy of Natural Sciences of Philadelphia, 10:116–117.

Bloch, J. I., E. D. Woodruff, A. R. Wood, A. F. Rincon, A. R. Harrington, G. S. Morgan, D. A. Foster et al. 2016. First North American fossil monkey and early Miocene tropical biotic interchange. Nature, DOI: 10.1038/nature17415.

Brodkorb, P. 1959. The Pleistocene avifauna of Arredondo, Florida. Bulletin of the Florida State Museum, 4:269–291.

Brown, R. W. 1959. A bat and some plants from the upper Oligocene of Oregon. Journal of Paleontology, 33:125–129.

Carlini, A. A., A. E. Zurita, and O. A. Aguilera. 2008. North American glyptodontines (Xenarthra, Mammalia) in the upper Pleistocene of northern South America. Paläontologische Zeitschrift, 82:125–138.

Carrillo, J. D., A. Forasiepi, C. Jaramillo, and M. R. Sánchez-Villagra. 2015. Neotropical mammal diversity and the Great American Biotic Interchange: spatial and temporal variation in South America's fossil record. Frontiers in Genetics, 5:1–11.

Cartelle, C., and Abuhid, V. S. 1994. Chiroptera do Pleistoceno final-Holoceno da Bahia. Acta Geologica Leopoldensia, 17:429–444.

Castro, M. C., F. C. Montefeltro, and M. C. Langer. 2014. The Quaternary vertebrate fauna of the limestone cave Gruta do Ioiô, northeastern Brazil. Quaternary International, 352:164–175.

Choate, J. R., and E. C. Birney. 1968. Sub-recent Insectivora and Chiroptera from Puerto Rico, with the description of a new bat of the genus Stenoderma. Journal of Mammalogy, 49:400–412.

Cione, A. L., G. M. Gasparini, E. Soibelzon, L. H. Soibelzon, and E. P. Tonni. 2015. The Great American Biotic Interchange: a South American perspective. Dordrecht: Springer Briefs in Earth System Sciences.

Clare E. L., H. R. Goerlitz, V. A. Drapeau, M. W. Holderied, A. M. Adams, J. Nagel, E. R. Dumont, P. D. N. Hebert, and M. B. Fenton. 2014. Trophic niche flexibility in Glossophaga soricina: how a nectar seeker sneaks an insect snack. Functional Ecology, 28:632–641.

Cockerell, T. D. A. 1930. An apparently extinct Euglandina from Texas. Proceedings of the Colorado Museum of Natural History, 9:52–53.

Cody, S., J. E. Richardson, V. Rull, C. Ellis, and R. T. Pennington. 2010. The Great American Biotic Interchange revisited. Ecography, 33:326–332.

Cushing, J. E., Jr. 1945. Quaternary rodents and lagomorphs of San Josecito Cave, Nuevo Leon, Mexico. Journal of Mammalogy, 26:182–185.

Czaplewski, N. J. 1990. Late Pleistocene (Lujanian) occurrence of Tonatia silvicola in the Talara Tar Seeps, Peru. Anais da Academia Brasileira de Ciências, 62:235–238.

Czaplewski, N. J. 1997. Chiroptera. Pp. 410–431 in: Vertebrate Paleontology in the Neotropics: The Miocene Fauna of La Venta, Colombia (R. F. Kay, R. H. Madden, R. L. Cifelli, and J. J. Flynn, eds.). Washington, DC: Smithsonian Institution Press.

Czaplewski, N. J. 2010. Bats (Mammalia: Chiroptera) from Gran Barranca (early Miocene, Colhuehuapian), Chubut Province, Argentina. Pp. 240–252 in: The Paleontology of Gran Barranca: Evolution and Environmental Change through the Middle Cenozoic of Patagonia (R. H. Madden, A. A. Carlini, M. G. Vucetich, and R. F. Kay, eds.). Cambridge: Cambridge University Press.

Czaplewski, N. J., and C. Cartelle. 1998. Pleistocene bats from cave deposits in Bahia, Brazil. Journal of Mammalogy, 79:784–803.

Czaplewski, N. J., J. Krejca, and T. E. Miller. 2003a. Late Quaternary bats from Cebada Cave, Chiquibul cave system, Belize. Caribbean Journal of Science, 39:23–33.

Czaplewski, N. J., G. S. Morgan, and S. A. McLeod. 2008. Chiroptera. Pp. 174–197 *in*: Evolution of Tertiary Mammals of North America, vol. 2: Small Mammals, Xenarthrans, and Marine Mammals (C. M. Janis, G. F. Gunnell, and M. D. Uhen, eds.). Cambridge: Cambridge University Press.

Czaplewski, N. J., and W. D. Peachey. 2003. Late Pleistocene bats from Arkenstone Cave, Arizona. Southwestern Naturalist, 48:597–609.

Czaplewski, N. J., A. D. Rincón, and G. S. Morgan. 2005. Fossil bat (Mammalia: Chiroptera) remains from Inciarte Tar Pit, Sierra de Perijá, Venezuela. Caribbean Journal of Science, 41:768–781.

Czaplewski, N. J., M. Takai, T. M. Naeher, N. Shigehara, and T. Setoguchi T. 2003b. Additional bats from the middle Miocene La Venta fauna of Colombia. Revista de la Academia Colombiana de Ciencias Exactas, Físicas y Naturales, 27: 263–282.

Dalquest, W. W., and E. Roth. 1970. Late Pleistocene mammals from a cave in Tamaulipas, Mexico. Southwestern Naturalist, 15:217–230.

Dávalos, L. M., and A. L. Russell. 2012. Deglaciation explains bat extinction in the Caribbean. Ecology and Evolution, 2:3045–3051.

Dávalos, L. M., and S. Turvey. 2012. West Indian mammals: the old, the new, and the recently extinct. Pp. 157–202 *in*: Bones, Clones, and Biomes: An Extended History of Recent Neotropical Mammals (B. D. Patterson and L. P. Acosta, eds.). Chicago: University of Chicago Press.

Dávalos L. M., P. M. Velazco, O. M. Warsi, P. D. Smits, and N. B. Simmons. 2014. Integrating incomplete fossils by isolating conflicting signal in saturated and non-independent morphological characters. Systematic Biology, 63:582–600.

Davis, W. B., and D. C. Carter. 1978. A review of the round-eared bats of the *Tonatia silvicola* complex, with descriptions of three new taxa. Occasional Papers, Museum of Texas Tech University, 53:1–12.

Desmarest, A. G. 1820. Mammalogie, ou description des espéces de mammifêres. Paris, 1:1–276.

Dumont, E. R., L. M. Dávalos, A. Goldberg, S. E. Santana, K. Rex, and C. C. Voigt. 2012. Morphological innovation, diversification and invasion of a new adaptive zone. Proceedings of the Royal Society B, 279:1797–1805.

Dumont, E. R., K. Samadevam, I. Grosse, O. M. Warsi, B. Baird, and L. M. Dávalos. 2014. Selection for mechanical advantage underlies multiple cranial optima in New World leaf-nosed bats. Evolution, 68:1436–1449.

Eiting, T. P., and G. F. Gunnell. 2009. Global completeness of the bat fossil record. Journal of Mammalian Evolution, 16: 151–173.

Evans, A., G. Wilson, M. Fortelius, and J. Jernvall. 2007. High-level similarity of dentitions in carnivorans and rodents. Nature, 445:78–81.

Ferrarezzi, H., and E. d. A. Gimenez. 1996. Systematic patterns and the evolution of feeding habits in Chiroptera (Archonta: Mammalia). Journal of Comparative Biology, 1:75–94.

Flynn, J. J., J. Guerrero, and C. C. Swisher III. 1997. Geochronology of the Honda Group. Pp. 44–59 *in*: Vertebrate Paleontology in the Neotropics: The Miocene Fauna of La Venta,

Colombia (R. F. Kay, R. H. Madden, R. L. Cifelli, and J. J. Flynn, eds.). Washington, DC: Smithsonian Institution Press.

Flynn, J. J., B. J. Kowallis, C. Nuñez, O. Carranza-Castaneda, W. E. Miller, C. C. Swisher III, and E. H. Lindsay. 2005. Geochronology of Hemphillian-Blancan aged strata, Guanajuato, Mexico, and implications for timing of the Great American Biotic Interchange. Journal of Geology, 113:287–307.

Fracasso, M. P. A., and L. O. Salles. 2005. Diversity of Quaternary bats from Serra da Mesa (State of Goiás). Zootaxa, 817:1–19.

Freeman, P. W. 1984. Functional cranial analysis of large animalivorous bats (Microchiroptera). Biological Journal of the Linnean Society, 21:387–408.

Freeman, P. W. 1988. Frugivorous and animalivorous bats (Microchiroptera): dental and cranial adaptations. Biological Journal of the Linnean Society, 33:249–272.

Freeman, P. W. 1995. Nectarivorous feeding mechanisms in bats. Biological Journal of the Linnean Society, 56:439–463.

Freeman, P. W. 1998. Form, function, and evolution in skulls and teeth of bats. Pp. 140–156 *in*: Bat Biology and Conservation (T. H. Kunz and P. A. Racey, eds.). Washington, DC: Smithsonian Institution Press.

Freeman, P. W. 2000. Macroevolution in Microchiroptera: recoupling morphology and ecology with phylogeny. Evolutionary Ecology Research, 2:317–335.

Geoffroy St. Hilaire, É. 1803. Catalogue des mammifères du Muséum National d'Histoire Naturelle. Muséum National d'Histoire Naturelle, Paris.

Geoffroy St. Hilaire, É. 1810. Sur les Phyllostomes et les Mégadermes. Annales du Museum d'Histoire Naturelle, Paris, 15:157–198.

Geoffroy St. Hilaire, É. 1818. Sur de nouvelles chauve — souris, sous le nom de Glossophages. Mémoires de Muséum National d'Histoire Naturelle, Paris, 4:411–418.

Gervais, P. 1856. Documents zoologiques pour server à la monographic des Cheiroptères Sud-Americaines. In Expédition dans les parties centrales de l'Amérique du Sud sous la direction du Compte Francis de Castelnau, pt. VII. Zoologie (tome II), Mammifères, 1855:25–88.

Gradstein, F. M., J. G. Ogg, M. D. Schmitz, and G. M. Ogg. 2012. The Geologic Time Scale 2012. Boston: Elsevier.

Grady, F. V., J. Arroyo-Cabrales, and E. R. Garton. 2002. The northernmost occurrence of the Pleistocene vampire bat *Desmodus stocki* (Chiroptera, Phyllostomidae, Desmodontinae) in eastern North America. Smithsonian Contributions to Paleobiology, 93:73–75.

Gray, J. E. 1833. Characters of a new genus of bats (*Brachyphylla*) obtained by the Society from the collection of the late Rev. Lansdown Guilding. Proceedings of the Zoological Society, London, 1833: Part II, 122–123.

Gray, J. E. 1838. A revision of the genera of bats (Vespertilionidae), and the description of some new genera and species. Magazine of Zoology and Botany, 2:483–505.

Gray, J. E. 1839. Description of some Mammalia discovered in Cuba by W. S. Macleay, Esq. Annals of Natural History, 4:1–7.

Gray, J. E. 1842. Descriptions of some new genera and fifty unrecorded species of Mammalia. Annals and Magazine of Natural History, ser. 1, 10:255–267.

Gray, J. E. 1843. [Letter from, on two new species of bats and a porcupine (*Hystrix subspinosus*, Lichtenstein) in the British Museum, and a new *Manis*.]. Proceedings of the Zoological Society, London, 1843:20–22.

Guerrero, J. 1997. Stratigraphy, sedimentary environments, and the Miocene uplift of the Colombian Andes. Pp. 15–43 *in*: Vertebrate Paleontology in the Neotropics: The Miocene Fauna of La Venta, Colombia (R. F. Kay, R. H. Madden, R. L. Cifelli, and J. J. Flynn, eds.). Washington, DC: Smithsonian Institution Press.

Gunnell, G. F., P. M. Butler, M. Greenwood, and N. B. Simmons. 2015. Bats (Chiroptera) from Olduvai Gorge, Early Pleistocene, Bed I (Tanzania). American Museum Novitates, 3846:1–36.

Gunnell, G. F., and M. O. Woodburne. In press. Toward an Evolutionary History of Southern Mammals: Background and Perspective from the North. *In*: Origins and Evolution of South American Mammals (A. R. Rosenberger and M. Tejedor, eds.). New York: Springer.

Gut, H. J. 1959. A Pleistocene vampire bat from Florida. Journal of Mammalogy, 40:534–538.

Guthrie, D. A. 1980. Analysis of avifaunal and bat remains from midden sites on San Miguel Island. Pp. 689–702 *in*: The California Channel Islands: Proceedings of a Multidisciplinary Symposium (D. M. Power, ed.). Santa Barbara, CA: Santa Barbara Museum of Natural History.

Habersetzer, J., and G. Storch, G. 1987. Klassifikation und funktionelle Flügelmorphologie paläogener Fledermäuse (Mammalia, Chiroptera). Courier Forschungsinstitut Senckenberg, 91:117–150.

Hadler, P., J. Ferigolo, and A. M. Ribeiro. 2010. Chiroptera (Mammalia) from the Holocene of Rio Grande do Sul state, Brazil. Acta Chiropterologica, 12:19–27.

Hand, S. J. 1985. New Miocene megadermatids (Chiroptera: Megadermatidae) from Australia with comments on megadermatid phylogenetics. Australian Mammalogy, 8:5–43.

Hand, S. J., B. Sigé, M. Archer, G. F. Gunnell, and N. B. Simmons. 2015. A new early Eocene (Ypresian) bat from Pourcy, Paris Basin, France, with comments on patterns of diversity in the earliest chiropterans. Journal of Mammalian Evolution, 22:343–354.

Harris, A. H. 1993. Quaternary vertebrates of New Mexico. New Mexico Museum of Natural History and Science Bulletin, 2:179–197.

Harris, A. H. 2016. Pleistocene/Holocene faunas from the Trans-Pecos. Pp. 157–175 *in*: Contributions in Natural History: A Memorial Volume in Honor of Clyde Jones (R. W. Manning, J. R. Goetze, and F. D. Yancey III, eds.). Special Publications Museum of Texas Tech University, 65.

Hatt, R. T., H. I. Fisher, D. A. Langebartel, and G. W. Brainerd. 1953. Faunal and archaeological researches in Yucatan caves. Bulletin of the Cranbrook Institute of Science, 33:1–119.

Head, J. J., A. F. Rincón, C. Suarez, C. Montes, and C. Jaramillo. 2012. Fossil evidence for earliest Neogene American faunal interchange: *Boa* (Serpentes, Boinae) from the early Miocene of Panama. Journal of Vertebrate Paleontology, 32:1328–1334.

Hutchison, J. H. 1967. A Pleistocene vampire bat (*Desmodus stocki*) from Potter Creek Cave, Shasta County, California. PaleoBios, 3:1–6.

Hutchison, J. H., and E. H. Lindsay. 1974. The Hemingfordian mammal fauna of the Vedder Locality, Branch Canyon Formation, Santa Barbara County, California. Part I: Insectivora, Chiroptera, Lagomorpha and Rodentia (Sciuridae). PaleoBios, 15:1–19.

James, G. T. 1963. Paleontology and nonmarine stratigraphy of the Cuyama Valley Badlands, California. Part I: Geology, faunal interpretations, and systematic descriptions of Chiroptera, Insectivora, and Rodentia. University of California Publications in Geological Sciences, 45:1–154.

Jaramillo, C., F. Moreno, A. J. Hendy, M. R. Sánchez-Villagra, and D. Marty. 2015. Preface: La Guajira, Colombia: a new window into the Cenozoic Neotropical biodiversity and the Great American Biotic Interchange. Swiss Journal of Paleontology, 134:1–4.

Jones, J., Jr. 1958. Pleistocene bats from San Josecito Cave, Neuvo Leon, Mexico. University of Kansas Publication, Museum of Natural History, 9:389–396.

Kalka, M., and E. K. V. Kalko. 2006. Gleaning bats as underestimated predators of herbivorous insects: diet of *Micronycteris microtis* (Phyllostomidae) in Panama. Journal of Tropical Ecology, 22:1–10.

Kay, R. F., R. H. Madden, R. L. Cifelli, and J. J. Flynn (editors). 1997. Vertebrate Paleontology in the Neotropics—the Miocene Fauna of La Venta, Colombia. Washington, DC: Smithsonian Institution Press.

Koopman, K. F. 1951. Fossil bats from the Bahamas. Journal of Mammalogy, 32:229.

Koopman, K. F. 1955. A new subspecies of *Chilonycteris* from the West Indies and a discussion of the mammals of La Gonave. Journal of Mammalogy, 36:109–113.

Koopman, K. F. 1958. A fossil vampire bat from Cuba. Breviora, 90:1–4.

Koopman, K. F., M. K. Hecht, and E. Ledecky-Janecek. 1957. Notes on the mammals of the Bahamas with special reference to the bats. Journal of Mammalogy, 38:164–174.

Koopman, K. F., and P. S. Martin. 1959. Subfossil mammals from the Gómez Farías region and the tropical gradient of eastern México. Journal of Mammalogy, 40:1–12.

Koopman, K. F., and R. Ruibal. 1955. Cave-fossil vertebrates from Camaguey, Cuba. Breviora, 46:1–8.

Koopman, K. F., and E. E. Williams. 1951. Fossil Chiroptera collected by H. E. Anthony in Jamaica, 1919–1920. American Museum Novitates, 1519:1–29.

Leach, W. E. 1821. The characters of seven genera of bats with foliaceous appendages to the nose. Transactions of the Linnean Society, London, 13:73–82.

Leigh, E. G., A. O'Dea, and G. J. Vermeij. 2014. Historical biogeography of the Isthmus of Panama. Biological Reviews, 89:148–172.

Linares, O. J. 1968. Quirópteros subfosiles encontrados en las cuevas Venezolanas. Parte 1. Boletín de la Sociedad Venezolana de Espeleología, 1:119–145.

Linares, O. J. 1970. Quirópteros subfosiles encontrados en las cuevas Venezolanas. Parte III. *Desmodus rotundus* en la

Cueva de la Brujula (Mi. 1) Miranda. Boletín de la Sociedad Venezolana de Espeleología, 3:33–36.

Linares, O. J. 1998. Mamíferos de Venezuela. Sociedad Conservacionista Audubon de Venezuela and British Petroleum, Caracas.

Linnaeus, C. 1758. Systema naturae per regna tria naturae, secundum classes, ordines, genera, species, cum characteribus, differentiis, synonymis, locis. Vol. 1: Regnum animale. Editio decima, reformata. Stockholm: Laurentii Salvii.

MacFadden, B. J., J. I. Bloch, H. Evans, D. A. Foster, G. S. Morgan, A. F. Rincon, and A. R. Wood. 2014. Temporal calibration and biochronology of the Centenario fauna, early Miocene of Panama. Journal of Geology, 122:113–135.

Maitre, E. 2014. Western European middle Eocene to early Oligocene Chiroptera: systematics, phylogeny and palaeoecology based on new material from the Quercy (France). Swiss Journal of Palaeontology, 133:141–242.

Mancina, C. A., and L. García-Rivera. 2005. New genus and species of fossil bat (Chiroptera: Phyllostomidae) from Cuba. Caribbean Journal of Science, 41:22.27.

Marivaux, L., S. Adnet, A. J. Altamirano-Sierra, F. Pujos, A. Ramdarshan, R. Salas-Gismondi, J. V. Tejada-Lara, and P. O. Antoine. 2016. Dental remains of cebid platyrrhines from the earliest late Miocene of western Amazonia, Peru: macroevolutionary implications on the extant capuchin and marmoset lineages. American Journal of Physical Anthropology, 161:478–493.

Martin, R. A. 1972. Synopsis of late Pliocene and Pleistocene bats of North America and the Antilles. American Midland Naturalist, 87:326–335.

Martínez, L., and B. Villa-R. 1940. Segunda contribución al conocimiento de los murciélagos mexicanos. II. Estado de Guerrero. Anales del Instituto de Biología Universidad Nacional Autónoma de México, 11:291–361.

Mayo, N. A. 1970. La fauna vertebrada de Punta Judas. Pp. 1–45 in: Sistema subterráneo de Punta Judas (A. G. Gonzalez and J. I. Bordon, eds.). Academia de Ciencias de Cuba, Serie Espeleológica y Carsológica, 30.

McDonald, H. G., and O. Carranza-Castañeda. 2019. Increased xenarthran diversity of the Great American Biotic Interchange: a new genus and species of ground sloth (Mammalia, Xenarthra, Megalonychidae) from the Hemphillian (late Miocene) of Jalisco, Mexico. Journal of Paleontology, 91:1069–1082.

McFarlane, D. A., J. Lundberg, and A. G. Fincham. 2002. A late Quaternary paleoecological record from caves of southern Jamaica, West Indies. Journal of Cave and Karst Studies, 64:117–125.

Miller, G. S., Jr. 1898. Descriptions of five new phyllostome bats. Proceedings of the Academy of Natural Sciences, Philadelphia, 50:326–337.

Miller, G. S. 1899. Two new glossophagines bats from the West Indies. Proceedings of the Biological Society of Washington, 8:33–37.

Miller, G. S., Jr. 1900a. The bats of the genus Monophyllus. Proceedings of the Washington Academy of Science, 2:31–38.

Miller, G. S., Jr. 1900b. Three new bats from the island of Curaçao. Proceedings of the Biological Society of Washington, 13:123–127.

Miller, G. S., Jr. 1902. Twenty new American bats. Proceedings of the Academy of Natural Sciences, Philadelphia, 54:389–412.

Miller, G. S., Jr. 1929. A second collection of mammals from caves near St. Michel, Haiti. Smithsonian Miscellaneous Collections, 81:1–30.

Monteiro, L. R., and M. R. Nogueira. 2011. Evolutionary patterns and processes in the radiation of phyllostomid bats. BMC Evolutionary Biology, 11:137.

Montes, C., A. Cardona, C. Jaramillo, A. Pardo, J. C. Silva, V. Valencia, C. Ayala et al.. 2015. Middle Miocene closure of the Central American seaway. Science, 348:226–229.

Morgan, G. S. 1989. Fossil Chiroptera and Rodentia from the Bahamas, and the historical biogeography of the Bahamian mammal fauna. Pp. 685–740 in: Biogeography of the West Indies (C. A. Woods, ed.). Gainesville, FL: Sandhill Crane Press.

Morgan, G. S. 1991. Neotropical Chiroptera from the Pliocene and Pleistocene of Florida. Bulletin of the American Museum of Natural History, 206:176–213.

Morgan, G. S. 1994. Late Quaternary fossil vertebrates from the Cayman Islands. Pp. 465–508 in: The Cayman Islands: Natural History and Biogeography (M. A. Brunt and J. E. Davies, eds.). Dordrecht: Kluwer Academic Publishers.

Morgan, G. S. 2001. Patterns of extinction in West Indian bats. Pp. 369–407 in: Biogeography of the West Indies: Patterns and Perspectives (C. A. Woods and F. E. Sergile, eds.). Boca Raton, FL: CRC Press.

Morgan, G. S. 2002. Late Rancholabrean mammals from southernmost Florida, and the Neotropical influence in Florida Pleistocene faunas. Smithsonian Contributions to Paleobiology, 93:15–38.

Morgan, G. S., and N. J. Czaplewski. 2012. Evolutionary history of the Neotropical Chiroptera: the fossil record. Pp. 105–161 in: Evolutionary History of Bats: Fossils, Molecules and Morphology (G. F. Gunnell and N. B. Simmons, eds.). Cambridge: Cambridge University Press.

Morgan, G. S., N. J. Czaplewski, A. F. Rincón, A. R Wood, and B. J. MacFadden. 2013. An early Miocene bat (Chiroptera: Phyllostomidae) from Panama and mid Cenozoic chiropteran dispersals between the Americas. Journal of Vertebrate Paleontology, Program and Abstracts, 2013:180.

Morgan, G. S., and S. D. Emslie. 2010. Tropical and western influences in vertebrate faunas from the Pliocene and Pleistocene of Florida. Quaternary International, 217:143–158.

Morgan, G. S., O. J. Linares, and C. E. Ray. 1988. New species of fossil vampire bats (Mammalia: Chiroptera: Desmodontidae) from Florida and Venezuela. Proceedings of the Biological Society of Washington, 101:912–928.

Norberg, U. M., and M. B. Fenton. 1988. Carnivorous bats? Biological Journal of the Linnean Society, 33:383–394.

O'Dea. A., H. A. Lessios, A. G. Coates, R. I. Eytan, S. A. Restrepo-Moreno, A. L. Cione, L. S. Collins et al. 2016. Formation of the Isthmus of Panama. Science Advances, 2:e1600883.

O'Leary, M. A., J. I. Bloch, J. J. Flynn, T. J. Gaudin, A. Giallombardo, N. P. Giannini, S. L. Goldberg et al. The placental mammal ancestor and the post-K-Pg radiation of placentals. Science, 339:662–667.

O'Leary M. A., and S. G. Kaufman. 2012. MorphoBank 3.0: Web application for morphological phylogenetics and taxonomy. Available from http://morphobank.org.

Olfers, I. von. 1818. Bemerkungen zu Illiger's Ueberblick der Säugthiere nach ihrer Vertheilung überdie Welttheile, rücksichtich der Südamericanischen Arten (Species). Abhandlung X. In W. L. von Eschwege, Journal von Brasilien, odor vermischte Nachrichten aus Brasilien, auf wissenschaftlichen Reisen gesammelt. Heft 2 In F. Bertuch (Ed.), Neue Bibliothek der wichtigsten Reisebeschreibungen zur Erweiterung der Erd- und Völkerkunde, 15:192–237.

Olsen, S. J. 1960. Additional remains of Florida's Pleistocene vampire. Journal of Mammalogy, 41:458–462.

Olson, S. L., and G. K. Pregill. 1982. Introduction to the paleontology of Bahaman vertebrates. Pp. 1–7 in: Fossil Vertebrates from the Bahamas (S. L. Olson, ed.). Smithsonian Contributions to Paleobiology, 48. Washington, DC: Smithsonian Institution Press.

Oprea, M., and D. E. Wilson. 2008. Chiroderma doriae (Chiroptera: Phyllostomidae). Mammalian Species, 816:1–7.

Orihuela, J. 2011. Skull variation of the vampire bat Desmodus rotundus (Chiroptera: Phyllostomidae): taxonomic implications for the Cuban fossil vampire bat Desmodus puntajudensis. Chiroptera Neotropical, 17:863–876.

Pallas, P. S. 1766. Miscellanea zoologica. Comitum: Hagae.

Pallas, P. S. 1767. Spicilegia zoologica, quibus novae et obscurae animalium species iconibus, descriptionibus atque commentariis illustrantur. Fasciculus tertius. Berlin: Gottl, August, Lange.

Pardiñas, U. F., and E. P. Tonni. 2000. A giant vampire (Mammalia, Chiroptera) in the late Holocene from the Argentinean pampas: paleoenvironmental significance. Palaeogeography, Palaeoclimatology, Palaeoecology, 160:213–221.

Paula Couto, C. de. 1953. Paleontología Brasileira. Mamíferos. Ministerio da Educacao e Saúde, Biblioteca Cientifica Brasileira, serie A-1:1–516.

Peters, W. C. H. 1856. Über die Chiropterengattungen Mormops und Phyllostoma. Abhandlungen der Königlichen Akademie der Wissenschaften zu Berlin, 1856:287–310.

Peters, W. C. H. 1860. Eine neue Gattung von Flederthieren, Chiroderma villosum, aus Brasilien. Monatsberichte der Königlich Preussischen Akademie der Wissenschaften zu Berlin, 1860:747–754.

Peters, W. C. H. 1861. Eine neue von Hern Dr. Gundlach beschriebene Gattung von Flederthieren aus Cuba, Phyllonycteris. Monatsberichte der Königlich Preussischen Akademie der Wissenschaften zu Berlin, 1861 (for 1860): 817–819.

Peters, W. C. H. 1866. Machte eine Mittheilung uber neue oder ungenugend bekannte Flerthiere (Vampyrops, Uroderma, Chiroderma, Ametrida, Tylostoma, Vespertilio, Vesperugo) und Nager (Tylomys, Lasiomys). Monatsberichte der Königlich Preussischen Akademie der Wissenschaften zu Berlin, 1866:392–397.

Peters, W. C. H. 1882. Ueber Sphaeronycteris toxophyllum, eine neue Gattung und Art der Fruigivoren Blattnasigen Flederthiere, aus dem Tropischen America. Sitzungsberichte der Königlich Preussischen Akademie der Wissenschaften zu Berlin, 45:987–990.

Peterson, O. A. 1917. Report upon the fossil material collected in 1913 by the Messrs. Link in a cave in the Isle of Pines. Annals Carnegie Museum, 11:359–361.

Pinto-Sánchez, N. R., R. Ibáñez, S. Madriñán, O. I. Sanjur, E. Birmingham, and A. J. Crawford. 2012. The Great American Biotic Interchange in frogs: multiple and early colonization of Central America by the South American genus Pristimantis (Anura: Craugastoridae). Molecular Phylogenetics and Evolution, 62:954–972.

Polaco, Ó. J., and L. Butrón M. 1997. Mamíferos pleistocénicos de la cueva La Presita, San Luis Potosí, México. Pp. 279–376 in: Homenaje al Profesor Ticul Alvarez (J. Arroyo-Cabrales and Ó. J. Polaco, eds.). México: Colección Científica, Instituto Nacional de Antropología e Historia.

Pregill, G. K., and S. L. Olson. 1981. Zoogeography of West Indian vertebrates in relation to Pleistocene climatic cycles. Annual Review of Ecological Systems, 12:75–98.

Pregill, G. K., D. W. Steadman, S. L. Olson, and F. V. Grady. 1988. Late Holocene fossil vertebrates from Burma Quarry, Antigua, Lesser Antilles. Smithsonian Contributions to Zoology, 463:1–27.

Ravel, A., M. Adaci, M. Bensalah, M. Mahboubi, F. Mebrouk, E. M. Essid, W. Marzougui et al. 2014. New philisids (Mammalia, Chiroptera) from the early-middle Eocene of Algeria and Tunisia: new insight into the phylogeny, palaeobiogeography and palaeoecology of the Philisidae. Journal of Systematic Palaeontology, 13:691–709.

Ray, C. E., O. J. Linares, and G. S. Morgan. 1988. Paleontology. Pp. 19–30 in: Natural history of vampire bats (A. M. Greenhall and U. Schmidt, eds.). Boca Raton, FL: CRC Press.

Ray, C. E., and D. E. Wilson. 1979. Evidence for Macrotus californicus from Terlingua, Texas. Occasional Papers, Museum of Texas Tech University, 57:1–10.

Rex, K., B. I. Czaczkes, R. Michener, T. H. Kunz, and C. C. Voigt. 2010. Specialization and omnivory in diverse mammalian assemblages. Ecosciences, 17:37–46.

Reynolds, T. E., K. F. Koopman, and E. E. Williams. 1953. A cave faunule from western Puerto Rico with a discussion of the genus Isolobodon. Breviora, 12:1–8.

Rincón, A. D. 2001. Quirópteros subfósiles presentes en los depósitos de guano de la Cueva de Los Murciélagos, Isla de Toas, Estado Zulia, Venezuela. Anartia, 13:1–13.

Rojas, D., O. M. Warsi, and L. M. Dávalos. 2016. Bats (Chiroptera: Noctilionoidea) challenge a recent origin of extant Neotropical diversity. Systematic Biology, 65:432–448.

Salles, L. O., J. Arroyo-Cabrales, A. C. Do Monte Lima, W. Lanzelotti, F. A. Perini, P. M. Velazco, and N. B. Simmons. 2014. Quaternary bats from the Impossível-Ioiô Cave System (Chapada Diamantina, Brazil): humeral remains and the first fossil record of Noctilio leporinus (Chiroptera, Noctilionidae) from South America. American Museum Novitates, 3798: 1–31.

Santana, S. E., I. Geipel, E. R. Dumont, M. B. Kalka, and E. K.

Kalko. 2011. All you can eat: high performance capacity and plasticity in the common big-eared bat, *Micronycteris microtis* (Chiroptera: Phyllostomidae). PLOS ONE, 6:e28584.

Santana, S. E., S. Strait, and E. R. Dumont. 2011. The better to eat you with: functional correlates of tooth structure in bats. Functional Ecology, 25:839–847.

Saussure, H. 1860. Note sur quelques mammifères du Mexique. Revue et Magazin de Zoologie, Paris, 2 (12): 479–494.

Savage, D. E. 1951. A Miocene phyllostomatid bat from Colombia, South America. University of California Publications in Geological Sciences, 28:357–366.

Schinz, H. R. 1821. Das Thierreich eingetheilt nach dem Bau de Thiere als Grundlage ihrer Naturgeschichte und der vergleichenden Anatomie von dem Herrn Ritter von Cuvier, Band 1. Säugethiere un Vögel, 1821:1–894.

Shockey, B. J., R. Salas-Gismondi, P. Baby, J.-L. Guyot, M. C. Baltazar, L. Huamán, A. Clack et al. 2009. New Pleistocene cave faunas of the Andes of central Peru: radiocarbon ages and the survival of low latitude, Pleistocene DNA. Palaeontologia Electronica, 12.3.15A.

Silva Taboada, G. 1974. Fossil Chiroptera from cave deposits in central Cuba, with description of two new species (Genera *Pteronotus* and *Mormoops*) and the first West Indian record of *Mormoops megalophylla*. Acta Zoologica Cracoviensia, 19:33–73.

Silva Taboada, G., and B. W. Wołoszyn. 1975. *Phyllops vetus* (Mammalia: Chiroptera) en Isla de Pinos. Miscelania Zoologica (Cuba) 1:3.

Simmons, N. B. 2005. Order Chiroptera. Pp. 312–529 *in*: Mammal Species of the World: A Taxonomic and Geographic Reference (D. E. Wilson and D. M. Reeder, eds.). Baltimore: Johns Hopkins University Press.

Simmons, N. B., and J. H. Geisler. 1998. Phylogenetic relationships of *Icaronycteris*, *Archaeonycteris*, *Hassianycteris*, and *Palaeochiropteryx* to extant bat lineages, with comments on the evolution of echolocation and foraging strategies in Microchiroptera. Bulletin of the American Museum of Natural History, 235:1–182.

Simmons, N. B., E. R. Seiffert, and G. F. Gunnell. 2016. A new family of large omnivorous bats (Mammalia, Chiroptera) from the Late Eocene of the Fayum Depression, Egypt, with comments on use of the name "Eochiroptera." American Museum Novitates, 3857:1–43.

Smith, T., J. Habersetzer, N. B. Simmons, and G. F. Gunnell. 2012. Systematics and paleobiogeography of early bats. Pp. 23–66 *in*: Evolutionary History of Bats: Fossils, Molecules and Morphology (G. F. Gunnell and N. B. Simmons, eds.). Cambridge: Cambridge University Press.

Smith, T., R. S. Rana, P. Missiaen, K. D. Rose, A. Sahni, H. Singh, and L. Singh. 2007. High bat (Chiroptera) diversity in the early Eocene of India. Naturwissenschaften, 94:1003–1009.

Spix, J. B. von. 1823. Simiarum et Vespertilionum Brasiliensium species novae, ou histoire naturelle des especes nouvelles de singes et de chauves-souris observes et recueillis pendant le voyage dans l'interieur du Brasil execute par order de S. M. le Roi de Baviere dans les anees 1817, 1818, 1819, 1820. Monachii, 4:1–72.

Suárez, W. 2005. Taxonomic status of the Cuban Vampire bat (Chiroptera: Phyllostomidae: Desmodontine: *Desmodus*). Caribbean Journal of Science, 41:761–767.

Suárez, W., and S. Díaz-Franco. 2003. A new fossil bat (Chiroptera: Phyllostomidae) from a Quaternary cave deposit in Cuba. Caribbean Journal of Science, 39:371–377.

Teeling, E. C., M. S. Springer, O. Madsen, P. Bates, S. J. O'Brien, and W. J. Murphy. 2005. A molecular phylogeny for bats illuminates biogeography and the fossil record. Science, 307:580–584.

Thomas, O. 1891. Notes on *Chiroderma villosum*, Peters, with description of a new species of the genus. Annali del Museuo Civico di Storia Naturale di Genova, ser. 2, 10:881–883.

Thomas, O. 1892. Description of a new bat of the genus *Artibeus* from Trinidad. Annals and Magazine of Natural History, ser. 6, 10:408–410.

Thomas, O. 1903. Notes on South American monkeys, bats, carnivores, and rodents, with descriptions of new species. Annals and Magazine of Natural History, ser. 7, 12:455–464.

Thomas, O. 1913. A new genus of glossophagine bat from Colombia. Annals and Magazine of Natural History, ser. 8, 12:270–271.

Thomas, O. 1928. A new genus and species of glossophagine, with a subdivision of the genus *Choeronycteris*. Annals and Magazine of Natural History, ser. 10, 1:120–123.

Tomes, R. F. 1863. On a new genus and species of leaf-nosed bats in the museum at Fort Pitt. Proceedings of the Zoological Society of London, 1863:81–84.

Trajano, E., and M. De Vivo. 1991. *Desmodus draculae* Morgan, Linares, and Ray, 1988, reported for southeastern Brasil, with paleoecological comments (Phyllostomidae, Desmodontinae). Mammalia, 55:456–460.

Tschapka M. 2004. Energy density patterns of nectar resources permit coexistence within a guild of Neotropical flower-visiting bats. Journal of Zoology, 263:7–21.

Tschudi, J. J. 1844–1845. Untersuchungen über die Fauna Peruana. St. Gallen, pts. 1–6: 1–262.

Ungar, P. S. 2010. Mammal teeth: origin, evolution, and diversity. Baltimore: Johns Hopkins University Press.

Velazco, P. M., and R. Cadenillas. 2011. On the identity of *Lophostoma silvicolum occidentalis* (Davis and Carter, 1978) (Chiroptera, Phyllostomidae). Zootaxa, 2962:1–20.

Velazco, P. M., H. O'Neill, G. F. Gunnell, R. Rimoli, A. L. Rosenberger, and N. B. Simmons. 2013. Quaternary bat diversity in the Dominican Republic. American Museum Novitates, 3779:1–20.

Wagner, J. A. 1843. Diagnosen neuer Arten brasilischer Handflügler. Wiegmanns Archiv für Naturgeschichte, 9:265–368.

Webb, S, D. 1974. Chronology of Florida Pleistocene mammals. Pp. 5–31 *in*: Pleistocene Mammals of Florida (S. D. Webb, ed.). University Presses of Florida, Gainesville.

Webb, S. D. 2006. The Great American Biotic Interchange: patterns and processes. Annals of the Missouri Botanical Garden, 93:245–257.

Webb, S. D., and K. T. Wilkins. 1984. Historical biogeography of Florida Pleistocene mammals. Pp. 370–383 *in*: Contributions in Quaternary Paleontology: A Volume in Memorial to

John E. Guilday (H. H. Genoways and M. R. Dawson, eds.). Carnegie Museum of Natural History, Special Publication, 8. Pittsburgh: Carnegie Museum of Natural History.

Weir, J. T., E. Bermingham, and D. Schluter. 2009. The Great American Biotic Interchange in birds. Proceedings of the National Academy of Sciences the United States of America, 106:21737–21742.

Wetterer, A. L., M. V. Rockman, and N. B. Simmons. 2000. Phylogeny of phyllostomid bats (Mammalia: Chiroptera): data from diverse morphological systems, sex chromosomes, and restriction sites. Bulletin of the American Museum of Natural History, 248:1–200.

Williams, E. E. 1952. Additional notes on fossil and subfossil bats from Jamaica. Journal of Mammalogy, 33:171–179.

Wing, E. S., C. A. Hoffman, Jr., and C. E. Ray. 1968. Vertebrate remains from Indian sites on Antigua, West Indies. Caribbean Journal of Science, 8:123–139.

Winge, H. 1893. Jordfundne og nulevende Flagermus (Chiroptera) fra Lagoa Santa, Minas Geraes, Brasilien. Med Udsigt over Flagermusenes indbyrdes Slaegtskab. E Museo Lundii, 2:1–65.

Wołoszyn, B. W., and N. A. Mayo. 1974. Postglacial remains of a vampire bat (Chiroptera: *Desmodus*) from Cuba. Acta Zoologica Cracoviensia, 19:253–265.

Wołoszyn, B. W., and G. Silva Taboada. 1977. Nueva especies fósil de *Artibeus* (Mammalia: Chiroptera) de Cuba, y tipificación preliminar de los depositos fosilíferos Cubanos contentivos de mamíferos terrestres. Poeyana, 161:1–17.

Woodburne, M. O. 2010. The Great American Biotic Interchange: dispersals, tectonics, climate, sea level and holding pens. Journal of Mammalian Evolution, 17:245–264.

Woodburne, M. O., A. L. Cione, and E. P. Tonni. 2006. Central American provincialism and the Great American Biotic Interchange. Pp. 73–101 *in*: Advances in Late Tertiary Vertebrate Paleontology in Mexico and the Great American Biotic Interchange (Ó. Carranza-Castañeda and E. H. Lindsay, eds.). Universidad Nacional Autónoma de México, Instituto de geología and Centro de Geociencias, Publicación Especial 4. México, City: UNAM, Instituto de Geología, Centro de Geociencias.

Yohe, L. R., P. M. Velazco, B. Gerstner, N. B. Simmons, and L. M. Dávalos. 2015. Bayesian hierarchical models suggest earliest known plant-visiting bat was omnivorous. Biological Letters, 11:20150501.0

Liliana M. Dávalos,
Paúl M. Velazco,
and Danny Rojas

6

Phylogenetics and Historical Biogeography

Introduction

The single largest change in the number of phyllostomid subfamilies took place in a 293-day span between 1 March and 19 December 2000. The first date marks the publication of Wetterer et al. (2000) comprehensive character-based phylogeny based primarily on morphological variation among genera and recognizing only seven subfamilies. The second corresponds to the first analyses based on DNA sequences from as many genera as could be sampled (Baker et al. 2000) that would eventually lead to subfamily classification for 11 clades (Baker et al. 2003). Together, both phylogenies stand as bookends signaling the end of one set of assumptions about the evolution, biogeography and adaptation of bats in the phyllostomid radiation and the beginning of another, whose full implications, reviewed here, have not yet been fully explored.

The shock inflicted by molecular phylogenies to the established phyllostomid phylogeny cannot be overstated. To understand the repercussions of those first well-resolved family-level analyses requires reviewing the culmination of multiple prior analyses and character sets in Wetterer et al. (2000). A preview of Andrea Cirranello's dissertation, Wetterer et al. (2000) assembled the largest data set to date to resolve phyllostomid phylogeny. With new morphological characters scored from the vast specimen holdings at the American Museum of Natural History and other North American collections, Wetterer et al. (2000) were able to include genera neglected by most previous efforts, from the monotypic nectar-feeding *Platalina* and *Scleronycteris*, to the diverse fruit-feeding *Artibeus* and *Platyrrhinus*. The character compilation was equally ambitious, compiling various sources to code, and in some cases describe for the first time, 150 characters from the skull, skeleton, soft tissues, fur, skin, and published accounts of restriction sites and karyotypes. By amassing such a large character matrix and including all the known genera, Wetterer et al. (2000) sought to generate a robust estimate of the phyllostomid phylogeny, comprehensive in the scope of both the characters and lineages included. The main result showed taxa that shared feeding specializations formed clades resulting in the support for and recognition of the seven traditionally recognized subfamilies.

The first indication that long-standing relationships within phyllostomids would be upended by analyses of DNA sequences arrived with the phylogeny based on the autosomal locus encoding for the recombination-activating protein 2 or RAG2 by Baker et al. (2000). While some of the phylogenetic relationships uncovered were unsurprising and in line with previously debated clades (e.g., the lack of monophyly among nectarivorous genera; Griffiths 1982, 1983), others were breathtakingly novel. Instead of being a "phyllostomine," *Macrotus*—and not the desmodontines—emerged as the sister to all other phyllostomids. *Lonchorhina*, a morphologically distinctive genus of primarily insectivorous bats, was sister to the nectarivorous *Lonchophylla* and *Lionycteris*. Instead of being sister taxa, the phenotypically and ecologically similar *Carollia* and *Rhinophylla* were paraphyletic, and *Carollia* was more closely related to the gleaning insectivores *Glyphonycteris* and *Trinycteris*. And the latter two genera, formerly part of a monophyletic "*Micronycteris*," were now only distant relatives rendering the genus polyphyletic.

Rejecting three paraphyletic subfamilies long thought to be monophyletic—"Phyllostominae," "Glossophaginae," and "Carolliinae"—the Baker et al. (2000) phylogeny was groundbreaking. But with poor support for relationships among newly redefined subfamilies, and its reliance on a single locus, the comparative implications of this phylogeny were not immediately grasped. A single gene tree can fail to reflect the species tree through incomplete lineage sorting of ancestral polymorphisms (Rosenberg and Nordborg 2002), lateral gene transfer (Sjöstrand et al. 2014), introgression (Litsios and Salamin 2014), paralogy (Roy 2009), or ecologically adaptive convergence (Liu et al. 2010). Instead of reflecting the evolutionary history of phyllostomids, the single-locus gene tree might reflect some idiosyncratic features of the RAG2 protein or its history.

With relationships among higher clades in flux, the historical biogeography of the family was not a priority for research. Nevertheless, the implications of the change from the morphology-based phylogeny to one based on sequences were noted early on. A possible North American origin for the family based on the position of *Macrotus* was first highlighted by Dávalos (2006), and the DNA-based phylogeny was adopted from the start in analyses of ecological biogeography by Stevens (2006). Despite several phylogenies for individual genera (e.g., Hoffmann and Baker 2001; Hoffmann et al. 2003; Porter et al. 2007), implementation of model-based biogeographic analyses has been relatively recent (Cuadrado-Ríos and Mantilla-Meluk 2016; Velazco and Patterson 2008, 2013). In short, the task of fleshing out the historical biogeography of phyllostomids has not yet been completed.

The purpose of this chapter is to review the literature since 2000, thereby summarizing the key changes between the two landmark studies of Wetterer et al. (2000) and Baker et al. (2000), as well as new discoveries published since then. In particular, we compare and contrast phylogenies introducing sources of phylogenetic characters or those including previously excluded lineages. In the process, we also identify outstanding questions on the phylogenetics and historical biogeography of the family and highlight areas of overlap between historical and ecological approaches.

Methods

The literature review by Wetterer et al. (2000) synthesized phylogenies, whether character or distance based, published until that point. Additionally, the character analyses they introduced directly related the systematics of the family to phylogenies through character changes defining nodes. This strictly phylogenetic approach had been applied to phyllostomids before (e.g., Baker et al. 1989; Griffiths 1982), but phylogenetic resolution resulting from those early analyses was poor. The taxonomic comprehensiveness and character-based approach of Wetterer et al. (2000) set a template for subsequent studies of relationships among subfamilies. The taxonomic scope of Wetterer et al. (2000) has guided our phylogenetic review of character-based analyses including multiple phyllostomid subfamilies. Additionally, we focus special attention on those studies introducing new sets of characters since 2000.

In contrast to phylogenetic analyses, most historical biogeographic analyses since 2000 have focused on particular genera with only recent interest in applying biogeographic models to the entire family. Some comprehensive biogeographic analyses tended to include historical signal incidentally, as part of studies of ecological biogeography analyzing diversity gradients and community structures. For these reasons, our biogeographic review has a broader taxonomic and thematic scope, including both genus-level historical biogeography studies and ecological biogeography analyses whose conclusions touch on historical aspects.

To visualize the differences between key phylogenies, we used the *cophylo* routine in the phytools v.0.5–38 R package (Revell 2012). The algorithm uses a matching table to present phylogenies with tips connected despite different input names or topologies. Nodes are then rotated as many times as needed to maximize the correspondence between different phylogenies. These analyses do not substitute for topology tests, which would reveal the statistical significance and source of conflict between resolutions (Dávalos et al. 2012). As these require reanalyses of all data sets, they are beyond the scope of this review.

For biogeographic analyses, we summarized the main findings of individual analyses, emerging common themes, and outstanding questions. Although there is an important body of work on classical phylogeography (Avise 2000) of certain lineages (e.g., Hoffmann and Baker 2001; Hoffmann et al. 2003; Larsen et al. 2007), our focus is on comparative, multi-species approaches. Hence we discuss phylogeographic studies only if they pertain to multiple species (e.g., Hoffmann and Baker 2003).

Results

Phylogenies of family-wide relationships published since the Wetterer et al. (2000) review are summarized in table 6.1. A comparison between the Wetterer et al. (2000) phylogeny and the first multilocus whole-family phylogeny based on DNA sequences by Baker

Table 6.1. Annotated phyllostomid phylogenies published since the comprehensive review by Wetterer et al. (2000)

Data type(s)	Loci	Analytical method	Notes	Source
DNA sequences	*rag2*	Parsimony	Phylogeny only	Baker et al. 2000
DNA sequences	*rag2*, mtr 12S, tRNA^val, 16S	Bayesian	Phylogeny and systematics	Baker et al. 2003
DNA sequences	*rag2*, mt *cytb*, 12S, tRNA^val, 16S	Likelihood, Bayesian	Dated phylogeny	Hoffmann et al. 2008
DNA sequences	*brca1*, *pepck*, *rag2*, *vwf*, mt *nd1*, *cox1*, *cytb*, 12S, tRNA^val, 16S	Parsimony, likelihood, Bayesian	Dated phylogeny	Datzmann et al. 2010
DNA sequences	*rag2*, mt 12S, tRNA^val, 16S, *cytb*	Likelihood, Bayesian	Dated phylogeny and ancestral traits	Rojas et al. 2011
DNA sequences	mt *cytb*	Bayesian	Dated phylogeny	Agnarsson et al. 2011
DNA sequences on fixed tree	*adra2b*, *rag1*, *rag2*, *vwf*, mt 12S, tRNA^val, 16S	Relative-rate fitting	Relative dates	Baker et al. 2012
DNA sequences	*rag2*, mt 12S, tRNA^val, 16S, *cox1*, *cytb*	Bayesian	Dated phylogeny and diversification	Dumont et al. 2012
DNA sequences, morphological characters	*rag2*, mt 12S, tRNA^val, 16S, *cox1*, *cytb*	Parsimony, likelihood, Bayesian	Phylogeny and method comparison	Dávalos et al. 2012
DNA sequences	*vwf*, *rag2*, mt genome	Likelihood, Bayesian	Phylogeny only	Botero-Castro et al. 2013
DNA sequences	*rag2*, mt 12S, tRNA^val, 16S, *cox1*, *cytb*, *nd2*	Bayesian	Dated phylogeny	Yu et al. 2014
DNA sequences, dental characters	*atp7a*, *bdnf*, *plcb4*, *rag2*, *stat5a*, *thy*, mt 12S, tRNA^val, 16S, *cox1*, *cytb*	Likelihood, Bayesian	Phylogeny only	Dávalos et al. 2014
DNA sequences	*atp7a*, *bdnf*, *plcb4*, *rag2*, *stat5a*, *thy*, mt 12S, tRNA^val, 16S, *cox1*, *cytb*	Bayesian	Dated phylogeny and selection analyses	Dumont et al. 2014
DNA sequences	*atp7a*, *bdnf*, *plcb4*, *rag2*, *stat5a*, *thy*, mt 12S, tRNA^val, 16S, *cox1*, *cytb*	Bayesian	Dated phylogeny and diversification	Rojas et al. 2016

Note: Abbreviations in column 1 above are as follows: adra2b, alpha-2B adrenergic receptor; *atp7*, X chromosome ATPase-7A gene; *bdnf*, brain-derived neurotrophic factor gene; *brca1*, breast cancer susceptibility gene 1; cox1, cytochrome oxidase 1 gene; *cytb*, cytochrome b gene; mtr, mitochondrial ribosomal RNAs 12S, tRNA^val and 16S; *nd1*, mitochondrial NADH dehydrogenase subunit 1 gene; *nd2*, mitochondrial NADH dehydrogenase subunit 2 gene; *pepck*, phosphoenolpyruvate carboxykinase gene; *plcb4*, phospholipase C beta 4 gene; *rag1*, recombination-activating gene 1; *rag2*, recombination-activating gene 2; *stat5a*, signal transducer and activator of 5A gene; *thy*, thyrotropin beta chain gene; *vwf*, von Willebrand factor gene.

et al. (2003) is shown in figure 6.1. Figure 6.2 compares two multilocus strictly molecular phylogenies (Botero-Castro et al. 2013; Datzmann et al. 2010), while figure 6.3 shows comparisons among recent phylogenies analyzing a combination of DNA sequences and morphological characters or DNA sequences alone (Dávalos et al. 2014; Rojas et al. 2016).

Rojas et al. (2016) provided model-based inferences of phyllostomid biogeography. Those analyses are based on the geographic distribution of extant taxa, fitting with rates for five biogeographic processes: vicariance, anagenetic dispersal, cladogenetic dispersal, founder-event speciation, and speciation in the same area. The rates are estimated based on branch lengths from the dated phylogeny, together with the observed and inferred geographic distribution of tips and nodes. The result is a phylogeny with a probabilistic inference of

the geographic range of each node of the tree, summarized in figure 6.4.

Discussion

Shaking Up the Bat Tree

Despite Wetterer et al.'s (2000) search for a fully resolved phylogeny, several phylogenetic challenges persisted. First, support for deep relationships was weak, with <50% bootstrap support for the earliest divergence separating desmodontine vampire bats from other phyllostomids, "Phyllostominae" from its sister taxon, and "Carolliinae" from Stenodermatinae (fig. 6.1). Second, the position of the enigmatic Antillean endemic genus *Brachyphylla* (Silva Taboada and Pine 1969) was unresolved relative to all other phyllosto-

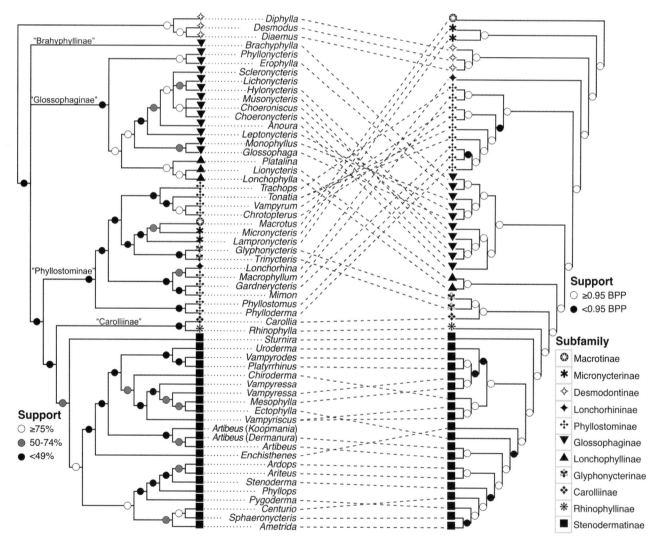

Figure 6.1. Co-phylogeny of the results of Wetterer et al. (2000, fig. 49) (*left*) and Baker et al. (2003, fig. 5) (*right*). Current subfamily classification and genus taxonomy follows Cirranello et al. (2016), former (traditional) subfamily classification is shown in quotations.

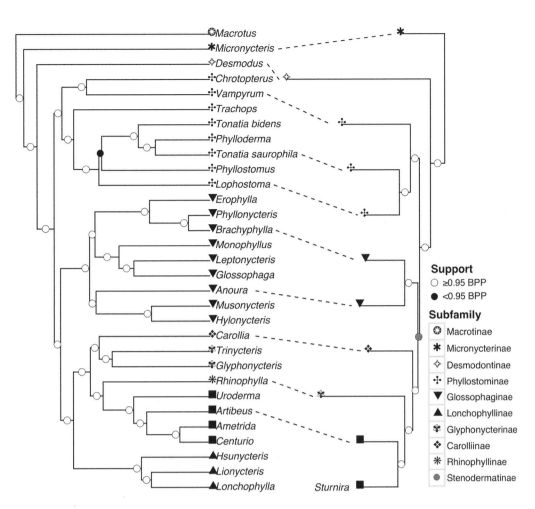

Figure 6.2. Co-phylogeny of the results of Datzmann et al. (2010, fig. 4) (*left*) and Botero-Castro et al. (2013, fig. 2) (*right*). Current subfamily classification and genus taxonomy follows Cirranello et al. (2016).

Support
○ ≥0.95 BPP
● <0.95 BPP

Subfamily
◎ Macrotinae
✳ Micronycterinae
◇ Desmodontinae
⁜ Phyllostominae
▼ Glossophaginae
▲ Lonchophyllinae
✺ Glyphonycterinae
❖ Carolliinae
✳ Rhinophyllinae
⬤ Stenodermatinae

mids except for desmodontines. Finally, most relationships among genera in the "Phyllostominae" clade were weakly supported, reflecting both the lack of shared derived character states and the conflict in signal from different suites of characters (Dávalos et al. 2012).

Strongly supported clades identified by Wetterer et al. (2000) also posed a challenge for comparative analyses, a subject discussed in the next chapter. With monophyletic clades corresponding to nectarivory ("Glossophaginae"), animalivory ("Phyllostominae"), and soft ("Carolliinae") and hard (Stenodermatinae) fruits specialists, ecological adaptation and phylogeny were perfectly correlated. As a result, comparisons between species accounting for the correlations in residuals arising from shared history lacked statistical power, yielding nonsignificant results (e.g., Cruz-Neto et al. 2001). The exceptions were cases in which many predictions were tested simultaneously, resulting in a series of matches validating an adaptive hypothesis for renal and digestive features (Schondube et al. 2001).

Ultimately, the character analyses of Wetterer et al.

(2000) were based on dozens of previous studies compiling mostly phenotypic characters, reflecting the underlying biases of those data sets. The most influential of these are the saturation of character states relative to steps in the phylogeny (Dávalos et al. 2012), a sign of exhaustion in phenotypes that can arise for structural or selective reasons (Wagner 2000; Wake 1991). Another indication of the correspondence between the Wetterer et al. (2000) phylogeny and the many phyllostomid phylogenies from which its characters were drawn is the congruence between the character-based results and the supertree of Jones et al. (2002). Although the analyses of Jones et al. (2002) differ from that of Wetterer et al. (2000), the two phylogenies match in the monophyly of all previously named subfamilies, differing only in the position of *Brachyphylla*, sister to a clade comprising *Phyllonycteris* and *Erophylla* in the supertree (fig. 6.1).

While the single-locus phylogeny of Baker et al. (2000) could be questioned as the result of lineage sorting of ancestral polymorphisms or the result of a

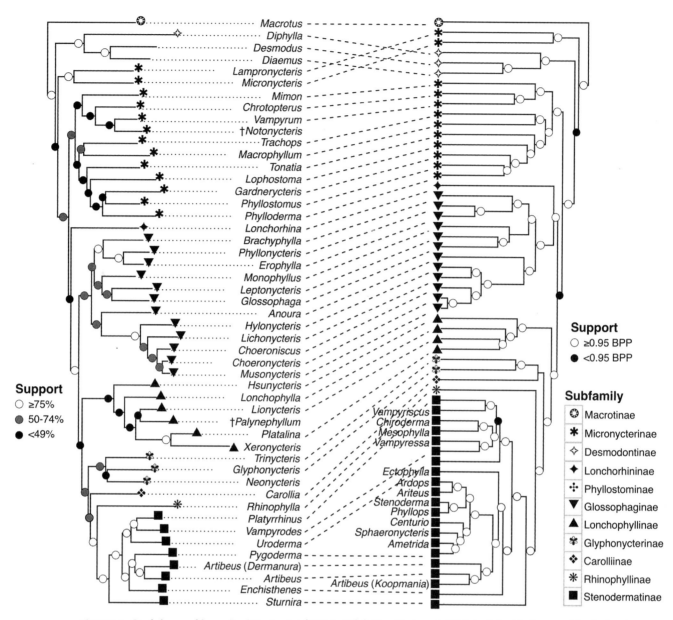

Figure 6.3. Co-phylogeny of the results of Dávalos et al. (2014, fig. 5B) (*left*) and Rojas et al. (2016, fig. 2) (*right*). Current subfamily classification and genus taxonomy follows Cirranello et al. (2016).

misleading gene tree, these explanations for phylogenetic conflict could no longer hold after the publication of Baker et al. (2003). Using a new Bayesian implementation of models of nucleotide evolution for two loci, RAG2 and the mitochondrial ribosomal RNAs 12S, tRNA^{val}, and 16S (mtr), their results confirmed the breakup of "Phyllostominae," "Glossophaginae," and "Carolliinae" and resolved the early phyllostomid divergences with posterior probabilities close to 1 (≥0.95, fig. 6.1). Many of the most important changes to the phyllostomid phylogeny first introduced by Baker et al. (2000) were confirmed including *Macrotus* as the earliest-diverging extant branch of phyllostomids, poly-

phyly of "Phyllostominae," polyphyly of "*Micronycteris*," paraphyly of "Glossophaginae," paraphyly of "Carolliinae," and the sister relationship between *Carollia* and Glyphonycterinae (*Glyphonycteris* and *Trinycteris*). One of the puzzling results from the earlier gene tree was dispelled, separating *Lonchorhina* as its own separate subfamily (fig. 6.1). Except for the stable Desmodontinae and Stenodermatinae, formerly monophyletic subfamilies with similar feeding ecologies were revealed to be collections of distant relatives united by ecologically convergent phenotypes.

The Baker et al. (2003) phylogeny represents a turning point in the phylogenetics and systematics of

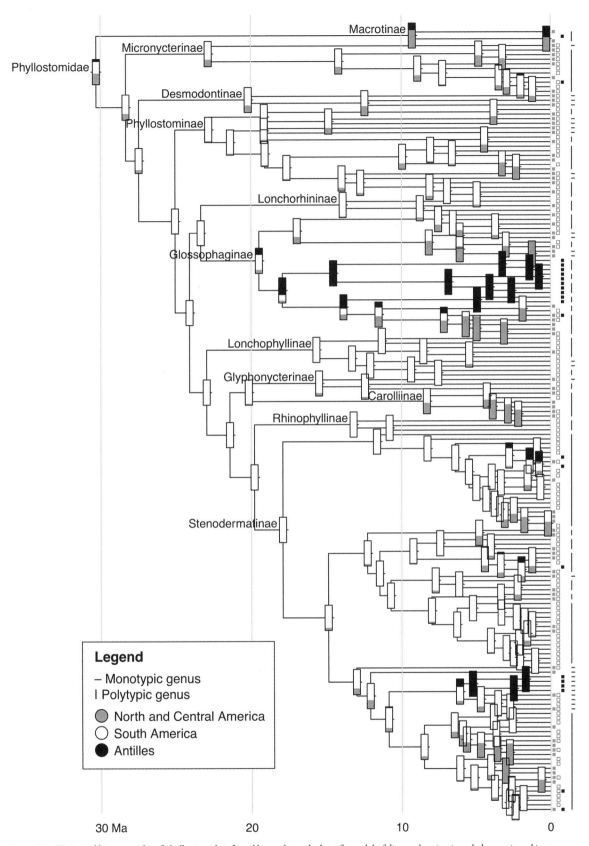

Figure 6.4. Historical biogeography of phyllostomids inferred by applying the best-fit model of dispersal-extinction-cladogenesis and jump speciation (Matzke 2013) on the dated phylogeny of Rojas et al. (2016). Subfamilies are shown along the branch defining the most recent common ancestor.

the family. Before that phylogeny, the traditional subfamily designations could be applied and the phylogeny might still uphold traditionally recognized subfamilies. After it, successive efforts to include new characters (e.g., Datzmann et al. 2010), or other lineages (e.g., Dávalos et al. 2014), left the contours of the phyllostomid phylogeny largely unchanged (figs. 6.2 and 6.3). These unchanging contours include the confirmation of Desmodontinae and Stenodermatinae as the only two monophyletic subfamilies inferred from both morphological data and DNA sequences (fig. 6.1); the separation of primarily nectar-feeding bats into the subfamilies Glossophaginae and Lonchophyllinae; the breakup of primarily insectivorous and carnivorous genera into the subfamilies Macrotinae, Micronycterinae, Lonchorhininae, Phyllostominae, and Glyphonycterinae; and the division of *Piper* and similar fruit specialists into the subfamilies Carolliinae and Rhinophyllinae (Cirranello et al. 2016).

Datzmann et al. (2010), in particular, helped strengthen support for the new phyllostomid phylogeny. In addition to the two loci used by Baker et al. (2003) and the published sequences encoding the mitochondrial cytochrome b *cytb* and cytochrome oxidase I *coi* genes, Datzmann et al. (2010) collected data for four other autosomal loci from fragments of the von Willebrand factor gene *vwf*, the breast cancer susceptibility gene 1 *brca1*, the 3′ untranslated region of the phospholipase C beta 4 gene *plcb4*, and an intron of the phosphoenolpyruvate carboxykinase gene *pepck*, as well as the mitochondrial NADH dehydrogenase subunit 1 gene *nd1* and its adjacent tRNA[leu]. With its focus on establishing relationships among nectarivorous lineages, the Datzmann et al. (2010) phylogeny also corroborated the sister relationship between a clade formed by *Brachyphylla* and *Erophylla*, and *Glossophaga*, *Leptonycteris* and *Monophyllus* (figs. 6.1 and 6.2), and the paraphyly of *Lonchophylla*, which had previously been reported based on the mitochondrial *cytochrome b* (*cytb*) gene tree (Dávalos and Jansa 2004). A paraphyletic *Tonatia* (fig. 6.2) is an uncorroborated result from Datzmann et al. (2010), as no subsequent study has inferred similar relationships despite including multiple *Tonatia* species (Dávalos et al. 2014). Outside of this single node, the relationships among subfamilies obtained in the multilocus analyses have been supported in subsequent analyses.

The first genomic analysis by Botero-Castro et al.

(2013) is similarly important in consolidating the new molecular phylogeny of phyllostomids (fig. 6.2). Based on the entire mitochondrial genome and applying second-generation sequencing techniques, the new data confirmed the subfamily relationships of Baker et al. (2003), while highlighting the relatively short internode uniting glossophagines, *Carollia*, *Rhinophylla* and stenodermatines, and separating that clade from phyllostomines (fig. 6.2). Despite >16 kb of mitochondrial DNA and 2.6 kb of the *rag2* and *vwf* genes, this branch was only moderately supported with a posterior probability of 0.91 and maximum likelihood bootstrap of 76%. This result highlighted the difficulties in resolving a few ancient phyllostomid nodes, analyzed in depth by Dávalos et al. (2012) and discussed in the next chapter.

Besides collecting new data, the Datzmann et al. (2010) analyses also applied relaxed molecular clock models to estimate divergence times among lineages. The relaxed clocks were calibrated at three nodes based on fossil phyllostomids from the Middle Miocene of La Venta in Colombia (Czaplewski et al. 2003) and a mormoopid fossil from the Early Oligocene of Florida and by constraining the divergence between Vespertilionidae and Molossidae. Applying relaxed molecular clocks with branch-specific rates of evolution drawn from an uncorrelated lognormal prior distribution and fitted to the entire alignment (Drummond et al. 2006), those analyses inferred substantially older divergence dates than alternative methods. For example, the divergence date between the vampire bats (Desmodontinae) and their sister group was estimated at 32 Ma (95% high probability density = 28, 36), while previous estimates using the Thorne and Kishino (2002) model of uncorrelated rates across loci estimated this divergence at 26 Ma (95% high probability density = 21, 30) (Teeling et al. 2005). Applying a rate-smoothing method to obtain branch- and partition-specific rates from "absolute" dates at constrained nodes (Bininda-Emonds 2007), the same node corresponds to 26 Ma (95% confidence interval = 21–31) (Baker et al. 2012). The disparate results suggest modeling different genes or partitions by drawing their rates from different distributions accommodates the between-gene variance in rates and reduces estimates of divergence times. Considering the much faster rates of change at mitochondrial loci, ensuring a sufficiently large variance in rates among loci is indispensable for dating future phylogenies.

While the analyses of Datzmann et al. (2010) and Botero-Castro et al. (2013) focused on expanding the sequence data available to infer phyllostomid phylogeny, only two analyses introduced new morphological characters (Dávalos et al. 2012, 2014). The challenges of combining DNA sequences and morphological characters in phyllostomid phylogeny had been analyzed in depth before by using likelihood-based statistical analyses to compare the phylogenies and data of Wetterer et al. (2000) and Baker et al. (2003) (fig. 6.1). Briefly, excluding morphological characters for which there is evidence of convergent evolution reduced conflict between the results of Wetterer et al. (2000) and those of Baker et al. (2003) and produced better-supported phylogenies combining both character types (Dávalos et al. 2012). Based on that insight, and aiming to adequately model the statistical behavior of morphological characters, Dávalos et al. (2014) investigated both rates of change and patterns of correlated change in character state in the dental data. The morphological rate of change was higher than the median substitution rate of DNA sequences by an order of magnitude. Coupled with evidence of character-state exhaustion or saturation and evidence for excess correlated changes in the dental data, the results implied the need to model the high rate of change and exclude characters for which convergence could be demonstrated. Dávalos et al. (2014) then developed two new methods to combine characters: one constraining morphological analyses based on the posterior sample of molecular trees and the other by identifying and excluding morphological characters significantly supporting ecologically convergent nodes. Results from the latter approach are in line with previous DNA-based phylogenies and, therefore, are in conflict with phylogenies based exclusively on morphological characters (fig. 6.3).

The Dávalos et al. (2014) phylogenies were the first to include the La Venta Miocene fossils using characters from multiple subfamilies. The results confirmed the close relationship between *Notonycteris* and *Vampyrum* (Czaplewski et al. 2003) and placed *Palynephyllum* within the Lonchophyllinae (fig. 6.3). These analyses of the Miocene fossils provide the first data sets for applying tip dating methods to the phyllostomid phylogeny (e.g., Herrera and Dávalos 2016). Until now, all dating analyses have relied on node dating methods (table 6.1), constraining particular nodes based on taxonomy or presumed phylogenetic placement, but without including fossil taxa as tips or fitting models of evolution to morphological characters (Heath et al. 2014; Ronquist et al. 2012). Thus, tip dating remains an unexplored frontier in estimates of the phyllostomid tree.

In addition to dental characters, Dávalos et al. (2014) introduced new markers for resolving the phyllostomid phylogeny, including introns in the thyrotropin beta chain gene or *thy* and the signal transducer and activator of 5A gene or *stat5a*, as well as the autosomal exons brain-derived neurotrophic factor or *bdnf*, titin 6 or *ttn6*, and the X chromosome exon ATPase-7A or *atp7a*. The resulting phylogenies were similar to previous molecular analyses in defining new subfamilies but changed the positions of Micronycterinae and Lonchorhininae relative to other lineages (cf. figs. 6.2 and 6.3). The low support for defining the placement of Micronycterinae is also evident in the difference between the undated phylogeny of Dávalos et al. (2014) and the relaxed molecular clock, node-dated phylogeny of Rojas et al. (2016).

Although based on the same set of markers, Rojas et al. (2016) expanded taxonomic sampling using both new and published sequences and specifically set out to evaluate rates of taxonomic diversification across the phylogeny. Rojas et al. (2016) was the first to apply Bayesian mixture models of diversification to phyllostomids and close relatives, confirming Stenodermatinae as the clade with the highest rates of diversification both among phyllostomids (Dumont et al. 2012) and across all bats (Shi and Rabosky 2015). This well-corroborated result raises the question, discussed in detail in the next chapter, of what traits differentiate stenodermatines from other phyllostomids and how these traits might contribute to their elevated rates of diversification.

Historical Biogeography

Although several analyses with biogeographic implications (e.g., Stevens 2006), as well as historical biogeographic analyses of genera or certain groups had been published before (e.g., Velazco and Patterson 2013), the first comprehensive historical biogeographic analyses of phyllostomids were conducted by Rojas et al. (2016). There are three sets of results relevant to biogeography. First, Rojas et al. (2016) evaluated if Quaternary climate change disproportionately contributed to speciation across the radiation. The Pleistocene refugia hypothesis has often been invoked to explain the diversity of species in the lowland forests of the Amazon (e.g., Haffer 1969; Hooghiemstra and van der Hammen 1998, but see Colinvaux et al. 2000; Lessa et al. 2003), where phyllostomids reach their highest richness. Since its inception, the refugia hypothesis has been extended to all Neotropical lowland forests, including the Brazilian Atlantic forests and the forests of Mesoamerica (Carnaval and Moritz 2008; Willis and Whittaker 2000). Unsurprisingly, studies of phyllostomid biogeography have applied this general framework to explain divergences, finding some genetic evidence for geographic structuring within species linked to such refugia for *Desmodus* in the Atlantic forests and the Pantanal (Martins et al. 2009), *Chrotopterus* in the Amazon (Clare 2011), and *Carollia brevicauda* and *Artibeus obscurus*, both in Brazil (Ferreira et al. 2014; Pavan et al. 2011). In contrast, no such patterns were obtained in earlier surveys of genetic diversity in the Atlantic Forest for *Artibeus lituratus, Carollia perspicillata, Sturnira lilium*, or *Glossophaga soricina* (Ditchfield 2000), as well as Amazonian populations of *Desmodus, Micronycteris megalotis, Trachops, Uroderma bilobatum*, and small *Platyrrhinus* species (Clare 2011).

Nevertheless, testing Quaternary glacial refugia as centers of speciation requires examining both the pattern of geographic structure and the timing for divergence. To this end, Rojas et al. (2016) estimated divergence times between sister species. If Quaternary climate change had played a role in isolating populations ultimately resulting in speciation, then divergence times between sister species should cluster during this period. Analyses of speciation events did not support a Quaternary origin for the extant diversity of phyllostomids and, instead, indicated most of the sister species diverged much earlier. Additionally, because the number of divergence events across a phylogeny depends on the number of branches present at a given time, an excess of divergences between sister species (e.g., Garzón-Orduña et al. 2014) is not enough to demonstrate that a particular period disproportionately contributed to speciation. Therefore, Rojas et al. (2016) analyzed diversification rates, or the net difference between speciation and extinction rates, across the phylogeny. If the Quaternary disproportionately contributed to speciation and background extinction rates had remained more or less constant, this period would show elevated diversification rates compared to previous eras. This prediction was refuted. Instead, the largest change in

diversification rates occurred in the ancestor to Stenodermatinae and was not concentrated in any particular period. In short, predictions from the hypothesis that Quaternary climate change played a substantial role in generating species-level diversity across phyllostomids were rejected.

The second contribution to the historical biogeography of phyllostomids by Rojas et al. (2016) is the model-based inference of ancestral areas for the entire family (fig. 6.4). Unlike earlier biogeographic analyses (e.g., Dávalos 2010), which tended to rely on equating areas with characters and optimizing using phylogenies, probabilistic model-based analyses explicitly include biogeographic processes that have no equivalent in character analyses and that account for branch lengths contributing to the probability of biogeographic events (Matzke 2013). Using one recently developed method called BioGeoBEARS, Rojas et al. (2016) encoded the three large and distinct areas of North and Central America, South America, and the Antilles. These analyses inferred South America as the ancestral area of most phyllostomid subfamilies (fig. 6.4).

Although these analyses were not designed to differentiate among habitats or between tropical and subtropical regions within continents, the South American—and specifically tropical Andean or lowland forest—origin of most subfamilies contributes to the latitudinal gradient of diversity in phyllostomids (Stevens 2006). Two historical mechanisms were proposed to explain the gradient: tropical niche conservatism and time for speciation (Stevens 2011). In the former, the high heritability of climate niche traits leads to speciation in areas with similar climate, with reduced dispersal to, and species accumulation in, climatically distinct areas resulting in fewer, shallower lineages (Buckley et al. 2010). Time for speciation, in contrast, results from the greater species accumulation in older areas (Wiens et al. 2006). Predictions for each of these mechanisms can only be distinguished by separating spatial and hence biogeographic gradients from climate/environmental variation. Unfortunately, biogeographic and climate variation within the range of phyllostomids is confounded, leading to inconclusive results and only a slight explanatory advantage for niche conservatism in regressions of phylogenetic characteristics as a function of spatial and environmental variables (Ramos Pereira and Palmeirim 2013; Stevens 2011).

Analyses of latitudinal gradients sometimes include

phylogenetic characteristics (e.g., branch lengths, genus to species ratios) (Arita et al. 2014; Ramos Pereira and Palmeirim 2013; Stevens 2006, 2011), without fully integrating the phylogeny. By directly integrating phylogenies with patterns of species coexistence, Villalobos et al. (2013) found support for the niche conservatism hypothesis to explain the phyllostomid diversity gradient. The disproportionate coexistence of phyllostomid species with close relatives across multiple phylogenetic scales is consistent with environmental niche conservation within the radiation. The gradient itself would result from the combination of conserved climate preferences coupled with higher speciation rates in the area of origin (Villalobos et al. 2013), although these rates were not modeled. Diversification analyses by Rojas et al. (2016) support the conclusions of Stevens (2011) and Villalobos et al. (2013) because increased speciation or decreased extinction rates are concentrated in stenodermatines, whose climate niche—along with biotic preferences—appear to be highly conserved (Dumont et al. 2014).

While South America has been the main center of diversification for most phyllostomids, North and Central America were both important for a few early divergences, as well as more recent divergences within subfamilies (fig. 6.4). The first finding is consistent with fossils showing early noctilionoids (Czaplewski and Morgan 2012), as well as an undescribed mormoopid (Morgan and Czaplewski 2012) from the Oligocene of Florida. Oligocene fossils in North America suggest that the ancestors of the family were present on this continent early on, well before any fossils found in South America (where the earliest findings correspond to the Miocene). The hypothesis of an early North American divergence or origin including both sets of Tertiary fossils, however, remains to be tested using model-based biogeographic methods. Additionally, the within-subfamily divergences are in line with analyses by Arita et al. (2014) who, based on latitudinal patterns of endemism and diversity across genera, identified the two continental landmasses as independent centers of diversification for phyllostomids. Additionally, Arita et al. (2014) proposed the absence of continuous desert and tropical dry forest habitats from Central America since the Miocene as the main ecological factor preventing *Macrotus*, *Choeronycteris*, and *Musonycteris* (but not *Leptonycteris*) from reaching South America from the north and the lonchophyllines *Dryadonycteris*,

Platalina, and *Xeronycteris* from reaching Central and North America from the south. This distinctly historical hypothesis also needs to be tested, for example, by combining phyllostomid phylogenies with inferred ancestral climate niche envelopes in phyloclimatic analyses (e.g., Yesson and Culham 2006).

The third important set of findings on phyllostomid biogeography from Rojas et al. (2016) centers on the role of the Antilles in the history of the clade. The descendants of the most recent common ancestor of *Glossophaga* and *Brachyphylla* (tribes Glossophagini and Brachyphyllini (Cirranello et al. 2016), along with the short-faced bats or Stenodermatina, have proposed origins in the Caribbean islands with subsequent colonization of the continent (Dávalos 2007, 2010). The biogeographic models supported reverse colonization only for the glossophagines (fig. 6.4), and not for the short-faced bats. This rejection of the Caribbean-origin hypothesis for short-faced bats, however, remains to be tested by including *Cubanycteris*, an extinct species from Cuba whose primitive morphology suggests it is an early branch in the subtribe (Mancina and Garcia-Rivera 2005), in character-based analyses and subsequent biogeographic models. Finally, dating analyses also supported a potential role for Miocene sea-level changes in facilitating colonization to and from the Antilles (Rojas et al. 2016).

Within phyllostomids, the historical biogeography of only four of the more than 50 phyllostomid genera has been analyzed to date (*Carollia*, *Platyrrhinus*, *Sturnira*, and *Uroderma*), all but one member of the subfamily Stenodermatinae (Cuadrado-Ríos and Mantilla-Meluk 2016; Hoffmann and Baker 2001; Velazco and Patterson 2008, 2013; table 6.2). But these genera are not all equally diverse. With 21 species each, *Platyrrhinus* and *Sturnira* are two of the most diverse phyllostomid genera, contributing to the high diversity of stenodermatines, with five species analyzed for *Carollia* (Hoffmann and Baker 2003), and the most liberal estimates of *Uroderma* diversity include only five species (Cuadrado-Ríos and Mantilla-Meluk 2016). Analyses of the only genus outside the subfamily Stenodermatinae, *Carollia*, included three species distributed in both Central and South America and two Central American endemics (Hoffmann and Baker 2003). Three biogeographic patterns were identified: (1) genetic continuity between the Chocó biodiversity hotspot west of the Andes in northern South America and Central America,

Table 6.2. Annotated historical biogeography analyses including phyllostomids.

Biogeographic method	Taxonomic scope	Source
Comparative phylogeography	*Carollia*	Hoffmann and Baker 2003
Reconciled area analyses	Caribbean mammals	Dávalos 2004
Polynomial regressions of richness, divergence, variance of divergence, clade age, and variance of clade age as functions of latitudinal gradient	Phyllostomidae	Stevens 2006
Shimodaira-Hasegawa tests of alternative phylogenies	Stenodermatina	Dávalos 2007
Regressions of richness against environmental variables	Lonchophyllinae	Mantilla-Meluk 2007
Dispersal-vicariance analysis, Lagrange	*Platyrrhinus*	Velazco and Patterson 2008
Analyses of assemblage composition and nestedness as functions of island characteristics	Caribbean chiroptera	Presley and Willig 2008
Parsimony optimizations of geographic distributions	Caribbean chiroptera	Dávalos 2010
Phylogenetic regressions	New World Noctilionoidea	Rojas et al. 2012
Generalized additive models of mean root distance as a function of latitude	New World chiroptera	Ramos Pereira and Palmeirim 2013
Dispersal-vicariance analyses, Lagrange	*Sturnira*	Velazco and Patterson 2013
Phylogenetic species variability, phylogenetic species clustering	Phyllostomidae	Villalobos et al. 2013
Phylogenetic distance, phylogenetic species variability, phylogenetic species clustering, and covariation among these	New World Noctilionoidea	Stevens and Tello 2014
Species/genus ratios, latitudinal richness analyses	New World chiroptera	Arita et al. 2014
Likelihood model comparisons	New World Noctilionoidea	Rojas et al. 2016
Dispersal-vicariance analysis and Bayesian binary MCMC method	*Uroderma*	Cuadrado-Ríos and Mantilla-Meluk 2016

(2) deep divergences within Amazonian populations of *brevicauda* and *castanea*, and (3) a very recent expansion of *perspicillata* and *sowelli* into the northwestern-most extent of the range in Yucatán. The similarities in faunas west of the Andes extending north into Central America was originally highlighted for birds (Chapman 1917; Haffer 1967), but few studies with bats have highlighted it, especially within species. The remaining patterns remain to be tested more generally, both in Amazonia and Mesoamerica.

The broad distribution of *Platyrrhinus* and *Sturnira* and radiation into a range of Neotropical biomes makes them great examples for testing biogeographic mechanisms at much finer scales. Current *Platyrrhinus* and *Sturnira* phylogenies include both mitochondrial and nuclear sequences (the mitochondrial *cytb*, NADH dehydrogenase subunit 2 *nd2* genes and the D-loop regulatory region and the nuclear *rag2* for both genera, and the recombination-activating protein 1 *rag1* for *Sturnira*), reducing the probability of inferring particular resolutions based on a single gene tree.

By defining several biogeographic areas within continents, analyses for both *Platyrrhinus* and *Sturnira* were able to test biogeographic hypotheses within South America (Velazco and Patterson 2008, 2013), and not just at the continental scale. Velazco and Patterson (2008) inferred the Brazilian Shield as the biogeographic area of origin for *Platyrrhinus* by applying model-based biogeographic analyses of dispersal, local extinction, and cladogenesis using Lagrange (Ree and Smith 2008). Ancestors of multiple species dispersed from the area of origin to the Amazon lowlands (most recent common ancestor of *P. brachycephalus* and *P. matapalensis*) and to the Andes (most recent common ancestor *P. albericoi* and *P. infuscus*), giving rise to the current distribution of species concentrated in the Andes. Individual lineages dispersed from the Andes to the Chocó or Pacific lowlands and then to Central America, explaining the distribution of *P. helleri*. Finally, dispersal from the Amazon to the Guianan Shield would be responsible for the widespread distribution of *P. aurarius* and *P. infuscus*. In summary, the results support dispersal as the primary biogeographic mechanism leading to the present-day distribution of *Platyrrhinus* and suggest that the Andes have been crucial for the diversification of the group, as was first proposed by Karl Koopman based on distributional analyses (Koopman 1978). This finding is also in line with previous studies for birds (Smith et al. 2014; Weir 2006), amphibians (Santos et al. 2009), and clearwing butterflies (Elias et al. 2009).

Velazco and Patterson (2013) applied the dispersal, local extinction, and cladogenesis model to infer the biogeographic history of *Sturnira*, crucially including endemic Antillean species. In contrast to *Platyrrhinus* and its colonization of the Andes, earliest-diverging lineages of *Sturnira* have exclusively Andean distributions, inferring the Andes as the area of origin for the genus. From the Andes, *Sturnira* lineages repeatedly dispersed to the Chocó and then to Central America. A single lineage dispersed to the Lesser Antilles and could have originated in one of several areas in northern South America (Chocó, northern Andes, or Caribbean lowlands). The timing of these events illuminates the links between earth history and the biogeography of phyllostomids. Dispersal into Central America followed the closing of the Isthmus of Panama, contributing to the Great American Biotic Interchange (GABI), a biogeographic mechanism first proposed by A. R. Wallace himself (Wallace 1876a, 1876b), with paleontological evidence accumulating throughout the twentieth century (Marshall 1988; Simpson 1980). More recent analyses grounded on phylogenetics have shown that *Sturnira* bats joined birds (Weir et al. 2009) and didelphid marsupials (Jansa et al. 2014) in the unusual south-to-north colonization associated with GABI. Arita et al. (2014) highlighted the main ecological factor contributing to GABI dispersal for *Sturnira* as well as for other frugivorous phyllostomids: continuous corridors of tropical wet forests connecting South America to southern Mexico. Similarly, the colonization of the Antilles dates to the Pleistocene (Velazco and Patterson 2013), when glaciations exposed large banks linking Grenada to Saint Vincent (Dávalos and Turvey 2012), facilitating dispersal from northern South America. Compared to dispersal to Central America or the Antilles, dispersal out of the Andes to other regions of South America has been the norm, giving rise to sympatric assemblages of both early and recently diverging species in the Andes.

Phylogenetic analyses for *Uroderma* have been restricted to sequences of the mitochondrial *cytb* orig-inally published by Hoffmann et al. (2003) to uncover the genetic diversity and divergence between chromosomal races in the genus. Given the known history of secondary hybridization in populations of *Uroderma bilobatum* (Baker 1981; Greenbaum 1981), using this exclusively matrilineal marker has the potential to produce a gene tree discordant with the species tree because of introgression. Nevertheless, Cuadrado-Ríos and Mantilla-Meluk (2016) used this best estimate of phylogeny to infer its biogeographic history by applying a statistical implementation of dispersal-vicariance analysis (Yu et al. 2015). As dispersal-vicariance analysis imposes no cost for vicariance (Ronquist 1997) and as no constraint was placed on the number of areas inferred for any node, all early nodes were inferred to be composites of multiple and sometimes discontinuous areas. Despite methodological challenges, a general pattern is evident in the biogeographic history of *Uroderma*. Both the Central Andes and Central America played important roles in the diversification of the genus since first emerging in the Miocene, with the northern Andes acting as a link between these two regions but not as a center of diversification.

Conclusions

Phylogenies based on molecular markers have the upended traditional understanding of evolutionary relationships among phyllostomids, and current phylogenies—whether based on several independent loci or entire mitochondrial genomes—differ from the original analyses of two loci by Baker et al. (2003) only in minor details. Although the monophyly of subfamilies is now settled, relationships among subfamilies—in particular, the phylogenetic positions of Micronycterinae relative to Desmodontinae (fig. 6.3)—and Lonchorhininae (cf. figs. 6.1 and 6.3) remain outstanding. Likewise, while phyllostomid Miocene fossils have been included in character-based analyses (fig. 6.3), support for those relationships is low, and future analyses should benefit from including several Oligocene close relatives of phyllostomids (Czaplewski and Morgan 2012; Morgan and Czaplewski 2012). Relaxed clock analyses accounting for the instability of the nodes offer a potential route forward (e.g., Herrera and Dávalos 2016) but have yet to be implemented, despite the publication of several new data sets (table 6.1).

Biogeographic analyses for the entire family have

rejected Quaternary glaciations as a primary species pump and have demonstrated many instances of dispersal at a continental scale (fig. 6.4). But analyses of individual genera have uncovered critical ecological factors necessary for both cladogenesis and dispersal, such as the continuity of habitats across continents, and have revealed direct connections between biogeographic events and earth history. The three genera analyzed to date have demonstrated the crucial role of the Andes in phyllostomid cladogenesis, as well as the importance of lowland rainforests in the long-term diversity of these stenodermatines. As family-wide analyses are forced to examine only large biogeographic regions, it is essential to extend genus or subfamily analyses to other clades to test these emerging patterns across the whole radiation. Much work remains to be completed to understand the historical biogeography of phyllostomids and the relative importance of geoclimatic events relative to adaptations to particular climatic niches in this family.

Acknowledgments

We thank A. Cirranello, T. Fleming, and two anonymous reviewers for comments on the manuscript. LMD was supported by the National Science Foundation (DEB-1442142). DR was supported by Foundation for Science and Technology, Portugal (www.fct.pt), fellowship SFRH/BPD/97707/2013. The Portuguese Foundation for Science and Technology supported CESAM RU (UID/AMB/50017) through national funds and FEDER funds, within the PT2020 Partnership Agreement.

References

Agnarsson, I., C. M. Zambrana-Torrelio, N. P. Flores-Saldana, and L. J. May-Collado. 2011. A time-calibrated species-level phylogeny of bats (Chiroptera, Mammalia). PLOS Currents 3:RRN1212. doi:10.1371/currents.RRN1212.

Arita, H. T., J. Vargas-Barón, and F. Villalobos. 2014. Latitudinal gradients of genus richness and endemism and the diversification of New World bats. Ecography, 37:1024–1033.

Avise, J. C. 2000. Phylogeography: The History and Formation of Species. Harvard University Press, Cambridge, MA.

Baker, R., C. Hood, and R. Honeycutt. 1989. Phylogenetic relationships and classification of the higher categories of the New World bat family Phyllostomidae. Systematic Zoology, 38:228–238.

Baker, R. J. 1981. Chromosome flow between chromosomally characterized taxa of a volant mammal, *Uroderma bilobatum* (Chiroptera: Phyllostomatidae). Evolution, 35:296–305.

Baker, R. J., O. R. P. Bininda-Emonds, H. Mantilla-Meluk, C. A. Porter, and R. Van Den Bussche. 2012. Molecular timescale of diversification of feeding strategy and morphology in New World leaf-nosed bats (Phyllostomidae): a phylogenetic perspective. Pp. 385–409 *in*: Evolutionary History of Bats: Fossils, Molecules and Morphology (G. F. Gunnell and N. B. Simmons, eds.). Cambridge Studies in Molecules and Morphology-New Evolutionary Paradigms. Cambridge University Press, Cambridge, MA.

Baker, R. J., C. A. Porter, S. R. Hoofer, and R. A. Van Den Bussche. 2003. Diversification among New World leaf-nosed bats: an evolutionary hypothesis and classification inferred from digenomic congruence of DNA sequence. Occasional Papers, Museum of Texas Tech University, 230:1–32.

Baker, R. J., C. A. Porter, J. C. Patton, and R. A. Van Den Bussche. 2000. Systematics of bats of the family Phyllostomidae based on RAG2 DNA sequences. Occasional Papers, Museum of Texas Tech University, 202:1–16.

Bininda-Emonds, O. R. P. 2007. Fast genes and slow clades: comparative rates of molecular evolution in mammals. Evolutionary Bioinformatics, 3:59–85.

Botero-Castro, F., M.-K. Tilak, F. Justy, F. Catzeflis, F. Delsuc, and E. J. P. Douzery. 2013. Next-generation sequencing and phylogenetic signal of complete mitochondrial genomes for resolving the evolutionary history of leaf-nosed bats (phyllostomidae). Molecular Phylogenetics and Evolution, 69:728–739.

Buckley, L. B., T. J. Davies, D. D. Ackerly, N. J. B. Kraft, S. P. Harrison, B. L. Anacker, H. V. Cornell et al. 2010. Phylogeny, niche conservatism and the latitudinal diversity gradient in mammals. Proceedings of the Royal Society B–Biological Sciences, 277:2131–2138.

Carnaval, A. C., and C. Moritz. 2008. Historical climate modelling predicts patterns of current biodiversity in the Brazilian Atlantic Forest. Journal of Biogeography, 35:1187–1201.

Chapman, F. M. 1917. The distribution of bird-life in Colombia: a contribution to a biological survey of South America. Bulletin of the American Museum of Natural History, 36:1–729.

Cirranello, A., N. B. Simmons, S. Solari, and R. J. Baker. 2016. Morphological diagnoses of higher-level phyllostomid taxa (Chiroptera: Phyllostomidae). Acta Chiropterologica, 18:39–71.

Clare, E. L. 2011. Cryptic species? Patterns of maternal and paternal gene flow in eight Neotropical bats. PLOS ONE, 6:e21460.

Colinvaux, P. A., P. E. De Oliveira, and M. B. Bush. 2000. Amazonian and Neotropical plant communities on glacial time-scales: the failure of the aridity and refuge hypotheses. Quaternary Science Reviews, 19:141–169.

Cruz-Neto, A. P., T. Garland, and A. S. Abe. 2001. Diet, phylogeny, and basal metabolic rate in phyllostomid bats. Zoology, 104:49–58.

Cuadrado-Ríos, S., and H. Mantilla-Meluk. 2016. Timing the evolutionary history of tent-making bats, genus *Uroderma* (Phyllostomidae): a biogeographic context. Mammalian Biology–Zeitschrift für Säugetierkunde, 81:579–586.

Czaplewski, N. J., and G. S. Morgan. 2012. New basal noctilion-oid bats (Mammalia: Chiroptera) from the Oligocene of subtropical North America. Pp. 162–209 in: Evolutionary History of Bats: Fossils, Molecules and Morphology (G. F. Gunnell and N. B. Simmons, eds.). Cambridge Studies in Morphology and Molecules: New Paradigms in Evolutionary Biology. Cambridge University Press, Cambridge.

Czaplewski, N. J., M. Takai, T. M. Naeher, N. Shigehara, and T. Setoguchi. 2003. Additional bats from the middle Miocene La Venta fauna of Colombia. Revista de la Academia Colombiana de Ciencias Físicas, Exactas y Naturales, 27:263–282.

Datzmann, T., O. von Helversen, and F. Mayer. 2010. Evolution of nectarivory in phyllostomid bats (Phyllostomidae Gray, 1825, Chiroptera: Mammalia). BMC Evolutionary Biology, 10:165.

Dávalos, L. M. 2004. Phylogeny and biogeography of Caribbean mammals. Biological Journal of the Linnean Society, 81:373–394.

Dávalos, L. M. 2006. The geography of diversification in the mormoopids (Chiroptera: Mormoopidae). Biological Journal of the Linnean Society, 88:101–118.

Dávalos, L. M. 2007. Short-faced bats (Phyllostomidae: Steno-dermatina): a Caribbean radiation of strict frugivores. Journal of Biogeography, 34:364–375.

Dávalos, L. M. 2010. Earth history and the evolution of Caribbean bats. Pp. 96–115 in: Island bats: Ecology, Evolution, and Conservation (T. H. Fleming and P. A. Racey, eds.). University of Chicago Press, Chicago.

Dávalos, L. M., A. L. Cirranello, J. H. Geisler, and N. B. Simmons. 2012. Understanding phylogenetic incongruence: lessons from phyllostomid bats. Biological Reviews, 87:991–1023.

Dávalos, L. M., and S. A. Jansa. 2004. Phylogeny of the Loncho-phyllini (Chiroptera: Phyllostomidae). Journal of Mammalogy, 85:404–413.

Dávalos, L. M., and S. Turvey. 2012. West Indian mammals: the old, the new, and the recently extinct. Pp. 157–202 in: Bones, Clones, and Biomes: An Extended History of Recent Neotropical Mammals (B. D. Patterson and L. P. Acosta, eds.). University of Chicago Press, Chicago.

Dávalos, L. M., P. M. Velazco, O. M. Warsi, P. Smits, and N. B. Simmons. 2014. Integrating incomplete fossils by isolating conflictive signal in saturated and non-independent morphological characters. Systematic Biology, 63:582–600.

Ditchfield, A. 2000. The comparative phylogeography of Neotropical mammals: patterns of intraspecific mitochondrial DNA variation among bats contrasted to nonvolant small mammals. Molecular Ecology, 9:1307–1318.

Drummond, A. J., S. Y. W. Ho, M. J. Phillips, and A. Rambaut. 2006. Relaxed phylogenetics and dating with confidence. PLOS Biology, 4:e88.

Dumont, E. R., L. M. Dávalos, A. Goldberg, C. C. Voigt, K. Rex, and S. E. Santana. 2012. Morphological innovation, diversification and the invasion of a new adaptive zone. Proceedings of the Royal Society B–Biological Sciences, 279:1797–1805.

Dumont, E. R., K. Samadevam, I. Grosse, O. M. Warsi, B. Baird, and L. M. Dávalos. 2014. Selection for mechanical advantage underlies multiple cranial optima in New World leaf-nosed bats. Evolution, 68:1436–1449.

Elias, M., M. Joron, K. Willmott, K. L. Silva-Brandão, V. Kaiser, C. F. Arias, L. M. G. Piñerez et al. 2009. Out of the Andes: patterns of diversification in clearwing butterflies. Molecular Ecology, 18:1716–1729.

Ferreira, W. A. S., B. d. N. Borges, S. Rodrigues-Antunes, F. A. G. d. Andrade, G. F. d. S. Aguiar, J. d. S. e. Silva-Junior, S. A. Marques-Aguiar et al. 2014. Phylogeography of the dark fruit-eating bat Artibeus obscurus in the Brazilian Amazon. Journal of Heredity, 105:48–59.

Garzón-Orduña, I. J., J. E. Benetti-Longhini, and A. V. Z. Brower. 2014. Timing the diversification of the Amazonian biota: butterfly divergences are consistent with Pleistocene refugia. Journal of Biogeography, 41:1631–1638.

Greenbaum, I. F. 1981. Genetic interactions between hybridizing cytotypes of the tent-making bat (Uroderma bilobatum). Evolution, 35:306–321.

Griffiths, T. A. 1982. Systematics of the New World nectar-feeding bats (Mammalia, Phyllostomidae), based on the morphology of the hyoid and lingual regions. American Museum Novitates, 2742:1–45.

Griffiths, T. A. 1983. On the phylogeny of the Glossophaginae and the proper use of outgroup analysis. Systematic Zoology, 32:283–285.

Haffer, J. 1967. Speciation in Colombian forest birds west of the Andes. American Museum Novitates, 2294:1–57.

Haffer, J. 1969. Speciation in Amazonian forest birds. Science, 165:131–137.

Heath, T. A., J. P. Huelsenbeck, and T. Stadler. 2014. The fossilized birth-death process for coherent calibration of divergence-time estimates. Proceedings of the National Academy of Sciences, USA, 111:E2957–E2966.

Herrera, J. P., and L. M. Dávalos. 2016. Phylogeny and divergence times of lemurs inferred with recent and ancient fossils in the tree. Systematic Biology, 65:772–791.

Hoffmann, F. G., and R. J. Baker. 2001. Systematics of bats of the genus Glossophaga (Chiroptera: Phyllostomidae) and phylogeography in G. soricina based on the cytochrome b gene. Journal of Mammalogy, 82:1092–1101.

Hoffmann, F. G., and R. J. Baker. 2003. Comparative phylo-geography of short-tailed bats (Carollia: Phyllostomidae). Molecular Ecology, 12:3403–3414.

Hoffmann, F. G., S. R. Hoofer, and R. J. Baker. 2008. Molecular dating of the diversification of Phyllostominae bats based on nuclear and mitochondrial DNA sequences. Molecular Phylogenetics and Evolution, 49:653–658.

Hoffmann, F. G., J. G. Owen, and R. J. Baker. 2003. mtDNA perspective of chromosomal diversification and hybridization in Peters' tent-making bat (Uroderma bilobatum: Phyllostomidae). Molecular Ecology, 12:2981–2993.

Hooghiemstra, H., and T. van der Hammen. 1998. Neogene and Quaternary development of the Neotropical rain forest: the forest refugia hypothesis, and a literature overview. Earth-Science Reviews, 44:147–183.

Jansa, S. A., F. K. Barker, and R. S. Voss. 2014. The early diversification history of didelphid marsupials: a window into South America's "splendid isolation." Evolution, 68:684–695.

Jones, K. E., A. Purvis, A. MacLarnon, O. R. P. Bininda-Emonds, and N. B. Simmons. 2002. A phylogenetic supertree of the bats (Mammalia: Chiroptera). Biological Reviews, 77: 223–259.

Koopman, K. F. 1978. Zoogeography of Peruvian bats, with special emphasis on the role of the Andes. American Museum Novitates, 2651:1–33.

Larsen, P. A., S. R. Hoofer, M. C. Bozeman, S. C. Pedersen, H. H. Genoways, C. J. Phillips, D. E. Pumo et al. 2007. Phylogenetics and phylogeography of the *Artibeus jamaicensis* complex based on cytochrome-*b* DNA sequences. Journal of Mammalogy, 88:712–727.

Lessa, E. P., J. A. Cook, and J. L. Patton. 2003. Genetic footprints of demographic expansion in North America, but not Amazonia, during the Late Quaternary. Proceedings of the National Academy of Sciences, USA, 100:10331–10334.

Litsios, G., and N. Salamin. 2014. Hybridisation and diversification in the adaptive radiation of clownfishes. BMC Evolutionary Biology, 14:1–9.

Liu, Y., J. A. Cotton, B. Shen, X. Han, S. J. Rossiter, and S. Zhang. 2010. Convergent sequence evolution between echolocating bats and dolphins. Current Biology, 20:R53–R54.

Mancina, C. A., and L. Garcia-Rivera. 2005. New genus and species of fossil bat (Chiroptera: Phyllostomidae) from Cuba. Caribbean Journal of Science, 41:22–27.

Mantilla-Meluk, H. 2007. Lonchophyllini, the Chocoan bats. Revista Institucional Universidad Tecnológica del Chocó, 26:49–57.

Marshall, L. G. 1988. Land mammals and the great American interchange. American Scientist, 76:380–388.

Martins, F. M., A. R. Templeton, A. C. Pavan, B. C. Kohlbach, and J. S. Morgante. 2009. Phylogeography of the common vampire bat (*Desmodus rotundus*): marked population structure, Neotropical Pleistocene vicariance and incongruence between nuclear and mtDNA markers. BMC Evolutionary Biology, 9:294.

Matzke, N. J. 2013. Probabilistic historical biogeography: new models for founder-event speciation, imperfect detection, and fossils allow improved accuracy and model-testing. Frontiers of Biogeography, vol. 5.

Morgan, G. S., and N. J. Czaplewski. 2012. Evolutionary history of the Neotropical Chiroptera: the fossil record. Pp. 105–161 *in*: Evolutionary History of Bats: Fossils, Molecules and Morphology (G. F. Gunnell and N. B. Simmons, eds.). Cambridge Studies in Morphology and Molecules: New Paradigms in Evolutionary Biology. Cambridge University Press, Cambridge.

Pavan, A. C., F. Martins, F. R. Santos, A. Ditchfield, and R. A. F. Redondo. 2011. Patterns of diversification in two species of short-tailed bats (Carollia Gray, 1838): the effects of historical fragmentation of Brazilian rainforests. Biological Journal of the Linnean Society, 102:527–539.

Porter, C. A., S. R. Hoofer, C. A. Cline, F. G. Hoffmann, and R. J. Baker. 2007. Molecular phylogenetics of the phyllostomid bat genus *Micronycteris* with descriptions of two new subgenera. Journal of Mammalogy, 88:1205–1215.

Presley, S. J., and M. R. Willig. 2008. Composition and structure of Caribbean bat (Chiroptera) assemblages: effects of inter-island distance, area, elevation and hurricane-induced disturbance. Global Ecology and Biogeography, 17:747–757.

Ramos Pereira, M. J., and J. M. Palmeirim. 2013. Latitudinal diversity gradients in New World bats: are they a consequence of niche conservatism? PLOS ONE, 8:e69245.

Ree, R. H., and S. A. Smith. 2008. Maximum likelihood inference of geographic range evolution by dispersal, local extinction, and cladogenesis. Systematic Biology, 57:4–14.

Revell, L. J. 2012. Phytools: an R package for phylogenetic comparative biology (and other things). Methods in Ecology and Evolution, 3:217–223.

Rojas, D., Á. Vale, V. Ferrero, and L. Navarro. 2011. When did plants become important to leaf-nosed bats? Diversification of feeding habits in the family Phyllostomidae. Molecular Ecology, 20:2217–2228.

Rojas, D., Á. Vale, V. Ferrero, and L. Navarro. 2012. The role of frugivory in the diversification of bats in the Neotropics. Journal of Biogeography, 39:1948–1960.

Rojas, D., O. M. Warsi, and L. M. Dávalos. 2016. Bats (Chiroptera: Noctilionoidea) challenge a recent origin of extant Neotropical diversity. Systematic Biology, 65:432–448.

Ronquist, F. 1997. Dispersal-vicariance analysis: a new approach to the quantification of historical biogeography. Systematic Biology, 46:195–203.

Ronquist, F., S. Klopfstein, L. Vilhelmsen, S. Schulmeister, D. L. Murray, and A. P. Rasnitsyn. 2012. A total-evidence approach to dating with fossils, applied to the early radiation of the hymenoptera. Systematic Biology, 61:973–999.

Rosenberg, N. A., and M. Nordborg. 2002. Genealogical trees, coalescent theory and the analysis of genetic polymorphisms. Nature Reviews Genetics, 3:380–390.

Roy, S. W. 2009. Phylogenomics: gene duplication, unrecognized paralogy and outgroup choice. PLOS ONE, 4:e4568.

Santos, J. C., L. A. Coloma, K. Summers, J. P. Caldwell, R. Ree, and D. C. Cannatella. 2009. Amazonian amphibian diversity is primarily derived from Late Miocene Andean lineages. PLOS Biology, 7:e1000056.

Schondube, J. E., L. G. Herrera M., and C. Martínez del Río. 2001. Diet and the evolution of digestion and renal function in phyllostomid bats. Zoology, 104:59–73.

Shi, J. J., and D. L. Rabosky. 2015. Speciation dynamics during the global radiation of extant bats. Evolution, 69:1528–1545.

Silva Taboada, G. S., and R. H. Pine. 1969. Morphological and behavioral evidence for the relationship between the bat genus *Brachyphylla* and the Phyllonycterinae. Biotropica, 1:10–19.

Simpson, G. G. 1980. Splendid isolation: the curious history of South American mammals. Yale University Press, New Haven, CT.

Sjöstrand, J., A. Tofigh, V. Daubin, L. Arvestad, B. Sennblad, and J. Lagergren. 2014. A Bayesian method for analyzing lateral gene transfer. Systematic Biology, 63:409–420.

Smith, B. T., J. E. McCormack, A. M. Cuervo, M. J. Hickerson, A. Aleixo, C. D. Cadena, J. Perez-Eman et al. 2014. The drivers of tropical speciation. Nature, 515:406–409.

Stevens, R. D. 2006. Historical processes enhance patterns of diversity along latitudinal gradients. Proceedings of the Royal Society B–Biological Sciences, 273:2283–2289.

Stevens, R. D. 2011. Relative effects of time for speciation and tropical niche conservatism on the latitudinal diversity gradient of phyllostomid bats. Proceedings of the Royal Society B–Biological Sciences, 278:2528–2536.

Stevens, R. D., and J. S. Tello. 2014. On the measurement of dimensionality of biodiversity. Global Ecology and Biogeography, 23:1115–1125.

Teeling, E. C., M. S. Springer, O. Madsen, P. Bates, S. J. O'Brien, and W. J. Murphy. 2005. A molecular phylogeny for bats illuminates biogeography and the fossil record. Science, 307:580–584.

Thorne, J. L., and H. Kishino. 2002. Divergence time and evolutionary rate estimation with multilocus data. Systematic Biology, 51:689–702.

Velazco, P., and B. D. Patterson. 2008. Phylogenetics and biogeography of the broad-nosed bats, genus *Platyrrhinus* (Chiroptera: Phyllostomidae). Molecular Phylogenetics and Evolution, 49:749–759.

Velazco, P. M., and B. D. Patterson. 2013. Diversification of the yellow-shouldered bats, genus *Sturnira* (Chiroptera, Phyllostomidae), in the New World tropics. Molecular Phylogenetics and Evolution, 68:683–698.

Villalobos, F., T. F. Rangel, and J. A. F. Diniz-Filho. 2013. Phylogenetic fields of species: cross-species patterns of phylogenetic structure and geographical coexistence. Proceedings of the Royal Society B–Biological Sciences, 280:1–9.

Wagner, P. J. 2000. Exhaustion of morphologic character states among fossil taxa. Evolution, 54:365–386.

Wake, D. B. 1991. Homoplasy: the result of natural selection, or evidence of design limitations? American Naturalist, 138:543–567.

Wallace, A. R. 1876a. The Geographical Distribution of Animals; with a Study of the Relations of Living and Extinct Faunas as Elucidating the Past Changes of the Earth's Surface, vol. 1. Harper & Brothers, New York.

Wallace, A. R. 1876b. The Geographical Distribution of Animals; with a Study of the Relations of Living and Extinct Faunas as Elucidating the Past Changes of the Earth's Surface, vol. 2. Harper & Brothers, New York.

Weir, J. T. 2006. Divergent timing and patterns of species accumulation in lowland and highland Neotropical birds. Evolution, 60:842–855.

Weir, J. T., E. Bermingham, and D. Schluter. 2009. The Great American Biotic Interchange in birds. Proceedings of the National Academy of Sciences, 106:21737–21742.

Wetterer, A. L., M. V. Rockman, and N. B. Simmons. 2000. Phylogeny of phyllostomid bats (Mammalia: Chiroptera): data from diverse morphological systems, sex chromosomes, and restriction sites. Bulletin of the American Museum of Natural History, 248:1–200.

Wiens, J. J., C. H. Graham, D. S. Moen, S. A. Smith, and T. W. Reeder. 2006. Evolutionary and ecological causes of the latitudinal diversity gradient in hylid frogs: treefrog trees unearth the roots of high tropical diversity. American Naturalist, 168:579–596.

Willis, K. J., and R. J. Whittaker. 2000. The refugial debate. Science, 287:1406.

Yesson, C., and A. Culham. 2006. A phyloclimatic study of *Cyclamen*. BMC Evolutionary Biology, 6:72.

Yu, W., Y. Wu, and G. Yang. 2014. Early diversification trend and Asian origin for extent bat lineages. Journal of Evolutionary Biology, 27:2204–2218.

Yu, Y., A. J. Harris, C. Blair, and X. He. 2015. RASP (Reconstruct Ancestral State in Phylogenies): a tool for historical biogeography. Molecular Phylogenetics and Evolution, 87:46–49.

7

Adapt or Live

Adaptation, Convergent Evolution, and Plesiomorphy

Liliana M. Dávalos,*
Andrea L. Cirranello,*
Elizabeth R. Dumont,
Stephen J. Rossiter,
and Danny Rojas

Introduction

Traditionally, the ecological theory of adaptive radiation relied on the fossil record to examine the relationship between increases in taxonomic diversity and the rates of evolution for, or emergence of, certain traits (Hunter and Jernvall 1995; Simpson 1953). A biological radiation was originally defined by observations of rapid diversification in taxonomic and trait diversity (Osborn 1902). Elaborating on this recurrent pattern, Simpson (1953) explained adaptive radiation as the result of high rates of diversification in species and traits after encountering ecological opportunity, followed by slowdowns in both rates as the opportunities were exhausted (Losos and Mahler 2010). The engine of diversity was therefore explicitly ecological, with new niches— whose sum Simpson called the "adaptive zone"—becoming available by reaching previously unexploited resources in new geographical areas, through the evolution of key innovations, or because of the extinction of competitors. As new species fill the niches in the new adaptive zone, their phenotypes diversify to match the diverse resources.

Based on these insights, and translating them to an explicitly phylogenetic framework, Schluter (2000) summarized the characteristics of adaptive radiation thus: (1) common ancestry, (2) correlated phenotype and ecology, (3) adaptive phenotype has higher fitness in new niche, and (4) rapid taxonomic and phenotypic evolution. Hence, ecological differentiation, competition, and opportunity are at the center of the theory, which remains the only general theory of how coupled taxonomic and ecological diversity evolve (Harmon et al. 2010; Rabosky et al. 2015). Schluter's (2000) framework translates into a series of predictions to be tested using phylogenies and comparative data. In addition to the relatively easily tested monophyly

---

* Equal contribution.

of proposed adaptive radiations, these predictions include bursts of diversification in response to ecological opportunities (Wagner et al. 2012; Yoder et al. 2010), diversity-dependent decreases in rates of taxonomic and trait diversification (Harmon et al. 2010; Rabosky and Glor 2010), and deterministic evolution of convergent phenotypes in response to ecological opportunities (Mahler et al. 2013; Reynolds et al. 2016).

Given that phyllostomids make up the only mammalian family to include species that feed on diets ranging from blood to figs, and the cranial anatomy of different species closely matches their ecological specialization (Dumont et al. 2014), the entire family has traditionally been assumed to be an adaptive radiation (Freeman 2000). Nevertheless, this is more often asserted than tested (Dumont et al. 2012) and has not been definitely established using current phylogenetic methods. Here we apply the phylogenetic framework of the theory of adaptive radiation to review comparative analyses of phyllostomid taxonomic and trait diversity. We also introduce new analyses of an index of trophic level, aiming to test for both distinct trophic adaptive zones and convergence in the evolution toward those zones across the family. Finally, we review a series of proposed adaptations, all associated with dietary specialization, discussing their implications and the questions emerging from them. This synthesis reveals both the wealth of evidence for increased fitness associated with derived features characterizing most plant-visiting lineages and the long-term persistence of ancestral dietary habits and traits across the family.

Methods

We applied new comparative methods to evaluate adaptation, ecological convergence, and plesiomorphy (or conservation of traits/phenotypes) as mechanisms underlying the diversity of trophic levels characteristic of phyllostomids. The response in the model was obtained from a continuous trophic-level metric developed based on the natural history literature (Rojas et al., in review). Briefly, the trophic-level index is given by $\log_{10}\{[(1 + \Sigma a_i + n_a)/n_a]/[(1 + \Sigma h_i + n_h)/n_h]\}$, in which Σa_i sums the relative importance of animal diets, while Σh_i is the counterpart for herbivory, with n_a quantifying the number of animal items in this category, and n_h the number of plant items. The index ranges from negative

values or low trophic level, to positive for high-trophic-level diets. Although this variable will not capture changes from, for example, 100% frugivory to 100% nectarivory, it will illuminate the transition from the ancestral chiropteran diet of primarily insects to both the carnivorous and frugivorous extremes that characterize the family (Freeman 2000).

The evolutionary pattern expected under adaptive radiation is open to debate. While all models emphasize that rapid taxonomic diversification is expected in every case, trait evolution can be modeled in different ways. An early burst of evolution followed by subsequent slowdown, for instance, is a poor fit to most adaptive radiations, including several classical examples such as Caribbean *Anolis*, Galapagos finches, or cichlids of the Great Lakes of Africa (Harmon et al. 2010). In all these cases, models specifying either adaptive optima or random walk fit the data better than early burst models, probably because both adaptive and partitioned Brownian motion models can capture aspects of directional evolution (Collar et al. 2009; Harmon et al. 2010). In practice, an adaptive model specifying multiple peaks can better fit adaptive radiations and simultaneously test for the convergence expected when adaptation to the ecology occurs (e.g., Mahler et al. 2013). Here, we focus on the adaptation expected under adaptive radiation and therefore use directional models with more than one peak in phylogenetic tests. To test for adaptive evolution we fitted the data to the comprehensive phylogeny of Phyllostomidae by Rojas et al. (2016) using an Ornstein-Uhlenbeck (OU) process of evolution (Butler and King 2004). The OU model enriches the Brownian motion model of evolution for continuous traits by two parameters, a rate of directional evolution often interpreted as adaptation α, and one or several adaptive optima or regimes with their values given by one or several θ (Hansen 1997).

To speed the analyses of the data, we applied a recently published lasso implementation of the phylogenetic OU model (Khabbazian et al. 2016). The lasso method penalizes the absolute value of the parameters leading to multiple coefficients estimated at zero (Tibshirani 1996) and thus overcomes the nonidentifiable nature of phylogenetic OU models without predetermined regimes (Bastide et al. 2016). As with previous methods to test nonpredetermined OU models (e.g., Ingram and Mahler 2013; Uyeda and Harmon 2014), the lasso method implemented in the R package llou (Khabbazian et al. 2016), examines shifts along all available branches. A newly developed phylogenetic Bayesian information criterion was applied to compare models (Khabbazian et al. 2016). Unlike commonly used criteria, such as the Akaike information criterion (AIC), the phylogenetic Bayesian information criterion accounts for phylogenetic correlation, and therefore its value is not strictly dependent on the number of shifts to different adaptive optima. Finally, to quantify the support for changes in adaptive zones along particular branches, we ran 100 bootstrap replicates. The method implemented in llou resamples the phylogenetic residuals of the regression of trait values against the inferred regimes (Khabbazian et al. 2016). To speed convergence and prevent parameter overfitting (Ho and Ané 2014), the ceiling on the number of regimes was set at the square root of the tips, or 13.

Additionally, we reviewed the literature on phyllostomid evolution since 2000, focusing on comparative analyses of either genes or continuous traits. In particular, we examined analyses in which specific adaptive or convergent hypotheses were tested, whether or not models such as the one applied here were used. Only analyses including at least three species were included; adaptations corresponding to single species are excluded from discussion as these are not comparative (e.g., ultraviolet-sensing cones in *Glossophaga soricina* Winter et al. [2003], or alternative splicing of genes linked to infrared sensation Gracheva et al. [2011]).

Results and Discussion

Parameters of the OU models fitted to the trophic-level data are summarized in table 7.1, and the inferred shifts

Table 7.1. Parameters inferred for Ornstein-Uhlenbeck (OU) models fitted to the trophic-level data on the phyllostomid phylogeny

Model	Number of regimes	pBIC
Ornstein-Uhlenbeck no convergence	6	–244.8
OU convergent evolution	5	–245.2

Note: Phylogenetic Bayesian information criterion, pBIC, lower values correspond to better-fitting models. A regime is a group of branches sharing a common adaptive optimum.

in adaptive optima are shown in figure 7.1. Five adaptive zones, two of them showing ecological convergence in trophic level, were identified. Comparative analyses of potential adaptations in quantitative traits are summarized in table 7.2. Comparative analyses finding potential genetic adaptation are summarized in table 7.3.

Evidence for Adaptive Radiation in Phyllostomids: Taxonomic and Trait Diversification

Analyses of traits for phyllostomid bats are relatively new compared to other classical examples of adaptive radiations, from fishes to anole lizards (e.g., Collar et al. 2014; Losos 2010; Near et al. 2013; Pinto et al. 2008), and consequently the literature to date has primarily focused on a few traits and even fewer genes. Nevertheless, several patterns emerge from both the new analyses of trophic-level evolution and the literature. In short: (1) a series of morphological, physiological and genetic adaptations to plant-based diets and frugivory in particular are well-documented; (2) part of the evidence of adaptation arises from the convergent evolution of traits associated with lower trophic levels, from olfactory receptor repertoires to the decrease in lower trophic level itself; (3) although convergent, the evolution of higher trophic level from plesiomorphic ancestors is associated with distinct, nonconvergent morphologies; and (4) the only trophic adaptation associated with higher rates of diversification is that of the fig-eating stenodermatines. We begin by summarizing and examining the results of the new analyses, before reviewing the evidence for emerging patterns.

Although greatly advanced since 2000, quantitative analyses among phyllostomids have focused almost exclusively on taxonomic diversification and traits associated with feeding ecology. A single shift toward higher taxonomic diversification in stenodermatines was first reported by Dumont et al. (2012), then confirmed by Rojas et al. (2016) and Shi and Rabosky (2015). None of these studies found diversity-dependent declines in taxonomic diversification rates, and only Dumont et al. (2012) correlated the shift to ecologically relevant traits. In particular, and in support of stenodermatines as an adaptive radiation, Dumont et al. (2012) found a correlation between phenotype (cranial morphology) and ecology (trophic level) with the phenotype evolv-

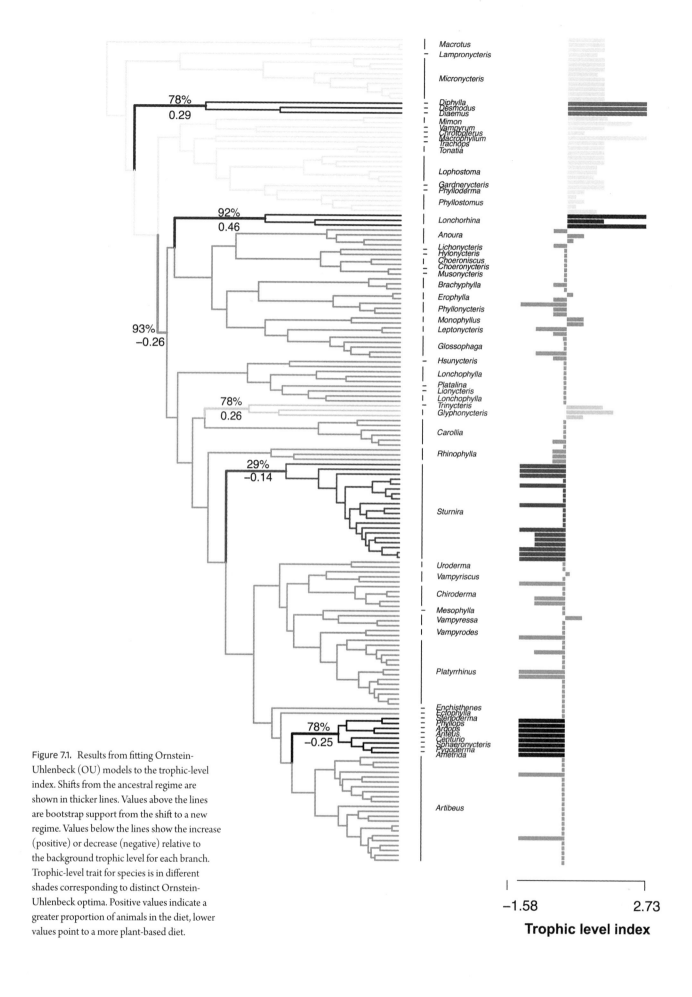

Figure 7.1. Results from fitting Ornstein-Uhlenbeck (OU) models to the trophic-level index. Shifts from the ancestral regime are shown in thicker lines. Values above the lines are bootstrap support from the shift to a new regime. Values below the lines show the increase (positive) or decrease (negative) relative to the background trophic level for each branch. Trophic-level trait for species is in different shades corresponding to distinct Ornstein-Uhlenbeck optima. Positive values indicate a greater proportion of animals in the diet, lower values point to a more plant-based diet.

Table 7.2. Annotated list of studies on (potential) trait adaptation, ecological convergence, or plesiomorphy in phyllostomids since 2000

Traits	Function	Taxonomic scope	Source
Skull proportions from linear measurements	Food acquisition	"Microchiroptera"	Freeman 2000
Intestinal sucrase, maltase and trehalase activity, relative medullary thickness of kidneys	Digesting nectar and fruit	Phyllostomidae	Schondube et al. 2001
Nephron and medullar morphology	Keep ions/concentrate urine	New World Chiroptera	Casotti et al. 2006
Intake response behavior for different sugars	Optimal nutrition	*Leptonycteris, Glossophaga, Artibeus*	Ayala-Berdon et al. 2008
Feeding habits	Food acquisition	Phyllostomidae	Rojas et al. 2011
Feeding habits and qualitative skull shape	Food acquisition	Phyllostomidae	Baker et al. 2012
Principal components of linear skull measurements, trophic level	Pierce through figs	Phyllostomidae	Dumont et al. 2012
Skull shape and loading behavior	Food acquisition	Phyllostomidae	Santana et al. 2012
Craniodental characters	Food acquisition	Phyllostomidae	Dávalos et al. 2012
Mechanical advantage and von Mises stress derived from engineering model of the skull	Food acquisition	Phyllostomidae	Dumont et al. 2014
Digestive capacity	Digestion	*Carollia, Sturnira, Artibeus*	Saldaña-Vázquez et al. 2015
Roost preference	Roosting	Amazonian Chiroptera	Voss et al. 2016

Table 7.3. Annotated list of studies on (potential) genetic adaptation and ecological convergence in phyllostomids since 2000

Traits	Function	Taxonomic scope	Source
Intestinal microbiome	Possibly digestion	New World Chiroptera	Phillips et al. 2012
Mitochondrial *cytochrome b gene*	Possibly optimize electron transport	Phyllostomidae	Dávalos et al. 2012
Olfactory receptor gene diversity analyzed as principal component of frequencies	Detect fruit	Chiroptera	Hayden et al. 2014
Intestinal microbiome	Possibly digestion	Phyllostomidae	Carrillo-Araujo et al. 2015
SLC2A2 promoter deletion	Increase glucose transport to liver	Chiroptera	Meng et al. 2016

ing more slowly within the group than outside it. As expected when new ecological opportunities become available (Yoder et al. 2010), the rates of trophic-level evolution in stenodermatines were higher than among other phyllostomids (Dumont et al. 2012). Finally, while the fitness of the new cranial architecture of stenodermatines was not directly assessed, the correlation between cranial morphology and bite force was a measure of fitness for an adaptive zone including hard fruits (figs). Applying the formal definition of adaptive radiation, especially the expectation of higher specia-

tion rates, the results of diversification analyses shift the focus from all phyllostomids (e.g., Baker et al. 2012; Freeman 2000) to the subfamily Stenodermatinae.

Olfactory Receptor Genes

Olfactory receptor (*OR*) genes are the largest family of protein-coding genes in mammals (Buck and Axel 1991), and extensive gains and losses of gene subfamilies underlie ecological shifts across mammals (Hayden et al. 2010; Niimura and Nei 2007). Each gene codes for one olfactory receptor protein expressed in olfactory epithelial cells, where odor perception occurs when volatile compounds bind to *OR*s in a combinatorial fashion (i.e., combinations of *OR*s recognize different odorants (Malnic et al. 1999). The composition and diversity of *OR* genes is therefore expected to reflect ecological adaptation across lineages, but, besides a handful of primate species, this had not been tested within a mammalian order (e.g., Matsui et al. 2010). To explore *OR* dynamics in an ecologically diverse clade, Hayden et al. (2014) sequenced paralogs within the gene family across bats. Focusing on hypotheses of exchange between sensory modalities, bat species were classified according to their use of laryngeal echolocation, as well as the presence of a functional vomeronasal organ. Additionally, the different species were also characterized by feeding ecology to investigate the interplay of sensory modalities and diet with *OR* genes.

Birth-death dynamics, in which genes easily duplicate and are lost even within populations (Nei and Rooney 2005), make it difficult to use standard measures of genetic adaptation in *OR* genes. With sequences from thousands of *OR* genes across 27 species of bats, Hayden et al. (2014) summarized the composition and diversity of genes by collapsing the data into phylogenetic principal components based on the relative frequencies across gene subfamilies and between putatively functional genes and pseudogenes—those with premature stop codons or shifts in the reading frame—as well as the proportion of pseudogenes found. After accounting for phylogeny, there were no significant differences between echolocating and nonecholocating bats for all *OR* genes, functional genes or pseudogenes, or the proportion of pseudogenes. Likewise, no differences were found between species with a functional vomeronasal organ versus those in which the vomeronasal organ is nonfunctional, for taxonomic groups such as Yinpterochiroptera versus Yangochiroptera, or between frugivorous and nonfrugivorous species across Chiroptera.

In contrast, examining *OR* gene diversity in frugivorous (Stenodermatinae) versus nonfrugivorous phyllostomids yielded significant results for all *OR* genes, functional genes, and pseudogenes. Relative to other phyllostomids, stenodermatines show increases in the relative diversity of *OR* gene subfamilies 1/3/7 and, to a lesser extent subfamilies 2/13. In non-stenodermatines, *OR* gene subfamilies 5/8/9 are better represented and

appear to be more important. Within Yinpterochiroptera, the *OR* gene subfamilies 1/3/7 and 2/13 lie on the main axis of differentiation between pteropodids and insectivorous Yinpterochiropteran bats for all genes and functional genes. In contrast to phyllostomids, significant differentiation in pseudogenes of Yinpterochiroptera lies on axes defined by *OR* gene subfamilies 11 and 52, separating pteropodids from other clades. In short, the greater representation of gene subfamilies 1/3/7 and 2/13 across clades of frugivorous bats evolving independently in the New World and Old World shows ecological convergence, while differences in patterns of pseudogenization lead to nonsignificant comparisons for all bats. In phyllostomids, similar patterns for all *OR* genes, functional genes, and pseudogenes may indicate adaptive changes linked to ecological specialization, while a similar pattern emerges in the frugivorous Pteropodidae but through greater pseudogenization in other gene subfamilies (Hayden et al. 2014). The results illuminate a clear genetic shift linked to the evolution of stenodermatines, and convergence with other frugivorous bats suggests adaptation to a specialized and previously unavailable diet.

Roosting Ecology

While most attention on phyllostomids in general and stenodermatines in particular has focused on diet and dietary diversity, Voss et al. (2016) analyzed the surprising diversity of roost types in the family. By assigning phyllostomid taxa to seven roosting guilds defined for Amazonia based on observations and data collected from the literature, Voss et al. (2016) found that roosts might be a limiting resource across lowland rainforest habitats, structuring communities to an extent hitherto unknown. Importantly, stenodermatines comprise almost all phyllostomid species known to roost in foliage and all species in the family known to make tents by modifying leaves. As the same skull architecture that enables stenodermatines to bite through figs despite their small size (Dumont et al. 2012) enables making leaf tents, the invasion into a new roost adaptive zone may be another factor contributing to the diversity of the subfamily. While the importance of these traits to adaptation or adaptive radiation remains unknown, a new roosting adaptive zone may contribute to increases in diversification rates at the base of the stenodermatine radiation.

Evidence for Adaptive Radiation in Phyllostomids: Multiple Adaptive Zones

Multiple adaptive zones corresponding to a variety of ecological resources are expected in adaptive radiations (Price and Hopkins 2015). As the Dumont et al. (2012) analyses focused on localizing shifts in taxonomic diversification and correlates of that shift, there was no exploration of the diversity of phenotypes known across phyllostomids. The first study to quantify diverse phyllostomid adaptive zones is the analysis of the evolution of functional cranial traits by Dumont et al. (2014). By fitting increasingly complex Ornstein-Uhlenbeck models to traits derived from engineering models of the skull, Dumont et al. (2014) tested whether these phenotypes corresponded to different adaptive zones for three traits: mechanical advantage and two types of biting stress. Using a set of predefined groups corresponding to (from longest to shorter skulls) nectarivores, other phyllostomids, stenodermatines except short-faced bats, and short-faced bats, Dumont et al. (2014) tested the fit of the data to increasingly complex models. A model with four adaptive zones was supported for mechanical advantage, although the adaptive optima of both other phyllostomids and most stenodermatines overlapped, supporting convergent evolution for this trait. In contrast, unilateral biting stress had optima corresponding to all stenodermatines, nectarivores, and other phyllostomids, and bilateral biting stress had two optima comprising nectarivores and all other phyllostomids (Dumont et al. 2014). There is, therefore, evidence of adaptation into divergent adaptive zones for mechanical advantage—a strong correlate of bite force—as expected if these traits contributed to the adaptive radiation of all phyllostomids.

The evolution of trophic level in phyllostomids has been intimately linked to changes in cranial morphology (Dumont et al. 2012; Freeman 2000) and traits associated with it, such as bite force (Dumont et al. 2014). But associated skeletal traits and their correlates are not the only ones relevant to shifts in trophic level. Santana et al. (2012) used Ornstein-Uhlenbeck models to relate dietary hardness to the evolution of the temporalis and masseter muscles and to relate phylogenetic regressions of loading behavior to muscles as well as skull shape. They found multiple distinct adaptive optima for both muscles and for observed mechanical advantage. Further, skull and muscle measurements best explained loading behaviors (a quantitative summary of the type of bite bats take for each mouthful) across phyllostomid species. These analyses highlight the tight association among form, function, and behavior, which results in adaptive phenotypes given diets of very different hardness. Additionally, support for multiple adaptive optima for muscle configuration across phyllostomids corroborates the adaptive nature of cranial morphology in the family (Santana et al. 2012).

Evidence for Adaptive Radiation in Phyllostomids: Within-Family Ecological Convergence

Deterministic convergent evolution is one of the predictions from the phylogenetic formulation of the ecological theory of adaptive radiation that remains untested. Through analyses that fit free—not predetermined—Ornstein-Uhlenbeck models to trophic-level data, our results provide the first quantitative evidence for convergence across phyllostomids (table 7.1, fig. 7.1). Despite the obstacles posed by previous phylogenetic hypotheses, several previous studies had either asserted or at least hinted at trait convergence (Freeman 2000; Rojas et al. 2011). Freeman (2000), for example, attributed the morphological diversity of phyllostomids to the ancestral inclusion of some plant food, resulting in the subsequent evolution of specialized cranial types associated with different plant-based diets. The data show a few species with reduced trophic level, reflecting occasional plant consumption, even among phyllostomines such as *Phyllostomus* (fig. 7.1). Seven trophic-level regimes or optima can be differentiated corresponding to the ancestral niche (1), the higher trophic level of vampire bats (2), *Lonchorhina* (3), a shared low trophic level for all nectar-feeding and most frugivorous lineages (4), with a secondary increase in level of glyphonycterines (5), and the reduction in level among *Sturnira* (6) and short-faced bats (7). Of these, the increases in animalivory of vampire bats and *Lonchorhina* are convergent. The secondary increase in trophic level in glyphonycterines is particularly relevant, as it upends the widely held assumption that no lineage reverts to insectivorous habits (Baker et al. 2012).

Another way to quantify these results is to compare the rate of directional evolution α to the Brownian motion rate of change σ, as the latter represents the background rate of random-walk change throughout the phylogeny. With a directional rate estimated at 0.32 units/million years, the adaptive rate of change is ~64 times higher than the background rate of change, indicating very strong directional tendencies to particular trophic-level optima. Hence, these first analyses of trophic-level evolution have shown there are shifts in trophic level across phyllostomids that have been shaped by natural selection, with two lineages converging on similarly high trophic levels. Thus, even in the absence of high taxonomic diversification rates during the early history of phyllostomids, the evolution of trophic level is consistent with adaptive trait diversification in diet for the family.

The analyses introduced here are the first to directly model convergence in an ecologically relevant trait, trophic level. These models, however, are not the first to propose ecological convergence to explain the evolution of morphological or even genetic traits. To address the persistent conflict between phylogenies based on morphology and those based on DNA sequences (see chap. 6, this vol.), Dávalos et al. (2012) examined methodological (e.g., taxonomic sampling, character sampling, methods of analysis) and biological (saturation, homoplasy, incongruence between gene trees, adaptive convergence) drivers of incongruence in the phyllostomid phylogeny. As the morphological data of Wetterer et al. (2000) supported clades composed of taxa that shared feeding behaviors strongly rejected by DNA sequences (Baker et al. 2003), there was already evidence for incongruence and conflict. Analyzing an updated set of 220 morphological characters for 80 species combined with roughly matching sequences, Dávalos et al. (2012) sought to minimize the effects of taxon and character sampling on incongruence by adding morphological characters and to reduce methodological conflict by using Bayesian algorithms for both the morphological and DNA sequence data. After removing these method-based sources of error, they found taxon choice produced some, but not all, of the significant phylogenetic conflict, and adding morphological characters and using the same algorithms had no effect on incongruence. In short, previously used methods and sets of taxa or characters were not responsible for the vast differences in phylogenetic results.

After ruling out methods as an explanation for phylogenetic conflict, Dávalos et al. (2012) searched for biological causes of incongruence. Here, it is important to consider two mechanisms underlying such conflicts: structural constraints imposed by development and natural selection causing ecological convergence (Wake 1991). Although reflecting very different processes, both mechanisms generate similar patterns of exhaustion in character states (Wagner 2000). Dávalos et al. (2012) found patterns of saturation in DNA substitutions and, more importantly, in morphological characters. But the pattern of morphological character-state evolution was not consistent with a hard ceiling on character states. Instead, the best fit to the character state to steps in tree curve was an ordered model, in which new character states only emerge from derived states (and not from other states). Subsequent analyses of exclusively dental characters revealed similar patterns, with the ordered model fitting the evolution of states for those characters (Dávalos et al. 2014). In short, the pattern of exhaustion of morphological character states is common to more than one data set collected by different teams. This suggests saturation in states may characterize phyllostomids more generally. Additionally, ordered states are also consistent with Ornstein-Uhlenbeck optima for cranial traits (Dumont et al. 2014): traits evolving toward an optimum are equivalent to derived states evolving from other derived states. Character-state saturation in phyllostomid characters, then, is consistent with the adaptive evolution toward a small number of optima. Additional analyses supported this interpretation, as morphological character changes supporting the monophyly of nectar-feeding phyllostomids or *Carollia* as sister to *Rhinophylla* were concentrated in suites of characters associated with feeding (Dávalos et al. 2012). To conclude, detailed analyses suggest that ecological convergence in morphological traits linked to feeding ecology influences the evolution of morphological character states and that adaptation helps explain conflict with phylogenies based on DNA sequences.

Natural selection is not the only process leading to exhaustion of character states in phyllostomids: developmental constraints also contribute to character-state exhaustion and phylogenetic conflict (Dávalos et al. 2014). Using distances between dental characters to quantify the similarity in state changes across characters, Dávalos et al. (2014) found an excess of similarities in change compared to similar analyses for DNA sequences. That is, there are more correlated changes in states than expected if the different dental characters were evolving independently from one another. Anatomical systems are tightly integrated, and correlated evolution in anatomical traits is to be expected (Voje et al. 2014, 2013). The result is convergence in form, even when natural selection is not involved. Phylogenetic analyses of morphological characters, however, assume the independent evolution of characters (O'Keefe and Wagner 2001). Although Dávalos et al. (2014) measured the expected excess of similarity, they did not exclude characters to assess their effects on phylogeny (e.g., Herrera and Dávalos 2016) or model the evolution of associated traits to determine the contribution of developmental constraint relative to natural selection in phyllostomid evolution (e.g., Bartoszek et al. 2012). The two latter approaches are promising avenues for disentangling the contributions of adaptation and constraint to evolution and uncovering the role of natural selection in trait diversity.

The pattern of ecological convergence among plant-visiting phyllostomids, and especially among nectar-feeding lineages, is not limited to morphological traits. Instead, there is evidence for convergent DNA substitutions in the mitochondrial *cytochrome b* gene (Dávalos et al. 2012). The first indication of ecological convergence emerged in the phylogenetic analyses of Dumont et al. (2012), which differed from the Baker et al. (2003) phylogeny in making nectar-feeding phyllostomids monophyletic (subfamilies Glossophaginae and Lonchophyllinae as sister taxa). To determine whether the relevant DNA substitutions were adaptive, Dávalos et al. (2012) fit models to estimate the ratio of nonsynonymous to synonymous (silent) substitutions at different branches and codons and ran simulations to determine if codons supporting the spurious nectar-feeding node underwent a shift toward more nonsynonymous changes. If codons with higher ratios tended to support the nectar-feeding node, it would be an indication of positive selection leading to convergent adaptation in both subfamilies. The results showed codons experiencing a ratio shift were massively overrepresented among codons supporting the nectar-feeding node. Further, the codons with shifting ratio were also concentrated in the carboxy-terminus of the protein, which is essential for correct assembly of the respiratory complex (di Rago et al. 1993). This study was the first to relate phylogenetic support for an ecologically convergent node to shifts in selection and particular protein regions in phyllostomids.

Evidence of Adaptation in Phyllostomids

The previous sections have focused on predictions arising from Simpson's (1953) ecological theory of adaptive radiation and, in particular, from the phylogenetic framework suggested by Schluter (2000) and developed subsequently (Mahler et al. 2013; Moen and Morlon 2014; Rabosky and Glor 2010; Yoder et al. 2010). However, many more studies have focused on adaptation in general, compared to explicit phylogenetic predictions of adaptive radiation (tables 7.1 and 7.2). Here, we summarize key analyses on adaptation across phyllostomids, seeking to place them within the framework of the ecological theory.

Comparative Physiology

In one of the first quantitative analyses in phyllostomids, Schondube et al. (2001) aimed to test if physiological changes in diet, digestion, and renal function were adaptations linked to changes from the ancestral diet. They collected detailed quantitative data from 16 phyllostomid species using the Wetterer et al. (2000) phylogeny, in which shifts to different diets from insectivory occurred once each (e.g., once to nectarivory, once to sanguivory). If the new diets resulted in physiological adaptations, then intestinal enzyme activity should match the corresponding diets after controlling for evolutionary relatedness. For example, there should be increasing sucrase and maltase activity for plant-based diets, increasing trehalase activity for insect-based diets—as trehalose is the primary insect storage sugar—and constant aminopeptidase N as a stand-in for protein digesting ability. Similarly, the ability to assimilate nutrients and meet metabolic demands would be reflected in the amount of nitrogenous waste and water that the kidney processes and in differences in kidney gross anatomy. Additionally, Schondube et al. (2001) examined the nitrogen isotopic composition of tissues, expecting nitrogen composition to reflect predominantly plant-based or animal-based diets. The data to test all predictions paired five shifts in diet against changes in the relative medullary thickness of the kidney, maltase, sucrose, trehalase, and aminopeptide-N activity.

In general, the adaptive hypothesis was supported

for most digestive traits, and the results for renal function were more nuanced (Schondube et al. 2001). First, shifts from insectivory to carnivory or sanguivory resulted in a large reduction in trehalase activity. While there was no detectable sucrase and maltase activity in the sanguivore, shifts from insectivory to nectarivory or frugivory resulted in increases in sucrase and maltase activity, as well as a decline in trehalase activity. Second, ^{15}N varied predictably with diet: species that ate a largely plant-based diet had lower values, while those that were omnivorous or had a largely animal-based diet had higher concentrations, reflecting the higher trophic level of their primary diets. Third, shifts from insectivory to nectarivory or frugivory were coupled with changes from the ancestral state of having well-developed papillae, and clearly differentiated outer and inner zones in the medulla (insectivory), to reduced medullary papillae and undifferentiated medullae (nectarivory, frugivory). And although no change was predicted to match the shift from insectivory to omnivory, a reduction in renal medullary thickness was also detected. Fourth, the lower water content of meat-based diets was not reflected in kidney anatomy, as there were no anatomical differences corresponding to shifts in diet from insectivory to sanguivory or carnivory. The changes in renal anatomy to reduced papillae and medullar thickness potentially reduce the ability to concentrate urine, and increase the ability to absorb electrolytes, matching the greater availability of water in diets containing nectar and fruit (Schondube et al. 2001). Finally, and despite not being originally highlighted, the ancestral traits linked to insectivory are themselves reflective of trophic specialization into this adaptive zone, to the point that including some plants results in increases in activity of maltase and sucrase and in reductions in medullar thickness. Although maintenance of ancestral phenotypes does not represent adaptation in phyllostomids, it does show how tightly optimized functional traits are to specific diets and how stable these traits have been through time across independent insectivore lineages such as *Macrotus* and *Micronycteris*. In short, there was systematic support for adaptive changes in digestive and renal function linked to the exploitation of new food sources, with the exception of certain kidney traits for lineages expanding their diet into higher trophic levels. This last result is not as surprising as first thought, given that the diets of carnivorous phyllostomids are just one end of

the gradient of animalivory and do not constitute a distinct ecological guild (Giannini and Kalko 2005).

To follow up on the groundbreaking work of Schondube et al. (2001), Casotti et al. (2006) delved into the renal morphology of 21 phyllostomid species, this time using stereology to assess the relative proportion of renal cortex, medulla, and vasculature and renal medullary thickness, as well as the proportions of nephron components and vasculature. Despite earlier findings of no changes in gross morphology in the transition from insectivory to omnivory, Casotti et al. (2006) expected changes in fine anatomy linked to these shifts and to those to frugivory or nectarivory. As the relative water content of the diet increases in each of these shifts, changes to smaller renal medullary surface and thickness and larger renal cortex were expected in each case. In contrast, and based on elements of nephron morphology conserved across vertebrates (Dantzler and Braun 1980), no changes to any nephron components or vasculature were expected with dietary shifts. Using a statistical approach based on Schondube et al. (2001), these predictions were tested using matrices of expected directional changes, one for cortex and medullary components and another for nephron components. The analyses test against a null probability of change in morphological elements at all transitions if all changes were independent and random.

As expected, shifts in diet away from ancestral insectivory to new diets were linked to reductions in the renal medullary surface and thickness and to enlargement of the renal cortex. Unexpectedly, and in line with previous results for gross anatomy by Schondube et al. (2001), no changes in these anatomical traits were found in the transition to omnivory. Also as expected, nephron elements were highly conserved and unrelated to shifts in diet. These results confirm at much finer scale the selective pressures of changes toward a primarily plant-based diet that result in adaptation, and the contrast with omnivory, which results in conserved morphology (Casotti et al. 2006).

The physiological effects of changing diets to include plenty of fruit and nectar are not limited to renal or digestive function. Instead, behavior of individuals should be integrated with their physiological adaptations to preserve metabolic homeostasis depending on relevant sources of energy. To test for this coupling of behavior and energy source, Ayala-Berdon et al. (2008) examined feeding responses in three species—the nectar-feeding *Leptonycteris curasoae* and *Glossophaga soricina* and the fruit-feeding *Artibeus jamaicensis*. Their experiments varied the composition—sucrose, or a mix of glucose and fructose, both hexoses, at 1:1 ratio—expecting preference for hexoses matching the composition of New World flower nectars. Experiments varying the concentration of artificial nectars aimed to show an "intake response" (Castle and Wunder 1995), in which intake decreases as a function of increases in energy. While all species lacked a preference for sucrose or the hexose mix, like nectar-feeding birds, bats showed the expected intake response: as the concentration of sugar in nectar increased, the amount of nectar consumed, or intake, decreased.

Despite the predominance of hexose-dominated nectars in New World bat-pollinated flowers, the three bat species that Ayala-Berdon et al. (2008) studied exhibited indistinguishable intake responses to the sucrose and the hexose solutions. The intake response of these species to both sucrose and hexose solutions demonstrates that neither sucrose hydrolysis nor uptake of hexoses is the limiting factor in food intake for these three species. Instead, as Ayala-Berdon et al. (2008) suggest, numerous factors—including digestion, osmoregulation, and metabolic processes—may limit the ability of individuals in these species to vary ingestion to maintain a constant sugar intake (compensatory feeding), and, when confronted with a low sugar concentration diet or high energetic demands, they may suffer potential impacts to fitness.

The prevalence of intake response is intriguing by itself because at low sugar concentrations and corresponding high intake, nectar-feeding bats would need to rapidly eliminate large amounts—from one to five times their body mass—of water. But, as outlined earlier, the kidneys of both nectar- and fruit-feeding phyllostomid bats are morphologically well-suited for eliminating water but not for recovering electrolytes. Among other adaptations to a diet of high water content are an undifferentiated medulla, lack of medullary papilla reducing the concentration of urine and increasing electrolyte recovery, and the enlarged renal cortex furthering the absorption of electrolytes and other solutes (Casotti et al. 2006; Schondube et al. 2001). These adaptations lead to an osmoregulatory dilemma, as the rate at which their kidneys can eliminate water may limit the amount of dilute nectar these species can ingest (Ayala-Berdon et al. 2008). How these bats re-

solve the conflict between osmoregulation and intake of dilute nectar remains to be discovered.

Previous physiological studies have lumped plant-visiting bats into a single category, but there are important differences between frugivorous and nectarivorous species. Saldaña-Vázquez et al. (2015) focused on testing the link between dietary diversity and factors such as body mass, gut nominal area, and digestive capacity—defined as the ability to obtain nutrients from foods of different nutrient quality. Digestive capacity is the negative of the slope of the relationship between volumetric intake and nutrient concentration in log-log scale, such that high digestive capacity indicates a strong intake response and low digestive capacity a weaker intake response. Using a sample of five representative species in three frugivorous genera (*Sturnira*—*S. ludovici*, and *S. lilium*, *Artibeus*—*A. jamaicensis* and *A. toltecus*, and *Carollia sowelli*), Saldaña-Vázquez et al. (2015) first tested for phylogenetic signal in digestive capacity, body mass, gut nominal area, and dietary diversity. These variables are not strongly constrained by phylogeny, demonstrating variability (and the potential for adaptation) within genera. Instead, digestive capacity was the sole factor predicting dietary diversity. Species with greater digestive capacity (e.g., *Sturnira lilium* and *Artibeus jamaicensis*) feed on a wider variety of plant food, including plants of lower quality, than congeners with smaller digestive capacity. In species with lower digestive capacity, diet was more restricted, with higher-quality fruits comprising much more of the total intake. In effect, 77% of the variation in dietary diversity in the species they considered could be explained by digestive capacity alone (Saldaña-Vázquez et al. 2015). Digestive capacity, then, has a direct bearing on the width of the dietary niche of frugivorous phyllostomids. As bats must limit gut size due to constraints imposed by flight, congeners may differ in gut physiology by changing reaction speed or the affinity of enzymes for their substrates, thus modifying digestive capacity without changing gut size or body mass (even though changes in body size occur relatively rapidly in some genera such as *Artibeus*).

Genetic Basis of Physiological Adaptation

The analyses summarized so far have focused on the evolution of adaptive traits in lineages with derived diets, and among these there is only one potential convergent genetic adaptation (Dávalos et al. 2012), without any functional tests. Only very recently have analyses integrating metabolic function, ecological characteristics, and functional genetics begun to include phyllostomid species. As the contemporary high-sugar diet of humans and resulting high-glucose levels in blood lead to type-2 diabetes, insulin resistance, and metabolic disease, among other diseases (Leturque et al. 2009), the search for biochemical adaptations to the derived plant-based diets of phyllostomids and other frugivorous bats has garnered interest beyond bat biology. Blood sugar regulation in bats with high-sugar diets is therefore a new focus of research in the quest for potential adaptations of value for future use in human treatments.

Meng et al. (2016) aimed to relate genotypes of the gene encoding glucose transporter 2 (*SLC2A2*), a transmembrane carrier protein responsible for bidirectional transport of glucose between the blood and liver (Shen et al. 2012), to the capacity to absorb glucose and maintain glucose homeostasis across bats. To this end, they began by exploring the relationship between blood glucose level, body size (measured by body mass or forearm length), and diet. Classifying 149 species of bats into frugivores, insectivores, and "others," they found the frugivore group has the highest body mass index, insectivores the lowest, with the "other" category intermediate between the two. For the 16 species for which blood glucose–level data were available, including the phyllostomid *Artibeus intermedius*, the relationship between this trait and body size was negative both across all bats and within diet categories. Hence, larger bats, corresponding largely to frugivorous species, tend to have lower blood glucose levels, but the slope of the relationship for the group is steeper, indicating a greater drop in blood glucose per size unit compared to other bats. Meng et al. (2016) interpreted these results to hypothesize that bats have adapted to eating fruit — high in sugar and poor in protein — by maintaining high glucose in blood, so energy is readily available for flight. If this were the case, however, frugivores should have higher glucose levels than expected for their body size, and there is no evidence to this effect. Indeed, a phylogenetic regression of blood glucose level against body mass index finds the best-fit model determined using the AIC fails to include diet as a predictor (table 7.4). With only a few independent observations and despite the apparent relationship, the data set is underpowered. Covariation between blood glucose level and body

Table 7.4. Comparison of phylogenetic regression models of blood glucose levels as a function of predictors

Predictor	Degrees of freedom	AIC
Body Mass Index (BMI)	2	60.59599
Two intercepts by diet and BMI	3	62.58643
Two BMI slopes by diet	3	61.94425
Two intercepts and BMI slopes by diet	4	62.46562

Note: The phylogeny of Shi and Rabosky (2015) was used to account for the phylogenetic structure of residuals. AIC = Akaike information criterion.

mass index is not significant ($R^2 = 0.16$, $F_{(1,14)} = 3.902$, $p = 0.068$). The relationship between diet and blood glucose levels across bats, therefore, remains to be assessed with more data and using appropriate statistical approaches.

Assuming that adaptation to the high-sugar diet further requires demonstrating enhanced function in transporting sugar, if frugivorous bats have evolved to cope with the sudden influx of sugar in their diet, then they should have better sugar transport. Meng et al. (2016) tested the intra-peritoneal glucose tolerance of the yinpterochiropterans *Rousettus leschenaulti* (a frugivore) and *Hipposideros armiger* (an insectivore). Testing blood glucose levels at 0, 10, 60, 90, and 120 minutes after the injection across multiple individuals in each species, they found both enhanced transfer from intestine to blood, and then from blood. Glucose blood concentration in the frugivorous bat peaked earlier and at a significantly lower value than in the insectivorous species and decreased more rapidly. This functional demonstration strengthens the case for adaptation in sugar transport in frugivorous bats, even if only for yinpterochiropterans.

The final piece of evidence regarding adaptation for glucose homeostasis in frugivorous bats lies in the proximal promoter sequence (275 bp) of *SLC2A2*, coupled with measures of expression of the gene. Although highly conserved across the 16 species they sequenced, and therefore subject to negative selection (and not adaptation), an important potentially functional difference emerged between plant-visiting and other bats (Meng et al. 2016, table 2). An 11 base pair deletion overlaps a portion of sequence thought to bind a transcriptional repressor (ZNF354C; a putative binding site found only in nonfrugivorous species), suggesting loss of the binding site causes increased transcription of *SLC2A2* in frugivores. The deletion, shared by phyllostomids (four plant-visiting bats and *Desmodus*) and Old World frugivorous bats, has evolved at least twice. The implications of the deletion in *Desmodus*, along with the evolution of the promoter across the diet shifts within Phyllostomidae, remain to be explored.

To close the chain of inference from the deletion to better sugar homeostasis for frugivorous bats, Meng et al. (2016) used quantitative PCR of *SLC2A2* mRNA to measure levels of the transporter in the liver. As expected if the deletion had functional effects, two insectivorous species lacking the deletion (*Hipposideros armiger, Myotis ricketti*) had much lower expression levels than the two frugivores (*Cynopterus sphinx, Rousettus leschenaulti*) with the 11-bp deletion (base level of 1 in the lowest values, 1.3 times higher in the other insectivore and 7.2 and 31 times higher in the frugivores). The convergent deletion, then, has substantial effects on transporter expression at least in yinpterochiropterans and is linked to better regulation in frugivorous species. Considering phyllostomids have evolved both nectarivory and frugivory more than once, as well as the intriguing presence of the high-expression deletion, there is great potential to explore this mechanism of adaptive glucose homeostasis in the family.

The Bacterial Microbiome

Gut bacterial communities are increasingly viewed as essential to host function because they provide nutrients to, regulate tissue development of, and interact with the immune system of their mammalian hosts (Eckburg et al. 2005). Although microbiome studies in humans have received the most attention (Kuczynski et al. 2012), comparative studies have grown in parallel to determine the ecological and evolutionary forces shaping these bacterial communities (Ley et al. 2008; Ochman et al. 2010). Sharing a similar method of sequencing the prokaryotic ribosomal RNA gene 16S, and based on samples from the New World, the two studies analyzing bat microflora have an excellent representation of phyllostomids (Carrillo-Araujo et al. 2015; Phillips et al. 2012). The pioneering analyses of Phillips et al. (2012) sampled seven families and focused on ecological and evolutionary covariates of bacterial diversity. Using phylogenetic distance (PD), or the total branch lengths contained in a sample for the overall 16S phylogeny, emballonurids have the most diverse

microbiomes, followed by vespertilionoids, and then noctilionoids, relating microbiome diversity to clade age. A non-phylogenetic regression of family PD against the time to the most recent common ancestor was significant, as were Kolmogorov-Smirnov tests comparing the PD subsample of each bat family to others. The sole exception to the latter pattern were noctilionoids, which shared similar values of PD. As measured by PD, the microfloral diversity of phyllostomids does not differ from that of mormoopids or noctilionoids, despite the much greater dietary diversity of the former.

Multiple factors seem to influence the bacterial diversity of the gut communities within phyllostomids. These factors, however, are not all related to diet; geographic locality, sex, sexual maturity, and reproductive condition also influenced the PD of various phyllostomids. Phylogenetic distance increased from sanguivores to insectivores, nectarivores, and frugivores, which had the highest PD values. This is consistent with the general mammalian pattern of increasing diversity from carnivory to omnivory to herbivory (Ley et al. 2008). Although not much can be said about the functional significance of particular bacterial lineages, the diversity of Lactobaccillales was higher in herbivorous phyllostomids, and bacterial species in this group could improve nutrient acquisition in herbivorous species (e.g., Famularo et al. 2005).

With hundreds of thousands of sequences clustered into hundreds of operational taxonomic units, the bacterial microbiome analyses of Phillips et al. (2012) defined the broad contours of diversity across diet specialists. New sequencing technology, however, quickly upended these first insights. Carrillo-Araujo et al. (2015) sequenced the same region of the prokaryotic 16S gene (hypervariable V4) and, in contrast with the longer-read method used by Phillips et al. (2012), sequenced using short-read technology (Illumina) for multiple individuals within species and at different regions of the intestine. With millions of sequences, grouped into thousands of operational taxonomic units, rarefaction analyses revealed lower bacterial diversity among the frugivorous *Carollia perspicillata* and *Artibeus jamaicensis* compared to the insectivorous *Macrotus waterhousii* and the sanguivorous *Desmodus rotundus*. This ordering of diversity was also reflected in analyses of phylogenetic diversity and cannot be explained by different measures of bacterial community richness. While intestinal region had no effect on bacterial diversity for most species, there were important differences for the frugivorous species. Gammaproteobacteria and firmicutes dominated the anterior intestine of *Carollia*, and cyanobacteria were prevalent in the posterior intestine. For *Artibeus*, tenericutes and firmicutes dominate the posterior intestine, with gammaproteobacteria prevalent in the anterior section. As in the previous microbiome study, phylogeny was a strong influence on the microbiome, with the effects of feeding strategy nested within the larger phylogenetic pattern. This is in line with analyses of the microbiomes of great apes, in which the gut microbiome recapitulates the host phylogeny (Ochman et al. 2010; Sanders et al. 2014). As the larger diversity of phyllostomids has not been sampled (e.g., Micronycterinae, Lonchorhininae, Lonchophyllinae, Rhinophyllinae), it is not yet possible to test if ecological convergence has reshaped the gut microbiome of ecologically similar mammals as it has in ant-eating lineages (Delsuc et al. 2014). As with the genomic basis of adaptation, much remains to be explored in the biology of phyllostomid microbiomes.

Conclusions

Since 2000, an important body of comparative work has focused on the adaptive radiation of phyllostomids and, in particular, on the diversity of adaptations linked to dietary specialization. Predictions from the ecological theory of adaptive radiation, however, have only been tested recently, and analyses of traits have focused mainly on feeding ecology and the morphology of the cranium. Using this framework, we find shifts in taxonomic diversification with the emergence of subfamily Stenodermatinae (Dumont et al. 2012; Shi and Rabosky 2015) and no diversity-dependent decrease in either this subfamily or phyllostomids as a whole (Rojas et al. 2016). Shifts in phenotype associated with the evolution in this subfamily include a new skull architecture leading to increased bite force (Dumont et al. 2012), a distinctive profile of olfactory receptor genes (Hayden et al. 2014), and a likely change in roosting behavior (Voss et al. 2016). Although there is no evidence of an increase in taxonomic diversification when phyllostomids first arose, several analyses reveal a diversity of adaptive zones across the family, including among cranial muscles and mechanical advantage (Santana et al. 2012) and in mechanical advantage inferred from engineering models (Dumont et al. 2014). Another line

of support for adaptive radiation in phyllostomids is the ecologically convergent evolution of phenotypes linked to particular adaptive zones. Quantitative analyses have documented convergence in trophic level (table 7.1, fig. 7.1), characters associated with feeding ecology for *Carollia* and *Rhinophylla* as well as nectar-feeding phyllostomids, and the possible adaptive evolution of the cytochrome *b* protein also in nectar-feeding lineages (Dávalos et al. 2012). The diversity of phyllostomid phenotypes corresponds to an adaptive radiation in traits and hence to the filling of various adaptive zones with distinctive niches, with stenodermatines comprising the fastest rate of taxonomic diversification in response to the completely new adaptive zone they occupy.

Evidence for adaptation to dietary specialization in phyllostomid subfamilies is more common but no less interesting: adaptations in feeding response to nectar concentration (Ayala-Berdon et al. 2008), digestive enzymatic activity and renal morphology—the latter especially in plant-visiting bats (Casotti et al. 2006; Schondube et al. 2001)—and digestive capacity defined by being able to extract nutrients from varying foods (Saldaña-Vázquez et al. 2015). All phyllostomids studied to date, including *Desmodus*, share a deletion in an 11-bp region of the promoter sequence for the *SLC2A2* gene that results in higher gene expression and faster sugar transport in Old World plant-visiting bats (Meng et al. 2016). The implications of this finding for phyllostomids remain to be fully explored. Similarly, variation in diversity of the microbiome suggests specialization whose functional significance remains unknown (Carrillo-Araujo et al. 2015; Phillips et al. 2012). Thus, the range of adaptations documented so far points to future lines of research, including exploring the limits some adaptations place on dietary intake, the regulation of sugar metabolism across lineages with different degrees of specialization, and determining the functional role of microbiome components in digestion, absorption, and metabolic regulation.

Acknowledgments

We thank N. Giannini and one anonymous reviewer for comments on the manuscript. LMD and SJR were supported in part by the National Science Foundation (DEB-1442142). DR was supported by Foundation for Science and Technology, Portugal (www.fct.pt), fellowship SFRH/BPD/97707/2013. The Portuguese Foundation for Science and Technology supported CESAM RU (UID/AMB/50017) through national funds and FEDER funds, within the PT2020 Partnership Agreement.

References

Ayala-Berdon, J., J. E. Schondube, K. E. Stoner, N. Rodriguez-Peña, and C. M. Del Río. 2008. The intake responses of three species of leaf-nosed Neotropical bats. Journal of Comparative Physiology B, 178:477–485.

Baker, R. J., O. R. P. Bininda-Emonds, H. Mantilla-Meluk, C. A. Porter, and R. Van Den Bussche. 2012. Molecular timescale of diversification of feeding strategy and morphology in New World leaf-nosed bats (Phyllostomidae): a phylogenetic perspective. Pp. 385–409 in: Evolutionary History of Bats: Fossils, Molecules and Morphology (G. F. Gunnell and N. B. Simmons, eds.). Cambridge Studies in Molecules and Morphology-New Evolutionary Paradigms. Cambridge University Press, Cambridge, MA.

Baker, R. J., C. A. Porter, S. R. Hoofer, and R. A. Van Den Bussche. 2003. Diversification among New World leaf-nosed bats: an evolutionary hypothesis and classification inferred from digenomic congruence of DNA sequence. Occasional Papers, Museum of Texas Tech University, 230:1–32.

Bartoszek, K., J. Pienaar, P. Mostad, S. Andersson, and T. F. Hansen. 2012. A phylogenetic comparative method for studying multivariate adaptation. Journal of Theoretical Biology, 314:204–215.

Bastide, P., M. Mariadassou, and S. Robin. 2016. Detection of adaptive shifts on phylogenies by using shifted stochastic processes on a tree. Journal of the Royal StatisticalSociety, ser. B: Statistical Methodology, 79:1067–1093.

Buck, L., and R. Axel. 1991. A novel multigene family may encode odorant receptors: a molecular basis for odor recognition. Cell, 65:175–187.

Butler, M. A., and A. A. King. 2004. Phylogenetic comparative analysis: a modeling approach for adaptive evolution. American Naturalist 164:683–695.

Carrillo-Araujo, M., N. Taş, R. J. Alcántara-Hernández, O. Gaona, J. E. Schondube, R. A. Medellín, J. K. Jansson et al. 2015. Phyllostomid bat microbiome composition is associated to host phylogeny and feeding strategies. Frontiers in Microbiology, 6:447.

Casotti, G., L. G. Herrera M., J. J. Flores M., C. A. Mancina, and E. J. Braun. 2006. Relationships between renal morphology and diet in 26 species of New World bats (suborder microchiroptera). Zoology, 109:196–207.

Castle, K. T., and B. A. Wunder. 1995. Limits to food intake and fiber utilization in the prairie vole, *Microtus ochrogaster*: effects of food quality and energy need. Journal of Comparative Physiology B, 164:609–617.

Collar, D. C., B. C. O'Meara, P. C. Wainwright, and T. J. Near. 2009. Piscivory limits diversification of feeding morphology in Centrarchid fishes. Evolution, 63:1557–1573.

Collar, D. C., J. S. Reece, M. E. Alfaro, P. C. Wainwright, and R. S. Mehta. 2014. Imperfect morphological convergence: variable changes in cranial structures underlie transitions to durophagy in moray eels. American Naturalist, 183:E168–E184.

Dantzler, W. H., and E. J. Braun. 1980. Comparative nephron function in reptiles, birds, and mammals. American Journal of Physiology—Regulatory, Integrative and Comparative Physiology, 239:R197.

Dávalos, L. M., A. L. Cirranello, J. H. Geisler, and N. B. Simmons. 2012. Understanding phylogenetic incongruence: lessons from Phyllostomid bats. Biological Reviews, 87:991–1023.

Dávalos, L. M., P. M. Velazco, O. M. Warsi, P. Smits, and N. B. Simmons. 2014. Integrating incomplete fossils by isolating conflictive signal in saturated and non-independent morphological characters. Systematic Biology, 63:582–600.

Delsuc, F., J. L. Metcalf, L. Wegener Parfrey, S. J. Song, A. González, and R. Knight. 2014. Convergence of gut microbiomes in myrmecophagous mammals. Molecular Ecology, 23:1301–1317.

di Rago, J.-P., C. Macadre, J. Lazowska, and P. P. Slonimski. 1993. The C-terminal domain of yeast cytochrome *b* is essential for a correct assembly of the mitochondrial cytochrome bc1 complex. FEBS Letters, 328:153–158.

Dumont, E. R., L. M. Dávalos, A. Goldberg, C. C. Voigt, K. Rex, and S. E. Santana. 2012. Morphological innovation, diversification and the invasion of a new adaptive zone. Proceedings of the Royal Society B–Biological Sciences, 279:1797–1805.

Dumont, E. R., K. Samadevam, I. Grosse, O. M. Warsi, B. Baird, and L. M. Dávalos. 2014. Selection for mechanical advantage underlies multiple cranial optima in New World leaf-nosed bats. Evolution, 68:1436–1449.

Eckburg, P. B., E. M. Bik, C. N. Bernstein, E. Purdom, L. Dethlefsen, M. Sargent, S. R. Gill et al. 2005. Diversity of the human intestinal microbial flora. Science, 308:1635–1638.

Famularo, G., C. De Simone, V. Pandey, A. R. Sahu, and G. Minisola. 2005. Probiotic lactobacilli: an innovative tool to correct the malabsorption syndrome of vegetarians? Medical Hypotheses, 65:1132–1135.

Freeman, P. W. 2000. Macroevolution in Microchiroptera: recoupling morphology and ecology with phylogeny. Evolutionary Ecology Research, 2:317–335.

Giannini, N. P., and E. K. V. Kalko. 2005. The guild structure of animalivorous leaf-nosed bats of Barro Colorado Island, Panama, revisited. Acta Chiropterologica, 7:131–146.

Gracheva, E. O., J. F. Cordero-Morales, J. A. Gonzalez-Carcacia, N. T. Ingolia, C. Manno, C. I. Aranguren, J. S. Weissman et al. 2011. Ganglion-specific splicing of TRPV1 underlies infrared sensation in vampire bats. Nature, 476:88–91.

Hansen, T. F. 1997. Stabilizing selection and the comparative analysis of adaptation. Evolution, 51:1341–1351.

Harmon, L. J., J. B. Losos, T. Jonathan Davies, R. G. Gillespie, J. L. Gittleman, W. Bryan Jennings, K. H. Kozak et al. 2010. Early bursts of body size and shape evolution are rare in comparative data. Evolution, 64:2385–2396.

Hayden, S., M. Bekaert, T. A. Crider, S. Mariani, W. J. Murphy, and E. C. Teeling. 2010. Ecological adaptation determines functional mammalian olfactory subgenomes. Genome Research, 20:1–9.

Hayden, S., M. Bekaert, A. Goodbla, W. J. Murphy, L. M. Dávalos, and E. C. Teeling. 2014. A cluster of olfactory receptor genes linked to frugivory in bats. Molecular Biology and Evolution, 31:917–927.

Herrera, J. P., and L. M. Dávalos. 2016. Phylogeny and divergence times of lemurs inferred with recent and ancient fossils in the tree. Systematic Biology, 65:772–791.

Ho, L. S. T., and C. Ané. 2014. Intrinsic inference difficulties for trait evolution with Ornstein-Uhlenbeck models. Methods in Ecology and Evolution, 5:1133–1146.

Hunter, J. P., and J. Jernvall. 1995. The hypocone as a key innovation in mammalian evolution. Proceedings of the National Academy of Sciences, USA, 92:10718–10722.

Ingram, T., and D. L. Mahler. 2013. SURFACE: detecting convergent evolution from comparative data by fitting Ornstein-Uhlenbeck models with stepwise Akaike Information Criterion. Methods in Ecology and Evolution, 4:416–425.

Khabbazian, M., R. Kriebel, K. Rohe, and C. Ané. 2016. Fast and accurate detection of evolutionary shifts in Ornstein-Uhlenbeck models. Methods in Ecology and Evolution, 7:811–824.

Kuczynski, J., C. L. Lauber, W. A. Walters, L. W. Parfrey, J. C. Clemente, D. Gevers, and R. Knight. 2012. Experimental and analytical tools for studying the human microbiome. Nature Reviews Genetics, 13:47–58.

Leturque, A., E. Brot-Laroche, and M. Le Gall. 2009. GLUT2 mutations, translocation, and receptor function in diet sugar managing. American Journal of Physiology—Endocrinology and Metabolism, 296:E985.

Ley, R. E., M. Hamady, C. Lozupone, P. J. Turnbaugh, R. R. Ramey, J. S. Bircher, M. L. Schlegel et al. 2008. Evolution of mammals and their gut microbes. Science, 320:1647–1651.

Losos, J. B. 2010. Adaptive radiation, ecological opportunity, and evolutionary determinism. American Naturalist, 175: 623–639.

Losos, J. B., and D. L. Mahler. 2010. Adaptive radiation: the interaction of ecological opportunity, adaptation, and speciation. Pp. 381–420 in: Evolution since Darwin: The First 150 Years (M. Bell, D. Futuyma, W. Eanes, and J. Levinton, eds.). Sinauer, Sunderland, MA.

Mahler, D. L., T. Ingram, L. J. Revell, and J. B. Losos. 2013. Exceptional convergence on the macroevolutionary landscape in island lizard radiations. Science, 341:292–295.

Malnic, B., J. Hirono, T. Sato, and L. B. Buck. 1999. Combinatorial receptor codes for odors. Cell, 96:713–723.

Matsui, A., Y. Go, and Y. Niimura. 2010. Degeneration of olfactory receptor gene repertoires in primates: no direct link to full trichromatic vision. Molecular Biology and Evolution, 27:1192–1200.

Meng, F., L. Zhu, W. Huang, D. M. Irwin, and S. Zhang. 2016. Bats: body mass index, forearm mass index, blood glucose levels and SLC2A2 genes for diabetes. Scientific Reports, 6:29960.

Moen, D., and H. Morlon. 2014. Why does diversification slow down? Trends in Ecology and Evolution, 29:190–197.

Near, T. J., A. Dornburg, R. I. Eytan, B. P. Keck, W. L. Smith, K. L. Kuhn, J. A. Moore et al. 2013. Phylogeny and tempo of diversification in the superradiation of spiny-rayed fishes. Proceedings of the National Academy of Sciences, USA, 110:12738–12743.

Nei, M., and A. P. Rooney. 2005. Concerted and birth-and-death evolution of multigene families. Annual Review of Genetics, 39:121–152.

Niimura, Y., and M. Nei. 2007. Extensive gains and losses of olfactory receptor genes in mammalian evolution. PLOS ONE, 2:e708.

Ochman, H., M. Worobey, C.-H. Kuo, J.-B. N. Ndjango, M. Peeters, B. H. Hahn, and P. Hugenholtz. 2010. Evolutionary relationships of wild hominids recapitulated by gut microbial communities. PLOS Biology, 8:e1000546.

O'Keefe, F. R., and P. J. Wagner. 2001. Inferring and testing hypotheses of cladistic character dependence by using character compatibility. Systematic Biology, 50:657–675.

Osborn, H. F. 1902. The law of adaptive radiation. American Naturalist, 36:353–363.

Phillips, C. D., G. Phelan, S. E. Dowd, M. M. McDonough, A. W. Ferguson, J. Delton Hanson, L. Siles et al. 2012. Microbiome analysis among bats describes influences of host phylogeny, life history, physiology and geography. Molecular Ecology, 21:2617–2627.

Pinto, G., D. L. Mahler, L. J. Harmon, and J. B. Losos. 2008. Testing the island effect in adaptive radiation: rates and patterns of morphological diversification in Caribbean and mainland *Anolis* lizards. Proceedings of the Royal Society B–Biological Sciences, 275:2749–2757.

Price, S. A., and S. S. B. Hopkins. 2015. The macroevolutionary relationship between diet and body mass across mammals. Biological Journal of the Linnean Society, 115:173–184.

Rabosky, D. L., and R. E. Glor. 2010. Equilibrium speciation dynamics in a model adaptive radiation of island lizards. Proceedings of the National Academy of Sciences, USA, 107:22178–22183.

Rabosky, D. L., and A. H. Hurlbert. 2015. Species richness at continental scales is dominated by ecological limits. American Naturalist, 185:572–583.

Reynolds, R. G., D. C. Collar, S. A. Pasachnik, M. L. Niemiller, A. R. Puente-Rolón, and L. J. Revell. 2016. Ecological specialization and morphological diversification in Greater Antillean boas. Evolution, 70:1882–1895.

Rojas, D., M. J. R. Pereira, C. Fonseca, and L. M. Dávalos. In review. Eating down the food chain: trophic specialization is not an evolutionary dead end. Ecology Letters.

Rojas, D., Á. Vale, V. Ferrero, and L. Navarro. 2011. When did plants become important to leaf-nosed bats? Diversification of feeding habits in the family Phyllostomidae. Molecular Ecology, 20:2217–2228.

Rojas, D., O. M. Warsi, and L. M. Dávalos. 2016. Bats (Chiroptera: Noctilionoidea) challenge a recent origin of extant Neotropical diversity. Systematic Biology, 65:432–448.

Saldaña-Vázquez, R. A., E. Ruiz-Sanchez, L. Herrera-Alsina, and J. E. Schondube. 2015. Digestive capacity predicts diet diversity in Neotropical frugivorous bats. Journal of Animal Ecology, 84:1396–1404.

Sanders, J. G., S. Powell, D. J. C. Kronauer, H. L. Vasconcelos,

M. E. Frederickson, and N. E. Pierce. 2014. Stability and phylogenetic correlation in gut microbiota: lessons from ants and apes. Molecular Ecology, 23:1268–1283.

Santana, S. E., I. R. Grosse, and E. R. Dumont. 2012. Dietary hardness, loading behavior, and the evolution of skull form in bats. Evolution, 66:2587–2598.

Schluter, D. 2000. The ecology of adaptive radiation. Oxford University Press, Oxford.

Schondube, J. E., L. G. Herrera M., and C. Martínez del Río. 2001. Diet and the evolution of digestion and renal function in phyllostomid bats. Zoology, 104:59–73.

Shen, B., X. Han, J. Zhang, S. J. Rossiter, and S. Zhang. 2012. Adaptive evolution in the glucose transporter 4 gene *Slc2a4* in Old World fruit bats (Family: Pteropodidae). PLOS ONE, 7:e33197.

Shi, J. J., and D. L. Rabosky. 2015. Speciation dynamics during the global radiation of extant bats. Evolution, 69:1528–1545.

Simpson, G. G. 1953. The Major Features of Evolution. Columbia University Press, New York.

Tibshirani, R. 1996. Regression shrinkage and selection via the lasso. Journal of the Royal Statistical Society, ser. B: Methodological, 58:267–288.

Uyeda, J. C., and L. J. Harmon. 2014. A novel Bayesian method for inferring and interpreting the dynamics of adaptive landscapes from phylogenetic comparative data. Systematic Biology, 63:902–918.

Voje, K. L., T. F. Hansen, C. K. Egset, G. H. Bolstad, and C. Pélabon. 2014. Allometric constraints and the evolution of allometry. Evolution, 68:866–885.

Voje, K. L., A. B. Mazzarella, T. F. Hansen, K. Østbye, T. Klepaker, A. Bass, A. Herland et al. 2013. Adaptation and constraint in a stickleback radiation. Journal of Evolutionary Biology, 26:2396–2414.

Voss, R. S., D. W. Fleck, R. E. Strauss, P. M. Velazco, and N. B. Simmons. 2016. Roosting ecology of Amazonian bats: evidence for guild structure in hyperdiverse mammalian communities. American Museum Novitates:1–43.

Wagner, C. E., L. J. Harmon, and O. Seehausen. 2012. Ecological opportunity and sexual selection together predict adaptive radiation. Nature, 487:366–369.

Wagner, P. J. 2000. Exhaustion of morphologic character states among fossil taxa. Evolution, 54:365–386.

Wake, D. B. 1991. Homoplasy: the result of natural selection, or evidence of design limitations? American Naturalist, 138:543–567.

Wetterer, A. L., M. V. Rockman, and N. B. Simmons. 2000. Phylogeny of phyllostomid bats (Mammalia: Chiroptera): data from diverse morphological systems, sex chromosomes, and restriction sites. Bulletin of the American Museum of Natural History, 248:1–200.

Winter, Y., J. Lopez, and O. von Helversen. 2003. Ultraviolet vision in a bat. Nature, 425:612–614.

Yoder, J. B., E. Clancey, S. Des Roches, J. M. Eastman, L. Gentry, W. Godsoe, T. J. Hagey et al. 2010. Ecological opportunity and the origin of adaptive radiations. Journal of Evolutionary Biology, 23:1581–1596.

8

The Evolution of Body Size in Noctilionoid Bats

Norberto P. Giannini,
Lucila I. Amador,
and R. Leticia Moyers Arévalo

Introduction

Body mass is likely the single most important variable affecting the functional biology of animals (MacNab 2007). Being flying vertebrates, size is critically important for bats. The body mass range observed in bats is ca. 2–1500 g, but most bats fall in a narrow 9–16 g interval of a significantly right-skewed distribution on a lognormal scale (Safi et al. 2013; Willig et al. 2003). This distribution differs from terrestrial mammals in a number of aspects (e.g., less dominant modal value, gradual decline toward smallest values, right-side tail less extended and less concave; Willig et al. 2003). Unlike some other diverse bat families with rather idiosyncratic body mass distributions (e.g., Emballonuridae, Vespertilionidae, Molossidae), phyllostomids exhibit a body mass distribution similar to that of bats as a group (Willig et al. 2003). In a wider phylogenetic framework, Noctilionoidea can be characterized as the clade of echolocating bats with the greatest evolutionary expression of size variation, for it includes some of the smallest bat species (thyropterids at ca. 3 g; Aguirre 2007) as well as one of the largest echolocating bats, the carnivorous spectral bat *Vampyrum spectrum* (Phyllostomidae: Phyllostominae) weighing up to 235 g (e.g., specimen USNM 335160, a pregnant female). This represents an impressive, near 80-fold size variation. Understanding how the extraordinary range and distribution of body mass in the group interacted with the noctilionoid evolution, arguably one of the greatest morphofunctional and ecological diversifications found in the mammalian evolutionary history (Dumont et al. 2011), becomes highly relevant.

Size is especially important for bats because important aspects of aerodynamics (e.g., Norberg 1994) and echolocation (e.g., Jones 1999), as well as many ecomorphological traits (e.g., Dumont et al. 2011), significantly interact with body size. Variables associated with flight variously depend on body mass. Aspect ratio—a key feature reflecting aerodynamic efficiency—is adimensional and so no specific relationship with size is expected (Norberg 1994). By contrast, wing loading (Pa) scales positively with body size (g) as expected and directly affects flight speed (m s^{-1}), home range (km^2), and commuting distance (km), among other key ecological traits (Norberg 1994). Wing beat frequency (Hz) scales inversely with body size (Bullen and McKenzie 2002) and may directly affect aspects of echolocation such as pulse repetition rate (Hz; Jones 1999; Speakman and Racey 1991). In addition, several other call parameters, such as peak frequency, frequency of harmonics, and call bandwidth (kHz), can all be expected to covary inversely with body size, just as other frequency parameters (e.g., Calder 1996). Thus, sophisticated laryngeal echolocation, in combination with powered flight, may impose severe physical constraints on body mass evolution in bats (e.g., Jones 1999). Regarding echolocation, phyllostomids are unusual in many ways; as in other gleaning bats, they tend to use diverse additional cues to detect food targets (e.g., rustling noises from insects, odor emanating from fruits or flowers) and use echolocation mainly for orientation in space (Schnitzler and Kalko 1998). As a consequence, phyllostomids emit rather uniform calls without distinct search, approach, and terminal phases as in aerial hawking bats, with few exceptions (notably, the trawling insectivore *Macrophyllum macrophyllum*; Weinbeer and Kalko 2007). These bats are also peculiar in terms of call intensity: phyllostomids are generally known as whispering bats because they produce faint calls of low intensity (dB), although recent research has reported the emission of calls up to 50% louder than previously recorded, suggesting that phyllostomids exhibit large output flexibility that may have ecological importance (see Brinkløv et al. 2009). Preliminary data suggest that maximum call intensity scales positively with body mass in phyllostomids (Brinkløv et al. 2009), but this and other aspects of the relationship between echolocation and body size remain largely unexplored in this group.

A number of purely ecological aspects is also expected to be significantly influenced by body size in noctilionoids. As a group, noctilionoids exploit an unparalleled diversity of resources, including insects, terrestrial vertebrates, fish, blood, nectar, pollen, fruits, and leaves (Dumont et al. 2011). Only a handful of species exhibit truly omnivorous habits (e.g., *Phyllostomus hastatus*; see Giannini and Kalko 2004, 2005); groups of species, in fact entire clades, specialize in just one, or a few, of these items: most frugivorous phyllostomids group in the Carolliinae and Stenodermatinae, sanguivorous species in Desmodontinae, nectarivorous species in the Glossophaginae and Lonchophyllinae, and animalivorous species in Macrotinae, Micronycterinae, Phyllostominae, and Lonchorhininae (see Baker et al. 2012). Body size may represent an integral functional component of the adaptation complex that enables a species to exploit such a diverse array of resources in a specialized fashion. For instance, carnivorous bats including phyllostomids and noctilionids tend to be larger than close relatives with more insectivorous diets (Giannini and Kalko 2005; Norberg and Fenton 1998); nectarivorous phyllostomids are species capable of hovering flight, which is greatly favored by small body mass (Norberg 1994).

Exactly how body size evolved in phyllostomids, and how the evolutionary dynamics of size are related to diet, to echolocation, or to any of several other highly relevant aspects of the biology of phyllostomids, remains unclear and invites exploration. Here we take a first step and interpret the evolutionary variation of body size in phyllostomids and close allies (Noctilionoidea) in light of their extraordinary ecological diversification. It has been shown that body size in bats as a group exhibits a strong phylogenetic signal, meaning that closely related species show a more similar body size than expected by chance (Safi et al. 2013). So here we tested phylogenetic signal in body size in noctilionoids and applied a detailed, node-by-node analysis of evolutionary changes as reconstructed in their phylogenetic tree. We hypothesized that the potential exists for substantial body size variation, significantly affecting the biology of ancestral noctilionoids and ancestral phyllostomids. In turn, crown clade phyllostomids are known to have expanded size to the documented upper boundary of weight in echolocating bats (see Jones 1999) and are presumed to have accumulated most of the body size variance of noctilionoids. We first examine general patterns of change and phylogenetic signal and the reconstructed changes along the phylogeny searching for the connection between body size change and the diversification of noctilionoids, with an emphasis on the largest groups, the Phyllostomidae and its subclades. We reveal a persistent stasis in roughly half the history of the group; a highly idiosyncratic, lineage-dependent pattern of increases and decreases that seemed to oscillate over time, adjusting body mass to specific ecological conditions; and trends of phyletic size increase sustained over millions of years, strongly suggestive of the relentless action of directional natural selection in specific lineages across the phyllostomid phylogenetic history. More intriguing, we interpret diverging trends of body size change as the result of a strong geographic imprint and the possible effect of past character displacement.

Phylogenetic Pattern of Body Mass Change

Data on mean body size, in grams, were compiled from diverse sources for a comprehensive sample of noctilionoid species (app. 8.1), coded as a single continuous character for each species, and optimized (see Goloboff et al. 2006) on our reference phylogeny from Rojas et al. (2016) using the program TNT (Goloboff et al. 2008). With this option, body size of internal nodes (ancestral species) can be reconstructed on the topology, changes on each branch (from ancestor to descendant) can be determined by subtraction, and net changes (common to all character reconstructions) can be distinguished from ambiguous changes (which depend on specific reconstructions; e.g., Amador and Giannini 2016). Some 42% of nodes remained stable over the course of noctilionoid evolution, including phyllostomids. Net decreases and increases exhibited

roughly similar frequencies (14–16% of nodes, respectively), whereas ambiguous nodes, that is, nodes with changes that vary with the specific reconstruction, represented ca. 28% of a total of 386 nodes (table 8.1). Overall amount of evolutionary change was 1,113 g (steps), but considerable levels of uncertainty were evident in the reconstruction, given that roughly one-quarter (26%) of total size change (or 292 g) was ambiguous. Increases clearly dominated over decreases, in terms of magnitude of change, and accounted for 55% of total size variation. Phyllostomids in particular accounted for 740 g of evolutionary body mass variation, or 66% of total changes, in spite of the group comprising 88% of terminals.

In phyllostomids especially, increases and decreases were balanced in terms of frequency of change (number of nodes of each kind); however, the magnitude of change was biased toward increases (546 g vs. 194 g of decreases), and the distribution of change varied greatly across selected monophyletic groups (subfamilies; table 8.1). In terms of magnitude of change (cumulative change in g), all changes were increases in Macrotinae, Desmodontinae, and Glyphonycterinae, or nearly so (>70% of nodes) in Phyllostominae, Glossophaginae, and Stenodermatinae, whereas all changes

Table 8.1. Summary of evolutionary body weight changes in subfamilies of Phyllostomidae

Subfamily	Total number of nodes	Increases		Decreases		Stasis		Increases		Decreases		Total
		Number	%	Number	%	Number	%	Number	%	Number	%	
Macrotinae	5	1	20.0	0	0.0	4	80.0	3	100.0	0	0.0	3
Micronycterinae	21	2	9.5	2	9.5	17	81.0	4	57.1	3	42.9	7
Desmodontinae	5	3	60.0	0	0.0	2	40.0	8	100.0	0	0.0	8
Phyllostominae	39	8	20.5	6	15.4	25	64.1	238	83.5	47	16.5	285
Lonchorhininae	5	0	0.0	1	20.0	4	80.0	0	0.0	6	100.0	6
Glossophaginae	53	10	18.9	9	17.0	34	64.2	51	68.9	23	31.1	74
Lonchophyllinae	17	1	5.9	4	23.5	12	70.6	4	26.7	11	73.3	15
Glyphonycterinae	5	1	20.0	0	0.0	4	80.0	6	100.0	0	0.0	6
Carolliinae	11	2	18.2	2	18.2	7	63.6	2	28.6	5	71.4	7
Rhynophyllinae	5	0	0.0	2	40.0	3	60.0	0	0.0	6	100.0	6
Stenodermatinae	165	27	16.4	25	15.2	113	68.5	228	71.0	93	29.0	321
Total	331	55	16.4	51	15.0	225	68.6	546	73.8	194	26.2	738

Note: Frequency and magnitude of change are classified in increases, decreases, and stasis (when applicable), and the respective quantity (in number of nodes or change in grams, g) expressed as actual value and as percentage.

were decreases in Lonchorhininae and Rhinophyllinae, or nearly so (>70% of nodes) in Lonchophyllinae and Carolliinae (table 8.1). That is, the distribution of change was highly idiosyncratic across subfamilies and, hence, ecological groups.

Gould and McFadden (2004) classified evolutionary change in body size in terms of nanism versus giantism and in terms of apomorphic change, or change occurring in a single branch, versus phyletic change, or change accumulating in successive branches. Amador and Giannini (2016) used a simple metric, the order, to account for the extent of phyletic change in terms of the number of successive nodes changing in the same direction; for example, successive increases along n branches represented a case of phyletic giantism of order n. Here we cross-classified evolutionary change in increases or decreases of apomorphic or phyletic nature (table 8.2), the latter described also in terms of its order (see specific cases below). For phyllostomids, 62% of changes were accommodated in phyletic series, and slightly more changes (ca. 57%) in phyletic increases as compared with phyletic decreases (table 8.2). In contrast, apomorphic decreases were slightly more common (55%) than apomorphic increases. Again the pattern of change varied greatly across subfamilies, from groups dominated by phyletic increases (e.g., Stenodermatinae), to balanced groups (e.g., Phyllostominae) with change distributed in several categories (table 8.2).

In addition to body size optimization, we applied a phylogenetic comparative method designed to detect signal in tree partitions (clades in a rooted tree) that correspond with major underlying data structure of the variable of interest (here body mass) and that estimate the fraction of variation significantly explained by those phylogenetic partitions. This method, Canonical Phylogenetic Ordination (Giannini 2003), uses simulations to test the significance of individual tree partitions and composes a linear model with the nonredundant tree partitions through a forward stepwise selection procedure. We applied the univariate case of Canonical Phylogenetic Ordination (see Giannini 2003) to our body size data (app. 8.1) using the program CANOCO v. 4.0 (ter Braak and Šmilauer 1998; main options RDA (redundancy analysis), no transformations, 9,999 unrestricted Monte Carlo permutations in each test, alpha level set to 0.01). Total variation nonredundantly explained by the four clades (out of a total of 198 nodes) that met one criterion of clade inclusion (that each clade should contribute at least 5% of total variation) was 63.2% (table 8.3). A more inclusive model that

Table 8.2. Summary of evolutionary body weight changes in subfamilies of Phyllostomidae cross-classified in apomorphic or phyletic increases or decreases, and the respective quantity in number of nodes and as percentage

Subfamily	Total changes	Phyletic					Apomorphic				
		Increases		Decreases		Total	Increases		Decreases		Total
		Number	%	Number	%		Number	%	Number	%	
Macrotinae	1	0	...	0	...	0	1	100	0	...	1
Micronycterinae	4	0	...	0	...	0	2	50	2	50	4
Desmodontinae	3	3	100	0	0	3	0	...	0	...	0
Phyllostominae	14	6	75	2	25	8	2	33.3	4	66.7	6
Lonchorhininae	1	0	...	0	...	0	0	0	1	100	1
Glossophaginae	19	7	53.8	6	46.2	13	3	50	3	50	6
Lonchophyllinae	5	0	...	4	100	4	1	100	0	0	1
Glyphonycterinae	1	0	...	0	...	0	1	100	0	0	1
Carolliinae	4	2	100	0	0	2	0	0	2	100	2
Rhynophyllinae	2	0	...	2	100	2	0	...	0	...	0
Stenodermatinae	52	19	55.9	15	44.1	34	8	44.4	10	55.6	18
Total	106	37	56.7	29	43.3	66	18	45	22	55	40

Table 8.3. Results of the phylogenetic comparative analysis applying Canonical Phylogenetic Ordination (CPO; Giannini 2003)

Tree Partition (Clade)	Ref. in Figs. 8.1–8.4	F-value	P-value	Variation Explained (%)	Cumulative Variation Explained (%)
Chrotopterus + Vampyrum	I	99.53	0.0002	34.9	34.9
Largest *Artibeus* (subgenus *Artibeus*)	II	52.55	0.0001	14.4	49.3
Phylloderma + Phyllostomus	III	33.83	0.0001	7.8	57.1
Large *Platyrrhinus*	IV	29.85	0.0001	6.1	63.2
Noctilionidae	V	12.99	0.0048	2.4	65.6
"Inner" Phyllostomidae*	VII	11.84	0.0007	2.1	67.7
Platyrrhinus albericoi + P. vittatus	VII	11.65	0.0055	2.0	69.7
Brachyphylla	VIII	10.61	0.0106	1.7	71.4
Predominantly phytophagous phyllostomids**	IX	9.27	0.0030	1.4	72.8
Largest *Phyllostomus* (*P. elongatus + P. hastatus*)	X	8.94	0.0130	1.3	74.1
Stenodermatinae	XI	8.61	0.0039	1.2	75.3
Subgenus *Artibeus* (all species)	. . .	5.56	0.0374	0.9	Not included

Note: Analyses were done using CANOCO v. 4.0 (ter Braak and Šmilauer 1998). Tree partitions were tested using 9,999 unrestricted Monte Carlo permutations shuffling clade membership and appear from top to bottom in the order they were included in the final reduced model. These groups appear numbered in figs. 8.1–8.4 with Roman numerals and indicated with dots in the corresponding branch. The observed *F*-statistic, its associated (randomized) probability (*P*-value), and the variation explained (as percentage of total variation) of a given clade correspond to values calculated with all previous groups included in the model. Cumulative variation explained is the partial summation of the contribution of each clade and all previous clades in the model. The first four clades correspond to the result of forward selection procedure stopped when variation explained by the candidate clade was <5% (variation explained 63.2%). Next clades correspond to those added to meet the criterion of selection until the predefined *P*-value was obtained. Here we set alpha = 0.01 as a conservative compromise between the conventional 5% alpha-value and a formal Bonferroni correction. Here variation explained was 75.3%, and 12 clades were accepted in the final nonredundant model.

* Phyllostomidae to the exclusion of basal Macrotinae and Micronycterinae.

** Predominantly phytophagous phyllostomids are members of a group containing Glossophaginae, Lonchophyllinae, Carolliinae, Rhynophyllinae, and Stenodermatinae, which also includes the Lonchorhininae and Glyphonycterinae in Rojas et al. (2016).

selected all 12 statistically significant clades explained 75.3% of total variation (see table 8.3). These tree partitions represented the most distinctive body weight differences between members (binary coded 1) and outsiders (binary coded 0) of a clade and roughly corresponded with major transitions in body size identified by optimization (see details below). Remarkably, these clades included highly specialized bats, particularly carnivorous, frugivorous, and omnivorous phyllostomids of large size such as *Chrotopterus* and *Vampyrum* and Noctilionidae among the former, large *Artibeus* and *Platyrrhinus* among the frugivores, and *Brachyphylla* and *Phyllostomus* among the omnivores (see detail of groups in table 8.3). Large, higher-level clades were also marginally important: "inner" phyllostomids (i.e., excluding macrotines and micronycterines) and Stenodermatinae (table 8.3).

Body Mass Evolution in Non-Phyllostomid Noctilionoids

Noctilionoids provide the frame for understanding in depth the evolutionary origins of phyllostomids (see Velazco and Giannini, chap. 3, this vol.). Here the ancestral body mass of the superfamily was estimated in a narrow range of 9–12 g (fig. 8.1), which is smaller but overlaps with the ancestral body mass range for crown clade Chiroptera, estimated at 12–14 g (see Giannini et al. 2012). The myzopodid and mystacinid branches did not show any change, as the single species included per family, *Myzopoda aurita* at ca. 9 g and *Mystacina tuberculata* at ca. 12 g, overlapped with the reconstructed ancestral range of noctilionoids, which they helped to define. Eventual inclusion of the recently discovered *Myzopoda schliemanni* would not have changed the

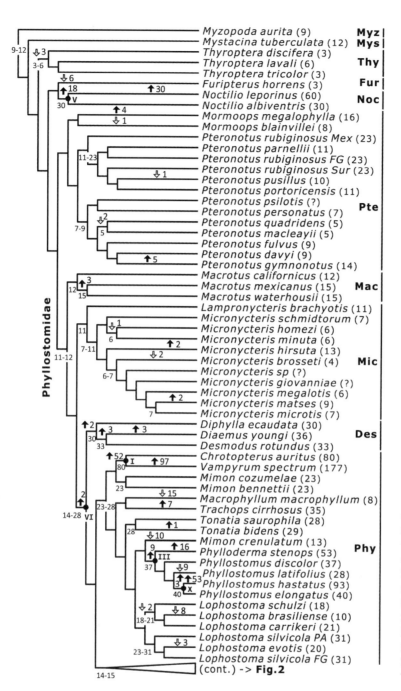

Figure 8.1. Reconstruction of body size evolution (see text for details) in noctilionoids and basal phyllostomids on our reference phylogeny (Rojas et al. 2016). Numbers below branches are the body weight values assigned to the corresponding node; branches without these numbers indicate stasis. Numbers above branches are net changes in body weight and are accompanied by a solid arrow pointing up (increases) or an outlined arrow pointing down (decreases). Dots and accompanying Roman numerals refer to tree partitions (clades in a rooted tree) selected as significantly associated with body size by the Canonical Phylogenetic Ordination analysis (results in table 8.3). Numbers in parentheses in each terminal are the average body weight (see sources in app. 8.1). Abbreviations: Des—Desmodontinae; Fur—Furipteridae; Mac—Macrotinae; Mic—Micronycterinae; Mor—Mormoopidae; Myt—Mystacinidae; Myz—Myzopodidae; Noc—Noctilionidae; Phy—Phyllostominae; Thy—Thyropteridae.

myzopodid branch, given that the reported size range is similar to the type species *M. aurita* at 8.6 ± 0.85 g (*n* = 26 in Goodman et al. 2006). Daniel (1979) estimated weight of the likely extinct *Mystacina robusta* to be 25–35 g (from extrapolation of forearm length in *M. tuberculata*); this means that inclusion of *M. robusta* in the analysis as sister to *M. tuberculata* would have had the effect of greatly increasing size variation in this family; however, subsequent estimates indicated a range of only 14–15 g for *Mystacina robusta* (Lloyd 2001), so only a modest additional change would be expected in the fam-

ily by including the species in the analysis. The ancestral node of the Neotropical noctilionoid subtree remained without change at 9–12 g. The first split, Thyropteridae, represented the first instance of an abrupt reduction in body mass with a net decrease of 3 g. No internal body mass change was reconstructed inside the family.

The noctilionoid backbone assignment remained stable for the next two nodes before diverging, that is, the furipterid—noctilionid clade—and the mormoopid—phyllostomid clade (fig. 8.1). The former clade exhibited a diverging pattern of change, with furipterids decreasing 6 g and noctilionids increasing as much as 18 g. Inclusion of *Amorphochilus schnablii*, the sole other member of Furipteridae, weighing ca. 3 g, would have placed the body size change in the ancestral node of the family. The result of size reduction seen in thyropterids and furipterids likely represents a change approaching a critically small value in bat evolution given that some members of these families are among the smallest bats. The noctilionid branch, on the contrary, experienced a drastic increase of 2.5 times the upper range of the ancestral size. Based on the fossil record, *Noctilio* entered a trawling insectivore niche ca. 16 Ma or earlier (Czaplewksi and Cartelle 1998). This ecological change was accompanied by the large size increase that we detected, but this change does not seem to be an ecological requirement as other trawling bats, including the phyllostomid *Macrophyllum* (see below), remained small during its evolution. However, this change may have facilitated access to large aquatic prey, so the size trend continued in the branch of the greater bulldog bat, *N. leporinus*, with a much greater net increase to reach an average of 60 g during the last ca. 4 Ma since the two *Noctilio* species last shared a common ancestor (Amador et al. 2016).

The mormoopid branch showed no internal backbone change, and only a few increases and decreases occurred inside the groups (fig. 8.1). The most relevant changes occurred in *Mormoops* with a diverging pattern; while *M. megalophylla* is twice as large as *M. blainvillei* (16 and 8 g, respectively), the net changes reconstructed were less marked, with the former species increasing 4 g with respect to the upper range of the mormoopid ancestor of 12 g, and the latter species decreasing only 1 g (with respect to the lower boundary of the range at 9 g). With regard to *Pteronotus*, there was considerable ambiguity expressed in the widening of the reconstructed intervals (11–23 g in the *P. parnellii* species complex, subgenus *Phyllodia*, and 7–9 g in its sister clade with the remaining species; fig. 8.1). Only three net changes were found, small (1–2 g) decreases in *P. pusillus* and subgenus *Chilonycteris* (*P. quadridens* + *P. macleayi*), and a +5 g change in *P. gymnonotus* (fig. 8.1).

The Size of the Ancestral Phyllostomid and Basal Animalivores

The ancestral phyllostomid likely shared an animalivorous diet with members of the first branches of the fam-

ily (Baker et al. 2012) and other noctilionoids, although most significantly, the hunting mode switched from aerial hawking to gleaning (Schnitzler and Kalko 1998). Based on optimization, the root of the phyllostomid subtree was assigned the narrow body mass range of 11–12 g. This represents no net change with respect to the backbone of the noctilionoid tree, only a shrinking of the reconstructed interval; that is, from 9–12 g to just 11–12 g—no steps counted. This has the interesting evolutionary implication of no required change in size during the transition to gleaning behavior and the origination of the phyllostomid clade, which remains in line with the stasis along the noctilionoid backbone (see above).

The first two phyllostomid branches, macrotines and micronycterines, showed only limited changes of 1–3 g in both directions (increases and decreases). However, given the small size of these bats (range 4–15 g), these changes likely are biologically significant; for instance, a 2 g decrease in the *Micronycteris brosseti* branch (see fig. 8.1) represents a one-third size reduction for this bat.

The phyllostomid backbone showed only one net increase of just 2 g in the next node up, which contains desmodontines as sister to the remaining phyllostomids. Desmodontines can be safely classified among the animalivorous phyllostomids in the sense that vampires appear nested among other animalivores (see Rojas et al. 2016) and that the protein-rich diet that characterizes animalivores is accentuated in these highly specialized blood feeders. In the clade of vampires, three phyletic increases of 2–3 g explain the final size of the three extant species (30–36 g). These changes represent roughly 10% increases with respect to successive ancestors; however, some of the fossil and subfossil vampires were larger: *Desmodus draculae* was ca. 30% greater than the extant *Desmodus* (the common vampire bat, *D. rotundus*) in linear dimensions (skull length; Morgan 1988), which translates to considerably larger net increases in (tridimensional) body weight occurring in this clade during the Pleistocene-to-Recent transition (see Pardiñas and Tonni 2000; Trajano and De Vivo 1991).

Body mass change in phyllostomines exhibited most of what has been reconstructed for the entire superfamily. Three major patterns were observed. First, massive, phyletic size increases occurred in two clades: two successive increases of 52 and 97 g in the vampyrine leaf-nosed bats *Chrotopterus* and *Vampyrum*, respectively, which split ca. 15 Ma, and a total net increase of 56 g in the younger branch (dated ca. 6 Ma; Amador et al. 2016), leading to *Phyllostomus hastatus*. In these cases and in the branch leading to *Noctilio leporinus*, which involved the acquisition of vertebrate prey hunting, the (uniform) rate of body mass change was very high at +6–8 g per Ma. Second, decreases in the range of 2–8 g characterized the entire *Lophostoma* clade, which, as in the case of micronycterines, may represent major changes for bats averaging 10–31 g. Third, three cases of diverging trends in left and right descendants (sister branches) were reconstructed (fig. 8.1): −15 g in *Macrophyllum* versus +7 g in *Trachops*; −10 g in *Gardnerycteris*

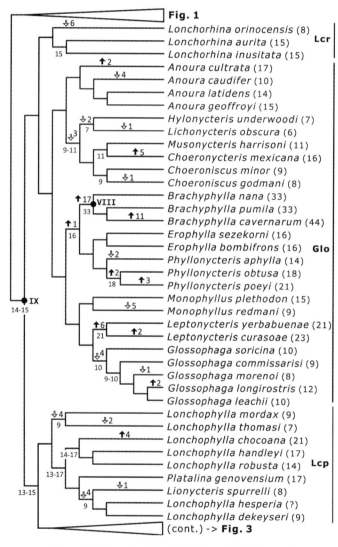

Figure 8.2. Reconstruction of body size evolution (see text for details) in nectarivorous phyllostomids and allies on our reference phylogeny (Rojas et al. 2016). Numbers and sources as in figure 8.1. Abbreviations: Glo—Glossophaginae; Lcp—Lonchophyllinae; Lcr—Lonchorhininae.

crenulatum versus +9 g in *Phyllostomus* + *Phylloderma*; and −9 g in *Phyllostomus latifolius* versus +3 g in *Phyllostomus hastatus* + *P. elongatus*. We discuss the implications of the character change reconstructed in the first case below.

Two groups of animalivores are more closely related to phytophagous phyllostomids than to other animalivores. One is Lonchorhininae (fig. 8.2), recovered as sister to Glossophaginae in Rojas et al. (2016; but see an alternative, more conservative position as sister to Phyllostominae in Amador et al. 2016). In this group, only one size change was reconstructed (a remarkable −6 g in the branch leading to the small *Lonchorhina orinocensis*). The second group, Glyphonycterinae (fig. 8.3), is deeply nested inside clades of phytophagous phyllostomids. This strongly suggests a reversal to an insectivorous diet (cf. Baker et al. 2012), although incomplete in the sense that these species may also include fruit in their diets (e.g., ca. 12% fruit in Panamanian *Trinycteris nicefori*; Giannini and Kalko 2004). Here, only one increase was reconstructed, specifically +6 g in the branch leading to the relatively large (average 21 g) *Glyphonycteris daviesi* (fig. 8.3).

Phytophagous Phyllostomids

The pattern of stasis along the backbone of the noctilionoid tree continued to the root of all the various subtrees of phytophagous phyllostomids (figs. 8.2–8.4), with just a single change of only +1 g and a few (costless) shifts in the interval of assigned body size. This is remarkable from an evolutionary perspective because essentially no body size change accompanied major ecological changes, such as the shift from a protein-based diet to a carbohydrate-based diet, and inside the latter, the exploitation of flowers versus fruits, which are functionally and energetically different. All changes were reconstructed inside well-defined groups, which we examine in the following.

Nectarivores

Several distinct patterns of body mass change were reconstructed in Glossophaginae. First, there were isolated increases and decreases in the range of 1–5 g (e.g., +2 g in the branch leading to *Anoura cultrata*, −5 g in *Monophyllus redmani*; fig. 8.2); second, phyletic decrease in the clade of the small (6–7 g) *Hylonycteris* and *Lichonycteris*; third, an abrupt phyletic increase (cumulative 29 g) to reach an average weight of 44 g in *Brachyphylla cavernarum*; fourth, diverging trends in left and right descendants in two clades: inside *Phyllonycteris*, and in *Leptonycteris* versus *Glossophaga* (fig. 8.2).

Not all these trends were replicated in Lonchophyllinae, likely because this clade is smaller than Glossophaginae. Phyletic decreases of low order (maximum order 2) were reconstructed independently in the branches leading to the small *Hsunycteris* (=*Lonchophylla*) thomasi (7 g) and *Lionycteris spurrelli* (8 g), as well as an isolated increase in *Lonchophylla chocoana* (fig. 8.2). The decreasing trends seen in both Glossophaginae and Lonchophyllinae can be related to specialization on small flowers with low nectar reward and the ability to feed successfully on them by hovering (e.g., Norberg 1994). The increasing trends are of two different kinds; on the one hand, mass increase in *Brachyphylla* may be related to the incorporation of fruit in the diet (Rodríguez and Dávalos 2008; Swanepoel and Genoways 1983 and citations therein); on the other hand, mass increase in the independent cases of *Choeronycteris* and *Leptonycteris* may be due to the fact that increased mass favors commuting and migration. These species use seasonally available flower resources in the Mexican desert and commute long distances daily from their roosting sites (they are obligate cave dwellers) to their shifting feeding patches; they also migrate following the availability of these resources (Fleming and Eby 2003). Both the increase in mass and the reduction in aerofoil surface area (small uropatagium as in *Leptonycteris*) contribute in combination to a faster flight that reduces commuting time (thus making resources available during the period of foraging activity, usually one night) and migration time (e.g., Norberg 1994). An anonymous reviewer also noted that larger size also permits greater storage of energetic reserve tissues. Body mass thus played a crucial role in the evolution of these habits for bats that are on the edge of an energetically viable budget (Tschapka 2004).

Frugivores

No major backbone change was reconstructed in the clades of frugivorous phyllostomids (figs. 8.3 and 8.4). *Carollia* is considered a *Piper* (Piperaceae) fruit special-

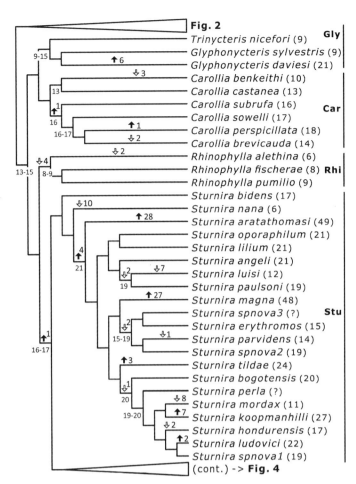

Figure 8.3. Reconstruction of body size evolution (see text for details) in basal frugivorous phyllostomids and allies on our reference phylogeny (Rojas et al. 2016). Numbers and sources as in figure 8.1. Abbreviations: Car—Carolliinae; Gly—Glyphonycterinae; Rhi—Rhinophyllinae; Stu—Sturnirini.

ist (Fleming 1986, 1988). In Carolliinae, evolutionary mass variation was low, with 1–3 g net apomorphic increases and decreases and one modest diverging trend in *Carollia perspicillata* (+1 g) versus *C. brevicauda* (−2 g). This variation, however low, has been associated with the degree of specialization in *Piper* fruits (the smaller *C. bevicauda* being more specialized; see Giannini and Kalko 2004). In Rhinophyllinae, a pattern of mass decrease was reconstructed, with the result of observed low mass (6–9 g) in extant species. Among phyllostomids, species of *Rhinophylla* are the main consumers of spikes of *Philodendron*, *Evodianthus*, *Anthurium*, and other Araceae (see review in Sánchez and Giannini, submitted), for which a small body mass may be advantageous given that these plants are root and tree-trunk climbing vines and epiphytes.

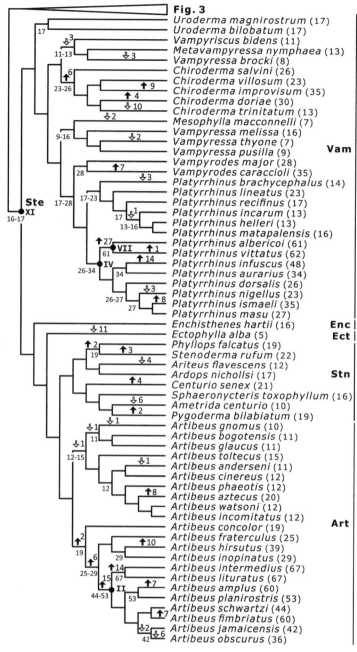

Fig. 3

Uroderma magnirostrum (17)
Uroderma bilobatum (17)
Vampyriscus bidens (11)
Metavampyressa nymphaea (13)
Vampyressa brocki (8)
Chiroderma salvini (26)
Chiroderma villosum (23)
Chiroderma improvisum (35)
Chiroderma doriae (30)
Chiroderma trinitatum (13)
Mesophylla macconnelli (7)
Vampyressa melissa (16)
Vampyressa thyone (7)
Vampyressa pusilla (9)
Vampyrodes major (28)
Vampyrodes caraccioli (35)
Platyrrhinus brachycephalus (14)
Platyrrhinus lineatus (23)
Platyrrhinus recifinus (17)
Platyrrhinus incarum (13)
Platyrrhinus helleri (13)
Platyrrhinus matapalensis (16)
Platyrrhinus albericoi (61)
Platyrrhinus vittatus (62)
Platyrrhinus infuscus (48)
Platyrrhinus aurarius (34)
Platyrrhinus dorsalis (26)
Platyrrhinus nigellus (23)
Platyrrhinus ismaeli (35)
Platyrrhinus masu (27)
Enchisthenes hartii (16)
Ectophylla alba (5)
Phyllops falcatus (19)
Stenoderma rufum (22)
Ariteus flavescens (12)
Ardops nichollsi (17)
Centurio senex (21)
Sphaeronycteris toxophyllum (16)
Ametrida centurio (10)
Pygoderma bilabiatum (19)
Artibeus gnomus (10)
Artibeus bogotensis (11)
Artibeus glaucus (11)
Artibeus toltecus (15)
Artibeus anderseni (11)
Artibeus cinereus (12)
Artibeus phaeotis (12)
Artibeus aztecus (20)
Artibeus watsoni (12)
Artibeus incomitatus (12)
Artibeus concolor (19)
Artibeus fraterculus (25)
Artibeus hirsutus (39)
Artibeus inopinatus (29)
Artibeus intermedius (67)
Artibeus lituratus (67)
Artibeus amplus (60)
Artibeus planirostris (53)
Artibeus schwartzi (44)
Artibeus fimbriatus (60)
Artibeus jamaicensis (42)
Artibeus obscurus (36)

Vam
Enc
Ect
Stn
Art
Ste

Figure 8.4. Reconstruction of body size evolution (see text for details) in derived frugivorous phyllostomids on our reference phylogeny (Rojas et al. 2016). Numbers and sources as in figure 8.1. Abbreviations: Art—Artibeina; Ect—Ectophyllina; Enc—Enchisthenina; Ste—Stenodermatini; Stn—Stenodermatina; Vam—Vampyressina.

Just a negligible 1 g increase was reconstructed in the node representing the major subfamily of frugivores, Stenodermatinae. However, changes inside the subfamily were numerous and some were of great magnitude (figs. 8.3 and 8.4). Sturnirini is sister to the other stenodermatines and contains the speciose genus *Sturnira* (Velazco and Patterson 2013). Within

this group, body mass variation was considerable and changes were reconstructed in most (14 out of 20) inner nodes (fig. 8.3). These changes included both small (−2 g) and large (−10 g) decreases, as well as increases that included two large independent changes (+27 and +28 g in *S. aratathomasi* and *S. magna*, respectively). Just one diverging trend was found (in the clade joining the small *S. mordax* and the large *S. koopmanhilli*), and one low-order (order 2) phyletic increase in the branch leading to *S. luisi* (fig. 8.3). Body mass changes in *Sturnira* thus showed a pattern of many isolated (apomorphic) changes of varying magnitude. No specific pattern reflecting the Andean versus lowland biogeographic imprint of the group (see Velazco and Patterson 2013) was detected in the variation of body mass.

Vampyressina was also reconstructed with many changes (fig. 8.4). Increases covered a wide range, from +1 g up to +27 g, as well as decreases from −1 g to −10 g, again without long phyletic trends (just one order-2 phyletic increase in *Platyrrhinus*). Two diverging trends of considerable magnitude of change were located in the group of *Chiroderma* (+6 g) and its vampyressine sister clade (−3 g), and inside *Chiroderma* (+4 g in *C. doriae* and −10 g in *C. trinitatum*). All the other, numerous changes in the Vampyressina were apomorphic (both increases and decreases), and thus the group closely resembles Sturnirini in this respect.

Two successive, isolated branches with one genus and species each, Enchisthenina including *Enchistenes hartii*, and Ectophyllina including *Ectophylla alba*, showed contrasting patterns of no change in the former, and one of the largest apomorphic decreases in the entire noctilionoid tree (−11 g, second only to *Macrophyllum* at −15 g) in the latter (fig. 8.4). This reduction in size resulted in *Ectophylla* being one of the smallest phyllostomids, and the smallest frugivore, at 5 g or just 33% of the mass of the reconstructed ancestor. The ecology of *Ectophylla* may reflect its minute size in at least three aspects: consumption of small, understory *Ficus* (Moraceae; Rodríguez and Pineda 2015); tent making and group living in relatively fragile *Heliconia* (Heliconiaceae) leaves (Brooke 1990; Rodríguez et al. 2011); and physiological advantage of heat preservation living in tents (Rodríguez et al. 2016).

In the group of short-faced bats, Stenodermatina, with eight species, all but one of the changes occurred at terminal branches, and all were balanced in terms of increases, decreases, and their corresponding magnitude (fig. 8.4). One diverging trend was detected in the clade of *Ametrida* (−6 g) and *Pygoderma* (+2 g), as well as a short (order-2) trend of increasing size leading to the largest short-faced bat, *Stenoderma rufum* at 22 g (fig. 8.4). No specific connection of body mass and the high specialization of these bats (e.g., of cranial morphology and diet; e.g., Dumont et al. 2011) was detected.

The last group of phyllostomids, Artibeina, comprises the single genus *Artibeus* and the distinct subgenera *Dermanura* and *Artibeus*. Here the pattern of body mass change was very conspicuously different from previous groups, with a predominance of long phyletic trends of both the decreasing and increasing types, in *Dermanura* and *Artibeus*, respectively (fig. 8.4). Decreases in *Dermanura* were modest (−1 g) but consistent in four branches, three of which were successive in a phyletic trend that may have begun 4 Ma (see Amador et al. 2016). An isolated, relatively large increase for the group (+8 g) was reconstructed in the branch leading to *Artibeus* (*Dermanura*) *aztecus*, a highland, cloud forest specialist (Solari 2016). In large *Artibeus*, continuous and discontinuous increasing phyletic trends occurred, from an estimated ancestor at 16–17 g, to the largest frugivores, such as *A. lituratus* at 67 g (fig. 8.4). Successive ancestral *Artibeus* species increased size almost fourfold during the ca. 6.8 Ma of their evolution (dated in Amador et al. 2016), with a rate of change at ca. 7 g per Ma—a rate comparable to that of large animalivores (see above). That is, large frugivores are smaller than large animalivores only because they are younger (ca. 7 Ma vs. ca. 15 Ma, respectively; see Amador et al. 2016). Fig-eating bats as a group (Stenodermatini) span a wide body size range (6–67 g; fig. 8.4; app. 8.1) but ecologically they have remained faithful to the consumption of fruits from the same core plant taxon, Moraceae, and more particularly *Ficus* (Fleming 1986, Giannini and Kalko 2004, Lobova et al. 2009; Mello et al. 2011; Saldaña-Vázquez 2014, Saldaña-Vázquez et al. 2013). This has been in order for the past 15 Ma (Fleming and Kress 2011, Rojas et al. 2011, Sánchez and Giannini, submitted). Specialization in Moraceae, especially on a wide size range of *Ficus* "fruits" (syconia), may have driven size evolution in *Artibeus sensu lato*, with small *Dermanura* species using small syconia, and large *Artibeus* species using medium to large syconia (Giannini and Kalko 2004; Kalko et al. 1996).

Emergent Patterns of Size Change and Evolutionary Implications

Stasis

Crown clade Chiroptera has been assigned an ancestral body mass of 12–14 g (or 19 g, in Safi et al. 2013), which roughly corresponds to the median value of body mass at ca. 14 g for a large sample of extant bats (Giannini et al. 2012). Just two nodes separate the root node of crown clade Chiroptera (dated 63 ± 6 Ma) from crown clade Noctilionoidea (dated 49 ± 3 Ma; Amador et al. 2016). A smaller but similarly narrow, overlapping range of 9–12 g was reconstructed here for Noctilionoidea, implying no net change of body weight over the 14 Ma elapsed across these key nodes. Another 14 Ma elapsed to the origin of Phyllostomidae (dated 35 ± 2.5 Ma; Amador et al. 2016), in the latest Eocene. These nodes occurred across the chiropteran tree, each resulting in major, superfamily or family-level cladogenetic events without any net changes in body weight. Inside Phyllostomidae, only a modest, even negligible, change of +2 g was assigned to one backbone node ("inner" phyllostomids). This stasis is remarkable for body weight considered as an evolutionary character. Stasis has been defined as the lack of evolutionary change over extended periods of geological time, and probable causes include lack of variation in the character and strong stabilizing selection maintaining the character around a central value (Futuyma 2005). In the next section, we show that variation in body weight was widespread in the phyllostomid subtree as well as in other noctilionoid groups, so stabilizing selection emerges as a plausible explanation for stasis in body weight during a period of ca. 28 Ma. Weight of most bats is concentrated in the narrow 9–16 g interval (Safi et al. 2013), which is shown here to include most backbone variation for almost the entire first half of bat evolutionary history. Echolocation constraints may provide one key mechanism for backbone stasis (Giannini et al. 2012). Frequencies vary inversely with size, so high frequencies of various call parameters required for the tasks involved in prey detection and capture, particularly in aerial hawking bats, may have kept body size within a given range (see Jones 1999).

The second half of the bat age, from latest Eocene to Recent, saw the diversification of the phyllostomid clade, which most significantly switched to a gleaning foraging mode (Schnitzler and Kalko 1998). During this time, the global climate deteriorated and habitat heterogeneity at the regional scale greatly increased, chiefly through the Andean uplift and the Amazon drainage basin remodeling (Hoorn et al. 2010a, 2010b; Jaramillo et al. 2010; Mora et al. 2010; Wesselingh et al. 2010), likely favoring speciation and creating new opportunities for specialization. Phyllostomids, in particular, but also noctilionids, greatly increased the body size space across the phylogeny. Also, more uncertainty was apparent in the shifting values of the phyllostomid backbone, and there was a slight trend to increase the range of assigned weights (figs. 8.2–8.4). However, in terms of magnitude of change, size range expansion was concentrated in specialized clades, as discussed below.

Ecology of Apomorphic Changes versus Phyletic Trends

Apomorphic changes in body size occurred in all groups, with a rather balanced frequency of increases and decreases. Some of these changes were of high magnitude, up to an observed 28 g in some *Sturnira* species (see above), which represented a twofold increase of size with respect to immediate ancestors. It is possible that these large changes are reconstructed this way because of extensive ghost lineages (extinction of intermediate branches; e.g., Simmons and Conway 2003). Most changes were small-magnitude variations in both directions (increases and decreases), which nonetheless demonstrated that variation existed to fuel the evolution of size in noctilionoids (see above). Overall, isolated apomorphic changes may reflect frequent reaccommodation of size to fit specific conditions faced by species in a changing environment (see Baker et al. 2015).

Phyletic trends, by contrast, indicate the presence of evolutionary pressure in one particular direction (increase or decrease) sustained over millions of years. These changes appeared concatenated in the phylogeny so that the size we sample for a given terminal appears as the result of adding up, or subtracting, the branch values along the series of nodes involved. For instance, weight of the largest frugivore, *Artibeus lituratus* (aver-

age mass of 67 g), resulted from a fourth-order, fourfold phyletic increase in size of this terminal with respect to the size of the genus ancestor. So in evolutionary terms (i.e., traversing the tree upwards), this terminal value emerges from adding the reconstructed changes (+2 + 6 + 15 + 14) to the upper boundary of the ancestral size (reconstructed at 16–17 g; fig. 8.4); yet some value (the missing 13 g of the summation) is "lost" to ambiguity meaning that the allocation of this amount varies with reconstructions and cannot be assigned with certainty to any of the involved nodes in all reconstructions.

These trends were distributed in several groups. Intense decreasing trends (phyletic nanism in terms of Gould and MacFadden [2004]), were reconstructed in the smallest animalivores (*Lophostoma*; fig. 8.1), nectarivores (e.g., *Lichonycteris, Lionycteris*; fig. 8.2), and frugivores (subgenus *Dermanura*; fig. 8.4). Intense increasing trends (phyletic giantism in terms of Gould and MacFadden 2004) were reconstructed in the largest animalivores (*Noctilio, Chrotopterus,* and *Vampyrum*; fig. 8.1), omnivores (*Phyllostomus hastatus* in fig. 8.1; *Brachyphylla* in fig. 8.2), and frugivores (subgenus *Artibeus* in fig. 8.4). These trends indicate size increase in different directions of the noctilionoid ecomorphospace (see below), most of which occurred in the phyllostomid clade and corresponded with dietary specialization (discussed above; see also Dumont et al. 2011).

Diverging Trends

Our analysis reconstructed 13 diverging trends of body size evolution, each mentioned above with the corresponding group. These trends indicated opposing changes in the character value of body size in the descendants. These diverging changes are most significant in the context of the broad sympatry of the species pairs or its lack thereof, because sympatry is a required condition for character displacement to be accepted as an evolutionary explanation for the divergence (see below).

Four diverging trends were recorded between allopatric descendants, so causes other than character displacement need to be invoked to account for these patterns. In two cases, the size difference between descendants likely is attributable to continent versus island, or small versus large island, effects. For instance, larger *Mormoops megalophylla* (continent) versus smaller *M. blainvillei* (Greater Antilles); smaller *Phyllonycteris aphylla* (small island; Jamaica) versus larger

P. poeyi and *P. obtusa* (large island; Cuba and Hispaniola, respectively). The other two cases are less clear geographically (*Chiroderma trinitatum* vs. *C. doriae*; *Sturnira mordax* vs. *S. koopmanhilli*) as they both occur on the continent, but they might be explained by a large versus small distribution-area effect. Still, all four cases can be attributed to purely geographic influences on body size.

The other nine species pairs with diverging trends occurred between sympatric species. Although this pattern may be the result of post-speciation dispersal, most likely the overlapping distribution reflects sympatric speciation mediated by character changes, such as the change in body size and associated characters. These cases include Furipteridae versus Noctilionidae, *Trachops* versus *Macrophyllum*, and *Gardnerycteris* versus *Phylloderma* + *Phyllostomus* (fig. 8.1), *Leptonycteris* versus *Glossophaga* (fig. 8.2), *Carollia perspicillata* versus *C. brevicauda* and *Sturnira nana* versus other *Sturnira* species (fig. 8.3), *Chiroderma* versus *Vampyriscus* and allies, *Ametrida* versus *Pygoderma*, and *Artibeus* subgenus *Dermanura* versus subgenus *Artibeus* (fig. 8.4).

Perhaps the most remarkable case of a diverging trend is that of the *Trachops-Macrophyllum* pair, which has an unusually deep time of divergence, estimated at 17 ± 2.5 Ma (Amador et al. 2016). This species pair includes the only two phyllostomids adapted to aquatic habitats in rainforest environments; in this pair, *Macrophyllum* decreased in size and specialized on the capture of aquatic insects (see Weinbeer and Kalko 2007), while its sympatric sister *Trachops* evolved in the opposite direction toward a larger size and inclusion of aquatic vertebrates in the diet (see Tuttle and Ryan 1981). The evolutionary rate in body size was considerably slower in *Trachops* (ca. 0.4 g per million years, or only 7 g of increase during the 17 Ma since the split with *Macrophyllum*) as compared with other partially or completely carnivorous lineages (see above). An explanation for the likely slow evolution of size in *Trachops* is a limitation that may exist in association with hunting small amphibian prey, unless the species entered the carnivorous guild very recently. If the rate of change of other vertebrate hunters (ca. 7 g per million years, see above) is applied to *Trachops*, then the species entered this guild during the last million years (see above).

We speculate that sympatric diverging trends may represent a present-day manifestation of past, even ancient, character displacement. At the microevolutionary timescale, character displacement involves sympatric populations of the same or closely related species differing more in a given character value, presumably affecting fitness, than populations of the same species in allopatry (Futuyma 2005). The sympatric speciation in the past may have been associated with character displacement and, once established, the difference was inherited and even accentuated in the corresponding descendants that remained in sympatry to the present. This may provide an explanation for the diverging trend in species pairs such as *Trachops cirrhosis* and *Macrophyllum macrophyllum*. These two species are broadly sympatric in the Neotropics; with the sympatry condition met, we suggest that the diverging pattern we see in body weight and hunting strategies is the result of ancient character displacement in phyllostomids associated with water. Another case is *Artibeus* subgenus *Dermanura* versus subgenus *Artibeus*, also broadly sympatric, each of which marked the beginning of extended decreasing and increasing phyletic trends, respectively, and were clearly associated with consumption of small versus large figs (see above; Kalko et al. 1996). No doubt the evolution of size divergence in *Artibeus* allowed the exploitation of the whole spectrum of chiropterochorous figs available in Neotropical rainforests by this lineage (Kalko et al. 1996). This is more remarkable considering that the presence of figs in Neotropical rainforests predated the origination of fig-eating bat clades (see review in Sánchez and Giannini, submitted).

Conclusions

Body size is crucially important for animals and more so for bats, which face the double constraints of flight and echolocation. Phylogeny explains much (up to 75%) of body size variation in noctilionoids, as in bats in general (see Safi et al. 2013). One of the most remarkable findings of this detailed (node-by-node) exploration of body size evolution was the stability of the noctilionoid backbone. Nothing relevant happened with body mass variation along roughly the first half of noctilionoid evolution; we suggest that body mass was kept nearly constant by virtue of stabilizing selection of echolocation call parameters acting as constraints on body mass. Stasis may have resulted in the fact that size per se never imposed a barrier to ecological diversification and indeed initial small size may have facilitated ecological transitions in noctilionoids. Perhaps the clearest example of this is glossophagines and lonchophyllines, which are highly specialized phyllostomids, but no body size changes were reconstructed in the transitional nodes leading to specialized nectarivory. The same argument is applicable to frugivores. However, a great potential for size change existed in the group. This potential triggered phyletic trends of high-rate body mass evolution in both directions (increases and decreases) that greatly expanded size space across the phyllostomid phylogeny. Likely, this allowed occupation of previously vacant niches along specialization gradients in carnivorous, omnivorous, and some frugivorous species (phyletic increases), and insectivorous, nectarivorous, and some frugivorous species (phyletic decreases). Finally, diverging trends that occurred in sympatry may represent the phylogenetic signal of past character displacement at a macroevolutionary scale that is apparent today in the phylogenetic patterns of size variation in noctilionoid bats.

Appendix 8.1. Body Size Data as Intraspecific Average in Grams (g) and Data Sources

Genus	Species	Body Mass (g)	Sources
Phyllostomidae			
Ametrida	centurio	10.15	Bobrowiec and Gribel 2010; dos Reis et al. 2007; Goodwin and Greenhall 1964; Jones et al. 2009; Mendes 2011; Smith et al. 2003
Anoura	caudifer	10.55	Aguirre 2007; Bárquez et al. 1993; Cruz-Neto et al. 2001; dos Reis et al. 2007; Jones et al. 2009; Mendes 2011; Ratcliffe 2009; Smith et al. 2003
A.	cultrata	17.42	Aguirre 2007; Carter et al. 1966; Hosken et al. 2002; Jones et al. 2009; Mendes 2011; Smith et al. 2003
A.	geoffroyi	15	Aguirre 2007; dos Reis et al. 2007; Hosken et al. 2002; Jones et al. 2009; Magalhães and Costa 2009; Mendes 2011; Ratcliffe 2009; Smith et al. 2003
A.	latidens	14.71	Jones et al. 2009; Mendes 2011; Smith et al. 2003; Solari et al. 1999; Soriano et al. 2002
Ardops	nichollsi	17.63	Hosken et al. 2002; Jones et al. 2009; Mendes 2011; Smith et al. 2003
Ariteus	flavescens	12.3	Genoways et al. 2005; Howe 1974; Mendes 2011
Artibeus	amplus	60.35	Jones et al. 2009; Mendes 2011; Smith et al. 2003
A.	anderseni	11.77	Aguirre 2007; Arias 2008; Jones et al. 2009; Patterson et al. 1996; Smith et al. 2003
A.	aztecus	20.75	Jones 1964; Jones et al. 2009; Mendes 2011; Smith et al. 2003
A.	bogotensis	11.5	Lim et al. 2005
A.	cinereus	12.19	Bobrowiec and Gribel 2010; dos Reis et al. 2007; Jones et al. 2009; Mendes 2011; Patterson et al. 1996; Ratcliffe 2009; Smith et al. 2003
A.	concolor	19.24	Arias 2008; Cruz-Neto et al. 2001; dos Reis et al. 2007; Jones et al. 2009; Mendes 2011; Ratcliffe 2009; Smith et al. 2003
A.	fimbriatus	60.93	Bárquez et al. 1993; Cruz-Neto et al. 2001; dos Reis et al. 2007; Jones et al. 2009; Mendes 2011; Smith et al. 2003
A.	fraterculus	25.54	Jones et al. 2009
A.	glauca	11.63	Jones et al. 2009; Mendes 2011; Ortega et al. 2015; Smith et al. 2003
A.	gnomus	10.86	Arias 2008; Bobrowiec and Gribel 2010; Lim et al. 2005; Mendes 2011
A.	hirsutus	39.36	Jones et al. 2009; Mendes 2011; Smith et al. 2003; Webster and Jones 1983
A.	incomitatus*	12.32	Albrecht et al. 2007; Chaverri and Kunz 2010; García-García et al. 2014; Kalko et al. 1996
A.	inopinatus	29.3	Webster and Jones 1983
A.	intermedius	67.65	Aguirre 2007; Arias 2008; Bárquez et al. 1993; Bobrowiec and Gribel 2010; dos Reis et al. 2007; Jones et al. 2009; Magalhães and Costa 2009; Norberg 1981; Ratcliffe 2009; Smith et al. 2003

Appendix 8.1. Continued

Genus	Species	Body Mass (g)	Sources
Phyllostomidae (cont.)			
A.	jamaicensis	42.56	Harvey et al. 2011; Jones et al. 2009; Lawlor 1973; Magalhães and Costa 2009; Mancina et al. 2004; Norberg 1981; Ratcliffe 2009; Silva Taboada 1979; Smith et al. 2003; Swartz et al. 2003
A.	lituratus	67.65	Aguirre 2007; Arias 2008; Bárquez et al. 1993; Bobrowiec and Gribel 2010; dos Reis et al. 2007; Jones et al. 2009; Magalhães and Costa 2009; Norberg 1981; Ratcliffe 2009; Smith et al. 2003
A.	obscurus	36.86	Aguirre 2007; Arias 2008; Bobrowiec and Gribel 2010; dos Reis et al. 2007; Jones et al. 2009; Mendes 2011; Smith et al. 2003
A.	phaeotis	12.45	García-García et al. 2014
A.	planirostris	53.38	Arias 2008; Beguelini et al. 2013; Bobrowiec and Gribel 2010; Lim et al. 2005
A.	schwartzi	44.45	Genoways et al. 2010
A.	toltecus	15.01	García-García et al. 2014; Jones 1964; Jones et al. 2009; Mendes 2011; Smith et al. 2003
A.	watsoni	12.32	Albrecht et al. 2007; Chaverri and Kunz 2010; García-García et al. 2014; Kalko et al. 1996
Brachyphylla	cavernarum	45.5	Gannon et al. 2005; Mendes 2011; Jones et al. 2009; Ratcliffe 2009; Smith et al. 2003; Swanepoel and Genoways 1983
B.	nana	33.44	Mancina et al. 2004; NMNH[**]; Jones et al. 2009; Smith et al. 2003
B.	pumila*	33.44	Mancina et al. 2004; NMNH[**]; Jones et al. 2009; Smith et al. 2003
Carollia	benkeithi	10.53	Arias 2008
C.	brevicauda	14.33	Arias 2008; Ávila Flores and Medellín 2004; Jones et al. 2009; Mendes 2011; Pacheco et al. 2004; Smith et al. 2003
C.	castanea	13.17	Jones et al. 2009; Mendes 2011; NMNH[**]; Ratcliffe 2009; Smith et al. 2003
C.	perspicillata	18.21	Ávila Flores and Medellín 2004; dos Reis et al. 2007; Jones et al. 2009; Lawlor 1973; Magalhães and Costa 2009; Norberg 1981; Ratcliffe 2009; Smith et al. 2003; Swartz et al. 2003
C.	sowelli	17.63	García-García et al. 2014; Voigt and Holderied 2011
C.	subrufa	16.31	García-García et al. 2014; Jones et al. 2009; Mendes 2011; Smith et al. 2003
Centurio	senex	21.56	García-García et al. 2014; Hosken et al. 2002; Jones et al. 2009; Kalko et al. 1996; Mendes 2011; Smith et al. 2003
Chiroderma	doriae	30.18	Oprea and Wilson 2008
C.	improvisum	35.13	Jones et al. 2009; Mendes 2011; Smith et al. 2003
C.	salvini	26.58	García-García et al. 2014; Jones et al. 2009; Mendes 2011; Ratcliffe 2009; Smith et al. 2003
C.	trinitatum	13.88	Arias 2008; dos Reis et al. 2007; Goodwin and Greenhall 1964; Jones et al. 2009; Mendes 2011; Ratcliffe 2009; Smith et al. 2003
C.	villosum	22.99	Arias 2008; dos Reis et al. 2007; Jones et al. 2009; Lawlor 1973; Mendes 2011; Norberg 1981; Ratcliffe 2009; Smith et al. 2003
Choeroniscus	godmani	8.63	Carter et al. 1966; Cruz-Neto et al. 2001; Jones et al. 2009; Lawlor 1973; Norberg 1981; Smith et al. 2003; Soriano et al. 2002
Choeronycteris	mexicana	16.19	Ávila Flores and Medellín 2004; Davis et al. 1964; Harvey et al. 2011; Mendes 2011; Jones et al. 2009; Smith et al. 2003
Choeroniscus	minor	9.23	Aguirre 2007; dos Reis et al. 2007; Jones et al. 2009; Mendes 2011; Patterson et al. 1996; Ratcliffe 2009; Smith et al. 2003

(continued)

Appendix 8.1. Continued

Genus	Species	Body Mass (g)	Sources
Phyllostomidae (*cont.*)			
Chrotopterus	*auritus*	80.31	Aguirre 2007; Bárquez et al. 1993; dos Reis et al. 2007; Jones et al. 2009; Kalko et al. 1996; Lawlor 1973; Mendes 2011; Norberg 1981; Smith et al. 2003
Desmodus	*rotundus*	33.46	Bárquez et al. 1993; Bobrowiec and Gribel 2010; dos Reis et al. 2007; Jones et al. 2009; Lawlor 1973; Magalhaẽs and Costa 2009; Norberg 1981; Ratcliffe 2009; Smith et al. 2003; Swartz et al. 2003
Diaemus	*youngi*	36.59	Bárquez et al. 1993; Cruz-Neto et al. 2001; dos Reis et al. 2007; Jones et al. 2009; Mendes 2011; Smith et al. 2003
Diphylla	*ecaudata*	30.05	Cruz-Neto et al. 2001; dos Reis et al. 2007; Harvey et al. 2011; Jones et al. 2009; Magalhaẽs and Costa 2009; Mendes 2011; Ratcliffe 2009; Smith et al. 2003
Ectophylla	*alba*	5.6	Chaverri and Kunz 2010; Hosken et al. 2002; Jones et al. 2009; Mendes 2011; Smith et al. 2003
Enchistenes	*hartii*	16.42	dos Reis et al. 2007; Jones et al. 2009; Mendes 2011; Ratcliffe 2009; Smith et al. 2003
Erophylla	*bombifrons*	16.19	Cruz-Neto et al. 2001; Mendes 2011
E.	*sezekorni*	16.41	Gannon et al. 2005; Harvey et al. 2011; Mancina et al. 2004; Jones et al. 2009; Mendes 2011; Silva Taboada 1979; Smith et al. 2003
Gardnerycteris	*crenulatum*	13.56	Aguirre 2007; Bobrowiec and Gribel 2010; dos Reis et al. 2007; Jones et al. 2009; Mendes 2011; Oliveira et al. 2013; Ratcliffe 2009; Smith et al. 2003
Glossophaga	*commissarisi*	9.05	dos Reis et al. 2007; García-García et al. 2014; Jones et al. 2009; Kalko et al. 1996; Mendes 2011; Patterson et al. 1996; Smith et al. 2003
G.	*leachii*	10.58	García-García et al. 2014; Jones et al. 2009; Mendes 2011; Smith et al. 2003
G.	*longirostris*	12.76	Cruz-Neto et al. 2001; dos Reis et al. 2007; Jones et al. 2009; Mendes 2011; Ratcliffe 2009; Smith et al. 2003; Soriano et al. 2002
G.	*morenoi*	8.91	García-García et al. 2014; Hosken et al. 2002; Jones et al. 2009; Mendes 2011; Smith et al. 2003
G.	*soricina*	10.01	Aguirre 2007; Ávila Flores and Medellín 2004; Bárquez et al. 1993; Bobrowiec and Gribel 2010; Cruz-Neto et al. 2001; dos Reis et al. 2007; Jones et al. 2009; Magalhaẽs and Costa 2009; Norberg 1981; Smith et al. 2003
Glyphonycteris	*daviesi*	21.29	Aguirre 2007; dos Reis et al. 2007; Gregorin and Rossi 2005; Jones et al. 2009; Mendes 2011; Smith et al. 2003; Solari et al. 1999
G.	*sylvestris*	9.44	dos Reis et al. 2007; Goodwin and Greenhall 1964; Jones et al. 2009; Mendes 2011; Smith et al. 2003
Hsunycteris	*thomasi*	7.22	Aguirre 2007; dos Reis et al. 2007; Jones et al. 2009; Louzada et al. 2015; Mendes 2011; Ratcliffe 2009; Smith et al. 2003
Hylonycteris	*underwoodi*	7.53	Gardner et al. 1970; Jones 1964; Jones et al. 2009; Mendes 2011; Smith et al. 2003
Lampronycteris	*brachyotis*	11.89	Aguirre 2007; dos Reis et al. 2007; Jones et al. 2009; Mendes 2011; Ratcliffe 2009; Smith et al. 2003; Solari et al. 1999; Tirira et al. 2010
Leptonycteris	*curasoae*	23.54	Ávila Flores and Medellín 2004; Cruz-Neto et al. 2001; Jones et al. 2009; Magalhaẽs and Costa 2009; Ratcliffe 2009; Smith et al. 2003; Swartz et al. 2003
L.	*yerbabuenae*	21.2	Harvey et al. 2011; Mendes 2011; Muijres et al. 2014
Lichonycteris	*obscura*	6.63	Carter et al. 1966; dos Reis et al. 2007; Jones et al. 2009; Lawlor 1973; Mendes 2011; Norberg 1981; Smith et al. 2003; Solari et al. 1999
Lionycteris	*spurrelli*	8.39	Aguirre 2007; dos Reis et al. 2007; Jones et al. 2009; Mendes 2011; Ratcliffe 2009; Smith et al. 2003; Woodman and Timm 2006

Appendix 8.1. Continued

Genus	Species	Body Mass (g)	Sources
Phyllostomidae (*cont.*)			
Lonchophylla	*chocoana*	21	Dávalos 2004; Woodman and Timm 2006
L.	*dekeyseri*	9	dos Reis et al. 2007
L.	*handleyi*	17.33	Dávalos 2004; Solari et al. 1999; Woodman and Timm 2006
L.	*hesperia*	...	...
L.	*mordax*	9.76	dos Reis et al. 2007; Jones et al. 2009; Ratcliffe 2009; Smith et al. 2003; Woodman 2007; Woodman and Timm 2006
L.	*robusta*	13.96	Goodman and Timm 2006; Jones et al. 2009; Mendes 2011; Smith et al. 2003; Solari et al. 1999; Tschapka et al. 2015
Lonchorhina	*aurita*	15.33	dos Reis et al. 2007; Jones et al. 2009; Louzada et al. 2015; Mendes 2011; Smith et al. 2003; Solari et al. 1999
L.	*inusitata*	15.25	dos Reis et al. 2007
L.	*orinocensis*	8.88	Jones et al. 2009; Mendes 2011; Smith et al. 2003
Lophostoma	*brasiliense*	10.18	Aguirre 2007; dos Reis et al. 2007; Jones et al. 2009; Mendes 2011; Patterson et al. 1996; Smith et al. 2003
L.	*carrikeri*	21.72	Aguirre 2007; Jones et al. 2009; Lim et al. 1999; Mendes 2011; Smith et al. 2003; Zortéa et al. 2009
L.	*evotis*	20.05	Jones et al. 2009; Medellín and Arita 1989; Mendes 2011; Smith et al. 2003
L.	*schulzi*	18.31	dos Reis et al. 2007; Jones et al. 2009; Mendes 2011; Smith et al. 2003; Zanella 2013
L.	*silvicolum centralis*	31.06	Aguirre 2007; Bobrowiec and Gribel 2010; dos Reis et al. 2007; Jones et al. 2009; Kalko et al. 1996; Medellín and Arita 1989; Mendes 2011; Smith et al. 2003
L.	*silvicolum laephotis*	31.06	Aguirre 2007; Bobrowiec and Gribel 2010; dos Reis et al. 2007; Jones et al. 2009; Kalko et al. 1996; Medellín and Arita 1989; Mendes 2011; Smith et al. 2003
Macrophyllum	*macrophyllum*	8.03	Aguirre 2007; Bárquez et al. 1993; dos Reis et al. 2007; Jones et al. 2009; Kalko et al. 1996; Mendes 2011; Ratcliffe 2009; Smith et al. 2003
Macrotus	*californicus*	12.13	Cruz-Neto et al. 2001; Harvey et al. 2011; Magalhães and Costa 2009; Jones et al. 2009; Mendes 2011; Smith et al. 2003
M.	*mexicanus**	15.18	Ávila Flores and Medellín 2004; Jones et al. 2009; Klingener et al. 1978; Magalhães and Costa 2009; Mancina et al. 2004; Mendes 2011; Smith et al. 2003
M.	*waterhousii*	15.18	Ávila Flores and Medellín 2004; Jones et al. 2009; Klingener et al. 1978; Magalhães and Costa 2009; Mancina et al. 2004; Mendes 2011; Smith et al. 2003
Mesophylla	*macconnelli*	7.24	Aguirre 2007; Arias 2008; Bobrowiec and Gribel 2010; dos Reis et al. 2007; Jones et al. 2009; Mendes 2011; Ratcliffe 2009; Smith et al. 2003
Micronycteris	*brosseti*	4.72	dos Reis et al. 2007; Lim et al. 1999; Porter et al. 2007
M.	*giovanniae*	...	...
M.	*hirsuta*	13.21	dos Reis et al. 2007; Gardner et al. 1970; Jones et al. 2009; Kalko et al. 1996; Mendes 2011; Porter et al. 2007; Smith et al. 2003
M.	*homezi*	6.38	dos Reis et al. 2007; Lim and Engstrom 2001; Ochoa and Sánchez 2005
M.	*matses*	9.05	Bernard 2001; Simmons et al. 2002
M.	*megalotis*	6.14	Aguirre 2007; Ávila Flores and Medellín 2004; Bobrowiec and Gribel. 2009; dos Reis et al. 2007; Jones et al. 2009; Moras et al. 2015; Ratcliffe 2009; Smith et al. 2003

(continued)

Appendix 8.1. Continued

Genus	Species	Body Mass (g)	Sources
Phyllostomidae *(cont.)*			
M.	microtis	7.91	dos Reis et al. 2007; Lim et al. 1999; Louzada et al. 2015; Moras et al. 2015
M.	minuta	6.9	Anderson et al. 1982; dos Reis et al. 2007; Jones et al. 2009; Mendes 2011; Ratcliffe 2009; Smith et al. 2003
M.	schmidtorum	7.63	dos Reis et al. 2007; Jones et al. 2009; Louzada et al. 2015; Mendes 2011; Ratcliffe 2009; Smith et al. 2003
Mimon	bennettii	23	Esberárd 2009; Crarvalho et al. 2008
M.	cozumelae	23.35	Carter et al. 1966; Fenton et al. 2001; Ortega and Arita 1997; Valdez and LaVal 1971
Monophyllus	plethodon	15.27	Hosken et al. 2002; Jones et al. 2009; Mendes 2011; Pedersen et al. 1996; Ratcliffe 2009; Smith et al. 2003
M.	redmani	9.16	Cruz-Neto et al. 2001; Gannon et al. 2005; Jones et al. 2009; Mancina et al. 2004; Mendes 2011; Silva Taboada 1979; Smith et al. 2003; Soriano et al. 2002
Musonycteris	harrisoni	11.68	Mendes 2011; Tschapka et al. 2008
Phylloderma	stenops	53.84	Aguirre 2007; Bobrowiec and Gribel 2010; dos Reis et al. 2007; Jones et al. 2009; Kalko et al. 1996; Mendes 2011; Ratcliffe 2009; Smith et al. 2003
Phyllonycteris	aphylla	14.28	Howe 1974; Mendes 2011
P.	obtusa	17.96	NMNH**
P.	poeyi	21.43	Mancina et al. 2004; Silva Taboada 1979
Phyllops	falcatus	19.5	da Cunha et al. 2008; Harvey et al. 2011; Mancina et al. 2004; Mendes 2011
Phyllostomus	discolor	37.79	Aguirre 2007; Cruz-Neto et al. 2001; dos Reis et al. 2007; García-García et al. 2014; Jones et al. 2009; Lawlor 1973; Magalhaẽs and Costa 2009; Norberg 1981; Ratcliffe 2009; Smith et al. 2003
P.	elongatus	40.37	Aguirre 2007; Bobrowiec and Gribel 2010; Cruz-Neto et al. 2001; dos Reis et al. 2007; Jones et al. 2009; Mendes 2011; Ratcliffe 2009; Smith et al. 2003
P.	hastatus	93.52	Cruz-Neto et al. 2001; dos Reis et al. 2007; Jones et al. 2009; Lawlor 1973; Magalhaẽs and Costa 2009; Norberg 1981; Ratcliffe 2009; Smith et al. 2003
P.	latifolius	28.25	dos Reis et al. 2007; Lim et al. 2005
Platalina	genovensium	17.8	Woodman and Timm 2006; Zamora Meza et al. 2013
Platyrrhinus	albericoi	61.5	Velazco 2005
P.	aurarius	34.51	Jones et al. 2009; Mendes 2011; Smith et al. 2003
P.	brachycephalus	14.05	Arias 2008; dos Reis et al. 2007; Jones et al. 2009; Louzada et al. 2015; Mendes 2011; Ratcliffe 2009; Smith et al. 2003
P.	dorsalis	26	Jones et al. 2009; Mendes 2011; Smith et al. 2003
P.	helleri	13.66	Arias 2008; Bobrowiec and Gribel 2010; dos Reis et al. 2007; Jones et al. 2009; Ratcliffe 2009; Smith et al. 2003; Velazco 2005
P.	incarum	13.58	NMNH**
P.	infuscus	48.07	Arias 2008; Jones et al. 2009; Mendes 2011; Ratcliffe 2009; Smith et al. 2003
P.	ismaeli	35.34	Rengifo et al. 2011; Velazco 2005
P.	lineatus	23.81	Bárquez et al. 1993; Cruz-Neto et al. 2001; dos Reis et al. 2007; Jones et al. 2009; Magalhaẽs and Costa 2009; Mendes 2011; Ratcliffe 2009; Smith et al. 2003; Soriano et al. 2002

Appendix 8.1. Continued

Genus	Species	Body Mass (g)	Sources
Phyllostomidae (*cont.*)			
P.	*masu*	27	Velazco 2005
P.	*matapalensis*	16	Velazco 2005
P.	*nigellus*	23.05	Aguirre 2007; Velazco 2005
P.	*recifinus*	17.7	Velazco 2005
P.	*vittatus*	62.3	Velazco 2005
Pygoderma	*bilabiatum*	19.3	Aguirre 2007; Bárquez et al. 1993; Jones et al. 2009; Mendes 2011; Smith et al. 2003
Rhinophylla	*alethina*	6.91	Jones et al. 2009
R.	*fischerae*	8.43	Arias 2008; dos Reis et al. 2007; Mendes 2011
R.	*pumilio*	9.49	Arias 2008; Cruz-Neto et al. 2001; dos Reis et al. 2007; Jones et al. 2009; Mendes 2011; Ratcliffe 2009; Smith et al. 2003
Sphaeronyc-teris	*toxophyllum*	16.28	dos Reis et al. 2007; Jones et al. 2009; Mendes 2011; Patterson et al. 1996; Ratcliffe 2009; Smith et al. 2003
Stenoderma	*rufum*	22	Gannon et al. 2005; Jones et al. 2009; Mendes 2011; Smith et al. 2003
Sturnira	*angeli*	21.1	Vaughan Jennings et al. 2004
S.	*aratathomasi*	48.96	Jones et al. 2009; Mendes 2011; Smith et al. 2003
S.	*bakeri*	19.9	Velazco and Patterson 2014
S.	*bidens*	17.44	Hosker et al. 2002; Jones et al. 2009; Mendes 2011; NMNH**; Smith et al. 2003
S.	*bogotensis*	20.51	Jones et al. 2009; Mendes 2011; Smith et al. 2003
S.	*burtonlimi*	19	Velazco and Patterson 2014
S.	*erythromos*	15.7	Aguirre 2007; Bárquez et al. 1993; Giannini and Bárquez 2003; Jones et al. 2009; Lee et al. 2008; Mendes 2011; Smith et al. 2003; Soriano et al. 2002
S.	*hondurensis*	17.2	NMNH**
S.	*koopmanhilli*	27.75	McCarthy et al. 2006
S.	*lilium*	21.17	Aguirre 2007; Arias 2008; Bobrowiec and Gribel 2010; dos Reis et al. 2007; Magalhães and Costa 2009; NMNH**; Ratcliffe 2009
S.	*ludovici*	22.43	García-García et al. 2014; NMNH**; Jones et al. 2009; Ratcliffe 2009; Smith et al. 2003
S.	*luisi*	11.99	Jones et al. 2009*
S.	*magna*	48.29	Aguirre 2007; Arias 2008; dos Reis et al. 2007; Mendes 2011
S.	*mordax*	11.79	Mendes 2011
S.	*nana*	6.64	Jones et al. 2009*
S.	*oporaphilum*	21.37	Aguirre 2007; Bárquez et al. 1993; Lee et al. 2008; Mendes 2011
S.	*parvidens*	14.85	Genoways and Timm 2005
S.	*paulsoni*	19.4	Vaughan Jennings et al. 2004
S.	*perla*	...	...
S.	*spnova3*	...	...
S.	*tildae*	24.13	Aguirre 2007; Arias 2008; Cruz-Neto et al. 2001; Jones et al. 2009; Mendes 2011; Ratcliffe 2009; Smith et al. 2003
Tonatia	*bidens*	29.29	Bárquez et al. 1993; Cruz-Neto et al. 2001; dos Reis et al. 2007; Jones et al. 2009; Mendes 2011; Patterson et al. 1996; Ratcliffe 2009; Smith et al. 2003

(continued)

Appendix 8.1. Continued

Genus	Species	Body Mass (g)	Sources
Phyllostomidae (*cont.*)			
Trachops	*cirrhosus*	35.63	Aguirre 2007; Bobrowiec and Gribel 2010; Carter et al. 1966; dos Reis et al. 2007; Jones et al. 2009; Mendes 2011; Ratcliffe 2009; Smith et al. 2003
Trinycteris	*nicefori*	8.95	Jones et al. 2009; Kalko et al. 1996; Mendes 2011; Smith et al. 2003; Zanella 2013
Uroderma	*bilobatum*	17.08	Arias 2008; Bobrowiec and Gribel 2010; Cruz-Neto et al. 2001; dos Reis et al. 2007; Jones et al. 2009; Norberg 1981; Ratcliffe 2009; Smith et al. 2003
Uroderma	*magnirostrum*	17.71	Arias 2008; dos Reis et al. 2007; Jones et al. 2009; Mendes 2011; Patterson et al. 1996; Smith et al. 2003
Vampyressa	*melissa*	16.39	Jones et al. 2009; Mendes 2011; Smith et al. 2003
V.	*pusilla*	9.08	Bárquez et al. 1993; Cruz-Neto et al. 2001; dos Reis et al. 2007; Jones et al. 2009; Lim and Engstrom 2001; Mendes 2011; Ratcliffe 2009; Smith et al. 2003
V.	*thyone*	7.3	Aguirre 2007; Arias 2008; Mendes 2011; Parker-Shames and Rodriguez-Herrera 2013
Vampyriscus	*bidens*	11.89	Aguirre 2007; Arias 2008; Bobrowiec and Gribel 2010; dos Reis et al. 2007; Jones et al. 2009; Mendes 2011; Patterson et al. 1996; Smith et al. 2003
V.	*brocki*	8.57	Arias 2008; dos Reis et al. 2007; Lim et al. 2005
V.	*nymphaea*	13.9	Gardner et al. 1970; Kalko et al. 1996; Patterson et al. 1996
Vampyrodes	*caraccioli*	35.12	Aguirre 2007; Arias 2008; dos Reis et al. 2007; Jones et al. 2009; Lawlor 1973; Mendes 2011; Norberg 1981; Ratcliffe 2009; Smith et al. 2003
V.	*major*	28.04	NMNH[**]
Vampyrum	*spectrum*	177.29	Aguirre 2007; dos Reis et al. 2007; Hosker et al. 2002; Jones et al. 2009; Mendes 2011; Ratcliffe 2009; Smith et al. 2003
Furipteridae			
Furipterus	*horrens*	3.14	dos Reis et al. 2007; Jones et al. 2009; Mendes 2011; Smith et al. 2003
Mormoopidae			
Mormoops	*blainvillii*	8.60	Gannon et al. 2005; Jones et al. 2009; Mancina et al. 2004; Mendes 2011; Smith et al. 2003
M.	*megalophylla*	16.12	Harvey et al. 2011; Jones et al. 2009; Mendes 2011; Smith et al. 2003; Soriano et al. 2002
Pteronotus	*davyi*	9.35	dos Reis et al. 2007; Jones et al. 2009; Mendes 2011; Smith et al. 2003; Soriano et al. 2002
P.	*fulvus*	9.35	dos Reis et al. 2007; Jones et al. 2009; Mendes 2011; Smith et al. 2003; Soriano et al. 2002
P.	*gymnonotus*	14.00	Aguirre 2007; dos Reis et al. 2007; Jones et al. 2009; Mendes 2011; Smith et al. 2003
P.	*macleayii*	5.21	Mancina et al. 2004; Mendes 2011
P.	*parnellii*	11.65	Genoways et al. 2005; NMNH[**]
P.	*personatus*	8.08	Aguirre 2007; dos Reis et al. 2007; Jones et al. 2009; Mendes 2011; Smith et al. 2003
P.	*portoricensis*	11.30	Vaughan Jennings et al. 2004
P.	*psilotis*	. . .	. . .
P.	*pusillus*	10.35	NMNH[**]
P.	*quadridens*	5.38	Gannon et al. 2005; Jones et al. 2009; Mancina et al. 2004; Mendes 2011; Smith et al. 2003

Appendix 8.1. Continued

Genus	Species	Body Mass (g)	Sources
Mormoopidae (*cont.*)			
P.	*rubiginosusMex*	23.78	Gutiérrez and Molinari 2008; NMNH[**]
P.	*rubiginosusSur*	23.78	Gutiérrez and Molinari 2008; NMNH[**]
P.	*rubiginosusFG*	23.78	Gutiérrez and Molinari 2008; NMNH[**]
Mystacinidae			
Mystacina	*tuberculata*	12.57	Daniel 1979; Jones et al. 2009; Mendes 2011; Smith et al. 2003
Myzopodidae			
Myzopoda	*aurita*	9.18	Garbutt 2007; Mendes 2011
Noctilionidae			
Noctilio	*albiventris*	30.03	Aguirre 2007; Bárquez et al. 1993; dos Reis et al. 2007; Jones et al. 2009; Lawlor 1973; Mendes 2011; Norberg 1981; Smith et al. 2003
N.	*leporinus*	60.49	Aguirre 2007; Bárquez et al. 1993; dos Reis et al. 2007; Gannon et al. 2005; Jones et al. 2009; Lawlor 1973; Magalhães and Costa 2009; Mendes 2011; Norberg 1981; Smith et al. 2003
Thyropteridae			
Thyroptera	*discifera*	3.43	Aguirre 2007; Mendes 2011; Pacheco et al. 2007; Jones et al. 2009; Smith et al. 2003
T.	*lavali*	6.00	Díaz 2011; Solari et al. 1999
T.	*tricolor*	3.88	Aguirre 2007; dos Reis et al. 2007; Lawlor 1973; Norberg 1981; Jones et al. 2009; Smith et al. 2003

[*] These terminals as such lack body size data but are considered as junior synonyms in Simmons (2005); therefore we assigned them the same body size as the valid species in Simmons (2005).

[**] NMNH: National Museum of Natural History, Washington, DC.

Acknowledgments

We thank Theodore Fleming, Liliana Dávalos, and Marco Mello for their kind invitation to contribute this chapter. Unidad Ejecutora Lillo (CONICET-FML, Tucumán, Argentina) provided working space and significant resources. We thank the American Museum of Natural History for granting access to off-site electronic and other resources. Two anonymous reviewers provided insightful comments and edits that greatly contributed to the final version of this chapter. We thank grant FONCYT PICT-2016–2389 for support.

References

Aguirre, L. F., ed. 2007. Historia Natural, Distribución y Conservación de los Murciélagos de Bolivia. Editorial: Centro de Ecología y Difusión Simón I. Patiño, Santa Cruz, Bolivia.

Albrecht, L., C. F. Meyer, and E. K. Kalko. 2007. Differential mobility in two small phyllostomid bats, *Artibeus watsoni* and *Micronycteris microtis*, in a fragmented Neotropical landscape. Acta Theriologica, 52 (2): 141–149.

Amador, L. I., and N. P. Giannini. 2016. Phylogeny and evolution of body mass in didelphid marsupials (Marsupialia: Didelphimorphia: Didelphidae). Organisms Diversity and Evolution, 16:641–657.

Amador, L. I., R. L. Moyers Arévalo, F. C. Almeida, S. A. Catalano, and N. P. Giannini. 2016. Bat systematics in the light of unconstrained analyses of a comprehensive molecular supermatrix. Journal of Mammalian Evolution, 1–34.

Anderson, S., G. K. Creighton, and K. F. Koopman. 1982. Bats of Bolivia: an annotated checklist. American Museum Novitates, no. 2750. American Museum of Natural History, New York.

Arias, L. C. 2008. Ecomorphological structure of an Amazonian phyllostomid bat assemblage. Master's thesis, Texas Tech University, Lubbock.

Avila-Flores, R., and R. A. Medellín. 2004. Ecological, taxonomic, and physiological correlates of cave use by Mexican bats. Journal of Mammalogy, 85 (4): 675–687.

Baker, R. J., O. R. Bininda-Emonds, H. Mantilla-Meluk, C. A. Porter, and R. A. Van Den Bussche. 2012. Molecular timescale of diversification of feeding strategy and morphology in New World leaf-nosed bats (Phyllostomidae): a phylogenetic perspective. Pp: 385–409 *in*: Evolutionary History of Bats: Fossils, Molecules and Morphology (G. F. Gunnell and N. B. Simmons, eds.). Cambridge University Press, Cambridge.

Baker, J., A. Meade, M. Pagel, and C. Venditti. 2015. Adaptive evolution toward larger size in mammals. Proceedings of the National Academy of Science 112 (16): 5093–5098.

Bárquez, R. M., N. P. Giannini, and M. A. Mares. 1993. Guide to the Bats of Argentina. Oklahoma Museum of Natural History, University of Oklahoma, Norman.

Beguelini, M. R., C. C. Puga, S. R. Taboga, and E. Morielle-Versute. 2013. Annual reproductive cycle of males of the flat-faced fruit-eating bat, *Artibeus planirostris* (Chiroptera: Phyllostomidae). General and Comparative Endocrinology, 185:80–89.

Bernard, E. 2001. Species list of bats (Mammalia, Chiroptera) of Santarém area, Pará state, Brazil. Revista brasileira de Zoologia, 18 (2): 455–463.

Bobrowiec, P. E. D., and R. Gribel. 2010. Effects of different secondary vegetation types on bat community composition in Central Amazonia, Brazil. Animal Conservation, 13 (2): 204–216.

Brinkløv, S., E. K. Kalko, and A. Surlykke. 2009. Intense echolocation calls from two "whispering" bats, *Artibeus jamaicensis* and *Macrophyllum macrophyllum* (Phyllostomidae). Journal of Experimental Biology, 212 (1): 11–20.

Brooke, A. P. 1990. Tent selection, roosting ecology and social organization of the tent-making bat, *Ectophylla alba*, in Costa Rica. Journal of Zoology (London), 221:11–19.

Bullen, R. D., and N. L. McKenzie. 2002. Scaling bat wingbeat frequency and amplitude. Journal of Experimental Biology, 205 (17): 2615–2626.

Calder, W. A. 1996. Size, Function, and Life History. Dover Publications, New York.

Carter, D. C., R. H. Pine, and W. B. Davis. 1966. Notes on Middle American bats. Southwestern Naturalist, 11 (4): 488–499.

Carvalho, F., A. P. da Cruz-Neto, and J. J. Zocche. 2008. Ampliação da distribuição e descrição da dieta de *Mimon bennettii* (Phyllostomidae, Phyllostominae) no sul do Brasil. Chiroptera Neotropical, 14 (2): 403–408.

Chaverri, G., and T. H. Kunz. 2010. Ecological determinants of social systems: perspectives on the functional role of roosting ecology in the social behavior of tent-roosting bats. Advances in the Study of Behavior, 42:275–318.

Cruz-Neto, A. P., T. Garland, and A. S. Abe. 2001. Diet, phylogeny, and basal metabolic rate in phyllostomid bats. Zoology, 104 (1): 49–58.

Czaplewksi, N. J., and C. Cartelle. 1998. Pleistocene bats from cave deposits in Bahia, Brazil. Journal of Mammalogy, 79:784–803.

da Cunha Tavares, V., and C. A. Mancina. 2008. *Phyllops falcatus* (Chiroptera: Phyllostomidae). Mammalian Species, 811:1–7.

Daniel, M. J. 1979. The New Zealand short-tailed bat, *Mystacina tuberculata*; a review of present knowledge. New Zealand Journal of Zoology, 6 (2): 357–370.

Dávalos, L. M. 2004. A new chocoan species of *Lonchophylla* (Chiroptera: Phyllostomidae). American Museum Novitates, no. 3426. American Museum of Natural History, New York.

Davis, W. B., D. C. Carter, and R. H. Pine. 1964. Noteworthy records of Mexican and Central American bats. Journal of Mammalogy, 45 (3): 375–387.

Díaz, M. M. 2011. New records of bats from the northern region of the Peruvian Amazon. Zoological Research, 32 (2): 168–178.

dos Reis, N. R., A. L. Peracchi, W. A. Pedro, and I. P. de Lima (Eds.) 2007. Morcegos do Brasil. [Ediçâo dos Editores], Londrina.

Dumont, E. R., L. M. Dávalos, A. Goldberg, S. E. Santana, K. Rex, and C. C. Voigt. 2011. Morphological innovation, diversification and invasion of a new adaptive zone. Proceedings of the Royal Society of London B–Biological Sciences, 279:1797–1805.

Esberárd, C. E. L. 2009. Observações preliminares sobre a atração intra-específica de fêmeas por jovens morcegos. Chiroptera Neotropical, 15 (2): 466–468.

Fenton, M. B., E. Bernard, S. Bouchard, L. Hollis, D. S. Johnston, C. L. Lausen, J. M. Ratcliffe, D. K. Riskin, J. R. Taylor, and J. Zigouris. 2001. The bat fauna of Lamanai, Belize: roosts and trophic roles. Journal of Tropical Ecology, 17 (4): 511–524.

Fleming, T. H. 1986. Opportunism versus specialization: the evolution of feeding strategies in frugivorous bats. Pp. 105–118 *in*: Frugivores and Seed Dispersal (A. Estrada and T. H. Fleming, eds.). Dr. W Junk, Dordrecht.

Fleming, T. H. 1988. The Short-Tailed Fruit Bat: A Study in Plant-Animal Interactions. University of Chicago Press, Chicago.

Fleming, T. H., and P. Eby. 2003. Ecology of bat migration. Pp. 156–208 *in*: Bat Ecology (T. H. Kunz and M. B. Fenton, eds.). University of Chicago Press, Chicago.

Fleming, T. H., and W. J. Kress. 2011. A brief history of fruits and frugivores. Acta Oecologica, 37 (6): 521–530.

Futuyma, D. J. 2005. Evolution. 3rd ed. Sinauer Associates, Sunderland, MA.

Gannon, M. R., A. Kurta, A. Rodríguez-Durán, and M. R. Willig. 2005. Bats of Puerto Rico: An Island Focus and a Caribbean Perspective. Texas Tech University Press, Lubbock.

Garbutt, N. 2007. Mammals of Madagascar: A Complete Guide. Yale University Press, New Haven, CT.

García-García, J. L., and A. Santos-Moreno. 2014. Efectos de la estructura del paisaje y de la vegetación en la diversidad de murciélagos filostómidos (Chiroptera: Phyllostomidae) en Oaxaca, México. Revista de Biología Tropical, 62 (1): 217–239.

Gardner, A. L., R. K. LaVal, and D. E. Wilson. 1970. The distributional status of some Costa Rican bats. Journal of Mammalogy, 51 (4): 712–729.

Genoways, H. H., J. W. Bickham, R. J. Baker, and C. J. Phillips. 2005. Bats of Jamaica. Special Publications/Museum of Texas Tech University, no. 48. Museum of Texas Tech University, Lubbock.

Genoways, H. H., G. G. Kwiecinski, P. A. Larsen, S. C. Pedersen, R. J. Larsen, J. D. Hoffman, M. de Silva, C. J. Phillips, and R. J. Baker. 2010. Bats of the Grenadine Islands, West Indies, and placement of Koopman's line. Chiroptera Neotropical, 16 (1): 529–549.

Giannini, N. P. 2003. Canonical Phylogenetic Ordination. Systematic Biology, 52 (5): 684–695.

Giannini, N. P., and R. M. Barquez. 2009. *Sturnira erythromos.* Mammalian Species, 729:1–5.

Giannini, N. P., G. F. Gunnell, J. Habersetzer, and N. B. Simmons. 2012. Early evolution of body size in bats. Pp: 530–555 *in:* Evolutionary History of Bats: Fossils, Molecules, and Morphology (G. F. Gunnell and N. B. Simmons, eds.). Cambridge University Press. Cambridge.

Giannini, N. P., and E. K. Kalko. 2004. Trophic structure in a large assemblage of phyllostomid bats in Panama. Oikos, 105 (2): 209–220.

Giannini, N. P., and E. K. Kalko. 2005. The guild structure of animalivorous leaf-nosed bats of Barro Colorado Island, Panama, revisited. Acta Chiropterologica, 7 (1): 131–146.

Goloboff, P. A., J. S. Farris, and K. C. Nixon. 2008. TNT, a free program for phylogenetic analysis. Cladistics, 24 (5): 774–786.

Goloboff, P. A., C. I. Mattoni, and A. S. Quinteros. 2006. Continuous characters analyzed as such. Cladistics, 22 (6): 589–601.

Goodman, S. M., F. Rakotondraparany, and A. Kofoky. 2006. The description of a new species of *Myzopoda* (Myzopodidae: Chiroptera) from western Madagascar. Mammalian Biology–Zeitschrift für Säugetierkunde, 72 (2): 65–81.

Goodwin, G. G., and A. M. Greenhall. 1964. New records of bats from Trinidad and comments on the status of *Molossus trinitatus* Goodwin. American Museum Novitates, no. 2195. American Museum of Natural History, New York.

Gould, G. C., and B. J. MacFadden. 2004. Gigantism, dwarfism, and Cope's rule: nothing in evolution makes sense without a phylogeny. Bulletin of the American Museum of Natural History, 285:219–237.

Gregorin, R., and R. V. Rossi. 2005. *Glyphonycteris daviesi* (Hill, 1964), a rare Central American and Amazonian bat recorded for Eastern Brazilian Atlantic Forest (Chiroptera, Phyllostomidae). Mammalia, 69 (3–4): 427–430.

Gutiérrez, E. E., and J. Molinari. 2008. Morphometrics and taxonomy of bats of the genus *Pteronotus* (subgenus *Phyllodia*) in Venezuela. Journal of Mammalogy, 89 (2): 292–305.

Harvey, M. J., J. S. Altenbach, and T. L. Best. 2011. Bats of the United States and Canada. Johns Hopkins University Press, Baltimore.

Hoorn, C., F. P. Wesseling, J. Hovikoski, and J. Guerrero. 2010a. The development of the Amazonian mega-wetland (Miocene; Brazil, Colombia, Peru, Bolivia). Pp. 123–142 *in:* Amazonia: Landscape and Species Evolution: A Look into the Past (C. Hoorn and F. P. Wesselingh, eds.). Wiley-Blackwell Publishing Ltd., Oxford.

Hoorn, C., F. P. Wesselingh, H. ter Steege, M. A. Bermudez, A. Mora, J. Sevink, I. Sanmartín et al. 2010b. Amazonia through time: Andean uplift, climate change, landscape evolution, and biodiversity. Science 330:927–931. DOI: 10.1126/science.1194585

Hosken, D., K. Jones, K. Chipperfield, and A. Dixson. 2002. Is the bat os penis sexually selected? Behavioral Ecology and Sociobiology, 51 (3): 302–307.

Howe, H. F. 1974. Additional records of *Phyllonycteris aphylla* and *Ariteus flavescens* from Jamaica. Journal of Mammalogy, 55 (3): 662–663.

Jaramillo, C., C. Hoorn, S. A. F. Silva, F. Leite, F. Herrera, L. Quiroz, R. Dino, and L. Antonioli. 2010. The origin of the modern Amazon rainforest: implications of the palynological and palaeobotanical record. Pp. 317–334 *in:* Amazonia: Landscape and Species Evolution: A Look into the Past (C. Hoorn and F. P. Wesselingh, eds.). Wiley-Blackwell Publishing Ltd., Oxford.

Jones, G. 1999. Scaling of echolocation call parameters in bats. Journal of Experimental Biology, 202 (23): 3359–3367.

Jones, J. K. 1964. Bats from western and southern Mexico. Transactions of the Kansas Academy of Science (1903–), 67 (3): 509–516.

Jones, K. E., J. Bielby, M. Cardillo, S. A. Fritz, J. O'Dell, C. D. L. Orme, K. Safi et al. 2009. PanTHERIA: a species-level database of life history, ecology, and geography of extant and recently extinct mammals. Ecology, 90:2648.

Kalko, E. K., E. A. Herre, and C. O. Handley. 1996. Relation of fig fruit characteristics to fruit-eating bats in the New and Old World tropics. Journal of Biogeography, 23 (4): 565–576.

Klingener, D., H. H. Genoways, and R. J. Baker. 1978. Bats from Southern Haiti. Annals of Carnegie Museum, vol. 47, article 5. Carnegie Museum of Natural History, Pittsburgh.

Lawlor, T. E. 1973. Aerodynamic characteristics of some Neotropical bats. Journal of Mammalogy, 54 (1): 71–78.

Lee, T. E., S. F. Burneo, M. R. Marchán, S. A. Roussos, and R. S. Vizcarra-Vásconez. 2008. The mammals of the temperate forests of Volcán Sumaco, Ecuador. Occasional papers/Museum of Texas Tech University, no. 276. Museum of Texas Tech University, Lubbock.

Lim, B. K., and M. D. Engstrom. 2001. Species diversity of bats (Mammalia: Chiroptera) in Iwokrama Forest, Guyana, and the Guianan subregion: implications for conservation. Biodiversity and Conservation, 10 (4): 613–657.

Lim, B. K., M. D. Engstrom, H. H. Genoways, F. M. Catzeflis, K. A. Fitzgerald, S. L. Peters, M. Djosetro, S. Brandon, and S. Mitro. 2005. Results of the Alcoa Foundation–Suriname expeditions. XIV. Mammals of Brownsberg Nature Park, Suriname. Annals of Carnegie Museum, 74 (4): 225–274.

Lloyd, B. D. 2001. Advances in New Zealand mammalogy 1990–2000: short-tailed bats. Journal of the Royal Society of New Zealand, 31 (1): 59–81.

Lobova, T. A., C. K. Geiselman, S. A. Mori. 2009. Seed Dispersal by Bats in the Neotropics. Memoir of the New York Botanical Garden, vol. 101. New York Botanical Garden, Bronx, NY.

Louzada, N. S. V., A. C. do Monte Lima, L. M. Pessôa, J. L. P. Cordeiro, and L. F. B. Oliveira. 2015. New records of phyllostomid bats for the state of Mato Grosso and for the Cerrado of Midwestern Brazil (Mammalia: Chiroptera). Check List, 11 (3): 1644.

Magalhaēs, J.P. and J. Costa. 2009. A database of vertebrate longevity records and their relation to other life-history traits. Journal of Evolutionary Biology, 22:1770–1774.

Mancina, C. A., R. Borroto-Páez, and L. García-Rivera. 2004. Tamaño relativo del cerebro en murciélagos cubanos. Orsis: organismes i sistemes, 19: 007-19.

McCarthy, T. J., V. L. Albuja, and M. S. Alberico. 2006. A new species of chocoan *Sturnira* (Chiroptera: Phyllostomidae: Stenodermatinae) from western Ecuador and Colombia. Annals of Carnegie Museum, 75 (2): 97–110.

McNab, B. K. 2007. The evolution of energetics in birds and mammals. Pp. 67–110 *in*: The Quintessential Naturalist: Honoring the Life and Legacy of Oliver P. Pearson (D. A. Kelt, E. P. Lessa, J. Salazar-Bravo, and J. L. Patton, eds.). University of California Publications in Zoology, vol. 134. University of California Press, Berkeley.

Medellín, R. A., and H. T. Arita. 1989. *Tonatia evotis* and *Tonatia silvicola*. Mammalian Species Archive, 334:1–5.

Mello, M. A. R., F. M. D. Marquitti, P. R. Guimaraes Jr., E. K. V. Kalko, P. Jordano, and M. A. Martinez de Aguiar. 2011. The missing part of seed dispersal networks: structure and robustness of bat-fruit interactions. PLOS ONE 6:e17395 DOI: 10.1371/journal.pone.001739

Mendes, P. 2011. Prioridades globais para a aconservação e características biológicas associadas ao risco de extinção em morcegos (Chiroptera: Mammalia). Master's thesis, Universidade Federal de Goiás, Goiânia.

Mora, A., P. Baby, M. Roddaz, M. Parra, S. Brusset, W. Hermoza, and N. Espurt. 2010. Tectonic history of the Andes and sub-Andean zones: implications for the development of the Amazon drainage basin. Pp. 38–60 *in*: Amazonia: Landscape and Species Evolution: A Look into the Past (C. Hoorn and F. P. Wesselingh, eds.). Wiley-Blackwell Publishing Ltd., Oxford.

Moras, L. M., and V. da Cunha Tavares. 2015. Distribution and taxonomy of the common big-eared bat *Micronycteris microtis* (Chiroptera: Phyllostomidae) in South America. Mammalia, 79 (4): 439–447.

Morgan, G. S. 1988. New species of fossil vampire bats (Mammalia, Chiroptera, Desmodontidae) from Florida and Venzuela. Proceedings of the Biological Society of Washington, 101: 912–928.

Muijres, F. T., L. C. Johansson, Y. Winter, and A. Hedenström. 2014. Leading edge vortices in lesser long-nosed bats occurring at slow but not fast flight speeds. Bioinspiration and Biomimetics, 9 (2): 025006.

Norberg, U. M. 1981. Allometry of bat wings and legs and comparison with bird wings. Philosophical Transactions of the Royal Society of London B–Biological Sciences, 292 (1061): 359–398.

Norberg, U. M. 1994. Wing design, flight performance, and habitat use in bats. Pp. 205–239 *in*: Ecological Morphology: Integrative Organismal Biology (P. C. Wainwright and S. M. Reilly, eds.). University of Chicago Press, Chicago.

Norberg, U. M., and M. B. Fenton. 1998. Carnivorous bats? Biological Journal of the Linnean Society, 33:383–394.

Ochoa, J. G., and J. H. Sánchez. 2005. Taxonomic status of *Micronycteris homezi* (Chiroptera, Phyllostomidae). Mammalia, 69 (3–4): 323–335.

Oliveira, G. L. 2013. Discriminando múltiplos fatores determinantes da partição de nicho em pequenos mamífero sul-americanos. Master's thesis, Universidade Federal do Rio Grande do Sul, Porto Alegre.

Oprea, M., and D. E. Wilson. 2008. *Chiroderma doriae* (Chiroptera: Phyllostomidae). Mammalian Species Archive, 816:1–7.

Ortega, J., and H. T. Arita. 1997. *Mimon bennettii*. Mammalian Species Archive, 549:1–4.

Ortega, J., J. Arroyo-Cabrales, N. Martínez-Mendez, M. Del

Real-Monroy, D. Moreno-Santillán, and P. M. Velazco. 2015. *Artibeus glaucus* (Chiroptera: Phyllostomidae). Mammalian Species, 47 (928): 107–111.

Pacheco, V., E. Salas, L. Cairampoma, M. Noblecilla, H. Quintana, F. Ortiz, and R. Ledesma. 2007. Contribución al conocimiento de la diversidad y conservación de los mamíferos en la cuenca del río Apurímac, Perú. Revista peruana de Biología, 14 (2): 169–180.

Pacheco, V., S. Solari, and P. Velazco. 2004. A new species of *Carollia* (Chiroptera: Phyllostomidae) from the Andes of Peru and Bolivia. Occasional Papers, Museum of Texas Tech University 236:1–15.

Pardiñas, U. F. J., and E. P. Tonni. 2000. A giant vampire (Mammalia, Chiroptera) in the Late Holocene from the Argentinean pampas: paleoenvironmental significance. Palaeogeography, Palaeoclimatology, Palaeoecology 160 (3–4): 213–221.

Parker-Shames, P., and B. Rodríguez-Herrera. 2013. Maximum weight capacity of leaves used by tent-roosting bats: implications for social structure. Chiroptera Neotropical, 19 (3): 36–43.

Patterson, B. D., V. Pacheco, and S. Solari. 1996. Distribution of bats along an elevational gradient in the Andes of south-eastern Peru. Journal of Zoology, 240 (4): 637–658.

Porter, C. A., S. R. Hoofer, C. A. Cline, F. G. Hoffmann, and R. J. Baker. 2007. Molecular phylogenetics of the phyllostomid bat genus *Micronycteris* with descriptions of two new subgenera. Journal of Mammalogy, 88 (5): 1205–1215.

Ratcliffe, J. M. 2009. Neuroecology and diet selection in phyllostomid bats. Behavioural Processes, 80 (3): 247–251.

Rengifo, E. M., V. Pacheco, and E. Salas. 2011. An additional record of *Platyrrhinus ismaeli* Velazco, 2005 on the western slope of Peru, with taxonomic comments. Chiroptera Neotropical, 17 (1): 903–907.

Rodríguez, A., and L. Dávalos. 2008. *Brachyphylla cavernarum*. The IUCN Red List of Threatened Species 2008: e.T2982A9528532. Downloaded on 27 November 2016.

Rodríguez, B., and W. Pineda. 2015. *Ectophylla alba*. The IUCN Red List of Threatened Species 2015: e.T7030A22027138. Downloaded on 27 November 2016.

Rodríguez-Herrera, B., G. Ceballos, and R. A. Medellín. 2011. Ecological aspects of the tent building process by *Ectophylla alba* (Chiroptera: Phyllostomidae). Acta Chiropterologica, 13 (2): 365–372.

Rodríguez-Herrera, B., L. Víquez-R, E. Cordero-Schmidt, J. M. Sandoval, and A. Rodríguez-Durán. 2016. Energetics of tent roosting in bats: the case of *Ectophylla alba* and *Uroderma bilobatum* (Chiroptera: Phyllostomidae). Journal of Mammalogy, 97 (1): 246–252.

Rojas, D., A. Vale, V. Ferrero, and L. Navarro. 2011. When did plants become important to leaf-nosed bats? Diversification of feeding habits in the family Phyllostomidae. Molecular Ecology, 20:2217–2228.

Rojas, D., O. M. Warsi, and L. M. Dávalos. 2016. Bats (Chiroptera: Noctilionoidea) challenge a recent origin of extant Neotropical diversity. Systematic Biology, 65 (3): 432–448.

Safi, K., S. Meiri, and K. E. Jones. 2013. Evolution of size in bats. Pp. 95–115 *in*: Animal Body Size: Linking Pattern and Process across Space, Time, and Taxonomic Group (F. A. Smith and S. K. Lyons, eds.). University of Chicago Press, Chicago.

Saldaña-Vázquez, R. A. 2014. Intrinsic and extrinsic factors affecting dietary specialization in Neotropical frugivorous bats. Mammal Review, 44 (3–4): 1–10.

Saldaña-Vázquez, R. A., V. J. Sosa, L. I. Iñíguez Dávalos, and J. E. Schondube. 2013. The role of extrinsic and intrinsic factors in Neotropical fruit bat–plant interactions. Journal of Mammalogy, 94 (3): 632–639.

Sánchez, M. S., N. P. Giannini. Submitted. Trophic structure of frugivorous bats (Phyllostomidae) in the Neotropics: emergent historical patterns. Journal of Animal Ecology.

Schnitzler, H. U., and E. K. Kalko. 1998. How echolocating bats search and find food. Pp: 183–196 in: Bat Biology and Conservation (T. H. Kunz and P. A. Racey, eds.). Smithsonian Institution Press, Washington, DC.

Silva Taboada, G. 1979. Los murciélagos de Cuba. Editorial Academia, Havana, Cuba.

Simmons, N. B., and T. M. Conway. 2003. Evolution of ecological diversity in bats. Pp. 493–535 in: Bat Ecology (T. H. Kunz and M. B. Fenton, eds.). University of Chicago Press, Chicago.

Simmons, N. B., R. S. Voss, and D. W. Fleck. 2002. A new Amazonian species of Micronycteris (Chiroptera, Phyllostomidae) with notes on the roosting behavior of sympatric congeners. American Museum Novitates, no. 3358. American Museum of Natural History, New York.

Smith, F. A., S. K. Lyons, S. K. M. Ernest, K. E. Jones, D. M. Kaufman, T. Dayan, P. A. Marquet, J. H. Brown, and J. P. Haskell. 2003. Body mass of Late Quaternary mammals. Ecology, 84 (12): 3403.

Solari, S. 2016. Dermanura azteca. The IUCN Red List of Threatened Species 2016: e.T2123A22000362. Downloaded on 27 November 2016.

Solari, S., V. Pacheco, and E. Vivar. 1999. New distribution records of Peruvian bats. Revista Peruana de Biología, 6 (2): 152–159.

Soriano, P. J., A. Ruiz, and A. Arends. 2002. Physiological responses to ambient temperature manipulation by three species of bats from Andean cloud forests. Journal of Mammalogy, 83 (2): 445–457.

Swanepoel, P., and H. H. Genoways. 1983. Brachyphylla cavernarum. Mammalian Species, 205:1–6.

Swartz, S. M., P. W. Freeman, and E. F. Stockwell. 2003. Ecomorphology of bats: comparative and experimental approaches relating structural design to ecology. Pp. 257–300 in: Bat Ecology (T. H. Kunz and M. B. Fenton, eds.). University of Chicago Press, Chicago.

Speakman, J. R., and P. A. Racey. 1991. No cost of echolocation for bats in flight. Nature, 350 (6317): 421–423.

Swanepoel, P., and H. H. Genoways. 1983. Brachyphylla cavernarum. Mammalian Species, (205): 1–6.

Ter Braak, C. J. F., and P. Šmilauer. 1998. CANOCO 4. CANOCO reference manual and user's guide to Canoco for Windows. Centre of Biometry, Wageningen.

Tirira, D. G., C. E. Boada, and S. F. Burneo. 2010. Mammalia, Chiroptera, Phyllostomidae, Lampronycteris brachyotis (Dobson, 1879): first confirmed record for Ecuador. Check List: 6 (2): 237–238.

Trajano, E., and M. De Vivo. 1991. Desmodus draculae Morgan, Linares, and Ray, 1988, reported for southeastern Brasil, with paleoecological comments (Phyllostomidae, Desmodontinae). Mammalia, 55 (3): 456–459.

Tschapka, M. 2004. Energy density patterns of nectar resources permit coexistence within a guild of Neotropical flower visiting bats. Journal of Zoology, 263:7–21.

Tschapka, M., T. P. Gonzalez-Terrazas, and M. Knörnschild. 2015. Nectar uptake in bats using a pumping-tongue mechanism. Science Advances, 1 (8): e1500525.

Tschapka, M., E. B. Sperr, L. A. Caballero-Martínez, and R. A. Medellín. 2008. Diet and cranial morphology of Musonycteris harrisoni, a highly specialized nectar-feeding bat in western Mexico. Journal of Mammalogy, 89 (4): 924–932.

Tuttle, M. D., and M. J. Ryan. 1981. Bat predation and the evolution of frog vocalizations in the Neotropics. Science, 214:677–678.

Valdez, R., and R. K. LaVal. 1971. Records of bats from Honduras and Nicaragua. Journal of Mammalogy, 52 (1): 247–250.

Vaughan Jennings, N., S. Parsons, K. E. Barlow, and M. R. Gannon. 2004. Echolocation calls and wing morphology of bats from the West Indies. Acta Chiropterologica, 6 (1): 75–90.

Velazco, P. M. 2005. Morphological phylogeny of the bat genus Platyrrhinus Saussure, 1860 (Chiroptera: Phyllostomidae) with the description of four new species. Fieldiana Zoology, 2005 (105): 1–53.

Velazco, P. M., and B. D. Patterson. 2013. Diversification of the yellow-shouldered bats, genus Sturnira (Chiroptera, Phyllostomidae) in the New World tropics. Molecular Phylogenetics and Evolution, 68 (3): 683–698.

Velazco, P., and B. Patterson. 2014. Two new species of yellow-shouldered bats, genus Sturnira Gray, 1842 (Chiroptera, Phyllostomidae) from Costa Rica, Panama and western Ecuador. ZooKeys, 402:43–66.

Voigt, C. C., and M. W. Holderied. 2011. High manoeuvring costs force narrow-winged molossid bats to forage in open space. Journal of Comparative Physiology B, 182 (3): 415–424.

Webster, W. D., and J. K. Jones. 1983. Artibeus hirsutus and Artibeus inopinatus. Mammalian Species Archive, 199:1–3.

Weinbeer, M., and E. K. Kalko. 2007. Ecological niche and phylogeny: the highly complex echolocation behavior of the trawling long-legged bat, Macrophyllum macrophyllum. Behavioral Ecology and Sociobiology, 61 (9): 1337–1348.

Wesselingh, F. P., C. Hoorn, S. B. Kroonenberg, A. Antonelli, J. G. Lundberg, H. B. Vonhof, and H. Hooghiemstra. 2010. On the origin of Amazonian landscapes and biodiversity: a synthesis. Pp. 421–431 in: Amazonia: Landscape and Species Evolution: A Look into the Past (C. Hoorn and F. P. Wesselingh, eds.). Wiley-Blackwell Publishing Ltd., Oxford.

Willig, M. R., B. D. Patterson, and R. D. Stevens. 2003. Patterns of range size, richness, and body size in the Chiroptera. Pp. 580–621 in: Bat Ecology (T. H. Kunz and M. B. Fenton, eds.). Chicago University Press, Chicago.

Woodman, N. 2007. A new species of nectar-feeding bat, genus Lonchophylla, from western Colombia and western Ecuador (Mammalia: Chiroptera: Phyllostomidae). Proceedings of the Biological Society of Washington, 120 (3): 340–358.

Woodman, N., and R. M. Timm. 2006. Characters and phylogenetic relationships of nectar-feeding bats, with descriptions of new *Lonchophylla* from western South America (Mammalia: Chiroptera: Phyllostomidae: Lonchophyllini). Proceedings of the Biological Society of Washington, 119 (4): 437–476.

Zamora Meza, H. T., C. T. Sahley, C. E. M. Pacheco, Y. E. A. Miranda, A. C. E. Montes, and A. P. Chipana. 2013. The Peruvian long-snouted Bat, *Platalina genovensium* Thomas, 1928 (Phyllostomidae, Lonchophyllinae), in the area of influence of the Peru LNG Gas Pipeline: population status and recommendations for conservation. Pp. 110–123 *in*:

Monitoring Biodiversity on a Trans-Andean Megaproject (A. Alonso, F. Dallmeier, and G. P. Servat, eds.). Smithsonian Institution Scholarly Press, Washington, DC.

Zanella F. F. 2013. Bat species vulnerability to forest fragmentation in the Central Amazon. Master's thesis, Universidade de Lisboa, Lisboa.

Zortéa, M., Z. D. Rocha, H. G. Carvalho, G. C. Oliveira, and P. S. Mata. 2009. First record of the Carriker's round-eared bat (*Lophostoma carrikeri*; Phyllostominae) in the Cerrado of central Brazil. Chiroptera Neotropical, 15 (1): 446–449.

Contemporary Biology

9

Structure and Function of Bat Wings

A View from the Phyllostomidae

*Sharon M. Swartz
and Justine J. Allen*

Introduction

Bats possess great diversity in morphology and physiology that allows a broad range of functional capabilities throughout the body. The respiratory, cardiovascular, sensory, and feeding systems, as well as the flight apparatus, are specialized and diverse. The bat flight apparatus supports performance capabilities as varied as hovering (Hermanson 1997; Hermanson and Altenbach 1983, 1985; Voigt and Winter 1999), the fastest recorded powered flights among vertebrates (McCracken et al. 2016), long-distance migration (Fleming and Eby 2003; Weller et al. 2016), acrobatic on-the-wing capture of elusive aerial prey (Corcoran and Conner 2016), and carrying young (Norberg and Rayner 1987). Today, methodological advances are opening doors to improved understanding of the biomechanical and physiological mechanisms underlying many aspects of animal flight, and powerful techniques adapted from experimental and computational motion analysis and aerodynamics are addressing long-standing questions in the biology of flight in bats, birds, insects, and gliders (Altshuler et al. 2015; Hedrick et al. 2015; Socha et al. 2015; Swartz 2015; Swartz and Konow 2015). Similarly, traditional methods of studying animal structure at gross and microscopic levels are now often complemented by methods that provide new information concerning morphology and its structural significance, such as micro–computed tomography (Curtis and Simmons 2017; Gignac et al. 2016) and cross polarized light imaging (Cheney et al. 2017; Lee and Simons 2015). As research advances deepen our understanding of the links between structure and function in bat wings, the phyllostomids play a special role: Phyllostomidae is a diverse and speciose family, thus this group provides a particularly rich setting for comparative study.

Phyllostomids have played central roles in studies of bat flight (Bergou et al. 2015; Cheney et al. 2015; Hedenstrom et al. 2007; Hermanson and Altenbach 1985; Konow et al. 2015, 2017; Muijres et al. 2014; Riskin et al. 2009; von Busse et al. 2012, 2014; Wolf et al. 2010) and of the functional morphology of bat wings (Altenbach and

Hermanson 1987; Hermanson and Altenbach 1985; Hermanson et al. 1993, 1998; Hermanson and Foehring 1988; Quinn and Baumel 1993; Strickler 1978; Swartz and Middleton 2008). Biomechanicians and comparative physiologists have begun to take advantage of the diversity in ecological niche and the resulting variety in wing morphology represented in the Phyllostomidae to form hypotheses about the structural and functional requirements for mammalian flight and about the phylogenetic relationships and evolutionary specialization of wings in this family. The many different feeding modes found among phyllostomids are naturally associated with a wide range of flight behaviors. For example, species that consume a substantial amount of nectar, such as glossophagines and lonchophyllines, must hover effectively. Blood-feeding taxa, for which a single meal can comprise more than 50% of body mass (Wimsatt 1969), must fly with a significant load, as do mothers carrying pups in any species. The diverse demands on the flight apparatus of phyllostomids make them ideal models to study and compare the requirements and adaptations of mammalian flight.

Much of the research on flight and on wing structure in phyllostomids has been carried out on species from the derived nectar- and fruit-eating Stenodermatinae, Carolliinae, and Glossophaginae, largely because they are common in laboratory colonies and adapt well to flight research. Insect- and vertebrate-eating lineages and all basal groups are much less well-studied. Here, we review how interpretation of the structure of wings and dynamics of flight have been shaped by studies of phyllostomids and the current understanding of ways in which phyllostomids may be distinctive among bats.

Muscles of the Wing Skin

Wings are the structural hallmark of all flight, and in every flying lineage, a part of the integument is modified by evolution as a portion of these crucial structures. In this way, the outermost layer of the limb takes on a novel function in animals that fly: interacting with the surrounding air to generate aerodynamic forces. Wings' surfaces interact directly with the air, and each time that powered, flapping flight has evolved—in insects, pterosaurs, birds, and bats—the integument has become specialized in a different way. The dissimilar properties of the material of which the wing is composed results in differences in the way the three-dimensional shape of

wings interacts with the surrounding air and, therefore, in differences in flight performance and aerodynamics. For example, the stiff, unjointed chitinous wings of insects and the feathered wings of birds interact with air in ways that are more passive than the dynamic, flexible, and complex interplay of stretching skin and airflow that occurs during bat flight. Although three-dimensional wing shape always determines aerodynamic force production, bats are unique in that they possess the capacity for active modulation and control of their aerodynamic surface.

Wing membrane skin, the primary aerodynamic surface for bats, is modified from the ancestral mammalian condition in a variety of ways and, in addition to flight, supports protection from pathogens (Pannkuk et al. 2014), temperature regulation (e.g., Muñoz-Garcia et al. 2012; Reichard et al. 2010); and sensation (e.g., Sterbing-D'Angelo et al. 2011, 2017). The wing skin of phyllostomids is, in many ways, similar to that of other bats. In particular, the skin of the wing is specialized to confer exceptional compliance or "stretchiness" and anisotropy (unequal properties along different axes) (Cheney et al. 2015; Swartz et al. 1996; Wegst and Ashby 2004). The extraordinary compliance of bat wing skin makes it possible for the wing to dramatically change shape during flight not only because of the motions of the highly jointed skeleton and the structural framework of the wing membrane but also because of deformations of the skin itself when it is subjected to dynamically changing aerodynamic forces (fig. 9.1A).

The skin of bat wings is also unique in that it encloses groups of muscles embedded within the dermis of the plagiopatagium or armwing in all bats (fig. 9.1B–D). The presence of muscles in the wings has long been recognized (e.g., Norberg 1972a; Schöbl 1871; Schumacher 1932), but there have been few comparisons among taxa. This is at least in part because a lack of clear and consistent nomenclature has obscured patterns and increased the difficulty of carrying out interspecific comparisons of morphology. Recently, we have aimed to describe patterns of attachment of major groups of intramembrous wing muscles and their distribution among bats, which allows us to begin to understand how bat lineages differ in wing muscle structure (Cheney et al. 2017).

Of the five major groups of intramembranous muscles, the membrane muscle (mm) plagiopatagiales proprii (hereafter, "plagiopatagiales"; singular, "pla-

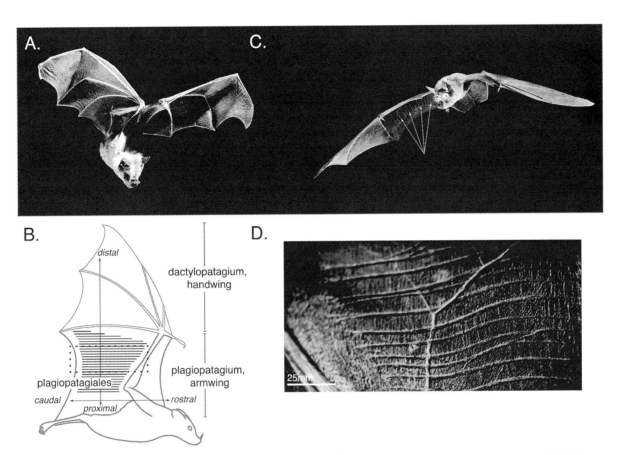

Figure 9.1. (*A*) The many-jointed wing skeleton combined with deformations of skin between wing segments produces wing shapes of high 3-D complexity as illustrated in the early downstroke of a *Dermanura phaeotis*. (*B*) Schematic of approximate anatomical location and orientation of plagiopatagiales muscles in lateral view of a flying bat at beginning of downstroke. *Dashed rectangle* indicates location of inset in *D*. (*C*) Plagiopatagiales, some indicated by *white arrows*, can be seen running in the craniocaudal or chordwise direction in the downstroke posture of the mormoopid bat *Pteronotus parnellii*. (*D*) Close-up photograph of plagiopatagiales (*right*) from a large pteropodid (*Eidolon helvum*). *A* and *C*: Photographs by Neil Boyle. *B* and *D*: adapted from Cheney, Konow et al. (2014).

giopatagialis") have been a particular focus of study to date (e.g., Cheney, Konow et al. 2014). These muscles are oriented craniocaudally, approximately perpendicular to the predominant orientation of the extensive network of bundles of connective tissue (hereafter, designated "elastin bundles"; see below, "Connective Tissues of Wing Skin," for more details on bundle composition). They both originate and insert within the armwing, neither attaching to bones nor spanning joints (fig. 9.1*B–D*). Individual plagiopatagiales that comprise an array are generally quite long, extending half or more of the wing chord (the distance from the anterior border or leading edge to the posterior border or trailing edge of the wing). Although this allows each muscle to undergo substantial shortening, the number of individual fibers in each of these muscles, and thus the physiological cross-sectional area of each muscle, is quite limited, hence arrays of plagiopatagiales generate relatively little force, even when all of the individual

muscles contract synchronously. For example, in *Carollia perspicillata*, a series of 9–13 muscle bellies are each composed of approximately 11–60 muscle fibers, and in *Artibeus lituratus*, an array of approximately 26–29 muscle bellies are each composed of approximately 35–79 fibers (JJA, unpublished data). Because the plagiopatagiales originate and insert within the wing membrane instead of attaching to bones or spanning joints, the effect of their contraction on the shape of the wing does not arise by directly causing movement of bones at joints but by modulation of tension of the wing membrane that, in turn, directly influences aerodynamic force production (Cheney, Konow et al. 2014; Cheney et al. 2015). The effects of activation of intramembranous muscles on three-dimensional wing shape are potentially very complex; bats may be able to activate numerous combinations of muscles in conjunction with activity in the muscles that directly control bone motion, and intramembranous muscle activity can, itself, potentially have complex, nonlinear effects on skin stiffness. Although we have yet to attain detailed understanding of muscle activity patterns and their relationship to wing function, the presence and orientation of these muscles suggest that bats can actively fine-tune the shape of their wings and, thus, the way their airfoils interact with the surrounding air and control wing function dynamically throughout the wingbeat cycle and in varying flight conditions.

The distribution of muscles across the skin of the wing in a highly patterned fashion could also be well-suited to sensory or sensorimotor roles. Because the plagiopatagiales are distributed along and across the chord of the armwing, they are potentially stretched when the wing experiences aerodynamic forces. The aerodynamic performance of compliant wings is highly sensitive to membrane stiffness (Rojratsirikul et al. 2010; Timpe et al. 2013), hence it is critical for bats to both sense and modulate this stiffness. For this reason, bats could benefit from a high density of muscle spindles in these muscles. Presence of these ubiquitous mammalian stretch-activated structures could provide effective signals to the motor system to initiate reflex responses in the plagiopatagiales (Bewick and Banks 2015; Dimitriou and Edin 2010). It has been suggested that muscles' spindle density or "spindle index," the number of spindles per gram of wet muscle weight, can indicate the degree to which a muscle performs a sensory function; muscles that provide fine-scale or "vernier" control of posture or movement should possess high spindle density (Botterman et al. 1978; Meyers and Hermanson 2006; Richmond and Abrahams 1975). To test this hypothesis, we searched for spindles in samples of the plagiopatagiales and compared the muscle tissue to that from two wing skeletal muscles, the deltoideus and biceps brachii in *C. perspicillata*. After detailed and exhaustive examination, we observed no muscle spindles in histological preparations of plagiopatagiales muscles from *C. perspicillata*. However, we were able to readily identify spindles in the deltoid and biceps brachii and found that they exhibited morphology typical of that observed in spindles in muscle from other mammals (e.g., Lionikas et al. 2013; Meyers and Hermanson 2006). We therefore suggest that it is unlikely that plagiopatagiales serve a primarily proprioceptive function and that, if they do, they must employ a mechanism that does not use spindles to detect changes in muscle length.

The morphology of the arrays of muscles varies substantially among families but is relatively consistent within families (Cheney et al. 2017; fig. 9.2). Phyllostomids possess a low to moderate number of plagiopatagiales muscles; in a recent comparison of 70 species from 16 bat families, molossids averaged over 160 muscles per wing, vespertilionids averaged fewer than 4, thyropterids 3, and phyllostomids ca. 19 (Price-Waldman and Cheney, unpublished). It is difficult to interpret the functional significance of this kind of variation at present. Using *A. jamaicensis* as a model species, it has recently been possible to explore the activity of the plagiopatagiales during flight (Cheney, Konow et al. 2014), a key step toward gaining mechanistic understanding of intramembranous muscle morphology. Direct measurement with electromyography (EMG) of plagiopatagiales activation over the course of the wingbeat cycle, during steady, level flight at low and moderate speeds shows that these muscles are recruited in every wingbeat (Cheney, Konow et al. 2014; fig. 9.3A). The intensity of muscle activity increases during the downstroke, particularly mid to late downstroke. As flight speed increases, this activity increases in magnitude and begins slightly earlier in downstroke (fig. 9.3B). Both increases and decreases in activity of individual muscle elements are synchronous across the array of plagiopatagiales. This pattern is consistent with the expectations for a muscle system tuned to reduce passive deflection and/or flutter and supports the hypothesis that the plagiopatagiales are recruited to dynamically modulate the stiffness of the wing skin during flight.

We can use these results to suggest functional hypotheses to explain the diversity in muscle morphology found among bat families. Species with a greater muscularization of the wing membrane—which could occur either via increased muscle cross-sectional area relative to body mass or by investiture of a greater proportion of the wing area with intramembranous muscle—might have greater needs for reduction of flutter or control of passive deformation. What factors could lead to such needs? Animals that employ a broader range of flight speeds, particularly higher speeds, may benefit preferentially from very fine control of skin tension or from control across a wider dynamic range. For example, *Tadarida brasiliensis*, which have been observed to fly at exceptionally high speeds (Davis et al. 1962; McCracken et al. 2016), possess a

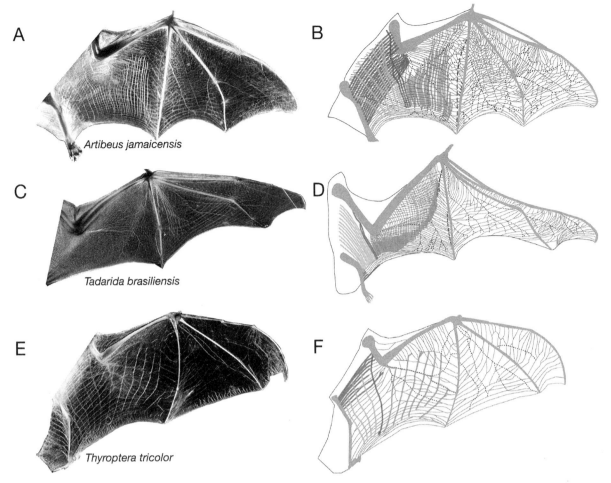

Figure 9.2. Diversity in wing membrane architecture, illustrating characteristic patterns for Phyllostomidae (represented by *Artibeus jamaicensis*), Molossidae (represented by *Tadarida brasiliensis*), and Thyropteridae (represented by *Thyroptera tricolor*, pattern similar to that in Vespertilionidae). *Left (A, C, E)*: Photographs taken with cross-polarized light. *Right (B, D, F)*: Schematics showing elastin bundles (*fine lines*), muscle arrays (*heavy lines*), and neurovasculature (*light dashed lines*) and collagenous fiber bundles (*heavy dashed lines*). Schematics were developed using multiple cross-polarized light images. Muscle arrays include tibiopatagiales, dorsopatagiales, coracopatagiales, plagiopatagiales proprii, and cubitopatagiales. Adapted from Cheney et al. (2107); see Cheney et al. (2017) for full detail.

large number of plagiopatagiales (>70), and among the 17 vespertilionid genera we have sampled to date, only species that regularly employ migratory behavior (*Lasiurus borealis*, *Lasiurus cinereus*, *Miniopterus schreibersii*) possess more than the family mean of five plagiopatagiales (Price-Waldman et al. in prep.). Similarly, those that regularly encounter turbulent flow conditions or pursue relatively massive evasive prey on the wing may have enhanced demand for control of airfoil properties, particularly the tuning of wing stiffness; we have observed that the number and cross-sectional area of plagiopatagiales are elevated in *Noctilio* (both *leporinus* and *albiventris*) relative to the condition in *Thyroptera*, *Mystacina*, and most mormoopids and phyllostomids (Price-Waldman et al. in prep.). If particular species possess wing membrane skin that has substantially

lower intrinsic stiffness, as some comparative studies suggest (Studier 1972; Swartz et al. 1996), greater muscular control may provide mechanisms that afford active compensation for otherwise passive characteristics. Concomitantly, the flight ecology of those species with reduced intramembranous musculature may impose reduced demands for modulating membrane stiffness through activity of the plagiopatagiales, or these species may achieve the mechanical and physiological goals attained by use of plagiopatagiales in some taxa by alternative mechanisms. Tension of the armwing skin could be controlled by contributions from multiple sources, including the plagiopatagiales but also other intramembranous muscles (dorsopatagiales, tibiopatagiales, and cubitopatagiales; Cheney et al. 2017) and muscles that contribute to skin tension by controlling the three-

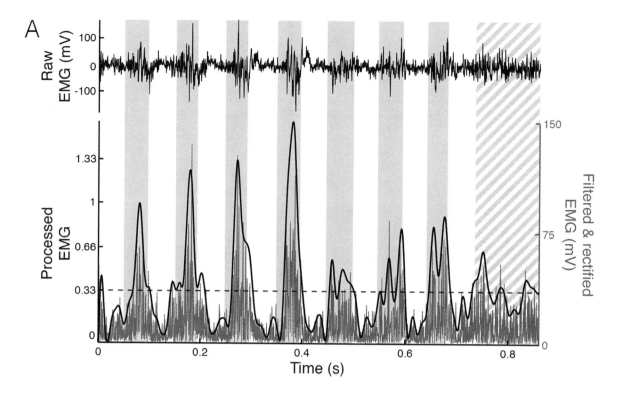

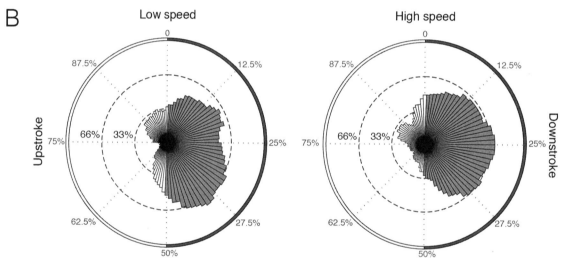

Figure 9.3. Activity of the plagiopatagiales from steady level flight in *Artibeus jamaicensis*. (*A*), *top strip*, EMG (electromyography) signal amplified 2,000x and with the bandpass left open (5–3,000 Hz) but with a 60 Hz notch filter engaged (raw EMG); *bottom strip*, signal was further filtered and rectified (filtered/rectified EMG) before it was low-pass filtered and normalized to produce the EMG envelope (processed EMG). Plagiopatagialis activity is elevated during the downstroke (*gray bands*) of all wingbeats during forward flight but not during landing (*diagonally striped band*). *B*, Rose plot of plagiopatagialis activity distribution over the wingbeat cycle at low-speed (*left*) and high-speed (*right*) flight (mean 2.2 and 5.5 ms⁻¹). Column height indicates percentage of wingbeats that displayed activity during that portion of the wingbeat cycle. *Gray columns* indicate downstroke. Adapted from Cheney, Konow et al. (2014).

dimensional spatial position of the humerus, radius, fifth digit, and hindlimb by direct attachments. Indeed, direct evidence indicates that some bats use movements of the hindlimb to actively modulate tension in the skin of the armwing (Cheney, Ton et al. 2014). We anticipate fruitful studies that further explore the capacity to

control wing camber through modulating skin stiffness. We predict that the ability of bats to employ fine-scale motion at joints will prove to interact with control of changes in skin tension during complex maneuvers and responses to flight perturbations. With greater understanding of these phenomena, it will be possible to assess the diversity of performance among phyllostomids and between phyllostomids and bats in other groups. The contrasts in morphology between phyllostomids and molossids on the one hand and vespertilionids and *Thyroptera* on the other (fig. 9.2) indicate that this is a promising area for future investigations.

Connective Tissues of the Wing Skin

Bat wing skin possesses exceptional physical performance, particularly high compliance and anisotropy, which arises from its distinctive structure, which extends beyond the presence of intramembranous muscles (Cheney et al. 2015; Skulborstad et al. 2015; Swartz et al. 1996). Detailed study, using *Carollia perspicillata* as a model species, shows that the complex mechanical dynamics of wing skin as a material arise from interactions between the relatively stiff matrix provided by the dermis and the large, organized, soft elastin bundles that are embedded within this matrix (Cheney et al. 2015, 2017; Crowley and Hall 1994; Holbrook and Odland 1978; Schöbl 1871; fig. 9.4C, E, H, J). When stress in the

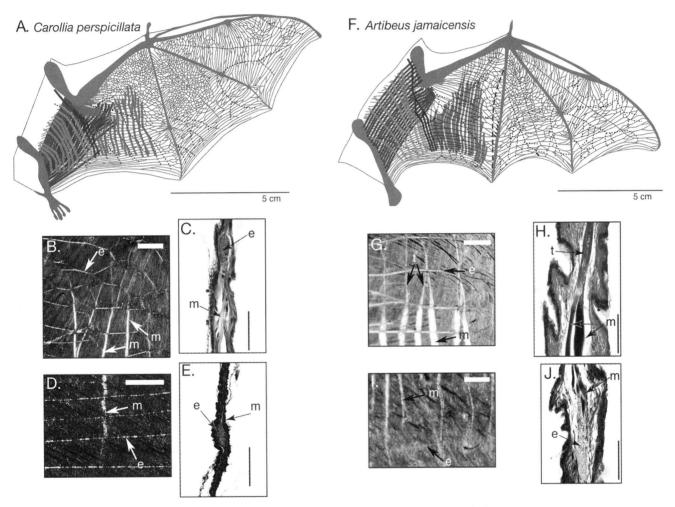

Figure 9.4. Architecture of plagiopatagiales muscles and elastin bundles in two phyllostomids, *Carollia perspicillata* (*left*), and *Artibeus jamaicensis* (*right*). *Top row* (A, F): Wing skin composition schematics, with elastin bundles (*fine lines*), muscle arrays (*heavy lines: light gray* = plagiopatagiales, *darker gray* = dorsopatagiales, tibiopatagiales, coracopatagiales, and cubitopatagiales), and neurovasculature (*light dashed lines*) and collagenous fiber bundles (*heavy dashed lines*). Schematics were developed using multiple cross-polarized light images. *Middle row* (B, G): Cross-polarized light photograph of ventral wing surface highlighting cranial or rostral attachment points of plagiopatagiales; (C, H): Longitudinal sections of cranial (rostral) attachments (stained with modified Verhoeff's [Garvey at al. 1991] and [C] Mallory's triple connective tissue stains [Humason 1962] or [H] van Gieson's stain [Garvey et al. 1991]; 6μm thick). In *C. perspicillata*, most muscle fibers attach cranially to spanwise-running elastin bundles and a few muscle fibers attach via endomysium to the collagenous layers of the skin. In *A. jamaicensis*, muscles attach to tendons composed of elastin bundles. *Bottom row* (D, I): Cross-polarized light photograph of highlighting caudal attachment points; (E, J): Longitudinal sections of caudal attachments (stained with modified Verhoeff's [Garvey et al. 1991] and Mallory's triple connective tissue stain [Humason 1962]). In both species, plagiopatagiales attach caudally via endomysium to spanwise-running elastin bundles. Scale bars: white, 2 mm; black, 100 μm. Abbreviations: e = elastin bundle; m = plagiopatagialis muscle; t = tendon.

membrane is low, as it is when the wings are furled or during portions of the wingbeat cycle in which they experience little aerodynamic force, the large connective tissue bundles are highly organized and generate highly anisotropic mechanical behavior. In this configuration, the skin is much more extensible proximodistally, parallel to the elastin bundles, than craniocaudally, perpendicular to the fiber bundles. As stress in the skin increases, as would occur when wings experience peak forces, the load-bearing role of the compliant fibers becomes less important, the stiffer matrix that surrounds the fiber bundles becomes functionally dominant, and the skin as a whole structure resists stretch equally well in all directions (Cheney et al. 2015).

These connective tissue bundles within the skin are themselves composite connective structures, composed in *T. brasiliensis* of approximately 50% elastin (Holbrook and Odland 1978) and in *C. perspicillata* and *A. lituratus* of at least 75% elastin (JJA unpublished data). These highly extensible elastin bundles, surrounded by and held together with collagen, are substantially more compliant than the tissue around them (Cheney et al. 2015; Holbrook and Odland 1978). In contrast, in the basal condition for mammals, elastin is distributed as an unorganized sheet of fibrils (Meyer et al. 1994). As in the case of the intramembranous muscles, the geometry of elastin bundle networks displays characteristic patterns in each species. Although patterns within families tend to be similar, this is not always the case (Cheney et al. 2017). Most fiber bundles run in the mediolateral or spanwise direction, along the axis of wing folding and unfolding, but there are variations in this pattern, even among closely related species, that are especially evident in particular regions of the wing (Cheney et al. 2017). Within Phyllostomidae, elastin fiber bundle architecture has been most carefully studied in *C. perspicillata* and *A. jamaicensis* (Cheney et al. 2015, 2017). To improve our understanding of the function of these fiber bundles, we have recently taken a histological approach to investigate the details of the attachments of intramembranous muscles within the skin of the wing in several species, including the phyllostomids *C. perspicillata* and *A. lituratus* (JJA, unpublished data).

Carollia perspicillata and *A. lituratus* share a similar pattern of attachment at the caudal (trailing-edge end) of the plagiopatagiales muscles. At the caudal attachment region in both species, the muscle fibers

are directly connected to a spanwise-running bundle of elastin (fig. 9.4). This connection is mediated by collagen, which, among other functions, surrounds the individual muscle fibers and elastin fibrils. In contrast, at the rostral end of each muscle belly, the plagiopatagiales of *C. perspicillata* and *A. lituratus* differ in their mode of attachment. In the former species, most muscle fibers attach via their endomysia to elastin bundles that comprise a honeycomb-like connective tissue network in the skin of the plagiopatagium near the radius. A few muscle fibers, however, take their attachment more cranially, beyond the elastin bundle closest to the cranial end of the muscle belly, to insert within the bilayered, collagenous extracellular matrix of the wing, also via their endomysia. At the cranial end of the plagiopatagiales in *A. lituratus*, many fibers in each muscle attach to spanwise-running bundles of elastin as at the caudal end, but those that attach most rostrally do so on an elastin bundle that lies parallel, not perpendicular, to the muscle belly.

Our studies of the details of connections among the primary constituents of phyllostomid wing skin help provide insight into how derived characteristics of the plagiopatagium and their variations contribute to flight. It has been suggested that the collagen-elastin bundle network plays an important role in wing folding by contributing to recoil or shortening in membrane skin as aerodynamic forces decline. This passive behavior likely helps maintain tension in the wing skin during upstroke, to counter what might otherwise be a detrimental tendency toward undesirable deflection and flutter (Cheney et al. 2015; Swartz et al. 1996; Timpe et al. 2013). The collagenous and muscular components of wing membrane skin can improve its strength and ability to resist tearing, as well as allow for the active modulation of wing shape, resulting in improved aerial agility (Holbrook and Odland 1978; Timpe et al. 2013).

The caudalmost or trailing edge of the plagiopatagium is also a functionally significant element of wing architecture. Based on its anatomical location, the trailing edge may be able to disproportionately influence the aerodynamic performance of bat wings and their resistance to flutter, unfavorable oscillations, or vibrations in a solid structure that arise from interaction with fluid dynamic forces (Hubner and Hicks 2011). Classic anatomical studies have suggested the possibility of muscle (Holbrook and Odland 1978) or muscle plus elastic fiber (Vaughan 1970) reinforcement in this region, but this work investigated only molossids. We investigated the histological morphology of the phyllostomid plagiopatagium trailing edge, using *C. perspicillata* as a model species.

We observed both muscle fibers and a large elastin bundle in the trailing edge of the *C. perspicillata* plagiopatagium (fig. 9.5). The most proximal portion of this region of wing is well reinforced with muscle fibers. These fibers originate from the tibia and astragulus, and the most proximal portion is oriented rostrocaudally, parallel to the tibia where the wing membrane skin attaches to the ankle. Just distal to the origin, the fibers bend approximately 90° to take on the proximodistal orientation, parallel to the trailing edge, that characterizes the great majority of their length (fig. 9.5A–C). This muscle bundle is relatively robust proximally, comprising 300–400 fibers, but decreases progressively in fiber number and hence cross-sectional area distally (fig. 9.5A, D–H). At the most distal extremity of the trailing edge of the plagiopatagium and at the distalmost portion of the distal phalanx V, we observed only a few muscle fibers in the skin. Thus, overall, the muscle is strongly tapered between the hindlimb and fifth digit attachments, with most of the muscle fibers ending in the proximal third of the arm wing.

An elastin bundle also reinforces the trailing edge in *C. perspicillata*, and its structure is intertwined with that of the muscle in a complex fashion. This bundle originates in the ankle region near the base of the tibia and forms a loop around the base of the calcar, tibia, and fibula proximally. Just distal to the ankle, it extends into the most caudal portion of the skin of the plagiopata-

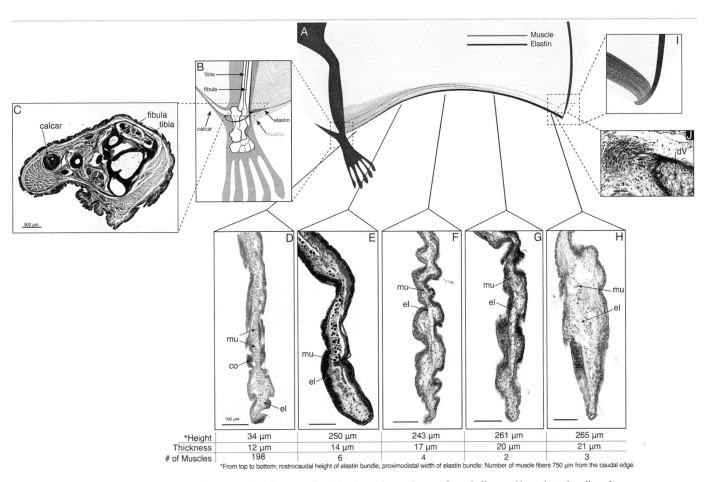

*Height	34 μm	250 μm	243 μm	261 μm	265 μm
Thickness	12 μm	14 μm	17 μm	20 μm	21 μm
# of Muscles	198	6	4	2	3

*From top to bottom: rostrocaudal height of elastin bundle; proximodistal width of elastin bundle; Number of muscle fibers 750 μm from the caudal edge.

Figure 9.5. Structure of trailing edge of the wing of *Carollia perspicillata*. This shows the attachment of muscle fibers and large elastin bundle at the ankle. (*A*) Diagram of tissue arrangement at trailing edge. (*B*) Diagram of attachment on and around bones of the ankle. Muscle fibers of the trailing edge insert directly onto the tibia and tarsals. Elastin bundle wraps around the ankle. (*C*) Cross-section through ankle illustrating bone structure with surrounding muscle and elastin. (*D–H*) Parasagittal sections through trailing-edge tissue. Rostrocaudal height and proximodistal width of elastin bundle increases distally. Number of muscle fibers decreases distally. Rostral is towards the top of the image, and caudal is towards the bottom. (*I*) Diagram of attachment of large elastin bundle at distal phalanx of digit V, ventral view. (*J*) Cross-section through distal phalanx of digit V showing detail of attachment of elastin fibers to collagen surrounding cartilaginous phalanx. Abbreviations: co = collagen; el = elastin; mu = muscle; nu = nuclei, dV = digit V distal phalanx.

gium (fig. 9.5A–C). Just distal to the ankle, it extends into the most caudal portion of the skin of the plagiopatagium (fig. 9.5A–C). Along the span of the armwing, a single large elastin bundle gradually increases in diameter from proximal to distal (fig. 9.5D–H), to take distal attachment on the unmineralized, distalmost extremity of the distal phalanx of digit V (fig. 9.5I, J). Approximately 2 mm distal to the ankle, these coalesce to form a single large elastin bundle that passes along the span of the armwing, gradually increasing in diameter from proximal to distal, to take distal attachment on the unmineralized, distalmost extremity of the distal phalanx of digit V (fig. 9.5I, J).

The Role of Tendons

Tendons, the structures that connect muscles to bones, have received much less attention than either bones or muscles in the literature concerning the wing morphology and function in bats. Although both classic and more recent studies have described the tendinous connections of wing muscles to the skeleton (e.g., Humphry 1869; Norberg 1968, 1972b; Vaughan 1959), we have only recently begun to consider tendons in the wing in the light of growing appreciation for the springlike function of tendons. These connective tissue structures function as shock absorbers and elements of "biological catapults" or power amplifiers to allow animals such as toads to feed via ballistic tongue projection and kangaroos to locomote by hopping (Alexander and Vernon 1975; Roberts and Azizi 2011). In large walking and running animals, impact with the ground can stretch tendons of major muscles in the limbs, such as the Achilles tendon that connects the gastrocnemius, plantaris, and soleus muscles to the calcaneus. Subsequent tendon recoil can contribute to force or power production (Biewener et al. 1998; Wilson et al. 2003).

The role of elastic effects in small animals has been ambiguous (Bullimore and Burn 2005; Ker et al. 1988). Further, the fundamental nature of motion during flight and swimming—movement through fluid—differs from substrate-to-limb impacts produced during terrestrial locomotion; the nature of elastic mechanisms under these conditions is even less well understood (see Askew and Marsh 2002; Blickhan and Cheng 1994; Tobalske and Biewener 2008). Understanding the details of the interplay of tendon and muscle mechanics

in bat flight requires the ability to independently measure the distinct contributions of length changes in the tendinous versus muscle fiber components of an intact, whole muscle, or more accurately, a muscle-tendon unit, as the length from origin to insertion changes over a locomotor cycle (Camp et al. 2016; Konow et al. 2015). This challenging technical requirement can now be met by application of fluoromicrometry, a technique in which minute metal spheres, as small as 0.5 to 1.0 mm in diameter, are surgically implanted into muscles and tendons and then tracked in 3-D by dual-camera high-resolution, high-speed X-ray videography (Camp et al. 2016). This approach allows investigators to simultaneously track muscle and tendon length changes in three dimensions, in synchrony with skeletal kinematics. With this method, we have learned that a major tendon in the elbow of C. perspicillata, that of the monoarticular lateral head of the triceps brachii, is stretched by elbow flexion at the end of downstroke, as the wing folds in preparation for upstroke (Konow et al. 2015). The elbow remains flexed during the upstroke portion of the wingbeat cycle, but at the end of this phase it extends as the wing begins to open, increasing aerodynamic force production through the downstroke. The length changes in the triceps tendon, in relation to joint movements, muscle fiber length changes, and muscle activity patterns, show that elbow extension is partly effected by tendon recoil; the fibers of the muscle undergo >20% shorter length changes than they would without the elastic compliance of the tendon. Because of the elastic action of this tendon, the elbow extends at high speed, even when triceps shortening is relatively slow (Konow et al. 2015).

At present, direct evidence for wing tendons contributing to the dynamics of flight comes from only a single species of phyllostomid, C. perspicillata, and therefore it is not yet possible to assess the degree to which phyllostomids are either specialized or more typical with respect to this feature. The long tendons of the muscles of the wing digits also merit much greater attention; thin tendons are particularly likely to display elastic effects in which motion of muscle fibers and movement of joints are decoupled (Roberts and Azizi 2011; Wilson et al. 2003), and stretch and recoil of these elastic elements is likely to have a significant influence on dynamics of the handwing. Analysis of tendon mechanics from living animals is time and labor intensive, but the potential importance of this element of wing structure for flight performance is great. Further study will surely contribute substantially toward elucidating the role of tendons in differences in flight performance among groups of bats.

Wing Function: Phyllostomids in Flight

A diverse array of measurement approaches has helped quantify the flight abilities of bats, including phyllostomids. Methods that have been applied to study of various aspects of flight in phyllostomids include high-speed cine or videography (e.g., Lindhe Norberg and Winter 2006; Sterbing-D'Angelo et al. 2011; von Busse et al. 2012; Wolf et al. 2010), measurement of muscle activity patterns with electromyography (Cheney, Konow et al. 2014; Hermanson 1997; Hermanson and Altenbach 1983, 1985; Konow et al. 2017), metabolic measurements of flight energetics (Voigt et al. 2006; Voigt and Winter 1999; von Busse et al. 2013), remotely monitoring flight behavior in natural environments (O'Mara et al. 2014; Sapir et al. 2014), and experimental aerodynamics (Hedenstrom et al. 2007; Muijres et al. 2008; von Busse et al. 2014). Together, these techniques contribute information required to fully understand the mechanistic basis of flight performance and its variation among flying animals (Swartz 2015).

Several phyllostomid species have been recent subjects of particular attention in analyses of the aeromechanics of bat flight, particularly *Glossophaga soricina* (Johansson et al. 2008; Muijres et al. 2008, 2011; Suarez et al. 2009), *Leptonycteris yerbabuenae* (Muijres et al. 2012; von Busse et al. 2012), *C. perspicillata* (Konow et al. 2015; Riskin et al. 2009; Sterbing-D'Angelo et al. 2011; Voigt and Lewanzik 2011; von Busse et al. 2013, 2014), and *A. jamaicensis* (Albertani et al. 2011; Carter and Adams 2015; Cheney, Konow et al. 2014; Stockwell 2001). Much of this body of work shows that flight in phyllostomids is similar to that in other bats at our current level of resolution and understanding. In the kinematics of all bats studied to date, angle of attack, the inclination of the wing relative to the oncoming air, and wingbeat frequency both tend to decrease with increasing flight speed (Swartz and Konow 2015). Similarly, as flight speed increases, all bats adjust motion at multiple joints to modulate the

aerodynamic force-producing area of the wing, to varying degrees. In some respects, however, no single pattern of motion describes all bats. Stroke plane angle— the angle at which the wing tip is swept relative to the horizontal—increases with speed in the phyllostomids that have been studied, as well as in pteropodids, but this pattern is not seen in the single molossid (*Tadarida brasiliensis*) and vespertilionid (*Myotis velifer*) species examined in depth (Hubel et al. 2016). Measurements of the activity of flight muscles are not yet extensive, but a recent comparison shows contrasting patterns of recruitment intensity (quantified as normalized average burst EMG, a measure of integrated EMG) in the pectoralis major and biceps brachii, the most important muscles powering downstroke motion at the shoulder and elbow, in the phyllostomid *C. perspicillata* compared to *Eptesicus fuscus*, a vespertilionid of comparable body mass and similar wing shape. In *C. perspicillata*, average muscle recruitment intensity is modulated in a U-shaped or linearly increasing relationship with speed, but in *E. fuscus*, muscle recruitment is independent of speed (Konow et al. 2017).

Particule image velocimetry, a technique adapted from experimental fluid dynamics, is a relatively new approach to analyzing animal flight and has revolutionized animal aerodynamics in the last 15 years (Bomphrey 2012). This method makes wakes, the "footprints" that describe the aerodynamic forces produced during flight, visible and accessible for quantification and analysis (Spedding et al. 2003). This approach is particularly valuable for improving understanding of animal flight because over the last 40 years recognition has steadily grown that the aerodynamics of flapping flight in animals differs in important respects from that of large, fast rigid fixed winged aircraft (Chin and Lentink 2016; Ellington et al. 1996; Rayner 1979a, 1979b). The vortex approach to animal flight, in contrast to an approach based on simplified airplane aerodynamics, does not assume that aerodynamic forces that lift, drag, and thrust are direct functions of the square of forward velocity and a lift or drag coefficient. Further, the vortex approach does not assume that aerodynamic forces operate at constant, steady-state levels or that time-varying aspects of force generation can be ignored (Dickinson 1996). This more nuanced view of aerodynamic force production also includes attention to specific unsteady effects, including delayed or dynamic stall, added mass, rotational circulation, and wing-wake interactions (Chin and Lentink 2016; Dickinson et al. 2000; Shyy et al. 2013). Unsteady effects can produce more lift than would be generated by steady flow over a given airfoil at comparable net flight speeds. Stable leading edge vortices—vortex structures that develop parallel to the rostral-most (leading) edge of animal wings and in which airflow spirals from the body toward the wingtip, creating a region of low air pressure along the front of the wing—appear to have evolved independently in insects, birds, and bats as mechanisms to augment lift during mid-downstroke and mid-upstroke (Chin and Lentink 2016). Evaluating the presence and relative importance of unsteady aerodynamic effects, such as leading edge vortices, requires careful experimental study of flow dynamics from animals and realistic animal-inspired models.

Particule image velocimetry has been applied to better understand flight in a diversity of insects, birds, and bats (Bomphrey et al. 2016; Chin and Lentink 2016; Henningsson et al. 2008, 2011; Hubel et al. 2016). Most studies of bat wakes that have employed particule image velocimetry have used phyllostomids as subjects, particularly *G. soricina*, *L. yerbabuenae*, and *C. perspicillata*, hence much of the emerging understanding of aerodynamics of bats over the last 10 years has been influenced strongly, perhaps disproportionately, by work on these species, particularly glossophagines. Particule image velocimetry studies demonstrate that leading edge vortices make a considerable contribution to net lift during hovering and slow flight in *G. soricina* and *L. yerbabuenae* (Chin and Lentink 2016; Muijres et al. 2008, 2011, 2014). These attached leading edge vortices can also help prevent stall when wings operate at high angles of attack. Other characteristic features of bat wakes, such as wingtip vortices, a pair of vortices shed from the junction of the body and the base of the wing, and vorticity shed at the end of upstroke, are observed in phyllostomids but also in the other bats studied with this approach, including *Cynopterus brachyotis* (Pteropodidae) and, to a lesser extent, in *Tadarida brasiliensis*, *Myotis velifer*, and *Plecotus auritus* (Håkansson et al. 2015; Hubel et al. 2010, 2012, 2016; Johansson et al. 2016). Indeed, bat species that differ in flight ecology and wing morphology may be very similar aerodynamically (Hubel et al. 2016). Such results suggest that identifying the morphological traits

of particular relevance for generating and controlling aerodynamic forces may be a more complex task than once envisioned, one that requires explicit attention to functional performance (Koehl 1996).

Conclusions and Future Directions

By applying conceptual frameworks and the experimental methods of physical and mathematical sciences, scientists studying flying animals have gained great insight into the structural and mechanistic basis of wing function and flight performance (e.g., Dudley 2000; Shyy et al. 2013; Vogel 2013; Wainwright et al. 1976). Experimental fluid dynamics approaches are proving instrumental to identifying physical mechanisms that produce and control flight in bats and birds and are critical for improving our understanding of how physical constraints shape the evolution of flying animals (Bomphrey 2012; Chin and Lentink 2016; Dickinson et al. 2000). The animal flight research community seeks to discern both commonalities and the distinctive traits in flight in bats, birds, and insects (Altshuler et al. 2015; Chin and Lentink 2016; Hedrick et al. 2015; Swartz 2015; Swartz and Konow 2015). As we work to identify where research may most fruitfully focus, it is clear that we have much to learn about flight in bats.

Placing phyllostomids into this growing framework presents multiple challenges. Methods such as particule image velocimetry require intensive laboratory-based work and applications for natural conditions are developing only slowly and are so far possible only for aquatic and not aerial flow measurements (Katija and Dabiri 2008; Wang et al. 2013). Species are typically selected for these studies according to pragmatic rather than scientific considerations; animals that thrive in laboratory conditions, that are well-suited to a particular experimental configuration because they will use artificial feeders, that are a convenient size (wingspan) for analysis, that have been studied previously, or that show extreme performance (Bomphrey 2012) are selected as focal species, with little attention to ecology or phylogeny. Labor- and time-intensive experimental methods often preclude phylogenetically well-informed comparative studies due to the relatively small sample sizes that can be processed in a reasonable amount of time. As students of animal flight aim to expand our range of data, careful attention to selection of study species, with an eye to meaningful biological questions as well as other concerns, could advance our knowledge of the evolution of flight. Much of the research on flight and wing structure in phyllostomids has been carried out in a very limited number of specialized species, almost all from three genera (*Artibeus*, *Carollia*, and *Glossophaga*). This limits our understanding of multiple facets of the biology of flight. However, some aspects of wing structure that are only recently receiving attention, such as retention of intramembranous muscle in the plagiopatagium and conservation of elastin bundle structure (Cheney et al 2017), and probably trailing-edge stabilization and propatagium control effected in part by the occipitopollicalis, are similar in most bats, across species with diverse ecologies and phylogenetic affinities. Other traits, such as details of attachment of plagiopatagiales to the elastin bundle network, may differ between species that share many similarities in life history, are closely related, and have no major differences in flight performance, as we have observed in comparisons of *Artibeus lituratus* and *Carollia perspicillata*. The role of body size is yet another factor whose influence is likely important but has not been clearly distinguished; many more data will be required to resolve this kind of question.

Effectively linking discoveries concerning structure and function with in-depth understanding of bats' interactions with each other and their environments — that is, integrating biomechanics and ecology into the emerging field of ecomechanics (Carrington 2002; Denny and Gaylord 2010; Denny and Helmuth 2009; Higham et al. 2015) — has advanced more slowly than basic insight into flight mechanisms. However, if we are to understand how and why bats evolved particular patterns of wing architecture, and how wings are employed to execute specific flight behaviors, we suggest that this integration of disciplines is essential. The wealth of impressive technologies now available enables us to collect and analyze rich and novel information about structure and function, and we look forward to employing new, innovative, and powerful tools to further reveal the mysteries of this relationship.

To date, few studies directly link factors that contribute specifically to locomotor performance, such as musculoskeletal morphology, limb kinematics, or locomotor forces, to quantitative observations of any locomotor behavior in natural environments (but see,

e.g., Combes et al. 2012; Ray et al. 2016). Quantifying the flight abilities of bats in nature is particularly difficult, but lab-based flight research often focuses on the behaviors most effectively controlled by investigators, not necessarily those of greatest ecological significance to the model species. Natural behaviors are notoriously difficult to elicit in lab settings (Corcoran and Conner 2016; Warrick et al. 2016) and might thus be neglected if we limit our research to laboratory studies.

Linking variation in wing architecture to flight performance in a clear, unambiguous fashion is also a major challenge (Bomphrey et al. 2016). As a result, the mechanistic basis of locomotor performance in naturally occurring behavioral interactions has rarely been treated quantitatively (but see Shelton et al. 2014; Wilson et al. 2013), and it remains difficult to make strong connections between our growing mechanistic understanding of bat flight and the demands placed on bats during relevant and ecologically meaningful performance in natural environments. It is now possible to envision uniting mechanistic understanding of the morphological, kinematic, and aerodynamic basis of bat flight with quantitative analyses of flight performance, thanks to advances for studying flight behavior in the field (Conner and Corcoran 2012; Theriault et al. 2014) and in the lab (Bergou et al. 2015; Corcoran and Conner 2016; Iriarte-Díaz et al. 2011). The rich ecological, phylogenetic, and morphological diversity represented by the phyllostomids makes them an intuitive and promising choice for integrative studies of this kind.

Acknowledgments

We are grateful to NSF (IOS 1145549 and CMMI 1426338) and AFOSR (FA9550-12-1-0210, monitored by Dr. Douglas Smith and FA9550-12-1-0301, monitored by Dr. Patrick Bradshaw) for supporting the work discussed here. This work depended on contributions of many members of our lab: Kenny Breuer, Nickolay Hristov, Tatjana Hubel, Elissa Johnson, Nicolai Konow, and Dan Riskin. Jorn Cheney made tremendous contributions to our thinking on all topics discussed here. Ted Fleming and Gary Kwiecinski provided suggestions that greatly improved this text, and we thank John Hermanson for his especially thoughtful and insightful comments. We thank Brock Fenton and Nancy Simmons for their generous leadership of research retreats in Belize, which nourish collaboration, creativity, and community around the study of bats and which have contributed to the ongoing scientific excellence of many who study phyllostomids.

References

Albertani, R., T. Hubel, S. M. Swartz, K. S. Breuer, and J. Evers. 2011. In-flight wing-membrane strain measurements on bats. Pp. 437–455 *in*: Experimental and Applied Mechanics. (T. Proulx, ed.). Conference Proceedings of the Society for Experimental Mechanics Series. Springer, New York.

Alexander, R. M., and A. Vernon. 1975. The mechanics of hopping by kangaroos (Macropodidae). Journal of Zoology 177:265–303.

Altenbach, J. S., and J. W. Hermanson. 1987. Bat flight muscle function and the scapulo-humeral lock. Pp. 100–118 *in*: Recent Advances in the Study of Bats (M. B. Fenton, P. Racey, and J. V. M. Rayner, eds.). Cambridge University Press, Cambridge.

Altshuler, D. L., J. W. Bahlman, R. Dakin, A. H. Gaede, B. Goller, D. Lentink, P. S. Segre et al. 2015. The biophysics of bird flight: functional relationships integrate aerodynamics, morphology, kinematics, muscles, and sensors. Canadian Journal of Zoology 93:961–975.

Askew, G. N., and R. L. Marsh. 2002. Muscle designed for maximum short-term power output: quail flight muscle. Journal of Experimental Biology 205:2153–2160.

Bergou, A. J., S. M. Swartz, H. Vejdani, D. K. Riskin, L. Reimnitz, G. Taubin, and K. S. Breuer. 2015. Falling with style: bats perform complex aerial rotations by adjusting wing inertia. PLOS Biology 13:e1002297.

Bewick, G. S., and R. W. Banks. 2015. Mechanotransduction in the muscle spindle. Pflügers Archiv–European Journal of Physiology 467:175–190.

Biewener, A. A., D. D. Konieczynski, and R. V. Baudinette. 1998. *In vivo* muscle force-length behavior during steady-speed hopping in tammar wallabies. Journal of Experimental Biology 201:1681–1694.

Blickhan, R., and J. Y. Cheng. 1994. Energy-storage by elastic mechanisms in the tail of large swimmers—a reevaluation. Journal of Theoretical Biology 168:315–321.

Bomphrey, R. J. 2012. Advances in animal flight aerodynamics through flow measurement. Evolutionary Biology 39:1–11.

Bomphrey, R. J., T. Nakata, P. Henningsson, and H. T. Lin. 2016. Flight of the dragonflies and damselflies. Philosophical Transactions of the Royal Society B–Biological Sciences 371.

Botterman, B. R., M. D. Binder, and D. G. Stuart. 1978. Functional anatomy of association between motor units and muscle receptors. American Zoologist 18:135–152.

Bullimore, S. R., and J. F. Burn. 2005. Scaling of elastic energy storage in mammalian limb tendons: do small mammals really lose out? Biology Letters 1:57–59.

Camp, A. L., H. C. Astley, A. M. Horner, T. J. Roberts, and E. L.

Brainerd. 2016. Fluoromicrometry: a method for measuring muscle length dynamics with biplanar videofluoroscopy. Journal of Experimental Zoology Part A: Ecological Genetics and Physiology 325:399–408.

Carrington, E. 2002. The ecomechanics of mussel attachment: from molecules to ecosystems. Integrative and Comparative Biology 42:846–852.

Carter, R. T., and R. A. Adams. 2015. Postnatal ontogeny of the cochlea and flight ability in Jamaican fruit bats (Phyllostomidae) with implications for the evolution of echolocation. Journal of Anatomy 226:301–308.

Cheney, J. A., J. J. Allen, and S. M. Swartz. 2017. Diversity in the organization of elastin bundles and intramembranous muscles in bat wings. Journal of Anatomy 230:510–523.

Cheney, J. A., N. Konow, A. Bearnot, and S. M. Swartz. 2015. A wrinkle in flight: the role of elastin fibres in the mechanical behaviour of bat wing membranes. Journal of the Royal Society Interface 12:20141286.

Cheney, J. A., N. Konow, K. M. Middleton, K. S. Breuer, T. J. Roberts, E. L. Giblin, and S. M. Swartz. 2014. Membrane muscle function in the compliant wings of bats. Bioinspiration and Biomimetics 9:025007.

Cheney, J. A., D. Ton, N. Konow, D. K. Riskin, K. S. Breuer, and S. M. Swartz. 2014. Hindlimb motion during steady flight of the lesser dog-faced fruit bat, *Cynopterus brachyotis*. PLOS ONE 9:e98093.

Chin, D. D., and D. Lentink. 2016. Flapping wing aerodynamics: from insects to vertebrates. Journal of Experimental Biology 219:920–932.

Combes, S. A., D. E. Rundle, J. M. Iwasaki, and J. D. Crall. 2012. Linking biomechanics and ecology through predator-prey interactions: flight performance of dragonflies and their prey. Journal of Experimental Biology 215:903–913.

Conner, W. E., and A. J. Corcoran. 2012. Sound strategies: the 65-million-year-old battle between bats and insects. Annual Review of Entomology 57:21–39.

Corcoran, A. J., and W. E. Conner. 2016. How moths escape bats: predicting outcomes of predator-prey interactions. Journal of Experimental Biology 219:2704–2715.

Crowley, G. V., and L. S. Hall. 1994. Histological observations on the wing of the grey-headed flying fox (*Pteropus poliocephalus*) (Chiroptera:Pteropodidae). Australian Journal of Zoology 42:215–236.

Curtis, A. A., and N. B. Simmons. 2017. Unique turbinal morphology in horseshoe bats (Chiroptera: Rhinolophidae). Anatomical Record 300:309–325.

Davis, R. B., C. F. Herreid, and H. L. Short. 1962. Mexican free-tailed bats in Texas. Ecological Monographs 32:311–346.

Denny, M. W., and B. Gaylord. 2010. Marine ecomechanics. Annual Review of Marine Science 2:89–114.

Denny, M., and B. Helmuth. 2009. Confronting the physiological bottleneck: a challenge from ecomechanics. Integrative and Comparative Biology 49:197–201.

Dickinson, M. H. 1996. Unsteady mechanisms of force generation in aquatic and aerial locomotion. American Zoologist 36:537–554.

Dickinson, M. H., C. T. Farley, R. J. Full, M. A. R. Koehl,

R. Kram, and S. Lehman. 2000. How animals move: an integrative view. Science 288:100–106.

Dimitriou, M., and B. B. Edin. 2010. Human muscle spindles act as forward sensory models. Current Biology 20:1763–1767.

Dudley, R. 2000. The Biomechanics of Insect Flight: Form, Function, Evolution. Princeton University Press, Princeton, NJ.

Ellington, C. P., C. van den Berg, A. P. Willmott, and A. L. R. Thomas. 1996. Leading-edge vortices in insect flight. Nature 384:626–630.

Fleming, T. H., and P. Eby. 2003. Ecology of bat migration. Pp. 156–208 in Bat Ecology (T. H. Kunz and M. B. Fenton, eds.). University of Chicago Press, Chicago.

Garvey, W. C. Jimenez, and B. Carpenter. 1991. A modified Verhoeff elastic–van Gieson stain. Journal of Histotechnology 14:113–115.

Gignac, P. M., N. J. Kley, J. A. Clarke, M. W. Colbert, A. C. Morhardt, D. Cerio, I. N. Cost et al. 2016. Diffusible iodine-based contrast-enhanced computed tomography (diceCT): an emerging tool for rapid, high-resolution, 3-D imaging of metazoan soft tissues. Journal of Anatomy 228:889–909.

Håkansson, J., A. Hedenström, Y. Winter, and L. C. Johansson. 2015. The wake of hovering flight in bats. Journal of the Royal Society Interface 12:20150357.

Hedenström, A., L. C. Johansson, M. Wolf, R. von Busse, Y. Winter, and G. R. Spedding. 2007. Bat flight generates complex aerodynamic tracks. Science 316:894–897.

Hedrick, T. L., S. A. Combes, and L. A. Miller. 2015. Recent developments in the study of insect flight. Canadian Journal of Zoology 93:925–943.

Henningsson, P., F. T. Muijres, and A. Hedenström. 2011. Time-resolved vortex wake of a common swift flying over a range of flight speeds. Journal of the Royal Society Interface 8:807–816.

Henningsson, P., G. R. Spedding, and A. Hedenström. 2008. Vortex wake and flight kinematics of a swift in cruising flight in a wind tunnel. Journal of Experimental Biology 211: 717–730.

Hermanson, J. W. 1997. Chiropteran muscle biology: a perspective from molecules to function. Pp. 127–139 in Bats: Phylogeny, Morphology, Echolocation, and Conservation Biology. (T. H. Kunz and P. A. Racey, eds.). Smithsonian Institution Press, Washington, DC.

Hermanson, J. W., and J. S. Altenbach. 1983. The functional anatomy of the shoulder of the pallid bat, Antrozous pallidus. Journal of Mammalogy 64:62–75.

Hermanson, J. W., and J. S. Altenbach. 1985. Functional anatomy of the shoulder and arm of the fruit-eating bat Artibeus jamaicensis. Journal of Zoology 205:157–177.

Hermanson, J. W., M. A. Cobb, W. A. Schutt, F. Muradali, and J. M. Ryan. 1993. Histochemical and myosin composition of vampire bat (Desmodus rotundus) pectoralis muscle targets a unique locomotory niche. Journal of Morphology. 217: 347–356.

Hermanson, J. W., and R. C. Foehring. 1988. Histochemistry of flight muscles in the Jamaican fruit bat, Artibeus jamaicensis—implications for motor control. Journal of Morphology 196: 353–362.

Hermanson, J. W., J. H. Ryan, M. A. Cobb, J. Bentley, and W. A. Schutt. 1998. Histochemical and electrophoretic analysis of the primary flight muscle of several phyllostomid bats. Canadian Journal of Zoology 76:1983–1992.

Higham, T. E., W. J. Stewart, and P. C. Wainwright. 2015. Turbulence, temperature, and turbidity: the ecomechanics of predator-prey interactions in fishes. Integrative and Comparative Biology 55:6–20.

Holbrook, K., and G. F. Odland. 1978. A collagen and elastic network in the wing of the bat. Journal of Anatomy 126:21–36.

Hubel, T. Y., N. I. Hristov, S. M. Swartz, and K. S. Breuer. 2012. Changes in kinematics and aerodynamics over a range of speeds in Tadarida brasiliensis, the Brazilian free-tailed bat. Journal of the Royal Society Interface 9:1120–1130.

Hubel, T. Y., N. I. Hristov, S. M. Swartz, and K. S. Breuer. 2016. Wake structure and kinematics in two insectivorous bats. Philosophical Transactions of the Royal Society B–Biological Sciences 371:20150385.

Hubel, T. Y., D. K. Riskin, S. M. Swartz, and K. S. Breuer. 2010. Wake structure and wing kinematics: the flight of the lesser dog-faced fruit bat, Cynopterus brachyotis. Journal of Experimental Biology 213:3427–3440.

Hubner, J. P., and T. Hicks. 2011. Trailing-edge scalloping effect on flat-plate membrane wing performance. Aerospace Science and Technology 15:670–680.

Humason, G. L. 1962. Animal Tissue Techniques. 2nd ed. W. H. Freeman & Co., San Francisco.

Humphry, G. M. 1869. The myology of the limbs of Pteropus. Journal of Anatomy and Physiology 3:289–319.

Iriarte-Díaz, J., D. K. Riskin, D. J. Willis, K. S. Breuer, and S. M. Swartz. 2011. Whole-body kinematics of a fruit bat reveal the influence of wing inertia on body accelerations. Journal of Experimental Biology 214:1546–1553.

Johansson, L. C., J. Håkansson, L. Jakobsen, and A. Hedenström. 2016. Ear-body lift and a novel thrust generating mechanism revealed by the complex wake of brown long-eared bats (Plecotus auritus). Scientific Reports 6:24886.

Johansson, L. C., M. Wolf, R. von Busse, Y. Winter, G. R. Spedding, and A. Hedenström. 2008. The near and far wake of Pallas' long tongued bat (Glossophaga soricina). Journal of Experimental Biology 211:2909–2918.

Katija, K., and J. O. Dabiri. 2008. In situ field measurements of aquatic animal-fluid interactions using a self-contained underwater velocimetry apparatus (SCUVA). Limnology and Oceanography: Methods 6:162–171.

Ker, R. F., R. M. Alexander, and M. B. Bennett. 1988. Why are mammalian tendons so thick? Journal of Zoology 216: 309–324.

Koehl, M. A. R. 1996. When does morphology matter? Annual Review of Ecology and Systematics 27:501–542.

Konow, N., J. A. Cheney, T. J. Roberts, J. Iriarte-Díaz, J. R. S. Waldman, and S. M. Swartz. 2017. Speed dependent modulation of wing muscle recruitment intensity and kinematics in two bat species. Journal of Experimental Biology 220:1820–1829.

Konow, N., J. A. Cheney, T. J. Roberts, R. J. R. S. Waldman, and S. M. Swartz. 2015. Spring or string: does tendon elastic action influence wing muscle mechanics in bat flight? Proceedings of the Royal Society of London B, 282:20151832.

Lee, A. H., and E. L. R. Simons. 2015. Wing bone laminarity is not an adaptation for torsional resistance in bats. PeerJ 3:e823.

Lindhe Norberg, U. M., and Y. Winter. 2006. Wing beat kinematics of a nectar-feeding bat, *Glossophaga soricina*, flying at different flight speeds and Strouhal numbers. Journal of Experimental Biology 209:3887–3897.

Lionikas, A., C. J. Smith, T. L. Smith, L. Bunger, R. W. Banks, and G. S. Bewick. 2013. Analyses of muscle spindles in the soleus of six inbred mouse strains. Journal of Anatomy 223:289–296.

McCracken, G. F., K. Safi, T. H. Kunz, D. K. N. Dechmann, S. M. Swartz, and M. Wikelski. 2016. Airplane tracking documents the fastest flight speeds recorded for bats. Royal Society Open Science 3:160398.

Meyer, W., K. Neurand, R. Schwarz, T. Bartels, and H. Althoff. 1994. Arrangement of elastic fibers in the integument of domesticated mammals. Scanning Microscopy 8:375–391.

Meyers, R. A., and J. W. Hermanson. 2006. Horse soleus muscle: Postural sensor or vestigial structure? Anatomical Record Part A: Discoveries in Molecular Cellular and Evolutionary Biology 288a:1068–1076.

Muijres, F. T., P. Henningsson, M. Stuiver, and A. Hedenstrom. 2012. Aerodynamic flight performance in flap-gliding birds and bats. Journal of Theoretical Biology 306:120–128.

Muijres, F. T., L. C. Johansson, R. Barfield, M. Wolf, G. R. Spedding, and A. Hedenström. 2008. Leading-edge vortex improves lift in slow-flying bats. Science 319:1250–1253.

Muijres, F. T., L. C. Johansson, Y. Winter, and A. Hedenström. 2011. Comparative aerodynamic performance of flapping flight in two bat species using time-resolved wake visualization. Journal of the Royal Society Interface 8:1418–1428.

Muijres, F. T., L. C. Johansson, Y. Winter, and A. Hedenström. 2014. Leading edge vortices in lesser long-nosed bats occurring at slow but not fast flight speeds. Bioinspiration and Biomimetics 9:025006.

Muñoz-Garcia, A., J. Ro, J. D. Reichard, T. H. Kunz, and J. B. Williams. 2012. Cutaneous water loss and lipids of the stratum corneum in two syntopic species of bats. Comparative Biochemistry and Physiology A: Molecular and Integrative Physiology 161:208–215.

Norberg, U. M. 1968. Functional osteology and myology of the wing of *Plecotus auritus* Linnaeus (Chiroptera). Arkiv für Zoologi 22:483–543.

Norberg, U. M. 1972a. Functional osteology and myology of the wing of the dog-faced bat *Rousettus aegypticus* (É. Geoffroy) (Pteropodidae). Zeitschrift für Morphologie der Tiere 73:1–44.

Norberg, U. M. 1972b. Functional osteology and myology of the wing of dog-faced bat *Rousettus aegyptiacus* (É. Geoffroy) (Mammalia, Chiroptera). Zeitschrift für Morphologie der Tiere 73:1–44.

Norberg, U. M., and J. M. V. Rayner. 1987. Ecological morphology and flight in bats (Mammalia; Chiroptera): wing adaptations, flight performance, foraging strategy and echolocation. Philosophical Transactions of the Royal Society of London B 316:335–427.

O'Mara, M. T., M. Wikelski, and D. K. N. Dechmann. 2014. 50 years of bat tracking: device attachment and future directions. Methods in Ecology and Evolution 5:311–319.

Pannkuk, E. L., N. W. Fuller, P. R. Moore, D. F. Gilmore, B. J. Savary, and T. S. Risch. 2014. Fatty acid methyl ester profiles of bat wing surface lipids. Lipids 49:1143–1150.

Quinn, T., and J. Baumel. 1993. Chiropteran tendon locking mechanism. Journal of Morphology 216:197–208.

Ray, R. P., T. Nakata, P. Henningsson, and R. J. Bomphrey. 2016. Enhanced flight performance by genetic manipulation of wing shape in *Drosophila*. Nature Communications 7:10851.

Rayner, J. M. V. 1979a. A vortex theory of animal flight. Part 1. The vortex wake of a hovering animal. Journal of Fluid Mechanics 91:697–730.

Rayner, J. M. V. 1979b. A vortex theory of animal flight. Part 2. The forward flight of birds. Journal of Fluid Mechanics 91:731–763.

Reichard, J. D., S. I. Prajapati, S. N. Austad, C. Keller, and T. H. Kunz. 2010. Thermal windows on Brazilian free-tailed bats facilitate thermoregulation during prolonged flight. Integrative and Comparative Biology 50:358–370.

Richmond, F. J., and V. C. Abrahams. 1975. Morphology and distribution of muscle spindles in dorsal muscles of the cat neck. Journal of Neurophysiology 38:1322–1339.

Riskin, D. K., J. W. Bahlman, T. Y. Hubel, J. M. Ratcliffe, T. H. Kunz, and S. M. Swartz. 2009. Bats go head-under-heels: the biomechanics of landing on a ceiling. Journal of Experimental Biology 212:945–953.

Roberts, T. J., and E. Azizi. 2011. Flexible mechanisms: the diverse roles of biological springs in vertebrate movement. Journal of Experimental Biology 214:353–361.

Rojratsirikul, P., Z. Wang, and I. Gursul. 2010. Effect of prestrain and excess length on unsteady fluid-structure interactions of membrane airfoils. Journal of Fluids and Structures 26: 359–376.

Sapir, N., N. Horvitz, D. K. N. Dechmann, J. Fahr, and M. Wikelski. 2014. Commuting fruit bats beneficially modulate their flight in relation to wind. Proceedings of the Royal Society B–Biological Sciences 281:20140018.

Schöbl, J. 1871. Die flughaut der Fledermäuse, namentlich die Endigung ihrer Nerven. Archiv für Mikroskopische Anatomie 7:1–31.

Schumacher, S. 1932. Muskeln und Nerven der Fledermausflughaut. Nach Untersuchungen an *Pteropus*. Anatomy and Embryology 97:610–621.

Shelton, R. M., B. E. Jackson, and T. L. Hedrick. 2014. The mechanics and behavior of cliff swallows during tandem flights. Journal of Experimental Biology 217:2717–2725.

Shyy, W., H. Aono, C.-K. Kang, and H. Liu. 2013. An Introduction to Flapping Wing Aerodynamics: Cambridge Aerospace Series. Cambridge University Press, Cambridge.

Skulborstad, A. J., S. M. Swartz, and N. C. Goulbourne. 2015. Biaxial mechanical characterization of bat wing skin. Bioinspiration and Biomimetics 10:036004.

Socha, J. J., F. Jafari, Y. Munk, and G. Byrnes. 2015. How animals glide: from trajectory to morphology. Canadian Journal of Zoology 93:901–924.

Spedding, G. R., A. Hedenström, and M. Rosen. 2003. A family of vortex wakes generated by a thrush nightingale in free flight over its entire range of flight speeds. Journal of Experimental Biology 206:2313–2344.

Sterbing-D'Angelo, S., M. Chadha, C. Chiu, B. Falk, W. Xian, J. Barcelo, J. M. Zook, and C. F. Moss. 2011. Bat wing sensors support flight control. Proceedings of the National Academy of Sciences, USA 108:11291–11296.

Sterbing-D'Angelo, S. J., M. Chadha, K. L. Marshall, and C. F. Moss. 2017. Functional role of airflow-sensing hairs on the bat wing. Journal of Neurophysiology 117:705–712.

Stockwell, E. F. 2001. Morphology and flight manoeuvrability in New World leaf-nosed bats (Chiroptera : Phyllostomidae). Journal of Zoology 254:505–514.

Strickler, T. L. 1978. Functional osteology and myology of the shoulder in the Chiroptera. Pp. 1–198 in: Contributions to Vertebrate Evolution, vol. 4. (M. K. Hecht and F. S. Szalay, eds.). S. Karger, New York.

Studier, E. H. 1972. Some physical properties of the wing membranes of bats. Journal of Mammalogy 53:623–625.

Suarez, R. K., K. C. Welch, S. K. Hanna, and M. L. G. Herrera. 2009. Flight muscle enzymes and metabolic flux rates during hovering flight of the nectar bat, Glossophaga soricina: further evidence of convergence with hummingbirds. Comparative Biochemistry and Physiology A: Molecular and Integrative Physiology 153:136–140.

Swartz, S. M. 2015. Advances in animal flight studies. Canadian Journal of Zoology 93:v–vi.

Swartz, S. M., M. S. Groves, H. D. Kim, and W. R. Walsh. 1996. Mechanical properties of bat wing membrane skin. Journal of Zoology 239:357–378.

Swartz, S. M., and N. Konow. 2015. Advances in the study of bat flight: the wing and the wind. Canadian Journal of Zoology 93:977–990.

Swartz, S. M., and K. M. Middleton. 2008. Biomechanics of the bat limb skeleton: scaling, material properties and mechanics. Cells Tissues Organs 187:59–84.

Theriault, D. H., N. W. Fuller, B. E. Jackson, E. Bluhm, D. Evangelista, Z. Wu, M. Betke et al. 2014. A protocol and calibration method for accurate multi-camera field videography. Journal of Experimental Biology 217:1843–1848.

Timpe, A., Z. Zhang, J. Hubner, and L. Ukeiley. 2013. Passive flow control by membrane wings for aerodynamic benefit. Experiments in Fluids 54:1–23.

Tobalske, B. W., and A. A. Biewener. 2008. Contractile properties of the pigeon supracoracoideus during different modes of flight. Journal of Experimental Biology 211:170–179.

Vaughan, T. A. 1959. Functional morphology of three bats: Eumops, Myotis, Macrotus. University of Kansas Publications, Museum of Natural History 12:1–153.

Vaughan, T. A. 1970. The muscular system. Pp. 140–194 in: The Biology of Bats. (W. A. Wimsatt, ed). Academic Press, New York.

Vogel, S. 2013, Comparative Biomechanics: Life's Physical World. Princeton University Press, Princeton, NJ.

Voigt, C. C., D. H. Kelm, and G. H. Visser. 2006. Field metabolic rates of phytophagous bats: do pollination strategies of plants make life of nectar-feeders spin faster? Journal of Comparative Physiology B: Biochemical Systemic and Environmental Physiology 176:213–222.

Voigt, C. C., and D. Lewanzik. 2011. Trapped in the darkness of the night: thermal and energetic constraints of daylight flight in bats. Proceedings of the Royal Society B–Biological Sciences 278:2311–2317.

Voigt, C. C., and Y. Winter. 1999. Energetic cost of hovering flight in nectar-feeding bats (Phyllostomidae: Glossophaginae) and its scaling in moths, birds and bats. Journal of Comparative Physiology B:Biochemical Systemic and Environmental Physiology 169:38–48.

von Busse, R., A. Hedenstrom, Y. Winter, and L. C. Johansson. 2012. Kinematics and wing shape across flight speed in the bat, Leptonycteris yerbabuenae. Biology Open 1:1226–1238.

von Busse, R., S. M. Swartz, and C. C. Voigt. 2013. Flight metabolism in relation to speed in Chiroptera: testing the U-shape paradigm in the short-tailed fruit bat Carollia perspicillata. Journal of Experimental Biology 216:2073–2080.

von Busse, R., R. M. Waldman, S. M. Swartz, C. C. Voigt, and K. S. Breuer. 2014. The aerodynamic cost of flight in the short-tailed fruit bat (Carollia perspicillata): comparing theory with measurement. Journal of the Royal Society Interface 11:20140147.

Wainwright, S. A., W. D. Biggs, J. D. Currey, and J. M. Gosline. 1976. Mechanical Design In Organisms. Edward Arnod, London.

Wang, B. B., Q. Liao, J. N. Xiao, and H. A. Bootsma. 2013. A free-floating PIV system: measurements of small-scale turbulence under the wind wave surface. Journal of Atmospheric and Oceanic Technology 30:1494–1510.

Warrick, D. R., T. L. Hedrick, A. A. Biewener, K. E. Crandell, and B. W. Tobalske. 2016. Foraging at the edge of the world: low-altitude, high-speed manoeuvering in barn swallows. Philosophical Transactions of the Royal Society B–Biological Sciences, 371.

Wegst, U. G. K., and M. F. Ashby. 2004. The mechanical efficiency of natural materials. Philosophical Magazine 84:2167–2181.

Weller, T. J., K. T. Castle, F. Liechti, C. D. Hein, M. R. Schirmacher, and P. M. Cryan. 2016. First direct evidence of long-distance seasonal movements and hibernation in a migratory bat. Scientific Reports 6:34585.

Wilson, A. M., J. C. Lowe, K. Roskilly, P. E. Hudson, K. A. Golabek, and J. W. McNutt. 2013. Locomotion dynamics of hunting in wild cheetahs. Nature 498:185–189.

Wilson, A. M., J. C. Watson, and G. A. Lichtwark. 2003. Biomechanics: a catapult action for rapid limb protraction. Nature 421:35–36.

Wimsatt, W. A. 1969. Transient behavior nocturnal activity patterns and feeding efficiency of vampire bats (Desmodus rotundus) under natural conditions. Journal of Mammalogy 50:233–244.

Wolf, M., L. C. Johansson, R. von Busse, Y. Winter, and A. Hedenstrom. 2010. Kinematics of flight and the relationship to the vortex wake of a Pallas' long tongued bat (Glossophaga soricina). Journal of Experimental Biology 213:2142–2153.

10

The Relationship between Physiology and Diet

*Ariovaldo P. Cruz-Neto
and L. Gerardo Herrera M.*

Introduction

What an animal eats is one of the most important factors underlying the diversification of ecological, behavioral, morphological, and physiological traits of organisms (Karasov and Douglas 2013). A classic example of correlated evolution between diet and several life-history traits are bats of the New World family Phyllostomidae. The most distinctive feature of this family is the extraordinary diversity of food items that its members consume. Fruit pulp, pollen grains, flower nectar, seeds, leaves, arthropods, small vertebrates, and blood encompass the range of food products used by these bats. This diversity is considered to be the main driver of the adaptive radiation experienced by phyllostomids as they evolved from arthropod-eating ancestors (Baker et al. 2012). Feeding on this array of food options implies adapting to contrasting ecological conditions in order to prey upon them. This has served as inspiration to several studies linking diet diversification to the morphological diversity present in these bats (Bolzan et al. 2015; Monteiro and Nogueira 2011; Rojas et al. 2012). This approach has proven useful for understanding how phyllostomid morphology allowed increased access to a plethora of food resources available during the early stages of ecological diversification. However, understanding how animals interact with their environment during this diversification process also requires knowledge of the physiological processes that determine their use of food resources (Karasov and Martínez del Río 2007). This principle is particularly important for phyllostomids due to the dissimilar nature of the items included in their diet. Flower nectar and fruit pulp are watery products composed mainly of sugars (i.e., glucose, fructose, and sucrose; Baker and Baker 1982), and pollen grains contain important amounts of amino acids, though access is limited by their tough outer wall (Stanley and Linskens 1974). Animal products eaten by phyllostomids are rich in amino acids, but in the case of blood, its ingestion involves processing large volumes of water. Some aspects of the digestive and excretory physiology of phyllostomids have been examined previously in relation to diet, but these analyses lack a strict phylogenetic approach (Cassoti et al. 2006; Schondube et al. 2001).

Like any other mammal, including bats from other families, the rate of energy expenditure by phyllostomids is mostly dictated by body mass (Cruz-Neto and Jones 2006; Cruz-Neto et al. 2001; McNab 2003). However, when this allometric effect is accounted for, there is still significant residual variability. The food-habit hypothesis (Cruz-Neto and Bozinovic 2004) posits that certain aspects of the diet, such as its quality, availability, and predictability, affect this residual variability, but attempts to test this hypothesis in phyllostomids have been controversial and have elicited acrimonious debate in the literature (Cruz-Neto et al. 2001; McNab 2003). Much of this controversy revolves around the effects of phylogeny and its colinearity with the broad dietary classification used in prior analyses (see also Speakman and Thomas 2003).

In this chapter, we complement previous attempts to link dietary and morphological diversity in phyllostomids by adding a physiological dimension. We analyzed the extent to which the dietary diversification observed in phyllostomids parallels diversification in whole-organism physiological traits that are essential for understanding the way in which these bats interact with their food. In particular, we focused our analysis on the interplay between diet and digestive, excretory, and metabolic physiology and explored some of its ecological consequences. We used a strict phylogenetic approach to test the relationship between several traits and diet when sufficient information for phyllostomids with different feeding habits was available: activity of intestinal enzymes, urine concentration, and basal metabolic rates. We also included kidney morphology in our analyses because this trait affects the function of the excretory system. This relationship has been examined previously but in a less strict phylogenetic approach (Cassoti et al. 2006; Schondube et al. 2001). For physiological traits for which data are scant or restricted to some trophic guilds, we review the information available in the literature to explore the relationship between functional patterns and diet.

Data Handling and Analysis

We reanalyzed the relationship between diet and several physiological traits in phyllostomid bats using both conventional and phylogenetic-informed analyses. The following traits were analyzed: intestinal activities of sucrase ($n = 15$ species), trehalase ($n = 14$), and aminopeptidase ($n = 14$); kidney relative medullary thickness (RMT, $n = 20$), relative cortex volume (RCV, $n = 21$). and relative medullary volume (RMV, $n = 21$); urine concentration capacity (U_{osm}, $n = 27$); and basal metabolic rate (BMR; $n = 33$). Enzyme activity was standardized per intestine measurement units following Schondube et al. (2001), but we used intestine mass rather than intestine area. Data for these parameters, along with body mass (M_b), were compiled from the literature and are available as supplementary files.

Dietary classification in comparative analyses has always been challenging, especially for clades with a high diversity of food habits as occurs in phyllostomid bats. Usually, a monotypic qualitative categorization is used for this purpose, but this approach is questionable because, among many other criticisms, there has been no standardization about the criteria used to make these categorizations and, most importantly, because it does not communicate the diversity of food used (Pineda-Muñoz and Alroy 2014). These constraints can be severe and lead to several problems and conflicting results when one is trying to correlate diversification of diet with the diversification of physiological traits (for an example of this problem with phyllostomid bats, see Cruz-Neto et al. [2001] and McNab [2003]). A possible solution to this caveat is to use a quantitative description of the diet (Pineda-Munoz and Alroy 2014). In this chapter, we used this approach following methodology described by Monteiro and Nogueira (2011) in their analysis of the association between diet diversification and the evolution of cranial morphology in phyllostomids. Briefly, we first generated a general matrix that categorized the dietary breadth of phyllostomid bats. This matrix was based on the relative usage (0—absent, 1—complementary, 2—predominant, 3—strict) of five main dietary categories (insectivory, carnivory, frugivory, nectarivory, and sanguivory). The rank categories followed Wetterer et al. (2000), with suggested modifications made by Monteiro and Nogueira (2011) and by M. Nogueira (personal communication). We then used Principal Component Analysis (PCA) to calculate the PCs of a correlation matrix among the diet variables. The diet PC scores, which reflected the relative usage of a given food item, were then used as independent variables in the analysis described below. The results from this analysis showed that the three first components of the PCA explained 96.9% of the variance in diet. The first diet PC (PC1) explained 36.6% of the variance and was positively correlated with insectivory (0.60) and carnivory (0.74), and negatively correlated with frugivory (−0.61) and nectarivory (−0.71). Thus, PC1 discriminates the diet of phyllostomids into two main groups. The first group includes animalivorous species that feed predominantly on diets with a high protein content but, except for vampire bats, with relatively low water levels (insects or vertebrates), without any distinction based on the nature of complementary dietary items. The second group includes species that feed predominantly on items with high water and carbohydrate content (fruits or nectar), with insects being complementary to the diet. The second diet PC (PC2) explained 32.4% of the variance and was positively correlated with sanguivory (0.96) and negatively correlated with frugivory (−0.45), nectarivory (−0.12), carnivory (−0.24), and insectivory (−0.65). Thus, PC2 basically discriminates bats that feed exclusively on blood. The third diet PC (PC3) explained 17.9% of the variance and discriminates between frugivory (0.64) and nectarivory (−0.61). These dietary trends, especially as revealed by the first two PC scores, are very similar to what was observed by Monteiro and Nogueira (2011) in their analysis of the coevolution between phyllostomid cranial morphology and diet.

For each of the physiological traits analyzed, we first tested for allometric trends by conventional analysis using ordinary least square regression (OLS) and, when deemed necessary to account for phylogenetic effects (see below), also by phylogenetic generalized least squares regression (PGLS; Rohlf 2001). Data for each trait (except for data on enzymatic activities) and the associated M_b value were log-transformed (base 10) before these analyses. For traits where we found a significant allometric effect (by OLS and PGLS), we used an F-test, as developed by Withers et al. (2006), to test the extent to which correction for the phylogenetic effect actually reduces the residual variability of the allometric effect. Where an allometric trend was detected, we performed a regression analysis using conventional and phylogenetically corrected residuals against the three diet PC scores to test for putative associations between the traits and diet. When no allometric effect

was evident, such regressions were undertaken using the conventional raw data and phylogenetically transformed data.

All phylogenetically informed analyses were carried out using a visual-basic (V6) program provided by Dr. Phil Withers (University of Western Australia) and a phylogenetic tree based on Rojas et al. (2016). We first used the k^*-statistic (Blomberg et al. 2003) to calculate the phylogenetic signal for each individual trait. Under a Brownian model of evolution, $k^* = 0$ indicates no phylogenetic signal, $k^* = 1$ indicates a perfect match between the trait value and the phylogeny, and $k^* > 1$ indicates that closely related species are more similar to each other than would be expected by Brownian motion. The significance of a particular k^* value was then assessed by a randomization test (Blomberg et al. 2003). All physiological traits, except those related to aminopeptidase activity ($k^* = 0.14$; $p = 0.86$), had a significant phylogenetic signal. The phylogenetic signal was lower than expected for BMR ($k^* = 0.83$; $p < 0.01$) and RMT ($k^* = 0.67$; $p = 0.02$), and higher than expected for U_{osm} ($k^* = 1.64$; $p < 0.001$), RMV ($k^* = 1.10$; $p < 0.01$), RCV ($k^* = 1.12$; $p < 0.001$), and trehalase ($k^* = 1.26$; $p < 0.01$) and sucrase ($k^* = 1.15$; $p < 0.01$) activities. M_b had a significant and lower-than-expected phylogenetic signal ($k^* = 0.86$; $p < 0.001$). All the diet PC scores had significant ($p < 0.001$ in all cases) and higher-than-expected phylogenetic signals (PC1: $k^* = 2.40$; PC2: $k^* = 1.96$; PC3: $k^* = 1.98$). The diet matrix and PC scores, as well as the distance matrix used, are available as supplementary files.

Digestion, Intestinal Absorption, and Oxidative Metabolism

Digestion of Food Items

Phyllostomid bats belonging to different trophic guilds are characterized by feeding on a particular type of food supplemented with other items. Fruit pulp, flower nectar, pollen, and arthropods are consumed by all phyllostomid trophic guilds except vampire bats, but the efficiency with which they are used as food has rarely been explored in a comparative framework (Becker et al. 2010; Herrera and Martínez del Río 1998; Saldaña-Vazquez and Schondube 2013).

Fruits consumed by phyllostomid bats are watery

items mainly composed of sugars with low amounts of protein and lipids (Dinerstein 1986). Frugivorous phyllostomids usually do not ingest the whole fruit, but eject varying proportions of the fibrous, indigestible material and swallow the juice. For example, *Artibeus jamaicensis* ingests 25–30% of metabolizable calories present in *Ficus* fruits, which are assimilated almost completely (Morrison 1980), whereas *Carollia brevicauda* ingests a high proportion (70–80%) of fruit biomass from *Vismia macrophylla* and *Piper auritum* and assimilates a high fraction (70–90%) of sugars (defecated sugar content relative to fruit content; Becker et al. 2010). Like most phyllostomid frugivores, *Chiroderma doria* and *C. villosum* extract and ingest the juice and spit out a first pellet containing dry fibrous material (Nogueira and Peracchi 2003). However, these bats do not ingest the seeds, which are crushed and spit in a second pellet (Nogueira and Peracchi 2003). Sugar assimilation (defecated sugar content relative to fruit content) by *C. villosum* using this feeding strategy to handle *Ficus* fruits is nearly complete (>98%; Wagner et al. 2015). Nectarivorous bats handle fruits in a different way than frugivores, resulting in a lower feeding efficiency (fruit mass eaten per second; Becker et al. 2010). For example, fruits of *P. auritum* and *V. macrophylla* are chewed by the nectarivorous *Glossophaga commissarisi* ingesting mostly juice, whereas the frugivorous *C. brevicauda* chews and swallows most fruit biomass (Becker et al. 2010). Assimilation of sugars (defecated sugar content relative to fruit content) contained in fruits of *P. auritum* and *V. macrophylla* by nectarivores can be high (80–90%) but depends on the extent to which they can avoid ingesting indigestible material (Becker et al. 2010). For example, *G. commissarisi* licks the juice of *P. hispidum* infructescence but swallows large amounts of pulp and seeds (~50% of fruit pulp biomass; Kelm et al. 2008). The energy digestion efficiency (energy retention in relation to energy uptake) of *G. commissarisi* using this feeding strategy is ~50% and decreases to only ~25% when energy contained in waste fruit parts is included (Kelm et al. 2008). In any case, fruit eating by most phyllostomid nectarivores is not considered to be a viable substitute for nectar feeding (Kelm et al. 2008).

Frugivorous phyllostomids have low nitrogen requirements and are able to maintain nitrogen balance on a fruit-only diet (Delorme and Thomas 1996; 1999; Herrera et al. 2011). Unlike most phyllostomids,

C. doria and *C. villosum* use seeds as a source of nitrogen (Nogueira and Peracchi 2003). *Chiroderma villosum* thoroughly chews fruit seeds and is able to extract higher proportions (nutrient content in chewed seeds relative to intact seeds) of soluble protein (~85%) and nitrogen (~58%) than from fruit pulp (soluble protein: ~33%, nitrogen: ~22%), presumably because the bat invests twice as much time chewing seeds than pulp (Wagner et al. 2015). *Chiroderma villosum* assimilates a high proportion (>98%) of ingested protein (defecated protein content relative to fruit content) and a lower proportion of nitrogen (~70%; Wagner et al. 2015). When offered protein in liquid diets, nectarivorous phyllostomids also have low nitrogen requirements (Herrera et al. 2006), but their capacity to maintain nitrogen balance on a fruit-only diet might be lower than that of frugivores due to their lower fruit-feeding efficiency. Further work is needed to determine whether nectarivorous bats are capable of maintaining a positive nitrogen balance on fruit-only diets. Digestive processing of fruit lipids by phyllostomids has been examined in only one study (Wagner et al. 2015). *Artibeus watsoni* and *C. villosum* extract 27–29% of lipids from pulp, and *C. villosum* extracts up to 89% from seeds (Wagner et al. 2015). Lipid assimilation (defecated lipid content relative to fruit content) by *C. villosum* is very high (>98%; Wagner et al. 2015).

Nectar is one of the simplest foods found in nature: it is basically a sugary solution with minute amounts of salts and amino acids (Baker and Baker 1982). Nectarivorous phyllostomids are capable of assimilating nearly all of the energy contained in nectar (~99%; Kelm et al. 2008). Under natural conditions, phyllostomid bats feed on nectars with widely differing sugar concentrations (5–29% weight/weight; Winter and von Helversen 2001) and may therefore be expected to adjust their volumetric food intake to maintain a constant sugar ingestion. Intake response to varying sugar densities is not uniform among phyllostomids. When nectarivorous bats are offered sugar solutions ranging from 5 to 49 g sugar 100 ml^{-1}, energy intake is lower at the lowest concentration in *Leptonycteris yerbabuenae* and *Glossophaga soricina* whereas *Choeronycteris mexicana* and *L. nivalis* maintain a constant caloric input (Ayala-Berdon and Schondube 2011; Ayala-Berdon et al. 2013; Herrera and Mancina 2008; Ramírez et al. 2005). Energy intake is also limited as sugar density decreases in frugivorous phyllostomids, but the extent of this constraint appears to be related to diet. Thus, *Sturnira ludovici* eats fruits with high sugar concentration, and its caloric intake is more limited at lower sugar densities than that of *A. jamaicensis*, a bat that appears to include fruits with lower sugar content in its diet (Saldaña-Vazquez and Schondube 2013). In summary, sugar density might affect the rate at which phyllostomid bats ingest food, which in turn affects ability to meet their caloric requirements. An additional aspect that has not been examined for phyllostomids is the effect that sugar density has on the efficiency with which sugars are assimilated as has been found in some avian nectarivores (Mancina and Herrera 2016 and references therein).

Pollen is a highly nutritious source of protein, nitrogen, amino acids, starch, sterols, and lipids (Roulston and Cane 2000). However, animals must penetrate the exine and intine that cover the pollen grains to gain access to the nutritious contents. These layers are highly resistant to degradation (Stanley and Linskens 1974), and several mechanisms have been proposed for animals to deal with the coat that covers pollen grains, including mechanical rupture, initiation of pollen germination, rupture via osmotic shock, and enzymatic digestion (Johnson and Nicolson 2001). Use of pollen as food by phyllostomid bats occurs mainly in Glossophaginae, but it is also reported for Stenodermatinae, Lonchophyllinae, Micronycterinae, Carolliinae, and Phyllostominae (Coelho and Marinho-Filho 2002; Giannini and Kalko 2004). Although pollen digestive processing has only been studied in a few phyllostomids, it appears that the efficiency with which it is used is related to the frequency with which it is included in the diet (Herrera and Martínez del Río 1998; Mancina et al. 2005). For example, nectarivorous bats (*Anoura geoffroyi*, *L. yerbabuenae*, and *Brachyphylla nana*) have higher digestive extraction efficiency of pollen grains collected from several bat-visited flowers than do frugivorous bats (*A. jamaicensis* and *Sturnira lilium*). The digestive mechanism by which phyllostomid bats extract nutrients from pollen grain is unknown, although pollen germination in the stomach, urea degradation of pollen proteins, and enzymatic degradation of the pollen wall have been proposed (Herrera and Martínez del Río 1998; Howell 1974). In any case, high extraction efficiency of pollen grain contents allows some nectarivorous bats (e.g., *L. yerbabuenae*) to maintain a positive nitrogen balance with pollen as their only source of proteins (Howell 1974). The extent to which pollen

grains satisfy the nitrogen requirements of frugivorous and insectivorous phyllostomids is unknown.

Arthropods are a rich source of protein and lipids (Bell 1990), but digestive efficiency has never been evaluated for phyllostomid bats. An interesting proximal approach to estimate the ability of phyllostomids to use arthropods as food is to measure chitinase activity. Chitinase is an enzyme whose activity has been measured in the intestine and stomach of some non-phyllostomid insectivores (Strobel et al. 2013; Whitaker et al. 2004). Chitinase hydrolyzes chitin, a long-chain polymer of *N*-acetylglucosamine that forms insect exoskeleton. One would expect higher chitinase activity in insectivorous phyllostomids than in their plant-feeding relatives.

Sugar Digestion and Intestinal Absorption

Sugars in food consumed by phyllostomids are mainly present in the forms of sucrose, glucose, and fructose in flower nectar and fruit pulp and as trehalose in insects. Sucrose is a disaccharide composed of glucose and fructose, whereas trehalose is a disaccharide formed by two units of glucose. Sucrose and trehalose are broken down by sucrase and trehalase, respectively, two enzymes located in the brush border of the small intestine (Martínez del Río and Karasov 1990; Sacktor 1968). Based on conventional analysis, we found no allometric trends related to the activity of sucrase ($r^2 = 0.02$; $F_{1,14} = 0.30$; $p = 0.59$) or trehalase ($r^2 = 0.006$; $F_{1,13} = 0.07$; $p = 0.80$) among phyllostomid bats. Correcting for phylogeny did not improve the variability in this relationship for any of the enzymes (trehalase: $F_{13,13} = 0.08$; $p > 0.05$; sucrase: $F_{14,14} = 0.09$; $p > 0.05$). Trehalase activity increased with an increase in PC1 scores either before ($r^2 = 0.25$; $F_{1,13} = 5.02$; $p = 0.04$; fig. 10.1*A*) or after phylogenetic correction ($r^2 = 0.23$; $F_{1,13} = 5.0$; $p = 0.04$; fig. 10.1*B*). This suggests that, independently of their phylogenetic relationships, animalivorous phyllostomids have higher trehalase activity than do phyllostomids that feed primarily on fruit and nectar. We found no difference in trehalase activity between frugivorous and nectarivorous phyllostomids ($r^2 = 0.01$; $F_{1,13} = 0.16$; $p = 0.69$). These patterns did not change after correction for phylogeny ($r^2 = 0.01$; $F_{1,13} = 0.15$; $p = 0.70$). Sucrase activity was negatively related to diet PC1 scores both before ($r^2 = 0.50$; $F_{1,14} = 14.1$; $p = 0.01$; fig. 10.2*A*) and after phylogenetic correction

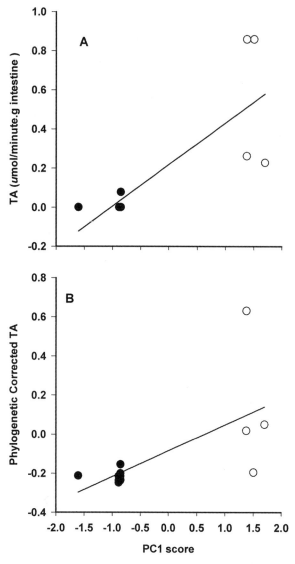

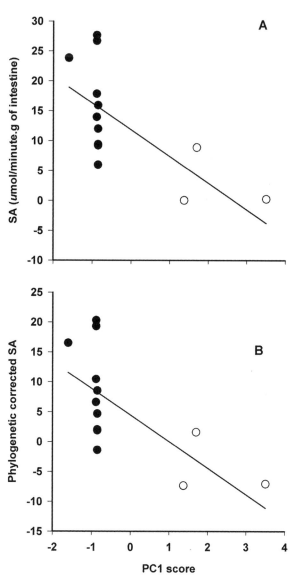

Figure 10.1. Relationship between the activity of the enzyme trehalase (TA) and diet of phyllostomid bats. Variability in the diet between these bats is described by the first score (PC1 score) of a Principal Component Analysis (PCA): positive values are associated with a diet composed primarily of insects or vertebrates, while negative values reflect a diet composed primarily of fruits and nectar. (*A*) uncorrected TA values. (*B*) Phylogenetically corrected (by PGLS) TA values. (●) Plant-eating species; (○) animalivorous species. See text for further details.

Figure 10.2. Relationship between the activity of the enzyme sucrase (SA) and diet of phyllostomid bats. (*A*) uncorrected SA values. (*B*) Phylogenetically corrected (by PGLS) SA values. See figure 10.1 for additional information and symbols.

$(r^2 = 0.49; F_{1,14} = 13.9; p < 0.01;$ fig. 10.2*B*), implying that phyllostomid bats whose diets are composed primarily of nectar or fruit have a higher activity of this enzyme compared with animalivorous phyllostomids. No difference in sucrase activity between frugivorous and nectarivorous phyllostmids was observed either before $(r^2 = 0.02; F_{1,14} = 0.32; p = 0.58)$ or after $(r^2 = 0.02; F_{1,14} = 0.32; p = 0.58)$ phylogenetic correction. Since we had only one data point for sucrose and trehalase activities for sanguinivores, we were not able to test for the asso-

ciation of the activity of any of these enzymes with PC2 scores.

Glucose and fructose are absorbed in the vertebrate intestine via transcellular and paracellular mechanisms. During transcellular transport, hexoses move across the brush-border membrane with the mediation of the sodium-dependent glucose-galactose transporter SGLT1, the facilitated glucose transporter GLUT2, and the facilitated fructose transporter GLUT5 (Cheeseman 2002; Kellett et al. 2008). In contrast, during paracellular transport, hexoses cross the epithelium through the intercellular space between cells (McWhorter et al. 2006). Hexose absorption by mammals occurs predominantly via transcellular transport (Fasulo et al. 2013; Lane et al. 1999; Schwartz et al. 1995), but bats appear to rely heavily on paracellular transport to compensate for reduced gut size associated with their flying habit (Price, Brun, Caviedes-Vidal et al. 2015).

Testing the relationship between the extent of glucose paracellular transport and feeding habits in phyllostomid bats is limited by the paucity of appropriate studies. Most intestinal glucose absorption occurs via paracellular transport in frugivores (*Carollia perspicillata*, *Artibeus lituratus*, and *S. lilium*; Brun et al. 2014; Caviedes-Vidal et al. 2008), nectarivores (*L. yerbabuenae*; Rodriguez-Peña et al. 2016), and vampire bats (*Desmodus rotundus*; Price, Brun, Gontero-Fourcade et al. 2015).Paracellular intestinal transport of glucose might also predominate in phyllostomid insectivores as found in non-phyllostomid insectivores (*Tadarida brasiliensis* and *Myotis lucifugus*; Fasulo et al. 2012; Price et al. 2013, 2014). Interestingly, although the extent to which bats use glucose paracellular transport is not related to diet, plant-eating phyllostomids have higher paracellular-mediated nominal clearance rates than vampires and non-phyllostomid insectivores, probably compensating for their shorter intestines (Price, Brun, Caviedes-Vidal et al. 2015; Price, Brun, Gontero-Fourcade et al. 2015).

Protein Digestion and Intestinal Absorption

In contrast to sugar digestion, protein digestion is a more complex process. Protein digestion begins in the stomach with the action of pepsin that breaks peptide bonds into polypeptides (chains of amino acids), which are then hydrolyzed in the small intestine by pancreatic and brush-border enzymes. Protein digestion was compared among phyllostomid bats using aminopeptidase N as a proxy. Aminopeptidase N is located in the brush-border membrane of enterocytes and is responsible for the last stages of protein hydrolysis. There was no allometric trend related to the activity of aminopeptidase N in phyllostomid bats, using either conventional analyses ($r^2 = 0.0003$; $F_{1,13} = 0.004$; $p = 0.98$) or correcting for phylogeny ($F_{13,13} = 0.03$; $p > 0.05$). We found no difference in aminopeptidase-N activity between animalivorous and plant-eating phyllostomids, both before ($r^2 = 0.01$; $F_{1,13} = 0.19$; $p = 0.67$) and after phylogenetic correction ($r^2 = 0.002$; $F_{1,13} = 0.03$; $p = 0.86$). There was also no difference in the activity of this enzyme between

frugivorous and nectarivorous phyllostomid bats before ($r^2 = 0.14$; $F_{1,13} = 3.15$; $p = 0.10$) or after phylogenetic correction ($r^2 = 0.1$; $F_{1,13} = 1.36$; $p = 0.27$). Similar to hexoses, amino acids can be absorbed in the intestine by paracellular and transcellular pathways. To date there are no measurements of amino acid absorption rates for phyllostomid bats, but they probably rely heavily on paracellular mechanisms as found in non-phyllostomid insectivorous bats (Fasulo et al. 2012; Price et al. 2013, 2014).

Intestinal Water Absorption

Flower nectar, fruit pulp, and blood are watery foods whose ingestion requires the processing of large volumes of water. For example, nightly intake of dietary preformed water when nectarivorous bats (*G. soricina*) feed on dilute nectar can be up to four times their body mass (Herrera and Mancina 2008). Some nectarivorous birds avoid processing large amounts of dietary water in their kidneys by reducing intestinal water absorption and eliminating most water through the intestinal tract (McWhorter et al. 2003; Purchase et al. 2013). In contrast to these birds, fractional water absorption by the gastrointestinal tract of *G. soricina* is not affected by water intake rate, resulting in high rates of water flux and body water turnover when feeding on dilute nectar (Hartman Bakken et al. 2008). Therefore, phyllostomid bats that feed on watery diets must have high rates of intestinal water absorption via transcellular and paracellular pathways. High rates of intestinal water absorption occur in vampire (*D. rotundus*) and nectarivorous (*G. soricina*) bats, with lower rates in frugivores (*A. lituratus*, *S. lilium*, and *C. perspicillata*; Price, Brun, Gontero-Fourcade et al. 2015). Intestinal absorption of water has not being measured in phyllostomid insectivores but, similar to non-phyllostomid insectivores (e.g., *T. brasiliensis* and *M. lucifugus*), they probably have lower rates than plant-eating bats and vampires (Price, Brun, Gontero-Fourcade et al. 2015).

Oxidative Metabolism

Recent work using stable isotope analysis of breath samples has shown that phyllostomid nectarivores and frugivores are capable of oxidizing recently ingested carbohydrates to sustain metabolic costs. Voigt and Speakman (2007) estimated that restrained individuals of *G. soricina* fueled most of their metabolism with recently ingested solutions of sucrose (77%), glucose (95%), and fructose (82%). In these bats, the incorporation of sugars into the pool of metabolized substrates occurs very rapidly: 50% of carbon atoms are exchanged in CO_2 expired by bats within 9-14 minutes after sugar ingestion (Voigt and Speakman 2007). Frugivorous phyllostomids are also able to use exogenous carbohydrates to sustain oxidative metabolism at a high rate: 50% of carbon atoms are exchanged into CO_2 by *C. perpicillata* after 11 minutes of being fed a hexose solution (Voigt et al. 2008a). The rate at which ingested food is processed to fuel oxidation in animalivorous phyllostomids has been examined only for vampire bats. Captive *D. rotundus* rapidly incorporates blood nutrients into the pool of metabolized substrate: 50% of carbon atoms are exchanged in the CO_2 of the bats 30 minutes after blood ingestion (Voigt et al. 2008b). The rate at which vampire bats incorporated recently ingested blood into oxidative metabolism is lower than for plant-eating bats, probably because proteins are more difficult to digest than simple sugars (Voigt et al. 2008b). Fuel oxidation rate has not been measured for insectivorous phyllostomids, but it might be similar to what has been found for non-phyllostomid insectivores. For example, 50% of carbon atoms are exchanged into CO_2 by *Noctilio albiventris* (Noctilionidae) about 27 minutes after being fed mealworms (Voigt et al. 2010).

Bats are characterized by flapping flight, and their digestive physiology appears to provide the necessary means to fulfill the high energy requirements of their expensive aerial lifestyle. For example, using a combination of breath stable isotope analysis and indirect calorimetry, Welch et al. (2008) estimated that *G. soricina* is able to use recently ingested sucrose to provide most (~78%) of the energy required during hovering flight. The respiratory exchange ratio (i.e., the ratio of CO_2 produced to O_2 consumed while food is metabolized) of these bats approaches a value of 1 within 30 minutes after their first feeding, indicating a nearly exclusive reliance on recently ingested sugars to sustain oxidative metabolism during hovering flight (Welch et al. 2008). The high rate at which nectarivorous phyllostomids use dietary sugars to sustain exercise metabolism is unusual among mammals, but it converges with the high capacity found in hummingbirds (Welch and Suarez 2007; Welch et al. 2006). The use of recently ingested nutrients to support the metabolic cost of flight has not

being directly measured in any other phyllostomid or in any non-phyllostomid bats. However, it is likely that a high capacity to rapidly mobilize ingested nutrients is a common feature in bats, including phyllostomids, to sustain metabolism during flight. For example, stable isotope analysis of breath samples of non-phyllostomid insectivores indicate that they oxidize recently ingested food, especially exogenous proteins and carbohydrates, while flying (Voigt et al. 2010; Voigt et al. 2012).

Solute Excretion and Water Processing

Excretory System

Phyllostomid bats ingest variable amounts of water and solutes from the food they eat, depending on their predominant feeding habit. For example, intake of fruit pulp and flower nectar represents the ingestion of large volumes of water with low amounts of protein and salts. The ingestion of blood implies processing large amounts of water but with high protein and salt content, and the ingestion of animal prey requires dealing with moderate amounts of water but high amounts of protein and salts. Therefore, the excretory systems of phyllostomids must accommodate contrasting dietary conditions to dispose of salts and nitrogenous wastes, recover filtered metabolites, and conserve water when it is scarce or excrete it when it is in excess. As in other mammals (Withers et al. 2016), bat kidneys excrete nitrogenous wastes, as well as salts and excess water, in the form of urine. The structure and function of the kidneys appear to correspond to the diet that predominates in the different trophic guilds. The macrostructure of bat kidneys can be divided into medullary and cortex regions. The cortex is the outer portion of the kidney, and it is responsible for filtering blood and removing unwanted substances from the body. The medulla is the innermost part of the kidney, and it contains structures responsible for concentrating urine. The three morphometric indexes used to describe kidney morphology of phyllostomid bats—relative medullary thickness (RMT), relative cortex volume (RCV), and relative medullary volume (RMV)—showed a significant allometric trend:

$$\text{Log}_{10}\text{RMT} = 0.47\ (\pm 0.1) + 0.21\ (\pm 0.07)$$
$$\text{Log}_{10}\text{M}_b\ (r^2 = 0.32;\ F_{1,19} = 7.9;\ p = 0.01;$$
$$\text{fig. 10.3}A),$$

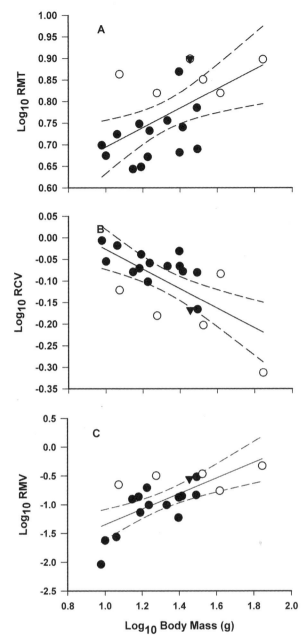

Figure 10.3. Allometric relationship of (*A*) relative medullary thickness (RMT), (*B*) relative cortex volume (RCV), and (*C*) relative medullary volume (RMV) in phyllostomid bats. Regressions lines (and 95% confidence limits—*dashed lines*) were fitted by ordinary least squares regression (OLS) in $\log_{10}$-transformed data without phylogenetic corrections. (●) Plant-eating species; (○) animalivorous species; (▼) hematofagous species.

$$\log_{10} RCV = 0.08 \, (\pm 0.009) - 0.14 \, (\pm 0.06)$$
$$\log_{10} M_b \, (r^2 = 0.19; \, F_{1,20} = 4.51; \, p = 0.04;$$
$$\text{fig. } 10.3B), \text{ and}$$

$$\log_{10} RMV = -2.15 \, (\pm 0.48) + 0.97 \, (\pm 0.36)$$
$$\log_{10} M_b \, (r^2 = 0.28; \, F_{1,20} = 7.20; \, p = 0.01;$$
$$\text{fig. } 10.3C).$$

Phylogenetic correction did not decrease the variability of these traits beyond that already explained by M_b (RMT: $F_{19,19} = 0.86$, $p > 0.05$; RCV: $F_{20,20} = 0.47$, $p > 0.05$; RMV: $F_{20,20} = 0.62$, $p > 0.05$), and allometric effects remained significant after correction for phylogeny (RMT: $r^2 = 0.34$, $F_{1,19} = 10.1$, $p = 0.005$; RCV: $r^2 = 0.32$, $F_{1,20} = 10.7$, $p < 0.01$; RMV: $r^2 = 0.44$, $F_{1,20} = 15.0$, $p < 0.01$). The slopes of the phylogenetically corrected regressions (RMT: slope = 0.19 ± 0.07; RCV: slope = -0.11 ± 0.05; RMV: slope = 0.81 ± 0.30) fell within the 95% confidence interval of the equivalent uncorrected slopes (RMT: 0.05–0.37; RCV: −0.28–0.002; RMV: 0.21–1.72).

Mass-corrected RMT was correlated with diet PC1, both before ($r^2 = 0.27$, $F_{1,19} = 7.9$, $p = 0.01$; fig. 10.4*A*) and after phylogenetic correction ($r^2 = 0.3$, $F_{1,19} = 9.1$, $p < 0.01$; fig. 10.4*B*). No association was observed between mass-corrected RMT and diet PC3, both before ($r^2 = 0.004$, $F_{1,19} = 0.08$, $p = 0.78$) and after phylogenetic correction ($r^2 = 0.008$, $F_{1,19} = 0.15$, $p = 0.71$). There was a significant and negative association between mass-corrected RCV and diet PC1, both before ($r^2 = 0.48$, $F_{1,20} = 17.6$, $p < 0.001$; fig. 10.4*C*) and after phylogenetic correction ($r^2 = 0.52$, $F_{1,20} = 21.5$, $p < 0.001$; fig. 10. 4*D*). No significant association was observed between mass-corrected RCV with diet PC3 before ($r^2 = 0.05$, $F_{1,20} = 0.98$, $p = 0.33$) and after phylogenetic correction ($r^2 = 0.07$, $F_{1,20} = 1.46$, $p = 0.24$). Finally, the association between mass-corrected RMV and diet PC1 was positive and significant, both before ($r^2 = 0.19$, $F_{1,20} = 5.68$, $p = 0.03$; fig. 10.4*E*) and after phylogenetic correction ($r^2 = 0.23$, $F_{1,20} = 6.85$, $p = 0.02$; fig. 10.4*F*). No difference in mass-corrected RMV between frugivorous and nectarivorous phyllostomids was observed both before ($r^2 = 0.08$, $F_{1,20} = 2.47$, $p = 0.13$) and after phylogenetic correction ($r^2 = 0.14$, $F_{1,20} = 3.14$, $p = 0.09$). Because we had only one data point for sanguivores, we could not test for correlations between any of these traits with diet PC2. Overall, these results suggest that animalivorous phyllostomids had higher mass-corrected and phylogenetically corrected RMT, RCT and RCV than do plant-eating phyllostomids. Likewise, the absence of an association between mass-corrected and phylogenetically corrected residuals of these traits with diet suggests that frugivorous and nectarivorous phyllostomids have kidneys with the same general morphology, irrespective of their M_b and phylogenetic relationships. In contrast to kidney macrostructure, renal micro-

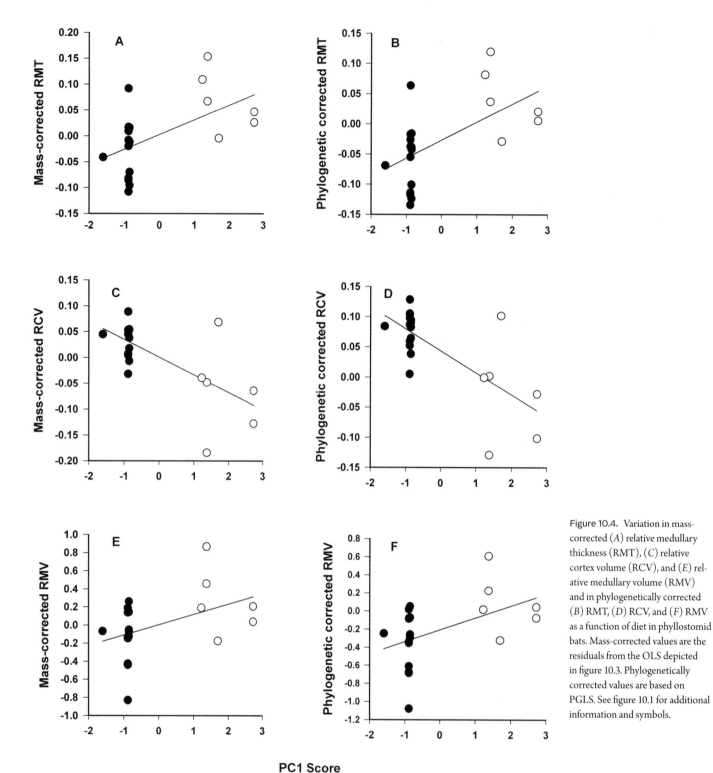

Figure 10.4. Variation in mass-corrected (A) relative medullary thickness (RMT), (C) relative cortex volume (RCV), and (E) relative medullary volume (RMV) and in phylogenetically corrected (B) RMT, (D) RCV, and (F) RMV as a function of diet in phyllostomid bats. Mass-corrected values are the residuals from the OLS depicted in figure 10.3. Phylogenetically corrected values are based on PGLS. See figure 10.1 for additional information and symbols.

PC1 Score

structure does not vary with diet in phyllostomids. No changes associated with diet were found in nephron components (collecting tubes, thick and thin limbs of Henle, and proximal tubules) within the cortex and medulla of phyllostomids (Cassoti et al. 2006).

Contrasting functional performance of the excretory system was evident after comparing the solute concentration of urine produced by phyllostomid bats with different feeding habits. Urine produced by nectarivorous and frugivorous bats under natural conditions is dilute in contrast to blood-, insect- and vertebrate-eating bats. Conventional regression showed no significant rela-

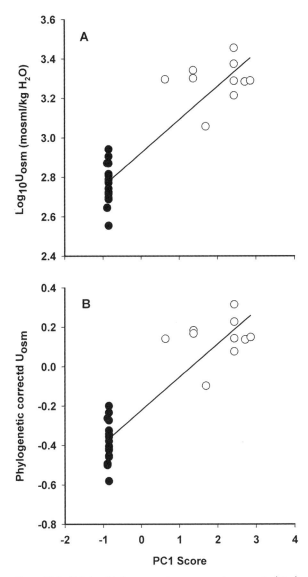

Figure 10.5. Relationship between urine concentration capacity (U_{osm}) and diet of plant-eating (●) and animalivorous (○) phyllostomid bats. (*A*) uncorrected, $\log_{10}$-transformed U_{osm} values. (*B*) Phylogenetically corrected (by PGLS) U_{osm} values. See figure 10.1 for additional information.

tionship between U_{osm} and M_b ($r^2 = 0.04$; $F_{1,26} = 1.12$; $p = 0.30$). However, correcting for phylogeny greatly improved this allometric relationship ($F_{25,25} = 39.4$; $p < 0.01$). After accounting for phylogenetic effects, we found a positive relationship between U_{osm} and M_b ($r^2 = 0.57$; $F_{1,26} = 36.2$; $p < 0.01$). Phyllostomid species that feed on nectar and fruits had significantly lower U_{osm} than those that feed on insects or vertebrates (U_{osm} vs. diet PC1: $r^2 = 0.81$; $F_{1,26} = 104.1$; $p < 0.01$; fig. 10.5*A*), a trend also observed after phylogenetic correction ($r^2 = 0.80$; $F_{1,26} = 102.7$; $p < 0.01$; fig. 10.5*B*). We found no difference in U_{osm} between nectar and fruit-eating phyllostomid bats before ($r^2 = 0.02$; $F_{1,26} = 0.47$; $p = 0.49$)

or after phylogenetic correction ($r^2 = 0.02$; $F_{1,26} = 0.49$; $p = 0.49$). Since we had only one data point on U_{osm} for sanguivores, we did not test for the correlation between U_{osm} and diet PC2.

The urine-concentrating ability of plant-eating phyllostomids is limited even when they are osmotically challenged. For example, maximum osmotic pressure of urine collected from *A. jamaicensis* (837 mOsm kg⁻¹; Studier et al. 1983) and *C. perspicillata* (1,090 mOsm kg⁻¹; Studier and Wilson 1983) without access to food or water, and from *A. jamaicensis* (866 mOsm kg⁻¹; Studier et al. 1983) and *G. soricina* (578 mOsm kg⁻¹; Hartman Bakken et al. 2008) forced to ingest salt-loaded solutions, was much lower than for urine collected from 11 species of water-deprived non-phyllostomid insectivores (3,350–5,000 mOsm kg⁻¹; Geluso 1978). During their natural daytime fast, nectarivorous bats (*G. soricina*) reduce urinary water loss by decreasing glomerular filtration rates to values 90% lower than when they are feeding (Hartman Bakken et al. 2008). Due to the watery nature of their food, the excretory system of plant-eating phyllostomids must deal with overhydration rather than dehydration. As mentioned above, bats do not regulate water intestinal absorption, so their excretory system must manage dietary water overloads. Nectarivorous birds that do not regulate water intestinal absorption deal with excessive water intake by reducing fractional renal water reabsorption and increasing glomerular filtration rates (Hartman Bakken and Sabat 2007). In nectarivorous bats (*G. soricina*), glomerular filtration rates are unresponsive to water loads, and excess water ingestion is managed by reducing fractional renal water reabsorption along with an increase in the rate of total evaporative water loss (Hartman Bakken et al. 2008). When feeding on watery food, excretion of high volumes of water entails electrolyte conservation, and under such circumstances, avian nectarivores are able to produce highly dilute cloacal fluids (Fleming and Nicolson 2003). Although nectarivorous bats are able to produce highly dilute urine when electrolyte concentration of food decreases, their capacity to recover filtered electrolytes does not appear exceptional among mammals, and it is certainly lower than for nectarivorous birds. For example, when *G. soricina* feeds on sugar solutions with no salt, urine osmolarity is much higher (31.0 mOsm kg⁻¹; Hartman Bakken et al. 2008) than that found in the cloacal fluid of the whitebellied sunbird *Nectarinia talatala* fed highly dilute nectar (6.2 mOsm kg⁻¹; Fleming and Nicolson 2003). The low capacity to maintain electrolyte balance when fed dilute sugar solutions is probably related to limits in daily intake rate as sugar diets get more dilute. For example, *G. soricina* can maintain a relatively constant daily energy intake on diets with sugar contents ranging from 10% to 40% weight per volume, but energy intake is significantly reduced when they are fed 5% sugar solutions (Herrera and Mancina 2008; Ramírez et al. 2005). The daily energy intake rate of avian nectarivores increases significantly when salt is added to watery sugar solutions (Mancina and Herrera 2016), and the same is probably true in plant-eating phyllostomids.

Energetics

The two most common energetic parameters that have been measured for bats are the rates of basal energy expenditure (basal metabolic rate, BMR) and field metabolic rate (FMR). Basal metabolic rate is defined as the minimum rate of energy expenditure measured within the thermoneutral zone and during the inactivity phase of the circadian cycle for an adult, postabsorptive, normothermic individual (Withers et al. 2016). Field metabolic rate is an integrative measure of the daily energy

expenditure of free-ranging animals, often measured using the doubly labeled water method (Withers et al. 2016). What an animal eats has been regarded as one of the main potential factors that could explain the residual variability in BMR and FMR in mammals (Withers et al. 2016). The effect of diet on mass-residual BMR and FMR is encapsulated by the food-habit hypothesis. A particular aspect of this hypothesis is related to the effects of diet quality. It is assumed that certain properties of the diet, such as digestibility, presence of secondary compounds, and/or energy content, are correlated with mass-corrected rates of energy expenditure (Cruz-Neto and Bozinovic 2004).

Although the effects of diet quality on mass-corrected FMR have been analyzed for mammals as a whole (Nagy 2005; Speakman 2000), the small sample size available for phyllostomid bats (*n* = 8; Speakman and Krol 2010), and the limited dietary breadth covered by these studies, precludes a rigorous analysis. However, a comparison between the FMRs of a nectar-feeding phyllostomid (*G. commissarisi*) and a fruit-eating phyllostomid bat (*C. brevicauda*) suggests that diet may play an important role in shaping rates of field energy expenditure (Voigt et al. 2006). Voigt et al. (2006) found that the mass-specific FMR of *G. commissarisi* exceeded that of *C. brevicauda* by a factor of two. The high mass-specific FMR of *G. commissarisi* is not directly related to the energetic content of its diet per se, as originally postulated, but reflects diet-specific effects on the activity budget since the foraging costs per energy reward are higher than for fruit-eating bats.

The effect of diet quality on BMR, in contrast, has been extensively studied in bats. In fact, this idea was first made clear by McNab (1969; see also McNab 1982) for bats: species that feed on vertebrates, pollen, and nectar usually have a high mass-corrected BMR, while those that feed on invertebrates, on blood and, to some extent, on fruits have intermediate to lower mass-corrected BMRs. In our reanalysis, the results from conventional analysis showed a significant relationship between BMR and M_b: $\text{Log}_{10}\text{BMR}$ (kJ.h⁻¹) $= -1.12$ $(\pm 0.08) + 0.71\ (\pm 0.06)M_b(g)$ ($r^2 = 0.82$; $F_{1,32} = 125.5$; $p < 0.01$; fig. 10.6). The effect of M_b on BMR remained significant after phylogenetic correction ($r^2 = 0.88$; $F_{1,32} = 257.5$; $p < 0.01$), with a slope (0.75 ± 0.06) that fell within the 95% confidence limit reported for the conventional slope ($0.57–0.83$). The slopes observed in our reanalysis (from OLS and PGLS) were similar to the

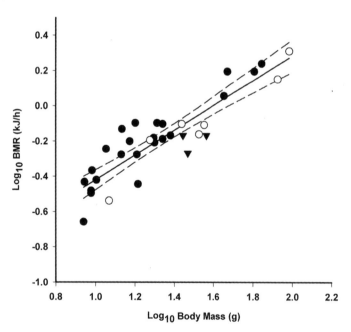

Figure 10.6. Allometric variation in basal metabolic rate (BMR) of phyllostomid bats. Regressions lines (and 95% confidence limits—*dashed lines*) fitted by ordinary least squares regression (OLS) in log₁₀-transformed values without phylogenetic corrections. Symbols as in figure 10.3.

slopes observed in previous analyses of phyllostomid bats (Cruz-Neto et al 2001; McNab 2003) and with all bats (Cruz-Neto and Jones 2006; McNab 2008; Speakman and Thomas 2003). Phylogenetic correction did not reduce the variability associated with the allometric relationship between BMR and M_b ($F_{32,32} = 0.89$; $p > 0.05$).

The BMR residuals from conventional analyses were negatively related to diet PC1 ($r^2 = 0.16$; $F_{1,32} = 6.04$; $p = 0.02$), suggesting that phyllostomids that feed on fruits and nectar have higher mass-corrected BMRs than animalivorous phyllostomids (insectivory + carnivory; fig. 10.7A). We also found a negative correlation between conventional residuals and diet PC2 ($r^2 = 0.12$; $F_{1,32} = 4.32$; $p = 0.04$; fig. 10.7B), suggesting that vampire bats have lower mass-corrected BMRs than the other phyllostomids. We found no correlation between conventional residuals and diet PC3 ($r^2 = 0.03$; $F_{1,32} = 0.88$; $p = 0.36$), suggesting no difference in mass-corrected BMR between nectarivorous and frugivorous phyllostomid bats. These trends remained unchanged after correction for phylogeny (PGLS residuals vs. PC1: $r^2 =$ 0.21; $F_{1,32} = 8.44$; $p < 0.01$; fig. 10.7C; PGLS residuals vs. PC2: $r^2 = 0.14$; $F_{1,32} = 4.47$; $p = 0.04$; fig. 10.7D; PGLS residuals vs. PC3: $r^2 = 0.01$; $F_{1,32} = 0.44$; $p = 0.51$).

Conclusions

All of the physiological traits that we reanalyzed for phyllostomid bats, except those associated with aminopeptidase activity, showed a significant phylogenetic signal. For BMR, such a pattern has been described before for phyllostomid bats (Blomberg et al. 2003) but not for the other traits, to the best of our knowledge. Nevertheless, the general pattern (a strong phylogenetic signal) we found for the traits associated with excretory physiology of phyllostomid bats was not different from what has been reported for rodents (al-Kahtani et al. 2004; Diaz et al. 2006). We are not aware of any studies for nonflying mammals that have analyzed the strength of the phylogenetic signal for enzymatic activity. For some of these traits (BMR, U_{osm}, RMT, RCV, and RMV), we also found a significant allometric effect, a pattern that has also been observed for other mam-

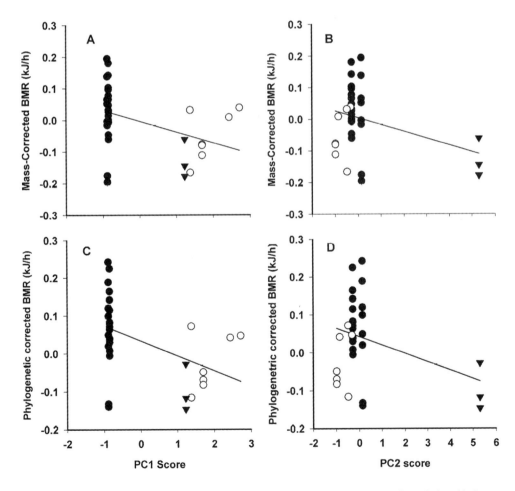

Figure 10.7. Variation in basal metabolic rate (BMR) for mass-corrected (A) PC1 and (B) PC2 scores and in phylogenetically corrected (C) PC1 and (D) PC2 scores as a function of diet in phyllostomid bats. Mass-corrected values are the residuals from the OLS depicted in figure 10.6. Phylogenetically corrected values are based on PGLS. See figure 10.1 for description of PC1 scores. PC2 scores separate bats with a hematofagous diet (high [> 5] positive scores) from the other species. Symbols as in figure 10.3.

mals (Withers et al. 2016), including phyllostomid bats (BMR—Cruz-Neto et al. 2001). Although a phylogenetic signal is still present in these traits after correcting for the effects of body mass, it seems that allometric effects prevail over phylogenetic effects in determining the magnitude of these traits for phyllostomid bats. This suggestion is based on the fact that, for all of these traits, correction for phylogeny did not increase residual variability beyond that already explained by body mass (Withers et al. 2006).

In a previous study, Cruz-Neto et al. (2001) found no correlation between diet and mass-corrected BMR for phyllostomid bats, after controlling for phylogeny. One plausible explanation for this lack of correlation stems from the way diet was classified and the impact of this classification on the analysis. Discrete, monotypic, dietary classification, such as used by Cruz-Neto et al. (2001), not only fails to encompass the diversity of food types in a bat's diet but also creates a potential problem of colinearity between phylogeny and diet, which reduces the power to detect a significant association between BMR and diet. Use of continuous dietary classification to analyze the relationship between diet and BMR tends to minimize this problem (Muñoz-Garcia and Williams 2005). In fact, in our reanalysis, mass-corrected BMR for phyllostomid bats was indeed correlated with diet, even after controlling for phylogeny. Although this general finding supports the food-habit hypothesis, the specifics of the association between diet and BMR differs from what has been previously suggested for bats in general, and for phyllostomid bats in particular (McNab 1982, 2003, 2008). For example, it has been postulated that phyllostomid bats that feed on insects and blood should have lower mass-corrected BMRs, while those that feed on vertebrates and nectar should have higher mass-corrected BMRs (McNab 1982). Fruit eaters should have intermediate levels of mass-corrected BMR (higher than insect-eating bats but lower than nectar- and vertebrate-eating bats). Our results showed that blood-feeding bats indeed have a lower BMR than do other phyllostomid bats. However, our results suggest that BMR of phyllostomid bats increases as these bats incorporate more plant material into their diets. Contrary to what has been suggested (McNab 2003), we did not find any difference in BMR between frugivores and nectarivores. Whatever the reasons behind these subtle differences between the association of diet and BMR compared

to previous findings (McNab 2003), it is clear that the dietary diversification observed for phyllostomid bats is correlated with changes in the level of BMR.

The first issue that arises from our literature review of digestive and excretory physiology is that virtually all experimental work has concentrated on nectarivores, frugivores, and vampires. Work that examines the digestive and excretory physiology of vertebrate- and insect-eating phyllostomids is probably limited by logistic constraints associated with their capture and maintenance in captivity. With this restriction in mind, we assume that findings of experimental work conducted with non-phyllostomid insectivores are probably representative of the functioning of vertebrate- and insect-eating phyllostomids. Accordingly, our review reveals that there are physiological traits that appear to be associated with feeding habits, whereas other traits are common to bats regardless of diet (see below). Aside from the need for experimental work that examines the digestive and excretory physiology of animal-eating phyllostomids, some aspects have been barely examined. For example, experimental work examining osmoregulatory physiology is scant (Hartman Bakken et al. 2008; McFarland and Wimsatt 1969). Plasticity of digestive physiology should also be considered in relation to migratory behavior and changes in ambient conditions, feeding habits, and reproductive state. For example, food intake rate by *G. soricina* is higher in the dry/cool season than in the wet/warm season (Ayala-Berdon et al. 2008). Future work with this bat family might also benefit from methodological approaches recently used to study other vertebrates. For example, use of isotopically labeled tracers can be used to examine the oxidative kinetics of different classes of macromolecules (e.g., protein, sugar, lipids; McCue et al. 2015) and even specific types of these macromolecules (e.g., essential vs. non-essential amino acids, hexoses vs. disaccharides, saturated vs. unsaturated fatty acids; McCue et al. 2010). This approach can be extended to examine the form in which different nutrients are used for other functions, such as tissue development (Muñoz-Garcia et al. 2012). Finally, the relationship between gut microbiota and dietary habits in phyllostomid bats was recognized several decades ago (Klite 1965), and recent development of massive DNA sequencing has sparked the interest in this relationship (Carrillo-Araujo et al. 2015). Gut microbiota modulates several aspects of host physiology (Sommer and Bäckhead 2013), and its interaction with phyllostomid physiology is a promising unexplored research area.

In summary, our phylogenetic comparisons indicate that trophic radiation in phyllostomids has been accompanied by changes in the intestinal activity of trehalase and sucrase, in the macrostructure of the kidney (RMT, RCV, and RMV), in urine concentration, and in basal metabolic rate. For traits that were not formally compared due to paucity of information, it appears that dietary diversification is related to the efficiency with which food types are processed, the rates at which glucose and water are absorbed by the intestine, the rates at which nutrients are incorporated into oxidative metabolism, and the maximum capacity to concentrate urine. In contrast, some traits appear to be unchanged regardless of diet, such as intestinal activity of aminopeptidases, kidney microstructure, and the extent to which the animal depends on the paracellular pathway for glucose intestinal absorption.

Acknowledgments

LGHM and APCN were supported by a grant from the Mexico–Brazil Bilateral Program 2014–2016 (Consejo Nacional de Ciencia y Tecnología and Conselho Nacional de Desenvolvimento Científico e Tecnológico — CNPq) during part of the development of the manuscript. APCN would like to thank State University of São Paulo, Fundação de Amparo a Pesquisa do Estado de São Paulo, and CNPq for logistic and financial support. We thank L. R. Monteiro and M. Nogueira for help with dietary analyses and P. Withers for help with phylogenetic-based statistical analyses. C. Korine and P. Withers kindly provided helpful comments on the manuscript.

References

al-Kahtani, M. A., C. Zuleta, E. Caviedes-Vidal, and T. Garland Jr. 2004. Kidney mass and relative medullary thickness of rodents in relation to habitat, body size, and phylogeny. Physiological and Biochemical Zoology, 77:346–365.

Ayala-Berdon, J., R. Galicia, C. Flores-Ortíz, R. A. Medellín, and J. E. Schondube. 2013. Digestive capacities allow the Mexican long-nosed bat (*Leptonycteris nivalis*) to live in cold environments. Comparative Biochemistry and Physiology A, 164:622–628.

Ayala-Berdon, J., and J. E. Schondube. 2011. A physiological perspective on nectar-feeding adaptation in phyllostomid bats. Physiological and Biochemical Zoology, 84:458–466.

Ayala-Berdon, J., J. E. Schondube, and K. S. Stoner. 2008. Seasonal intake responses in the nectar-feeding bat *Glossophaga soricina*. Journal of Comparative Physiology B, 179:553–562.

Baker, H. G., and I. Baker. 1982. Chemical constituents of nectar in relation to pollination mechanisms and phylogeny. Pp. 131–171 *in*: Biochemical Aspects of Evolutionary Biology (M. H. Nitecki, ed.). University of Chicago Press, Chicago.

Baker, R. J., O. R. P. Bininda-Emonds, H. Mantilla-Meluk, C. A. Porter, and R. A. van den Bussche. 2012. Molecular timescale of diversification of feeding strategy and morphology in new world leaf-nosed bats (Phyllostomidae): a phylogenetic perspective. Pp. 385–409 *in*: Evolutionary History of Bats: Fossils, Molecules and Morphology (G. F. Gunnel and N. B. Simmons, eds.). Cambridge University Press, Cambridge.

Becker, N. I., C. Rothenwöhrer, and M. Tschapka. 2010. Dynamic feeding habits: efficiency of frugivory in a nectarivorous bat. Canadian Journal of Zoology, 88:764–773.

Bell, G. P. 1990. Birds and mammals on an insect diet: a primer on diet composition analysis in relation to ecological energetics. Pp. 416–422 *in*: Avian Foraging: Theory, Methodology and Applications (M. L. Morrison, C. J. Ralph, J. Verner, and J. R. Jehl Jr., eds.), special issue of Studies in Avian Biology, vol. 13.

Blomberg, S. P., T. Garland, and A. R. Ives. 2003. Testing for phylogenetic signal in comparative data: behavioural traits are more labile. Evolution, 57:717–745.

Bolzan, D. P., L. M. Pessôa, A. L. Peracchi, and R. E. Strauss. 2015. Allometric patterns and evolution in Neotropical nectar-feeding bats (Chiroptera, Phyllostomidae). Acta Chiropterologica, 17:59–73.

Brun, A., E. R. Price, M. N. Gontero-Fourcade, G. Fernandez-Marinone, A. P. Cruz-Neto, W. H. Karasov, and E. Caviedes-Vidal. 2014. High paracellular nutrient absorption in intact bats is associated with high paracellular permeability in perfused intestinal segments. Journal of Experimental Biology, 217:3311–3317.

Carrillo-Araujo, M., N. Tas, R. J. Alcántara-Hernández, O. Gaona, J. E. Schondube, R. A. Medellín, J. K. Jansson, and L. I. Falcón. 2015. Phyllostomid bat microbiome composition is associated to host phylogeny and feeding strategies. Frontiers in Microbiology, 6:447.

Cassoti, G., L. G. Herrera M., J. J. Flores, C. A. Mancina, and E. J. Braun. 2006. Relationships between renal morphology and diet in 26 species of New World bats (suborder Microchiroptera). Zoology, 109:196–207.

Caviedes-Vidal, E., W. H. Karasov, J. G. Chediack, V. Fasulo, A. P. Cruz-Neto, and L. Otani. 2008. Paracellular absorption: a bat breaks the mammal paradigm. PLOS ONE, 3:e1425.

Cheeseman, C. I. 2002. Intestinal hexose absorption: transcellular or paracellular fluxes. Journal of Physiology, 544:336.

Coelho, D. C., and J. Marinho-Filho. 2002. Diet and activity of *Lonchophylla dekeyseri* (Chiroptera, Phyllostomidae) in the Federal District, Brazil. Mammalia, 66:319–330.

Cruz-Neto, A. P., and F. Bozinovic. 2004. The relationship between diet quality and basal metabolic rate in endotherms: insights from intraspecific analysis. Physiological and Biochemical Zoology, 77:877–889.

Cruz-Neto, A. P., T. Garland Jr., and A. S. Abe. 2001. Diet, phylogeny, and basal metabolic rate in phyllostomid bats. Zoology, 104:49–58.

Cruz-Neto, A. P., and K. E. Jones. 2006. Exploring the evolution of the basal metabolic rate in bats. Pp. 58–69 *in*: Functional and Evolutionary Ecology of Bats (A. Zubaid, G. F. Mc-Cracken, and T. H. Kunz, eds.). Oxford University Press, New York.

Delorme, M., and D. W. Thomas. 1996. Nitrogen and energy requirements of the short-tailed fruit bat (*Carollia perspicillata*): fruit bats are not nitrogen constrained. Journal of Comparative Physiology B, 166:427–434.

Delorme, M., and D. W. Thomas. 1999. Comparative analysis of the digestive efficiency and nitrogen requirements of the phyllostomid fruit bat (*Artibeus jamaicensis*) and the pteropodid fruit bat (*Rousettus aegyptiacus*). Journal of Comparative Physiology B, 169:123–132.

Diaz, G. B., R. A. Ojeda, and E. L. Rezende. 2006. Renal morphology, phylogenetic history and desert adaptation of South American hystricognath rodents. Functional Ecology, 20:606–620.

Dinerstein, E. 1986. Reproductive ecology of fruit bats and the seasonality of fruit in a Costa Rican cloud forest. Biotropica, 18:307–318.

Fasulo, V., Z. Zhang, E. R. Price, J. G. Chediack, W. H. Karasov, and E. Caviedes-Vidal. 2013. Paracellular absorption in laboratory mice: molecule size-dependent but low capacity. Comparative Biochemistry and Physiology A, 164:71–76.

Fasulo, V., Z. ZhiQiang, J. G. Chediack, F. D. Cid, W. H. Karasov, and E. Caviedes-Vidal. 2012. The capacity for paracellular absorption in the insectivorous bat *Tadarida brasiliensis*. Journal of Comparative Physiology, 183:289–296.

Fleming, P. A., and S. W. Nicolson. 2003. Osmoregulation in an avian nectarivore, the whitebellied sunbird *Nectarinia talatala*: response to extremes of diet concentration. Journal of Experimental Biology, 206:1845–1854.

Geluso, K. N. 1978. Urine concentrating ability and renal structure of insectivorous bats. Journal of Mammalogy, 59:312–323.

Giannini, N. P., and E. K. V. Kalko. 2004. Trophic structure in a large assemblage of phyllostomid bats in Panama. Oikos, 105:209–220.

Hartman Bakken, B., L. G. Herrera M., R. M. Carroll, J. Ayala-Berdón, J. E. Schondube, and C. Martínez del Río. 2008. A nectar-feeding mammal avoids body fluid disturbances by varying renal function. American Journal of Physiology–Renal Physiology, 295:F1855–F1863.

Hartman Bakken, B., and P. Sabat. 2007. Evaporative water loss and dehydration during the night in hummingbirds. Revista Chilena de Historia Natural, 80:267–273.

Herrera M., L. G., and C. A. Mancina G. 2008. Sucrose hydrolysis does not limit food intake by Pallas's long-tongued bats. Physiological and Biochemical Zoology, 81:119–124.

Herrera M., L. G., and C. Martínez del Río. 1998. Pollen digestion by New World bats: the effects of processing time and feeding habits. Ecology, 79:2828–2838.

Herrera M., L. G., J. Osorio M., and C. A. Mancina G. 2011 Ammonotely in a Neotropical frugivorous bat as energy intake decreases. Journal of Experimental Biology, 214:3775–3781.

Herrera M., L. G., N. Ramírez, and L. Mirón M. 2006. Ammonia excretion increased and urea excretion decreased in urine of a New World Nectarivorous bat with decreased nitrogen intake. Physiological and Biochemical Zoology, 79:801–809.

Howell, D. J. 1974. Bats and pollen: physiological aspects of the syndrome of chiropterophily. Comparative Biochemistry and Physiology A, 48:263–276.

Johnson, S. A., and S. W. Nicolson. 2001. Pollen digestion by flower-feeding Scarabidae: protea beetles (Cetoniini) and monkey beetles (Hopliini). Journal of Insect Physiology, 47:725–733.

Karasov, W. H., and A. E. Douglas. 2013. Comparative digestive physiology. Comprehensive Physiology, 3:741–783.

Karasov, W. H., and C. Martínez del Río. 2007. Physiological Ecology: How Animals Process Energy, Nutrients, and Toxins. Princeton University Press, Princeton, NJ.

Kellett, G. L., E. Brot-Laroche, O. J. Mace, and A. Leturque. 2008. Sugar absorption in the intestine: the role of GLUT2. Annual Review of Nutrition, 28:35–54.

Kelm, D. H., J. Schaer, S. Ortmann, G. Wibbelt, J. R. Speakman, and C. C. Voigt 2008. Efficiency of facultative frugivory in the nectar-feeding bat *Glossophaga commissarisi*: the quality of fruits as an alternative food source. Journal of Comparative Physiology B, 178:985–996.

Klite, P. D. 1965. Intestinal bacterial flora and transit time of three Neotropical bat species. Journal of Bacteriology, 90:375–379.

Lane, J. S., E. E. Whang, D. A. Rigberg, O. J. Hines, D. Kwan, M. J. Zinner, D. W. McFadden, J. Diamond, and S. W. Ashley. 1999. Paracellular glucose transport plays a minor role in the unanesthetized dog. American Journal of Physiology—Gastrointestinal and Liver Physiology, 276: G789–G794.

Mancina, C. A., F. Balseiro, and L. G. Herrera M. 2005. Pollen digestion by nectarivorous and frugivorous Antillean bats. Mammalian Biology, 70:282–290.

Mancina, C. A., and L. G. Herrera M. 2016. The effect of salt content on nectar intake of a New World generalist avian nectarivore (*Cyanerpes cyaneus*: Thraupidae). Auk, 133:52–58.

Martínez del Río, C., and W. H. Karasov. 1990. Digestion strategies in nectar- and fruit-eating birds and the sugar composition of plant rewards. American Naturalist, 136:618–637.

McCue, M. D., R. M. Guzman, and C. A. Passement. 2015. Digesting pythons quickly oxidize the proteins in their meals and save the lipids for later. Journal of Experimental Biology, 218:2089–2096.

McCue, M. D., O. Sivan, S. R. McWilliams, and B. Pinshow. 2010. Tracking the oxidative kinetics of carbohydrates, amino acids and fatty acids in the house sparrow using exhaled $^{13}CO_2$. Journal of Experimental Biology, 213:782–789.

McFarland, W. N., and W. A. Wimsatt. 1969. Renal function and its relation to the ecology of the vampire bat, *Desmodus rotundus*. Comparative Biochemistry and Physiology, 28:98–1006.

McNab, B. K. 1969. The economics of temperature regulation in Neotropical bats. Comparative Biochemistry and Physiology, 31:227–268.

McNab, B. K. 1982. Evolutionary alternatives in the physiological ecology of bats. Pp. 151–200 *in*: Ecology of Bats (T. H. Kunz, ed.). Academic Press, New York.

McNab, B. K. 2003. Standard energetics of phyllostomid bats: the inadequacies of phylogenetic-corrected analyses. Comparative Biochemistry and Physiology A, 135:357–368.

McNab, B. K. 2008. An analysis of the factors that influence the level and scaling of mammalian BMR. Comparative Biochemistry and Physiology A, 151:5–28.

McWhorter, T. J., B. Hartman Bakken, W. H. Karasov, and C. Martínez del Río. 2006. Hummingbirds rely on both para-cellular and carrier-mediated intestinal glucose absorption to fuel high metabolism. Biology Letters, 2:131–134.

McWhorter, T. J., C. Martínez del Río, and B. Pinshow. 2003. Modulation of ingested water absorption by Palestine sun-birds: evidence for adaptive regulation. Journal of Experimental Biology, 206:659–666.

Monteiro, L. R., and M. R. Nogueira. 2011. Evolutionary patterns and processes in the radiation of phyllostomid bats. BMC Evolutionary Biology, 11:137.

Morrison, D. W. 1980. Efficiency of food utilization by fruit bats. Oecologia, 45:270–273.

Muñoz-Garcia, A., S. E. Aamidor, M. D. McCue, S. R. Mc-Williams, and B. Pinshow. 2012. Allocation of endogenous and dietary protein in the reconstitution of the gastro-intestinal tract in migratory blackcaps at stopover sites. Journal of Experimental Biology, 215:1069–1075.

Muñoz-Garcia, A., and J. B. Williams. 2005. Basal metabolic rate in carnivores is associated with diet after controlling for phylogeny. Physiological and Biochemical Zoology, 78:1039–1056.

Nagy, K. 2005. Field metabolic rate and body size. Journal of Experimental Biology, 208:1621–1625.

Nogueira, M. R., and A. L. Peracchi. 2003. Fig-seed predation by two species of Chiroderma: discovery of a new feeding strategy in bats. Journal of Mammalogy, 84:225–233.

Pineda-Munoz, S., and J. Alroy. 2014. Dietary characterization of terrestrial mammals. Proceedings of the Royal Society of London B, 281:20141173.

Price, E. R., A. Brun, E. Caviedes-Vidal, and W. H. Karasov. 2015. Digestive adaptations of aerial lifestyles. Physiology 30, 69–78.

Price, E. R., A. Brun, V. Fasulo, W. H. Karasov, and E. Caviedes-Vidal. 2013. Intestinal perfusion indicates high reliance on paracellular nutrient absorption in an insectivorous bat Tadarida brasiliensis. Comparative Biochemistry and Physiology A, 164:351–355.

Price, E. R., A. Brun, M. Gontero-Fourcade, G. Fernández-Marinone, A. P. Cruz-Neto, W. H. Karasov, and E. Caviedes-Vidal. 2015. Intestinal water absorption varies with expected dietary water load among bats but does not drive paracellular nutrient absorption. Physiological and Biochemical Zoology, 88:680–684.

Price, E. R., K. H. Rott, E. Caviedes-Vidal, and W. H. Karasov. 2014. Paracellular nutrient absorption is higher in bats than rodents: integrating from intact animals to the molecular level. Journal of Experimental Biology, 217: 3483–3492.

Purchase, C., K. R. Napier, S. W. Nicolson, T. J. McWhorter, and P. A. Fleming. 2013. Gastrointestinal and renal responses to variable water intake in whitebellied sunbirds and New Holland honeyeaters. Journal of Experimental Biology, 216:1537–1545.

Ramírez N., L. G. Herrera M., and L. Mirón M. 2005. Physiological constraint to food ingestion in a New World nectarivorous bat. Physiological and Biochemical Zoology 78:1032–1038.

Rodriguez-Peña, N., E. R. Price, E. Caviedes-Vidal, C. M. Flores-Ortiz, and W. H. Karasov. 2016. Intestinal paracellular absorption is necessary to support the sugar oxidation cascade in nectarivorous bats. Journal of Experimental Biology, 219:779–782.

Rohlf, F. J. 2001.Comparative methods for the analysis of continuous variables: geometric intepretations. Evolution, 55:2143–2160

Rojas, D., A. Vale, V. Ferrero, and L. Navarro. 2012. The role of frugivory in the diversification of bats in the Neotropics. Journal of Biogeography, 39:1948–1960.

Rojas, D., O. M. Warsi, and L. M. Dávalos. 2016. Bats (Chiroptera: Noctilionoidea) challenge a recent origin of extant Neotropical diversity. Systematic Biology, 65:432–448.

Roulston, T. H., and J. H. Cane. 2000. Pollen nutritional content and digestibility for animals. Plant Systematics and Evolution, 222:187–209.

Sacktor, B. 1968. Trehalase and the transport of glucose in the mammalian kidney and intestine. Proceedings of the National Academy of Sciences, USA, 60:1007–1014.

Saldaña-Vazquez, R. A., and J. E. Schondube. 2013. Food intake changes in relation to food quality in the Neotropical frugivorous bat Sturnira ludovici. Acta Chiropterologica, 15:69–75.

Schondube, J. E., L. G. Herrera M., and C. Martínez del Río. 2001. Diet and the evolution of digestion and renal function in phyllostomid bats. Zoology, 104:59–73.

Schwartz, R. M., J. K. Furne, and M. D. Levitt. 1995. Paracellular intestinal transport of six-carbon sugars is negligible in the rat. Gastroenterology, 109:1206–1213.

Sommer, F., and F. Bäckhead. 2013. The gut microbiota: masters of host development and physiology. Nature Reviews, Microbiology, 11:228–238.

Speakman, J. R. 2000. The cost of living: field metabolic rates of small mammals. Advances in Ecological Research, 30: 177–297.

Speakman, J. R., and E. Krol. 2010. Maximal heat dissipation capacity and hyperthermia risk: neglected key factors in the ecology of endotherms. Journal of Animal Ecology, 79: 726–746.

Speakman, J. R., and D. W. Thomas. 2003. Physiological ecology and energetics of bats. Pp. 430–490 in: Bat Ecology (T. H. Kunz and M. B. Fenton, eds.). University of Chicago Press, Chicago.

Stanley, R. G., and H. F. Linskens. 1974. Pollen: Biology, Biochemistry, Management. Springer-Verlag, Berlin.

Strobel, S., A. Roswag, N. I. Becker, T. E. Trenczek, and J. A. Encarnação. 2013. Insectivorous bats digest chitin in the stomach using acidic mammalian chitinase. PLOS ONE, 8:e72770.

Studier, E. H., B. C. Boyd, A. T. Feldman, R. W. Dapson, and D. E. Wilson. 1983. Renal function in the Neotropical bat, Artibeus jamaicensis. Comparative Biochemistry and Physiology A, 74:199–209.

Studier, E. H., and D. E. Wilson. 1983. Natural urine concentrations and composition in Neotropical bats. Comparative and Biochemical Physiology A, 75:509–515.

Voigt, C. C., P. Grasse, K. Rex, S. K. Hetz, and J. R. Speakman. 2008b. Bat breath reveals metabolic substrate use in free-ranging vampires. Journal of Comparative Physiology B, 178:9–16.

Voigt, C. C., D. H. Kelm, and G. H. Visser. 2006. Field metabolic rates of phytophagous bats: do pollination strategies of plants make life of nectar-feeders spin faster? Journal of Comparative Physiology B, 176:213–222.

Voigt, C. C., K. Rex, R. H. Michener, and J. R. Speakman. 2008a. Nutrient routing in omnivorous animals tracked by stable carbon isotopes in tissue and exhaled breath. Oecologia, 157:31–40.

Voigt, C. C., K. Sörgel, and D. K. Dechmann. 2010. Refueling while flying: foraging bats combust food rapidly and directly to power flight. Ecology, 91:2908–2917.

Voigt, C. C., K. Sörgel, J. Šuba, O. Keišs, and G. Pētersons. 2012. The insectivorous bat *Pipistrellus nathusii* uses a mixed-fuel strategy to power autumn migration. Proceedings of the Royal Society of London B, 279:3772–3778.

Voigt, C. C., and J. R. Speakman. 2007. A mammal nectar specialist fuels its metabolically expensive nocturnal life directly and almost exclusively with exogenous carbohydrates. Functional Ecology, 21:913–921.

Wagner, I., J. U. Ganzhorn, E. K. V. Kalko, and M. Tschapka. 2015. Cheating on the mutualistic contract: nutritional gain through seed predation in the frugivorous bat *Chiroderma villosum* (Phyllostomidae). Journal of Experimental Biology, 218:1016–1021.

Welch, K. C., Jr, B. H. Bakken, C. Martínez del Río, and R. K. Suarez. 2006. Hummingbirds fuel hovering flight with newly ingested sugar. Physiological and Biochemical Zoology, 79:1082–1087.

Welch, K. C., Jr, L. G. Herrera M., and R. K. Suarez. 2008. Dietary sugar as a direct fuel for flight in the nectarivorous bat *Glossophaga soricina*. Journal of Experimental Biology, 211:310–316.

Welch, K. C., Jr, and R. K. Suarez. 2007. Oxidation rate and turnover of ingested sugar in hovering Anna's (*Calypte anna*) and rufous (*Selasphorus rufus*) hummingbirds. Journal of Experimental Biology, 210:2154–2162.

Wetterer, A. L., M. V. Rockman, and N. B. Simmons. 2000. Phylogeny of phyllostomid bats (Mammalia: Chiroptera): data from diverse morphological systems, sex chromosomes, and restriction sites. Bulletin of the American Museum of Natural History, 248:1–200.

Whitaker, J. O., Jr, H. K. Dannelly, and D. A. Prentice. 2004. Chitinase in insectivorous bats. Journal of Mammalogy, 85:15–18.

Withers, P. C., C. E. Cooper, and A. N. Larcombe. 2006. Environmental correlates of physiological variables in marsupials. Physiological and Biochemical Zoology, 79:437–453.

Withers, P. C., C. E. Cooper, S. N. Maloney, F. Bozinovic, and A. P. Cruz-Neto. 2016. Ecological and Environmental Physiology of Mammals. Oxford University Press, Oxford.

Winter, Y., and O. von Helversen. 2001. Bats as pollinators: foraging energetics and floral adaptations. Pp. 148–176 *in*: Cognitive Ecology and Pollination: Animal Behavior and Floral Evolution (L. Chittka and J. D. Thompson, eds.). Cambridge University Press, Cambridge.

11

Sensory and Cognitive Ecology

Jeneni Thiagavel,
Signe Brinkløv, Inga Geipel,
and John M. Ratcliffe

Introduction

A vertebrate's foraging ecology entails different sensory requirements (Geva-Sagiv et al. 2015) and can often be deduced from its morphological characteristics, which have been shaped by the species' environment and nutritional requirements (Safi and Dechmann 2005). Most vertebrates build real-time (and sometimes stored) neural representations of their surroundings and augment decision making based on external stimuli transduced by several different sensory systems (Dusenbury 1992; Shettleworth 2010). The brain and its components are thus under constant evolutionary pressure to improve such sensory acquisition and information storage (Dukas and Ratcliffe 2009). Brain regions are often specialized for different forms of information processing and reflect species-specific reliance on different sensory systems (Baron et al. 1996).

Combinations of three distance-sensing systems (visual, olfactory, auditory) as well as spatial memory have been studied extensively in bats (e.g., Eisenberg and Wilson 1978; Hutcheon et al. 2002; Jolicoeur and Baron 1980; Jolicoeur et al. 1984; Ratcliffe 2009a; Safi and Dechmann 2005; Thiagavel et al. 2018; Yao et al. 2012). The superior colliculi, inferior colliculi, and olfactory bulb are involved in processing visual, auditory, and olfactory stimuli, respectively (Baron et al. 1996; Frahm 1981). The hippocampus plays an important role in spatial memory and spatial information processing (Healy et al. 2005). Over evolutionary time, foraging ecology and diet can profoundly impact these sensory systems and the corresponding brain regions (Baron et al. 1996; Rojas et al. 2013). Because neural tissue is expensive to build and maintain and braincase volume is limited, enhancement of one or more brain regions may be accompanied by reduction of other relatively less important regions (Jolicoeur and Baron 1980). Mass-limited, flying vertebrates may be especially prone to such compensations (McGuire and Ratcliffe 2011).

Most bats are laryngeal echolocators and obligate predators, with the vast ma-

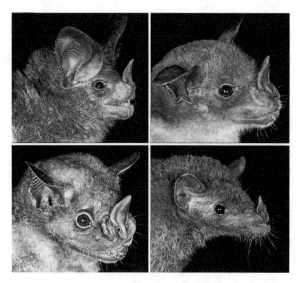

Figure 11.1. Headshots of phyllostomid species representing four diet categories. *Clockwise from top left*: *Tonatia saurophila*, a predatory species; *Phyllostomus discolor*, an omnivorous species; *Anoura geoffroyi*, a nectarivorous species; *Artibeus jamaicensis*, a frugivorous species. Note the relatively elongated snout of *A. geoffroyi* and relatively large ears of *T. saurophilia*, species-specific morphological traits presumably associated with taking nectar from flowers and listening to prey-generated cues, respectively. Photographs by Marco Tschapka, used with permission. Figure adapted from Ratcliffe 2009a.

jority of species feeding almost exclusively on insects and other arthropod prey. However, frugivory and nectarivory are commonly observed among the phyllostomids and the Old World fruit bats (family Pteropodidae). Pteropodids do not echolocate (with the exception of the genus *Rousettus*, which uses tongue clicks rather than laryngeal echolocation [Kulzer 1956]). Whether this family lost or never had the ability to echolocate remains unknown (Teeling 2009). Instead, pteropodids rely primarily on vision and olfaction to detect and localize food, and their brains have been described as being visual and olfactory (Eisenberg and Wilson 1978), having enlarged brain regions associated with these two senses (Thiagavel et al. 2018). Their auditory areas are small relative to laryngeal echolocating bats, most of which, in turn, have reduced visual and olfactory centers. Interestingly, this apparent trade-off is not so starkly observed in the phyllostomids (Jolicoeur and Baron 1980), which are characterized by the broadest range of dietary niches (e.g., Kalko, Handley et al. 1996) and greatest interspecific behavioral plasticity of any bat family (figs. 11.1 and 11.2). Instead, they have relatively larger brains than other laryngeal echolocating bat families (fig. 11.2).

As we alluded to above, some phyllostomids have diets that converge upon those of the phylogenetically distant pteropodids (Baron and Jolicoeur 1980). In particular, the frugivorous and nectarivorous (i.e., phytophagous) phyllostomids have expanded the breadth of their diets from once strict insectivory (the ancestral diet [Freeman 2000]) to now include fruits and nectar. Today, over 75% of phyllostomids consume plant matter (Fleming et al. 2009). The exploitation of plants by an ancestral phyllostomid species may have initiated the evolution of larger overall brains (Sol et al. 2005). Rather than taking animal prey opportunistically, vegetarian phyllostomids must use visual, olfactory, and spatial cues to localize spatially distributed fruits and flowers, effectively handle these nonanimal foods, and all the while assess their quality (e.g., ripeness, toxicity), before consumption (Pirlot and Stephan 1970; Ratcliffe 2009a). Thus, like the pteropodids, phytophagous phyllostomids experience greater demands on their senses of sight and smell than most bats. Correspondingly, their visual and olfactory brain regions are relatively large. This goes some way in explaining the relatively large brains we see in these two families (Baron et al. 1996; fig. 11.2). However, the phyllostomids are laryngeal echolocators, like all other bats except the pteropodids, and thus are also characterized by enhanced auditory regions similar in size to obligate predatory bats (deWinter and Oxnard 2001). Relative to both pteropodids and solely predatory laryngeal echolocating bat families, the relative size of brain regions associated with visual, olfactory, and auditory modalities are more balanced in many phyllostomid brains. Because many phyllostomids rely on the integration of inputs from multiple sensory systems and a well-developed capacity for spatial memory for successful foraging, they are an excellent model family for the study of sensory and cognitive ecology in bats.

Although bat sensory systems and associated brain regions have been studied for decades, early analyses did not account for the shared ancestry between species and were thus often based on incorrect assumptions on the independence of species (e.g., Eisenberg and Wilson 1978). Others have used comparative methods to control for common ancestry, albeit with outdated or incorrect phylogenies and foraging categorizations (e.g., Hutcheon et al. 2002). Even so, some general trends observed between sensory systems and brain development in bats have withstood the test of time. This chapter provides an overview of these trends and dis-

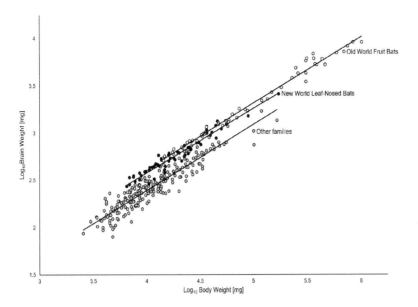

Figure 11.2. Brain mass versus body mass in bats. Relative to body mass, the phyllostomids and pteropodids have larger brains than do the remaining bat families (data from Baron et al. 1996).

cusses the influence of behavior and foraging ecology on phyllostomid sensory systems and the development of corresponding brain regions. We begin by exploring the visual, olfactory, and auditory systems, as well as the role of spatial memory in this family, and have included four case studies to further elucidate the extent to which different phyllostomids rely on different senses. We close by discussing whether overall brain size has increased over evolutionary time in this lineage.

The Visual System

Echolocation is the primary spatial sensory modality for many, perhaps most, bats, while the nonecholocating pteropodids, like most other mammals, depend primarily on vision. However, among laryngeal echolocating bats, vision is particularly important for the phyllostomids (Baron and Jolicoeur 1980). Indeed, phyllostomids—regardless of diet—are assumed to have good night vision (Bell and Fenton 1986). There are limitations to any sensory modality, and although vision can be used over greater distances than echolocation, it often has lower acuity (Suthers 1970). Despite this, when presented with conflicting cues, some phyllostomids (e.g., *Anoura geoffroyi*) prioritize visual over auditory inputs (Altringham and Fenton 2003).

Obligate animal-eating phyllostomids may not possess a visual system as developed as that of vegetarian species. However, predatory *Macrotus californicus* apparently relies solely on vision to glean terrestrial prey when light levels permit (Bell 1985). Indeed, *M. califor-*

nicus's vision has been demonstrated to be as good as its frugivorous and nectarivorous counterparts, which may be still more reliant upon it. Some phyllostomids, such as the frugivorous *Carollia perspicillata*, can visually discriminate between different colors, shapes, and sizes (Suthers 1969). Other bats, such as the omnivorous *Phyllostomus hastatus*, have been shown to use vision for long-range orientation. When *P. hastatus* is displaced from its roost and fitted with transparent goggles, the bat is more likely to home successfully than when blindfolded (Williams et al. 1966). Vespertilionid bats, in contrast, do not seem to accrue equivalent benefits from vision (Griffin 1958).

Because bats forage at night, their retinae were thought to have the rods required for night vision but not the cones required to discriminate colors (Müller et al. 2009). However, studies with the mainly vegetarian phyllostomids *Glossophaga soricina* and *Carollia perspicillata* have shown that, in addition to having a significant number of cones, they are also sensitive to ultraviolet (UV) light (Müller et al. 2009). Being able to detect UV light is advantageous for nectar-loving bats, such as some phyllostomids, as many bat-pollinated flowers are capable of reflecting UV light (Burr et al. 1995). Specifically, the atmospheric scattering at twilight results in a light spectrum that consists primarily of short wavelengths (von Helversen and Winter 2003), giving UV light–sensitive bats like *G. soricina* and *C. perspicillata* an advantage when foraging. In fact, these species are among the few mammals with the ability to see into the ultraviolet range (Müller et al. 2009). Whether functional UV vision has evolved independently or is characteristic of frugivorous and nectarivorous sister clades is not known.

Both vegetarian and animal-eating phyllostomids are thought to have eyes that are larger than the strictly insectivorous non-phyllostomid bats (Thiagavel et al. 2018; von Helversen and Winter 2003). They do appear to have more developed superficial layers of the superior colliculus compared to other laryngeal echolocating bats (von Helversen and Winter 2003), and both the phyllostomids and the highly visual pteropodids have the largest superior colliculi volumes among bats (Hu et al. 2006). Similarly, the nerves that project to the lateral geniculate nucleus are larger in frugivorous phyllostomids than in most insectivorous species (Eklöf 2003). Selection on the size and complexity of these visual centers in the brain (including the neocortex) may contribute substantially to the increase in overall brain volume seen in these two families (Barton et al. 1995).

The Olfactory System

Bats exhibit huge variation in the extent to which individual species rely on olfactory and chemical information (Eiting et al. 2014). Vegetarian species rely heavily on their sense of smell for efficient foraging, while insectivorous bats do not (Laska 1990). Frugivorous species assess the quality of fruit by smell, and nectarivorous bats can be guided by flower odors alone (Altringham and Fenton 2003). Congruently, phytophagous phyllostomids (and pteropodids) have the largest and most developed olfactory structures (Fleming 1988). Phytophagous bats also have larger nasal epithelia (von Helversen et al. 2000) and higher olfactory acuity than do insectivorous bats (Bhatnagar and Kallen 1974). Further, a well-developed vomeronasal organ is found in phyllostomids but is absent in all other bats, including the pteropodids (Cooper and Bhatnagar 1976). A recent study of the enlarged olfactory recess (a posterior region of the nasal fossa) in phyllostomid bats has shown that it improves airflow, and thus olfaction, by allowing odor molecules to linger in the airspace (Eiting et al. 2014).

Olfaction plays a crucial role for detecting and localizing fruits for many frugivorous phyllostomids (Kalko, Herre et al. 1996). For example, *Carollia perspicillata* can detect and recognize minute concentrations of odors even after the smell has been removed or chemically masked (Laska 1990). Social transmission of food preferences via chemical cues on a conspecific's breath has also been observed in this species (Ratcliffe and ter Hofstede 2005). Sensitivity to chemical compounds is often directly related to a species' diet. Nutritional specializations result in an abundance of receptor types that are sensitive to some substances and not others (Davies 1971). The common vampire bat *Desmodus rotundus* is highly sensitive to butyric acids associated with animal metabolism, while the frugivore *C. perspicillata* is highly sensitive to esters found in fruit (Laska 1990). Both of these phyllostomids have much higher sensitivity for such acids and alcohols than do insectivorous vespertilionids such as *Myotis myotis* (Laska 1990). Some phyllostomids rely primarily on their sense of smell to find and assess food. When presented with an odor that smelled of ripe bananas and one that was a cocktail of odors including ripe bananas, *C. perspicillata* always preferred the pure smell to the cocktail (Laska 1990). Surprisingly, this preference was seen even when components of the cocktail had concentrations that exceeded those of pure banana. This suggests that frugivorous bats, such as *C. perspicillata*, use their keen sense of smell not only to assess a fruit's quality based on its ripeness (nutritional benefits tend to increase with ripeness) but also to discriminate between safe and over- (or under-) ripe fruit (Laska 1990). More than this, *C. perspicillata* can also use olfactory cues and airborne chemical gradients for spatial orientation (Laska 1990). The following case study considers *Artibeus jamaicensis*, another frugivorous phyllostomid that depends on its sense of smell.

Case Study 1: Olfaction in the Jamaican Fruit Bat (*Artibeus jamaicensis*)

Artibeus jamaicensis is a medium-sized frugivorous phyllostomid (Stenodermatinae) that uses vision and olfaction to detect figs as well as brightly colored, odorous fruits (Handley et al. 1991). While olfaction is often considered to be a secondary sensory modality for bats, *A. jamaicensis* relies heavily on its sense of smell for the detection and coarse localization of fruit, perhaps more so than on vision and echolocation (Kalko, Herre et al. 1996). Fig trees fruit irregularly, and these bats are known to travel for several kilometers presumably guided by the scent of ripe figs alone (Morrison, 1978).

The olfactory structures and corresponding brain regions in *A. jamaicensis* are well developed. Controlling for body size, *A. jamaicensis* has an olfactory bulb double the size of the insectivorous vespertilionid *Myotis lucifugus* (Bhatnagar and Kallen 1974). Also, unlike in *M. lucifugus*, the olfactory and respiratory portions of *A. jamaicensis*'s nasal cavity are separated, increasing airstream control (Bhatnagar and Kallen 1974). As in all phyllostomids, *A. jamaicensis* has a vomeronasal organ and a nasal epiglottis that maximize the flow of odorous air, increasing olfactory acuity. *Artibeus jamaicensis* also has pronounced nasal glands and nerve bundles along with a thick olfactory epithelium with a large surface area and roughly twice as many receptors as vespertilionids. Its nostril movements have also been shown to direct odor molecules toward the nasal ethmoturbinates, further increasing its olfactory acuity (Bhatnagar and Kallen 1974). This species' reliance on olfaction may well impose selection pressure on fig trees to increase the strength of their odor to help attract seed-dispersing bats (Kalko, Herre et al. 1996).

While *A. jamaicensis* predominantly relies on olfaction for fruit finding, olfaction alone is not sufficient. For example, when discriminating individual figs, *A. jamaicensis* uses its presumably narrow biosonar sound beam to precisely locate the position of a fruit before picking it up with its mouth (Brinkløv et al. 2009, 2011). Also, *A. jamaicensis* has a well-developed visual system with large eyes, high visual acuity (Heffner et al. 2001), and large information-processing visual nuclei (Hope and Bhatnagar 1979). Surprisingly, Heffner and colleagues (2001) found that this bat's visual acuity is even higher than that of the tongue clicking (i.e., echolocating) pteropodid *Rousettus aegyptiacus*. As in most phyllostomids, eye size and olfactory sensitivity scaled positively in this species, where larger-eyed individuals also had a better sense of smell (Bhatnagar and Kallen 1974).

Echolocation and the Auditory System

Most bats rely on echolocation as their primary sensory modality for orientation and foraging (Griffin 1958). Phyllostomids, despite using vision and olfaction to a greater extent than other laryngeal echolocating bats, are no exception. Unsurprisingly, phyllostomids have relatively large auditory brain regions, such as inferior colliculi (which integrates incoming auditory information) and auditory nuclei (comprising the cochlear nuclei and superior olivary complexes), compared to similarly sized nonecholocating mammals (Baron et al. 1996).

Phyllostomids tend to emit short biosonar signals (<10 ms) that are well-suited for detecting food or prey with little overlap of echoes from background foliage (i.e., clutter echoes) (Schnitzler and Kalko 1998). Their close association with vegetation may be why these bats are constrained to biosonar signals with remarkably similar underlying templates across species (Ratcliffe 2009b). In most species, their frequency-modulated (FM) signal is high-pass filtered in the bat's vocal tract to dampen the fundamental and emphasize the signal's energy at higher frequencies in the second and third harmonics (Brinkløv et al. 2010; fig. 11.3). Energy at these frequencies can either be spread evenly across multiple harmonics to increase the bandwidth of its echolocation beam or be shifted between harmonics by the bat to accentuate lower or higher frequencies (Brinkløv et al. 2010). Through relatively short FM sweeps containing multiple high-energy harmonics, the signal covers a broad range of frequencies, providing clear time stamps and increased resolution for texture discrimination (Kalko and Schnitzler 1993; Schnitzler and Kalko 2001).

Donald Griffin, who discovered echolocation in bats, originally estimated sound pressure levels (SPL) of phyllostomid calls to be around 75 decibels sound pressure level (dB SPL) at 10 cm from the bat's mouth (Griffin 1958). Phyllostomids apparently often produce relatively faint echolocation calls and were long considered to be "whispering bats" (Schnitzler and Kalko 1998). They are difficult to detect acoustically in the field even when they are flying close to ultrasonic bat detectors. While phyllostomids rarely approach the source levels of most open-space aerial hawkers and trawlers, such as molossids, emballonurids, vespertilionids, or even other noctilionoid bats, such as *Noctilio leporinus*, that emit calls averaging 120–140 dB SPL (Surlykke and Kalko 2008), recent studies reveal that phyllostomids are not necessarily "whispering" (Brinkløv et al. 2009). Subdued relative to hawkers and trawlers, phyllostomids still emit calls well above 100 dB SPL (Nørum et al. 2012). The combined effects of volume adjustment and a highly directional (i.e., narrow) sound beam may be why phyllostomids often evade acoustic monitoring, as signals are only de-

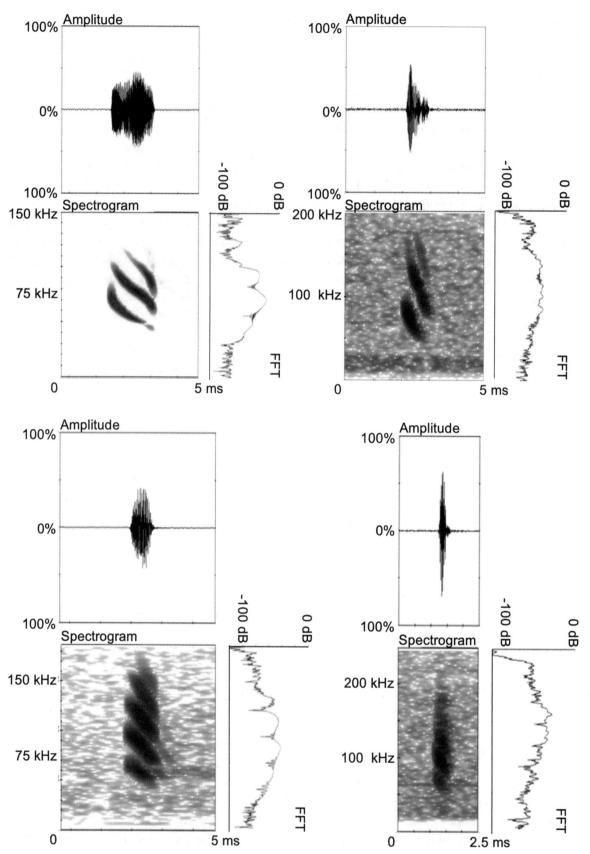

Figure 11.3. Echolocation calls of four phyllostomid bats with different foraging strategies. For each call we show an oscillogram (*top*), a spectrogram (*bottom left*), and a power spectrum (fast Fourier transform [FFT], *bottom right*). All recordings have been high-pass filtered (~15 kHz). Note that the spectrogram axes differ between species. *Clockwise from top left*: *Artibeus jamaicensis*, a frugivorous species; *Glossophaga soricina*, a nectarivorous species; *Micronycteris microtis*, a gleaning species; *Macrophyllum macrophyllum*, a trawling species.

tectable when the bat is on-axis with the microphone (Brinkløv et al. 2011). While this is inconvenient for the researcher, it may be a useful strategy for a forest-dwelling bat trying to decrease the echo load from surrounding vegetation (clutter echoes) that would otherwise unnecessarily tax its auditory system (Jakobsen, Brinkløv et al. 2013).

As bats forage in different soundscapes, the spatial energy distribution of their sonar beam is under active control, and sound waves can be focused in the forward direction by adjusting the emitter size and the intensity and frequency content of the call (Jakobsen, Brinkløv et al. 2013). Output intensity directly affects the range over which echoes can be detected and used to find and localize food. High-intensity calls travel farther and maximize range, to the benefit of most bats (including phyllostomids) when foraging or commuting in open space. Emitting calls of lower intensity is advantageous to bats foraging in and around vegetation, as this helps the bat reject clutter echoes and increases the relative strength of target echoes (Jakobsen, Brinkløv et al. 2013). Decreasing the signal's intensity may also allow predatory phyllostomids to sneak up undetected on eared, ultrasound-sensitive prey (Bell 1985; Surlykke 1988). This tactic allows the phyllostomid *G. soricina*, for example, to remain undetected during prey pursuit, supplementing its primarily nectarivorous diet with insects (Clare et al. 2014). Alongside the ability to find food by listening for the echoes of their signals (active sensing), some phyllostomids can rely on plants to make themselves acoustically more conspicuous. Chiropterophilous (bat-loving) flowers are often bell shaped and can optimize echo reflectance (Ross and Holderied 2013). Some plants have even evolved structures, such as dish-shaped leaves above their flowers, which act as acoustic beacons for pollinating bats. In other words, these plants increase the likelihood of their detection by vegetarian phyllostomids by acoustically attracting and guiding them (Simon et al. 2011). From these reflections, nectarivorous bats can gauge the intensity as well as the spectral and temporal patterns of specific flower echoes (Ross and Holderied 2013). The most surprising example may be from the legume *Mucuna holtonii*, which explodes when first visited by a bat. Virgin flowers have concave vexillums (i.e., large uppermost petals) that maximize the echo's reflected energy (von Helversen and von Helversen 1999). As a bat attempts to access nectar, the keel of the plant bursts, launching pollen onto the bat's rump. The emphasis on virgin flowers is notable in that most of the nectar (~100 µl) is delivered upon the first visit, while subsequent visits yield only 10–20 µl. Nectarivorous bats can distinguish between the two configurations. The importance of echoes from the vexillum is evident, as removing the vexillum decreases the number of visits with or without olfactory cues (von Helversen and von Helversen 1999). Another example is *Marcgravia evenia*, a vine with disc-shaped leaves that sit upright above the inflorescences and act as echo beacons for pollinating bats. Experiments with *G. soricina* have shown that the strong, multidirectional echoes reflected from these leaves substantially increase foraging efficiency. Echoes reflected off these disc-shaped leaves have constant echo signatures that make the plant detectable, as bats recognize the leaves as the only invariant objects (Simon et al. 2011). Similarly, it has been suggested that the cephalia (i.e., flower-bearing outgrowth) of glossophagine-pollinated cacti maximize echo reflectance through its flowers while simultaneously protecting the wings of hovering bats (von Helversen and Winter 2003).

Phyllostomids may also exert dynamic control over call intensity to optimize the information they acquire from the environment (Brinkløv et al. 2010, 2011). Specifically, field recordings of frugivorous *A. jamaicensis* have demonstrated an increase in mean signal intensity from 96 dB SPL in a cluttered flight room to 113 dB SPL when leaving their roost at dusk in the wild (Brinkløv 2010). All else being equal, the higher the frequency, the more directional (i.e., narrower) the sound beam (Jakobsen, Ratcliffe et al. 2013). While active call-to-call control of beam width has not been demonstrated in phyllostomids, insect-trawling *Macrophyllum macrophyllum* (Phyllostominae), the subject of our next case study, has been shown to increase maximum signal energy from the second harmonic in open space to higher harmonics in clutter (Brinkløv et al. 2010). This illustrates the potential for phyllostomid bats to affect beam shape by altering the frequency of their calls. In the following case study, we explore the echolocation behavior of *M. macrophyllum*, which uses calls typical of phyllostomids but is one of few species in this family to emit high repetition feeding buzzes, characteristic of echolocating bats that track and catch airborne prey (Ratcliffe et al. 2013), and is the only phyllostomid known to trawl insects from water surfaces.

Case Study 2: Echolocation in the Long-Legged Bat (*Macrophyllum macrophyllum*)

Macrophyllum macrophyllum (Phyllostominae) is a small insectivorous phyllostomid that is distinguished by its very long legs and large tail membrane. It uses calls that are typical of other phyllostomids (i.e., short duration, high pitched, broadband, frequency modulated). With respect to foraging strategies, however, it is unique among phyllostomids, being the only species that trawls insects from water surfaces (Weinbeer et al. 2013). In doing so, *M. macrophyllum* might experience less clutter than a gleaning phyllostomid, as water surfaces may act like mirrors by reflecting sound away from the surface while increasing target echo strength (Hulgard and Ratcliffe 2016; Siemers et al. 2005).

Several phyllostomids emit calls with shorter pulse intervals prior to hawking airborne prey (Clare et al. 2014) or gleaning silent prey (Geipel, Jung et al. 2013) or when taking fruits and nectar (Gonzales-Terrazas et al. 2016). In studies using multi-microphone array recordings to reconstruct flight paths of phyllostomids and compare source levels of bats in different clutter scenarios, it has been found that both *M. macrophyllum* and the frugivore *A. jamaicensis* emit echolocation signals with average source levels of ~100 dB SPL, despite having different foraging strategies (Brinkløv et al. 2011). Similarly, another frugivorous gleaner, *Carollia perspicillata*, calls at ~100 dB SPL (Brinkløv et al. 2009). This is about 16 times louder than the 75 dB initially estimated by Griffin for phyllostomid bats (Griffin 1958), indicating that phyllostomid bats do not whisper as quietly as once thought.

Macrophyllum macrophyllum's potential flexibility in call intensity was evaluated across habitats representing three different degrees of clutter. *Macrophyllum macrophyllum* nearly doubled its mean signal intensity from 100 dB SPL in the highly cluttered confines of a flight room to 105 dB SPL at a partially cluttered field site. Mean source levels yet again nearly doubled when the bats were recorded foraging in more open space (111 dB SPL; Brinkløv et al. 2010). In addition to increasing its signal intensity, *M. macrophyllum* also lowers call peak frequency and increases call duration, increasing sonar range when moving into more open space. *Macrophyllum macrophyllum*, unlike other phyllostomine clade members, also produces a feeding buzz (calls at rates >100/s) just prior to gaffing insects from lakes and ponds, much like aerial hawking and trawling bats from other laryngeally echolocating families (Ratcliffe et al. 2013; Weinbeer et al. 2013).

Noseleaves

Phyllostomids, commonly known as New World or American leaf-nosed bats, are characterized in part by leaflike fleshy facial structures. The noseleaf of most phyllostomids consists of a horseshoe-shaped structure sweeping below the nostrils and a bladelike dorsal lancet of highly variable appearance above them (Arita 1990; fig. 11.1). Though noseleaves are present in a handful of bat families (e.g., Rhinolophidae, Hipposideridae, Megadermatidae, Rhinopomatidae), they have been suggested to be a uniquely derived feature of the phyllostomids (Simmons 1998). Wetterer and colleagues (2000) suggest that the ancestor of this lineage had a small noseleaf with a short spear and a horseshoe that was undifferentiated from the upper lip (Wetterer et al. 2000). Over evolutionary time, the noseleaf then increased in complexity within each subfamily. Specifically, spears got longer and horseshoes more defined. Huge variation in size and shape of noseleaf morphology is observed in extant phyllostomids, from the extraordinarily long lancets of *Lonchorhina aurita*, to very short ones in the nectarivores, to the seemingly rudimentary structures found in *D. rotundus* and *Centurio senex* (Arita 1990; fig. 11.1). Noseleaf morphology tends to be a unique and diagnostic feature of each phyllostomid subfamily (Wetterer et al. 2000).

While the exact function of the phyllostomid noseleaf is not entirely resolved, it has been demonstrated to provide directional effects on sound emissions. In restrained *C. perspicillata*, the noseleaf affects the vertical pattern of the echolocation beam (Hartley and Suthers 1987). Indeed, moving the lancet back and forth apparently allows the bat to steer its beam (Vanderelst et al. 2010). Hartley and Suthers (1987) found that blocking one of the two nostrils doubles the horizontal beam width while the vertical dimension remains unaffected. Specifically, high horizontal directionality has been proposed to be the result of the overlap between nostril emissions. Removing the lancet, in contrast, results in a pronounced broadening of the sound beam in the vertical, but not the horizontal, dimension (Hartley and Suthers 1987).

Two nose-leafed phyllostomids, the frugivore *C. perspicillata* and the frog-eating *Trachops cirrhosus*, produce narrower beam widths than non-phyllostomid hawkers without noseleaves (Ghose and Moss 2003; Jakobsen, Brinkløv et al. 2013). It has been presumed that successful nasal emission of ultrasound is contingent upon the presence of a noseleaf, emphasizing the potential importance of these structures to nostril-emitting bats. By focusing sound energy into a narrow cone, the noseleaf may increase the likelihood of detecting a weak target echo in the focal area while minimizing potentially distracting (off-axis) clutter echoes. In other words, noseleaves may help to improve the signal-to-noise ratio along the axis, thus increasing the strength of focal echoes (Vanderelst et al. 2010).

This effect seems to vary in species with different ecological requirements (Vanderelst et al. 2010) and the morphology and function of noseleaves may relate directly to the foraging behavior and sensory specialization of species (Arita 1990; Bogdanowicz et al. 1997). Phyllostomids that are less reliant on echolocation have smaller nostrils and less elaborate noseleaves than those that rely heavily on sound to localize and evaluate their prey. For example, the predatory phyllostomines, the most echolocation-dependent phyllostomids, also have the most prominent noseleaves with the greatest variation in size and shape (Bogdanowicz et al. 1997). Corroborating this, removing the lancet in omnivorous *Phyllostomus discolor* was not as detrimental to spatial sensitivity as it was when removed from the gleaning predator *Micronycteris microtis*. This may be because *P. discolor* is an opportunistic generalist, consuming fruits, flowers, nectar, and some insects, relying on visual and olfactory cues in addition to echolocation. *Micronycteris microtis*, conversely, relies solely on echolocation to forage and has a longer noseleaf than *P. discolor* relative to the wavelength of its calls (Vanderelst et al. 2010). However, there is scant evidence that the noseleaf plays a crucial role in echolocation call emission in phyllostomids, even the extent to which these bats emit their calls through their nostrils as opposed to their mouths is not well understood (Jakobsen, Brinkløv et al. 2013); however, the role of prey-generated sounds is.

Passive Listening

Bats have morphological adaptations and hunting strategies that optimize foraging in extreme habitats. Open-space aerial hawkers detect and track airborne prey in flight based solely on echolocation. Gleaners, such as the typical phyllostomid, capture prey directly off the ground or foliage (Norberg and Rayner 1987), most often using sensory cues provided by the prey itself. For example, potential prey may make sounds as they move about or for intraspecific communication (Tuttle and Ryan 1981). Gleaning bats listen for these prey-generated sounds (passive listening), in addition to their own echoes, for prey detection (Ratcliffe and Dawson 2003). Because these bats are sensitive to broadband incidental sounds, they are able to localize noisy prey among clutter (Page and Ryan 2008). Impressively, gleaners can even acquire information about the size of potential prey and the substrate on which it is found using prey-generated rustling noises alone (Goerlitz et al. 2008).

Some eared prey become silent when they detect a bat, and in some instances bats may then use echolocation to help localize them (Page et al. 2012). For example, the fringe-lipped bat, *Trachops cirrhosus*, is classified as a passive listener and has evolved to be sensitive to the mating calls of its anuran prey, the túngara frog (*Engystomops pustulosus*) (Tuttle and Ryan 1981). This bat continues to echolocate while listening for prey sounds (Barclay et al. 1981). Even if the frog becomes silent after having been detected by an approaching bat, *T. cirrhosus* can sometimes capture it by detecting water ripples made by the frog's air sac (Halfwerk et al. 2014). Bats like *T. cirrhosus*, which use both active and passive listening to locate their prey, seem to have two frequency-sensitive ranges: one for high-frequency echolocation, one for detection of low-frequency prey sounds (Reep and Bhatnagar 2000; see chap. 14). Unsurprisingly, bats such as these are assumed to experience greater auditory demands and do have relatively large inferior colliculi (Baron et al. 1996).

Trachops cirrhosus prefers complex male túngara frog calls to simple ones (Ryan and Tuttle 1982). Complex calls are easier for the bat to localize, as they have broader bandwidths, longer durations, and sharper on and off sets. This bias towards complex calls has also been argued to have a learned component, as frogs often use complex calls when conspecifics are in the vicinity (Page and Ryan 2005). And so it may be that bats associate complex calls with higher prey density and greater capture success. Further, *T. cirrhosus* quickly associates high-quality prey with particular acoustic cues and can discriminate the calls of poisonous toads from palatable frogs (Page and Ryan 2006; Tuttle and Ryan 1981). However, evidence for a perceptual or sensory bias for complex calls has been mixed. When presented with populations of Peter's dwarf frogs (*Engystomops petersi*) that use both complex and simple calls, one study reported that *T. cirrhosus* was biased towards complex calls (Trillo et al. 2012). However, another study (Fugère et al. 2015) failed to find evidence for this existing sensory bias, suggesting that more work needs to be done to better understand how *T. cirrhosus* selects prey. For more information about the biology of the fringe-lipped bat, see Hemingway et al., chapter 14 (this vol.).

Bats that rely heavily on passive listening are also known to abort attacks entirely if the prey is silenced (ter Hofstede et al. 2008). In fact, it was once considered impossible for bats to glean silent prey in acoustically complex environments using echolocation alone (Arlettaz et al. 2001). However, one particular phyllostomid, *Micronycteris microtis*, the subject of the next case study, has very recently been show to defy this assumption (Geipel, Jung et al. 2013).

Case Study 3: Echolocation in the Common Big-Eared Bat (*Micronycteris microtis*)

Micronycteris microtis (subfamily Micronycterinae), like almost all animal-eating phyllostomids, is a gleaner. Weighing ~7g, it is among the smallest of carnivorous bats and feeds on a range of insects (e.g., dragonflies), spiders, and small vertebrates that it hunts within the acoustic clutter of dense rainforest understory (Kalka and Kalko 2006); it relies solely on echolocation to detect and acquire prey (Geipel, Jung et al. 2013). Hunting for prey in clutter is acoustically challenging because of the large number of echoes that will be generated from the surrounding foliage: weaker prey echoes can be masked by stronger background echoes (backward masking), while the call and returning echoes can overlap with one another (forward masking) (Neuweiler 1989). To reduce the effects of acoustic clutter, bats often adjust call parameters such as bandwidth, duration, pulse interval, and intensity (Schnitzler and Kalko 2001). However, until the discovery of *M. microtis*, it was thought that locating still and silent prey with echolocation alone was impossible (Arlettaz et al. 2001).

Surprisingly, *M. microtis* does not appear to use prey-generated sounds as it can glean silent, motionless (e.g., dragonflies), and even camouflaged (e.g., walking sticks) prey (Kalka and Kalko 2006). It can distinguish between real and fake dragonflies, discriminating small differences in target shape, structure, and texture (Geipel, Jung et al. 2013). Furthermore, after classifying an object as a prey item, the bat exhibits very precise localization, attacking the prey at the thorax before carrying it to its roost by the head (IG, pers. obs.). This precision is necessary, as many of this bat's prey are larger than it (e.g., small *Anolis* lizards) and often well-defended (e.g., strong jaws and mandibles). *Micronycteris microtis* has the greatest bite force for its size among the phyllostomids (Santana et al. 2011) and consumes between 61% and 84% of its body mass per night (Kalka and Kalko 2006).

Micronycteris microtis's slow and maneuverable foraging flight is interrupted by nearly stationary three-dimensional hovering phases upon detection of an object of interest. During hovering flight, the bats echolocate while changing their relative angles to the prey and leaf surface, reducing the masking of prey echoes by background echoes (Geipel, Jung et al. 2013). Prey echoes tend to be stronger on smooth surfaces (if the sound is emitted from an oblique angle), as clutter echoes are mainly reflected away from the echolocating bat (Siemers et al. 2005). When approaching a target, *M. microtis* uses short, broadband, FM, multiharmonic calls with high frequencies to increase resolution and prevent call-echo overlap at short range. The combination of flight, three-dimensional hovering, and echolocation call design provides *M. microtis* with a detailed acoustic image of its would-be targets (Geipel, Jung et al. 2013).

Micronycteris microtis may be unique among bats in that it gleans motionless and silent prey using echolocation. *Micronycteris microtis* is also the first bat species observed to feed its pups with solid prey, up to and beyond five months after the pups have been weaned (Geipel, Kalko et al. 2013). In addition to nutritional benefits, provisioning pups with solid food may have informational benefits as well. That is, this species' difficult and unusual foraging strategy of locating and capturing silent, motionless prey might require gradual mastery by the young bats while mothers continue to provision them. Through maternal prey transfers, the pups can also safely practice the handling of large, well-defended prey and perhaps even learn the echo-

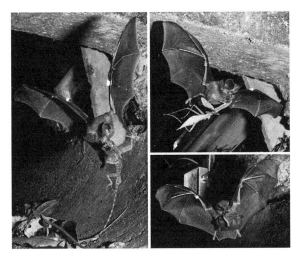

Figure 11.4. *Micronycteris microtis* and its various prey. *Clockwise from top left:* adult *M. microtis* with a lizard in its mouth; adult *M. microtis* with mantid in its mouth; adult female *M. microtis* with caterpillar in its mouth (and its pup attached to its underside).

acoustic images of different prey items. As pups are weaned, they start hunting alone, and return to their natal roost with their prey. As pup's prey-capture rate increases, there is a parallel drop in maternal provisioning until it stops altogether (Geipel, Kalko et al. 2013).

As we have seen, phyllostomids rely heavily on visual, olfactory, and auditory cues, resulting in a larger brain relative to other echolocating bats. However, in addition to these sensory systems, phyllostomid bats also depend heavily on memory and spatial navigation abilities for foraging and have the largest relative hippocampus volumes among laryngeally echolocating bats (Fleming et al. 1977; Thiagavel et al. 2018).

Spatial Memory

The hippocampus is part of the limbic system and is associated with spatial navigation (Healy et al. 2005). Hippocampus size has been shown to correspond to spatial memory in a range of mammals and birds (Krebs et al. 1989). The large hippocampus volumes found in phyllostomids (Fleming et al. 1977) may reflect their reliance on memory to return to stationary food resources (Thiele and Winter 2005). Bats in the Stenodermatine subfamily (frugivores that often hunt in the canopy) have the largest overall brain size among phyllostomids. Their large brains have been primarily attributed to their excellent spatial memories and explorative tendencies (Baron et al. 1996). The nectar-feeding glossophagines also have well-developed spatial memories and are ca-

pable of recalling rewarding sites long after the rewards have been removed (Carter et al. 2010; Stich and Winter 2006). Most bat-pollinated plants secrete nectar at a steady rate and flower at the same locations over a number of nights, making spatial memory one of the most important mechanisms with which nectarivorous bats locate flowers. The importance of memory for nectarivores has been demonstrated in *Glossophaga soricina* (von Helversen and Winter 2003). In that study, the bats were presented with a number of feeders of varying shapes, one of which was rewarding. When the feeders were rearranged, the bats preferred unrewarding feeders that were at the location where they had previously been rewarded. Most insectivorous bats, conversely, are opportunistic hunters that apparently do not rely as heavily on their memories to find food (Hulgard and Ratcliffe 2014; Stich and Winter 2006).

Multimodal Sensory Integration

While some phyllostomids may have sensory systems that they heavily rely on, multimodal flexibility with respect to sensory acquisition is advantageous, as certain signals and cues are better suited for some purposes than they are for others. The ability to switch between different sensory modalities means that when a prey cue is masked to one sensory system, it may remain detectable by another (Gomes et al. 2016).

Phyllostomids are generally considered to be quite flexible and often switch between different modes to optimize sensory acquisition and foraging (Altringham and Fenton 2003). For example, *A. jamaicensis* relies primarily on its sense of smell to find fruit but switches to vision and echolocation when olfaction is no longer sufficient. Like *A. jamaicensis*, which travels far in search of ripened figs, *Leptonycteris yerbabuenae* flies far in search of nectar-rewarding cactus flowers. Having different means of acquiring environmental information makes these bats well equipped for their long journeys (Gonzalez-Terrazas et al. 2016). Other examples include the primarily frugivorous *C. perspicillata*, which switches from using olfactory cues to echolocation when detecting and removing fruit and the omnivorous *P. hastatus* that uses shape cues provided by echolocation to detect fruit and olfactory cues to assess the fruit's ripeness (Kalko and Condon 1998; Thies et al. 1998). While *P. discolor* relies so strongly on spatial memory that it will try to land on perches that have long been removed, it predominantly uses echolocation and vision during prey pursuit (Höller and Schmidt 1996).

The obligate blood-feeding common vampire bat *D. rotundus* can acquire host information via thermoperception (Altringham and Fenton 2003), having four infrared sensors between the nasal pad and noseleaf (Kurten and Schmidt 1982). By maintaining these organs at a cooler temperature than the rest of the face, *D. rotundus* can detect the warmer temperatures of its would-be hosts and optimize its bite site choice. In addition to this exceptional ability, *D. rotundus* has large eyes, echolocates, has a well-developed olfactory bulb and vomeronasal organ (Baron et al. 1996; Jolicoeur and Baron 1980), and can locate sleeping prey by listening for their breathing noises (Schmidt et al. 1991; for more information see chap. 15, this vol.).

Case Study 4: Multisensory Integration in Pallas's Long-Tongued Bat (*Glossophaga soricina*)

Excellent visual, olfactory, and auditory information processing, complemented by a well-developed spatial memory, characterizes the glossophagine subfamily. Among the primarily nectar-loving bats, *G. soricina* has been studied most extensively. In addition to nectar, *G. soricina* also consumes fruits and insects. With excellent vision, it is also sensitive to flowers that reflect ultraviolet light (Müller et al. 2009). In addition to vision, *G. soricina* also relies on its sense of smell when foraging. It has been shown to have large olfactory bulbs that are comparable in size to nonecholocating fruit bats and has a strong preference for the smell of sulfur. Indeed, even when a sugar water reward was paired with flowers with nonsulfuric scents, *G. soricina* preferred those flowers that lacked a reward but smelled like sulfur. In fact, these bats were attracted to the sulfuric scent prior to any exposure or positive reinforcements, suggesting that this preference may be innate in primarily nectarivorous phyllostomid bats (von Helversen et al. 2000).

Unlike pteropodids, phyllostomids such as *G. soricina* use echolocation during foraging. Like *M. macrophyllum*, *G. soricina* is also one of the few phyllostomids to emit high repetition feeding buzzes, typical of bats that track and catch moving prey. *Glossophaga soricina* uses an acoustic strategy that is apparently unique among phyllostomids. It has been shown

to stealth hunt, which is a strategy that is typical of the insectivorous vespertilionid *Barbastella barbastellus* (Goerlitz et al. 2010). This strategy means that *G. soricina* is able to opportunistically supplement its diet with insects. Its high-frequency, low-intensity calls may allow it to remain undetected, thus not triggering an escape response in insects with bat-detecting ears (Clare et al. 2014).

In addition to using visual, olfactory, and auditory cues, *G. soricina* also has excellent spatial memory. This species can recall which flowers it has visited and which it yet needs to visit (Thiele and Winter 2005). However, studies have shown that *G. soricina* prefers spatial cues over visual, olfactory, and auditory cues, even when those spatial cues are unreliable and misleading. In fact, it struggles to associate food with location when given only shape or scent cues (Carter et al. 2010). Being able to recall the distribution of flowers and nectar over periods of time is highly advantageous for nectarivores since many bat-pollinated plants flower multiple times throughout the year at the same locations (Hopkins 1984). We have now seen how a variety of phyllostomids rely on environmental information and individual learning for effective foraging.

Overall Brain Size

Phytophagous phyllostomids and pteropodids rely on more sensory modalities than do obligate predatory bats and thus experience greater demands on sensory information processing and integration. Bats in these two families also have large overall brain volumes relative to other bats (Eisenberg and Wilson 1978; fig. 11.2). Relative brain region size helps deduce those sensory modalities with which the bat primarily acquires information. Pteropodids and phyllostomids have large visual and olfactory regions, as well as large neocortices (higher-order cognition, perception of complex stimuli) and hippocampi (spatial memory and orientation) (Thiagavel et al. 2018). All contribute to larger brains.

Aerial insectivores, which tend to forage in more open spaces than pteropodids and phyllostomids, have smaller brains relative to facultative gleaners that hunt in clutter (Safi and Dechmann 2005). There have been several explanations for this positive relationship between brain size and habitat complexity. While large brains are generally advantageous for survival (Jones et al. 2013), they are especially valuable to gleaning bats, which may often find themselves in close proximity to both terrestrial (Jones et al. 2013) and aerial (Fenton 1995) predators. Larger forebrains are often favored, as they increase survival, improve decision making, and allow for behavioral flexibility in fluctuating environments (Sol et al. 2005). Also, because gleaners dwell in cluttered habitats and must detect prey against complex backgrounds (Safi and Dechmann 2005), they experience greater demands on their visual, olfactory, and auditory systems than do open space bats. Similarly, the hippocampal volumes of bats that glean prey have been shown to be relatively larger than those of hawkers, as gleaners experience more obstacles in their constantly changing environments (Safi and Dechmann 2005). The cerebellar nuclei (involved in flight control and locomotory behavior) may also be more developed in gleaning species, particularly in hovering bats (Baron et al. 1996).

However, larger brains may not be advantageous for all bats. Insectivorous bats such as vespertilionids and molossids, for example, may be under pressure to maintain a reduced brain size, as smaller brains (and body mass, in general) may facilitate the high speeds and long-distance flights observed in these families (McGuire and Ratcliffe 2011). Unlike most phyllostomids (and pteropodids), these bats must react quickly to opportunistically catch their airborne prey (Geberl et al. 2015), and they rely less on spatial and temporal distribution of their food source or on visual and olfactory stimuli. It may be that relying primarily on echolocation has allowed these bats to evolve relatively small brains (Thiagavel et al. 2018).

It is generally assumed that the ancestral bat was a small-brained insectivore (Safi et al. 2005). Thus, the expansion of the diet to include fruit and nectar may have initiated the evolution of larger brains. However, the ancestral state of the common noctilionoid ancestor has not been estimated and the move into new niches (e.g., fishing, gleaning vertebrates) may, like phytophagy, also be associated with relatively large brains. The sensory needs of insectivorous hawkers, conversely, may have stayed focused on audition, allowing brains to remain small. In fact, phylogenetic reconstruction of the ancestral brain size using fossils has supported the idea of an increase in brain size over evolutionary time in bats (Yao et al. 2012). In other words, the brains of insectivorous bats may simply be small because they did not need to expand from the presumably small ancestral brain size. As a side note, insectivorous bats with the most sophisticated calls have the smallest brains (i.e., vespertilionids and rhinolophids) while the call designs of the phyllostomids are apparently conserved (Ratcliffe 2009a).

Conclusions

The evolution of brain regions with respect to their ecology has long been studied in bats and is still being explored today. Despite now defunct phylogenies, most of the same general trends remain clear. For instance, that phyllostomids and nonecholocating pteropodids have larger brains after controlling for their larger size is widely accepted. With updated molecular phylogenies and modern phylogenetic methods, we can now, better than ever before, validate these and other relationships between foraging behavior and information use and storage. Bats in general have proven to be an excellent model system with which to carry out brain studies. Families show extreme differences in their sensory-system specializations, reflective of the ecological pressures they face. Perhaps due to the limitations of flight, different brain regions tend to be enhanced or reduced to optimize sensory acquisition and efficiency.

Non-laryngeal echolocating, frugivorous pteropodids have reduced auditory and enhanced visual and olfactory centers, while the opposite trade-offs are obvious in echolocating predatory bats. Phyllostomids, conversely, have a variety of means with which they can acquire sensory information and showcase balanced representations of visual, olfactory, and auditory modalities in the brain (Baron et al. 1996). As sensory generalists, phyllostomids exploit a wide spectrum of feeding ecologies, make use of multiple sensory systems, and have diversified into one of the most species-rich bat families today (Jones et al. 2002). The species- and niche-rich nature of the New World leaf-nosed bats makes them unique among bats, and a group from which we will continue to gain a better understanding of sensory specializations, cognitive ecology, and the evolution of mammalian brains.

Acknowledgments

We thank the editors for inviting us to contribute a chapter to this volume and Marco Tschapka for letting us use his photographs of phyllostomid bats. Our research has been funded by the Danish Research Council, the Natural Sciences and Engineering Research Council of Canada, and the Smithsonian Tropical Research Institute.

References

Altringham, J. D., and M. B. Fenton. 2003. Sensory ecology and communication in the Chiroptera. Pp. 90–127 *in*: Bat Ecology (T. H. Kunz and M.B. Fenton, eds.). University of Chicago Press, Chicago.

Arita, H. T. 1990. Noseleaf morphology and ecological correlates in phyllostomid bats. Journal of Mammalogy, 71:36–47.

Arlettaz, R., G. Jones, and P. A. Racey. 2001. Effect of acoustic clutter on prey detection by bats. Nature, 414:742–745.

Barclay, R. M. R., M. B. Fenton, M. D. Tuttle, and M. J. Ryan. 1981. Echolocation calls produced by *Trachops cirrhosus* (Chiroptera: Phyllostomatidae) while hunting for frogs. Canadian Journal of Zoology, 59:750–753.

Baron, G., and P. Jolicoeur. 1980. Brain structure in Chiroptera: some multivariate trends. Evolution, 34:386–393.

Baron, G., H. Stephan, and H. D. Frahm. 1996. Comparative Neurobiology in Chiroptera. Vol 1. Birkhäuser Verlag, Basel.

Barton, R. A., A. Purvis, and P. H. Harvey. 1995. Evolutionary radiation of visual and olfactory brain systems in primates, bats and insectivores. Philosophical Transactions of the Royal Society of London B, 348:381–392.

Bell, G. P. 1985. The sensory basis of prey location by the California leaf-nosed bat *Macrotus californicus* (Chiroptera: Phyllostomidae). Behavioral Ecology and Sociobiology, 16:343–347.

Bell, G. P., and M. B. Fenton. 1986. Visual acuity, sensitivity and binocularity in a gleaning insectivorous bat, *Macrotus californicus* (Chiroptera: Phyllostomidae). Animal Behavior, 34:409–414.

Bhatnagar, K. P., and F. C. Kallen. 1974. Morphology of the nasal cavities and associated structures in *Artibeus jamaicensis* and *Myotis lucifugus*. American Journal of Anatomy, 139:167–190.

Bogdanowicz, W., R. D. Csada, and M. B. Fenton. 1997. Structure of noseleaf, echolocation and foraging behavior in the Phyllostomidae (Chiroptera). Journal of Mammalogy, 78:942–953.

Brinkløv, S. 2010. Echolocation call intensity and beam pattern of Neotropical leaf-nosed bats (Phyllostomidae). PhD thesis, University of Southern Denmark, Odense.

Brinkløv, S., L. Jakobsen, J. M. Ratcliffe, E. K. V. Kalko, and A. Surlykke. 2011. Echolocation call intensity and directionality in flying short-tailed fruit bats, *Carollia perspicillata* (Phyllostomidae). Journal of the Acoustical Society of America, 129:427–435.

Brinkløv, S., E. K. V. Kalko, and A. Surlykke. 2009. Intense echolocation calls from two "whispering" bats, *Artibeus jamaicensis* and *Macrophyllum* (Phyllostomidae). Journal of Experimental Biology, 212:11–20.

Brinkløv, S., E. K. V. Kalko, and A. Surlykke. 2010. Dynamic adjustment of biosonar intensity to habitat clutter in the bat

Macrophyllum macrophyllum (Phyllostomidae). Behavioral Ecology and Sociobiology, 64:1867–1874.

Burr, B., D. Rosen, and W. Barthlott. 1995. Untersuchungen zur Ultraviolettreflexion von Angiospermenblüten III. Dilleniidae and Asteridae s. I. Tropische und Subtropische Pflanzenwelt, 93:1–185.

Carter, G. G., J. M. Ratcliffe, and B. G. Galef. 2010. Flower bats (*Glossophaga soricina*) and fruit bats (*Carollia perpsicillata*) rely on spatial cues over shapes and scents when relocating food. PLOS ONE, 5:1–6.

Clare, E. L., H. R. Goerlitz, V. A. Drapeau, M. W. Holderied, A. M. Adams, J. Nagel, E. R. Dumont, P. D. N. Hebert, and M. B. Fenton. 2014. Trophic niche flexibility in *Glossophaga soricina*: how a nectar seeker sneaks an insect snack. Functional Ecology, 28:632–641.

Cooper, J. G., and K. P. Bhatnagar. 1976. Comparative anatomy of the vomeronasal organ complex in bats. Journal of Anatomy, 122:571–601.

Davies, J. T. 1971. Olfactory theories. Pp. 322–350 *in*: Handbook of Sensory Physiology. Vol. 4, pt. 1 (L. M. Beidler, ed.). Springer-Verlag, New York.

deWinter, W., and C. E. Oxnard. 2001. Evolutionary radiations and convergences in the structural organization of mammalian brains. Nature, 409:710–714.

Dukas, R., and J. M. Ratcliffe. 2009. Prospects. Pp. 298–300 *in*: Cognitive Ecology II (R. Dukas and J. M. Ratcliffe, eds). University of Chicago Press, Chicago.

Dusenbery, D. B. 1992. Sensory Ecology: How Organisms Acquire and Respond to Information. W. H. Freeman, New York.

Eisenberg, J. F., and D. E. Wilson. 1978. Relative brain size and feeding strategies in the Chiroptera. Evolution, 32:740–751.

Eiting, T. P., T. D. Smith, and E. R. Dumont. 2014. Olfactory epithelium in the olfactory recess: a case study in New World leaf-nosed bats. Anatomical Record, 297:2105–2112.

Eklöf, J. 2003. Vision in echolocating bats. PhD thesis, Göteborg University, Göteborg, Sweden.

Fenton, M. B. 1995. Natural history and biosonar signals. Pp. 37–86 *in*: Hearing in Bats (A. N. Popper and R. R. Fay, eds.). Springer, New York.

Fleming, T. H. 1988. The Short-Tailed Fruit Bat: A Study in Plant-Animal Interactions. University of Chicago Press, Chicago.

Fleming, T. H., C. Geiselman, and W. J. Kress. 2009. The evolution of bat pollination: a phylogenetic perspective. Annals of Botany, 104:1017–1043.

Fleming, T. H., E. R. Heithaus, and W. B. Sawyer. 1977. An experimental analysis of the food location behavior of frugivorous bats. Ecology, 58:619–627.

Frahm, H. D. 1981. Volumetric comparison of the accessory olfactory bulb in bats. Acta Anatomica, 109:173–183.

Freeman, P. W. 2000. Macroevolution in Microchiroptera: recoupling morphology and ecology with phylogeny. Evolutionary Ecology Research, 2:317–335.

Fugère, V., M. T. O'Mara, and R. A. Page. 2015. Perceptual bias does not explain preference for prey call adornment in the frog-eating bat. Behavioral Ecology and Sociobiology, 69:1353–1364.

Geberl, C., S. Brinkløv, L. Wiegrebe, and A. Surlykke. 2015. Fast sensory-motor reactions in echolocating bats to sudden changes during the final buzz and prey intercept. Proceedings of the National Academy of Sciences, USA, 112:4122–4127.

Geipel, I., K. Jung, and E. K. V. Kalko. 2013. Perception of silent and motionless prey on vegetation by echolocation in the gleaning bat *Micronycteris microtis*. Proceedings of the Royal Society B, 280:1754.

Geipel, I., E. K. V. Kalko, K. Wallmeyer, and M. Knörnschild. 2013. Postweaning maternal food provisioning in a bat with a complex hunting strategy. Animal Behaviour, 85:1435–1441.

Geva-Sagiv, M., L. Las, Y. Yovel, and N. Ulanovsky. 2015. Spatial cognition in bats and rats: from sensory acquisition to multiscale maps and navigation. Nature Reviews Neuroscience, 16:94–108.

Ghose, K., and C. F. Moss. 2003. The sonar beam pattern of a flying bat as it tracks tethered insects. Journal of the Acoustical Society of America, 114:1120–1131.

Goerlitz, H. R., S. Greif, and B. M. Siemers. 2008. Cues for acoustic detection of prey: insect rustling sounds and the influence of walking substrate. Journal of Experimental Biology, 211:2799–2806.

Goerlitz, H. R., H. M. ter Hofstede, M. R. K. Zeale, G. Jones, and M. W. Holderied. 2010. An aerial-hawking bat uses stealth echolocation to counter moth hearing. Current Biology, 20:1568–1572.

Gomes, D. G. E., R. A. Page, I. Geipel, R. C. Taylor, M. J. Ryan, and W. Halfwerk. 2016. Bats perceptually weight prey cues across sensory systems when hunting in noise. Science, 353:1277–1280.

Gonzalez-Terrazas, T. P., C. Martel, P. Milet-Pinheiro, M. Ayasse, E. K. V. Kalko, and M. Tschapka. 2016. Finding flowers in the dark: nectar-feeding bats integrate olfaction and echolocation while foraging for nectar. Royal Society Open Science, 3:160199.

Griffin, D. R. 1958. Listening in the Dark: The Acoustic Orientation of Bats and Men. Yale University Press, New Haven, CT.

Halfwerk, W., P. L. Jones, R. C. Taylor, M. J. Ryan, and R. A. Page. 2014. Risky ripples allow bats and frogs to eavesdrop on a multisensory sexual display. Science, 343:413–416.

Handley, C. O., Jr., D. E. Wilson, and A. L. Gardner. 1991. Demography and natural history of the common fruit bat, *Artibeus jamaicensis*, on Barro Colorado Island, Panamá. Smithsonian Contributions to Zoology, 511:1–173.

Hartley, D. J., and R. A. Suthers. 1987. The sound emission pattern and the acoustical role of the noseleaf in the echolocating bat, *Carollia perspicillata*. Journal of the Acoustical Society of America, 82:1892–1900.

Healy, S. D., S. R. de Kort, and N. S. Clayton. 2005. The hippocampus, spatial memory and food hoarding: a puzzle revisited. Trends in Ecology and Evolution, 20:17–22.

Heffner, R. S., G. Koay, and H. E. Heffner. 2001. Sound localization in a new-world frugivorous bat, *Artibeus jamaicensis*: acuity, use of binaural cues, and relationship to vision. Journal of the Acoustical Society of America, 109:412–421.

Höller, P., and U. Schmidt. 1996. The orientation behavior of the lesser spearnosed bat, *Phyllostomus discolor* (Chiroptera) in a model roost. Journal of Comparative Physiology A, 179:245–254.

Hope, G. M., and K. P. Bhatnagar. 1979. Effect of light adaptation on electrical responses of the retinas of four species of bats. Experientia, 35:1191–1193.

Hopkins, H. C. 1984. Floral biology and pollination ecology of the Neotropical species of *Parkia*. Journal of Ecology, 72:1–23.

Hu, K., Y. Li, X. Gu, H. Lei, and S. Zhang. 2006. Brain structures of echolocating and noncholocating bats, derived *in vivo* from magnetic resonance images. Neuroreport, 17:1743–1746.

Hulgard, K., and J. M. Ratcliffe. 2014. Niche-specific cognitive strategies: object memory interferes with spatial memory in the predatory bat, *Myotis nattereri*. Journal of Experimental Biology, 217:3293–3300.

Hulgard, K., and J. M. Ratcliffe. 2016. Sonar sound groups and increased terminal buzz duration reflect task complexity in hunting bats. Scientific Reports, 6:21500.

Hutcheon, J. M., J. W. Kirsch, and T. Garland Jr. 2002. A comparative analysis of brain size in relation to foraging ecology and phylogeny in the Chiroptera. Brain, Behavior and Evolution, 60:165–180.

Jakobsen, L., S. Brinkløv, and A. Surlykke. 2013. Intensity and directionality of bat echolocation signals. Frontiers in Physiology, 4:89.

Jakobsen, L., J. M. Ratcliffe, and A. Surlykke. 2013. Convergent acoustic field of view in echolocating bats. Nature, 493: 93–96.

Jolicoeur, P., and G. Baron. 1980. Brain center correlations among Chiroptera. Brain, Behavior and Evolution, 17:419–431.

Jolicoeur, P., P. Prilot, G. Baron, and H. Stephan. 1984. Brain structure and correlation patterns in Insectivora, Chiroptera, and Primates. Systematic Zoology, 33:14–29.

Jones, K. E., A. Purvis, A. Maclarnon, O. R. P. Bininda-Emonds, and N. B. Simmons. 2002. A phylogenetic supertree of the bats (Mammalia: Chiroptera). Biological Reviews, 77: 223–259.

Jones, P. L., M. J. Ryan, V. Flores, and R. A. Page. 2013. When to approach novel prey cues? Social learning strategies in frog-eating bats. Proceedings of the Royal Society of London B, 280:20132330.

Kalka, M., and E. K. V. Kalko. 2006. Gleaning bats as underestimated predators of herbivorous insects: diet of *Micronycteris microtis* (Phyllostomidae) in Panama. Journal of Tropical Ecology, 22:1–10.

Kalko, E. K. V., and M. A. Condon. 1998. Echolocation, olfaction, and fruit display: how bats find fruit of flagellichorous cucurbits. Functional Ecology, 12:364–372.

Kalko, E. K. V., C. O. Handley Jr., and D. Handley. 1996. Organization, diversity, and long-term dynamics of a Neotropical bat community. Pp. 503–553 *in*: Long-Term Studies in Vertebrate Communities (M. Cody and J. Smallwood, eds). Academic Press, Los Angeles.

Kalko, E. K. V., E. A. Herre, and C. O. Handley Jr. 1996. Relation of fig fruit characteristics to fruit-eating bats in the New and Old World tropics. Journal of Biogeography, 23:565–576.

Kalko, E. K. V., and H.-U. Schnitzler. 1993. Plasticity in echolocation signals of European pipistrelle bats in search flight: implications for habitat use and prey detection. Behavioral Ecology and Sociobiology, 33:415–428.

Krebs, J. R., D. F. Sherry, S. D. Healy, V. H. Perry, and A. L. Vaccarino. 1989. Hippocampal specialization of food-storing birds. Proceedings of the National Academy of Sciences, USA, 86:1388–1392.

Kulzer, E. 1956. Flughunde erzeugen Orientierungslaute durch Zungenschlag. Naturwissenschaften, 43:117–118.

Kurten, L., and U. Schmidt. 1982. Thermoperception in the common vampire bat (*Desmodus rotundus*). Journal of Comparative Physiology A, 146:223–228.

Laska, M. 1990. Olfactory discrimination ability in short-tailed fruit bat, *Carollia perspicillata* (Chiroptera: Phyllostomidae). Journal of Chemical Ecology, 16:3291–3299.

McGuire, L. P., and J. M. Ratcliffe. 2011. Light enough to travel: migratory bats have smaller brains, but not larger hippocampi, than sedentary species. Biology Letters, 7:233–236.

Morrison, D. W. 1978. Foraging ecology and energetics of the frugivorous bat, *Artibeus jamaicensis*. Ecology, 59:716–723.

Müller, B., M. Glösmann, L. Peichl, G. C. Knop, C. Hagemann, and J. Ammermüller. 2009. Bat eyes have ultraviolet-sensitive cone photoreceptors. PLOS One, 4:e6390.

Neuweiler, G. 1989. Foraging ecology and audition in echolocating bats. Trends in Ecology and Evolution, 4:160–166.

Norberg, U. M., and J. M. V. Rayner. 1987. Ecological morphology and flight in bats (Mammalia; Chiroptera): wing adaptations, flight performance, foraging strategy and echolocation. Philosophical Transactions of the Royal Society B, 316:335–427.

Nørum, U., S. Brinkløv, and A. Surlykke A. 2012. New model for gain control of signal intensity to object distance in echolocating bats. Journal of Experimental Biology, 215: 3045–3054.

Page, R. A., and M. J. Ryan. 2005. Flexibility in assessment of prey cues: frog-eating bats and frog calls. Proceedings of the Royal Society of London B, 272:841–847.

Page, R. A., and M. J. Ryan. 2006. Social transmission of novel foraging behavior in bats: frog calls and their referents. Current Biology, 16:1201–1205.

Page, R. A., and M. J. Ryan. 2008. The effect of signal complexity on localization performance in bats that localize frog calls. Animal Behaviour, 76:761–769.

Page, R. A., T. Schnelle, E. K. V. Kalko, T. Bunge, and X. E. Bernal. 2012. Sequential assessment of prey through the use of multiple sensory cues by an eavesdropping bat. Naturwissenschaften, 99:505–509.

Pirlot, P., and H. Stephan. 1970. Encephalization in Chiroptera. Canadian Journal of Zoology, 48:433–444.

Ratcliffe, J. M. 2009a. Neuroecology and diet selection in phyllostomid bats. Behavioral Processes, 80:247–251.

Ratcliffe, J. M. 2009b. Predator-prey interaction in an auditory world. Pp. 201–225 *in*: Cognitive Ecology II (R. Dukas and J. M. Ratcliffe, eds). University of Chicago Press, Chicago.

Ratcliffe, J. M., and J. W. Dawson. 2003. Behavioral flexibility: the little brown bat, *Myotis lucifugus*, and the northern long-eared bat, *M. septentrionalis*, both glean and hawk prey. Animal Behaviour, 66:847–856.

Ratcliffe, J. M., C. P. H. Elemans, L. Jakobsen, and A. Surlykke. 2013. How the bat got its buzz. Biology Letters, 9:1–5.

Ratcliffe, J. M., and H. M. ter Hofstede. 2005. Roosts as social information centers: social learning of food preferences in bats. Biology Letters, 1:72–74.

Reep, R. L., and K. P. Bhatnagar. 2000. Brain ontogeny and ecomorphology in bats. Pp. 93–136 *in*: Ontogeny, Functional Ecology, and Evolution of Bats (R. A. Adams and S. C. Pedersen, eds.). Cambridge University Press, Cambridge.

Rojas, D., C. A. Mancina, J. J. Flores-Martines, and L. Navarro. 2013. Phylogenetic signal, feeding behavior and brain volume in Neotropical bats. Journal of Evolutionary Biology, 26:1925–1933.

Ross, G., and M. W. Holderied. 2013. Learning and memory in bats: a case study on object discrimination in flower-visiting bats. Pp 207–224 *in*: Bat Evolution, Ecology and Conservation (R. Adams and S. C. Pedersen, eds.). Springer, New York.

Ryan, M. J., and M. D. Tuttle. 1982. Bat predation and sexual advertisement in a Neotropical anuran. American Naturalist, 119:136–139.

Safi, K., and D. K. N. Dechmann. 2005. Adaptation of brain regions to habitat complexity: a comparative analysis in bats (Chiroptera). Proceedings of the Royal Society of London B, 272:179–186.

Safi, K., M. A. Seid, and D. K. N. Dechmann. 2005. Bigger is not always better: when brains get smaller. Biology Letters, 1:283–286.

Santana, S. E., I. Geipel, E. R. Dumont, M. B. Kalka, and E. K. V. Kalko. 2011. All you can eat: high performance capacity and plasticity in the common big-eared bat, *Micronycteris microtis* (Chiroptera: Phyllostomidae). PLOS ONE, 6:e28584.

Schmidt, U., P. Schlegel, H. Schweizer, and G. Neuweiler. 1991. Audition in vampire bats, *Desmotus rotundus*. Journal of Comparative Physiology, 168:45.

Schnitzler, H.-U., and E. K. V. Kalko. 1998. How echolocating bats search and find food. Pp 183–196 *in*: Bat Biology and Conservation (T. H. Kunz and P. A. Racey, eds.). Smithsonian Institution Press, Washington, DC.

Schnitzler, H.-U., and E. K. V. Kalko. 2001. Echolocation by insect-eating bats. BioScience, 51:557–569.

Shettleworth, S. J. 2010. Cognition, Evolution and Behaviour. Oxford University Press, Oxford.

Siemers, B. M., E. Baur, and H.-U. Schnitzler. 2005. Acoustic mirror effect increases prey detection distance in trawling bats. Naturwissenschaften, 92:272–276.

Simmons, N. B. 1998. A reappraisal of interfamilial relationships of bats. Pp. 3–26 *in*: Bat Biology and Conservation (T. H. Kunz and P. A. Racey, eds.). Smithsonian Institution Press, Washington, DC.

Simon, R., M. W. Holderied, C. U. Koch, and O. von Helversen. 2011. Floral acoustics: conspicuous echoes of a dish-shaped leaf attract bat pollinators. Science, 333:631–633.

Sol, D., R. P. Duncan, T. M. Blackburn, P. Cassey, and L. Lefebvre. 2005. Big brains, enhanced cognition, and response of birds to novel environments. Proceedings of the National Academy of Sciences, USA, 102:5460–5465.

Stich, K. P., and Y. Winter. 2006. Lack of generalization of object discrimination between spatial contexts by a bat. Journal of Experimental Biology, 209:4802–4808.

Surlykke, A. 1988. Interaction between echolocating bats and their prey. Pp. 551–566 *in*: Animal Sonar: Processes and

Performance (P. E. Nachtigall and P. W. B. Moore, eds.). Plenum Press, New York.

Surlykke, A., and E. K. V. Kalko. 2008. Echolocating bats cry out loud to detect their prey. PLOS ONE 3:e2036.

Suthers, R. A. 1969. Visual form discrimination by echolocating bats. Biology Bulletin, 137:535–546.

Suthers, R. A. 1970. Vision, olfaction and taste: Pp. 265–281 *in*: Biology of Bats. Vol. 2. (W. A. Wimsatt, ed.). Academic Press, New York.

Teeling, E. C. 2009. Hear, hear: the convergent evolution of echolocation in bats? Trends in Ecology and Evolution, 24:351–354.

ter Hofstede, H. M., J. M. Ratcliffe, and J. H. Fullard. 2008. The effectiveness of katydid (*Neoconocephalus ensiger*) song cessation as antipredator defence against the gleaning bat *Myotis septentrionalis*. Behavioral Ecology and Sociobiology, 63:217–226.

Thiagavel, J., C. Céchetto, S. Santana, L. Jakobsen, E. J. Warrant and J. M. Ratcliffe. 2018. Auditory opportunity and visual constraint enabled the evolution of echolocation in bats. Nature Communications, 9:98.

Thiele, J., and Y. Winter. 2005. Hierarchical strategy for relocating food targets in flower bats: spatial memory versus cue-directed search. Animal Behaviour, 69:315–327.

Thies, W., E. K. V. Kalko, and H.-U. Schnitzler. 1998. The roles of echolocation and olfaction in two Neotropical fruit-eating bats, *Carollia perspicillata* and *C. castanea*, feeding on *Piper*. Behavioral Ecology and Sociobiology, 42:397–409.

Trillo, P. A., K. A. Athanas, D. H. Goldhill, K. L. Hoke, and W. C. Funk. 2012. The influence of geographic heterogeneity in predation pressure on sexual signal divergence in an Amazonian frog species complex. Journal of Experimental Biology, 26:216–222.

Tuttle, M. D., and M. J. Ryan. 1981. Bat predation and the evolution of frog vocalizations in the Neotropics. Science, 214:677–678.

Vanderelst, D., F. D. Mey, H. Peremans, I. Geipel, E. K. V. Kalko, and U. Fizlaff. 2010. What noseleaves do for FM bats depends on their degree of sensory specialization. PLOS ONE, 5:e11893.

von Helversen, D., and O. von Helversen. 1999. Acoustic guide in bat-pollinated flower. Nature, 398:759–760.

von Helversen, O., L. Winkler, and H. J. Bestmann. 2000. Sulphur-containing "perfumes" attract flower-visiting bats. Journal of Comparative Physiology A, 186:143–153.

von Helversen, O., and Y. Winter. 2003. Glossophagine bats and their flowers: costs and benefits for plants and pollinators. Pp. 346–397 *in*: Ecology of Bats (T. H. Kunz and M. B. Fenton, eds.). University of Chicago Press, Chicago.

Weinbeer, M., E. K. V. Kalko, and K. Jung. 2013. Behavioral flexibility of the trawling long-legged bat, *Macrophyllum macrophyllum* (Phyllostomidae). Frontiers in Physiology, 4:1–11.

Wetterer, A. L., M. V. Rockman, and N. B. Simmons. 2000. Phylogeny of phyllostomid bats (Mammalia: Chiroptera): data from diverse morphological systems sex chromosomes and restriction sites. Bulletin of the American Museum of Natural History, 248:1–200.

Williams, C. F., J. M. Williams, and D. R. Griffin. 1966. The homing ability of the Neotropical bat, *Phyllostomus hastatus*, with evidence for visual orientation. Animal Behaviour, 14:468–473.

Yao, L., J. P. Brown, M. Stampanoni, F. Marone, K. Isler, and R. D. Martin. 2012. Evolutionary change in the brain size of bats. Brain, Behavior and Evolution, 80:15–25.

12

Reproduction and Life Histories

Robert M. R. Barclay
and Theodore H. Fleming

Introduction

Bats are unusual mammals in various ways, including their reproduction and life histories. Mammalian life histories are viewed as falling along a "fast-slow continuum" (e.g., Promislow and Harvey 1990; Read and Harvey 1989). This reflects species such as small rodents and shrews that mature quickly, produce litters with many young several times a year, and live short lives, compared with other generally larger species, such as cetaceans and primates, that mature slowly, produce litters with few young, and live long lives. Bats fall at the slow end of the continuum, something that is unusual for such small mammals (Barclay and Harder 2003). Explanations for the evolution of the slow life histories of bats revolve around the evolution of flight. Flight is associated with a reduced risk of extrinsic mortality (predation; Healy et al. 2014; Holmes and Austad 1994; Pomeroy 1990), which favors investment in maintenance and relatively large parental investment in few offspring that increases their chance of survival. Flight is also costly and requires that young attain near full adult size, especially in terms of skeletal development, before flight and weaning is possible (Barclay1994). Each offspring is thus energetically and nutritionally expensive to produce, constraining the number that can be raised at one time. Except in one family (Vespertilionidae), all bats produce litters with only a single young.

Among families of bats, the Phyllostomidae is arguably the most diverse in many respects. They range in body size from under 10 g to the largest microchiropteran (*Vampyrum spectrum*, 170 g). They occupy diverse environments that vary in climate, seasonality, and vegetation, and have diverse feeding niches, including nectarivory, frugivory, insectivory, carnivory, and sanguinivory (see chap. 2, this vol.). We might expect that this diversity in ecology and behavior would have resulted in the evolution of diverse reproductive patterns and life histories (i.e., traits that affect the schedule of births and deaths of an organism; Stearns 1992). Despite this, however, one key life-history trait, litter size, is invariable among phyllostomids; all species have a litter size of one.

Our goal in this chapter is to analyze available data to assess hypotheses regarding the evolution of variation in phyllostomid reproductive and life-history characteristics other than litter size. We summarize the data and test predictions based on life-history theory (e.g., Stearns 1992) using data compiled from the primary literature. These data are somewhat limited. Of the 216 currently recognized species of phyllostomids (chap. 4, this vol.), we were able to find at least some data on only 71 species (33%), and for many traits such as longevity, size of young at birth or weaning, and duration of gestation and lactation, sample sizes are less than 25. This obviously limits our ability to test predictions but also presents interesting avenues for future research.

We begin by reviewing the aspect of phyllostomid reproduction for which there is the most information, the timing and number of reproductive cycles per year. Given the lack of variation in litter size, the number of litters per year is one of the few variables that natural selection on reproductive effort (output) could act upon. Variation in annual reproductive output may also be associated with variation in other life-history traits, and we investigate this in the second part of the chapter. Our emphasis is on gross, rather than microscopic, aspects of phyllostomid reproduction. Many papers by John Rasweiler and colleagues (e.g., Badwaik and Rasweiler 2000; Rasweiler and Badwaik 2000; Rasweiler et al. 2010, 2011) describe in detail various aspects of the developmental and reproductive biology, including the occurrence of menstruation, of a well-studied phyllostomid, *Carollia perspicillata*. We end by outlining some of the questions that remain unanswered regarding phyllostomid life histories that we feel future research should address.

Reproductive Patterns

Unlike temperate-zone bats that are faced with severe seasonal climatic conditions that profoundly restrict their reproductive seasons, phyllostomids and their close relatives live in less rigorous subtropical and tropical climates that potentially permit broader reproductive seasons. As has been widely discussed (e.g., Heideman 2000; Racey 1982; Racey and Entwistle 2000), the reproductive phenology of tropical bats is driven more by seasonal rainfall patterns and their effect on plant phenology than by seasonal changes in temperature. As a result, many more tropical bats are polyestrous than monestrous, the predominant reproductive pattern in temperate bats. Supporting this, Bernard and Cumming (1997) identified ten different reproductive patterns in eight families of African bats, most of which live in tropical or subtropical climates; a number of these patterns involve polyestry, especially in pteropodid and molossid bats. They identified four different patterns in African pteropodid bats, which are the ecological analogues of nectar- and fruit-eating phyllostomids. These include restricted seasonal monestry (found in *Eidolon helvum*), extended seasonal monestry (in *Rousettus aegyptiacus*), continuous bimodal polyestry with a postpartum estrus (in *Epomophorus anurus*), and aseasonal polyestry (in *E. wahlbergi*).

As in African pteropodids, several different reproductive patterns, including both monestry and polyestry, occur in phyllostomids and their close relatives (table 12.1). Noctilionids and mormoopids are mostly monestrous. Monestry also occurs in three phyllostomid clades: Macrotinae, Phyllostominae, and Glossophaginae (table 12.1). Seasonal bimodal polyestry is common in many plant visitors, including species of Glossophaginae, Carolliinae, and Stenodermatinae. In most of these species a postpartum estrus occurs shortly after the first pregnancy of the year. Apparently only vampires (e.g., *Desmodus rotundus*), which have a gestation period of about seven months, exhibit nonseasonal polyestry in this family.

What intrinsic and extrinsic factors are associated with these different reproductive patterns? The most obvious intrinsic factor, which is also important in pteropodid bats, is body size (Racey and Entwistle 2000). Two large phyllostomines (80–170 g), *Vampyrum spectrum* and *Phyllostomus hastatus*, are monestrous. Similarly, large pteropodids weighing up to 1,500 g (e.g., species of *Pteropus, Acerodon, Eidolon*, etc.) are monestrous whereas many smaller species (e.g., species of *Cynopterus, Haplonycteris, Macroglossus*, and *Syconycteris*, etc.) are bimodally polyestrous (e.g., Hall and Richards 2000; Heideman 1995). Among 66 species of phyllostomids for which we could find data, the mean body mass of monestrous species is significantly larger than that of polyestrous species (32.6 g vs. 18.9 g, respectively; $t = 2.07$, $p = 0.046$).

Another important intrinsic factor influencing reproductive patterns is migratory behavior. Long-

Table 12.1. Summary of the reproductive modes in Phyllostomidae and its relatives

Family or subfamily	Reproductive Mode		
		Polyestrus	
	Monestrus	Seasonal	Nonseasonal
Noctilionidae	*Noctilio labialis*	*N. leporinus*	...
Mormoopidae	*Mormoops blainvillei, Pteronotus parnellii*	...	...
Macrotinae	*Macrotus californicus**	...	...
Micronycterinae	...	*Micronycteris hirsuta*	...
Desmodontinae	...	...	*Desmodus rotundus*
Phyllostominae	*Mimon cozumelae?, Phyllostomus discolor?, P. hastatus, Vampyrum spectrum*	*Lophostoma brasiliense, L. silvicolum, Tonatia bidens?*	...
Glossophaginae	*Anoura geoffroyi, Brachyphylla nana, Choeronycteris mexicana, Erophylla* spp, *Glossophaga longirostris* (Curaçao), *Leptonycteris* spp, *Monophyllus redmani* (Jamaica), *Musonycteris harrisoni, Phyllonycteris poei*	*B. cavernarum, Glossophaga* spp., *Hylonycteris underwoodi, Monophyllus redmani* (Puerto Rico), *Phyllops falcatus?*	...
Lonchorhininae	...	...	...
Lonchophyllinae	...	...	...
Glyphonycterinae	...	...	...
Carolliinae	...	*Carollia perspicillata** + other *Carollia*	...
Rhinophyllinae	...	...	...
Stenodermatinae	...	*Artibeus bogotensis, A. cinereus, A. jamaicensis*, A. lituratus, A. phaeotis, A. planirostris, A. watsoni, Chiroderma salvini?, Ectophylla alba, Enchisthenes hartii, Platyrrhinus dorsalis, P. helleri, P. lineatus, P. umbratus, Stenoderma rufum, Stunira* spp., *Uroderma bilobatum, U. magnirostrum?, Vampyressa dorsalis, V. pusilla, Vampyriscus nymphaea, Vampyrodes caraccioli*	...

Note: Question marks (?) indicate species in which information is suggestive but not complete. Data from: Anderson and Wimsatt 1963; Baker et al. 1978; Beguelini et al. 2013; Bernard 2002; Bleier 1975; Bradshaw 1962; Ceballos et al. 1997; Chaverri and Kunz 2006; Costa et al. 2007; Dechmann et al. 2005; Dinerstein 1986; Durant et al. 2013; Fleming 1971, 1988; Fleming and Nassar 2002; Fleming et al. 1972, 2009; Galindo-Galindo et al. 2000; Gannon et al. 2005; Genoways et al. 2005; Heideman et al. 1992; Krutzsch and Nellis 2006; Kwiecinski 2006; Mancina et al. 2007; Mello et al. 2004, 2009; Molinari and Soriano 2014; Montiel et al. 2011; Petit 1997; Porter and Wilkinson 2001; Ramos-Pereira et al. 2010; Rasweiler and Badwaik 1997; Silva Taboada 1979; Sosa and Soriano 1996; Sperr et al. 2011; Stoner 2001; Taft and Handley 1991; Tschapka 2005; Willig 1985; Wilson 1979; Zortea 2003.

* Delayed development occurs in this species.

distance migrations (i.e., >500 km) are uncommon in phyllostomid bats and are best known in several glossophagines, including three species of *Leptonycteris* and *Choeronycteris mexicana* (Fleming and Eby 2003; Fleming and Nassar 2002). Three of these species, *L. yerbabuenae, L. nivalis,* and *Ch. mexicana,* migrate from Mexico into the southwestern United States to give birth in the spring. The fourth species, *L. curasoae,* occurs in arid regions of northern South America and adjacent Caribbean islands and migrates seasonally among different Andean and lowland habitats, including islands (Newton et al. 2003; Simal et al. 2015). All of these species are monestrous.

Extrinsic factors influencing phyllostomid reproductive patterns include latitude, occurrence on islands, and, most importantly, the type of and seasonal changes in resource availability, especially flowers and fruit. In addition to large body size, relatively high latitude, and island life are associated with monestrous reproductive patterns. *Macrotus californicus,* a nonmigratory member of the most basal clade in the Phyllostomidae, has the most northern distribution; its range includes the southwestern United States and northwestern Mexico and Baja California. It is monestrous (Bleier 1975) and has a reproductive schedule reminiscent of that of temperate-zone vespertilionids, with mating occurring in October and births occurring the following June. But, unlike temperate-zone vespertilionids, ovulation and fertilization follow directly after mating. Implantation begins in October but is not complete until February; embryonic growth and development then accelerate in March, and the single young is born in June after a gestation period of about eight months (Bleier 1975).

Macrotus californicus is also insectivorous, and diet appears to be associated with variation in reproductive patterns. Among 51 species of insectivorous and frugivorous/nectarivorous bats on the mainland for which we could find data, insectivores make up only 5.9% (two of 34 species; *Lophostoma silvicola* and *Micronycteris megalotis*) of those with a polyestrous reproductive pattern, while 41.2% (seven of 17 species) of monestrous species are insectivorous. This is a significant difference (Fisher's exact test, $p = 0.004$) and does not appear to be related to differences in body mass as insect and plant feeders do not differ significantly in body mass (14.1 g vs. 20.5 g, respectively; $t = 1.72$, $p = 0.09$). Instead, it may reflect the longer gestation periods of animalivorous species of phyllostomids (see below).

Piscivorous/insectivorous noctilionoids and insectivorous mormoopids are strictly tropical in distribution, and most species are also monestrous (table 12.1). Except for *Noctilio leporinus*, which weighs up to 78 g, members of these two families are small to medium in size, ranging from 10 to 37 g, which again runs counter to the trend of monestry being associated with large size. Mormoopids are especially common in highly seasonal tropical dry forests (chap. 2, this vol.) and throughout the Caribbean. Their seasonal environments apparently favor monestry.

The widespread occurrence of monestry in noctilionids, mormoopids, and basal phyllostomids has at least two important evolutionary implications: (1) monestry, not polyestry, is likely to be the ancestral reproductive mode in this group and (2) the ancestral home of these bats is likely to have been at the northern end of these families' current ranges (see chap. 6, this vol.). If monestry is the ancestral mode in these bats, this runs counter to Bernard and Cummings's (1997) proposal that seasonal polyestry is the ancestral reproductive mode in (all?) tropical bats. It also suggests that ancestral noctilionoids evolved in (highly) seasonal, perhaps subtropical, habitats in which insect food resources underwent strong seasonal fluctuations.

In addition to mormoopids, a number of other island-dwelling phyllostomids, including glossophagines and stenodermatines, are monestrous, or populations vary in their reproductive cycles. These include members of a clade of old endemic Greater Antillean glossophagines: species in the genera *Brachyphylla*, *Phyllonycteris*, and *Erophylla* (Dávalos 2009; table 12.1). Not all West Indian glossophagines are monestrous, however. The large and widespread *Brachyphylla cavernarum*, for example, is polyestrous, whereas the smaller and geographically restricted *B. nana* is monestrous. *Monophyllus redmani* is polyestrous on Cuba and Puerto Rico but is monestrous on Jamaica (Gannon et al. 2005; Genoways et al. 2005; Silva Taboada 1979). *Glossophaga longirostris* is monestrous on Curaçao whereas it is polyestrous in northern South America (Petit 1997; Sosa and Soriano 1996). The stenodermatine *Artibeus jamaicensis* is a recent colonist of the Greater Antilles (Fleming et al. 2009), where it retains the bimodal polyestrous pattern found on the Neotropical mainland. Its heavy reliance on fig (*Ficus* sp.) fruits, which are generally available year-round throughout the tropics, probably favors polyestry throughout its range. Another West Indian stenodermatine, *Stenoderma rufum*, is also polyestrous on Puerto Rico (Gannon et al. 2005). *Artibeus lituratus* and *Phyllostomus discolor* also appear to vary geographically in terms of numbers of litters per year (Kwiecinski 2006; Willig 1985; Wilson 1979). Such species present interesting subjects in terms of whether other life-history traits also differ geographically.

In general, monestry in many island phyllostomids likely reflects a more seasonal resource base (insects, flowers, fruit, etc.) on islands compared with mainland habitats. For example, two columnar cacti, *Stenocereus griseus* and *Subpilocereus repandus*, are important sources of nectar, pollen, and fruit for *G. longirostris* (and monestrous *L. curasoae*) on both Curaçao and Venezuela, but their flowering and fruiting seasons overlap less and are more extended on the mainland than on the island (Petit 1997; Sosa and Soriano 1996). Greater year-round availability of these and other plant resources allows *G. longirostris* to have a longer reproductive season on the mainland than on the adjacent islands, and thus two rather than one litter per year.

Occurring throughout the arid regions of Mexico and, seasonally, in southern Arizona, *Leptonycteris yerbabuenae* is monestrous, but its reproductive sched-

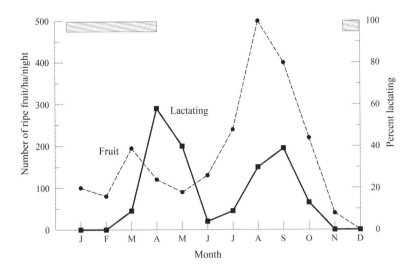

Figure 12.1. The relationship between reproductive (*solid line*) and resource seasonality (*dashed line*) in *Carollia perspicillata* in Costa Rican tropical dry forest. Data come from Fleming (1988). The *hatched bars* indicate the dry season. Number of ripe fruits represents the collective abundance for 16 species.

ule appears to be unique in the Phyllostomidae. Like the African hipposiderid *Hipposideros cafer*, which has two different reproductive schedules in adjacent locations near the equator (Brosset 1968; Brosset and Saint-Girons 1980), *L. yerbabuenae* has at least two reproductive demes (Fleming and Nassar 2002). In one deme, mating occurs in November–December in southwestern Mexico, and births occur primarily in the Sonoran Desert in May. In the other deme, mating occurs in May–June in central Mexico, and births occur in December–January in southern Mexico. The northern birth and lactation periods occur when Sonoran Desert columnar cacti are in peak bloom. The southern birth and lactation periods occur when bat-pollinated trees and shrubs are in peak bloom in tropical dry forest. Both females and males apparently participate in only one of the two breeding seasons each year (e.g., Ceballos et al. 1997). Although *L. yerbabuenae* is acting like two reproductively isolated species, genetic analyses indicate that the two reproductive demes are not genetically isolated from each other (Newton 2002). Unlike *L. yerbabuenae*, the other two species of *Leptonycteris* are monestrous with only one birth and lactation schedule, which is the ancestral condition in these bats.

Seasonal changes in rainfall and hence plant productivity undoubtedly drive patterns of seasonal bimodal polyestry in phyllostomids, as they do in pteropodids (Racey and Entwistle 2000). The typical pattern, first described by Fleming et al. (1972), is for the first birth of the year to occur near the end of the dry season (March–April in Panama) followed by a postpartum estrus and a second birth occurring in the middle of the wet season (July–August in Panama). In northwestern Costa Rica, birth periods are shifted about one month earlier, reflecting the earlier onset of the dry and wet seasons there (Fleming 1988). Geographic variation in the timing of birth periods is likely to be common in geographically widespread phyllostomids, including *T. cirrhosus*, *P. discolor*, and *A. lituratus* (Wilson 1979). A detailed study of the population biology of *Carollia perspicillata* in northwestern Costa Rica (Fleming 1988) clearly shows the relationship between birth periods and fruit resources in this species (fig. 12.1). Peak lactation periods include April–May and August–September, with births occurring during or just before the dry and wet season fruiting peaks, respectively. Weaning occurs about 1.5 months after birth, and most newborns are therefore volant for several months before the period of low fruit availability at the beginning of the dry season. In *C. perspicillata* and other seasonally polyestrous phyllostomids, most adult females undergo two pregnancies per year (e.g., Fleming 1988; Taft and Handley 1991; Tschapka 2005).

Although seasonal changes in resource availability are clearly important drivers of reproductive seasonality in phyllostomids, two other extrinsic factors, temperature and photoperiod, have also been implicated in this seasonality. Mello et al. (2009), for example, reported that the bimodal polyestrous system in a montane population of *Sturnira lilium* in Brazil was more strongly correlated with seasonal changes in ambient temperature than with changes in fruit availability. Seasonal photoperiodic changes, though small in the tropics, have also been implicated in the seasonal reproductive periodicity in females of the Neotropical emballonurid *Saccopteryx bilineata* (Greiner et al. 2011) and in males of the Paleotropical pteropodid *Cynopterus brachyotis* (Bumrungsri at al. 2007). In contrast, experimental studies have shown that the seasonal testicular cycles of the phyllostomid *Anoura geoffroyi* are not entrained by photoperiod (Heideman and Bronson 1994). Clearly, more work needs to be done in this area of tropical bat reproduction, especially in relation to climate change.

The degree of reproductive (birth) synchrony is variable in phyllostomid bats. For example, it is high in *P. hastatus* and *A. jamaicensis* both in the field and in the lab, and in *Monophyllus redmani* on Jamaica (Genoways et al. 2005; Porter and Wilkinson 2001; Taft and Handley 1991). Births are less synchronous in *Desmodus rotundus*, *C. perspicillata*, and *Stenoderma rufum* in the field (Fleming 1988; Gannon et al. 2005; Wilson 1979). In their detailed study of factors influencing reproduction in *Phyllostomus hastatus* in the field and laboratory, Porter and Wilkinson (2001) reported a high degree of birth synchrony within female groups (harems) in particular caves in Trinidad, and different birth peaks in different caves. Between-cave differences were correlated with differences in rainfall patterns. In captivity, a high degree of within-group birth synchrony persisted in the absence of environmental cues, and females that moved between groups synchronized reproductively with their new group, which suggests that social factors are likely involved in this synchrony. Porter and Wilkinson (2001) suggested that birth synchrony is advantageous in this species because it allows young to receive alloparental care (which is not particularly common in phyllostomids) and female group members can share food information during the common lactation period. Factors influencing degree of birth synchrony or lack thereof in other phyllostomids have not been studied, but in species with low birth synchrony, extended periods of high food availability are likely to be involved.

Male reproductive cycles have been much less studied than those of females in the Phyllostomidae. Available data seem to indicate that while some adult males remain spermatogenic throughout the year, testis size in most males fluctuates seasonally in rhythm with female cycles, with peaks occurring during mating periods (e.g., Ceballos et al. 1997; Fleming 1988; Fleming et al. 1972; Tschapka 2005).

Three kinds of reproductive delays are found in bats: (1) delayed fertilization with sperm storage (found in many temperate-zone bats); (2) delayed implantation (found in the pteropodids *Eidolon helvum* and three species of *Cynopterus*); and (3) delayed embryonic development (Racey and Entwistle 2000). Delayed development occurs in at least three species of pteropodids (*Haplonycteris fischeri*, *Otopterus carilagonodus*, and *Ptenochirus jagori*) (Kofron 2008). It also occurs in three species of phyllostomids: *M. californicus* (described above), *C. perspicillata*, and *A. jamaicensis*. Rasweiler and Badwiak (1997) reported that, in captivity, some *C. perspicillata* had gestation periods as long as 230 days compared with a normal gestation of 120 days. They also noted that this species sometimes has unusually long gestation periods on Trinidad, although Fleming et al. (1972) and Fleming (1988) did not find any evidence of delayed development in this species in Panama and Costa Rica. Rasweiler and Badwiak (1997) attributed delayed development in the lab, and possibly also in the field, to be the result of stress. Delayed development occurs in *A. jamaicensis* both in the field and in captivity (Fleming 1971; Taft and Handley 1991). In this species, a postpartum estrus likely occurs shortly after the second pregnancy of the year, and in Panama, the embryo implants in late August or early September. These embryos do not begin continuous development, however, until mid-November (a period of 2.5 months) and are born in March or April, after a gestation period of about 7 months compared to a non-delayed gestation of about 4 months. Why *A. jamaicensis* undergoes delayed development while other polyestrous stenodermatines do not is not yet clear. Finally, to judge from its 7-month gestation period, delayed development is likely to occur in the migratory Mexican glossophagine, *L. yerbabuenae*.

In summary, reproductive output in phyllostomids is low and ranges from one to two offspring per year. Monestry is likely to be ancestral in this family (and its superfamily) and is often associated with large body size, migratory behavior, island life, and an insectivorous or carnivorous diet. Polyestry is usually associated with diets that include nectar, pollen, and fruit. Delayed development is uncommon in this family; it occurs regularly only in *M. californicus* (a northern insectivore), *A. jamaicensis* (a widespread frugivore), and possibly in *L. yerbabuenae* (a migratory nectarivore).

Life Histories

An organism's life history involves traits that influence its schedule of growth and reproduction, and its ultimate death. For mammals, these traits include timing variables such as the length of gestation and lactation, the age of sexual maturity, and longevity. They also involve counts of events such as the number of litters per year and the number of young per litter, and size traits such as size at birth and weaning. As previously mentioned, bats are unusual in being small mammals with slow reproduction (few young per year, along with slow growth and sexual maturity) and long lives (Barclay and Harder 2003). In general terms, phyllostomids follow this pattern (table 12.2), but does the morphological, ecological, and behavioral diversity within the family correlate with variation in life-history traits?

Typical litter size in all phyllostomids that have been assessed is one. Although occasional twinning has been reported in some species (e.g., *A. lituratus*), it is rare (Barlow and Tamsitt 1968; Stevens 2001). In contrast, of the species we found data for, 60.6% (*n* = 43/71) have at least some populations that reproduce twice (or more) each year. However, while at a population level such species have at least two reproductive cycles a year, whether each female produces more than one litter in a year is difficult to assess. Nonetheless, simultaneously lactating and pregnant females have been found in some species (e.g., *A. jamaicensis*, *C. perspicillata*, *Uroderma bilobatum*; Fleming et al. 1972), clearly indicating that at least some females are able to produce two litters per year.

There are relatively few data on the age of sexual maturity in phyllostomids. As in many other bats (Barclay and Harder 2003), females of *M. californicus*, *A. jamaicensis*, and *C. perspicillata* reach sexual maturity in their first year, while males only do so in their second year (Bradshaw 1962; Cloutier and Thomas 1992; Fleming 1988). Later maturity (16 months) of female *Chrotopterus auritus* (Esbérard et al. 2006) could be related to this species' large size, requiring a long period of development before reaching maturity. In the migratory nectar-feeder *L. yerbabuenae*, females first reproduce as 2-year-olds, which is also likely in males (Fleming and Nassar 2002).

Other than litter size and number of litters per year, the life-history trait for which there are the most data is size at birth. As in other bats, phyllostomids give birth to relatively large babies, averaging 28% of their mother's nonpregnant mass (*n* = 23; table 12.2). There is considerable variation, however. Some give birth to pups that are less than 20% of their mother's mass

Table 12.2. Means and ranges for various life history traits of phyllostomid bats

Trait	*n*	Mean	Range
Litter size	97	1	1
Litters/year	71	1.58	1–2
Birth mass/female mass	23	0.281	0.145–0.413
Birth forearm/ female forearm	19	0.513	0.416–0.642
Weaning mass/ female mass	6	0.828	0.612–0.972
Weaning forearm/ female forearm	5	0.960	0.881–1.022
Gestation length (days)	19	134.4	90–240
Lactation length (days)	12	89.5	30–278
Female age at maturity (months)	7	8.04	3–16
Male age at maturity (months)	5	12.7	9–15.5

Note: Data for particular species are available in supplementary information.

(e.g., *A. lituratus*; Kurta and Kunz 1987), while others produce pups that are over 35% of their mother's mass (e.g., *S. rufum*; Kurta and Kunz 1987). Our ability to tease apart potential factors contributing to this variation is limited by the small sample size and by the fact that there are data for only one insectivorous species. Birth mass scales with female mass, as one might expect, but larger species give birth to relatively smaller pups (log birth mass = 0.75 log adult mass—0.21; r^2 = 0.81; fig. 12.2). For example, the 11.6 g *Artibeus watsoni* produces pups averaging 4 g (34.5% of adult mass; Chaverri and Kunz 2006), whereas the 75 g *Phylloderma stenops* produces 15 g pups (20%; Esbérard 2012). Taking this into account, there is no difference in relative birth mass between species of animalivorous (carnivorous, omnivorous, insectivorous, and sanguinivorous) species (*n* = 6) and fruit or nectar feeders (*n* = 17; *F* = 0.01, *p* > 0.9).

Length of gestation varies among phyllostomids, and this variation is related to several factors. As noted earlier, delayed development occurs in *M. californicus*, *C. perspicillata*, *A. jamaicensis*, and possibly *L. yerbabuenae* such that gestation is prolonged. Data exist for only 18 other species, and for some of them, gestation is described in vague terms (e.g., approximately 3 months, or between 3 and 4 months). Regardless, gestation length varies between approximately 3 months in species such as *G. longirostris* (Sosa and Soriano 1996) and *A. watsoni* (Chaverri and Kunz 2006), to over 5 months for *Desmodus rotundus* (Delpietro and Russo 2002; Greenhall et al. 1983) and 7 months for *Chrotopterus auritus*

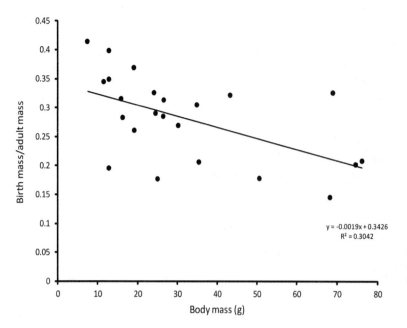

Figure 12.2. Relationship between adult body mass and relative mass at birth (i.e., birth mass/adult mass) among phyllostomid bats. Larger species give birth to relatively smaller young.

and *L. yerbabuenae* (Esbérard et al. 2006; Fleming and Nassar 2002). This variation is due partly to increased gestation length with increased adult mass, but there is considerable variation around that relationship (r^2 = 0.25). Some of that variation is due to a significant difference in the mass-specific gestation length with diet (ANCOVA, $F = 8.82$, $p = 0.01$). At a particular adult body mass, animalivorous phyllostomids ($n = 5$) have longer gestation times than do frugivorous and nectarivorous species ($n = 13$; fig. 12.3). For example, the 26.5 g *Platyrrhinus lineatus*, a frugivore, has a gestation duration of approximately 3.5 months (Willig and Hollander 1987), while the similarly sized vampire *Diphylla ecaudata* is pregnant for about 5 months (Delpietro and Russo 2002). This is despite the fact that both species produce neonates that are approximately 30% of the adult female mass (31.3% vs. 28.4%, respectively; Delpietro and Russo 2002; Willig and Hollander 1987).

Duration of the lactation period and size of juveniles at weaning is known for only a few phyllostomids. In *C. perspicillata* and *A. jamaicensis*, lactation lasts 5–6 weeks, and newly volant young are 32–40 days old (Fleming 1988; Taft and Handley 1991). As in other bats (Barclay and Harder 2003), juvenile phyllostomids attain almost full adult forearm size before they begin to fly (table 12.2), a feature common to bats and birds and related to the need for fully ossified wing bones to handle the stresses placed on them during flight (Barclay 1994). For example, first flight in *P. hastatus* occurs when juveniles have reached 94% of adult fore-

arm length (Stern et al. 1997), and in *A. watsoni* at 99% (Chaverri and Kunz 2006). Mass of pups at weaning is not as great (relative to adult mass), but for six species, the mean is 82.8% of adult female mass (table 12.2).

Given that phyllostomids vary within and among species in terms of annual reproductive output (numbers of litters), we expect that there are corresponding variations in other life-history traits, based on life-history theory (e.g., Stearns 1992). For example, within the fast-slow perspective of mammalian life histories, monestrous species of phyllostomids are on the slower end of the continuum, whereas polyestrous species are more towards the fast end. If polyestrous species have evolved to reproduce at a higher rate, and assuming populations are stable, then survival should be lower than for monestrous species (see also Lentini et al. 2015).

Unfortunately, survival data are almost nonexistent for phyllostomids. Annual mortality in *C. perspicillata* over the first 2 years of life was estimated to be 53%, and 22% in subsequent years (Fleming 1988). This resulted in a life expectancy in females of only 2.6 years at birth and a maximum expected longevity of 10 years, although, subsequently, the maximum longevity in captivity was recorded as 17 years (Weigl 2005). Long-term studies on *A. jamaicensis* estimated annual survival of only 58% across all ages (Leigh and Handley 1991). Mortality among pre-weaning juvenile *P. hastatus* in the wild was estimated at 40% (Stern and Kunz 1998). Maximum longevity records for phyllostomids

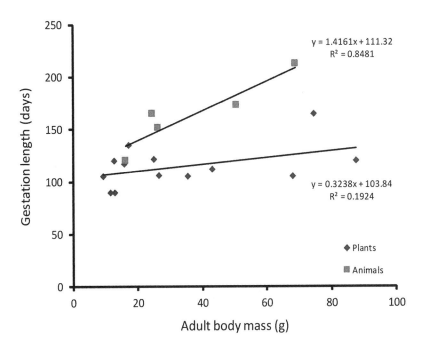

Figure 12.3. Relationship between adult mass of phyllostomid bats and length of gestation. Species that feed on fruit and/or nectar have significantly shorter gestation lengths for their size than do species that feed on animals.

are scarce ($n = 15$). They also come from studies on captive and wild individuals, which provide different environments and mortality risks, making them difficult to compare. Longevity records range from 8 years in the wild for *D. ecaudata* (Martino et al. 2006), to 29.2 years in captivity for *D. rotundus* (Weigl 2005). The mean longevity for five studies on wild individuals is 11.3 years, and from 10 studies on captive populations it is 15.9 years. Overall, the mean longevity across all 15 species is 14.4 years, which is significantly less than for all other species of bats (mean = 19.5 years, $n = 59$; data from Healy et al. 2014; $t = 2.29$, $p = 0.025$). This might reflect a real difference, for example, associated with longer lifespans of hibernating species in temperate zones (Wilkinson and South 2002). It may also, however, reflect differences in study durations, as many species of temperate-zone bats were banded and observed for many decades, especially while in hibernation (e.g., Keen and Hitchcock 1980).

Among mammals in general, longevity is correlated with adult body mass (Lindstedt and Calder 1981), if for no other reason than that predation risk scales with body mass. However, there is no such relationship among the 15 species of phyllostomids for which there are data (slope = 0.90, $r^2 = 0.129$, $p = 0.19$). Limited data and the combination of studies on captive and wild individuals may explain this. There is also no significant difference in maximum longevity between plant and animal feeders, despite the fact that various hypotheses

propose that diet should have an effect on survival, due, for example, to increased intake of antioxidants by frugivorous species (Schneeberger et al. 2014). Lastly, our prediction that monestrous species would have greater longevity than polyestrous species is not supported. Monestrous species have an average maximum longevity of 14.8 years ($n = 7$) and polyestrous species have a mean of 14.3 years ($n = 7$; $t = 0.16$, $p > 0.88$). Again the small sample size and combination of wild and captive studies hinders this analysis and its interpretation.

If polyestry in phyllostomids represents one trait in a relatively fast life history, we would predict that these females would invest less in each offspring, assuming that there is a trade-off between such investment and the ability to reproduce a second time in the same year. Investment might be measured in terms of size at birth and weaning, or length of gestation and lactation. For 12 monestrous and 10 polyestrous species, however, relative mass at birth does not differ between the two groups (mean relative mass = 0.260 vs. 0.308, respectively; $t = 1.52$, $p = 0.14$).

Relative mass at weaning would be a better measure of investment by females in their pups than size at birth. Unfortunately, size at weaning is difficult to determine, and size at first flight is sometimes substituted (e.g., Stern and Kunz 1998). For three monestrous species (*P. stenops*, *P. hastatus*, and *L. yerbabuenae*), "weaning" mass averaged 75.1% of adult mass, while for three species of polyestrous phyllostomids (*A. watsoni*, *C. per-*

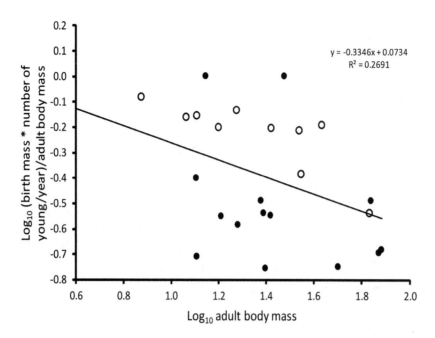

Figure 12.4. Relationship between adult body mass and mass-specific annual productivity (i.e., total mass of neonates) for phyllostomid bats. Larger species (*closed circles*) produce, on average, a smaller amount of neonatal tissue per year relative to their body mass. Species with two litters per year (*open circles*) have relatively high productivity.

spicillata, and *P. discolor*), the average was 91.4%. The sample size is too small to analyze statistically, although the trend is in the opposite direction to our prediction.

Sibly and Brown (2007) analyzed the annual reproductive output of mammals and related its variation to body size and "lifestyle" (including ecological, physiological, and anatomical traits). In general, mass-specific annual productivity (total mass of offspring produced) declines with adult body mass across all mammalian groups. They specifically predicted that bats (and some other taxa) would have reduced productivity associated with their reduced extrinsic mortality. The data they used indicated that bats had reduced productivity, as predicted, but also a particularly shallow negative slope to the relationship with body mass. For all bats, the slope was −0.085, the lowest of any mammalian group they analyzed. In other words, there was less of an effect of adult size on relative annual productivity than expected.

Using our data from 22 species of phyllostomids, we tested whether this family differed in the relationship between body size and mass-specific productivity compared to the relationship Sibly and Brown (2007) found for all bats. There was a significant negative relationship between productivity and body mass among the phyllostomids ($r^2 = 0.216$; $t = 2.35$, $p = 0.029$; fig. 12.4), with a slope of −0.361. This is a much steeper slope than Sibly and Brown (2007) found for all bats and may be due to several factors. First, there is significant variation around the slope for phyllostomids

caused primarily by the difference between productivity per year in monestrous versus polyestrous species (fig. 12.4). Polyestrous species have greater productivity than do monestrous species of the same size. In addition, however, polyestrous species are on average smaller than monestrous species, and this would result in a steeper slope to the regression. There also appear to be errors in the data used by Sibly and Brown (2007) to compare productivity rates among mammals. There are some extremely small productivity rates for bats, apparently due to unusually small neonate masses. For example, in PanTHERIA (Jones et al. 2009), the source for Sibly and Brown's (2007) data, *E. sezekorni* is listed as producing newborns weighing 0.9 g (5.6% of adult mass), yet a value of 4.6 g (28.2%) is given in Soto-Centeno and Kurta (2003) and is more in line with relative neonate masses of other phyllostomids and other bats (Barclay and Harder 2003; Kurta and Kunz 1987). The low value for this species in Sibly and Brown (2007), and other unusually low productivity rates associated with small neonate masses in relatively small species of bats, would result in a shallower slope than if these values were removed. The slope we found for phyllostomids (−0.361) is in line with those of other small mammals listed by Sibly and Brown (2007) (e.g., Insectivora −0.361, Rodentia −0.378). Phyllostomids, as in other bats, produce few but relatively large young (at birth) per year, and given the large size at weaning, overall productivity and reproductive effort is high.

Are Vampires Unique?

Perhaps the most unusual phyllostomid bats in terms of ecology and behavior are the three species of vampires (*D. rotundus*, *D. ecaudata*, and *Diaemus youngi*; chap. 15, this vol.). Relatively little is known about the latter two in terms of reproduction and life history. *Diphylla ecaudata* is monestrous and has the second longest lactation period (217 days; Delpietro and Russo 2002) of any phyllostomid. *Diaemus youngi* is polyestrous and has been kept in captivity for up to 20 years (Animal Diversity Web 2016).

Much more is known about *D. rotundus*, and in some respects it is an outlier in terms of its reproduction and life history. The length of gestation is one of the longest (173 days; Delpietro and Russo 2002; Greenhall et al. 1983), despite *D. rotundus* giving birth to relatively small pups (under 20% of adult body mass; Kurta and Kunz 1987). The small size at birth might be an adaptation associated with the terrestrial locomotion of adult females and the need to be able to fly from the ground after feeding while carrying both the fetus and a blood meal. Lactation lasts longer in *D. rotundus* (278 days; Delpietro and Russo 2002; Greenhall et al. 1983) than in any other phyllostomid for which data are available. Females also feed their young regurgitated blood for up to one year (Delpietro and Russo 2002; Greenhall et al. 1983). Female *D. ecaudata* also regurgitate blood to their young for approximately 7.5 months (Delpietro and Russo 2002). These are the only species of bat we know of that supplement lactation, and this behavior may allow a longer period of dependency compared to other species. Lastly, *D. rotundus* has the greatest known longevity (29.2 years in captivity) of any phyllostomid (Weigl 2005).

In many respects, the data regarding vampire bats suggest particularly slow life histories. Whether this is associated with their diet (blood), or is unique to *Desmodus* alone, requires much more data on all three species. It is, however, one of many interesting questions remaining to be answered about the evolution of life histories in this diverse family.

Conclusions and Opportunities for Future Research

The variation in size, ecology, and behavior found among Phyllostomidae is mirrored in at least some aspects of their reproduction and behavior, and correlations are evident despite the relatively low proportion of species for which data exist. Rainfall and its seasonal pattern clearly influences reproduction among phyllostomids due to its influence on food resources (plant and animal). The degree and timing of seasonal patterns varies across the range of the family and results in among- and within-species variation in the timing of reproduction, its synchrony, and whether a population is monestrous, seasonally polyestrous, or aseasonal. Monestry appears to be the ancestral state, suggesting that a northern origin for the family is likely.

In terms of life-history traits, other than litter size (uniformly one), there is variation within the family, although limited data hinder in-depth analyses of potential patterns. As in other bats, phyllostomids mature and reproduce relatively slowly but live a long time. Body size influences some traits. As expected, gestation duration and time to weaning scale positively with size. However, relative size of pups at birth is lower in larger species than smaller ones. Diet also appears to be related to some variation in traits. For example, monestry is more common among insectivorous species than in frugivorous and nectarivorous species, and animalivorous species have longer gestation durations for their size than do frugivores and nectarivores. Perhaps unexpectedly, there is no difference in the relative birth mass of insectivorous versus frugivorous/nectarivorous species, despite the fact that carrying a large embryo might be expected to hinder foraging in insectivores in particular.

We suggest that the following would be interesting research questions to address to allow a better understanding of the linkages between the ecological variation and reproductive variation in this family and, perhaps by extension, across families of bats.

1. How do life histories differ in populations within a species that vary in terms of numbers of litters per year? Life-history theory would predict that monestrous populations would have greater survival and perhaps greater investment in each pup. What factors would result in greater survival? *Glossophaga longirostris* in mainland Venezuela (polyestrous) and Curaçao (monestrous) would make an interesting comparison, for example.

2. Among species, does survival rate differ between

those that are monestrous versus those that are polyestrous and how does that difference come about? For example, does the degree of coloniality or type of roost influence survival and thus favor the evolution of different reproductive patterns?

3. Given the apparent difference in numbers of litters per year between frugivorous/nectarivorous species and insectivorous species, do other aspects of the life histories of insectivorous phyllostomids differ from those of frugivores and nectarivores, and if so why? Are survival rates different, or growth rate of young?

4. Does flight style influence reproduction and life history? For example, do species that depend on greater maneuverability or the ability to hover (aerial insectivores and nectarivores) produce smaller neonates because of the importance of lower wingloading for their foraging success?

Acknowledgments

RMRB's research on bats has been supported primarily by grants from the Natural Sciences and Engineering Research Council of Canada, Bat Conservation International, the governments of Alberta, British Columbia, Yukon, Northwest Territories, and Manitoba, and the Universities of Calgary and Manitoba. THF's research on phyllostomid bats was supported primarily by the Smithsonian Institution, the U.S. National Science Foundation, the National Geographic Society, and the Universities of Missouri–Saint Louis and Miami. We thank M. Mello and two anonymous reviewers for their valuable suggestions on drafts of this chapter.

References

Anderson, J. W., and W. A. Wimsatt. 1963. Placentation and fetal membranes of the Central American noctilionid bat, *Noctilio labialis minor*. Developmental Dynamics, 112:181–201.

Animal Diversity Web. 2016. http://animaldiversity.org/ Accessed 1 December 2016.

Badwaik, N. K., and Rasweiler IV, J. J. 2000. Pregnancy. Pp. 221–293 *in*: Reproductive Biology of Bats (E. G. Crichton and P. H. Krutzsch, eds.). Academic Press, San Diego.

Baker, R. I., H. H. Genoways, and C. J. Patten. 1978. Bats of Guadeloupe. Occasional Papers, Museum of Texas Tech University, 50:1–16.

Barclay, R. M. R. 1994. Constraints on reproduction by flying vertebrates: energy and calcium. American Naturalist, 144:1021–1031.

Barclay, R. M. R., and L. D. Harder. 2003. Life histories of bats: life in the slow lane. Pp. 209–253 *in*: Bat Ecology (T. H. Kunz and M. B. Fenton, eds.). University of Chicago Press, Chicago.

Barlow, J. C., and J. R. Tamsitt. 1968. Twinning in American leaf-nosed bats (Chiroptera: Phyllostomatidae). Canadian Journal of Zoology, 46:290–292.

Beguelini, M. R., C. C. I. Puga, S. R. Taboga, and E. Morielle-Versute. 2013. Annual reproductive cycle of males of the flat-faced fruit-eating bat, *Artibeus planirostris* (Chiroptera: Phyllostomidae). General and Comparative Endocrinology, 185:80–89.

Bernard, E. 2002. Diet, activity and reproduction of bat species (Mammalia, Chiroptera) in Central Amazonia, Brazil. Revista Brasileira de Zoologia, 19:173–188.

Bernard, R. T. F., and G. S. Cumming. 1997. African bats: evolution of reproductive patterns and delays. Quarterly Review of Biology, 72:253–274.

Bleier, W. J. 1975. Early embryology and implantation in the California leaf-nosed bat, *Macrotus californicus*. Anatomical Record, 182:237–253.

Bradshaw, G. V. R. 1962. Reproductive cycle of the California leaf-nosed bat, *Macrotus californicus*. Science, 136:645–646.

Brosset, A. 1968. La permutation du cycle sexuel saisonnier chez Chiroptere *Hipposideros caffer*, as voisinage de l'equateur. Biologica Gabonica, 4:325–341.

Brosset, A., and H. Saint-Girons. 1980. Cycles de reproduction des microchiropteres troglophiles du nord-est du Gabon. Mammalia, 44:225–232.

Bumrungsri, S., W. Bumrungsri, and P. A. Racey. 2007. Reproduction in the short-nosed fruit bat in relation to environmental factors. Journal of Zoology, 272:73–81.

Ceballos, G., T. H. Fleming, C. Chavez, and J. Nassar. 1997. Population dynamics of *Leptonycteris curasoae* (Chiroptera: Phyllostomidae) in Jalisco, Mexico. Journal of Mammalogy, 78:1220–1230.

Chaverri, G., and T. H. Kunz. 2006. Reproductive biology and postnatal development in the tentmaking bat *Artibeus watsoni* (Chiroptera: Phyllostomidae). Journal of Zoology 270:650–656.

Cloutier, D., and D. W. Thomas. 1992. *Carollia perspicillata*. Mammalian Species, 417:1–9.

Costa, L. M., J. C. Almeida, and C. E. L. Esbérard. 2007. Dados de reprodução de *Platyrrhinus lineatus* em estudo de longo prazo no Estado do Rio de Janeiro (Mammalia, Chiroptera, Phyllostomidae). Iheringia Série Zoologia, 97:152–156.

Dávalos, L. M. 2009. Earth history and the evolution of Caribbean bats. Pp. 96–115 *in*: Island Bats: Evolution, Ecology, and Conservation (T. H. Fleming and P. A. Racey, eds.). University of Chicago Press, Chicago.

Dechmann, D. K. N., E. K. V. Kalko, B. Konig, and G. Kerth. 2005. Mating system of a Neotropical roost-making bat: the whitethroated, round-eared bat, *Lophostoma silvicolum* (Chiroptera: Phyllostomidate). Behavioral Ecology and Sociobiology, 53:316–325.

Delpietro, V. H. A., and R. G. Russo. 2002. Observations of the common vampire bat (*Desmodus rotundus*) and the hairy-legged vampire bat (*Diphylla ecaudata*) in captivity. Mammalian Biology, 67:65–78.

Dinerstein, E. 1986. Reproductive ecology of fruit bats and the seasonality of fruit production in a Costa Rican cloud forest. Biotropica, 18:307–318.

Durant, K. A., R. W. Hall, L. M. Cisneros, R. M. Hyland, and M. R. Willig. 2013. Reproductive phenologies of phyllostomid bats in Costa Rica. Journal of Mammalogy, 94:1438–1448.

Esbérard, C. E. L. 2012. Reproduction of *Phylloderma stenops* in captivity (Chiroptera, Phyllostomidae). Brazilian Journal of Biology, 72:171–174.

Esbérard, C. E. L., A. G. Motta, J. C. Almeida, L. C. S. Ferreira, and L. M. Costa. 2006. Reproduction of *Chrotopterus auritus* (Peters) in captivity (Chiroptera, Phyllostomidae). Brazilian Journal of Biology, 66:955–956.

Fleming, T. H. 1971. *Artibeus jamaicensis*: delayed embryonic development in a Neotropical bat. Science, 171:402–404.

Fleming, T. H. 1988. The Short-Tailed Fruit Bat: A Study in Plant-Animal Interactions. University of Chicago Press, Chicago.

Fleming, T. H., and P. Eby. 2003. Ecology of bat migration. Pp. 156–208 *in*: Bat Ecology (T. H. Kunz and M. B. Fenton, eds.). University of Chicago Press, Chicago.

Fleming, T. H., E. T. Hooper, and D. E. Wilson. 1972. Three Central American bat communities: structure, reproductive cycles, and movement patterns. Ecology 53:555–569.

Fleming, T. H., and J. Nassar. 2002. Population biology of the lesser long-nosed bat, *Leptonycteris curasoae*, in Mexico and northern South America. Pp. 283–305 *in*: Columnar Cacti and Their Mutualists: Evolution, Ecology, and Conservation (T. H. Fleming and A. Valiente-Banuet, eds.). University of Arizona Press, Tucson.

Fleming, T. H., K. L. Murray, and B. Carstens. 2009. Phylogeography and genetic structure of three evolutionary lineages of West Indian phyllostomid bats. Pp. *in*: Island Bats: Evolution, Ecology, and Conservation (T. H. Fleming and P. A. Racey, eds.). University of Chicago Press, Chicago.

Galindo-Galindo, C., A. Castro-Campillo, A. Salame-Méndez, and J. Ramírez-Pulido. 2000. Reproductive events and social organization in a colony of *Anoura geoffroyi* (Chiroptera: Phyllostomidae) from a temperate Mexican cave. Acta Zoológica Mexicana, 80:51–68.

Gannon, M. R., A. Kurta, A. Rodríguez-Durán, and M. R. Willig. 2005. Bats of Puerto Rico: An Island Focus and a Caribbean Perspective. Texas Tech Press, Lubbock.

Genoways, H. H., R. J. Baker, J. W. Bickham, and C. J. Phillips. 2005. Bats of Jamaica. Special Publications of the Museum of Texas Tech University, 48:1–155.

Greenhall, A. M., G. Joermann, and U. Schmidt. 1983. *Desmodus rotundus*. Mammalian Species, 202:1–6.

Greiner, S., F. Schwarzenberger, and C. C. Voigt. 2011. Predictable timing of oestrus in the tropical bat *Saccopteryx bilineata* living in a Costa Rican rain forest. Journal of Tropical Ecology, 27:121–131.

Hall, L. S., and G. Richards. 2000. Flying Foxes: Fruit and Blossom Bats of Australia. Krieger Publishing, Malabar, FL.

Healy, K., T. Guillerme, S. Finlay, A. Kane, S. B. A. Kelly, D. McClean, D. J. Kelly, I. Donohue, A. L. Jackson, and N. Cooper. 2014. Ecology and mode-of-life explain lifespan variation in birds and mammals. Proceedings of the Royal Society of London B, 281:20140298.

Heideman, P. D. 1995. Synchrony and seasonality of reproduction in tropical bats. Pages 151–165 *in*: Ecology, Evolution and Behaviour of Bats (P. Racey and S. M. Swift, eds.). Oxford Science Publications, Oxford.

Heideman, P. D. 2000. Environmental regulation of reproduction. Pp. 469–500 *in*: Reproductive Biology of Bats (E. G. Crichton and P. H. Krutzsch, eds.). Academic Press, San Diego.

Heideman, P. D., and F. H. Bronson. 1994. An endogenous circannual rhythm of reproduction in a tropical bat, *Anoura geoffroyi*, is not entrained by photoperiod. Biology of Reproduction, 50:607–614.

Heideman, P. D., P. Deoraj, and F. H. Bronson. 1992. Seasonal reproduction of a tropical bat, *Anoura geoffroyi*, in relation to photoperiod. Journal of Reproduction and Fertility, 96:765–773.

Holmes, D. J., and S. N. Austad. 1994. Fly now and die later: life-history correlates of gliding and flying in mammals. Journal of Mammalogy, 75:224–226.

Jones, K. E., J. Bielby, M. Cardillo, S. A. Fritz, J. O'Dell, C. D. L. Orme, Kamran Safi et al. 2009. PanTHERIA: a species-level database of life history, ecology, and geography of extant and recently extinct mammals. Ecology 90:2648.

Keen, R., and H. B. Hitchcock. 1980. Survival and longevity of the little brown bat (*Myotis lucifugus*) in southeastern Ontario. Journal of Mammalogy, 61:1–7.

Kofron, C. P. 2008. Reproduction of the long-tongued nectar bat *Macroglossus minimus* (Pteropodidae) in Brunei, Borneo. Acta Zoologica, 89:53–58.

Krutzsch, P. H., and D. W. Nellis. 2006. Reproductive anatomy and cyclicity of the male bat *Brachyphylla cavernarum* (Chiroptera: Phyllostomidae). Acta Chiropterologica, 8:497–507.

Kurta, A., and T. H. Kunz. 1987. Size of bats at birth and maternal investment during pregnancy. Symposium of the Zoological Society of London, 57:79–106.

Kwiecinski, G. 2006. *Phyllostomus discolor*. Mammalian Species, 801:1–11.

Leigh, E. G., Jr., and C. O. Handley Jr. 1991. Population estimates: demography and natural history of the common fruit bat, *Artibeus jamaicensis*, on Barro Colorado Island, Panama. Pp. 77–88, *in*: Demography and Natural History of the Common Fruit Bat, *Artibeus jamaicensis*, on Barro Colorado Island, Panama (C. O. Handley Jr., D. E. Wilson, and A. L. Gardner, eds.). Smithsonian Institution Press, Washington, DC.

Lentini, P. E., T. J. Bird, S. R. Griffiths, L. N. Godinho, and B. A. Wintle. 2015. A global synthesis of survival estimates for microbats. Biology Letters, 11:20150371.

Lindstedt, W. A., and S. L. Calder III. 1981. Body size, physiological time, and longevity of homeothermic animals. Quarterly Review of Biology, 56:1–16.

Mancina, C. A., L. García-Rivera, and R. T. Capote. 2007. Habitat use by phyllostomid bat assemblages in secondary forests of the "Sierra del Rosario" Biosphere Reserve, Cuba. Acta Chiropterologica, 9:203–218.

Martino, A. M. G., J. Aranguren, and A. Arends. 2006. New longevity records in South American microchiropterans. Mammalia, 70:166–167.

Mello, M. A. R., G. M. Schittini, P. Selig and H. G. Bergallo. 2004. A test of the effects of climate and fruiting of *Piper* species (Piperaceae) on reproductive patterns of the bat *Carollia perspicillata* (Phyllostomidae). Acta Chiropterologica, 6:309–318.

Mello, M. A. R., E. K. V. Kalko and W. R. Silva. 2009. Ambient temperature is more important than food availability in explaining reproductive timing of the bat *Sturnira lilium* (Mammalia: Chiroptera) in a montane Atlantic Forest. Canadian Journal of Zoology, 87:239–245.

Molinari, J., and P. J. Soriano. 2014. Breeding and age-structure seasonality in *Carollia brevicauda* and other frugivorous bats (Phyllostomidae) in cloud forests in the Venezuelan Andes. Therya, 5:81–109.

Montiel, S., A. Estrada, and P. León. 2011. Reproductive seasonality of fruit-eating bats in northwestern Yucatan, Mexico. Acta Chiropterologica, 13:139–145.

Newton, L. R. 2002. Mobility, reproduction, and population genetic structure of two phyllostomid bats. MS thesis, University of Miami, Coral Gables, FL.

Newton, L. R., J. M. Nassar, and T. H. Fleming. 2003. Genetic population structure and mobility of two nectar-feeding bats from Venezuelan deserts: inferences from mitochondrial DNA. Molecular Ecology, 12:3191–3198.

Petit, S. 1997. The diet and reproductive schedules of *Leptonycteris curasoae curasoae* and *Glossophaga longirostris elongata* (Chiroptera: Glossophaginae) on Curaçao. Biotropica, 29:214–223.

Pomeroy, D. 1990. Why fly? The possible benefits for lower mortality. Biological Journal of the Linnean Society, 40:53–65.

Porter, T. A., and G. S. Wilkinson. 2001. Birth synchrony in greater spear-nosed bats (*Phyllostomus hastatus*). Journal of Zoology, 253:383–390.

Promislow, D. E. L., and P. H. Harvey. 1990. Living fast and dying young: a comparative analysis of life-history variation among mammals. Journal of Zoology (London), 220:417–437.

Racey, P. A. 1982. Ecology of bat reproduction. Pp. 57–104 *in*: Ecology of Bats (T. H. Kunz, ed.). Plenum Press, New York.

Racey, P. A., and A. C. Entwistle. 2000. Life-history and reproductive strategies of bats. Pp. 363–404 *in*: Reproductive Biology of Bats. (E. G. Crichton and P. H. Krutzsch, eds.). Academic Press, San Diego.

Ramos Pereira, M. J., J. T. Marques, and J. M. Palmeirim. 2010. Ecological responses of frugivorous bats to seasonal fluctuation in fruit availability in Amazonian forests. Biotropica, 42:680–687.

Rasweiler, J. J., IV, and N. K. Badwaik. 1997. Delayed development in the short-tailed fruit bat, *Carollia perspicillata*. Journal of Reproduction and Fertility, 109:7–20.

Rasweiler, J. J., IV, and N. K. Badwaik. 2000. Anatomy and physiology of the female reproductive tract. Pp. 157–219 *in*: Reproductive Biology of Bats (E. G. Crichton and P. H. Krutzsch, eds.). Academic Press, San Diego.

Rasweiler, J. J., IV, N. K. Badwaik, and K. V. Mechineni. 2010. Selectivity in the transport of spermatozoa to oviductal reservoirs in the menstruating fruit bat, *Carollia perspicillata*. Reproduction, 140:743–757.

Rasweiler, J. J., IV, N. K. Badwaik, and K. V. Mechineni. 2011. Ovulation, fertilization, and early embryonic development in the menstruating fruit bat, *Carollia perspicillata*. Anatomical Record—Advances in Integrative Anatomy and Evolutionary Biology, 294:506–519.

Read, A. F., and Harvey, P. H. 1989. Life-history differences among the eutherian radiations. Journal of Zoology (London), 219:329–353.

Schneeberger, K., G. A. Czirják, and C. C. Voigt. 2014. Frugivory is associated with low measures of plasma oxidative stress and high antioxidant concentration in free-ranging bats. Naturwissenschaften, 101:285–290.

Sibly, R. M., and J. H. Brown. 2007. Effects of body size and lifestyle on evolution of mammal life histories. Proceedings of the National Academy of Sciences, USA, 104:17707–17712.

Silva Taboada, G. 1979. Los Murcielagos de Cuba. Editora de la Academia de Ciencias de Cuba, Havana.

Simal, F., C. de Lannoy, L. García-Smith, O. Doest, J. A. de Freitas, F. Franken, I. Zaandam et al. 2015. Island-island and island-mainland movements of the Curacaoan long-nosed bat, *Leptonycteris curasoae*. Journal of Mammalogy, 96:579–590.

Sosa, M., and P. J. Soriano. 1996. Resource availability, diet and reproduction in *Glossophaga longirostris* (Mammalia: Chiroptera) in an arid zone of the Venezuelan Andes. Journal of Tropical Ecology, 12:805–818.

Soto-Centano, J. A., and A. Kurta. 2003. Description of fetal and newborn brown flower bats, *Erophylla sezekorni* (Phyllostomidae). Caribbean Journal of Science, 39:233–234.

Sperr, E. B., L. A. Caballero-Martínez, R. A. Medellín, and M. Tschapka. 2011. Seasonal changes in species composition, resource use and reproductive patterns within a guild of nectar-feeding bats in a west Mexican dry forest. Journal of Tropical Ecology, 27:133—145.

Stearns, S. C. 1992. The Evolution of Life Histories. Oxford University Press, Oxford.

Stern, A. A., and T. H. Kunz. 1998. Intraspecific variation in postnatal growth in the greater spear-nosed bat. Journal of Mammalogy, 79:755–763.

Stern, A. A., T. H. Kunz, and S. S. Bhatt. 1997. Seasonal wing loading and the ontogeny of flight in *Phyllostomus hastatus* (Chiroptera: Phyllostomidae). Journal of Mammalogy, 78:1199–1209.

Stevens, R. D. 2001. Twinning in the big fruit-eating bat *Artibeus lituratus* (Chiroptera: Phyllostomidae) from eastern Paraguay. Mammalian Biology, 66:178–180.

Stoner, K. E. 2001. Differential habitat use and reproductive patterns of frugivorous and nectarivorous bats in tropical dry forest of northwestern Costa Rica. Canadian Journal of Zoology 79:1626–1633.

Taft, L. K., and C. O. Handley Jr. 1991. Reproduction in a captive colony. Pp. 19–41, *in*: Demography and Natural History of the Common Fruit Bat, *Artibeus jamaicensis*, on Barro Colorado Island, Panama (C. O. Handley Jr., D. E. Wilson, and A. L. Gardner, eds.). Smithsonian Institution Press, Washington, DC.

Tschapka, M. 2005. Reproduction of the bat *Glossophaga commissarisi* (Phyllostomidae: Glossophaginae) in the Costa Rican rain forest during frugivorous and nectarivorous periods. Biotropica, 37:408–414.

Weigl, R. 2005. Longevity of mammals in captivity: from the living collections of the world. Kleine Senckenberg-Reihe, no. 48. Schweizerbart Science Publishers, Stuttgart.

Wilkinson, G. S., and J. M. South. 2002. Life history, ecology and longevity in bats. Aging Cell, 1:124–131.

Willig, M. R. 1985. Reproductive patterns of bats from Caatingas and Cerrado biomes in Northeast Brazil. Journal of Mammalogy, 66:668–681.

Willig, M. R., and R. R. Hollander. 1987. *Vampyrops lineatus*. Mammalian Species, 275:1–4.

Wilson, D. E. 1979. Reproductive patterns. Pp. 317–378 *in*: Biology of Bats of the New World Family Phyllostomatidae, Part 3 (R. J. Baker, J. K. Jones Jr., and D. C. Carter, eds.). Texas Tech Press, Lubbock.

Zortéa, M. 2003. Reproductive patterns and feeding habits of three nectarivorous bats (Phyllostomidae: Glossophaginae) from the Brazilian Cerrado. Brazilian Journal of Biology, 63:159–168.

13

Patterns of Sexual Dimorphism and Mating Systems

Danielle M. Adams,
Christopher Nicolay,
and Gerald S. Wilkinson

Introduction

Bats exhibit a diverse range of mating systems from monogamy to extreme polygyny (McCracken and Wilkinson 2000). Describing a species' mating system typically requires a long-term study to determine the spatial distributions and behavioral interactions among males and females, in addition to assigning parentage. Relatively few phyllostomid species have been studied in such detail, but the available evidence suggests that species in the family exhibit much of the mating system variation present in the order. In the absence of additional studies, patterns of mating behavior can be inferred by examining variation in traits likely to influence male mating success before and after mating.

Socioecological factors can offer insight into expected mating systems because they influence the spatial and temporal distribution of resources and the females that depend on them. When resources or females are spatially clumped and limiting, they become defensible, thus promoting resource or female defense polygyny (Emlen and Oring 1977). Bats form aggregations that range from a few to thousands of individuals (Kerth 2008) and that occupy an array of roost structures that vary in size and length of use (Kunz et al. 2003). Both roost abundance and permanence are known to affect social associations (Chaverri and Kunz 2010), so that species that roost in abundant but ephemeral roosts, such as foliage or leaf tents, tend to have more fluid social structures with short-term associations (Chaverri et al. 2007; Sagot and Stevens 2012). By contrast, species in less abundant but more permanent structures, such as caves or hollow trees, exhibit more stable social associations (Brooke 1997; McCracken and Bradbury 1981; Wilkinson 1985a), often amid a much larger assemblage of individuals.

When resources or females are defensible, males are expected to compete to control them (Emlen and Oring 1977). Because male mammals are largely liberated from the demands of parental care, they are free to invest in competition to maximize mating opportunities (Trivers 1972), and thus most mammals exhibit some form of polygyny (Clutton-Brock 1989). In both female and resource defense polygyny, selection typically favors large, aggressive males that can compete effectively to control access to females (Andersson 1994; Clutton-Brock et al. 1977; Plavcan and van Schaik 1997). Thus, precopulatory sexual selection has been inferred to be the primary cause of male-biased sexual size dimorphism (SSD) in mammals (Lindenfors et al. 2002, 2007; Plavcan and van Schaik 1997; Weckerly 1998). Alternative explanations based on ecological differences between the sexes have also been proposed (Isaac 2005; Ralls 1977) but have typically received less empirical support. Among bats, however, females are often larger than males (Ralls 1976). One explanation for why female bats are larger is to carry additional weight during and after pregnancy given that bat litters can approach 50% of maternal body mass at birth (Kunz and Kurta 1987). This idea is commonly referred to as the big-mother hypothesis (Ralls 1976; Stevens et al. 2013). Therefore, even modest male-biased sexual dimorphism may be indicative of strong sexual selection in bats.

In addition to body size, precopulatory sexual selection often promotes the development of weapons (Andersson 1994; Darwin 1871). Unlike large terrestrial mammals that wield obvious weapons, such as horns or antlers, bats are constrained by aerodynamics, given their need to fly, which leaves their canine teeth, and possibly their thumbs, as potential weapons. Primates and carnivores also use their canines as weapons, and the degree of sexual dimorphism in canine length is associated with differences in their mating behavior (Gittleman and Van Valkenburgh 1997; Kappeler 1996; Plavcan 2012; Plavcan and van Schaik 1992). Among carnivores, canine sexual dimorphism is greatest in polygynous species with single-male, multifemale groups (Gittleman and Van Valkenburgh 1997). A similar pattern is seen among many primates in which increasing canine dimorphism is correlated with increasing levels of intrasexual aggression (Plavcan and van Schaik 1992), although lemurs and lorises are an exception (Kappeler 1996). Therefore, sexual dimorphism in canine length can serve as an additional indicator of the strength of precopulatory sexual selection.

In situations where males cannot control female mating, precopulatory sexual selection can result from females choosing traits that reflect attributes of male quality other than fighting ability, such as the amount of carotenoid pigment (Blount et al. 2003) or the length of feathers (Andersson 1982). There is evidence that female choice occurs in some bat species and has resulted in sexually dimorphic traits used for signaling, such as the enlarged rostrum of male hammer-headed bats, *Hypsignathus monstrosus* (Bradbury 1977), the wing sacs of some male emballonurid bats (Bradbury and Vehrencamp 1977; Voigt and von Helversen 1999), and the complex songs produced by some male molossid bats (Smotherman et al. 2016). The role of female choice is largely unexplored among phyllostomid species; the presence of sexually dimorphic features that can act as signals may reveal candidates worthy of further study.

Reproductive success is not guaranteed by acquiring copulations, because multiple mating by females creates opportunities for postcopulatory sexual selection via sperm competition (Ginsberg and Huck 1989). In many taxa, including bats (Wilkinson and McCracken 2003), there is a strong positive correlation between the opportunity for female promiscuity and size of the testes (Harcourt et al. 1995; Moller and Briskie 1995; Stockley et al. 1997), as males with larger testes are able to produce more sperm (Moller 1988) and are thus more likely to successfully sire offspring. Given the challenges of observing copulations of bats in the wild, measures of relative testis size can provide insight into the degree of female promiscuity and the resulting sperm competition among species of phyllostomids.

Information on roosting habits, particularly aggregation sizes and the structures used for roosting, is more readily available than detailed observations of mating behavior. Therefore, the aim of this chapter is to examine how roosting habits may shape mating systems by influencing opportunities for precopulatory and postcopulatory sexual selection. We infer the strength of sexual selection from measures of sexual dimorphism and testis size using both museum collections and live, wild bats. Finding strong associations will improve our ability to predict mating behavior from simple observations of roosting behavior. We hypothesize that increasing aggregation size increases opportunities for male-male competition and thus promotes precopulatory sexual selection for larger, heavier males with longer canines. Additionally, we expect larger aggregations to facilitate opportunities for multiple mating by females, thus increasing postcopulatory selection for larger testes. Whether such postcopulatory selection occurs depends on whether males can control females within aggregations. When roosts are ephemeral, social groups are likely to be more labile, which may decrease direct competition among males but could increase sperm competition as females will have the opportunity to mate sequentially with multiple males as the group composition changes. Therefore, as roost permanence decreases, we expect sexual dimorphism in body mass and canine length to become less prominent and testes mass to increase. Because the Phyllostomidae include several groups of species that have undergone recent radiations (Rojas et al. 2016), we incorporate phylogenetic similarity (Pagel 1999) into our analyses to determine if relationships among traits or factors are due to recent selection or are the result of gradual evolutionary change that occurred in proportion to the time since a common ancestor. We further examine how the patterns uncovered in our analyses align with what is known about the subset of phyllostomid bats whose mating behavior has been studied.

Methods

Data Collection

To evaluate the role of roost permanence and aggregation size on precopulatory and postcopulatory sexual selection we utilized data from several sources. We downloaded 212,823 phyllostomid specimen records from VertNet (http://www.vertnet.org) and then added 29,721 records from the Smithsonian National Museum of Natural History (http://collections.nmnh.si.edu/search/mammals/). We used species names provided by Cirranello and Simmons (chap. 4, this vol.). From these records we extracted sex, life stage, forearm length, mass, testis size, and capture location whenever it was available. We supplemented these data with direct measurements of canine length or testes that we made on specimens at the National Museum of Natural History, the American Museum of Natural History, the University of Kansas Museum of Natural History, and the Carnegie Natural History Museum. For each species, we selected at least 10 adult skulls of each sex that showed little or no evidence of tooth wear to measure the length of the left canine to at least 0.05 mm using calipers. When available, we measured specimens from the same collecting excursion to a single country. We measured length and width of one testis either from fluid specimens or from live animals that either DMA or GSW captured in Trinidad, West Indies, or Costa Rica. After eliminating records without usable data or irreconcilable species names, our data set contained 60,338 specimen records on 154 species, including 149 phyllostomids, 2 species of *Noctilio*, and 3 mormoopids (table S13.1). We then examined the range of measurements for each trait and removed obvious outliers, that is, greater than ±3 SD from the mean, to ensure that data entry errors did not distort mean values. In table 13.1 we summarize the number of species and number of specimens for each character in the data set.

To measure sexual dimorphism in canine length and body size, we first perform a phylogenetic size correction (Revell 2009) because canine length and body mass are not independent of body size. Using phylogenetic generalized least squares (PGLS), as implemented in CAPER for R (Orme et al. 2013), we regressed species mean canine length on mean forearm length. Because PGLS operates on species means rather than sex-specific means, we used the resulting coefficients and the sex-specific trait means to calculate the sex-specific residuals. We then measured sexual dimorphism as the residual male trait minus the residual female trait divided by the average value of the trait multiplied by 100, so that each dimorphism measure would represent the percentage difference in the trait between the sexes independent of body size. We calculated percentage difference in mass similarly, after excluding pregnant females, except that we estimated residuals from the PGLS of log mass on log forearm to account for the nonlinear relationship between mass and forearm. All phylogenetic analyses are based on the phylogeny of Rojas and colleagues (2016).

As a measure of the intensity of postcopulatory sexual selection, we used natural log (combined testes mass), estimated as double the volume of a prolate spheroid. To compensate for the fact that testes regress during the nonbreeding season and expand during the breeding season, we used the 90% quantile of combined testes mass to represent an average breeding male for each species. This correction likely still underestimates maximum testes mass. For example, average combined testes mass for 174 *Phyllostomus discolor* was 0.597 g while the 90% quantile was 1.053 g and the maximum was 1.868 g. Because testes mass increases with body size, we also include male body mass (natural log-transformed) as a covariate in all models. Finally, we only used measures of dimorphism or testes in subsequent analyses if there were three or more measurements per sex per species.

We used information from the literature or from museum records to score each species with regard to the degree of permanence of a roosting site and the relative number of individuals typically found in a roosting site (table S13.1). For each species, we scored roost permanence on an ordinal scale with (1) foliage or roots, (2) tents, (3) hollow trees, logs, or excavated termite nests, and (4) caves, culverts, mines, or buildings, according to reports (Arita 1993; Eisenberg 1989; Kunz et al. 2003; Reid 1997; Tuttle 1976). We calculated the average roost score for species that have been observed in multiple types of roosts. We also scored aggregation size on an ordinal scale with (1) small or family groups less than 10, (2) groups containing 11–25 individuals, (3) small colonies of 25–100, (4) large colonies greater than 100 based on comments in Eisenberg (1989), Goodwin and Greenhall (1961), Reid (1997), or in a Mammalian Species Account (mspecies.oxfordjournals.org; see table S13.1 for references). In cases where sources differed, we again used the average of the ordinal scores.

Following McCracken and Wilkinson (2000), we also used information from the literature to characterize the mating system as either single male/single female (SM/SF), single male/multifemale (SM/MF), or multimale/multifemale (MM/MF). In addition, in cases where paternity studies have been conducted, we required harem male paternity to exceed 60% before characterizing a species as SM/MF. As a consequence, some species that were previously described as harem forming or SM/MF are now scored MM/MF here (table S13.1). We made this change because reduced paternity means that sperm competition is likely to be greater and precopulatory selection on body mass or canine length is likely to be lower in such species.

Statistical Analyses

To determine the extent to which sexual dimorphism for a trait in any extant species is due to phylogenetic history—that is, closely related species are more likely to exhibit similar degrees of dimorphism—we estimated Pagel's lambda (λ) using CAPER (Orme et al. 2013). This parameter ranges from 0 to 1, such that $\lambda = 0$ represents no phylogenetic signal and $\lambda = 1$ represents strong phylogenetic signal consistent with gradual evolution via a Brownian motion model (Harvey and Pagel 1991; Pagel 1999).

We used PGLS, implemented in CAPER (Orme et al. 2013) to examine the effects of aggregation size and roost permanence on measures of dimorphism and testes mass. As before, we used the recent noctilionoid tree by Rojas et al. (2016). In the context of PGLS, λ represents the degree to which the phylogeny influences the regression, which may differ from the phylogenetic signal of a particular trait (Symonds and Blomberg 2014). We used Akaike information criterion (AICc) for model selection to evaluate the candidate models (Burnham and Anderson 2002), such that the model with the lowest AICc is preferred and models with ΔAICc <2 are considered equivalent. Because the number of species for which we have data differs depending on which traits are considered, we use only those species for which we have complete data during model selection but then apply the selected model to all possible species.

Results

We find that different traits vary in the degree to which phylogenetic similarity has an effect (table 13.1). Average forearm length has a high lambda value, indicating it is highly influenced by phylogenetic relationships. By

Table 13.1 Phylogenetic signal and sample sizes for each trait

Character	No. Species (No. Specimens)	Pagel's λ	Confidence Interval
Forearm length	137 (17,999)	0.98	[0.92, 1.00]
Forearm dimorphism (% difference)	129 (17,974)	0.00	[0.00, 0.55]
Mass dimorphism[1]	110 (17,990)	0.60	[0.29, 0.83]
Canine dimorphism[2]	87 (2,111)	0.86	[0.56, 0.99]
Testes mass[3]	107 (6,339)	0.89	[0.72, 0.97]
Roost permanence	105	0.60	[0.33, 0.82]
Aggregation size	83	0.96	[0.85, 1.00]

[1] Percentage difference of sex-specific residuals from regression on ln(forearm length).

[2] Percentage difference of sex-specific residuals from regression on forearm length.

[3] ln(combined testes mass).

contrast, forearm sexual dimorphism has a low lambda value, indicating that SSD varies independently of phylogenetic relationships, that is, has evolved rapidly among phyllostomid bats. Sexual dimorphism in both mass and canine length exhibits moderate phylogenetic signal, while relative testes mass is also influenced by phylogeny, an observation consistent with the large family-level differences in relative testes mass reported by Wilkinson and McCracken (2003). In addition to these morphological traits, both roost permanence and aggregation size are influenced by phylogeny, with aggregation size having a lambda value not significantly different from 1. Similarities between related species could be due to genetic constraints or to similar patterns of selection; regardless, this finding highlights the need to control for phylogeny rather than assume species values represent independent observations in comparative analyses.

Patterns of Dimorphism

Phyllostomid bats vary greatly in body size as measured both by length and by sexual dimorphism of forearms. One of the smallest bats in the group, *Ametrida centurio*, exhibits the greatest female-biased sexual size dimorphism (SSD, male forearm [mean ± SD]: 25.56 ± 0.48 mm, female forearm: 31.95 ± 0.76 mm, % difference: −22.22%). By contrast, the largest bat, *Vampyrum spectrum*, exhibits only weak SSD (male forearm: 105.56 ± 2.98 mm, female forearm: 104.33 ± 3.02 mm, % difference: 1.17%). Rensch's rule predicts that among

species with male-biased SSD, larger species will show greater degrees of SSD, while among female-biased species, larger species will show less dimorphism (Rensch 1959). We did not find support for this predicted pattern among either female-biased species (PGLS: $F_{1,39}$ = 0.01, p = 0.93, λ = 0.00) or male-biased species (PGLS: $F_{1,87}$ = 3.39, p = 0.07, λ = 0.31; fig. 13.1).

Sexual dimorphism in mass ranges from extreme female bias in *Macrophyllum macrophyllum* (−20.76%), to minimal sex bias in *Diphylla ecaudata* (−0.01%), to extreme male bias in *Monophyllus redmani* (20.67%). Similarly, canine sexual dimorphism ranges from moderately female biased (*Centurio senex*: −7.51%) to strongly male biased (*Phyllonycteris poeyi*: 23.24%), with males possessing relatively longer canines than females in most species. Additionally, canine sexual dimorphism is positively associated with mass sexual dimorphism (PGLS: $F_{1,80}$ = 4.33, p = 0.04, λ = 0.66, R^2 = 0.05; fig. 13.2).

Although Phyllostomidae tend to have smaller testes than other bat families (Wilkinson and McCracken 2003), they still span a broad range from 0.07% of body mass (*Leptonycteris yerbabuenae*) to 3.67% of body mass (*Diaemus youngi*), indicating that postcopulatory sexual selection is likely important for many species. As expected, log testes mass increases with log body mass (PGLS: $F_{3,72}$ = 9.01, λ = 0.65, p < 0.01, R^2 = 0.27; coefficient ± SE: 0.61 ± 0.15, p < 0.01) but does not covary with measures of canine dimorphism (coefficient ± SE: −0.02 ± 0.02, p = 0.22) or mass dimorphism (coefficient ± SE: 0.02 ± 0.01, p = 0.23).

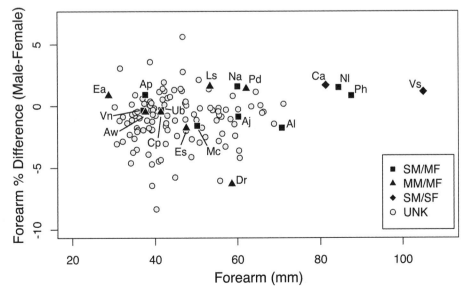

Figure 13.1. Relationship between sexual size dimorphism and body size for 129 species. Sexual size dimorphism is measured as the difference in forearm length between males and females expressed as a percentage of the species average. Symbols represent mating system types: single male/multifemale (SM/MF), multimale/mulitfemale (MM/MF), single male/single female (SM/SF), and unknown (UNK). Species labels indicate genus and species as follows: Aj—*Artibeus jamaicensis*, Ap—*A. phaeotis*, Aw—*A. watsoni*, Cp—*Carollia perspicillata*, Ca—*Chrotopterus auritus*, Dr—*Desmodus rotundus*, Ea—*Ectophylla alba*, Es—*Erophylla sezkorni*, Ls—*Lophostoma silvicolum*, Mc—*Macrotus californicus*, Nl—*Noctilio leporinus*, Na—*N. albiventris*, Pd—*Phyllostomus discolor*, Ph—*P. hastatus*, Ub—*Uroderma bilobatum*, Vn—*Vampyriscus nymphaea*, Vs—*Vampyrum spectrum*.

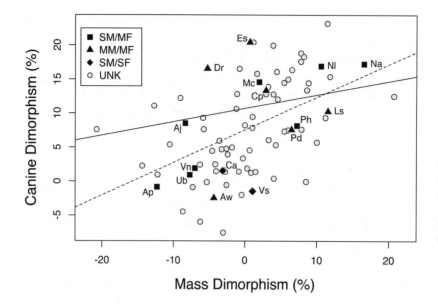

Figure 13.2. Relationship between measures of sexual dimorphism in canine length and mass among 82 species. Lines are fit by ordinary least squares (*dashed*) or phylogenetic generalized least squares (*solid*). Symbols represent mating system types and species labels indicate genus and species as in figure 13.1.

Effect of Roosting Ecology on Sexual Dimorphism and Testis Size

As expected, variation in both canine sexual dimorphism and mass sexual dimorphism is best explained by relative aggregation size (table S13.2), but measures of roost permanence do not improve model fits, as per AICc. Species that form large aggregations are more likely to exhibit male-biased mass dimorphism (PGLS: $F_{1,69} = 19.90$, $p < 0.001$, $R^2 = 0.22$, $\lambda = 0.00$, table 13.2, fig. 13.3). Similarly, canine dimorphism increases with aggregation size (PGLS: $F_{1,59} = 10.70$, $p = 0.002$, $R^2 = 0.15$, $\lambda = 0.63$, table 13.2, fig. 13.3).

The best-fit model (table S13.2), which explains variation in testes mass, includes a positive effect of body mass and a negative quadratic relationship with aggregation size (PGLS: $F_{3,61} = 10.77$, $p < 0.001$, $R^2 = 0.35$, $\lambda = 0.55$, table 13.2). The model including body mass plus quadratic effects of both aggregation size and roost permanence has a ΔAICc of 0.92 and is considered equivalent (PGLS: $F_{5,59} = 7.60$, $p < 0.001$, $R^2 = 0.39$, $\lambda = 0.51$, table S13.2). Upon examination of the effect sizes (table 13.2), it is clear that aggregation size has a stronger effect on relative testes mass than on roost permanence. Species with moderate aggregation sizes tend to have larger testes for their body size than species that form very small or very large aggregations (fig. 13.4). However, there is considerable variation among species

Table 13.2 Effect sizes of roost permanence, aggregation size, and ln(body mass) on measures of dimorphism and testes mass from four phylogenetic generalized least squares models

Model	Response	Predictor	Estimate ± SE	t	p
1	Mass dimorphism	Aggregation	3.67 ± 0.82	4.46	<0.001
2	Canine dimorphism	Aggregation	3.03 ± 0.93	3.27	0.002
3	Testes mass	Aggregation	1.10 ± 0.52	2.10	0.040
		Aggregation²	−0.28 ± 0.11	−2.57	0.012
		ln(Body mass)	0.64 ± 0.14	4.60	<0.001
4	Testes mass	Roost	1.48 ± 0.78	1.91	0.062
		Roost²	−0.28 ± 0.15	−1.86	0.068
		Aggregation	1.24 ± 0.55	2.26	0.028
		Aggregation²	−0.29 ± 0.11	−2.63	0.011
		ln(Body mass)	0.65 ± 0.14	4.74	<0.001

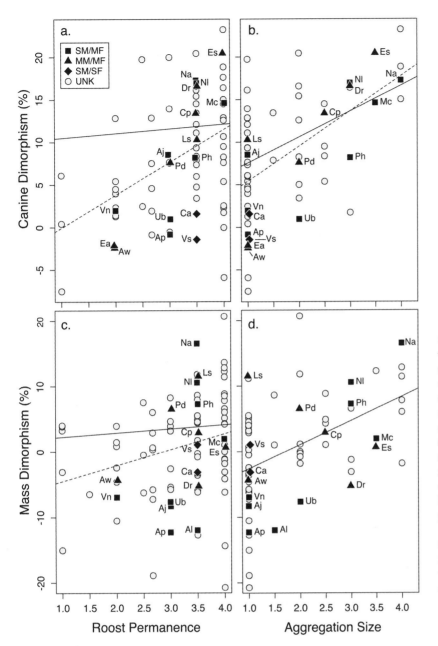

Figure 13.3. Sexual dimorphism for weaponry and weight increase with opportunity for male-male competition due to roost permanence and aggregation size. Canine dimorphism plotted against (a) degree of roost permanence for 73 species and (b) aggregation size category for 61 species. Mass dimorphism plotted against (c) degree of roost permanence for 90 species and (d) aggregation size category for 71 species. Lines are fit by ordinary least squares (dashed) or phylogenetic generalized least squares (solid) and are identical when lambda is zero. Model details available in table S13.3. Symbols represent mating system types and species labels indicate genus and species as in figure 13.1.

that roost in small groups, with combined testes mass ranging from 0.07% to 2.90% of body mass.

Discussion

Dimorphism as a Signature of Precopulatory Selection

We hypothesized that both larger aggregations and more permanent roosting structures would promote competition among males for access to reproductive females and thereby select for larger body mass and longer canines in males relative to females. We found that as roost aggregations increase in size, males become heavier and have longer canines for their size, thus supporting our hypothesis of greater competition in larger groups. We did not, however, find such support for the effect of the roost structure permanence.

For 18 of the 149 species included in our analyses, we have more detailed information on mating behavior and can thus examine where these species lie in the family-wide patterns we have found for sexual dimorphism. Two of the least dimorphic species are *V. spectrum* and *Chrotopterus auritus*, both of which roost in small groups in hollow trees or caves. *Vampyrum spectrum* is socially monogamous and the roosting group typically

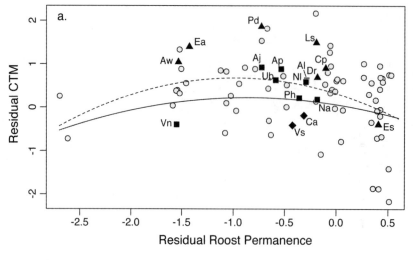

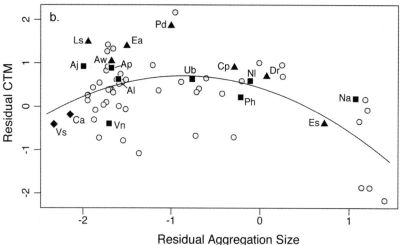

Figure 13.4. Effect of roost permanence (*a*) and aggregation size (*b*) on combined testes mass (CTM). To account for the effect of male body mass, ln(CTM), roost permanence, and aggregation size are each regressed on ln(body mass) via PGLS, and the residuals are used for plotting. (*a*) Residual testes mass plotted against residual roost permanence for 83 species. (*b*) Residual testes mass plotted against aggregation size for 65 species. Lines fit by ordinary least squares (*dashed*) and phylogenetic generalized least squares (*solid*) are identical when lambda is zero. Symbols represent mating system types and species labels indicate genus and species as in figure 13.1. Additional model details available in table S13.4.

consists of a single male and female along with recent offspring that have not yet dispersed (Vehrencamp et al. 1977). *Chrotopterus auritus* also appears to be socially monogamous, as accounts indicate roosting groups consist of family groups similar to those of *V. spectrum* (Reid 1997). How pairs form in either species is still unknown, but the lack of sexual dimorphism suggests very limited direct competition between males.

Several species show no sex bias or female bias in dimorphism, including *Uroderma bilobatum*, *Ectophylla alba*, and three of the four *Artibeus* species for which some information on mating system is available (*A. watsoni*, *A. lituratus*, *A. phaeotis*). These species all roost in small groups and construct leaf tents, except *A. lituratus*, which often roost in foliage or, occasionally, in hollow trees or caves. Leaf tents cannot accommodate the large aggregations found in more permanent roosting structures, such as hollow trees, caves, and buildings. Additionally, the limited life span of tents

requires movement between roosts, which may limit the stability of social groups (Sagot and Stevens 2012). Both of these attributes would limit opportunities for direct competition among males. However, precopulatory sexual selection may act on males if females are choosing a mate based on his tent. Kunz and McCracken (1996) suggest that tent roosts are a defendable resource and thus likely to be constructed by males to attract females. Observations of *A. watsoni* support this claim. Male *A. watsoni* construct and defend leaf tents, and roosting groups generally consist of a single male with multiple females, which suggests a mating system based on resource defense (Chaverri and Kunz 2006). Although males invest in tent construction, they do not restrict themselves to a single tent, and both males and females frequently switch among roost sites (Chaverri and Kunz 2006). By contrast, *E. alba* roost in mixed-sex groups (Brooke 1990), and both males and females engage in tent construction, with multiple individuals making modifications to a single tent (Rodríguez-Herrera et al. 2011). How group composition and tent construction influence individual mating success is still unknown, but the lack of male-biased dimorphism in mass or canine length among these species implies that small aggregations limit opportunities for direct male-male competition.

The other tent-roosting bats, *A. phaeotis*, *U. bilobatum*, *Vampyriscus nymphaea*, and occasionally *Artibeus jamaicensis*, appear to form small harem groups consisting of a single male with multiple females. These species exhibit female-biased mass dimorphism, but *V. nymphaea* and *A. jamaicensis* have male-biased canine dimorphism. Little is known about the details of *V. nymphaea*'s mating behavior, but *A. jamaicensis* has been well-studied (Kunz et al. 1983; Morrison 1979; Ortega and Arita 1999, 2000, 2002; Ortega and Maldonado 2006; Ortega et al. 2003). *Artibeus jamaicensis* roosts in a variety of structures ranging from leaf tents to caves, and as a result, aggregation sizes also vary (Kunz et al. 1983). Dominant males aggressively defend groups of females from other males (Ortega and Arita 2000), but the composition of the female groups is labile (Ortega and Arita 1999). In large harems, the dominant male tolerates the presence of a subordinate male, whose presence allows the dominant male to maintain control of the large female group in exchange for a fraction of the paternity (Ortega and Arita 2002; Ortega et al. 2003). The males' enlarged canines are presumably valuable for female defense. Our measures of dimorphism are based on species averages, but given the widespread geographic range of this species along with the diversity of roosting structures and aggregation sizes, a within-species examination of variation in mass and canine length could reveal interesting patterns. Population differences in relative testes mass have already been reported for this species (Wilkinson and McCracken 2003), and geographic variation in sexual dimorphism is known in two other phyllostomids (Willig and Hollander 1995).

Male *Lophostoma silvicolum* also build roosts to attract females, but rather than modify leaves, they excavate the underside of arboreal termite nests. The size of these roosts constrains aggregation size, but they are more permanent than most leaf tents. Roost construction appears to be under sexual selection, as only males perform the excavation and reproductive success is greater for males with roosts (Dechmann and Kerth 2008). Unlike tent-making bats, males are larger than females and have larger canines. Moreover, males with roosts are heavier than those without roosts, and females prefer to associate with larger roost holders (Dechmann and Kerth 2008). While these observations have been interpreted as a consequence of female choice, the patterns of dimorphism are consistent with a history of male-male competition.

Another species in which female choice may be important is *Erophylla sezekorni*, which forms labile mixed-sex groups with lek-like mating behavior. Males perform multimodal displays that involve visual wing flapping and display flights, vocalizations, acoustic wing buzzes, and olfactory signals (Murray and Fleming 2008). Similar to classic lekking species, *E. sezekorni* males often perform these displays in small aggregations within the cave. According to our analyses, *E. sezekorni* males and females have a similar body mass, but males have much larger canines (fig. 13.2). However, Murray and Fleming (2008) found that males are heavier and in better condition (i.e., heavier controlling for forearm length) than females. This difference may be due to population differences or seasonal variation. Regardless, their large aggregation size and male-biased canine dimorphism suggest the presence of precopulatory sexual selection, but how mass and canine size play a role in mating success remains to be determined.

In addition to *L. silvicolum*, the species with the greatest degree of mass dimorphism for which mating behavior has been reported are *Phyllostomus hastatus* and the two outgroup species, *Noctilio leporinus* and *N. albiventris*. All three of these species form harem groups, often within larger colony aggregations (Brooke 1997; McCracken and Bradbury 1981; Schad et al. 2012). In both *P. hastatus* and *N. leporinus*, female aggregations are remarkably stable over several years and remain intact despite turnover of the harem male. When in residence, the harem male fiercely defends the group by aggressively driving away any approaching males (Brooke 1997; McCracken and Bradbury 1981). This guarding behavior enables the harem male to secure most of the paternity within the group, thus making harem defense critical to reproductive success (McCracken and Bradbury 1977). In turn, competition among males to obtain harem status is expected to be fierce, and observations of *P. hastatus* males with injuries during the breeding season support this expectation (GSW, pers. obs.). Large body mass and long canines are thus a likely advantage in these competitive interactions. Less is known about the mating behavior of *N. albiventris*, but its similarity to both to *P. hastatus* and *N. leporinus* in dimorphism, testes mass, and roosting ecology suggests it is similar with respect to mating system.

Overall as aggregation size increases, sexual dimorphism becomes increasingly male biased. However, within a narrow range of aggregation sizes we still find variation in the degree of dimorphism, particularly among species with small aggregation sizes. Thus, even basic knowledge of roosting aggregations can provide insight into the degree of precopulatory selection a species may have experienced, and simple measures of sexual dimorphism can improve inferences. However, comprehensive understanding of the mating system and precopulatory selective pressures still requires long-term careful observation.

Testes Mass as a Signature of Postcopulatory Selection

We expected species that use more ephemeral roosts to have larger testes due to increased opportunities for multiple mating. Additionally, roosting in large aggregations might provide females access to more potential mates and make it more difficult for males to defend their mates, resulting in increased post-copulatory sexual selection. We found that aggregation size has a greater effect on testes mass than does roost permanence. In our analysis, moderate aggregations have higher levels of postcopulatory competition as indicated by the negative quadratic relationship between relative testes mass and increasing aggregation size. However, an additional pattern emerges when we examine the variation within each aggregation level. In doing so, we see that species known to have MM/MF groups have consistently larger testes than SM/MF groups, which in turn have larger testes than SM/SF species (fig. 13.4). This observation is consistent with previously reported patterns for the entire order (Wilkinson and McCracken 2003).

Of the species for which we have mating system information, some of the largest testes relative to body size are found in *E. alba*, *A. watsoni*, and *L. silvicolum*. As described above, these three species occur in small groups in roosts constructed from leaves or termite mounds. The relatively large testes of *E. alba* are consis-

tent with multiple mating by females and strong sperm competition, which may be expected in their mixed-sex roosting groups. Interestingly, after parturition, the composition of roosting groups changes; groups of nursing females associate only with a single adult male, while males form separate aggregations (Brooke 1990). To date, no studies have examined paternity in *E. alba*, so it is unknown what proportion of the pups in a group are sired by the resident male. Both *A. watsoni* and *L. silvicolum* form groups that appear to be harems (Chaverri et al. 2008; Dechmann et al. 2005); however, females do not remain loyal to a single male, and frequent roost switching creates opportunities for multiple mating. As a consequence, less than 30% of the pups in an *A. watsoni* tent are sired by the resident male (Chaverri et al. 2008). Similarly, less than half of the pups in an *L. silvicolum* termite roost have been sired by the resident male or by a male the female was known to roost with previously (Dechmann et al. 2005). Harems with labile female membership are also observed in *P. discolor* (GSW, pers. obs.) and *Carollia perspicillata* (Fleming 1988; Porter 1979), and males of both species also have relatively large testes. Additionally, *C. perspicillata*, bachelor males are able to successfully sneak copulations, further increasing the degree of sperm competition (Fasel et al. 2016). Flexible harems create opportunities for multiple mating and thus sperm competition; as a consequence, selection is expected to favor males with larger testes (fig. 13.4).

In contrast, *P. hastatus*, *N. leporinus*, and *N. albiventris* all form stable harem groups within permanent roosts (hollow trees and caves) and harem males aggressively defend their female groups. The relatively small testes of these males suggest their defensive behavior limits multiple mating while still monopolizing paternity. For example, *P. hastatus* harem males are able to successfully exclude most intruders and secure 70–100% paternity of pups born in their harems (McCracken and Bradbury 1977).

Both *E. sezekorni* and *Desmodus rotundus* form large multisex aggregations but appear to have small testes relative to the other multimale/multifemale species. Small testes may be expected in lekking species because females can exercise mate choice freely and the remating rate is expected to be low; however, studies of lekking birds reveal that females may mate multiply even when able to choose freely (Hess et al. 2012; Lank et al. 2002; Petrie et al. 1992). The testes of *E. sezekorni* are relatively small, which supports the hypothesis that remating rates are low when females are able to choose. However, paternity analyses reveal that reproductive skew is relatively weak, and thus this species does not conform to a classic lek (Murray and Fleming 2008). These observations do not preclude low female remating rates, but further research is needed. *Desmodus rotundus* also has moderately sized testes for a species with MM/MF groups. Unlike many bats, *D. rotundus* reproduction is asynchronous and year-round (Wilkinson 1985b). Increasing asynchrony may allow males to sequentially defend preferred females to reduce remating and limit sperm competition (Emlen and Oring 1977). This hypothesis is consistent with their substantial canine-length sexual dimorphism, although behavioral observations indicate that females sometimes remate or reject males (Wilkinson 1985b).

As expected, the two monogamous species, *V. spectrum* and *C. auritus*, have relatively small testes. Pair bonding and low population densities drastically reduce opportunities for sperm competition, thus selection should not favor large testes. Other species with small testes include several nectar feeders (*Anoura cultrata*, *Leptonycteris yerbabuenae*, and *Monophyllus redmani*) and *Centurio senex*. The three nectar feeders all roost within caves but with varying aggregation sizes. While little is known about their mating behavior, sexual dimorphism in canine length is substantially male biased in all three species (14.35%—*A. cultrata*, 15.22%—*L. yerbabuenae*, 12.27%—*M. redmani*), suggesting that males likely fight to control access to groups of females. *Centurio senex* roosts in small aggregations among the foliage. In addition to males having small testes, females are heavier and have relatively larger canines than males, suggesting weak sexual selection on males, but again little is known about their mating behavior.

In addition to variation in testis size, postcopulatory sexual selection could be influenced by the rate of sperm production and sperm morphology (Anderson et al. 2005; Tourmente et al. 2011). Bats are known to exhibit the full range of reproductive delays, from delayed fertilization (sperm storage) to delayed development after implantation (Orr and Zuk 2013; Racey and Entwistle 2000). Consequently, some of the variation in testis size could be related to species differences in sperm storage or fertilization. For example, *Noctilio albiventris* can delay fertilization (Badwaik and Ras-

weiler 2000; Rasweiler 1979), which should increase the opportunity for sperm competition. Similarly, *Glossophaga soricina* can delay implantation (Badwaik and Rasweiler 2000; Rasweiler 1979), which should allow females to bet-hedge by mating with multiple males before committing to a pregnancy. Finally, at least three phyllostomids (*A. jamaicensis* [Fleming 1971]; *C. perspicillata* [Rasweiler and Badwaik 1997; Roellig et al. 2011]; *Macrotus californicus* [Bleier 1975]) can delay development, which could allow females to compare developing embryos and only invest in the most successful one. This could favor multiple mating if each embryo was fathered by a different male. Given that females of these species typically give birth to a single pup, this scenario would also require selective embryo resorbtion, which has been reported in some bats (van der Merwe and Rautenbach 1987). The extent to which females use some form of reproductive delay to manipulate postcopulatory sexual selection is an interesting topic worthy of further investigation.

Other Consequences of Sexual Selection: Signaling and Mate Choice

Sexual selection can lead to some of the most elaborate ornaments and armaments (Darwin 1871), and while most bats are drab in color and their morphology is constrained by the demands of flight, several species possess sexually dimorphic traits and behaviors that may function as signals in agonistic interactions or for mate attraction. Vocalizations are a common sexual signal in many taxa, and several studies have documented vocalizations for courtship and territory defense in bats (Behr and von Helversen 2004; Bradbury 1977; Davidson and Wilkinson 2004). Among phyllostomids, two species known to use vocalizations are *E. sezekorni* and *C. perspicillata*, both of which perform multimodal displays. As mentioned previously, *E. sezekorni* males perform a visual display consisting of wing movements and short display flights, which are accompanied by acoustic signals produced both vocally and percussively by vibrating the wings. These signals are directed toward females, suggesting a role in mate attraction; however, the importance of these signals for acquiring mates is unclear because, as noted above, reproductive skew is low (Murray and Fleming 2008). *Carollia perspicillata* males also perform wing displays, which include poking with the wings and hovering flights, along with a courtship-specific trill vocalization. These trills exhibit acoustic differences among individuals (Knörnschild et al. 2014), but again, how these calls influence male reproductive success is still not fully understood. An additional set of vocalizations is used in aggressive contexts (Knörnschild et al. 2014), and playback experiments demonstrate that males are able to discriminate individuals by their aggressive calls, which may mitigate conflict between territorial neighbors (Fernandez et al. 2014; Porter 1979).

Olfactory signals are common among bats (Bloss 1999), but their role in mate defense or courtship is poorly understood. The best-studied example is the wing-sac odor of *Saccopteryx bilineata*, an emballonurid bat (Caspers et al. 2008; Voigt et al. 2005; Voigt and von Helversen 1999), but there is evidence that phyllostomid bats also employ olfaction in both competitive and courtship interactions. Some species have sexually dimorphic scent-producing structures, while others have monomorphic structures with dimorphic chemical profiles. Within the subfamily Phyllostominae, adult males of multiple species have a glandular throat sac that is absent or rudimentary in females (*P. discolor* [Holler and Schmidt 1993]; *P. hastatus* [McCracken and Bradbury 1981], *C. auritus* [Medellín 1989], *Phylloderma stenops* [Nowak 1994]). In both *P. discolor* and *P. hastatus*, males rub this gland on their roost site, and *P. hastatus* males also rub the gland on their harem females. Holler and Schmidt (1993) report that *P. discolor* males are able to discriminate their own scent from that of other males, which implies utility in territorial defense; furthermore, females can discriminate the scent of familiar and unfamiliar harem males, suggesting a potential role in mate recognition or choice. Similarly, the chemical composition of *P. hastatus* gland secretions indicates that males possess individually distinct scent profiles that could facilitate discrimination (Adams et al. 2018). The chemical profile also reflects variation in social status as either a bachelor or harem male, but further work is needed to determine how individuals respond to these signals.

Many chemical signals are associated with a visual component, such as swellings or hairs specialized for disseminating the odor (plate 2C, D). In two nectar-feeding bats, *Leptonycteris curasoae* and *L. yerbabuenae*, males create a visible patch between their shoulder blades by smearing secretions from various glands on their backs during the mating season (Muñoz-Romo and Kunz 2009; Nassar et al. 2008). In *L. curasoae*, the presence of this dorsal patch is correlated with larger testes, lower body condition, and fewer ectoparasites (Muñoz-Romo and Kunz 2009; Muñoz-Romo et al. 2012). The secretion mixture contains compounds that act as natural insecticides in other taxa, which might contribute to the lower parasite load, but this causal relationship is yet to be confirmed (Muñoz-Romo et al. 2012). Although little is known about the mating system of this species, this visual and chemical signal is expected to serve a role in either mate attraction or choice, as females preferentially associate with odors from males with a dorsal patch over odors from males without a patch (Muñoz-Romo et al. 2011). In both species, males are relatively heavier and possess longer canines than females but have small testes, with *L. yerbabuenae* having the smallest testes (0.07% of body mass) of the phyllostomid bats for which we have data. This implies strong precopulatory competition but weak postcopulatory competition among males despite large aggregations.

Males and females of several *Sturnira* species possess epaulettes, tufts of hair on the shoulders that are associated with underlying sebaceous glands (plate 2). There is no sexual dimorphism in the structure of these epaulettes (Scully et al. 2000), but there is significant sexual dimorphism in the bacterial colonies present in these glandular regions (González-Quiñonez et al. 2014). Bacteria play an important role in olfactory communication by altering chemical signals through the metabolism of secretory products, and as a result, odor can serve as a signal of condition or infection status (Penn and Potts 1998; Zala et al. 2004).

Some of the most bizarre sexually dimorphic traits in bats are found among the short-faced bats within the family Stenodermatinae. For example, male *Pygoderma bilabiatum* have glandular tissue around their eyes, under their chins, and on their wrists that undergo seasonal swelling (Tavares and Tejedor 2009) (plate 2C), during which time males emit a subtle musky odor (R. L. M. Novaes, personal communication). Their small testes and weak female-biased size dimorphism suggest that sexual selection is weak, but the function of these structures and the behavior of this species are largely unknown. Close relatives of *P. bilabiatum* also possess a variety of intriguing facial structures. Male *Sphaeronycteris toxophyllum*, appropriately named the visored bat, have a large sexually dimorphic outgrowth on their foreheads (plate 2*A*, *B*). In addition to the forehead protrusion, males also have a fleshy flap under their chins, which can be raised to cover their faces (Nowak 1994). Unfortunately, little is known about the behavior of this species, but we can infer there is some element of male-male competition due to the strong male bias in relative mass, despite weak sexual dimorphism in skeletal and canine size.

In addition to choosing males based on morphological features and behavioral displays, females may also select mates based on the quality of the roosts they construct. As described above, males of several species construct roosts from leaves or termite mounds, and these roosts may serve as an extended phenotype (Schaedelin and Taborsky 2009). Because roosts are an essential resource for females, they may prefer males that are able to build suitable roosts. Furthermore, roosts may signal male quality. However, this requires that the resident male roosts only in tents that he has constructed. While this is clearly the case for *L. silvicolum* males that excavate termite nests (Dechmann and Kerth 2008), it is unlikely to be true for many tent-making bats, which frequently switch between several roosts at any given time.

Conclusions

Even though females have larger forearms than males in 69% of phyllostomid species, males have relatively longer canines than do females in 85% of species, and males have relatively greater body mass than females in 57% of species. These patterns of sexual dimorphism are consistent with strong precopulatory sexual selection acting via male competition for access to mates. Moreover, these two measures of sexual dimorphism are correlated with each other, and both are better predicted by size of roosting aggregations than by the degree of permanence of the roosting site. These results indicate that proximity within a roost facilitates male competition and likely enables males to defend larger groups of females. Postcopulatory sexual selection appears to operate independently of precopulatory sexual selection, as testis size is uncorrelated with sexual dimorphism in either relative canine length or mass. In contrast to our predictions, postcopulatory sexual selection is not positively related with aggregation size. Instead, species that form relatively small to intermediate-sized aggregations exhibit some of the largest testes, although considerable

variation in testis size is unexplained. Variation in reproductive delay among species might explain some of that variation.

The patterns of sexual dimorphism we observed are consistent with what is known about the mating systems of phyllostomid bats. Species that form single-male/multifemale (or harem) groups exhibit some of the most extreme male-biased sexual dimorphism for canine length and body mass, while species that form single-male/single-female or multimale/multifemale groups typically show much less sexual dimorphism for these traits. In contrast, testis size tends to be largest in species that form multimale/multifemale groups in various sized aggregations. The sexually dimorphic traits of the stenodermatine species, such as extended brow ridges, swollen eye tissue, or glandular tissue in males, may be used by females for mate selection; unfortunately, too little is known about their mating systems, but this possibility is certainly worthy of further study. Despite the limited behavioral information available, morphological signatures of sexual selection and commonly collected information on roosting ecology reveal widespread patterns among the phyllostomids but also highlight the need for further research.

Acknowledgments

We thank Ted Fleming, Gary McCracken, and an anonymous reviewer for comments that helped improve the manuscript. Our work on phyllostomid bats has been supported by grants from the American Society of Mammalogists and the Society for the Study of Evolution to DMA and the National Science Foundation to GSW.

References

Adams, D. M., Y. Li, and G. S. Wilkinson. 2018. Male scent gland signals mating status in greater spear-nosed bats, *Phyllostomus hastatus*. Journal of Chemical Ecology, 44:975–986.

Anderson, M. J., J. Nyholt, and A. F. Dixson. 2005. Sperm competition and the evolution of sperm midpiece volume in mammals. Journal of Zoology, 267:135–142.

Andersson, M. 1982. Female choice selects for extreme tail length in a widowbird. Nature, 299:818–820.

Andersson, M. 1994. Sexual Selection. Princeton University Press, Princeton, NJ.

Arita, H. T. 1993. Conservation biology of the cave bats of Mexico. Journal of Mammalogy, 74:693–702.

Badwaik, N. K., and J. J. I. V. Rasweiler. 2000. Pregnancy.

Pp. 221–293 in: Reproductive Biology of Bats (E. G. Crichton and P. H. Krutzsch, eds.). Academic Press, San Diego.

Behr, O., and O. von Helversen. 2004. Bat serenades—complex courtship songs of the sac-winged bat (*Saccopteryx bilineata*). Behavioral Ecology and Sociobiology, 56:106–115.

Bleier, W. J. 1975. Fine structure of implantation and the corpus luteum in the California leaf-nosed bat, *Macrotus californicus*. PhD thesis, Texas Tech University.

Bloss, J. 1999. Olfaction and the use of chemical signals in bats. Acta Chiropterologica, 1:31–45.

Blount, J. D., N. B. Metcalfe, T. R. Birkhead, and P. F. Surai. 2003. Carotenoid modulation of immune function and sexual attractiveness in zebra finches. Science, 5616:125–127.

Bradbury, J. W. 1977. Lek mating behavior in the hammer-headed bat. Zeitschrift für Tierpsychologie—Journal of Comparative Ethology, 45:225–255.

Bradbury, J. W., and S. L. Vehrencamp. 1977. Social organization and foraging in emballonurid bats: 3. Mating systems. Behavioral Ecology and Sociobiology, 2:1–17.

Brooke, A. P. 1990. Tent selection, roosting ecology, and social organization of the tent-making bat, *Ectophylla alba*, in Costa Rica. Journal of Zoology, 221:11–19.

Brooke, A. P. 1997. Organization and foraging behaviour of the fishing bat, *Noctilio leporinus* (Chiroptera:Noctilionidae). Ethology, 103:421–436.

Burnham, K. P., and D. R. Anderson. 2002. Model selection and multimodel inference: a practical information-theoretic approach. Springer, New York.

Caspers, B., S. Franke, and C. C. Voigt. 2008. The wing-sac odour of male greater sac-winged bats *Saccopteryx bilineata* (Emballonuridae) as a composite trait: seasonal and individual differences. Pp. 151–160 in: Chemical Signals in Vertebrates 11 (J. L. Hurst, R. J. Beynon, S. C. Roberts, and T. D. Wyatt, eds.). Springer, New York.

Chaverri, G., and T. H. Kunz. 2006. Roosting ecology of the tent-roosting bat *Artibeus watsoni* (Chiroptera: Phyllostomidae) in southwestern Costa Rica. Biotropica, 38:77–84.

Chaverri, G., and T. H. Kunz. 2010. Ecological determinants of social systems: perspectives on the functional role of roosting ecology in the social behavior of tent-roosting bats. Pp. 275–318 in: Advances in the Study of Behavior: Behavioral Ecology of Tropical Animals (R. Macedo, ed.). Academic Press, Burlington, MA.

Chaverri, G., O. E. Quiros, M. Gamba-Rios, and T. H. Kunz. 2007. Ecological correlates of roost fidelity in the tent-making bat *Artibeus watsoni*. Ethology, 113:598–605.

Chaverri, G., C. J. Schneider, and T. H. Kunz. 2008. Mating system of the tent-making bat, *Artibeus watsoni* (Chiroptera: Phyllostomidae). Journal of Mammalogy, 89:1361–1371.

Clutton-Brock, T. H. 1989. Review lecture: mammalian mating systems. Proceedings of the Royal Society of London B–Biological Sciences, 236:339–372.

Clutton-Brock, T. H., P. H. Harvey, and B. Rudder. 1977. Sexual dimorphism, socionomic sex ratio and body weight in primates. Nature, 269:797–800.

Darwin, C. 1871. The Descent of Man, and Selection in Relation to Sex. John Murray, London.

Davidson, S. M., and G. S. Wilkinson. 2004. Function of male

song in the greater white-lined bat, *Saccopteryx bilineata*. Animal Behaviour, 67:883–891.

Dechmann, D. K. N., E. K. V. Kalko, B. König, and G. Kerth. 2005. Mating system of a Neotropical roost-making bat: the white-throated, round-eared bat, *Lophostoma silvicolum* (Chiroptera: Phyllostomidae). Behavioral Ecology and Sociobiology, 58:316–325.

Dechmann, D. K. N., and G. Kerth. 2008. My home is your castle: roost making is sexually selected in the bat *Lophostoma silvicolum*. Journal of Mammalogy, 89:1379–1390.

Eisenberg, J. F. 1989. Mammals of the Neotropics: The Northern Neotropics. University of Chicago Press, Chicago.

Emlen, S. T., and L. W. Oring. 1977. Ecology, sexual selection, and the evolution of mating systems. Science, 197:215–223.

Fasel, N., V. Saladin, and H. Richner. 2016. Alternative reproductive tactics and reproductive success in male *Carollia perspicillata* (Seba's short-tailed bat). Journal of Evolutionary Biology, 29:2242–2255.

Fernandez, A. A., N. Fasel, M. Knörnschild, and H. Richner. 2014. When bats are boxing: aggressive behaviour and communication in male Seba's short-tailed fruit bat. Animal Behaviour, 98:149–156.

Fleming, T. H. 1971. *Artibeus jamaicensis*—delayed development in a Neotropical bat. Science, 171:402–404.

Fleming, T. H. 1988. The Short-Tailed Fruit Bat: A Study in Plant-Animal Interactions. University of Chicago Press, Chicago.

Ginsberg, J. R., and U. W. Huck. 1989. Sperm competition in mammals. Trends in Ecology and Evolution, 4:74–79.

Gittleman, J. L., and B. Van Valkenburgh. 1997. Sexual dimorphism in the canines and skulls of carnivores: Effects of size, phylogeny, and behavioural ecology. Journal of Zoology, 242:97–117.

González-Quiñonez, N., G. Fermin, and M. Muñoz-Romo. 2014. Diversity of bacteria in the sexually selected epaulettes of the little yellow-shouldered bat *Sturnira lilium* (Chiroptera: Phyllostomidae). Interciencia, 39:882–889.

Goodwin, G. G., and A. M. Greenhall. 1961. A review of the bats of Trinidad and Tobago. Bulletin of the American Museum Natural History, 122:195–301.

Harcourt, A. H., A. Purvis, and L. Liles. 1995. Sperm competition: mating system, not breeding season, affects testes size of primates. Functional Ecology, 9:468–476.

Harvey, P. H., and M. D. Pagel. 1991. The Comparative Method in Evolutionary Biology. Oxford University Press, Oxford.

Hess, B. D., P. O. Dunn, and L. A. Whittingham. 2012. Females choose multiple mates in the lekking greater prairie chicken (*Tympanuchus cupido*). Auk, 129:133–139.

Holler, P., and U. Schmidt. 1993. Olfactory communication in the lesser spear-nosed bat *Phyllostomus discolor* (Chiroptera, Phyllostomidae). Zeitschrift für Saugetierkunde, 58:257–265.

Isaac, J. L. 2005. Potential causes and life-history consequences of sexual size dimorphism in mammals. Mammal Review, 35:101–115.

Kappeler, P. M. 1996. Intrasexual selection and phylogenetic constraints in the evolution of sexual canine dimorphism in strepsirhine primates. Journal of Evolutionary Biology, 9:43–65.

Kerth, G. 2008. Causes and consequences of sociality in bats. Bioscience, 58:737–746.

Knörnschild, M., M. Feifel, and E. K. V. Kalko. 2014. Male courtship displays and vocal communication in the polygynous bat *Carollia perspicillata*. Behaviour, 151:781–798.

Kunz, T. H., P. V. August, and C. D. Burnett. 1983. Harem social organization in cave roosting *Artibeus jamaicensis* (Chiroptera: Phyllostomidae). Biotropica, 15:133–138.

Kunz, T. H., and A. Kurta. 1987. Size of bats at birth and maternal investment during pregnancy. Pp. 79–106 *in*: Symposia of the Zoological Society of London, no. 57. Cambridge University Press, Cambridge.

Kunz, T. H., and L. F. Lumsden. 2003. Ecology of cavity and foliage roosting bats. Pp. 3–89 *in*: Bat Ecology (T. H. Kunz and M. B. Fenton, eds.). University of Chicago Press, Chicago.

Kunz, T. H., and G. F. McCracken. 1996. Tents and harems: apparent defence of foliage roosts by tent-making bats. Journal of Tropical Ecology, 12:121–137.

Lank, D. B., C. M. Smith, O. Hanotte, A. Ohtonen, S. Bailey, and T. Burke. 2002. High frequency of polyandry in a lek mating system. Behavioral Ecology, 13:209–215.

Lindenfors, P., J. L. Gittleman, and K. E. Jones. 2007. Sexual size dimorphism in mammals. Pp. 16–26 *in*: Sex, Size and Gender Roles: Evolutionary Studies of Sexual Size Dimorphism (D. J. Fairbairn, W. U. Blanckenhorn, and T. Székely, eds.). Oxford University Press, Oxford.

Lindenfors, P., B. S. Tullberg, and M. Biuw. 2002. Phylogenetic analyses of sexual selection and sexual size dimorphism in pinnipeds. Behavioral Ecology and Sociobiology, 52:188–193.

McCracken, G. F., and J. W. Bradbury. 1977. Paternity and genetic heterogeneity in the polygynous bat, *Phyllostomus hastatus*. Science, 198:303–306.

McCracken, G. F., and J. W. Bradbury. 1981. Social organization and kinship in the polygynous bat *Phyllostomus hastatus*. Behavioral Ecology and Sociobiology, 8:11–34.

McCracken, G. F., and G. S. Wilkinson. 2000. Bat mating systems. Pp. 321–362 *in*: Reproductive Biology of Bats (E. G. Crichton and P. H. Krutzsch, eds.). Academic Press, London.

Medellín, R. A. 1989. *Chrotopterus auritus*. Mammalian Species, 343:1–5.

Moller, A. P. 1988. Ejaculate quality, testes size, and sperm competition in primates. Journal of Human Evolution, 17:479–488.

Moller, A. P., and J. V. Briskie. 1995. Extra-pair paternity, sperm competition and the evolution of testis size in birds. Behavioral Ecology and Sociobiology, 36:357–365.

Morrison, D. W. 1979. Apparent male defense of tree hollows in the bat, *Artibeus jamaicensis*. Journal of Mammalogy, 60:11–15.

Muñoz-Romo, M., J. F. Burgos, and T. H. Kunz. 2011. Smearing behaviour of male *Leptonycteris curasoae* (Chiroptera) and female responses to the odour of dorsal patches. Behaviour, 148:461–483.

Muñoz-Romo, M., and T. H. Kunz. 2009. Dorsal patch and chemical signaling in males of the long-nosed bat, *Leptonycteris curasoae* (Chiroptera: Phyllostomidae). Journal of Mammalogy, 90:1139–1147.

Muñoz-Romo, M., L. T. Nielsen, J. M. Nassar, and T. H. Kunz. 2012. Chemical composition of the substances from dorsal patches of males of the curasoaen long-nosed bat, *Leptonycteris curasoae* (Phyllostomidae: Glossophaginae). Acta Chiropterologica, 14:213–224.

Murray, K. L., and T. H. Fleming. 2008. Social structure and mating system of the buffy flower bat, *Erophylla sezekorni* (Chiroptera, Phyllostomidae). Journal of Mammalogy, 89:1391–1400.

Nassar, J. M., M. V. Salazar, A. Quintero, K. E. Stoner, M. Gomez, A. Cabrera, and K. Jaffe. 2008. Seasonal sebaceous patch in the nectar-feeding bats *Leptonycteris curasoae* and *L. yerbabuenae* (Phyllostomidae : Glossophaginae): phenological, histological, and preliminary chemical characterization. Zoology, 111:363–376.

Nowak, R. M. 1994. Walker's Bats of the World. Johns Hopkins University Press, Baltimore, MD.

Orme, D., R. Freckleton, G. Thomas, T. Petzoldt, S. Fritz, N. Isaac, and W. Pearse. 2013. CAPER: comparative analyses of phylogenetics and evolution in R. Methods in Ecology and Evolution 3:145–151.

Orr, T. J., and M. Zuk. 2013. Does delayed fertilization facilitate sperm competition in bats? Behavioral Ecology and Sociobiology, 67:1903–1913.

Ortega, J., and H. T. Arita. 1999. Structure and social dynamics of harem groups in *Artibeus jamaicensis* (Chiroptera : Phyllostomidae). Journal of Mammalogy, 80:1173–1185.

Ortega, J., and H. T. Arita. 2000. Defence of females by dominant males of *Artibeus jamaicensis* (Chiroptera : Phyllostomidae). Ethology, 106:395–407.

Ortega, J., and H. T. Arita. 2002. Subordinate males in harem groups of Jamaican fruit-eating bats (*Artibeus jamaicensis*): satellites or sneaks? Ethology, 108:1077–1091.

Ortega, J., and J. E. Maldonado. 2006. Female interactions in harem groups of the Jamaican fruit-eating bat, *Artibeus jamaicensis* (Chiroptera : Phyllostomidae). Acta Chiropterologica, 8:485–495.

Ortega, J., J. E. Maldonado, G. S. Wilkinson, H. T. Arita, and R. C. Fleischer. 2003. Male dominance, paternity, and relatedness in the Jamaican fruit-eating bat (*Artibeus jamaicensis*). Molecular Ecology, 12:2409–2415.

Pagel, M. 1999. Inferring the historical patterns of biological evolution. Nature, 401:877–884.

Penn, D., and W. K. Potts. 1998. Chemical signals and parasite-mediated sexual selection. Trends in Ecology and Evolution, 13:391–396.

Petrie, M., M. Hall, T. Halliday, H. Budgey, and C. Pierpoint. 1992. Multiple mating in a lekking bird: why do peahens mate with more than one male and with the same male more than once? Behavioral Ecology and Sociobiology, 31:349–358.

Plavcan, J. M. 2012. Sexual size dimorphism, canine dimorphism, and male-male competition in primates: where do humans fit in? Human Nature—an Interdisciplinary Biosocial Perspective, 23:45–67.

Plavcan, J. M., and C. P. van Schaik. 1992. Intrasexual competition and canine dimorphism in anthropoid primates. American Journal of Physical Anthropology, 87:461–477.

Plavcan, J. M., and C. P. van Schaik. 1997. Intrasexual competition and body weight dimorphism in anthropoid primates. American Journal of Physical Anthropology, 103:37–67.

Porter, F. L. 1979. Social behavior in the leaf-nosed bat, *Carollia perspicillata*. 1. Social Organization. Journal of Comparative Ethology, 49:406–417.

Racey, P. A., and A. C. Entwistle. 2000. Life-history and reproductive strategies of bats. Pp. 636–414 *in*: Reproductive Biology of Bats (E. G. Crichton and P. H. Krutzsch, eds.). Academic Press, London.

Ralls, K. 1976. Mammals in which females are larger than males. Quarterly Review of Biology, 51:245–276.

Ralls, K. 1977. Sexual dimorphism in mammals: avian models and unanswered questions. American Naturalist, 111: 917–938.

Rasweiler, J. J. 1979. Early embryonic development and implantation in bats. Journal of Reproduction and Fertility, 56:403–416.

Rasweiler, J. J., and N. K. Badwaik. 1997. Delayed development in the short-tailed fruit bat, *Carollia perspicillata*. Journal of Reproduction and Fertility, 109:7–20.

Reid, F. 1997. A field guide to the mammals of Central America and Southeast Mexico. Oxford University Press, Oxford.

Rensch, B. 1959. Evolution above the Species Level. Columbia University Press, New York.

Revell, L. J. 2009. Size-correction and principal components for interspecific comparative studies. Evolution, 63:3258–3268.

Rodríguez-Herrera, B., G. Ceballos, and R. A. Medellín. 2011. Ecological aspects of the tent building process by *Ectophylla alba* (Chiroptera: Phyllostomidae). Acta Chiropterologica, 13:365–372.

Roellig, K., B. R. Menzies, T. B. Hildebrandt, and F. Goeritz. 2011. The concept of superfetation: a critical review on a "myth" in mammalian reproduction. Biological Reviews, 86:77–95.

Rojas, D., O. M. Warsi, and L. M. Dávalos. 2016. Bats (Chiroptera: Noctilionidae) challenge a recent origin of extant Neotropical diversity. Systematic Biology, 65:432–448.

Sagot, M., and R. D. Stevens. 2012. The evolution of group stability and roost lifespan: perspectives from tent-roosting bats. Biotropica, 44:90–97.

Schad, J., D. K. Dechmann, C. C. Voigt, and S. Sommer. 2012. Evidence for the "good genes" model: association of MHC class II DRB alleles with ectoparasitism and reproductive state in the Neotropical lesser bulldog bat, *Noctilio albiventris*. PLOS ONE, 7:e37101.

Schaedelin, F. C., and M. Taborsky. 2009. Extended phenotypes as signals. Biological Reviews, 84:293–313.

Scully, W. M. R., M. B. Fenton, and A. S. M. Saleuddin. 2000. A histological examination of the holding sacs and glandular scent organs of some bat species (Emballonuridae, Hipposideridae, Phyllostomidae, Vespertilionidae, and Molossidae). Canadian Journal of Zoology, 78:613–623.

Smotherman, M., M. Knörnschild, G. Smarsh, and K. Bohn. 2016. The origins and diversity of bat songs. Journal of Comparative Physiology A, 202:535–554.

Stevens, R. D., M. E. Johnson, and E. S. McCulloch. 2013.

Absolute and relative secondary-sexual dimorphism in wing morphology: a multivariate test of the "Big Mother" hypothesis. Acta Chiropterologica, 15:163–170.

Stockley, P., M. J. G. Gage, G. A. Parker, and A. P. Moller. 1997. Sperm competition in fishes: the evolution of testis size and ejaculate characteristics. American Naturalist, 149:933–954.

Symonds, M. R. E., and S. P. Blomberg. 2014. A primer on phylogenetic generalised least squares *in*: Modern Phylogenetic Comparative Methods and Their Application in Evolutionary Biology (L. Z. Garamszegi, ed.). Springer-Verlag, Berlin.

Tavares, V. d. C., and A. Tejedor. 2009. The forelimb swellings of *Pygoderma bilabiatum* (Chiroptera: Phyllostomidae). Chiroptera Neotropical, 15:411–416.

Tourmente, M., M. Gomendio, and E. R. S. Roldan. 2011. Sperm competition and the evolution of sperm design in mammals. BMC Evolutionary Biology, 11:12.

Trivers, R. L. 1972. Parental investment and sexual selection. Pp. 136–179 *in*: Sexual Selection and the Descent of Man (B. Campbell, ed.). Aldine, Chicago.

Tuttle, M. D. 1976. Collecting techniques. Pp. 71–88 *in*: Biology Bats of the New World Family Phyllostomatidae, Part 1 (R. J. Baker, J. K. Jones Jr., and D. C. Carter, eds.). Texas Tech Press, Lubbock.

van der Merwe, M., and I. L. Rautenbach. 1987. Reproduction in Schlieffen's bat, *Nycticeius schlieffenii*, in the eastern Transvaal lowveld, South Africa. Journal of Reproduction and Fertility, 81:41–50.

Vehrencamp, S. L., F. G. Stiles, and J. W. Bradbury. 1977. Observations on the foraging behavior and avian prey of the Neotropical carnivorous bat, *Vampyrum spectrum*. Journal of Mammalogy, 58:469–478.

Voigt, C. C., B. Caspers, and S. Speck. 2005. Bats, bacteria, and bat smell: sex-specific diversity of microbes in a sexually selected scent organ. Journal of Mammalogy, 86:745–749.

Voigt, C. C., and O. von Helversen. 1999. Storage and display of odour by male *Saccopteryx bilineata* (Chiroptera, Emballonuridae). Behavioral Ecology and Sociobiology, 47:29–40.

Weckerly, F. W. 1998. Sexual-size dimorphism: influence of mass and mating systems in the most dimorphic mammals. Journal of Mammalogy, 79:33–52.

Wilkinson, G. S. 1985a. The social organization of the common vampire bat. I. Pattern and cause of association. Behavioral Ecology and Sociobiology, 17:111–121.

Wilkinson, G. S. 1985b. The social organization of the common vampire bat. II. Mating system, genetic structure, and relatedness. Behavioral Ecology and Sociobiology, 17:123–134.

Wilkinson, G. S., and G. F. McCracken. 2003. Bats and balls: sexual selection and sperm competition in Chiroptera. Pp. 128–155 *in*: Bat Ecology (T. H. Kunz and M. B. Fenton, eds.). University of Chicago Press, Chicago.

Willig, M. R., and R. R. Hollander. 1995. Secondary sexual dimorphism and phylogenetic constraints in bats—a multivariate approach. Journal of Mammalogy, 76:981–992.

Zala, S. M., W. K. Potts, and D. J. Penn. 2004. Scent-marking displays provide honest signals of health and infection. Behavioral Ecology, 15:338–344.

Trophic Ecology

14

The Omnivore's Dilemma

The Paradox of the Generalist Predators

Claire T. Hemingway,
M. May Dixon,
and Rachel A. Page

Introduction

Phyllostomid bats are known for their spectacular radiation into distinct dietary niches. But not all phyllostomids are specialists; many phyllostomid species have quite generalized diets. Historically, the generalist phyllostomids fell within the subfamily Phyllostominae, a diverse group that primarily hunts insects and small vertebrates such as frogs, lizards, birds, rodents, and other species of bats. This now paraphyletic group consists of over 30 species. While this group is characterized by predators with broad, generalized diets, many of these species have distinct, specialized foraging strategies. In this chapter, we will discuss the apparent paradox of these generalist predators: bats that can be classified as dietary generalists, broadly overlapping with one another in the prey they consume, but often possessing quite specialized foraging adaptations and distinct behavioral strategies for prey finding. To untangle the puzzle of this understudied and historically elusive group of generalist predators, we examine the influences of diet breadth on hunting behavior, spatial distribution, conservation concern, and morphology.

Phyllostomids are the most ecologically diverse of any of the mammalian families (Freeman 2000; Gould 1976). Their diets range from insects to fish, frogs, lizards, rodents, bats, birds, fruits, pollen, nectar, and blood (Kalko et al. 1996). Phyllostomids offer stunning examples of extreme specialization, including the nectar-feeding bat, *Anoura fistulata*, with a tongue 150% the length of its body, perfectly matched to the *Centropogon nigricans* flowers it pollinates (Muchhala 2006; Muchhala and Tschapka, chap. 16, this vol.), or the vampire bat, *Desmodus rotundus*, an obligate blood feeder with heat-sensing pit organs on its face, which allow it to localize blood vessels from which to feed (Gracheva et al. 2011; Hermanson and Carter, chap. 15, this vol.; Kurten and Schmidt 1982).

But within phyllostomids, beyond the broad categories of frugivore, nectarivore, and sanguivore, finer-scale categorization becomes difficult. The remaining phyllostomid bats form a group we refer to as the "animalivorous phyllostomids." These are predators that consume a wide array of animal prey, from small insects to large vertebrates. This group includes a wide array of body sizes, from *Vampyrum spectrum*, the largest bat (~175 g) in the New World, that hunts birds as large as motmots (Vehrencamp et al. 1977), to *Micronycteris microtis* (~6 g), a tiny bat that consumes well-defended katydids as long as itself (Kalka and Kalko 2006). Despite these differences, there are clear characteristics that unite this group. A large portion of the diet of animalivorous phyllostomids consists of insects, small vertebrates, or both, with a strong trend towards broad, generalist diets. While no longer considered monophyletic, the animalivorous phyllostomids are currently phylogenetically clustered in three distinct clades, the subfamilies Phyllostominae, Micronycterinae, and Lonchorhininae (Rojas et al. 2016; plate 3).

A number of researchers have tried to make sense of the diversity of this group by categorizing its members into guilds, or functional groups of species sharing sim-

ilar adaptations (Denzinger and Schnitzler 2013; Kalko et al. 1996). Humphrey et al. (1983), for example, suggested that the generalist, animal-eating phyllostomids represent two distinct guilds, an insectivorous group, primarily consuming insects, and a carnivorous group, primarily consuming vertebrates. Currently, however, it is agreed that the animal-eating phyllostomids actually represent a single diverse guild that vary in the proportion of their diet that is insectivorous versus carnivorous (Giannini and Kalko 2005). Giannini and Kalko's (2005) dietary-structure study of these bats reveals a continuum, with carnivory gradually replacing insectivory as bat body size, and correspondingly, prey body size, increases (fig. 14.1). Indeed, Freeman (1984, 403) argues that "there is no distinct boundary where insectivory ends and carnivory begins," adding that "flesh-eating bats are often bigger versions" of smaller insectivorous bats.

Over the years, a wide range of names has been applied to this group of generalist phyllostomids, with different species included and excluded. Names for

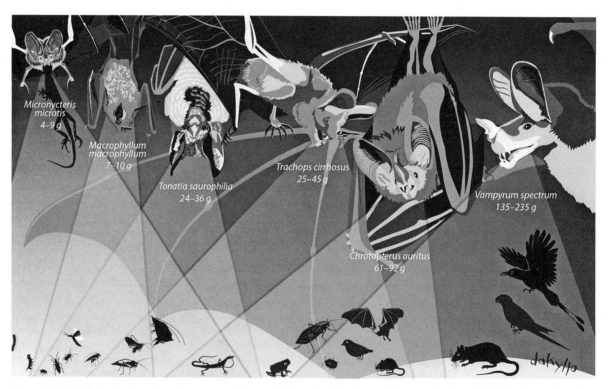

Figure 14.1. Scaling of predators and prey showing a trend for increased carnivory as a function of increased body size (Giannini and Kalko 2005). While animalivorous phyllostomids follow this trend, they also show a great deal of flexibility and variation, with the smallest species in this group, *Micronycteris microtis*, consuming reptiles such as anoles in addition to insects, and the largest species, *Vampyrum spectrum*, while almost completely carnivorous, also consuming insects such as beetles. In general, we see broad generalist diets across the animalivorous phyllostomids, with the larger bat species consuming larger prey items, and the proportion of carnivory increasing with increasing body size. Drawings by Damond Kyllo.

this group and proposed subgroups have included carnivores (Freeman 1984; Santana and Cheung 2016), gleaning carnivores (Bonaccorso 1979), insectivores (Arita 1993; Freeman 1984), insectivore/carnivores (Humphrey et al. 1983), omnivores (García-Morales et al. 2013), animalivores (Giannini and Kalko 2005), or combinations thereof. In this chapter, we argue that struggles to identify common characteristics of this group and lack of consensus in classifying the group as a single diverse guild have hindered attempts at comprehensive study. A primary goal of this chapter, therefore, is to unravel some of the puzzles associated with this elusive group, identify common features uniting it, and lay the groundwork for research to come. For the remainder of this chapter, we will refer to this continuum of generalist bats as a single, diverse guild: the animalivorous phyllostomids.

The animalivorous phyllostomids present something of a paradox. Most species in this group possess broad, generalized diets (albeit some more extreme than others) but highly derived, specialized hunting strategies. For example, species such as *Lophostoma silvicolum*, *Tonatia saurophila*, and *Trachops cirrhosus* (discussed in one of our case studies below), eavesdrop on the acoustic cues emitted by their prey (Falk et al. 2015). While all three species exhibit this specialized hunting strategy, the prey that they consume range from insects to frogs to small mammals (Giannini and Kalko 2005).

The majority of the animalivorous phyllostomids are substrate gleaners, picking their prey off of surfaces such as vegetation. However, the group also includes bats that hunt airborne prey and trawl prey from the water, such as *Macrophyllum macrophyllum*, which has a distinct foraging adaptation, specialized for its foraging niche. Convergent with aerial insectivores that use short echolocation calls at a high repetition rate at the final stage of attack to gather information on prey location, *M. macrophyllum* employs a "terminal buzz" to localize small insect prey on the water surface (by trawling) or just above the water surface (by aerial hawking) (Weinbeer and Kalko 2007). *Macrophyllum macrophyllum*'s specialized foraging technique has expanded the range of prey items available to it. Thus, like other animalivorous phyllostomids, *M. macrophyllum* has a relatively generalized diet but a specialized foraging strategy.

Another example of wide diet breadth but specialized foraging strategy is *M. microtis*. While it may occasionally use prey mating signals as its eavesdropping relatives do (Falk et al. 2015), its primary mode of prey localization is echolocation (Geipel, Jung et al. 2013). Through a series of elegantly designed experiments, Geipel, Jung, and colleagues (2013) discovered that *M. microtis* can use echolocation alone to pinpoint silent, motionless prey amidst the clutter of the forest understory. This feat was previously considered impossible, given the echo overlap between prey and the surrounding forest foliage (Neuweiler 1989; Schnitzler and Kalko 1998). *Micronycteris microtis* uses a distinct stereotypical hovering flight search behavior, coupled with the production of short, multiharmonic, broadband echolocation calls (Geipel, Jung et al. 2013). As with the examples outlined above, *M. microtis*'s specialized foraging behavior gives it access to a broad range of prey species, from dragonflies to spiders to caterpillars to anolis lizards (Santana et al. 2011). So while it is a highly specialized bat with a unique foraging specialization, its diet is characteristically broad.

The examples above illustrate the feature that we feel best characterizes the animalivorous phyllostomid group: bats that have broad, overlapping diets but distinctly specialized foraging strategies. Diet breadth also varies in this group (Giannini and Kalko 2005). *Phyllostomus hastatus* is at one end, consuming an extremely broad diet from pollen to insects to rodents (Gardner 1977; Santos et al. 2003); we will discuss this bat in detail in a case study later in the chapter. In contrast, other animalivorous phyllostomids have much narrower diets: *Lophostoma silvicolum*, for example, consumes exclusively insects, albeit from a diverse range of taxa (Giannini and Kalko 2005).

In the remainder of the chapter we will identify common attributes of animalivorous phyllostomid bats that unite this group behaviorally, morphologically, and ecologically. Our goal is to bring together key findings across this eclectic guild and identify avenues for future research.

Morphology

Body Size

In general, the larger the bat, the larger its prey (Norberg and Fenton 1988). In the animalivorous phyllostomids, large bats can subdue and consume large prey items such as birds and terrestrial vertebrates, while smaller bats tend to eat smaller vertebrates and more insects. The degree of carnivory in this family is strongly correlated with body size (Giannini and Kalko 2005). The largest of the Neotropical gleaners are phyllostomids and are almost exclusively carnivorous: *V. spectrum* and *Chrotopterus auritus*. *Vampyrum spectrum* weighs around 170 g and feeds primarily on rodents, birds, and other bat species (Bonato et al. 2004; Gardner 1977; Navarro and Wilson 1982). *Chrotopterus auritus* is about 75 g and eats rodents, opossums, bats, birds, lizards, and anurans (Medellín 1988). However, in keeping with the wide diet breadth characteristic of this animalivorous phyllostomid group, both *V. spectrum* and *C. auritus* also include insect prey in their diets (Coleoptera: Bonato et al. 2004). The smaller *T. cirrhosus*, approximately 30 g, has a diet consisting of around 20–40% vertebrates and 60–80% insects (Bonoto et al. 2004; Giannini and Kalko 2005).

There are also interesting deviations from this size-diet trend within animalivorous phyllostomids. For instance, *P. hastatus* is the one of the largest bat species in the Neotropics, weighing approximately 105 g, but it largely consumes fruits and insects, with only minor components of its diet consisting of vertebrates (Santos et al. 2003). Similarly, *L. silvicolum*, which is comparable in size to *T. cirrhosus*, is not known to consume any vertebrate prey (Giannini and Kalko 2005). Perhaps the most extreme example is *M. microtis*, which weighs only ~6 g but consumes both vertebrate and invertebrate prey, making it the world's smallest carnivorous bat (Santana et al. 2011). So while most of the roughly 30 species in the animalivorous phyllostomid guild exhibit increasing carnivory with increasing body size, we see exceptions on both ends of the spectrum: large bats that are largely insectivorous (or omnivorous, also consuming fruit), and small bats preying on vertebrates.

Functionally, body size in bats reflects at least one strong trade-off, especially for predatory bats (Stockwell 2001). Larger body sizes may increase dietary niche breadth but can also significantly decrease maneuverability (Fenton 1990; Giannini and Kalko 2005; Stockwell 2001). The costs of increased size on flight performance may necessitate detection of prey at greater distances (Rosenzweig 1968). Within the animalivorous phyllostomids, other aspects of morphol-ogy, sensory ecology, and behavior may help to offset these costs. We expand on each of these topics in the sections that follow.

Ears

Ears are perhaps the most conspicuous and varied features of bats (Obrist et al. 1993). Because all species of Yangochiropteran bats (the suborder to which the phyllostomid bats belong) interact with their environments primarily via sound, ears can have a significant effect on foraging performance in bats and often serve as reliable predictors of adaptive trade-offs between foraging and flying (Dechmann et al. 2006; Fenton and Bogdanowicz 2002; Gardiner et al. 2011).

For species that glean from substrates, echoacoustic information about prey on surfaces is typically severely degraded and difficult to detect due to echo clutter from the surrounding vegetation (Neuweiler 1983). Rather than relying on echolocation to locate prey, many gleaners listen for the sounds that prey generate while moving (Coles et al. 1989; Fiedler 1979) or signaling to mates (e.g., Tuttle and Ryan 1981). Large pinnae (external ears) help facilitate this departure from echolocation-based foraging by amplifying the faint, low-frequency sounds emanating from moving prey (Coles et al. 1989; Gardiner et al. 2011). Indeed, gleaning animalivores from disparate families have independently evolved spectacularly large ears. For example, megadermatids (Megadermatidae), which are more closely related to Old World fruit bats than to phyllostomids, have the largest pinnae among bats and look remarkably similar to the gleaning animalivorous phyllostomids (Neuweiler et al. 1984; fig. 14.2).

Gleaning animalivorous phyllostomids tend to have larger ears than their non-gleaning relatives (Gardiner et al. 2011; fig. 14.2). Because gleaning animalivores are not monophyletic within this family (plate 3), it is likely that evolution has favored large ears among the gleaning bats. For example, the common vampire bat, *Desmodus rotundus* (a strict sanguivore and not included in the animalivorous phyllostomid group), has small ears in relation to body size and is considered an outlier for ear size in a comparative study across 12 families and 98 species (Gardiner et al. 2011). Because the genus *Desmodus* has evolved within the group of animalivorous gleaners (plate 3), evidence suggests that

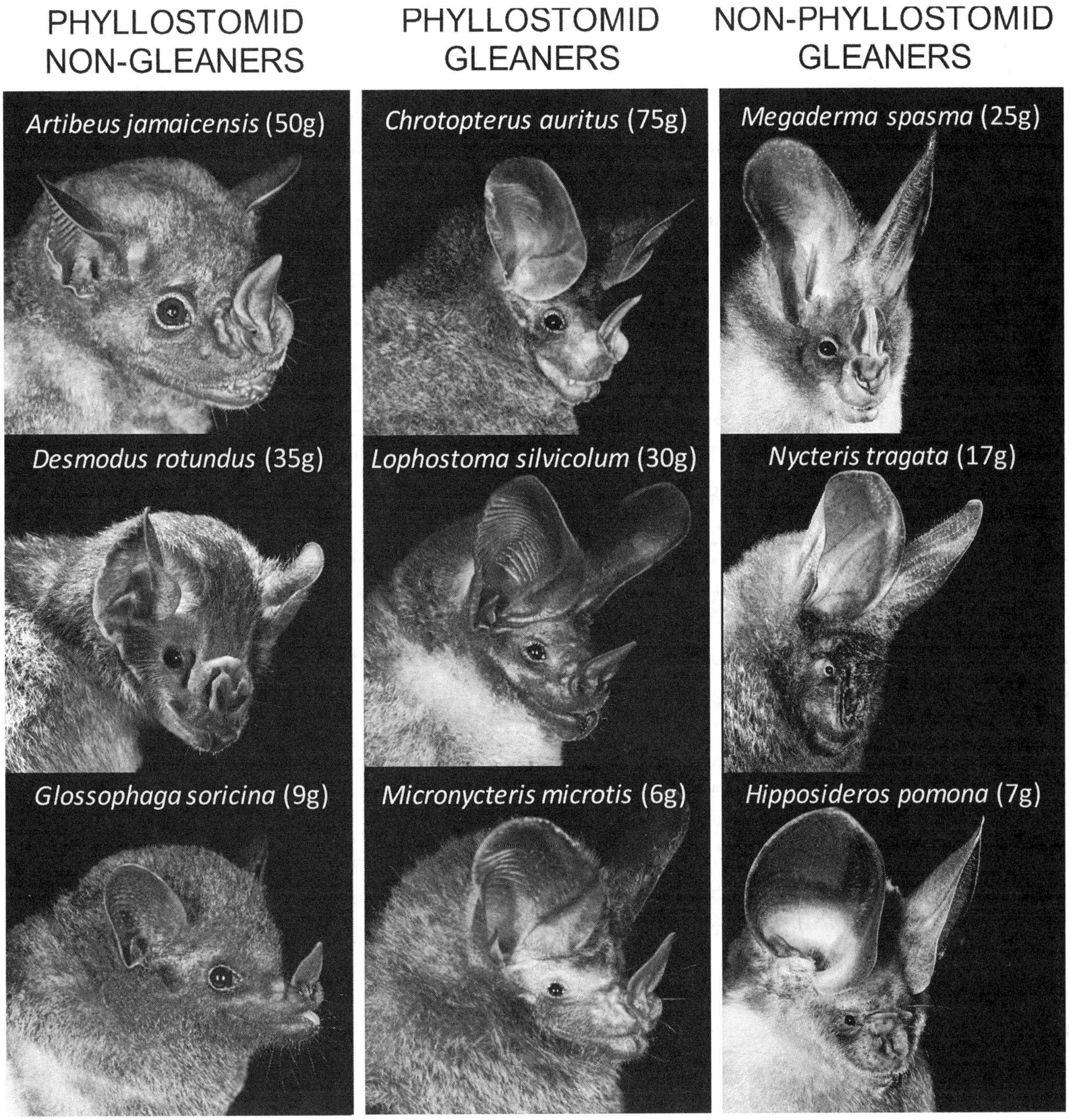

PHYLLOSTOMID NON-GLEANERS	PHYLLOSTOMID GLEANERS	NON-PHYLLOSTOMID GLEANERS
Artibeus jamaicensis (50g)	*Chrotopterus auritus* (75g)	*Megaderma spasma* (25g)
Desmodus rotundus (35g)	*Lophostoma silvicolum* (30g)	*Nycteris tragata* (17g)
Glossophaga soricina (9g)	*Micronycteris microtis* (6g)	*Hipposideros pomona* (7g)

Figure 14.2. Portraits illustrating the ear morphologies of non-gleaning phyllostomid bats, gleaning phyllostomid bats, and gleaning bats from other families. *Chrotopterus auritus, Lophostoma silvicolum, Micronycteris microtis* (all Phyllostomidae), and *Megaderma spasma, Nycteris tragata,* and *Hipposideros pomona* (from the phylogenetically distant families of Megadermatidae, Nycteridae, and Hipposideridae, respectively) all have large, rounded pinna, ideal for eavesdropping on prey-emitted sounds. The physical likeness among these species is striking, despite differences in body size and, for the three examples in the far right column, phylogenetic distance. As Freeman (1984) points out, the larger species of gleaning bats seem to simply be "bigger versions" of the smaller species. In contrast, the three non-gleaning phyllostomids, exemplified here with a frugivore (*Artibeus jamaicensis*), a sanguivore (*Desmodus rotundus*), and a nectarivore (*Glossophaga soricina*), though much more closely related to the gleaning phyllostomids, have very different facial morphologies with much smaller ears. Approximate body weights are noted in parentheses. Photos by Marco Tschapka and Merlin Tuttle.

ear size is not a phylogenetically conserved trait within this lineage but, rather, represents an adaptive feature related to foraging behavior.

Large ears incur costs to flight stability and agility (Gardiner et al. 2011); they increase drag during flight, with the longest-eared bats incurring the greatest ear aerodynamic cost, especially at higher flight speeds (Fenton 1972; Norberg 1976). Because flight is energetically costly and large ears impede flight, for gleaning phyllostomids this trade-off must pay off: large ears must be an effective adaptation for detecting prey-emitted sounds (Maina 2000).

Bite Force

Bite force in bats is an important predictor of what is included in the diets of different species and is largely predicted by cranial morphology (Aguirre et al. 2003; Freeman 1984). Increasing bite force has the potential to broaden available prey types (Aguirre et al. 2003). Bite force in bats is likely to be under extreme selection pressure relative to other mammals, as most bats are unable to manipulate and subdue prey items with their forelimbs (Vandoros and Dumont 2004).

The evolution of bite force can be shaped by changes in cranial shape that maximize mechanical advantage, muscle anatomy modifications that result in high force production, or both (Vandoros and Dumont 2004). Within phyllostomids, several key morphological changes have been shown to correspond with bite force and dietary specialization (Dumont 1997). For example, the false vampire bat (*V. spectrum*) has an immense skull equipped with sharp canine teeth and shearing molars, which aid with the crushing of bones and cutting of flesh (Feldhamer et al. 2015). In frugivores, the rostrum (or snout) has typically become shorter, whereas in nectarivores, it has lengthened (Dumont 1997). Within the animalivorous phyllostomids, skull morphology tends to be similar across animalivorous diets (Arbour et al. 2019).

The maximum bite force that different phyllostomid species can produce is correlated with the hardness of food items in their diets, with harder items corresponding to higher bite forces (Aguirre et al. 2003). This is particularly important for bats that eat hard insects and bony vertebrates. Bats that consume hard-shelled arthropods and vertebrates typically have wider skulls and stout jaws, which allow for more powerful bite force (Freeman 1984). While there is a positive correlation between bite force and shorter rostra in insectivorous and frugivorous phyllostomids, most animalivores have relatively long rostrums (Nogueria et al. 2009). Conversely, Giannini and Kalko (2005) have found that the gleaning insectivorous and carnivorous bats in this family do not have cranial morphology consistent with the consumption of hard-bodied prey, even though most have a large portion of Coleoptera in their diet (21–56%), which are among the hardest of the prey consumed in this family (Giannini and Kalko 2005).

Most mammals are also able to modulate their bite force through behavioral modifications, such as biting with posterior or anterior teeth (Santana and Dumont 2009). This interaction between behavior and performance may have played a large role in the diversity within this order (Jones et al. 2005). It is possible that variation in diet in animalivorous phyllostomids is due to selection for behavioral traits that overcome morphological limitations, allowing for further partitioning of resources (Giannini and Kalko 2005; Santana and Dumont 2009).

For example, *M. microtis* has an elevated bite force and performance plasticity that seems to have allowed it to expand its diet to include both invertebrates and vertebrates of varying degrees of hardness (Santana et al. 2011). Although it uses its molars when eating all prey types, *M. microtis* changes its feeding behavior in other ways, such as by changing the number of bites per prey item depending on prey hardness (Santana et al. 2011). Other animalivorous phyllostomids with narrower diets, such as *T. saurophila* and *Gardnernycteris crenulatum*, have been found to have lower plasticity in bite force (Santana et al. 2011).

Wing Size and Shape

One can tell a great deal about the types of environments that bats inhabit and the type of food that they eat by examining their wings (Aldridge and Rautenback 1987; Norberg and Rayner 1987). Bat that forage in highly cluttered habitats tend to have short, broad wings with low aspect ratio (wing span²/wing area) and low wing loading (weight/wing area) (Aldridge and Rautenbach 1987; Norberg and Rayner 1987). Bats that must achieve high flight speeds to forage on flying insects typically have long, slender wings with high aspect ratios and above-average wing loading (Norberg and Rayner 1987).

Most phyllostomids, regardless of their foraging guild, forage within the forest understory (Schnitzler and Kalko 1998). Neotropical forests are characterized by a high density of obstacles such as trees, lianas, branches, and other vegetation. As a result, most phyllostomids have low wing loading and low wing aspect ratios compared to other bat families (Norberg and Rayner 1987).

Within Phyllostomidae are subtler differences in

wing shape and loading that vary according to diet and are more readily predicted based on foraging guild. Animalivorous phyllostomids are characterized by lower wing loading than most other phyllostomids and have longer, pointed wing tips (Norberg and Fenton 1988; Norberg and Rayner 1987). This is especially true for the larger phyllostomids, such as *V. spectrum* and *C. auritus*, which need very slow and maneuverable flight for foraging among clutter and flying with heavy prey (Norberg and Fenton 1988; Norberg and Rayner 1987).

Gleaning insectivorous bats may also differ in wing morphology based on whether they are hover-gleaners that take prey items from vegetation in the understory or ground-gleaners that take prey from the ground. Hover-gleaning is common for most of the smaller insectivorous phyllostomids and can be characterized by longer wings with low wing loading and rounded wingtips (Norberg and Rayner 1987). These features help them to maneuver in dense foliage over vegetation to find and capture their prey. Ground-gleaning bats typically have the added challenge of needing to take off from the ground with heavy prey in their mouths, and low wing loading is necessary to assist in takeoff (Norberg and Rayner 1987). For example, *T. cirrhosus* has low wing loading and a low aspect ratio, which likely aids in the slow and maneuverable flight necessary to glean frogs and insects from water or foliage and then take off with them (Barclay et al. 1981; Norberg and Rayner 1987). This has also been shown experimentally for *Lophostoma silvicolum*, which forages almost exclusively on large orthopterans (Stockwell 2001).

Cognition and Behavior

Behavioral Flexibility versus Stereotypy

Reliance on learning can be costly. Energy and neural space are needed to process and store memories. It also may be risky to sample the environment in trial-and-error learning or to rely on potentially unreliable information in social learning (Burns et al. 2010; Dukas 1999; Safi and Dechmann 2005). Thus if a stereotyped, innate behavior can evolve for catching food, it may be less costly than reliance on learning (Dukas 1999). Stereotypy may be favored if all prey items have similar handling requirements and are consistently available

(Dukas 1998). Flexibility and learning, in contrast, are thought to be advantageous in heterogeneous and variable conditions, because animals can adapt to current conditions more quickly (Dukas 1998).

Animalivorous phyllostomids eat a wide variety of prey species and even within a prey type, such as katydids, different species vary radically in size and antipredator defenses (ter Hofstede et al. 2017). Given diets that include food items that may vary seasonally and involve very different handling approaches, we predict that animalivorous phyllostomids are likely to rely relatively more heavily on learning than on innate foraging behaviors. There is little empirical research on variation in propensity to learn in phyllostomids, but the question has been addressed in vespertilionids of the genus *Myotis* (Clarin et al. 2013). Clarin and colleagues (2013) found that species that forage in open space areas learned a complex maze task more slowly than species that hunt their prey by gleaning or foraging in the cluttered forest understory. Within animalivorous phyllostomids, we predict that species with more varied diets (e.g., *P. hastatus*, *T. cirrhosus*, *Mimon bennettii*) rely more on learning than those with narrower diets (e.g., *M. macrophyllum*, *Lonchorina aurita*) (fig. 14.1).

Another major factor that may affect the evolution of learning is whether there is mutualistic or antagonistic evolution between a consumer and its food. Many phytophagous phyllostomids provide essential pollination or seed dispersal services for their food plants, which puts selection pressure on the plants to be more easily found by bats (Chittka and Raine 2006; Kunz et al. 2011; Ross and Holderied 2013). Indeed, some plants provide echoacoustically conspicuous "guides" to aid detection and approach by their bat mutualists or even alter their echoacoustic properties to indicate that a flower has already been visited (Schöner et al. 2015; von Helversen and von Helversen 1999).

In contrast, predatory bats impose strong negative selection pressures on the species that they eat. Their prey may respond with increased crypsis or antipredator defenses (ter Hofstede and Ratcliffe 2016). For example, animalivorous phyllostomids such as *M. microtis*, *T. saurophila*, *L. silvicolum*, and *T. cirrhosus* prey on Neotropical katydids (Belwood and Morris 1987; Falk et al. 2015; Tuttle et al. 1985). In turn, katydids have evolved a suite of physical and behavioral defenses, including calling from protected perches, communicating with substrate-borne vibratory signals rather than calling, ceasing to call when they hear bat echolocation calls, and calling infrequently overall (see ter Hofstede et al. 2017 for review). Because animalivorous phyllostomids are confronted with prey that do not "want" to be eaten, they may be under strong selection for behavioral flexibility, which may enable these predators to overcome the many defenses of their prey.

Social Learning

Social learning may be more dictated by social structure than by foraging strategy (see section on *P. hastatus*'s vocal learning in the second case study in this chapter), but the type of foraging may also influence the extent to which species use social information (Lachmann et al. 2000). Like other kinds of learning, social learning is more likely when trial-and-error learning is risky and when prey is variable, difficult to detect, or hard to handle (Thorten and Clutton-Brock 2011). There is indirect evidence of vertical transmission of foraging information from mothers to their young in animalivorous phyllostomids. As previously mentioned, *M. microtis* has a complex foraging strategy wherein it gleans a wide variety of silent and motionless prey from vegetation (Geipel, Jung et al. 2013). These prey items are often well-defended and large, relative to the size of the bat. Mothers provision their young with pieces of food for months, which may teach juveniles prey-handling skills (Geipel, Kalko et al. 2013). *Trachops cirrhosus* juveniles seem to forage close to their mothers after they are volant, which may also allow them to learn foraging skills (Ripperger et al. 2016). It is unknown whether these animals can learn to hunt on their own, or whether this type of social learning is an essential element of growing up for the species. Captive rearing studies will be necessary to further understand these issues.

Horizontal transmission of information has also been found in animalivorous phyllostomids: *T. cirrhosus* learns a novel foraging task more quickly when exposed to experienced conspecifics (Page and Ryan 2006) but only attends to food cues from conspecifics when its own individually acquired information about food is unreliable (Jones et al. 2013). Animalivorous phyllostomids can acquire foraging information from heterospecifics as well (Patriquin et al. 2018). *Lophostoma silvicolum* and *T. cirrhosus* both attend to katydid calling songs to detect and locate prey (Falk et al. 2015). In a flight cage setting, *T. cirrhosus* learns novel acoustic cues by observing the predatory behavior of *L. silvicolum* as quickly as it learns from observing other *T. cirrhosus* (Patriquin et al. 2018). Attending to the behavior of heterospecifics may be advantageous when conspecifics are not present or when heterospecifics have familiarity with new prey types. How often such social learning takes place in nature, however, requires further investigation.

Niche-Specific Cognitive Strategies

Stich and Winter (2006) proposed the niche-specific cognitive strategies hypothesis to explain why nectar-feeding bats (e.g., *Glossophaga soricina*) prefer to use spatial cues over object cues in foraging experiments. This hypothesis posits that animals should have cognitive abilities associated with the senses that they rely on most to detect their prey. Thus, animals from the same ecological guild should have similar learning abilities in a given sensory modality and related animals should vary along a cognitive continuum based on their diets (Stich and Winter 2006). For example, bird species that depend on stored food tend to have stronger spatial memories in comparison to birds that are less reliant on food caches (Balda and Kamil 2006). From this hypothesis, we can make predictions about how animalivorous phyllostomids compare to frugivores and nectarivores in their ability to make associations with spatial, olfactory, visual, and acoustic cues. Because their prey animals are mobile and often unpredictable, it seems likely that most animalivorous phyllostomids may rely relatively less on spatial memory than do other phyllostomids. Bats that key in on the acoustic cues of their prey are likely to be relatively better at learning sounds than bats in other foraging guilds. There is currently some indirect evidence for this (Page and Ryan 2005), but this hypothesis has never been tested in a comparative framework. Bats that find prey by echoacoustic shape, such as *M. microtis*, may be relatively better at forming shape associations (Geipel, Kalko et al. 2013). The niche-specific cognitive strategies hypothesis has been tested with frugivorous and nectarivorous phyllostomids (Carter et al. 2010; Stich and Winter 2006), and in an Old World insectivore (*Myotis nattereri*, Hulgard and Ratcliffe 2014), but has not yet been tested comparatively with animalivorous phyllostomids.

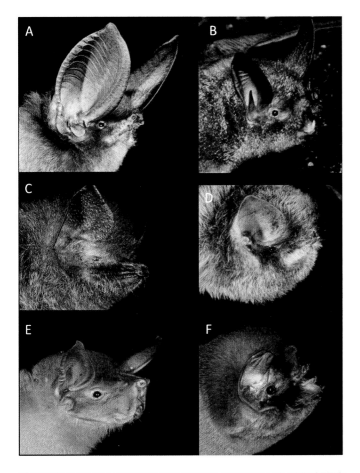

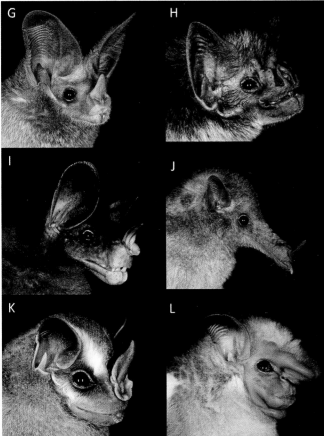

Plate 1a. Portrait gallery of non-phyllostomid noctilionoid bats. From top left: (*A*) Myzopodidae: *Myzopoda schliemanni*, holotype, (*B*) Mystacinidae: *Mystacina tuberculata*, (*C*) Thyropteridae: *Thyroptera discifera*, (*D*) Furipteridae: *Furipterus horrens*, (*E*) Noctilionidae: *Noctilio leporinus*, (*F*) Mormoopidae: *Mormoops megalophylla*,

Plate 1b. Phyllostomidae: (*G*) *Macrotus waterhousii*, (*H*) *Desmodus rotundus*, (*I*) *Vampyrum spectrum*, (*J*) *Musonycteris harrisoni*, (*K*) *Vampyressa nymphaea*, (*L*) *Sphaeronycteris toxophyllum*, male. Not to scale. Photo credits: Merlin Tuttle (*A*), Stuart Parsons (*B*), Marco Tschapka (*C–L*).

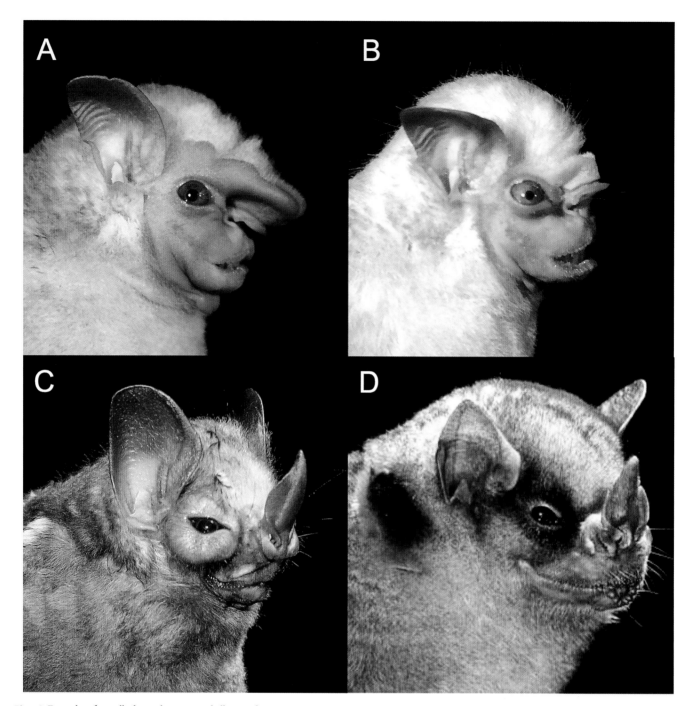

Plate 2. Examples of sexually dimorphic traits in phyllostomids.
(A) Male *Sphaeronycteris toxophyllum* with enlarged browridge.
(B) Female *Sphaeronycteris toxophyllum* without pronounced browridge. (C) Male *Pygoderma bilaboatum* with swollen tissue surrounding the eyes. Males also have swollen tissue around their wrists. (D) Male *Sturnira tildae* with shoulder epaulettes caused by glandular secretions. Photos courtesy of Rodrigo Medellín (A, B), Roberto L. M. Novaes (C), and Merlin Tuttle (D).

Plate 3. (*Opposite*) Phyllostomid phylogeny redrawn with permission from Rojas et al. (2016). Animalivorous phyllostomids are found on the shaded insert below and are indicated in bold font. *Colored circles* indicate the diets of these bat species: insects (*blue*), amphibians (*orange*), reptiles (*purple*), birds (*red*), mammals (*brown*), plants (*green*). We have indicated even trace components of the diet in this figure. For example, plant matter is a large component of the diet of *Phyllostomus hastatus* but is only found in very small amounts in the diet of *Trachops cirrhosus*. But because plants have been reported in the diets of both, we have designated each of these bat species with a *green circle*. Drawings by Damond Kyllo.

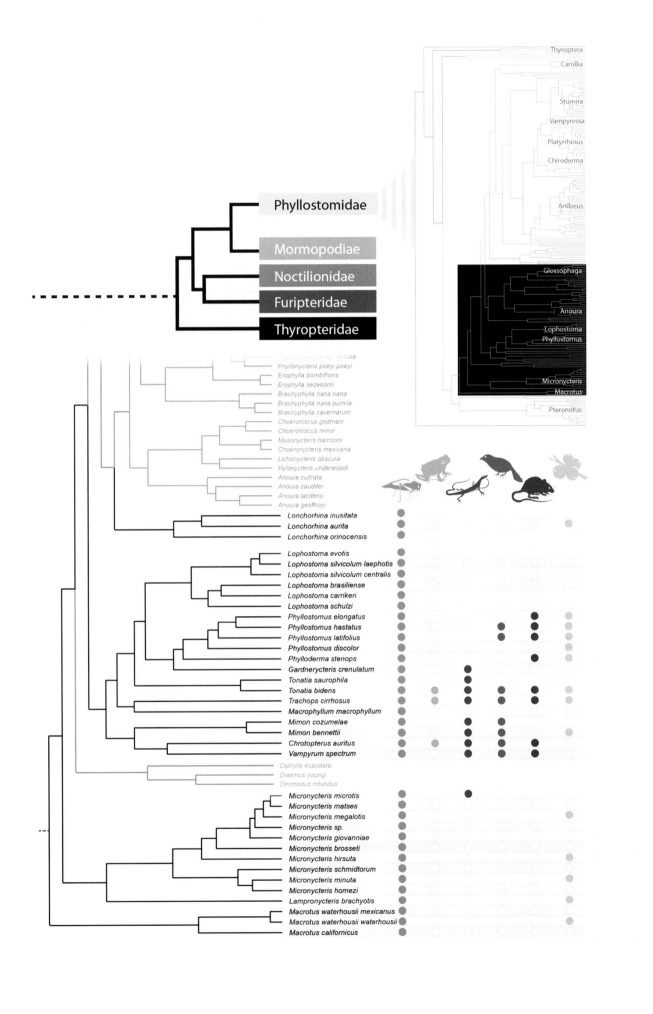

Phyllostomidae

Mormopodiae

Noctilionidae

Furipteridae

Thyropteridae

Thyroptera
Carollia
Sturnira
Vampyressa
Platyrrhinus
Chiroderma
Artibeus
Glossophaga
Anoura
Lophostoma
Phyllostomus
Micronycteris
Macrotus
Pteronotus

Phyllonycteris poeyi obtusa
Phyllonycteris poeyi poeyi
Erophylla bombifrons
Erophylla sezekorni
Brachyphylla nana nana
Brachyphylla nana pumila
Brachyphylla cavernarum
Choeroniscus godmani
Choeroniscus minor
Musonycteris harrisoni
Choeronycteris mexicana
Lichonycteris obscura
Hylonycteris underwoodi
Anoura cultrata
Anoura caudifer
Anoura latidens
Anoura geoffroyi
Lonchorhina inusitata
Lonchorhina aurita
Lonchorhina orinocensis
Lophostoma evotis
Lophostoma silvicolum laephotis
Lophostoma silvicolum centralis
Lophostoma brasiliense
Lophostoma carrikeri
Lophostoma schulzi
Phyllostomus elongatus
Phyllostomus hastatus
Phyllostomus latifolius
Phyllostomus discolor
Phylloderma stenops
Gardnerycteris crenulatum
Tonatia saurophila
Tonatia bidens
Trachops cirrhosus
Macrophyllum macrophyllum
Mimon cozumelae
Mimon bennettii
Chrotopterus auritus
Vampyrum spectrum
Diphylla ecaudata
Diaemus youngi
Desmodus rotundus
Micronycteris microtis
Micronycteris matses
Micronycteris megalotis
Micronycteris sp.
Micronycteris giovanniae
Micronycteris brosseti
Micronycteris hirsuta
Micronycteris schmidtorum
Micronycteris minuta
Micronycteris homezi
Lampronycteris brachyotis
Macrotus waterhousii mexicanus
Macrotus waterhousii waterhousii
Macrotus californicus

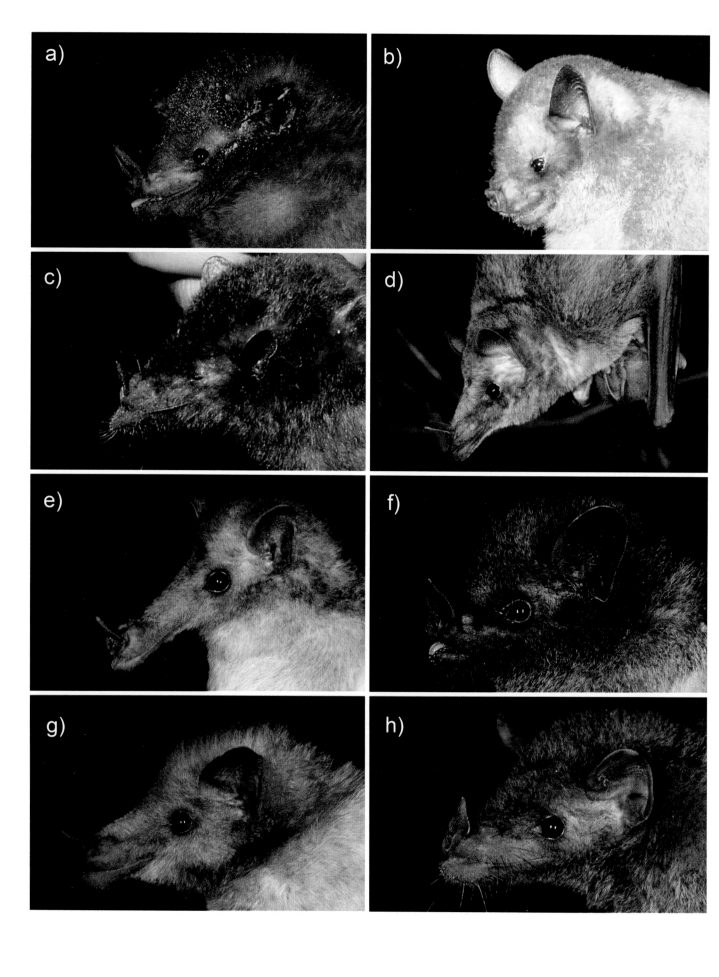

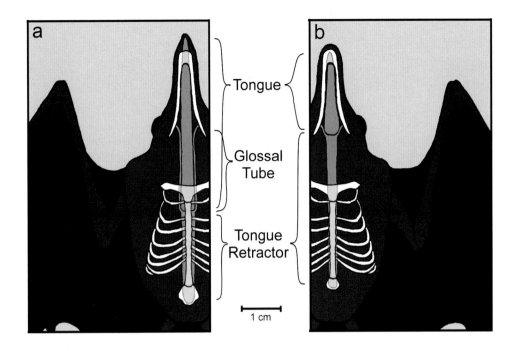

Tongue

Glossal
Tube

Tongue
Retractor

1 cm

Plate 5. Comparison of the tongue morphology of *Anoura fistulata* with that of a typical glossophagine. (*a*) Ventral view of *A. fistulata*, showing tongue (*pink*), glossal tube and tongue retractor muscle (*blue*), and skeletal elements (*white*). (*b*) Corresponding ventral view of a typical glossophagine bat. Adapted from Muchhala (2006a).

Plate 4. (*Opposite*) Nectar-feeding phyllostomids. (*a*) *Hsunycteris thomasi*, (*b*) *Brachyphylla cavernarum*, (*c*) *Anoura fistulata*, (*d*) *Xeronycteris vierai*, (*e*) *Musonycteris harrisoni*, (*f*) *Glossophaga soricina*, (*g*) *Platalina genovensium*, (*h*) *Anoura geoffroyi*

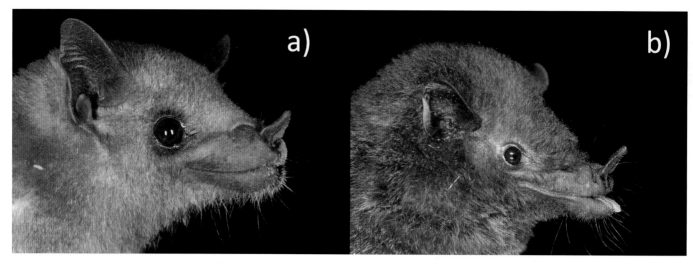

Plate 6. Relative eye size varies distinctly between (*a*) *Leptonycteris yerbabuenae*, which often flies in open desert habitats, and (*b*) *Hylonycteris underwoodi*, which inhabits rainforest understory and is not known to move seasonally between sites.

Plate 7. (*Opposite*) Nectar bat visits to bat-pollinated flowers. (*a*) *Macrocarpea* (Gentianaceae) visited by *Anoura caudifer*, (*b*) *Cleome anomala* (Capparaceae) visited by *A. geoffroy*, (*c*) *Symbolanthus* (Gentianaceae) visited by *A. geoffroyi*, (*d*) *Centropogon nigricans* (Lobeliaceae) visited by *A. fistulata*, (*e*) *Pachycereus pringlei* (Cactaceae) visited by *Leptonycteris yerbabuenae*, (*f*) *Trianea* (Solanaceae) and (*g*) *Burmeistera glabrata* (Lobeliaceae) visited by *A. cultrata*, and (*h*) *Werauhia* cf. *sanguinolenta* (Bromeliaceae) visited by *Lonchophylla robusta*.

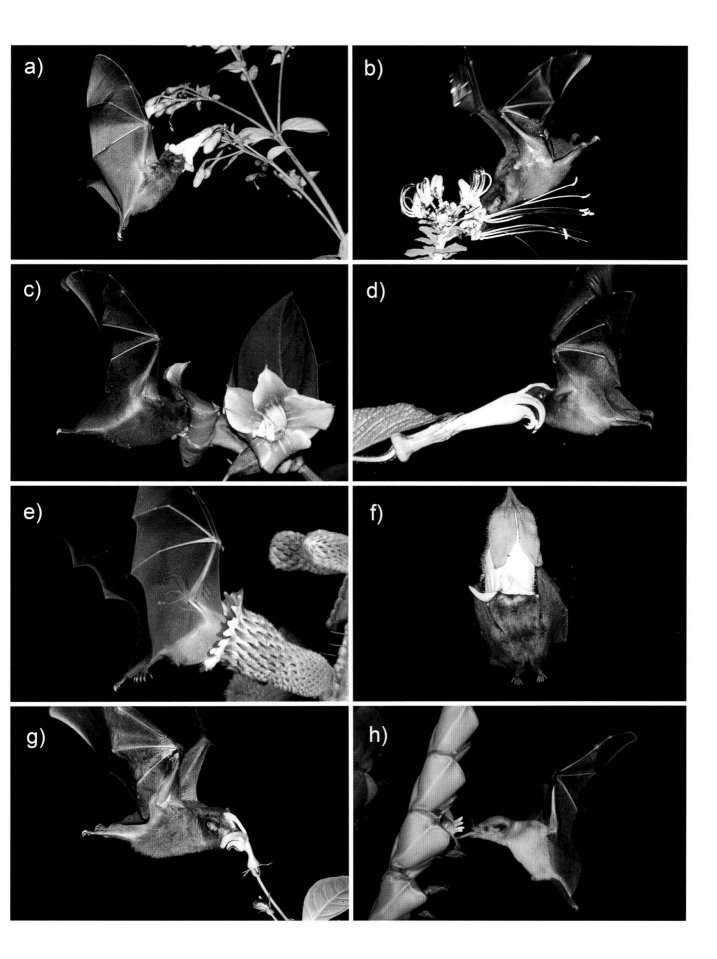

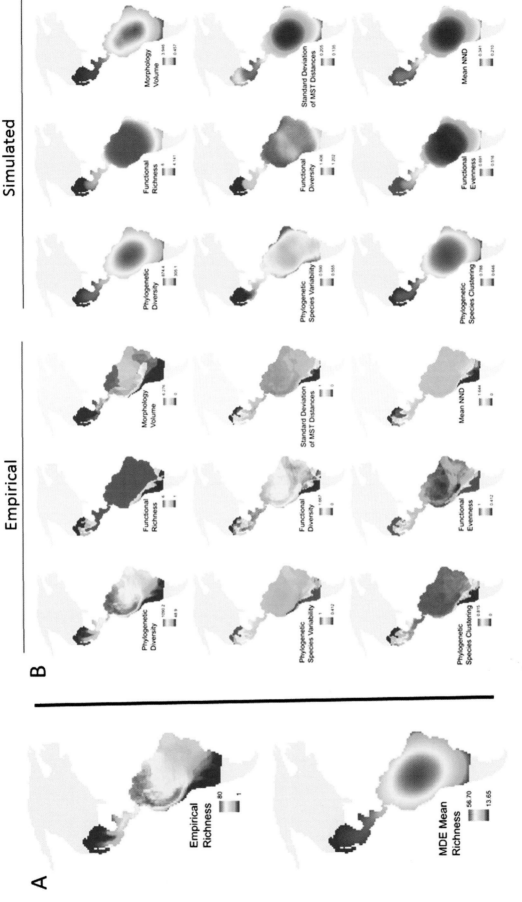

Plate 8. Empirical and simulated gradients of biodiversity of Noctilionoidea across the continental New World. Simulations produce gradients according to a mid-domain effect (MDE). (A) Gradients of species richness that are often compared to that produced by an MDE. (B) Gradients of phylogenetic, functional, and phenetic diversity that are also compared to gradients of like characteristics but determined by an MDE. The empirical gradient in species richness is not statistically different from that produced by the MDE. Nonetheless, gradients across all other dimensions of biodiversity are significantly different. The abbreviation MST refers to a minimum spanning tree, whereas NND refers to nearest-neighbor distance. Figure redrawn from Stevens et al. (2013).

Need for Comparative Studies

Many studies have demonstrated impressive rapid learning in different bat species (e.g., Clarin et al. 2013; O'Mara et al. 2014; Winter 2005), including a few documenting learning or behavioral flexibility in animalivorous phyllostomids (e.g., Page and Ryan 2005; Weinbeer and Kalko 2007). These studies generally hypothesize that the learning performance is due to some element of the animal's life history. To test these types of questions experimentally, however, it is necessary to directly compare the learning performance of animals with different life histories, which can only be done through comparative studies (Shettleworth 2013). Unfortunately, comparative cognition studies are notoriously difficult to do (but for examples in bats, see Ratcliffe et al. 2006 and Clarin et al. 2013).

Another difficulty is that comparative experiments should be conducted across multiple species focused on a putatively important trait, while taking phylogeny into account. This can be problematic when there is a high investment in time per animal in cognitive experiments, making testing multiple species time consuming and difficult. It can also be problematic to take phylogenetically independent contrasts into consideration, as foraging strategies are often largely confounded by clade (see Dávalos et al., chap. 7, this vol.; Felsenstein 1984; plate 3). Careful cross-species experiments need to be designed to test how animalivorous phyllostomids compare cognitively to other bats in the ways we outline above.

Case Studies

To characterize the attributes that define animalivorous phyllostomids as a group and to illustrate the scope of their diversity, in the case studies below we detail two species that we feel represent this group well: the fringe-lipped bat, *T. cirrhosus*, and the greater spear-nosed bat, *P. hastatus*.

Case Study 1: *Trachops cirrhosus*

We have argued that many of the generalist species in the animalivorous phyllostomid guild in fact have fairly specialized foraging strategies. This paradoxical relationship is exemplified in the fringe-lipped bat, *T. cirrhosus*. This understory gleaner has both generalist and specialist characteristics. While *T. cirrhosus* is well-known for feeding on frogs and for using frog mating calls to detect, locate, and assess its prey, it does not exclusively feed on frogs (Page and Jones 2016; Tuttle and Ryan 1981). Rather, it consumes a wide range of prey items, from birds to lizards to smaller species of bats (Bonato et al. 2004; Rodrigues et al. 2004). Small amounts of fruits are also occasionally found in its diet, but the bulk of what it feeds on is insects (Bonaccorso 1979; Bonato et al. 2004; Gardner 1977; Giannini and Kalko 2004, 2005). What is curious is not the breadth of *T. cirrhosus*'s diet, for as we have seen, broad diets are common across the animalivorous phyllostomids (plate 3 and fig. 14.1) and, indeed, can be found in other phyllostomid clades as well (an excellent example is the primarily nectarivorous *G. soricina*, that is known to also sometimes feed on insects [Clare et al. 2014]). What is striking about *T. cirrhosus* is that, despite its broad generalist diet, it possesses a number of sensory and cognitive adaptations that seem to make it highly specialized for frog predation. We discuss a few of these specializations below.

The first of these adaptations is its low-frequency hearing. *Trachops cirrhosus*, like other echolocating bats, is sensitive to the frequencies of its own echolocation calls. The echolocation calls of *T. cirrhosus* consist of multiharmonic downward frequency sweeps, ranging from approximately 100 to 50 kHz, with the main energy at about 75 kHz (Barclay et al. 1981; Surlykke et al. 2013). Sensitivity at lower frequencies has rarely been tested for in other bats, but what little work has been done suggests that bats that attend to low-frequency prey sounds actually have two peaks of auditory sensitivity: one high, in the range of their own echolocation calls, and one much lower, for attending to the lower frequency sounds of their prey (Baron et al. 1996; Reep and Bhatnagar 2000).

For *T. cirrhosus*, this second peak of auditory sensitivity is below 5 kHz, encompassing frequencies that coincide with the mating calls of its anuran prey (Ryan et al. 1983). Histological analysis reveals anatomical features consistent with its low-frequency hearing (Bruns et al. 1989). *Trachops cirrhosus* has a long basilar membrane and a large baso-apical stiffness difference, indicating sensitivity to a broad range of frequencies (Bruns et al. 1989). It also has both the highest number of cochlear neurons and the second highest cochlear neuron innervation reported for any mammal (Bruns et al. 1989). It has a peak of neural density in the apical portion of the cochlea, the portion of the cochlea sensitive to low-frequency sounds (Bruns et al. 1989; von Békésy 1960). These anatomical adaptations, enabling the detection of low-frequency sounds in the frequency range of frog mating calls, suggests that the auditory system of *T. cirrhosus* is indeed specialized for frog call detection.

For most of the over 1,300 extant bat species worldwide, hearing sensitivities and auditory physiology are still unknown. A thorough investigation of auditory sensitivities across a broad range of frequencies and wide number of eavesdropping predators would be a fruitful avenue for research. We suspect that many eavesdropping species, not just *T. cirrhosus*, have auditory systems tuned to the sounds of their prey.

Another of *T. cirrhosus*'s apparent adaptations for frog predation is its unusual submandibular salivary glands (Tandler et al. 1997). Phillips et al. (1987) conducted a study of the submandibular salivary glands of 38 genera of bats and found that only three bat species possessed submandibular salivary glands characterized by large, follicle-like structures: *T. cirrhosus*, *Megaderma lyra* (the greater false vampire bat), and *M. spasma* (the lesser false vampire bat). In a separate study, *Cardioderma cor* (the heart-nosed bat), was also found to possess these glands (Tandler et al. 1996). These four species are from three very different lineages on three different continents (*T. cirrhosus* from Latin America, *M. lyra* and *M. spasma* from Southeast Asia, and *C. cor* from East Africa). What links them is that all are frog eaters. The hypothesis is that these unusual submandibular salivary glands serve to neutralize toxins found in the skin of the frogs and toads eaten by these bats. The authors argue that this is a strong case of convergent evolution in the histology of these bats, enabling them to feed on frogs (Tandler et al. 1996).

Another hypothesized adaptation for frog eating in *T. cirrhosus* are the pronounced tubercles found on its chin and lips that give it its common name, the fringe-lipped bat. These tubercles may play a chemosensory role, allowing *T. cirrhosus* to quickly assess the palatability of its prey by lightly brushing the skin of a frog or toad (Page et al. 2012). In the dense rainforest, where numerous species of both palatable and highly poisonous anurans call, the ability to rapidly discern prey palatability would be advantageous. While behavioral studies have shown that *T. cirrhosus* does indeed

use chemical cues to reject unpalatable prey upon approach, it is still unclear what role, if any, its tubercles play in this process (Page et al. 2012).

In addition to these potential sensory and histological specializations, *T. cirrhosus* has a number of cognitive traits that seem to adapt it well to predation on frogs. For example, it is highly flexible in its ability to associate prey cues and prey quality, even to the point of reversing preexisting associations (Page and Ryan 2005). It can rapidly learn new acoustic associations by observing other bats, and it has long memories of up to four years for prey cue–prey quality associations (Dixon et al. in prep; Jones et al. 2013; Page and Ryan 2006; Patriquin et al. 2018).

In sum, despite *T. cirrhosus*'s wide diet breadth, it seems to have a number of specializations for frog predation. As such, *T. cirrhosus* exemplifies the enigma of the animalivorous phyllostomids.

Case Study 2: *Phyllostomus hastatus*

The greater spear-nosed bat, *P. hastatus*, is perhaps one of the best examples of a true phyllostomid generalist. It provides an interesting case study because, despite being one of the largest phyllostomids (>100 g), this species is infrequently carnivorous but is instead rather impressively omnivorous, feeding on a wide variety of insects, small vertebrates, fruit, flowers, nectar, and pollen (Dunn 1933; Gardner 1977; Nowak 1991; Santos et al. 2003). Diet analyses indicate that *P. hastatus* feeds on at least 28 families of insects (Giannini and Kalko 2005; Willig et al. 1993). Among plants, it feeds on over 20 tropical fruit species (Santos et al. 2003), and among vertebrates, it hunts bats, mice, and birds (Dunn 1933; Emmons and Feer 1997; Oprea et al. 2014).

Phyllostomus hastatus also differs from the other large animalivorous phyllostomids in terms of the morphological features associated with its diet. For instance, it does not have the large, rounded pinna that are disproportionally large for its body size but instead has curved, triangular ears that are allometrically proportional to its body, suggesting that it does not rely on prey-emitted sounds to find its food (Gardiner et al. 2011; Santos et al. 2003: fig. 14.2).

The extremity of *P. hastatus*'s generalist diet is also reflected in its teeth. While the other carnivores, such as *V. spectrum*, have molar patterns specialized for carnivory, *P. hastatus*'s dentition is not specialized (Freeman 1984). Additionally, *P. hastatus* has greater molar areas, which aid in crushing insect exoskeletons, than most of the primarily carnivorous phyllostomids (Freeman 1984).

Phyllostomus hastatus also has a wing aspect ratio much higher than most of the other phyllostomid gleaners that forage within the understory, making it a fast and agile flier (Norberg and Rayner 1987; Santos et al. 2003). This may facilitate flying long distances in foraging.

Cognitive adaptations associated with its diet are still largely unexplored, and we argue that *P. hastatus* provides an excellent opportunity to understand the suite of cognitive adaptations associated with being a broad dietary generalist. While dietary specialization is often thought to be associated with cognitive specializations, being a generalist may also impose a high cognitive demand on an animal (Dukas 1998; Stich and Winter 2006). Generalists must be behaviorally flexible to make decisions about the relative profitability of different possible food types (Dukas 1998). For example, the diet of *P. hastatus* changes dramatically based on season. In drier months (December to January), it eats mainly pollen and fruits, whereas from April to June its diet is characterized by insects and fruit (Wilkinson and Boughman 1998). Thus, it likely has to frequently revise its prey search image at different times of the year, based on what is seasonally available, an ecological demand that might require enhanced cognitive flexibility (Shettleworth 2010).

Additionally, we know that *P. hastatus*'s hunting strategy changes when eating plentiful resources that cannot be monopolized, such as trees full of mature balsa pollen and nectar. Female *P. hastatus* have been shown to fly in large groups and to emit group-specific screech calls to recruit roost mates to foraging sites when feeding on plentiful, highly abundant balsa flowers (Wilkinson and Boughman 1998). These calls not only recruit and coordinate foraging among group members, but they also allow feeding sites to be more effectively defended (Wilkinson and Boughman 1998). This behavior changes seasonally with diet; individuals do not emit screech calls when eating prey such as small vertebrates that are not usually shared (Boughman and Wilkinson 1998; Wilkinson and Boughman 1998).

While little is known about the cognitive abilities of *P. hastatus* in the context of foraging, we do have insights in the context of social vocal learning. Groups of unrelated females produce calls that are identical to others within the group but different between groups (Boughman 1997). When a new, unrelated adult with a different call joins, the group's calls converge until they are unified, indicating that *P. hastatus* can flexibly change its vocalizations (Boughman and Wilkinson 1998), which likely helps facilitate group foraging behavior. This form of vocal learning has rarely been seen in other mammals (Knörnschild 2014).

In sum, *P. hastatus* is a broad generalist. While other species in this group include small amounts of plant matter in their diets (e.g., *Lampronycteris brachyotis* [Weinbeer and Kalko 2004], *Micronycteris megalotis* [Whitaker and Findley 1980], *Tonatia bidens* [Felix et al. 2013], *T. cirrhosus* [Gardner 1977]; plate 3), the diet of *P. hastatus* consists primarily of plants during certain months of the year (Wilkinson and Boughman 1998). It has ear, wing, and teeth morphologies that differ from those characteristic of most animalivorous phyllostomids, reflecting differences in foraging behavior. While much work remains to be done to fully understand *P. hastatus*, this species is well-studied in comparison with the other *Phyllostomus* species: *P. elongatus*, *P. latifolius*, and *P. discolor*. For these other species, little is known beyond the most basic dietary records (Gardner 1977). Like *P. hastatus*, other species of *Phyllostomus* are broadly omnivorous, with diets that consist of fruit, nectar, pollen, and insects (Gardner 1977; plate 3). To fully understand the *Phyllostomus* species group and how it fits into animalivorous phyllostomids specifically, and the Neotropical bat community more generally, detailed studies of the morphology, foraging behavior, cognitive flexibility, and social structures of this genus are needed.

Ecological Implications

All other aspects being equal, dietary generalists are thought to have lower extinction risk than specialists in response to habitat loss and climate change (Devictor et al. 2008). Since specialist animals have a limited number of food options, if these are disturbed, then specialists will not have alternatives to fall back on. Generalists, in contrast, can shift their diets adaptively depending on what is available and are less likely to be tied to the success of any one food source (Devictor et al. 2008). Where animalivorous phyllostomids fall on this continuum, however, is unclear.

Predictions for other bat species have been mixed. Boyles and Storm (2007) found evidence in temperate vespertilionids in North America, Australia, and Europe that a narrower dietary breadth is correlated with higher IUCN concern levels. Conversely, Safi and Kerth (2004) found that diet specialization (as determined by fecal analysis) and extinction risk (as determined by IUCN Red Listing) are not correlated in temperate-zone bats. Instead, they found that extinction risk is predicted by wing morphology, with bats with rounder and broader wings more likely to be at risk. This suggests that habitat specialization is more critical than diet specialization, with bats that fly in and around foliage being at the most risk (Safi and Kerth 2004). Since most phyllostomid bats fly within the forest, this high-risk category is likely to be relevant to all phyllostomids (Kalko et al. 1996).

Roost specialization may also be more critical than diet specialization for determining extinction risks. For animals that roost in trees, loss of roost trees in degraded habitats may impose severe limits on the number of bats or even number of species that can persist in a patch of forest. Voss et al. (2016) argue that loss of roosts may be more important than reduced food for some bats, especially animalivores, during tropical habitat degradation (see chap. 19, this vol.).

Diet specialization has also not been a major predictor of sensitivity in the few studies on habitat fragmentation in phyllostomids. One such study compared the dispersal abilities and demographic response to fragmentation between mainland phyllostomids (*Uroderma bilobatum* and *Carollia perspicillata*) to those living in fragmented islands formed by the creation of the Panama Canal and did not find that dietary specialization was important (Meyer et al. 2009). Rather, decline in frequency in fragmented areas was linked to sensitivity to edge effects and, to a lesser degree, to natural abundance and mobility. One reason why diet may not be a good predictor of success in response to habitat change is that even most chiropteran "specialists" are more generalized than specialists in other orders (Fleming 1986; Meyer et al. 2008). Most bats considered to be specialists actually consume multiple species, which means that they may survive even if some of the species that they consume decline.

Animalivorous bats, particularly those in the former Phyllostominae, tend to decline in abundance in response to habitat fragmentation (Fenton et al. 1992; Gorresen and Willig 2004; Medellín et al. 2000; Meyer et al. 2008; but see Klingbeil and Willig 2009). These bats are more sensitive to deforestation than other guilds, which often reach highest species richness at intermediate levels of disturbance (Gorresen and Willig 2004; Klingbeil and Willig 2009). The broad wings that help these species maneuver in dense forests and capture prey on surfaces may reduce long-range mobility, especially across open spaces with higher predation risk (Fenton 1990; Safi and Kerth 2004). Thus, animalivorous phyllostomids may have a harder time crossing fragmented landscapes. This is thought to make gleaning animalivores highly edge sensitive (Meyer et al. 2008). Additionally, most animalivorous phyllostomids rely on tree cavities to roost in and are therefore affected by deforestation (Voss et al. 2016). Phyllostomine bats have been proposed to be good indicator species for habitat disturbance and should be the focus of conservation efforts (Fenton et al. 1992; Medellín et al. 2000).

Conclusions

Animalivorous phyllostomids are a fascinating and diverse group of bats. In some respects, this group consists of the species that remain when all the other phyllostomids are neatly categorized (e.g., this vol.: sanguivores [chap. 15], nectarivores [chap. 16], frugivores [chap. 17]). We argue that this eclectic group of leftovers can be characterized as a single diverse guild (following Giannini and Kalko 2005). The animalivorous phyllostomids are united by their broad, generalist diets, which reflect a continuum of increasing prey size as bat body size increases (fig. 14.1). Likewise, there is a trend across these species for carnivory gradually replacing insectivory as body size increases (fig. 14.1). Animalivorous phyllostomids are also united by the apparent paradox of having broad generalist diets, on the one hand, and often quite specialized foraging behaviors, on the other. As such, this group offers fascinating avenues for research. Indeed, some of the most exciting discoveries in bat biology have come from this group: bats that hunt both birds and other bats, bats that learn socially from conspecifics, bats that can perceive silent, motionless prey from clutter, and predatory bats that eavesdrop on prey mating signals (Boughman 1997;

Geipel, Jung et al. 2013; Jones et al. 2013; Page and Ryan 2006; Tuttle and Ryan 1981; Vehrencamp et al. 1977).

Animalivorous phyllostomids consume a broad range of animals, from moths to mammals, as well as plant material, to varying degrees. While the width of diet breadth across animalivorous phyllostomids is striking, we also see this trend in other phyllostomid guilds. The nectivorous bat, *G. soricina*, for example, eats beetles, flies, and moths, along with nectar, pollen, and fruit (Clare et al. 2014). This sort of opportunistic omnivory may be common among phyllostomids. There is likely to be a continuum of generalists and specialists within every guild.

Historically, categorizing the animalivorous phyllostomids has been difficult, and discrepancies among categorization systems have led to conflicting analyses on key behavioral and morphological characteristics, such as brain size, bite force, and learning abilities (e.g., Giannini and Kalko 2005; Jones et al. 2004; Nogueira et al. 2009; Rojas et al. 2013). While any generalization for this group will have exceptions, we hope future authors can agree upon a more consistent system to avoid this conflict. Our recommendation, as argued in this chapter, is that all animalivorous phyllostomids be considered together as one broad guild (following Giannini and Kalko 2005), which in turn can be subdivided into multiple distinct foraging strategies, such as passive gleaners (e.g., *T. cirrhosus, L. silvicolum, T. saurophila, C. auritus, V. spectrum*), active gleaners (e.g., *M. microtis*), "true" generalists (e.g., *P. hastatus, P. discolor*), and trawlers (*M. macrophyllum*). Crucially, animalivorous phyllostomids should not be grouped with aerial insectivores from other families (see chap. 2, this vol.). While both groups are technically animalivorous, consuming vast quantities of insect prey, animalivorous phyllostomids are distinct from aerial insectivores in almost every component of their ecology: behaviorally, morphologically, and otherwise.

The hallmark of phyllostomid bats is their dietary diversity. While the diets of some phyllostomids have been very well-studied (e.g., *Artibeus jamaicensis* [Ortega and Carto-Arellano 2001], *C. perspicillata* [Cloutier and Thomas 1992], *D. rotundus* [Greenhall and Schmidt 1988], and *G. soricina* [Clare et al. 2014]), for many of the animalivorous phyllostomids we have only the most basic dietary records (e.g., *Lonchorhina orinocensis, Lophostoma schulzi, Glyphonycteris daviesi,*

Gardnernycteris cozumelae, Phylloderma stenops, P. latifolius). The more we begin to fill in dietary information about the animalivorous phyllostomids, the more the trends of broad, generalist diets become apparent (e.g., *T. bidens*, which is now known to feed broadly across prey categories, including insects, frogs, geckos, birds, and bats [Esbérard and Bergallo 2004], plate 3). Foundational knowledge of diet is critical to inform more detailed studies to come. The next building blocks toward robust understanding of these animals are studies of sensory ecology, social structure, and cognition, which are currently sparse. Once the basic biology of more species has been well-characterized we will be able to conduct richer comparative studies. Currently, only a tiny subset of the animalivorous phyllostomids, such as the case studies we highlight here (*T. cirrhosus* and *P. hastatus*), have been studied thoroughly, and even for these bats, we are missing important aspects of their natural histories. Given the amount of diversity uncovered with just a small amount of study, what more might be out there to be discovered?

Acknowledgments

We thank the Smithsonian Tropical Research Institute for facilitating decades of Neotropical bat research and for their support during the writing of this chapter. MD was additionally supported by a National Science Foundation Graduate Research Fellowship. We are grateful to Damond Kyllo for his illustrations, and to Marco Tschapka and Merlin Tuttle for their photos. We thank Ted Fleming, Inga Geipel, John Ratcliffe, and Marco Tschapka for their helpful comments on drafts of this manuscript. We dedicate this chapter to the memory of Elisabeth Kalko, whose visionary work with phyllostomids continues to inspire Neotropical research.

References

Aguirre, L. F., A. Herrel, R. Van Damme, and E. Matthysen. 2003. The implications for food hardness for diet in bats. Functional Ecology, 17:201–212.

Aldridge, H. D. J. N., and I. L. Rautenback. 1987. Morphology, echolocation, and resource portioning in insectivorous bats. Journal of Animal Ecology, 56:763–778.

Arbour, J. H., A. A. Curtis, S. E. Santana. 2019. Signatures of echolocation and dietary ecology in the adaptive evolution of skull shape in bats. Nature Communications, 10:2036.

Arita, H. T. 1993. Rarity in Neotropical bats: correlations with phylogeny, diet, and body mass. Ecological Applications, 3:507–517.

Balda, R. P., and A. Kamil. 2006. Linking life zones, life history traits, ecology, and spatial cognition in four allopatric southwestern seed caching corvids. Papers in Behavior and Biological Sciences, 36:1–32.

Barclay, R. M. R., M. B. Fenton, M. D. Tuttle, and M. J. Ryan.1981. Echolocation calls produced by *Trachops cirrhosus* (Chiroptera: Phyllostomatidae) while hunting for frogs. Canadian Journal of Zoology, 59:750–753.

Baron, G., H. Stephan, and H. D. Frahm. 1996. Comparative Neurobiology in Chiroptera. Birkhäuser Verlag, Basel.

Belwood, J. J., and G. K. Morris. 1987. Bat predation and its influence on calling behavior in Neotropical katydids. Science, 238:64–67.

Bonaccorso F., 1979. Foraging and reproductive ecology in a community of bats in Panama. Bulletin of the Florida State Museum of Biological Sciences 24:359–408.

Bonato, V., K. G. Facure, and W. Uieda. 2004. Food habits of bats of subfamily vampyrinae in Brazil. Journal of Mammalogy, 85:708–713.

Boughman, J. W. 1997. Greater spear-nosed bats give group-distinctive calls. Behavioral Ecology and Sociobiology, 40:61:70.

Boughman, J. W., and G. S. Wilkinson. 1998. Greater spear-nosed bats discriminate group mates by vocalizations. Animal Behaviour, 55:1717–1732.

Boyles, J. G., and J. J. Storm. 2007. The perils of picky eating: dietary breadth is related to extinction risk in insectivorous bats. PLOS ONE, 2:e672.

Bruns, V., H. Burda, and M. J. Ryan. 1989. Ear morphology of the frog-eating bat (*Trachops cirrhosus*, family: Phyllostomidae): apparent specializations for low-frequency hearing. Journal of Morphology, 199:103–118.

Burns, J. G., J. Foucaud, and F. Mery. 2010. Costs of memory: lessons from "mini" brains. Proceedings of the Royal Society B–Biological Sciences, 278:923–929.

Carter, G. G., J. M. Ratcliffe, and B. Galef. 2010. Flower bats (*Glossophaga soricina*) and fruit bats (*Carollia perspicillata*) rely on spatial cues over shapes and scents when relocating food. PLOS ONE, 5:e10808.

Chittka, L., and N. E. Raine. 2006. Recognition of flowers by pollinators. Current Opinion in Plant Biology, 9:428–435.

Clare, E. L., H. R. Goerlitz, V. A. Drapeau, M. W. Holderied, A. M. Adams, J. Nagel, E. R. Dumont, P. D. N. Herbert, and B. Fenton. 2014. Tropic niche flexibility in *Glossophaga soricina*: how a nectar seeker sneaks an insect snack. Functional Ecology, 28:632–641.

Clarin, T. M. A., I. Ruczyński, R. A. Page, and B. M. Siemers. 2013. Foraging ecology predicts learning performance in insectivorous bats. PLOS ONE, 8:e64823.

Cloutier, D., and D. W. Thomas. 1992. *Carollia perspicillata*. Mammalian Species, 417:1–9.

Coles, R. B., A. Guppy, M. E. Anderson, and P. Schlegel. 1989. Frequency sensitivity and directional hearing in the gleaning bat, *Plecotus auritus* (Linnaeus 1758). Journal of Comparative Physiology A, 165:269–280.

Dechmann, D. K. N., K. Safi, and M. J. Vonhof. 2006. Matching morphology and diet in the disc-winged bat *Thyroptera tricolor* (Chiroptera). Journal of Mammalogy, 87:1013–1019.

Denzinger A., and H.-U. Schnitzler. 2013. Bat guilds, a concept to classify the highly diverse foraging and echolocation behaviors of microchiropteran bats. Frontiers in Physiology, 4:164.

Devictor, V., R. Julliard, and F. Jiguet. 2008. Distribution of specialist and generalist species along spatial gradients of habitat disturbance and fragmentation. Oikos, 117:507–514.

Dixon, M. M., P. L. Jones, S. Meneses, and R. A. Page. In prep. Frog-eating bats remember an associative learning task after four years in the wild.

Dukas, R. 1998. Evolutionary ecology of learning. Pp. 129–174 *in*: Cognitive Ecology: The Evolutionary Ecology of Information Processing and Decision Making (R. Dukas, ed.). University of Chicago Press, Chicago.

Dukas, R. 1999. Costs of memory: ideas and predictions. Journal of Theoretical Biology, 197:41–50.

Dumont, E. R. 1997. Cranial shape in fruit, nectar, and exudate feeders: implications for interpreting the fossil record. American Journal of Physical Anthropology, 102:187–202.

Dunn, L. W. 1933. Observations on the carnivorous habits of the spear-nosed bat, *Phyllostomus hastatus panamensis Allen*, in Panama. Journal of Mammalogy, 14:188–199.

Emmons, L. H., and F. Feer. 1997. Neotropical Rainforest Mammals: A Field Guide. University of Chicago Press, Chicago.

Esbérard, C. E. L., and H. G. Bergallo. 2004. Aspectos sobre a biologia de *Tonatia bidens* (Spix) no estado do Rio de Janeiro, sudeste do Brasil (Mammalia, Chiroptera, Phyllostomidae). Revista Brasileira de Zoologia, 21:253–259.

Falk, J. J., H. M. ter Hofstede, P. L. Jones, M. M. Dixon, P. A. Faure, E. K. V. Kalko, and R. A. Page. 2015. Sensory-based niche partitioning in a multiple predator–multiple prey community. Proceedings of the Royal Society B–Biological Sciences, 282:20150520. DOI: 10.1098/rspb.2015.0520.

Feldhamer, G. A., L. C. Drickhamer, S. H. Vessey, J. F. Merritt, and C. Krajewski. 2015. Mammalogy: Adaptation, Diversity, Ecology. 4th ed. Johns Hopkins University Press, Baltimore, MD.

Felix, S., R. L. M. Novaes, R. F. Souza, and R. T. Santori. 2013. Diet of *Tonatia bidens* (Chiroptera, Phyllostomidae) in an Atlantic Forest area, southeastern Brazil: first evidence for frugivory. Mammalia, 77:451–454. DOI:10.1515/mammalia-2012–0117.

Felsenstein, J. 1984. Distance methods for inferring phylogenies: a justification. Evolution, 38:16–24.

Fenton, M. B. 1972. The structure of aerial-feeding bat faunas as indicated by ears and wing elements. Canadian Journal of Zoology, 50:287–296.

Fenton, M. B. 1990. The foraging behaviour and ecology of animal-eating bats. Canadian Journal of Zoology, 68:411–422.

Fenton, M. B., L. Acharya, D. Audet, M. B. C. Hickey, C. Merriman, M. K. Obrist, D. M. Syme, and B. Adkins. 1992. Phyllostomid bats (Chiroptera: Phyllostomidae) as indicators of habitat disruption in the Neotropics. Biotropica, 24: 440–446.

Fenton, M. B., and W. Bogdanowicz. 2002. Relationship between external morphology and foraging behavior: bats in the genus *Myotis*. Canadian Journal Zoology, 80:1004–1013.

Fiedler, J. 1979. Prey catching with and without echolocation in the Indian False Vampire (*Megaderma lyra*). Behavioral Ecology and Sociobiology, 6:155–160.

Fleming, T. H. 1986. Opportunism versus specialization: the evolution of feeding strategies in frugivorous bats. Pp. 105–118 *in*: Frugivores and Seed Dispersal (A. Estrada and T. H. Fleming, eds.). Springer, Dordrecht.

Freeman, P. W. 1984. Functional cranial analysis of large animal-eating bats (Microchiroptera). Biological Journal of the Linnean Society, 21:387–408.

Freeman, P. W. 2000. Macroevolution in Microchiroptera: re-coupling morphology and ecology with phylogeny. Evolutionary Ecology Research, 2:317–333.

García-Morales, R., E. I. Badano, and C. E. Moreno. 2013. Response of Neotropical bat assemblages to human land use. Conservation Biology, 27:1096–1106.

Gardiner, J. D., J. R. Codd, and R. L. Nudds. 2011. An association between ear and tail morphologies of bats and their foraging style. Canadian Journal of Zoology, 89:90–99.

Gardner, A. L. 1977. Feeding habits. Pp. 293–350 *in*: Biology of Bats of the New World Family Phyllostomatidae, pt. 2 (R. J. Baker, J. K. Jones Jr., and D. C. Carter, eds.). Special Publications of the Museum Texas Tech University, no. 13. Texas Tech Press, Lubbock.

Geipel, I., K. Jung, and E. K. V. Kalko. 2013. Perception of silent and motionless prey on vegetation by echolocation in the gleaning bat *Micronycteris microtis*. Proceedings of the Royal Society B–Biological Sciences, 280:1754.

Geipel, I., K. K. V. Kalko, K. Wallmeyer, and M. Knörnschild. 2013. Postweaning maternal food provisioning in a bat with a complex hunting strategy. Animal Behaviour, 85:1435–1441.

Giannini, N. P., and E. K. V. Kalko. 2004. Trophic structure in a large assemblage of phyllostomid bats in Panama. Oikos, 105:209–220.

Giannini, N. P., and E. K. V. Kalko. 2005. The guild structure of animalivorous leaf-nosed bats of Barro Colorado Island, Panama, revisited. Acta Chiroterologica, 7:131–146.

Gorresen, P. M., and M. R. Willig. 2004. Landscape responses of bats to habitat fragmentation in Atlantic Forest of Paraguay. Journal of Mammalogy, 85:688–697.

Gould, E. 1976. Echolocation and communication. Pp. 247–292 *in*: Biology of the New World Family Phyllostomatidae, pt. 2 (R. J. Baker, J. K. Jones, Jr., and D. C. Carter, eds.). Special Publications of Museum Texas Tech University, no. 13. Texas Tech Press, Lubbock.

Gracheva, E. O., J. F. Cordero-Morales, J. A. Gonzalez-Carcacia, N. T. Ingolia, C. Manno, C. I. Aranguren, J. S. Weissman, and D. Julius. 2011. Ganglion-specific splicing of TRPV1 underlies infrared sensation in vampire bats. Nature, 476:88–91.

Greenhall, A. M., and U. Schmidt. 1988. Natural History of Vampire Bats. CRC Press, Boca Raton, FL.

Hulgard, K., and J. M. Ratcliffe. 2014. Niche-specific cognitive strategies: object memory interferes with spatial memory in the predatory bat *Myotis nattereri*. Journal of Experimental Biology, 217:3293–3300.

Humphrey S. R., F. J. Bonaccorso, and T. L. Zinn. 1983. Guild

structure of surface-gleaning bats in Panama. Ecology, 64:284–294.

Jones, K. E., O. R. P. Bininda-Emonds, and J. L. Gittleman. 2005. Bats, clocks, and rocks: diversification in patterns in Chiroptera. Evolution, 59:2243–2255.

Jones, K. E., A. M. MacLarnon, and A. E. M. Pagel. 2004. Affording larger brains: testing hypotheses of mammalian brain evolution on bats. American Naturalist, 164:20–31.

Jones, P. L., R. A. Page, and J. M. Ratcliffe. 2016. To scream or to listen? Prey detection and discrimination in animal-eating bats. Pp. 93–116 *in*: Bat Bioacoustics (M. B. Fenton, A. D. Grinnell, A. N. Popper, and R. R. Fay, eds.). Springer, New York.

Jones, P. L., M. J. Ryan, V. Flores, and R. A. Page. 2013. When to approach novel prey cues? Social learning strategies in frog-eating bats. Proceedings of the Royal Society B–Biological Sciences, 280:20132330.

Kalka, M., and E. K. V. Kalko. 2006. Gleaning bats as under-estimated predators of herbivorous insects: diet of *Micronycteris microtis* (Phyllostomidae) in Panama. Journal of Tropical Ecology, 22:1–10.

Kalko, E. K. V., C. O. Handley, and D. Handley. 1996. Organization, diversity, and long-term dynamics of a Neotropical bat community. Pp. 503–553 *in*: Long-Term Studies in Vertebrate Communities (M. L. Cody and J. A. Smallwood, eds.). Academic Press, San Diego.

Klingbeil, B. T., and M. R. Willig. 2009. Guild-specific responses of bats to landscape composition and configuration in fragmented Amazonian rainforest. Journal of Applied Ecology, 46:203–213.

Knörnschild, M. 2014. Vocal production learning in bats. Current Opinions in Neurobiology, 80. DOI:10.1016/j.conb.2014.06 .014.

Kunz, T. H., E. Braun de Torrez, D. Bauer, T. Lobova, and T. H. Fleming. 2011. Ecosystem services provided by bats. Annals of the New York Academy of Sciences, 1223:1–38.

Kurten, L., and U. Schmidt. 1982. Thermoperception in the common vampire bat (*Desmodus rotundus*). Journal of Comparative Physiology A, 146:223–228.

Lachmann, M., G. Sella, and E. Jablonka. 2000. On the advantages of information sharing. Proceedings of the Royal Society of London B–Biological Sciences, 267:1287–1293.

Maina, J. N. 2000. What it takes to fly: the structural and functional respiratory refinements in birds and bats. Journal of Experimental Biology, 203:3045–3064.

Medellín, R. A. 1988. Prey of *Chrotopterus auritas*, with notes on feeding behavior. Journal of Mammalogy, 69:841–844.

Medellín, R. A., M. Equihua, and M. A. Amin. 2000. Bat diversity and abundance as indicators of disturbance in Neotropical rainforests. Conservation Biology, 14:1666–1675.

Meyer, C. F. J., J. Fründ, W. P. Lizano, and E. K. V. Kalko. 2008. Ecological correlates of vulnerability to fragmentation in Neotropical bats. Journal of Applied Ecology, 45:381–391.

Meyer, C. F. J., E. K. V. Kalko, and G. Kerth. 2009. Small-scale fragmentation effects on local genetic diversity in two phyllostomid bats with different dispersal abilities in Panama. Biotropica, 41:95–102.

Muchhala, N. 2006. Nectar bat stows huge tongue in rib cage. Nature, 444:701–702.

Navarro, D. L., and D. E. Wilson. 1982. *Vampyrum spectrum.* Mammalian Species 184:1–4.

Neuweiler, G. 1983. Echolocation and adaptivity to ecological constraints. Pp. 280–302 *in*: Neuroethology and Behavioural Physiology (H. Markl and F. Huber, eds.). Springer-Verlag, Berlin.

Neuweiler, G. 1989. Foraging ecology and audition in echolocating bats. Trends in Ecology and Evolution, 4:160–166.

Neuweiler, G., S. Singh, and K. Sripathi. 1984. Audiograms of a south Indian bat community. Journal of Comparative Physiology A, 154:133–142.

Nogueira, M. R., A. L. Peracchi, and L. R. Monteiro. 2009. Morphological correlates of bite force and diet in the skull and mandible of phyllostomid bats. Functional Ecology, 23:715–723.

Norberg, U. M. 1976. Aerodynamics, kinematics, and energetics of horizontal flapping flight inthe long-eared bat *Plecotus auritus.* Journal of Experimental Biology, 65:179–212.

Norberg, U. M., and M. B. Fenton. 1988. Carnivorous bats? Biological Journal of the Linnean Society, 33:383–394.

Norberg, U. M., and J. M. V. Rayner. 1987. Ecological morphology and flight in bats (Mammalia; Chiroptera): wing adaptations, flight performance, foraging strategy and echolocation. Philosophical Transactions of the Royal Society of London B–Biological Sciences, 316:335–427.

Nowak, R. M. 1991. Walker's Mammals of the World. Johns Hopkins University Press, Baltimore, MD.

Obrist, M., M. B. Fenton, J. L. Eger, and P. Schlegel. 1993. What ears do for bats: a comparative study of pinna sound pressure transformation in Chiroptera. Journal of Experimental Biology, 180:119–152.

O'Mara, M. T., D. K. N. Dechmann, and R. A. Page. 2014. Frugivorous bats evaluate the quality of social information when choosing novel foods. Behavioral Ecology, 25:1233–1239.

Oprea, M. T. B. Vieira, V. T. Pimenta, P. Mendes, D. Brito, A. D. Ditchfield, L. V. Knegt, and C. E. L. Esbérard. 2014. Bat predation by *Phyllostomus hastatus*. Chiroptera Neotropical, 12:255–8.

Ortega, J., and I. Carto-Arellano. 2001. *Artibeus jamaicensis.* Mammalian Species, 662:1–9.

Page, R. A., and P. L. Jones. 2016. Overcoming sensory uncertainty: factors affecting foraging decisions in frog-eating bats. Pp. 285–312 *in*: Perception and Cognition in Animal Communication (M. A. Bee and C. T. Miller, eds.). Springer, New York.

Page, R. A., and M. J. Ryan. 2005. Flexibility in assessment of prey cues: frog-eating bats and frog calls. Proceedings of the Royal Society B–Biological Sciences, 272:841–847.

Page, R. A., and M. J. Ryan. 2006. Social transmission of novel foraging behavior in bats: frog calls and their referents. Current Biology, 16:1201–1205.

Page, R. A., T. Schnelle, E. K. V. Kalko, T. Bunge, and X. E. Bernal. 2012. Sequential assessment of prey through the use of multiple sensory cues by an eavesdropping bat. Naturwissenschaften, 99:505–509.

Patriquin, K. J., J. Kohles, R. A. Page, and J. R. Ratcliffe. 2018. Bats without borders: predators learn novel prey cues from other predatory species. Science Advances, 4:eaaq0579.

Phillips, C. J., B. Tandler, and C. A. Pinkstaff. 1987. Unique salivary glands in two genera of tropical microchiropteran bats an example of evolutionary convergence in histology and histochemistry. Journal of Mammalogy, 68:235–242.

Ratcliffe, J. M., M. B. Fenton, and S. J. Shettleworth. 2006. Behavioral flexibility positively correlated with relative brain volume in predatory bats. Brain, Behavior and Evolution, 67:165–176.

Reep, R. L., and K. P. Bhatnagar. 2000. Brain ontogeny and ecomorphology in bats. Pp. 93–136 *in*: Ontogeny, Functional Ecology, and Evolution of Bats (R. A. Adams and S. C. Pedersen, eds.). Cambridge University Press, Cambridge.

Ripperger, S., D. Josic, M. Hierold, A. Koelpin, R. Weigel, M. Hartmann, R. A. Page, and F. Mayer. 2016. Automated proximity sensing in small vertebrates: design of miniaturized sensor nodes and first field tests in bats. Ecology and Evolution, 6:2179–2189.

Rodrigues, F. H. G., M. L. Reis, and V. S. Braz. 2004. Food habits of the frog-eating bat, *Trachops cirrhosus*, in Atlantic Forest of northeastern Brazil. Chiroptera Neotropical, 10:180–182.

Rojas, D., C. A. Mancina, J. J. Flores-Martínez, and L. Navarro. 2013. Phylogenetic signal, feeding behaviour and brain volume in Neotropical bats. Journal of Evolutionary Biology, 26:1925–1933.

Rojas, D., O. M. Warsi, and L. M. Dávalos. 2016. Bats (Chiroptera: Noctilionoidea) challenge a recent origin of extant Neotropical diversity. Systematic Biology, 65:432–448.

Rosenzweig, M. 1968. The strategy of body size in mammalian carnivores. American Midland Naturalist, 80:299–315.

Ross, G., and M. W. Holderied. 2013. Learning and memory in bats: a case study on object discrimination in flower-visiting bats. Pp. 207–224 *in*: Bat Evolution, Ecology, and Conservation (R. A. Adams and S. C. Pedersen, eds.). Springer, New York.

Ryan, M. J., M. D. Tuttle, R. M. R. Barclay. 1983. Behavioral responses of the frog-eating bat, *Trachops cirrhosus*, to sonic frequencies. Journal of Comparative Physiology, 150: 413–418.

Safi, K., and D. K. Dechmann. 2005. Adaptation of brain regions to habitat complexity: a comparative analysis in bats (Chiroptera). Proceedings of the Royal Society of London B–Biological Sciences, 272:179–186.

Safi, K., and G. Kerth. 2004. A comparative analysis of specialization and extinction risk in temperate-zone bats. Conservation Biology, 18:1293–1303.

Santana, S. E., and E. Cheung. 2016. Go big or go fish: morphological specializations in carnivorous bats. Proceedings of the Royal Society B, 283:20160615.

Santana, S. E., and E. R. Dumont. 2009. Connecting behavior and performance: the evolution of biting behavior and bite performance in bats. Journal of Evolutionary Biology, 22:2131–2145.

Santana, S. E., I. Geipel, E. R. Dumont, M. B. Kalka, and E. K. V. Kalko. 2011. All you can eat: high performance capacity and plasticity in the common big-eared bat, *Micronycteris microtis* (Chiroptera: Phyllostomidae). PLOS ONE, 6:1–7.

Santos M., L. F. Aguirre, L. B. Vázquez, and J. Ortega. 2003. *Phyllostomus hastatus*. Mammalian Species, 722:1–6.

Schnitzler, H. U., and E. K. V. Kalko. 1998. How echolocating bats search and find food. Pp. 183–196 in: Bat Biology and Conservation (T. H. Kunz and P. A. Racey, eds.). Smithsonian Institution Press, Washington, DC.

Schöner, M. G., C. R. Schöner, R. Simon, T. U. Grafe, S. J. Puechmaille, L. L. Ji, and G. Kerth. 2015. Bats are acoustically attracted to mutualistic carnivorous plants. Current Biology, 25:1911–1916.

Shettleworth, S. J. 2010. Cognition, Evolution, and Behavior. 2nd ed. Oxford University Press, New York.

Shettleworth, S. J. 2013. Fundamentals of Comparative Cognition. 1st ed. Oxford University Press, New York.

Stich, K. P., and Y. Winter. 2006. Lack of generalization of object discrimination between spatial contexts by a bat. Journal of Experimental Biology, 209:4802–4808.

Stockwell, E. F. 2001. Morphology and flight maneuverability in New World leaf-nosed bats (Chiroptera: Phyllostomidae). Journal of Zoology London, 254:505–514.

Surlykke, A., L. Jakobsen, E. K. V. Kalko, and R. A. Page. 2013. Echolocation intensity and directionality of perching and flying fringe-lipped bats, Trachops cirrhosus (Phyllostomidae). Frontiers in Physiology, 4:143.

Tandler, B., T. Nagato, and C. J. Phillips. 1997. Ultrastructure of the unusual accessory submandibular gland in the fringe-lipped bat, Trachops cirrhosus. Anatomical Record, 248:164–175.

Tandler, B., C. J. Phillips, and T. Nagato. 1996. Histological convergent evolution of the accessory submandibular glands in four species of frog-eating bats. European Journal of Morphology, 34:163–168.

ter Hofstede, H., and J. M. Ratcliffe. 2016. Evolutionary escalation: the bat-moth arms race. Journal of Experimental Biology, 219:1589–1602.

ter Hofstede, H., S. Voigt-Heucke, A. Lang, H. Römer, R. Page, P. Faure, and D. Dechmann. 2017. Revisiting adaptations of Neotropical katydids (Orthoptera: Tettigonidae) to gleaning bat predation. Neotropical Biodiversity, 3:41–49.

Thornton, A., and T. Clutton-Brock. 2011. Social learning and the development of individual and group behaviour in mammal societies. Philosophical Transactions of the Royal Society B–Biological Sciences, 366:978–987.

Tuttle, M. D., and M. J. Ryan. 1981. Bat predation and the evolution of frog vocalizations in the Neotropics. Science, 214:677–678.

Tuttle M. D., M. J. Ryan, and J. J. Belwood. 1985. Acoustical resource partitioning by two species of Phyllostomid bats (Trachops cirrhosus and Tonatia sylvicola). Animal Behaviour, 33:1369–1371.

Vandoros, J. D., and E. R. Dumont. 2004. Use of the wings in manipulative and suspensory behaviors during feeding by frugivorous bats. Journal of Experimental Zoology A, 301A:361–366.

Vehrencamp, S. L., F. G. Stiles, and J. W. Bradbury. 1977. Observations on the foraging behavior and avian prey of the Neotropical carnivorous bat, Vampyrum spectrum. Journal of Mammalogy, 58:469–478.

von Békésy, G. 1960. Experiments in Hearing. McGraw-Hill, New York.

von Helversen, D., and O. von Helversen. 1999. Acoustic guide in bat-pollinated flower. Nature, 398:759–760.

Voss, R. S., D. W. Fleck, R. E. Strauss, P. M. Velazco, and N. B. Simmons. 2016. Roosting ecology of Amazonian bats: evidence for guild structure in hyperdiverse mammalian communities. American Museum Novitates, 3870:1–44.

Weinbeer, M., and E. K. V. Kalko. 2004. Morphological characteristics predict alternate foraging strategy and microhabitat selection in the orange-bellied bat, Lampronycteris brachyotis. Journal of Mammalogy, 85:1116–1123.

Weinbeer, M., and E. K. V. Kalko. 2007. Ecological niche and phylogeny: the highly complex echolocation behavior of the trawling long-legged bat, Macrophyllum macrophyllum. Behavioral Ecology and Sociobiology, 61:1337–1348.

Whitaker, J. O., and J. S. Findley. 1980. Foods eaten by some bats from Costa Rica and Panama. Journal of Mammalogy, 61:540–544.

Wilkinson, G. S., and J. M. Boughman. 1998. Social calls coordinate foraging in greater spear-nosed bats. Animal Behavior, 55:337–350.

Willig, M. R., G. R. Camilo, and S. J. Noble. 1993. Dietary overlap in frugivorous and insectivorous bats from Edaphic Cerrado habitats of Brazil. Journal of Mammalogy, 74:117–128.

Winter, Y. 2005. Foraging in a complex naturalistic environment: capacity of spatial working memory in flower bats. Journal of Experimental Biology, 208, 539–548.

15

John W. Hermanson
and Gerald G. Carter

Vampire Bats

Introduction

The dramatic adaptive radiation of phyllostomids led to the invasion of a novel niche for mammals: obligate parasitic blood feeding. There are three extant vampire bats that feed solely on blood: the common vampire bat *Desmodus rotundus*, its sister taxon the white-winged vampire bat *Diaemus youngi*, and the hairy-legged vampire bat *Diphylla ecaudata*. *Desmodus* is by far the most common and best studied. All three species primarily parasitize livestock and poultry. *Desmodus* feeds largely on mammalian blood, while *Diaemus* and especially *Diphylla* show preferences and adaptations for parasitizing birds.

The evolutionary origin of blood feeding has been the subject of much speculation (reviewed by Schutt 1998). Fossils give little insight into the origin of this clade because extinct vampire species are similar to *D. rotundus*. The extinct *Desmodus* include the 30% larger *D. draculae* from the Late Pleistocene and Holocene in Argentina, Venezuela, Brazil, and the Yucatan, Mexico; the 10–20% larger *D. stocki* from the Late Pleistocene and Holocene found as far north as Shasta County, California, and West Virginia, USA; and the similarly sized *D. praecursor* (or *D. archaeodaptes*) from the Early Pleistocene in Florida (Czaplewski et al. 2003; Czaplewski and Cartelle 1998; Morgan et al. 1988; Ray et al. 1988). No extinct species of *Diaemus* or *Diphylla* have yet been described.

Regardless of the evolutionary path taken to obligate sanguivory, the invasion of this dietary niche required adaptations to overcome key challenges, with far-reaching consequences for virtually all other aspects of vampire bat biology and ecology. In this chapter, we highlight adaptive changes to anatomy, physiology, behavior, and cognition linked to this unique sanguivorous lifestyle. We begin by considering the act of blood feeding itself.

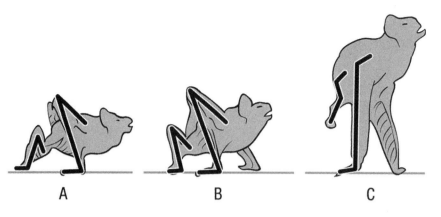

Figure 15.1. A flight-initiating jump sequence 8 ms from start to finish. The initial crouching posture of *Desmodus rotundus* in preparation (*A*). The bat's abdomen and thorax are held well above the substrate with its weight borne directly on the thumb pads and dorsum of the carpus. The orientation of the pollex (thumb) is directed laterally away from the sagittal plane. The bat proceeds through a sequence of adduction of the wings to propel the bat vertically (*B*). After the hindlimbs have become airborne, the thumbs remain in contact with the substrate and will provide a final vertically directed force finishing the jump (*C*). Figure modified from Schutt et al. (1997).

Feeding Behavior

Typical feeding by *Desmodus* proceeds as follows. First, the bat approaches the bite site on its host by walking on the ground or along the animal's body. Similar to blood-feeding insects, vampire bats can feed most efficiently on sedentary hosts. Landing directly on the animal can awaken it or incur swatting or biting, so an alternative approach is along the underside of a branch (especially common for *Diaemus* and *Diphylla*) or on the ground (especially common for *Desmodus*; Altenbach 1988; Greenhall 1972, 1988; Greenhall et al. 1969, 1971). *Desmodus* are unique among phyllostomids in their proficiency at walking and hopping on the ground (Altenbach 1979, 1988; Riskin and Hermanson 2005; Schutt et al. 1997), allowing them to follow hosts on the ground without taking flight or to escape quickly if detected by their hosts, many of which are several orders of magnitude larger in mass.

Jumping

To avoid being bitten or crushed by their host, *Desmodus* is capable of remarkably fast jumps away (Altenbach 1979) or into flight (Schutt et al. 1997). Using well-developed thoracic wing muscles, *Desmodus* are able to launch into a jump with a vertical force impulse of 0.06 N and peak vertical force of 9.5 times their body weight (Schutt et al. 1997). In under 30 ms, the bat's jump can accelerate it to a velocity of 2.4 m/s, which is near their horizontal flight speed (Altenbach 1979). *Desmodus* thumbs are key launching points and are unique among bats for their relatively large size and for possessing two metacarpal pads (Altenbach 1979). After the pectoralis muscle initiates the jump, elbow ex-

tension powered by the M. triceps brachii continues the push-off of the limb against the substrate, and a ventral flexion of the thumb provides a final push at the end of the jump (Schutt et al. 1997). Two thumb muscles, the M. abductor pollicis brevis and the M. flexor pollicis brevis, are well positioned to flex the thumb during a jump and to keep the digits flexed while walking (Altenbach 1979) (fig. 15.1).

Walking and Bounding

In addition to jumping, *Desmodus* shows a clear divergence from the pattern of poor walking and running seen in almost all other observed bats (Riskin et al. 2016). Many poorly walking bats hold the wings in an abducted and extended position when on the ground, essentially flapping as they ambulate on the ground (Riskin et al. 2016). In contrast, a walking or running *Desmodus* keeps the wings folded and holds the contact points with the ground (the carpus and thumbs) close to the sagittal plane of the body, which creates an upright posture similar to that of standing mammals and unlike that of sprawling reptiles. When *Desmodus* increases its walking velocity and transitions to a hopping gait (0.3–1 m/s), the thoracic limbs drive the motion while the hind limbs provide mere stability (Riskin and Hermanson 2005), similar to crutch walking in humans or knuckle walking in some primates (Tuttle 1975). Because the ancestors of vampire bats could not run, this bounding ability in vampire bats represents the novel evolution of running in bats. *Desmodus* can maintain a stride frequency that is lower than that of running mice, perhaps due to the elongate structure of the wings (Riskin and Hermanson 2005). Recent work attempting to quantify the metabolic cost of running suggests a relatively low oxygen consumption rate compared to other non-volant mammals (unpublished data, Hermanson et al. 2016) and this work is still ongoing.

Diaemus and *Diphylla* possess only a single metacarpal pad on the thumb and are not capable of the same degree of dexterous terrestrial locomotion seen in *Desmodus*. Specifically, *Diaemus* does not exhibit the running gait seen in *Desmodus* and is not capable of flight-initiating jumps at the levels seen in *Desmodus* (Schutt et al. 1997). *Diaemus* and *Diphylla* are, however, skilled at climbing on branches (Greenhall 1988; Schutt et al. 1997). *Diphylla* is unique in having a modified calcar that functions like a sixth digit while grasping branches (Schutt and Altenbach 1997). *Cheiromeles torquatus* (Molossidae) has an opposable hallux on the foot that also appears to facilitate grasping of branches and functions as a mechanism for tucking the distal wing elements into an axial pouch while performing terrestrial and arboreal locomotion (Schutt and Simmons, 2001).

A few non-vampire bats are also adept on the ground, such as the New Zealand short-tailed bat (*Mystacina tuberculata*, Mystacinidae), which feeds on myriad prey in the leaf litter of the forest (Daniel 1979). Although *Mystacina* has specialized morphology for scurrying in pursuit of terrestrial prey, it lacks the coordinated and efficient running gait of *Desmodus* (Riskin et al. 2006). The elbow and associated musculature of *Mystacina* is quite different from that of the desmodontines, which suggests that walking gaits evolved independently in these two groups (Hand et al. 2009). Riskin et al. (2016) found that *Pternotus parnellii* (Momoopidae) walked poorly and *Natalus tumidirostrus* (Natalidae) never walked or crawled in their experiments but could take flight from the ground with ease. Molossid bats are often cited as having an ability to scamper on the ground, for example, in *Cheiromeles* spp. (Schutt and Simmons 2001). Several non-vampire phyllostomid species, *Artibeus jamaicensis*, *Artibeus lituratus*, and *Carollia perspicilla* have shown little to no walking ability (JWH, pers. obs.).

Preparing the Bite

Scurrying about near prey and investigating possible bite sites can take anywhere from a few minutes to a half hour (Greenhall 1972). Using unique heat receptors on its noseleaf, *Desmodus* can detect infrared radiation as low as 5×10^{-5} Wcm^{-2} allowing detection of a warm-blooded bite site at a distance of 13–16 cm (Gracheva et al. 2011; Kürten and Schmidt 1982). *Desmodus* often appears to sniff the area before biting, a behavior that might indicate olfaction or heat perception. Next, *Desmodus* cautiously touches the site with vibrissae on its noseleaf that are linked to sensitive mechanoreceptors (Kürten 1985; Schmidt 1988b). Together with the jumping abilities described above, such tactile sensitivity allows the bat to feed above a large animal's hoof and sense the earliest instance of movement of the hoof so as to be able to quickly leap away (Schmidt 1988b). If necessary, the bat will take up to 40 min to prepare the bite site by shaving off hair or feathers, but this is unnecessary for reopening an old wound or for sites without thick fur such as on the flank and shoulders of some tropical breeds of cattle (Greenhall 1972).

Biting

After briefly licking the site, the bat uses its lower incisors to secure the skin while the razor-sharp and concave upper incisors cut a roughly 3-mm diamond-shaped crater in the skin. During this typical bite, the incisor tips remove a divot of skin, but there are other feeding bites that are rarely used, including a scratching bite with the upper incisors and even a rasping wound with the tongue (Greenhall 1972).

Dentition

The unique and specialized design of vampire bat teeth is a good illustration of the tight coupling of form and function (Davis et al. 2010). There is no need to chew, but acquiring adequate blood with minimal disturbance requires an effective bite. The dental formulae are reduced (*Diphylla*: i2/2, c1/1, p1/2, m2/2 X2 = 26; *Diaemus*: i1/2 c1/1 p1/2, m2/1 X2 = 22, *Desmodus*: i1/2, c1/1, p1/2, m1/1 X2 = 20; Koopman, 1988). The incisors are specialized as the primary wound openers, while the molariform teeth are reduced in size and number relative to other phyllostomids (fig. 15.2).

The very large incisors and elongate canine teeth require extra room within the oral cavity. For the incisors, prominent depressions, termed "mandibular pits" by Davis et al. (2010), penetrate into the bony structure of the mandible on the ventral aspect of the oral cavity, imme-

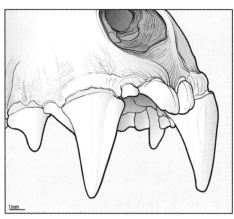 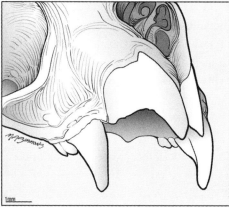

Figure 15.2. Oblique view of the upper rostral dentition of omnivore *Phyllostomus hastatus* (*left*) and the vampire *Desmodus rotundus* (*right*) showing divergence of the incisor teeth in terms of number, sharpness, breadth, and shape. The canine tooth of *P. hastatus* is large and conical, which is fairly representative of other bats and mammals. The canine of the vampire is smaller but extremely sharp-edged and nonconical. While the canine of *P. hastatus* may be suited to puncturing and crushing food items, the canine of *D. rotundus* is well-suited to slicing, a function exhibited even more by the upper incisors of the vampire.

diately caudal to the lower incisor teeth (on the lingual side). The upper canines fit into a space on the lateral side of the mandible immediately caudal to the lower canines. The mandibular pits are described as being most extreme in *Diphylla*, requiring divergent placement of the roots of the lower incisor teeth but allowing for secure fit of the relatively vertical upper incisors (which are less procumbent than seen in *Desmodus* or *Diaemus*) when the mouth is closed (Davis et al. 2010) (fig. 15.3).

In rodents and artiodactyls that grind fibrous foods, the interface of enamel with dentin is what creates a sharp cutting edge. Vampire bat incisors and canines, however, have been described as lacking enamel (Greenhall 1988; Phillips and Steinberg 1976). Microscopic examination reveals a razor-sharp edge in both young and old. The bats were thought to self-sharpen the teeth by scraping the superior and inferior teeth together (Greenhall 1988) or via abrasion using the rough tip of the tongue (Vierhaus 1983). The notion that vampire bats lack enamel was challenged by Bhatnagar and Wimsatt (1988) using hematoxylin and eosin staining and microscopy and by Lester et al. (1988) using scanning microscopy. More work is required to determine the composition of vampire bat teeth: do they have full enamel, very little, or none?

The supposed dearth of enamel begs a simple question. If the teeth are not constantly growing, does the lack or diminished amount of enamel limit the longevity of wild vampire bats? In the wild, where bats must bite to obtain blood, the longevity record is for a male *Desmodus* who was at least 17 years old (from a mark-recapture study with more than 13,000 captures; Delpietro et al. 2017). This old bat had missing and deteriorated canine and incisor teeth. In captivity, where bats

do not suffer predation risk but also do not use their teeth to feed, there are two records of *Desmodus* older than 29 years (unpublished data, GGC). The specialization of the upper incisors and the canines for sharpness invites further analysis regarding the tooth composition and how this sharp edge is maintained throughout the life of a vampire bat.

The bite force of *Desmodus* is relatively low compared to phyllostomid bats of comparable body mass. Two *Desmodus* with a mean body mass of 40.9 g demonstrated a mean maximum bite force of 8.60 N (Aguirre et al. 2002). For perspective, eight *Phyllostomus discolor* (Phyllostomidae), an omnivore with a mean body mass of 36.7 g, showed a mean maximum bite force of 21.6 N. The relatively low bite force of *Desmodus* is still effective because of the sharpness of the upper incisors.

Processing Blood

To drink, *Desmodus* extends its tongue through a space between its lower incisors along a groove in the lower lip. This diastema between the lower incisors is best developed in *Desmodus* and *Diaemus* and is not present in *Diphylla* (Davis et al. 2010). Unsurprisingly, the tongue is specialized for lapping blood (Griffiths and Criley 1989; Phillips et al.1986) and possibly grooming (Tandler et al. 1997). The ventrolateral surface of the tongue in *Desmodus* and *Diaemus* possess longitudinal grooves, which are absent in most bats but similar nonhomologous grooves are found in the nectar-feeding lonchophylline phyllostomids (Griffiths and Criley 1989). These grooves on the tongue's surface may facilitate the inflow of blood via capillary action. Near the

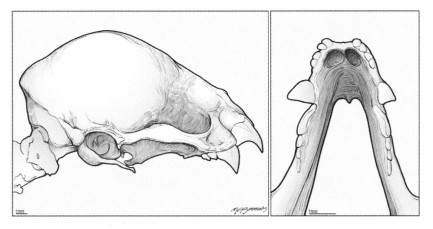

Figure 15.3. Lateral view of the skull and dorsal view of the rostral mandible of *Desmodus rotundus* (Phyllostomidae). Note the procumbent form of the upper incisors. These teeth are so sharp that there needs to be consideration of how to avoid self-inflicted wounds while handling the skull. Much of the crown of the incisor is exposed, as the bony boundary between the crown of the tooth and the premaxillary (incisive) bone is somewhat dorsally and caudally placed. The cutting edge of the incisor has an upside-down "V-shape," like a large serration on a bread knife. The upper canine is oriented more vertically than the incisor and has a sharpened caudal edge. On the mandible, the mandibular pits appear as depression on the mandible caudal to lower incisors 1 and 2. The upper incisors fit into these depressions when the mouth is closed.

tongue tip, there are large spiked papillae, which might sharpen the incisors (Vierhaus 1983) or be used in the tongue rasping bite (Greenhall 1972), described above.

To prevent the coagulation of blood while feeding, the saliva of *Desmodus* contains several components, including anticoagulant compounds (draculin [Apitz-Castro et al. 1995] and desmolaris [Ma et al. 2013]) and plasminogen activator to break down blood clots—first called desmokinase (Hawkey 1966) and later *Desmodus* salivary plasminogen activator (DSPA) and desmoteplase—which is now being tested for treatment of stroke (Lindsberg and Caso 2016; Medcalf 2012). The plasminogen activator compounds of *Desmodus* and *Diaemus* are specialized for mammalian and avian blood, respectively (Cartwright and Hawkey 1969).

In a single feeding, wild common vampire bats can consume roughly 15–20 ml of blood, about half of their body mass; less active captive bats consume less, 12–20 ml (Schmidt 1978; Wimsatt 1969; Wimsatt and Guerriere 1962). The digestive system of the vampire bat is highly specialized in its form and simplicity (Harlow and Braun 1997; Morton 1986). The morphology of the stomach is quite divergent from the simple stomach of mammals in that it has a large tubular extension, termed a "cecum." This part of the stomach extends out from the side of the stomach's body and may have extremely reduced areas of glandular distribution along the body and pylorus (Forman et al. 1979). This "cecum" is not to be confused with the cecum associated with the transition between the ileum and ascending colon of most domestic mammals (see Dyce et al. 2010). Ingested blood goes straight to the intestine and then overflows back into the stomach and this expansive cecal region (Mitchell and Tigner 1970). The shapes of the stomachs and the "cecal extensions" of the three vampire bats are quite different: it is tubular and smooth in *Desmodus*; it is obtund and enlarged in diameter away from the connection with the stomach in *Diaemus*; and it is highly pocketed in *Diphylla* (Forman et al. 1979). These morphologic specializations allow for rapid ingestion of the blood meal, and storage of blood in the cecum might facilitate regurgitation of blood to offspring (see below).

The renal system is also specialized (Busch 1988; Horst 1968; McFarland and Wimsatt 1969). Blood is about 90% water, and vampire bats rapidly expel this water while or after feeding to facilitate flight. After feeding has ended, the rate of water absorption from blood roughly equals the rate of urine production (Morton and Richards 1981). A vampire bat may drink for 8–40 min, but the entire process of preparing, biting, and feeding can take up to 2 h, depending on whether the host is asleep or awake (Greenhall 1972; Wimsatt 1969). Like many other bats, *Desmodus* may rest, hanging on a tree branch immediately after feeding, presumably to digest and urinate before returning to its roost (Goodwin and Greenhall 1961; Schmidt 1978). In the first few hours after eating, roughly 25% of the blood weight is passed in urine; increased excretion and concentrated urination continues for up to 20 h (Wimsatt and Guerriere 1962).

Social Behaviors While Feeding

Given the difficulty of making a bite compared to the ease of drinking from an open wound, some individuals save energy and reduce risk by exploiting the bites already made by others. For example, during 3 h of observation, seven bats were seen drinking from a single bite, including two at the same time (Greenhall et al. 1971; Sazima 1978). *Desmodus* will also defend wounds from conspecifics with agonistic behavior including screeching, hair bristling, lunging, and pushing by thumbs and wings (Schmidt 1978; Schmidt and van de Flierdt 1973).

When threatened, *Diaemus* will sometimes spray an odorous mist from a pair of 2–3 mm spherical glands, one in each corner of its open mouth (Goodwin and Greenhall 1961). These glands are exposed, and perhaps inflated, accompanied by loud hissing. This multimodal display occurs in response to handling by people and in agonistic encounters with conspecifics (Greenhall 1988, Schutt et al. 1999). The structure of these unique glands and the composition of the spray have not been studied in detail.

Loss of a Cognitive Trait

There is evidence that adaptive specializations for blood feeding extend even to cognitive traits that influence how animals learn. Taste aversion learning is the association of a novel flavor with gastrointestinal sickness based on a single experience and is a specialized form of learning found across diverse animal taxa, including many vertebrates, insects, and molluscs. The loss of taste aversion learning was predicted for vampire bats because blood from a live animal never spoils or becomes toxic. Indeed, *Desmodus* is the only known mammal to lack taste aversion learning (Ratcliffe et al. 2003).

Differences in Feeding by Diaemus and Diphylla

The feeding behavior of *Diaemus* is similar to that of *Desmodus* except that the bats will typically climb along the underside of a branch and feed from the exposed toes of birds, whereas when *Desmodus* feeds on birds, it more often walks along the top of branches as if on the ground (Greenhall 1988; Sazima and Uieda 1980). Although *Diaemus* is less agile than *Desmodus* on the ground, it will also feed from the ground or crawl on the animal itself, but chickens on the ground are less likely to be preyed upon by *Diaemus* than those off the ground (Greenhall 1988; Sazima and Uieda 1980).

To our knowledge, *Diphylla* has never been observed feeding from the ground and is specialized for climbing on branches (Schutt 1998). The calcar, which typically supports the tail membrane, is virtually absent in *Diaemus* and *Desmodus*, but the small calcar of *Diphylla* appears to function as an opposable thumb-like sixth digit to grasp branches (Schutt 1998; Schutt and Altenbach 1997). When feeding on chickens, *Diphylla* often attaches itself to the feathers near the cloaca (Hoyt and Altenbach 1981). When nestled into the feathers, it can be difficult to see the bat's body. Based on the morphology of the face, teeth, mouth, and tongue of *Diphylla*, Greenhall (1988) speculated that this species, unlike *Diaemus* and *Desmodus*, might literally suck blood from the wound; however, footage of *Diphylla* feeding from a chicken cloaca revealed the same basic licking method as seen in the other two vampires (Carter 2016).

Host Preferences

Desmodus has long been considered to prefer mammalian hosts while *Diaemus* and *Diphylla* were considered bird specialists. However, increasing evidence suggests that these preferences might vary by region. *Desmodus* feeds primarily on herds of cattle in agricultural sites throughout its range, but observations show that *Desmodus* will readily parasitize a wide variety of wild and domestic hosts, including horse, goat, chicken, dog, sea lion, pelican, cormorant, capybara, tapir, llama, armadillo, porcupine, rabbit, small rodent, crocodile, turtle, iguana, boa constrictor, coral snake, rattlesnake, rat snake, penguin, falcon, and human (Greenhall 1988; Luna-Jorquera and Culik 1995; Muela et al. 2011). Only a few hundred years ago, vampire bats switched from feeding on wildlife to almost exclusively feeding on the domestic livestock introduced by Europeans. Given early written reports by conquistadors of vampire bats parasitizing both horses and humans, it is likely that *Desmodus* often parasitized native people in villages and cities before the proliferation of livestock (Turner 1975). Today, feeding on humans is more likely in sites where livestock have been removed or in remote regions where livestock are absent and native wildlife populations have been reduced through deforestation and hunting (Caraballo 1996; Stoner-Duncan et al. 2014).

Greenhall (1970) developed the use of a precipitin test to identify hosts from stomach contents of *Desmo-*

dus. His analysis of 3,500 samples showed the following order of host preference in Trinidad: cattle, water buffalo, horses, mules, donkeys, goats, pigs, chicken, sheep, dogs, and people. In Mexico, 500 samples showed a preference of *Desmodus* for cattle, and one sample identified a squirrel (Greenhall 1988).

More sophisticated, efficient, and noninvasive testing of host preference is possible through amplifying and sequencing host DNA from feces (Bobrowiec et al. 2015; Carter et al. 2006). When applied to *Desmodus* and *Diaemus* caught in riverside villages in the Brazilian Amazon, chickens were both the most available and most attacked hosts, but *Desmodus* most preferred to prey on pigs, with no evidence of feeding on wildlife (Bobrowiec et al. 2015). Another method for assessing diet is stable isotope analysis, which has revealed a similar strong preference of *Desmodus* for livestock in some areas but also for wildlife hosts in areas with fewer livestock, such as sea lions along coastal habitats (Streicker and Allgeier 2016; Voigt and Kelm 2006).

Diaemus appears to favor domestic chickens where available but will also feed on mammals. Out of 23 *Diaemus* sampled in Trinidad, 13 fed on mammals, 8 fed on both mammals and birds, and 2 fed on birds (Greenhall 1988). In captivity, they preferred blood of goats, donkeys, and guinea pigs over that of cattle, but they will feed on cattle blood in captivity and in the wild. When tested for their ability to feed on various wild birds, *Diaemus* was able to successfully parasitize birds ranging in mass from 15 to 200 g, including doves, parrots, woodpeckers, woodcreepers, manakins, tyrant flycatchers, mockingbirds, thrushes, orioles, tanagers, and finches. The exception was a 7 g trochilid hummingbird, which was not approached due to the perch site being inaccessible (Greenhall 1988).

Compared to *Diaemus*, *Diphylla* is considered to have an even stronger preference for birds (Greenhall 1988). It has refused to feed on rat, rabbit, or bovine blood (Hoyt and Altenbach 1981), but there are a few reports of wild bats feeding on blood of mammals (Piccinini et al. 1991), including humans (Ito et al. 2016).

Digestive Specializations to Host-Specific Blood

Two adaptive traits mentioned above, locomotion and antihemostatic saliva, show specialization to mammals for *Desmodus* and to birds for *Diaemus* and *Diphylla*. In addition, there is evidence that the digestive systems of *Desmodus* and *Diaemus* are adapted to processing mammalian and avian blood, respectively. Coen (2002) fed captive *Desmodus* on chicken, bovine, or pig blood for week-long periods in digestion chambers and then measured the responses in survival, body mass, consumption, and nutrient retention before and after switching bats of each species from their preferred diet (bovine blood for *Desmodus* and chicken blood for *Diaemus*) to their nonpreferred diet. Chicken blood, compared with pig and bovine blood, is roughly 3 times lower in dry matter, 5 times lower in nitrogen, and 1.5 times lower in iron, but 2.5 times higher in sodium, and 2–6 times higher in phosphorus. *Desmodus* sometimes consumed twice as much of the more dilute chicken blood, drinking more than the equivalent of its own body mass of chicken blood (35–44 g), but only half its mass of pig or bovine blood (14–19 g). When *Diaemus* was switched from an avian diet to a mammalian diet, these bats likewise decreased their mean consumption from 51 to 27 g/day. Despite consuming more avian blood, both species retained less dry matter on their nonpreferred diet, and each species remained more efficient at retaining nutrients on its preferred diet. Moreover, 25% of the *Desmodus* fed on only avian blood died within 7 days. On the first day that *Diaemus* were switched to bovine blood, they did not reduce their intake sufficiently, which caused swelling of the abdomen beyond what is typically observed, because the food had congealed into a gelatinous mass. This abnormal condition lasted 7–11 days and seemed uncomfortable to the bats (Coen 2002). Hence, despite an ability to feed from a diversity of prey, these results suggest that each species might not be capable of sustained feeding on nonpreferred hosts.

Finding Prey

Very little is known about how vampire bats locate prey animals, but it is likely that they use spatial memory and a combination of several long-range senses, including passive listening (Heffner et al. 2013, 2015; Schmidt et al. 1991), echolocation (Joermann 1984; Schmidt and Schmidt 1977), vision (Schmidt 1988b), and olfaction (Bahlman and Kelt 2007). The use of vision by vampire bats was demonstrated by trials where flight paths of both *P. discolor* and *D. rotundus* were influenced by the visual illusion of a rewarded landing grid, although both species detected the illusion at 20–40 cm using echolocation (Joermann et al. 1988). Another experiment compared *Desmodus* and *Phyllostomus* on the relative salience of sensory cues by presenting rewarded combinations of stimuli that were acoustic (2–38 kHz broadband noise), visual (6 cm-diameter illuminated plate), and olfactory (grass odor) and showed that *Desmodus* used acoustic cues over visual ones, while *Phyllostomus* did the opposite in the same test (Schmidt et al. 1988). The role of olfaction in vampire bat prey selection is unclear. Hungry *Desmodus* are attracted to the smell of cattle (Bahlman and Kelt 2007) but are not lured to cattle with cattle blood applied to the cattle fur (Turner 1975).

Echolocation is relatively typical compared with other phyllostomids; however, their low-frequency hearing is more than an octave lower than the three other phyllostomids for which audiograms are available: *Phyllostomus hastatus*, *A. jamaicensis*, and *C. perspicillata* (Heffner et al. 2013). At 60 dB, *Desmodus* can hear sounds as low as 716 Hz, which shows better low-frequency hearing than the Virginia opossum and the nonecholocating flying fox *Eidolon helvum* (Heffner et al. 2013). Some authors have suggested, based on neurophysiology and psychoacoustics, that *Desmodus* uses passive low-frequency hearing to recognize the breathing sounds of different individual hosts (Gröger and Wiegrebe 2006; Schmidt et al. 1991). This intriguing hypothesis requires further study.

Foraging Behavior

The overall time spent outside the roosts each night varies from 0.5 to 4 h (Ripperger and Carter, unpublished data; Turner 1975; Wilkinson 1985a; Wimsatt 1969; Young 1971). Wilkinson (1985a, 1988) found that females spent on average 2 h outside the roost, while resident males spent consistently less time (1.5 h) outside. These estimates approximate the 2-h period that Greenhall (1972) observed is needed for a bat to feed, suggesting that these bats did not take much time to find their livestock hosts. At some sites, *Desmodus* appear to use rivers to commute to feeding sites, and cattle closer to rivers are bitten more frequently (Turner 1975). *Desmodus* are light phobic and reduce their flight activity when there is bright moonlight (Turner 1975). Heavy rain also reduces flight activity (Wimsatt 1969). When Young (1971) released marked bats at various distances from their roost to test homing ability, all 72

bats returned to their roost after being released 0.5–6 miles away: 6 of 12 bats flew back to their roost after being moved 10 miles; and none of 12 bats returned when displaced 18 miles.

The flight of vampire bats might be influenced by some design trade-offs with their extraordinary terrestrial locomotion. From takeoff, *Desmodus* quickly achieve forward velocity of about 2–2.5 m/s and accelerate to about 5 m/s during observed short flights (Altenbach 1979). Longer-distance flight speed is about 13.5 km/h (range: 7–27 km/h; *n* = 47; Sanchez-Hernandez et al. 2006). The relatively broad wings of vampire bats are suitable for lift production during slow but not highly maneuverable flight (Norberg and Rayner 1987). The wing stroke is not particularly different from that of other bats, and the neuromuscular control of the stroke is comparable to that reported in other species (Altenbach 1979; Altenbach and Hermanson 1987; Hermanson and Altenbach 1985). Vampire bats often fly low; 82% of 1,864 *Desmodus* captured in mist nets in Argentina were entangled in the bottom half of the net (Delpietro et al. 2017), an observation in agreement with our collective field experience in Belize, Costa Rica, Panama, and Trinidad.

Evidence for Social Foraging

Some authors report *Desmodus* arriving in small groups of two to six individuals flying near the ground and circling cattle (Crespo et al. 1974; Greenhall et al. 1969; Schmidt 1978). Results from multiple-year field tracking studies provide some evidence that experienced females may lead their volant offspring to prey sites (Wilkinson 1985a, 1988). Although Wilkinson (1985a) did not detect a correlation between co-roosting and foraging range overlap, he noted that recent female offspring shared 29% of their foraging ranges with their mothers, but only 4% with other members of their social group (Wilkinson 1985a, 1988). In contrast, none of the adult males showed any overlap in foraging range. Resident males always flew to the nearest pasture to feed and returned to their roost afterwards. Females from different social groups both roosted and foraged in different areas, but pairs of adult females spent more time in the same areas compared with pairs of resident males (Wilkinson 1985a). Radio-tracking data suggested that *Desmodus* flew an average of 3.4 km, which was roughly twice the round-trip distance necessary to find and feed upon prey (Wilkinson 1985a). This suggests that not all flights are for foraging; some likely serve social purposes. Improvements in tracking technology, including GPS (Cvikel et al. 2015) and proximity loggers (Ripperger et al. 2016), will undoubtedly shed new insight into vampire bat foraging.

Social Behavior

Desmodus females demonstrate fission-fusion social dynamics, where a larger social group consists of smaller subgroups that break apart and recombine in a nonrandom fashion. Fission-fusion dynamics have been observed in many forest-dwelling bats and allow individuals to maintain long-term relationships despite frequent roost switching (Carter and Wilkinson 2013a; Kerth et al. 2011). Wimsatt (1969) reported frequent roost switching by vampire bats among caves and tunnels in Mexico, and Schmidt (1978) noted the existence of stable social associations. These patterns were later confirmed and described in detail for *Desmodus* living in tree hollows in Costa Rica (Wilkinson 1985a). Here, females typically visited one to three other roosts shared by their social groups, and roost-switching decisions appeared to depend on the other females roosting there. Females maintained associations while switching among tree hollows that are each defended by a resident dominant male. As is typical in resource defense polygyny, these males defend territories from other males, and vicious fights among males to gain access to female groups can result in injuries (Wilkinson 1985a, 1988). Subordinate males might reside in the same hollow below the group, and nonresident males might roost in other trees alone or in small bachelor groups (Wilkinson 1985a). Nonresident males visit female roosts to seek mating opportunities. Wilkinson (1988) found that nonresident males often flew several kilometers (probably more than 10 km) in a single night along rivers in the direction of other roosts (Wilkinson 1988). At mist nets set in front of roosts in Belize and Panama at dusk, more males were caught entering the roost from the outside than males or females exiting (GGC, pers. obs.). In Argentina, more males, but not females, were observed leaving and entering maternity roosts by night than the number that remained inside the same roosts by day, and females were occasionally observed visiting bachelor roosts (Delpietro et al. 2017).

Larger stable groups with more than 50 bats usually contain smaller subgroups of 8 to 20 adult females with young (Wilkinson 1988). Large cave aggregations can exceed 2,000 individuals (Wilkinson 1988), and some studies report that vampire bats do not appear to switch roosts when living in caves or man-made structures (Delpietro et al. 2017; Streicker et al. 2012; Young 1971). It is therefore possible that the bats are switching among smaller roost sites within the cave. Studies tracking the roosting locations of individuals within larger cave roosts would clarify this issue.

Females typically stay in their natal social group, and one unrelated female joins the group on average every two years (Wilkinson 1984, 1985a). Despite female philopatry, average group relatedness is low, with estimates varying from 0.02 to 0.11 depending on the estimation method (Wilkinson 1984, 1985b, 1987, 1988). About 50% of adult pairs in a group have a kinship coefficient (r) of 0–0.05; about 40% have $r = 0.15$–0.25; and about 10% of pairs are mother-offspring pairs where r is 0.5 (Wilkinson 1984). Relatedness is low in part because the tenure of alpha males typically lasts less than 17 months: their mean maximum paternity is only 46%; and their aggression towards other males causes related young males to disperse (Wilkinson 1985a, 1985b).

Reproduction

Vampire bats possess a life-history strategy of high investment and slow but continuous reproduction. The gestation period of *Desmodus* has been estimated to be 165–220 days (Delpietro and Russo 2002; Schmidt 1978, 1988a), and sexual maturity can take up to 285 days (Delpietro and Russo 2002). Based on palpable pregnancy, the gestation period of *Diphylla* seems about 14 days shorter than that of *Desmodus* (Delpietro and Russo 2002). Female bats become sexually mature after their first year (Wilkinson 1985b), and a female can then produce one new offspring about every 10 months. There is often no strict reproductive season or synchrony (Schmidt 1988a), but in Argentina, there is clear reproductive seasonality (Delpietro et al. 2017).

Desmodus pups are born with open eyes and large feet and grow quickly from about ~6 g at birth to ~12 g in about 3 weeks, and ~24 g in 3 months (Schmidt 1988a). In captivity, mothers will carry their pups around while nursing them for 2 months, when the pups weigh roughly half their mother's mass. Although the bats are flying independently at 4 months, weaning occurs at 9–10 months of age (Delpietro and Russo 2002, Schmidt 1988a). Mothers gradually switch pups from milk to regurgitated blood. Pups are fed with regurgitated blood primarily by their mother through their first year, then increasingly by other groupmates, especially maternal kin (Carter and Wilkinson 2013c, 2015a; Schmidt 1988a; Wilkinson 1984). Mothers have been seen to regurgitate food to a pup just moments after birth (Wilkinson 1988). Bats as young as 4.5 months will regurgitate food to feed their unfed mother (Carter and Wilkinson 2013c). For female pups, maternal care extends into an enduring mother-daughter relationship that can last many years (see details of cooperation behavior below).

Vampire bat juveniles depend on their mothers for longer than any other bats. For example, in comparison with a weaning period of up to 10 months in *Desmodus*, the typical lactation period for both temperate and Neotropical vespertilionids is under 2 months (Jenness and Studier 1976). For instance, *Eptesicus fuscus* are fully weaned at 35 days and volant as early as 18 days (Hamilton and Barclay 1998); *Myotis lucifugus* are volant at 20–26 days (Powers et al. 1991); *Pipistrellus subflavus* are volant at about 21 days (Hoying and Kunz 1998). The molossid *Tadarida brasiliensis* is weaned and volant at about 42 days (Kunz and Robson 1995). Phyllostomids such as *Artibeus watsoni, C. perspicillata, Macrotus californicus, Leptonycteris* are also weaned at 1–2 months (Chaverri and Kunz 2006; Jenness and Studier 1976). Weaning is a bit longer, at 2 months for *Artibeus jamaicensis* and 3 months for *Phyllostomus discolor* (Kwiecinski et al. 2003).

Cooperation

Although not seen in field studies (Wilkinson 1988), allonursing has been observed in captivity in *Diphylla* (Delpietro and Russo 2002), *Diaemus* (GGC, pers. obs.), and *Desmodus* (Schmidt 1988a). One orphaned *Desmodus* pup was adopted by a non-lactating female that began lactating after a few days and raised the adopted bat successfully (Schmidt 1988a).

Vampire bats share food via regurgitation. *Desmodus* demonstrates a reduced ability to store energy reserves, such as liver glycogen or lipids, and are especially susceptible to starvation, showing severely low blood glucose and eventually death after two to three nights of fasting (Freitas et al. 2003, 2005, 2013). This ease of

starvation is probably a result of their extreme diet, but the details are not entirely clear. In addition, vampire bats, especially younger bats, often fail to feed. Wilkinson (1984) reported that 33% of 258 *Desmodus* younger than two years and 7% of 340 older bats failed to feed on a given night. If young bats fail to feed on a third of their nights and if they can starve to death after missing three consecutive meals, then almost all young bats are likely to starve to death in their first year of foraging without regurgitated food sharing from their mothers.

Not only do mothers regurgitate blood to feed their pups and volant young (Mills 1980; Schmidt and Manske 1973), but all three vampire bat species also perform adult-to-adult food sharing (Carter et al. 2008; Elizalde-Arellano et al. 2007; Wilkinson 1984). Adult food sharing is likely to have evolutionary origins in prolonged maternal care, and kin selection likely extended this care to closely related adults.

Several lines of evidence suggest that donors benefit from sharing food because it both aids the survival of close kin and helps maintain social bonds that lead to reciprocal helping when the donor is in need (Carter and Wilkinson 2013b, 2013c; Carter et al. 2017; Wilkinson 1984). First, the costs of giving food are low for a successful forager with a full belly, while the benefits for unfed bats are high (Wilkinson 1984). Female vampire bats form stable associations with both kin and nonkin, and adult regurgitations occur primarily between close female kin, and only between bats spending at least 60% of their roosting time in the same roost (Wilkinson 1984). Among captive adults, reciprocal help was about eight times more important than kinship for predicting donation rates (Carter and Wilkinson 2013b, 2013c). Nonkin food sharing occurred among bats that first encountered each other as adults. Each female formed her own fairly consistent network of food donors (Carter and Wilkinson 2013c). Moreover, when a primary female donor was removed, she received less food overall compared to when a female not in her donor network was removed, which shows that food-sharing partners in the social group are not interchangeable (Carter and Wilkinson 2015a). Females that invest in more nonkin bonds cope better with the removal of donors from their sharing network, so helping nonkin expands a female's network of possible donors beyond what would be possible if sharing was limited to close kin (Carter et al. 2017).

Vampire bat food-sharing seems to depend on enduring social relationships that are also maintained through other cooperative services. Vampires are the only bats known to engage in high levels of social grooming (Carter and Leffer 2015; Wilkinson 1986). Social grooming and food sharing are highly correlated among and within pairs (Carter and Wilkinson 2013c; Wilkinson 1986), and both are affected by oxytocin administration (Carter and Wilkinson 2015b).

Rabies Transmission

The social nature of vampire bats combined with their frequent biting of domestic livestock makes them an optimal vector for rabies virus. Rabies transmitted by *Desmodus* poses a serious threat to agricultural development and human health in Latin America; it costs about 30 million US dollars per year in livestock mortality (Blackwood et al. 2013). Even worse, lethal human rabies outbreaks are increasingly detected in remote villages in the Amazon (Batista-da-Costa et al. 1993; Blackwood et al. 2013; Da Rosa et al. 2006; Gilbert et al. 2012; Stoner-Duncan et al. 2014). A recent case included a person bitten repeatedly on different nights by vampire bats in Brazil (Filho et al. 2014), highlighting the potential of *Desmodus* to affect human populations.

In Latin America, vampire bats are often killed with a topical anticoagulant (e.g., vampiricide) to control populations. However, such culling appears to be ineffective and perhaps even counterproductive for rabies control (Blackwood et al. 2013; Streicker et al. 2012), for at least two reasons. First, rabies exposure in bats appears independent of the size of social groups (Streicker et al. 2012). In 1,086 serum samples collected in Peru, 10% of vampire bats possessed rabies virus-neutralizing antibodies, suggesting a past or recent exposure to rabies: in roosts with more than four bats, seroprevalence ranged from 3% to 29% (Streicker et al. 2012). Curiously, the main predictor of rabies antibodies was age class; bats of ages 2–9 months were more likely to be seropositive, suggesting higher rabies exposure during the first year of life (Streicker et al. 2012). The reason for this is unclear.

The second reason for why culling is ineffective for controlling rabies is that rabies persistence appears driven by the opposing frequencies of immunizing exposures and movements between roosts or contacts at feeding sites between bats from different colonies (Blackwood et al. 2013). Infected bats, especially males, immigrate into new areas, and rabies exposure only leads to lethal infection in a few vampire bats (~10%), such that small groups do not collapse (Blackwood et al. 2013). Even if culling were locally effective, viral persistence would therefore remain at the regional level. Culling might even increase bat movement by reducing competition or causing more dispersal, which could elevate rabies transmission rates (Blackwood et al. 2013). Despite evidence that culling does not reduce rabies transmission, farmers are likely to favor or even demand it, because vampire bat parasitism is believed to lead to anemia, secondary infections, and decreased milk production in cows (but see Thompson et al. 1977). As an alternative to culling, vampire bats can be vaccinated via intramuscular, scarification, oral, or aerosol administration of rabies vaccines (Aguilar-Setién et al. 2002; Mayen 2003; Sétien et al. 1998). Region-specific knowledge of their prey preferences, ecology, and demography will be crucial for mitigating the dangers of vampire bat–transmitted rabies.

Desmodus can host many other pathogens besides rabies. Brandão et al. (2008) detected a novel coronavirus in vampire bats. Wray et al. (2016) detected *Bartonella* spp. bacteria in 38% of individuals and also found rhabdovirus, adenovirus, and herpesvirus. These findings warrant further investigation to possibilities for spillover of other zoonotic pathogens to domestic livestock or humans.

Conclusions

Vampire bats are of great importance because of their role as rabies reservoir hosts and because of their many adaptations to a unique ecological niche. Vampire bats have both gained and lost many ancestral traits and serve as an illustrative case of the power of natural selection to reshape morphology, physiology, behavior, and cognition. The blood-feeding lifestyle has led to dramatic anatomical and physiological changes for approaching, parasitizing, and efficiently processing the blood of their hosts. The energetic constraints imposed by this difficult diet have likely set the stage for the evolution of cooperative food sharing, where some energy is stored not in the body but as a cooperative investment in kin and nonkin that reciprocate. Given their extensive range, it would be interesting to understand the extent to which disparate populations differ in physiology, ecology, and behavior. For example, regional variation in rates of rabies infections, livestock availability, and culling via socially transmitted poisons could all lead to regional differences in cooperative traits such as food-sharing propensity. Given their extreme specialization, vampire bats represent an excellent case study for assessing the relative roles of adaptation and phylogenetic constraints in the evolution of mammals.

Acknowledgments

GC is supported by a Smithsonian Postdoctoral Fellowship. We thank T. Fleming, R. Crisp, D. Becker, and D. Riskin for helpful comments on an earlier version of the manuscript. We would also thank Michael A. Simmons, MFA, who created the illustrations used in this chapter.

References

Aguilar-Setién, A., Y. L. Campos, E. T. Cruz, R. Kretschmer, B. Brochier, and P. Pastoret. 2002. Vaccination of vampire bats using recombinant vaccinia-rabies virus. Journal of Wildlife Diseases, 38:539–544.

Aguirre, L. F., A. Herrel, R. van Damme, and E. Matthysen. 2002. Ecomorphological analysis of trophic niche partitioning in a tropical savannah bat community. Proceedings of the Royal Society London B, 269:1271–1278. DOI: 10.1098/rspb.2002.2011.

Altenbach, J. S. 1979. Locomotor morphology of the vampire bat, Desmodus rotundus. Special Publications of the American Society of Mammalogists, 6:1–137.

Altenbach, J. S. 1988. Locomotion. Pp. 71–84 in: Natural History of Vampire Bats (A. M. Greenhall and U. Schmidt, eds.). CRC Press, Boca Raton, FL.

Altenbach, J. S., and J. W. Hermanson. 1987. Bat flight muscle function and the scapulo-humeral lock. Pp. 100–118 in: Recent Advances in the Study of Bats (M. B. Fenton, P. A. Racey, and J. M. Rayner, eds.). Cambridge University Press, Cambridge.

Apitz-Castro, R., S. Beguin, A. Tablante, F. Bartoli, J. C. Holt, and H. C. Hemker. 1995. Purification and partial characterization of draculin, the anticoagulant factor present in the saliva of vampire bats (Desmodus rotundus). Thrombosis and Haemostasis, 73:94–100.

Bahlman, J. W., and D. Kelt. 2007. Use of olfaction during prey location by the common vampire bat (Desmodus rotundus). Biotropica, 39:147–149.

Batista-da-Costa, M., R. F. Bonito, and S. A. Nishioka. 1993. An outbreak of vampire bat bite in a Brazilian village. Tropical Medicine and Parasitology, 44:219–20.

Bhatnagar, K. P., and W. A. Wimsatt. 1988. Vampire bats do have fully developed enamel. Anatomical Record 223:14A. (abstract)

Blackwood, J. C., D. G. Streicker, S. Altizer, and P. Rohani. 2013. Resolving the roles of immunity, pathogenesis, and immigration for rabies persistence in vampire bats. Proceedings of the National Academy of Sciences, USA, 110:20837–42. DOI: 10.1073/pnas.1308817110.

Bobrowiec, P. E. D., M. R. Lemes, and R. Gribel. 2015. Prey preference of the common vampire bat (Desmodus rotundus, Chiroptera) using molecular analysis. Journal of Mammalogy, 96:54–63.

Brandão, P. E., K. Scheffer, L. Y. Villarreal, S. Achkar, R. de Novaes Oliveira, W. de Oliveira Fahl, J. G. Castilho, I. Kotait, and L. J. Richtzenhain. 2008. A coronavirus detected in the vampire bat Desmodus rotundus. Brazilian Journal of Infectious Diseases 12:466–468.

Busch, C. 1988. Consumption of blood, renal function and utilization of free water by the vampire bat, Desmodus rotundus. Comparative Biochemistry and Physiology, 90A:141–146.

Caraballo, A. J. 1996. Outbreak of vampire biting in a Venezuelan village. Revista de Saúde Pública, 30:483.

Carter, G. 2016. Footage of feeding by the hairy-legged vampire bat Diphylla ecaudata refutes the hypothesis that this vampire bat sucks. Figshare. https://dx.doi.org/10.6084/m9.figshare.4490621.v1.

Carter, G. G., and L. L. Leffer. 2015. Social grooming in bats: are vampire bats exceptional? PLOS ONE, 10:e0138430. DOI: 10.1371/journal.pone.0138430.

Carter, G. G., and G. S. Wilkinson. 2013a. Cooperation and conflict in the social lives of bats. Pp. 225–242 in: Bat Evolution, Ecology, and Conservation (R. Adams and S. Pedersen eds.). Springer Science Press, New York.

Carter, G. G., and G. Wilkinson. 2013b. Does food sharing in vampire bats demonstrate reciprocity? Communicative and Integrative Biology, 6:e25783. DOI: 10.4161/cib.25783.

Carter, G. G., and G. S. Wilkinson. 2013c. Food sharing in vampire bats: reciprocal help predicts donations more than relatedness or harassment. Proceedings of the Royal Society B, 280:20122573.

Carter, G. G., and G. S. Wilkinson. 2015a. Social benefits of non-kin food sharing by female vampire bats. Proceedings of the Royal Society B, 282:20152524–20152524. DOI: 10.1098/rspb.2015.2524.

Carter, G. G., and G. S. Wilkinson. 2015b. Intranasal oxytocin increases social grooming and food sharing in the common vampire bat Desmodus rotundus. Hormones and Behavior, 75:150–153. DOI: 10.1016/j.yhbeh.2015.10.006.

Carter, G. G., C. E. Coen, L. M. Stenzler, and I. J. Lovette. 2006. Avian host DNA isolated from the feces of white-winged vampire bats (Diaemus youngi). Acta Chiropterologica, 8:255–258.

Carter, G. G., M. Skowronski, P. Faure, and M. B. Fenton. 2008. Antiphonal calling allows individual discrimination in white-winged vampire bats. Animal Behaviour, 76:1343–1355.

Carter, G. G., D. R. Farine, and G. S. Wilkinson. 2017. Social bet-hedging in vampire bats. Biology Letters, 13:20170112.

Cartwright, T., and C. Hawkey. 1969. Activation of the blood fibrinolytic mechanism in birds by saliva of the vampire bat (Diaemus youngi). Journal of Physiology, 201:45–46.

Chaverri, G., and T. H. Kunz. 2006. Reproductive biology and postnatal development in the tent-making bat Artibeus *watsoni* (Chiroptera: Phyllostomidae). Journal of Zoology, 270:650–656.

Coen, C. E. 2002. Comparative nutritional ecology of two genera of vampire bats: *Desmodus rotundus* and *Diaemus youngi*. PhD diss., Cornell University, Ithaca, NY.

Crespo, R. F., S. S. Fernández, R. J. Burns, and G. C. Mitchell. 1974. Observaciones sobre el comportamiento del vampiro común (*Desmodus rotundus*) al alimentarse en condiciones naturales. Revista Mexicana de Ciencias Pecuarias, 1:39.

Cvikel, N., K. E. Berg, E. Levin, E. Hurme, I. Borissov, A. Boonman, E. Amichai, and Y. Yovel. 2015. Bats aggregate to improve prey search but might be impaired when their density becomes too high. Current Biology, 24:2962–2967.

Czaplewski, N. J., and C. Cartelle. 1998. Pleistocene bats from cave deposits in Bahia, Brazil. Journal of Mammalogy 79:784–803.

Czaplewski, N. J., J. Krejca, and T. E. Miller. 2003. Late Quaternary bats from Cebada Cave, Chiquibul cave system, Belize. Caribbean Journal of Science, 39:23–33.

Da Rosa, E. S., I. Kotait, T. F. Barbosa, M. L. Carrieri, P. E. Brandão, A. S Pinheiro, A. L. Begot, M. Y. Wada, R. C. De Oliveira, and E. C. Grisard. 2006. Bat-transmitted human rabies outbreaks, Brazilian Amazon. Emerging Infectious Diseases, 12:1197–1202.

Daniel, M. J. 1979. The New Zealand short-tailed bat *Mystacina tuberculata*: a review of present knowledge. New Zealand Journal of Zoology, 6:357–370.

Davis, J. S., C. W. Nicolay, and S. H. Williams. 2010. A comparative study of incisor procumbency and mandibular morphology in vampire bats. Journal of Morphology, 271:853–862. DOI: 10.1002/jmor.10840.

Delpietro, H., and R. G. Russo. 2002. Observations of the common vampire bat (*Desmodus rotundus*) and the hairy-legged vampire bat (*Diphylla ecaudata*) in captivity. Zeitschrift für Säugetierkunde, 67:65–78.

Delpietro, H. A., R. G. Russo, G. G. Carter, R. D. Lord, and G. Delpietro. 2017. Reproductive seasonality, sex ratio, and philopatry in Argentina's common vampire bats. Royal Society Open Science, 4. DOI: 10.1098/rsos.160959

Dyce, K. M., W. O. Sack, and C. J. G. Wensing. 2010. Textbook of Veterinary Anatomy. Saunders Elsevier, Saint Louis.

Elizalde-Arellano, C., J. C. Lopez-Vidal, J. Arroyo-Cabrales, R. A. Medellín, and J. W. Laundre. 2007. Food sharing behavior in the hairy-legged vampire bat *Diphylla ecaudata*. Acta Chiropterologica, 9:314–319.

Filho, F. B., B. K. Kac, D. R. Azylay, G. Martines, J. A. da Costa Nery, G. S. Luchi, and L. Azylay-Abulafia. 2014. Multiple lesions by vampire bat bites in a patient in Niterói, Brazil— Case Report. Annals Brasil Dermatology, 89:340–343. DOI: http://dx.doi.org/10.1590/abd1806–4841.20142996.

Forman, G. L., C. J. Phillips, and C. S. Rouk. 1979. Alimentary tract. Pp. 205–227 *in*: Biology of Bats of the New World Family Phyllostomatidae. Pt. 3 (R. J. Baker, J. Knox Jones Jr., and D. C. Carter, eds.). Special Publications of the Museum, Texas Tech University, no. 16. Texas Tech Press, Lubbock.

Freitas, M. B., C. B. C. Passos, R. B. Vasconcelos, and E. C. Pinheiro. 2005. Effects of short-term fasting on energy reserves of vampire bats (*Desmodus rotundus*). Comparative

Biochemistry and Physiology Part B: Biochemistry and Molecular Biology, 140:59–62.

Freitas, M. B, J. F. Queiroz, C. I. D. Gomes, C. B. Collares-Buzato, H. C. Barbosa, A. C. Boschero, C. A. Gonçalves, and E. C. Pinheiro. 2013. Reduced insulin secretion and glucose intolerance are involved in the fasting susceptibility of common vampire bats. General and Comparative Endocrinology, 183:1–6.

Freitas, M. B., A. F. Welker, S. F. Millan, and E. C. Pinheiro. 2003. Metabolic responses induced by fasting in the common vampire bat *Desmodus rotundus*. Journal of Comparative Physiology B, 173:703–707.

Gilbert, A. T., B. W. Petersen, S. Recuenco, M. Niezgoda, J. Gomez, V. A. Laguna-Torres, and C. Rupprecht. 2012. Evidence of rabies virus exposure among humans in the Peruvian Amazon. American Journal of Tropical Medicine and Hygiene, 87:206–215.

Goodwin, G. G., and A. M. Greenhall. 1961. A review of the bats of Trinidad and Tobago: descriptions, rabies infections, and ecology. Bulletin of the American Museum of Natural History, 122:187–342.

Gracheva, E. O., J. F. Cordero-Morales, J. A. González-Carcacía, N. T. Ingolia, C. Manno, C. I. Aranguren, J. S. Weissman, and D. Julius. 2011. Ganglion-specific splicing of TRPV1 underlies infrared sensation in vampire bats. Nature, 476:88–91.

Greenhall, A M. 1970. The use of a precipitin test to determine host preferences of the vampire bats *Desmodus rotundus* and *Diaemus youngi*. Bijdragen tot de Dierkunde, 40:36–39.

Greenhall, A. M. 1972. The biting and feeding habits of the vampire bat *Desmodus rotundus*. Journal of Zoology (London), 168:451–461.

Greenhall, A. M. 1988. Feeding behavior. Pp. 111–131 *in*: Natural History of Vampire Bats (A. M. Greenhall and U. Schmidt, eds.). CRC Press, Boca Raton, FL.

Greenhall, A. M., U. Schmidt, and W. Lopez-Forment. 1969. Field observations on the mode of attack of the vampire bat (*Desmodus rotundus*) in Mexico. Anales del Instituto de Biologia UNAM Serie Zoologia, 40:245–252.

Greenhall, A. M., U. Schmidt, and W. Lopez-Forment. 1971. Attacking behaviour of the vampire bat, *Desmodus rotundus*, under field conditions in Mexico. Biotropica, 3:136–141.

Griffiths, T. A., and B. B. Criley. 1989. Comparative lingual anatomy of the bats *Desmodus rotundus* and *Lonchophylla robusta* (Chiroptera: Phyllostomidae). Journal of Mammalogy, 70:608–613.

Gröger, U., and L. Wiegrebe. 2006. Classification of human breathing sounds by the common vampire bat, *Desmodus rotundus*. BMC Biology, 4:18.

Hamilton, I. M., and R. M. R. Barclay. 1998. Ontogenetic influences on foraging mass accumulation by big brown bats (*Eptesicus fuscus*). Journal of Animal Ecology, 67:930–940.

Hand, S. J., V. Weisbecker, R. M. D. Beck, M. Archer, H. Godthelp, A. J. D. Tennyson, and T. H. Worthy. 2009. Bats that walk: a new evolutionary hypothesis for the terrestrial behaviour of New Zealand's endemic mystacinids. BMC Evolutionary Biology, 9:169. DOI: 10.1186/1471-21148-9-169.

Harlow, H. J., and E. J. Braun. 1997. Gastric Na+ K+ ATPase activity and intestinal urea hydrolysis of the common vampire bat, *Desmodus rotundus*. Comparative Biochemistry and Physiology Part A: Physiology, 118:665–669.

Hawkey, C. 1966. Plasminogen activator in saliva of the vampire bat *Desmodus rotundus*. Nature, 211:434–435.

Heffner, R. S., G. Koay, and H. E. Heffner. 2015. Sound localization in common vampire bats: acuity and use of the binaural time cue by a small mammal. Journal of the Acoustical Society of America, 137:42–52. DOI: 10.1121/1.4904529.

Heffner, R. S., G. Koay, and H. E. Heffner. 2013. Hearing in American leaf-nosed bats. IV: the common vampire bat, *Desmodus rotundus*. Hearing Research, 296:42–50. DOI: 10.1016/j.heares.2012.09.011.

Hermanson, J. W., and J. S. Altenbach. 1985. Functional anatomy of the shoulder and arm of the fruit-eating bat *Artibeus jamaicensis*. Journal of Zoology, 205:157–177.

Hermanson, J. W., Y. Dzal, T. Orr, J. York, Z. Czenze, and S. Parsons. 2016. The efficiency of bounding vampires. Annual Meeting of the Society of Integrative and Comparative Biology, Portand, Oregon. (abstract).

Horst, R. 1968. Observations on renal morphology and physiology in the vampire bat *Desmodus rotundus*. PhD diss., Cornell University, Ithaca, NY.

Hoying, K. M., and T. H. Kunz. 1998. Variation in size at birth and post-natal growth in the insectivorous bat *Pipistrellus subflavus* (Chiroptera:Vespertilionidae). Journal of Zoology (London), 245:15–27.

Hoyt, R. A., and J. S. Altenbach. 1981. Observations on *Diphylla ecaudata* in captivity. Journal of Mammalogy, 62:215–216.

Ito, F., E. Bernard, and R. A. Torres. 2016. What is for dinner? First report of human blood in the diet of the hairy-legged vampire bat *Diphylla ecaudata*. Acta Chiropterologica, 18:509–515.

Jenness, R., and E. H. Studier. 1976. Lactation and milk. Pp 201–218 *in*: Biology of Bats of the New World Family Phyllostomatidae. Pt. 1 (R. J. Baker, J. Knox Jones Jr., and D. C. Carter, eds.). Texas Tech Press, Lubbock.

Joermann, G. 1984. Recognition of spatial parameters by echolocation in the vampire bat, *Desmodus rotundus*. Journal of Comparative Physiology A, 155:67–74.

Joermann, G., U. Schmidt, and C. Schmidt. 1988. The mode of orientation during flight and approach to landing in two phyllostomid bats. Ethology, 78:332–340.

Kerth, G., N. Perony, and F. Schweitzer. 2011. Bats are able to maintain long-term social relationships despite the high fission-fusion dynamics of their groups. Proceedings of Royal Society London B, 278:2761–2767. DOI: 10.1098/rspb.2010.2718.

Koopman, K. F. 1988. Systematics and distribution. Pp. 7–17 *in*: Natural History of Vampire Bats (A. M. Greenhall and U. Schmidt, eds.). CRC Press, Boca Raton, FL.

Kunz, T. H., and S. K. Robson. 1995. Postnatal growth and development in the Mexican free-tailed bat (*Tadarida brasiliensis mexicana*): birth size, growth rates, and age estimation. Journal of Mammalogy, 76:769–783.

Kürten, L. 1985. Mechanoreceptors in the nose-leaf of the vampire bat *Desmodus rotundus*. Zeitschrift für Säugetierkunde, 50:26–35.

Kürten, L., and U. Schmidt. 1982. Thermoperception in the common vampire bat (*Desmodus rotundus*). Journal of Comparative Physiology, 146:223–228.

Kwiecinski, G. G., M. Falzone, and E. H. Studier, 2003. Milk concentration and postnatal accretion of minerals and nitrogen in two phyllostomid bats. Journal of Mammalogy, 84:926–936.

Lester, K. S., S. J. Hand, and F. Vincent. 1988. Adult phyllostomid (bat) enamel by scanning electron microscopy—with a note on dermopteran enamel. Scanning Microscopy, 2:371–383.

Lindsberg, P. J., and V. Caso. 2016. Desmoteplase after ischemic stroke in patients with occlusion or high-grade stenosis in major cerebral arteries. Stroke, 47:901–903.

Luna-Jorquera, G., and B.-M. Culik. 1995. Penguins bled by vampires. Journal für Ornithologie, 136:471–472.

Ma, D., D. M. Mizurini, T. C. F. Assumpção, Y. Li, Y. Qi, M. Kotsyfakis, J. M. C. Ribeiro, R. Q. Monteiro, and I. M. B. Francischetti. 2013. Desmolaris, a novel factor XIa anticoagulant from the salivary gland of the vampire bat (Desmodus *rotundus*) inhibits inflammation and thrombosis in vivo. Blood, 122:4094–4106.

Mayen, F. 2003. Haematophagous bats in Brazil, their role in rabies transmission, impact on public health, livestock industry and alternatives to an indiscriminate reduction of bat population. Journal of Veterinary Medicine B, 50:469–472.

Medcalf, R. L. 2012. Desmoteplase: discovery, insights and opportunities for ischaemic stroke. British Journal of Pharmacology, 165:75–89.

McFarland, W. N., and W. A. Wimsatt. 1969. Renal function and its relation to the ecology of the vampire bat, *Desmodus rotundus*. Comparative Biochemistry and Physiology, 28:985–1006.

Mills, R. S. 1980. Parturition and social interaction among captive vampire bats, *Desmodus rotundus*. Journal of Mammalogy 61:336–337.

Mitchell, G. C., and J. R. Tigner. 1970. The route of ingested blood in the vampire bat *Desmodus rotundus*. Journal of Mammalogy 51 (4): 814–817.

Morgan, G. S., O. J. Linares, and C. E. Ray. 1988. New species of fossil vampire bats (Mammalia, Chiroptera, Desmodontidae) from Florida, USA, and Venezuela. Proceedings of the Biological Society of Washington, 101:912–928.

Morton, D. 1986. The structure of the fundic cecum of the recently fed common vampire bat *Desmodus rotundus*. Anatomical Record, 214:89A.

Morton, D., and J. F. Richards. 1981. The flow of excess dietary water through the common vampire bat during feeding. Comparative Biochemistry and Physiology Part A, 69:511–515.

Muela, A., M. Curti, Y. Seminario, and A. Hernandez. 2011. An incubating orange-breasted falcon (*Falco deiroleucus*) as host for a vampire bat. Journal of Raptor Research, 45:277–279.

Norberg, U. M., and J. M.V. Rayner. 1987. Ecological morphology and flight in bats (Mammalia; Chiroptera): wing adaptations, flight performance, foraging strategy, and echolocation. Philosophical Transactions Royal Society B, 316:337–419. DOI: 10.1098/rstb.1987.0030.

Phillips, C. J., and B. Steinberg. 1976. Histological and scanning electron microscopic studies of the tooth structure and thegosis in the common vampire bat, *Desmodus rotundus*.

Occasional Papers, The Museum, Texas Tech University, 42:1–12.

Phillips, C. J., B. Tandler, and K. Toyoshima. 1986. Structure of the dorsal surface of the tongue of the vampire bat *Desmodus rotundus*. Anatomical Record, 214:101A–102A.

Piccinini, R. S., A. L. Peracchi, S. D. L. Raimundo, A. M. Tannure, J. C. P. de Souza, S. T. de Albuquerque, and L. L. Furtado. 1991. Observations on the food habits of *Diphylla ecaudata* Spix, 1923. Revista Brasileira de Medicina Veterinária, 13:8–10.

Powers, L. V., S. C. Kandarian, and T. H. Kunz. 1991. Ontogeny of flight in the little brown bat, *Myotis lucifugus*: behavior, morphology, and muscle histochemistry. Journal of Comparative Physiology A, 168:675–685.

Ratcliffe, J. M., M. B. Fenton, and B. G. Galef. 2003. An exception to the rule: common vampire bats do not learn taste aversions. Animal Behaviour, 65:385–389. DOI: 10.1006 /anbe.2003.2059.

Ray, C. E., O. J. Linares, and G. S. Morgan. 1988. Paleontology. Pp. 19–30 *in*: Natural History of Vampire Bats (A. M. Greenhall and U. Schmidt, eds.). CRC Press, Boca Raton, FL.

Ripperger, S., D. Josic, M. Hierold, A. Koelpin, R. Weigel, M. Hartmann, R. Page, and F. Mayer. 2016. Automated proximity sensing in small vertebrates: design of miniaturized sensor nodes and first field tests in bats. Ecology and Evolution, 6:2179–2189.

Riskin, D. K., and J. W. Hermanson. 2005. Independent evolution of running in vampire bats. Nature, 434 (7031): 292.

Riskin, D. K., S. Parsons, W. A. Schutt, G. G. Carter, and J. W. Hermanson. 2006. Terrestrial locomotion of the New Zealand short-tailed bat *Mystacina tuberculata* and the common vampire bat *Desmodus rotundus*. Journal of Experimental Biology, 209:1725–1736.

Riskin, D. K., J. E. A. Bertram, and J. W. Hermanson. 2016. The evolution of terrestrial locomotion in bats: the bad, the ugly, and the good. Pp. 307–323 *in*: Understanding Mammalian Locomotion (J. E. A. Bertram, ed.). John Wiley & Sons, Hoboken, NJ

Sanchez-Hernandez, C., M. D. L. Romero-Almarez, M. C. Wooten, and M. L. Kennedy. 2006. Speed in flight of common vampire bats (*Desmodus rotundus*). Southwestern Naturalist, 51:422–425.

Sazima, I. 1978. Aspects of the feeding behavior of the blood sucking bat *Desmodus rotundus* in the Region of Campinas State of Sao-Paulo Brazil. Boletim de Zoologia, 3:97–120.

Sazima, I., and W. Uieda. 1980. Feeding behavior of the white-winged vampire bat *Diaemus youngii* on poultry. Journal of Mammalogy, 61:102–103.

Schmidt, U. 1978. Vampirfledermäuse. Westarp Wissenschaften, Magdeburg.

Schmidt, C. 1988a. Reproduction. Pp. 99–110 *in*: Natural History of Vampire Bats (A. M. Greenhall and U. Schmidt, eds.). CRC Press, Boca Raton, FL.

Schmidt, U. 1988b. Orientation and sensory functions in *Desmodus rotundus*. Pp 143–166 *in*: Natural History of Vampire Bats (A. M. Greenhall and U. Schmidt, eds.). CRC Press, Boca Raton, FL.

Schmidt, U., and U. Manske. 1973. Die Jugendentwicklung der Vampirfledermaus (*Desmodus rotundus*). Zeitschrift für Säugetierkunde, 38:14–33.

Schmidt, U., and K. van de Flierdt. 1973. Intraspecific agonistic behavior of the vampire bat *Desmodus rotundus* at the feeding site. Zeitschrift für Tierpsychologie, 32:139–146.

Schmidt, U., and C. Schmidt. 1977. Echolocation performance of the vampire bat (*Desmodus rotundus*). Zeitschrift für Tierpsychologie, 45:349–358.

Schmidt, U., G. Joermann, and G. Rother. 1988. Acoustical vs. visual orientation in Neotropical bats. Pp. 589–593 *in*: Animal Sonar (P. E. Nachtigall and P. W. B. Moore, eds.). Springer, New York.

Schmidt, U., P. Schlegel, H. Schweizer, and G. Neuweiler. 1991. Audition in vampire bats *Desmodus rotundus*. Journal of Comparative Physiology A: Sensory Neural and Behavioral Physiology, 168:45–52.

Schutt, W. A., Jr. 1998. Chiropteran hindlimb morphology and the origin of blood feeding in bats. Pp. 157–168 *in*: Bat Biology and Conservation (T. H. Kunz and P. A. Racey eds.). Smithsonian Institution Press, Washington, DC.

Schutt, W. A., Jr., and J. S. Altenbach. 1997. A sixth digit in *Diphylla ecaudata*, the hairy legged vampire bat. Mammalia, 6:280–285.

Schutt, W. A., Jr., and N. B. Simmons. 2001. Morphological specializations of *Cheiromeles* (naked bulldog bats; Molossidae) and their possible role in quadrupedal locomotion. Acta Chiropterologica, 3:225–235.

Schutt, W. A, Jr., J. S. Altenbach, Y. H. Chang, D. M. Cullinane, J. W. Hermanson, F. Muradali, and J. E. A. Bertram. 1997. The dynamics of flight-initiating jumps in the common vampire bat *Desmodus rotundus*. Journal of Experimental Biology, 200:3003–3012.

Schutt, W. A., F. Muradali, N. Mondol, K. Joseph, and K. Brockmann. 1999. Behavior and maintenance of captive white-winged vampire bats, *Diaemus youngi*. Journal of Mammalogy, 80:71–81.

Sétien, A. A., B. Brochier, N. Tordo, O. De Paz, P. Desmettre, D. Péharpré, and P.-P. Pastoret. 1998. Experimental rabies infection and oral vaccination in vampire bats (*Desmodus rotundus*). Vaccine, 16:1122–1126.

Stoner-Duncan, B., D. G. Streicker, and C. M. Tedeschi. 2014. Vampire bats and rabies: toward an ecological solution to a public health problem. PLOS Neglected Tropical Diseases, 8:e2867.

Streicker, D. G., and J. E. Allgeier. 2016. Foraging choices of vampire bats in diverse landscapes: potential implications for land-use change and disease transmission. Journal of Applied Ecology, 53:1280–1288.

Streicker, D. G., S. Recuenco, W. Valderrama, J. G. Benavides, I. Vargas, V. Pacheco, R. E. C. Condori, J. Montgomery, C. E. Rupprecht, and P. Rohani. 2012. Ecological and anthropogenic drivers of rabies exposure in vampire bats: implications for transmission and control. Proceedings of the Royal Society B, 279 (1742): 3384–3392.

Tandler, B., K. Toyoshima, Y. Seta, and C. J. Phillips. 1997. Ultrastructure of the salivary glands in the midtongue of the common vampire bat, *Desmodus rotundus*. Anatomical Record, 249:196–205.

Thompson, R. D., D. J. Elias, and G. C. Mitchell. 1977. Effects of vampire bat control on bovine milk production. Journal of Wildlife Management, 41:736–739.

Turner, D. C. 1975. The Vampire Bat: A Field Study in Behavior and Ecology. Johns Hopkins University Press, Baltimore.

Tuttle, R. H. 1975. Knuckle-walking and knuckle-walkers: a commentary on some recent perspectives on Hominoid evolution. Pp. 203–212 in: Primate Functional Morphology and Evolution (R. H. Tuttle, ed.). Mouton Publishers, The Hague.

Vierhaus, V. H. 1983. Wie vampirfledermäuse (*Desmodus rotundus*) ihre zähne schärfen. Zeitschrift für Säugetierkunde, 48:269–277.

Voigt, C. C., and D. H. Kelm. 2006. Host preference of the common vampire bat (*Desmodus rotundus*; Chiroptera) assessed by stable isotopes. Journal of Mammalogy, 87:1–6.

Wilkinson, G. S. 1986. Social grooming in the common vampire bat, *Desmodus rotundus*. Animal Behaviour, 34:1880–1889.

Wilkinson, G. S. 1987. Altruism and cooperation in bats. Pp. 299–323 in: Recent Advances in the Study of Bats (P. A. Racey, M. B. Fenton, and J. M. V. Rayner, eds.). Cambridge University Press, Cambridge.

Wilkinson, G. S. 1984. Reciprocal food sharing in the vampire bat. Nature, 308:181–184.

Wilkinson, G. S. 1988. Social organization and behavior. Pp. 85–98 in: Natural History of Vampire Bats (A. M. Greenhall and U. Schmidt, eds.). CRC Press, Boca Raton, FL.

Wilkinson, G. S. 1985a. The social organization of the common vampire bat: I. pattern and cause of association. Behavioral Ecology and Sociobiology, 17:111–121.

Wilkinson, G. S. 1985b. The social organization of the common vampire bat: II. mating system, genetic structure, and relatedness. Behavioral Ecology and Sociobiology, 17:123–134.

Wimsatt, W. A. 1969. Transient behavior, nocturnal activity patterns, and feeding efficiency of vampire bats (*Desmodus rotundus*) under natural conditions. Journal of Mammalogy, 50:233–244.

Wimsatt, W. A., and A. Guerriere. 1962. Observations of feeding capacities and excretory functions of captive vampire bats. Journal of Mammalogy, 43:17–27.

Wray, A. K., K. J. Olival, D. Morán, M. Renee Lopez, D. Alvarez, I. Navarrete-Macias, E. Liang et al. 2016. Viral diversity, prey preference, and *Bartonella* prevalence in *Desmodus rotundus* in Guatemala. EcoHealth, 13:761–774.

Young, A. M. 1971. Foraging of vampire bats *Desmodus rotundus* in Atlantic wet lowland Costa Rica. Revista de Biologia Tropical, 18:73–88.

16

The Ecology and Evolution of Nectar Feeders

Nathan Muchhala
and Marco Tschapka

Introduction

As detailed in this book, New World leaf-nosed bats have undergone a rapid and extensive diversification, in terms of both species richness and the niches they occupy. One trophic niche particularly important to the ecology of New World ecosystems is the nectar-feeding specialists, which many plants have evolved to depend on for their pollination services (Fleming et al. 2009; Garibaldi et al. 2012). Nectar provides an especially carbohydrate-rich food source that is relatively easy to digest (Baker et al. 1998), and many phyllostomids occasionally include nectar in their diets. This is particularly common among primarily frugivorous genera such as *Artibeus*, *Sturnira*, and *Carollia* and omnivores such as *Phyllostomus* (Giannini and Kalko 2004; Heithaus et al. 1975; Sazima 1976). However, only a handful of phyllostomid genera have evolved adaptations for a diet composed primarily of nectar. In this chapter, we review the ecology and evolution of these bats, as well as the flowers with which they have coevolved.

Evolution of Nectarivory

Nectarivore specialists occur in the subfamilies Glossophaginae and Lonchophyllinae. Glossophaginae comprises 14 genera and 36 species, and Lonchophyllinae 5 genera and 20 species (table 16.1); thus, in total, nectarivores make up about one-quarter (26%) of the 216 known species of phyllostomids (Cirranello et al. 2016; chap. 4, this vol.). These species share a suite of traits that adapts them to their nectarivorous lifestyle, including reduced dentition, elongated snouts, long, extensible tongues with hairlike papillae, and the ability to hover in front of flowers while extracting nectar (Freeman 1995; Norberg et al. 1993; von Helversen and Winter 2003). Early phylogenetic work based largely on morphology suggested that the two subfamilies were sister clades (e.g., Slaughter 1970), which implies that nectarivory

Table 16.1. Classification scheme for the 52 known species of nectarivorous phyllostomid bats (following Cirranello and Simmons, chap. 4, this vol.)

Subfamily	Tribe	Subtribe	Genus	Species
Glossophaginae	Brachyphyllini	Brachyphyllina	*Brachyphylla*	*B. cavernarum, B. nana*
		Phyllonycterina	*Erophylla*	*E. bombifrons, E. sezekorni*
			Phyllonycteris	*P. aphylla, P. poeyi*
	Choeronycterini	Anourina	*Anoura*	*A. aequatoris, A. cadenai, A. carishina, A. caudifer, A. cultrata, A. fistulata, A. geoffroyi, A. latidens, A. luismanueli, A. peruana*
		Choeronycterina	*Choeroniscus*	*C. godmani, C. minor, C. periosus*
			Choeronycteris	*C. mexicana*
			Dryadonycteris	*D. capixaba*
			Hylonycteris	*H. underwoodi*
			Lichonycteris	*L. degener, L. obscura*
			Musonycteris	*M. harrisoni*
			Scleronycteris	*S. ega*
	Glossophagini	…	*Glossophaga*	*G. commissarisi, G. leachii, G. longirostris, G. morenoi, G. soricina*
			Leptonycteris	*L. curasoae, L. nivalis, L. yerbabuenae*
			Monophyllus	*M. plethodon, M. redmani*
Lonchophyllinae	Hsunycterini	…	*Hsunycteris*	*H. cadenai, H. dashe, H. pattoni, H. thomasi*
	Lonchophyllini	…	*Lionycteris*	*L. spurrelli*
			Lonchophylla	*L. bokermanni, L. chocoana, L. concava, L. dekeyseri, L. fornicata, L. handleyi, L. hesperia, L. inexpectata, L. mordax, L. orcesi, L. orienticollina, L. peracchii, L. robusta*
			Platalina	*P. genovensium*
			Xeronycteris	*X. vieirai*

evolved only once in the family. A detailed analysis of hyoid and lingual morphology, however, led Griffiths (1982) to suggest that lonchophyllines and glossophagines represented independent origins of nectarivory. This caused some controversy, with further analyses arguing for a single origin of nectarivory (Baker et al. 1981; Haiduk and Baker 1982; Smith and Hood 1984; Wetterer et al. 2000) or giving equivocal results (Carstens et al. 2002). The latest molecular work now strongly supports the conclusion of Griffiths (1982) that the Glossophaginae and Lonchophyllinae are not, in fact, sister clades (Baker et al. 2016; Datzmann et al. 2010; Rojas et al. 2011, 2016). Morphological and behavioral evidence further supports independent origins of nectarivory: while lonchophyllines and glossophagines are superficially similar in their elongated snouts and tongues and in their ability to hover, differences in tongue musculature, papillae arrangement, and feeding behavior suggest these similarities are examples of convergent evolution (Datzmann et al. 2010; Griffiths

1982; Tschapka et al. 2015; Winter and von Helversen 2003; see "Morphological Adaptations to Nectarivory" below). Interestingly, among these two nectar-feeding radiations, there appears to be one instance of a shift to a frugivorous diet, in the genus *Brachyphylla*. The two species in this genus possess extensible tongues with papillae, yet are more similar to frugivorous phyllostomids in skull (Freeman 1995) and tooth morphology (Griffiths 1985), and limited dietary studies suggest they do in fact feed primarily from fruits (Lenoble et al. 2014; Silva Taboada and Pine 1969; Swanepoel and Genoways 1983a, 1983b).

What ancestral feeding mode did the nectar-feeding phyllostomids evolve from? While initially it was thought that nectarivory evolved from frugivory, which in turn evolved from insectivory (Ferrarezzi and Gimenez 1996), more recent phylogenetic reconstruction of ancestral diets suggests that each plant-feeding specialization has evolved independently from primarily insectivorous ancestors (Baker et al. 2012; Rojas et al. 2011). The idea that insectivory could shift directly to nectarivory is supported by the fact that an insectivorous bat (*Antrozous pallidus*) from a highly insectivorous family (Vespertilionidae) was recently discovered to supplement its diet with nectar from cactus flowers (Frick et al. 2009, 2013). However, it can be difficult to reliably reconstruct ancestral states, particularly if there has been extensive extinction of less-specialized ancestral taxa. An alternate hypothesis is that the ancestral species that gave rise to the nectar-feeding clades were in fact omnivores that fed on insects, fruits, and nectar. In support of this scenario, the less-derived taxa among extant species of glossophagines tend to be particularly generalized (Fleming and Kress 2013), including taxa such as *Phyllonycteris*, *Erophylla*, *Brachyphylla*, and *Glossophaga soricina*, relative to other species of *Glossophaga*. In the same way that *Antrozous pallidus* suggests that shifts directly from insectivory to nectarivory may be possible, the unusual New Zealand endemic *Mystacina tuberculata* (Mystacinidae) may support the possibility of a shift through an omnivorous intermediate; it feeds on insects and fruits but also visits flowers and possesses distinct adaptations to nectarivory, such as a long tongue and reduced dentition (Arkins et al. 1999; Carter and Riskin 2006). Overall, given phylogenies and current diets, it seems likely that the ancestral nectarivores had a generalized diet that was supplemented at least by insects and possibly by fruits as well.

Dietary Breadth

While phyllostomid nectar-feeding bats possess specialized morphologies, as outlined above, the majority have relatively flexible diets in that they will feed occasionally on insects and/or fruits. The three bats with the most extreme adaptations to nectarivory include the glossophagines *Anoura fistulata* and *Musonycteris harrisoni* and the lonchophylline *Xeronycteris vieirai*. While most nectar bats can extend their tongues to approximately 60% of their body length, *A. fistulata* has a tongue extension of 150% of its body length, longer than any other mammal and exceeded only by chameleons among vertebrates (Muchhala 2006a). A glossal tube extends back from the jaw and houses the base of the tongue in the rib cage, allowing a much greater resting tongue length. Despite this specialized morphology, *A. fistulata* also feeds on insects (Muchhala, unpublished data). *Musonycteris harrisoni* also has a re-

markable tongue extension; it does not possess a glossal tube, so houses the resting tongue entirely in its mouth but has an extremely long snout (20 mm), longer than any other nectar bat (Gonzalez-Terrazas et al. 2012; Tellez and Ortega 1999). It feeds nearly exclusively on flowers from a wide variety of species, although insects have also been found in its feces (Tschapka et al. 2008). Finally, *Xeronycteris vieirai* is notable for having the greatest reduction in dentition (in terms of smaller teeth and larger gaps between them) among nectar bats. Given their increased reliance on a liquid diet, teeth are less important for nectar feeders and are much less robust than in other phyllostomids (Freeman 1995); *X. vieirai* takes this trend to an extreme, with many typical tooth structures absent or greatly reduced (Gregorin and Ditchfield 2005).

At the opposite end of the specialization spectrum, *G. soricina* is a widespread, abundant bat with a relatively short rostrum and a highly opportunistic diet. It is distributed from Mexico to Argentina, encompassing nearly the entire distribution of glossophagines (Alvarez et al. 1991), and is known to feed primarily on insects and fruit for large portions of the year in parts of its range (Clare et al. 2014; Hobson et al. 2001; Howell 1974b; Zortéa 2003). Dietary data on many other nectar bat species suggest they regularly supplement their floral diet with insects and fruit, which has led some authors to argue that they (and phyllostomids in general) should be considered opportunistic omnivores (Rex et al. 2010). For instance, *Glossophaga commissarisi* in a Costa Rican rainforest was found to switch to frugivory for several months of the year when nectar is scarce, with up to 30% of the individuals per month showing no pollen at all in the fur (Tschapka 2004). While species of *Anoura* in cloud forests are rarely found to consume fruits (Muchhala and Jarrín-V. 2002; Muchhala et al. 2005), in the Brazilian Cerrado, seeds are regularly found in their feces (Zortéa 2003). In fact, for *Anoura geoffroyi*, Zortéa (2003) estimated that pollen/nectar represented only 13% of their diet, while fruit parts represented 45% and insect parts represented 42%. Other records of fruit feeding among nectar bats include *Choeronycteris mexicana* and *Leptonycteris yerbabuenae* in Mexican deserts (Godinez-Alvarez and Valiente-Banuet 2000), *L. curasoae* and *Glossophaga longirostris* in Venezuelan shrublands (Sosa and Soriano 1993), and *Erophylla sezekorni* and *Monophyllus redmanii* in Puerto Rican moist forests (Soto-Centeno and Kurta 2006). Records of insectivory are even more common than those of frugivory; in fact, in all studies we have reviewed where feces of nectar bats were analyzed, at least some proportion of the samples contained insect parts (citations in this paragraph, and Barros et al. 2013; Goyret and Yuan 2015; Petit 1997; Rex et al. 2010; Sánchez and Medellín 2007; Sazima 1976; Sperr et al. 2011; Willig et al. 1993). The extent to which this represents aerial hunting remains unclear, as many of these insects may have been gleaned from flowers or other surfaces; however there are also records of nectar bats actively capturing flying insects in hawking bouts (Clare et al. 2014; Howell 1974b).

Morphological Adaptations to Nectarivory

The evolution of elongated mouthparts to facilitate access to nectar in flowers, arguably the most important physical adaptation to a nectar-feeding lifestyle, is common across the nectar-feeding members of various animal groups, including moths, flies, butterflies, and birds (Fenster et al. 2004). For phyllostomids, elongated jaws (see plate 4) allow bats to house longer tongues (Bolzan et al. 2015; Freeman 1995), which in turn allows for longer tongue extension and the ability to extract nectar from flowers with deep floral tubes (Gonzalez-Terrazas et al. 2012; Muchhala 2006a; Winter and von Helversen 2003). However, this does come with a trade-off: the longer the jaw, the greater the out-lever of the jaw system, and thus the weaker the bite force (Aguirre et al. 2002; Nogueira et al. 2009; Santana 2015; Santana et al. 2012), which may limit the ability to eat hard-bodied insects and tougher fruits or to defend against predators.

With an increasing degree of specialization, the jaw becomes less important as a tooth-bearing structure and more important for supporting the tongue (Freeman 1995). In correspondence with a predominantly fluid diet, there is a tendency toward reduction of the dentition within nectar-feeding phyllostomids. Bats with increased morphological specialization show a gap between the lower incisors (e.g., *Monophyllus* spp.) or have even have lost them entirely (e.g., *Anoura*; fig. 16.1), probably to maximize space for tongue movements (Carstens et al. 2002; Freeman 1995). In more derived species, there are also large gaps (diastemas) between the comparably small molars, as seen in *X. vieirai* (Gregorin and Ditchfield 2005) and *M. har-*

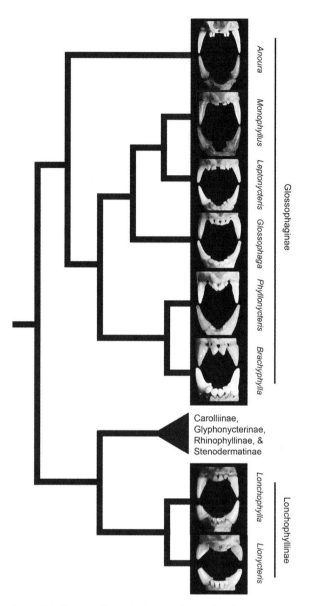

Figure 16.1. Upper and lower incisor morphology for select nectarivorous phyllostomids, showing the frontal gap through which the tongue is exserted, with phylogenetic relationships for the species. Adapted from Carstens et al. (2002), using the phylogeny of Rojas et al. (2016); images kindly provided by B. Carstens.

risoni (Tellez and Ortega 1999). Interestingly, canine teeth remain well developed across nectarivores (fig. 16.1), suggesting an important role. Freeman (1995) hypothesized that nectarivores brace upper and lower canines against each other to support the jaw during rapid tongue movements, citing patches of wear on the teeth where they contact each other. Canines may also function in "scraping" nectar off of the tongue's papillae as it moves past, or alternatively, they may primarily be used for defense or intraspecific aggression.

Tongue length is correlated with jaw length across

species of phyllostomid nectarivores (Winter and von Helversen 2003) and seems to depend on the degree of specialization on a nectar diet. In many species, animals maintain the tongue protruding slightly from the tip of the snout while not feeding. While less-derived nectarivores such as *Glossophaga* spp. can extend their tongues ca. 4 cm, more specialized species with particularly long jaws such as *C. mexicana* or *M. harrisoni* can extend them to 7–8 cm (Winter and von Helversen 2003). As in all mammals, tongues of nectar-feeding bats are muscular hydrostats, that is, they have a constant volume and are moved through complex bundles of muscles that allow elongation, shortening, bending, and torsion (Harper et al. 2013; Kier and Smith 1985). As in other mammals, the tongue initiates at the base of the oral cavity but differs in having a tongue retractor muscle that attaches much farther back, to the xiphoid process of the rib cage (plate 5*b*). *Anoura fistulata*, as mentioned above, provides a remarkable exception to this pattern in that the base of the tongue is shifted back into the thoracic cavity, close to the sternum (Muchhala 2006a; plate 5*a*). *Anoura* spp. are among the most insectivorous of the nectar-feeding bat species; it is possible that selection favored this solution for *A. fistulata* rather than elongation of the rostrum because it conserved the bite force necessary to feed successfully on insects.

Although Glossophaginae and Lonchophyllinae both evolved long tongues with papillae, closer inspection reveals profound differences in their morphologies, consistent with their independent evolution of nectarivory. While glossophagines have brush-like tongues with numerous filamentous papillae at the tip, lonchophyllines are characterized by papillae-lined grooves along the lateral sides of the tongue (Griffiths 1982; Howell and Hodgkin 1976). Recent studies of glossophagine feeding show that, as tongues approach maximum extension, the papillae on the tip become hydraulically erected through blood infusion and actively aid in collecting nectar (Harper et al. 2013). Glossophagine tongues perform sinusoidal lapping movements, alternating between dipping into the nectar and retracting into the mouth. In contrast, the extended lonchophylline tongue stays immersed in the nectar fluid throughout flower visits (Tschapka et al. 2015). Nectar rises along the canals into the mouth cavity, probably due to a combination of active peristaltic pumping and capillary forces in the small canals.

Interestingly, the efficiency of these two very different and independently evolved drinking methods are quite similar in terms of nectar extraction per unit of time, at least while feeding from artificial flowers (Tschapka et al. 2015; Muchhala, unpublished data). These differences in nectar extraction might offer advantages at flower types that differ in nectar amount and presentation: the lonchophylline pumping mechanism may work better at flowers with a larger nectar volume concentrated in one spot (e.g., *Ochroma pyramidale*—Malvaceae), while the glossophagine "nectar mop" may be better-adapted for the extraction of small and diffusely distributed droplets from smaller flowers (e.g., *Capanea grandiflora*—Gesneriaceae; *Merinthopodium neuranthum*—Solanaceae). Regardless, for both lonchophyllines and glossophagines, the amount of nectar they can extract within a certain time span (extraction efficiency) decreases sharply when they have to reach deeper into flowers, possibly due to greater time needed for tongue extension, increased leakage during tongue retraction, and/or decreased pumping power at longer extensions (Tschapka et al. 2015). Flower depth can vary widely across bat-visited flowers, from openly presented nectar in species such as *Cleome spinosa* (Capparaceae) to long-tubed species such as the columnar cactus *Pachycereus pecten-aboriginum*, with its 9-cm-deep corollas (Tschapka et al. 2008).

Another key to efficiently visiting flowers is nectar-feeding bats' ability for extended hovering flight. Extracting nectar without needing to support their weight on the flowers during visits allows bats to exploit flowers of less-robust plants, which likely expanded their possible dietary spectrum. In slow flight, and particularly while hovering, glossophagine bats reverse the hand wings during the upstroke to generate thrust and a small amount of lift (Håkansson et al. 2015; Norberg and Winter 2006). In contrast, hummingbirds rotate the entire wing around the shoulder joint and thus are able to produce even more lift during the upstroke (Warrick et al. 2005).

Physiological Adaptations to Nectarivory

Nectar is primarily an aqueous sugar solution that is very easy to digest. Stomachs of nectar-feeding bats are therefore relatively simple compared to those from other dietary habits. Bats match their feeding behavior to nectar sugar content, increasing food uptake in response to lowered sugar concentration (Ayala-Berdon and Schondube 2011). This ability to compensate likely has a lower limit due to the physiology of hydrolysis and water processing from extremely dilute sugar solutions (Ramírez et al. 2005). The enzyme sucrase in the digestive system of the glossophagine *Leptonycteris nivalis* was found to have a much higher affinity for sugars than that of its sister species *L. yerbabuenae*, which might translate into a more efficient energy uptake that may have permitted it to expand into higher, colder altitudes (Ayala-Berdon et al. 2013).

The high energetic needs of nectar-feeding bats, a product of a high basal metabolic rate and a feeding ecology requiring visitation to many flowers, translate into enormous food requirements. Small species such as *G. soricina* regularly consume up to 150% of their body weight per night when feeding on 17% sugar water (Winter and von Helversen 2003). This extremely high liquid uptake requires a particularly efficient water management physiology. Through regulation of glomeral filtration rate, urination is tightly coupled to liquid uptake, and excess water is shed by reducing renal water reabsorption. Nectar-feeding phyllostomids, like their frugivorous counterparts, need to excrete water while conserving electrolytes and they have a relatively well-developed renal cortex compared to their insectivorous ancestors (Casotti et al. 2006).

The constant influx of sugar may lead to extremely high glucose concentrations in the blood. In contrast to most other mammals that mobilize required energy from previously built-up fat stores, flight activity of glossophagine bats is largely fueled directly by their incoming carbohydrates, a physiological adaptation that they share with hummingbirds (Voigt and Speakman 2007). Although extended periods of high blood glucose concentration are detrimental to the health of most mammals, nectar-feeding bats do not appear to suffer such costs, either due to specialized physiological adaptations or because they are able to regulate high glucose levels through high physical activity (Kelm et al. 2011). This corresponds to experiments showing that glossophagines fly longer each night than necessary, based solely on their foraging needs (e.g., Horner et al. 1998). The consequence is a daily energy expenditure that is kept within a narrow range (von Helversen and Winter 2003). Energy turnover in glossophagines is extremely high, and in fact most of the body fat built up during a night of foraging (ca. 1 g for *G. soricina*) is consumed

over the course of the following day (von Helversen and Winter 2003). In situations where available resources do not cover daily energy requirements, glossophagines have been found to use diurnal torpor. In contrast to non-phyllostomid bats using torpor, glossophagines do not lower their body temperature close to ambient temperature but try to maintain the highest levels possible, given the energetic resources available to them (Kelm and von Helversen 2007). To date, torpor in nectar-feeding phyllostomids has only been investigated in laboratory experiments; the extent to which they actually use this strategy in nature remains unknown.

While in the majority of cases nectar bats visit flowers primarily for their nectar, after visits they will groom and consume the pollen that collects on their fur (Álvarez and González Quintero 1970; Herrera and Martínez del Río 1998; Howell 1974a). There are even occasional records of bats directly feeding on pollen during visits, for example, from *Pachira aquatica* (Hernández-Montero and Sosa 2015) or the nectarless, wind-pollinated species of *Cecropia* (Tschapka 2004). In captivity, they will actively consume dry pollen provided in dishes (NM and MT, pers. obs.). The exine of pollen grains is extremely tough and probably cannot be digested by bats. However, the stomach of nectar-feeding bats primarily contains a warm sugar solution that promotes the germination and emergence of the pollen tube from the exine shell and allows it to be digested (Howell 1974a; von Helversen 1995). Pollen is nearly always detected in nectar bat feces (e.g., Álvarez and González Quintero 1970), and, as might be expected, nectar bats that have been tested to date (*L. yerbabuenae*, *A. geoffroyi*, and *Brachyphylla nana*) are physiologically much more efficient at extracting the nutritious contents of pollen grains than other phyllostomids, even those known to opportunistically include pollen in their diets (Herrera and Martínez del Río 1998; Mancina and Balseiro 2005).

Sensory Ecology

Echolocation

While searching for flowers, nectar-feeding phyllostomids use a combination of active echolocation and passive perception of scent and other floral cues. Based on this foraging mode and their habitat, they can be classified as members of the "narrow space passive-/

active-gleaning" forager guild (Denzinger et al. 2016). Their echolocation calls correspond largely to the usual phyllostomid pattern, with very short (1–2 ms), multi-harmonic FM calls of low intensity. Traditionally, phyllostomids have been considered to be "whispering" bats because call recordings typically register low intensities. However, recent work shows their calls can be quite loud but are emitted with high directionality and thus can be difficult to record (Brinkløv et al. 2009; Kalko 2004). When flying in the open, the nectar-feeding bats *Phyllonycteris poeyi* and *Leptonycteris yerbabuenae* also use longer calls (up to 8 ms), with the highest energy in the first harmonic (Mora and Macías 2007; Gonzalez-Terrazas, pers. comm.). These open space calls are easily recorded from over 10 m away. The same species, however, emit typical phyllostomid calls when flying in more confined environments, such as a flight cage (Gonzalez-Terrazas, Koblitz et al. 2016), indicating a high flexibility in echolocation behavior.

Echolocation not only is used for orientation and obstacle avoidance during distance flight, but it also plays an important role for flower recognition and approach coordination. Glossophagine bats are able to recognize minute surface structures by echolocation, to a resolution of 0.38 mm (Simon et al. 2014). Additionally, they have been shown to recognize similar shapes despite varying sizes, demonstrating that they can generalize shape cues (von Helversen 2004). A few flowers have been shown to exploit these echolocation capabilities, and it is probable that even more will be recognized as acoustically conspicuous (see "Bat-Pollinated Flowers" for more details).

Olfaction

Scent is clearly an important sensory modality for phyllostomid nectar bats, as suggested by the fact that their olfactory bulbs and associated brain regions are relatively large (Bhatnagar and Kallen 1974; Stephan and Pirlot 1970). Accordingly, Neotropical bat-pollinated flowers typically produce strong, musky scents to attract their pollinators. Traditionally, scent has been viewed as a long-distance attractant that can guide bats to the general vicinity of bat-adapted plants before they switch to relying on echolocation (and perhaps vision) to locate specific flowers. However, recent flight cage experiments show that floral scent also plays a role at short distances (Gonzalez-Terrazas, Martel et al. 2016).

Scented flowers are consistently located faster than unscented flowers, and odor seems to be particularly important when locating flowers in structurally more complex environments (Muchhala and Serrano 2015).

As in most mammals, olfaction in nectar bats is important not only for locating food but also in interspecific interactions. Relatively few studies have explored mating systems of nectar bats, but those that have point to a central role for odor cues. For example, *Erophylla sezekornii* males produce garlic-scented secretions above their eyes, which they use along with wing movements to attract females to lek mating sites (Murray and Fleming 2008). Similarly, during its mating season, male *L. yerbabuenae* smear an odoriferous mixture of various body fluids on their backs, which may provide females with a cue to mate quality given that less-scented males have been found to have higher parasite loads (Muñoz-Romo et al. 2011, 2012; Rincón-Vargas et al. 2013). The similarity between these odors and those of bat-adapted flowers suggests a possible link, in that flowers may be exploiting a preference that originally evolved for interspecific communication.

Vision

The role of vision in nectar-feeding bats still seems to be rather unexplored. As in other phyllostomids, nectar feeders are relatively flexible and may switch between different sensory modalities for different activities. There are older reports that *A. geoffroyi* relied on visual over echolocation cues during escape behavior (Chase 1983), although bats of this species never stop echolocating throughout. Laboratory tests with *G. soricina* that have investigated its spectral sensitivity in choice experiments showed that it cannot distinguish between different parts of the color spectrum and therefore has to be considered color blind, at least for the scotopic (low-light) conditions used in this study (Winter et al. 2003). Interestingly, the same experiments also showed sensitivity at extremely short wavelengths, that is, in the ultraviolet range, which may be an adaptation to enhance visual contrast of ultraviolet-reflecting flowers that could increase their flower search efficiency. The physiological basis for this short wave sensitivity includes the fact that, unlike the visual systems of most diurnal mammals, both their cornea and lens transmit ultraviolet light (Müller et al. 2009), which can then be perceived by both rods (due to the beta-band in rod opsins;

Winter et al. 2003) and cones (due to shortwave-sensitive opsins; Müller et al. 2009). There are several night-blooming flowers, including chiropterophilous ones, that reflect short wavelengths (Biedinger and Barthlott 1993); however, the actual role of ultraviolet perception during foraging remains to be investigated.

Eye sizes vary distinctly between species of nectar bats. While rain forest species such as *Hylonycteris underwoodi* have small eyes, the desert-dwelling *L. yerbabuenae* has relatively large eyes (plate 6). Given the distinct differences in the structure of their habitats, this could indicate an adaptation to visual landmark orientation, for example, the outlines of mountains against the horizon during the long-distance flight of the latter in relatively open habitats, which probably plays a lesser role for most rain forest bats. So far, there are no experimental data on the use of vision during identification of flowers as food resources. It is conceivable that, in relatively open habitats such as dry forests or deserts, detection of the silhouettes of flowers may be important, especially at dusk or against a moonlit sky, while relatively closed habitats such as the understory of a rain forest may not provide sufficient light for this. Deserts also tend to have comparably bright flowers, such as those of columnar cacti (e.g., species of *Carnegiea*, *Pachycereus*, and *Stenocereus*), which may provide visual cues from short distances by maximizing contrast between flower and stem.

Spatial Memory

While most fruits or insects can be consumed only once, most flowers produce nectar over a period of several hours or even days, and nectar-feeding bats frequently visit the same floral resource repeatedly throughout the course of a night. In this sense, it is likely more energy-efficient for nectarivores to remember the precise location of past resources than to constantly search for new ones. Because their main nectar resources are renewable, selection likely favors high spatial memory abilities among nectarivores relative to frugivores and insectivores. Accordingly, spatial memory seems to be more important than other sensory cues for relocating previously visited flowers (Carter et al. 2010; Thiele and Winter 2005; Toelch et al. 2008; Winter and Stich 2005). Likely in response to such well-developed spatial memory, individual bat-adapted plants typically flower for months while presenting only one or few open flowers each night ("steady-state flowering," *sensu* Gentry [1974]) in certain habitats. For instance, plants of the bat-adapted rain forest species *M. neuranthum* will remain in bloom for up to 10 months (Tschapka 2004). Many species of the primarily bat-pollinated rain forest bromeliad genus *Werauhia* cue into bat's spatial memory at an even finer scale: flower buds originally orient outwards from two opposite sides of a flowering stalk, yet all reorient to face in the same direction just before opening, which results in an inflorescence that may be visited from the same side over its entire flowering duration of several weeks (Tschapka and von Helversen 2007). In contrast, bat-pollinated desert columnar cacti have shorter, cornucopia (*sensu* Gentry 1974) flowering seasons (Fleming 2002; Nassar et al. 1997) that could still favor well-developed spatial memory. Trapline foraging, that is, the repeated visitation of a circuit of flowers by the same individual, is likely and has been mentioned repeatedly as a strategy used by nectar-feeding bats (e.g., Fleming 1982; Sazima et al, 1999; von Helversen, 1993). However, actual observational proof of it in the field is still lacking. Advances in modern GPS-based tracking techniques may allow elucidation of such foraging patterns, particularly in species with large home ranges.

Nectar Bat Communities

Phyllostomid nectar bats form guilds (*sensu* Root 1967) that contain more specialized species than bat guilds in other parts of the world (Fleming and Muchhala 2008). Numbers of species in a guild are not easy to access from the literature. It can be difficult to differentiate rare but permanent members of the local guild, which will be shaped by more deterministic assembly processes, from the stochastic occurrence of transient species. The only way to differentiate these is by monitoring the community at a site over extended periods for several years. Species lists from field stations do not necessarily reflect the actual ecological communities, as they have a tendency to accumulate species over time, through stray individuals from neighboring habitats as well as through misidentification of species. The minute diagnostic differences in dentition among species of *Glossophaga*, for example, make nectar-feeding bats in this context a particularly problematic group.

Local guilds of nectar bats usually contain one to six species (Fleming and Muchhala 2008; Fleming et al. 2005). Additionally, often a smaller number of generalist phyllostomids feed from the locally available flower resources. These species, such as the omnivorous *Phyllostomus* spp. or the frugivorous *Carollia perspicillata* and *Artibeus jamaicensis*, are typically less capable at hovering and only use the few plant species that provide large amounts of easily accessible nectar, such as *Ochroma pyramidale* or *Ceiba pentandra* (Kays et al. 2012). Drier or more seasonal habitats also seem to promote nectar feeding among bats not specialized on flower visitation (MacSwiney G. et al. 2012; Tschapka and Dressler 2002). Most commonly, the core of lowland nectar-feeding guilds is formed by species of *Glossophaga*, although in South America these are sometimes replaced by small lonchophyllines such as *Hsunycteris thomasi*, while in montane habitats species of *Anoura* are often the most abundant nectar specialists (Fleming et al. 2005). Species that are more specialized morphologically, such as *M. harrisoni*, *H. underwoodi*, *A. fistulata*, and *Choeronycteris mexicana*, are in general far less abundant than the less-derived species, such as *G. soricina*.

Neotropical nectar bat communities are highly dynamic because the species depend on seasonally changing flower resources. Consequently, communities possess a core of resident species that are consistently present year-round, while other species fluctuate in numbers or are simply absent for part of the year (Sperr et al. 2011; Tschapka 2004). A seasonal presence at a location can be the consequence of long-distance migration or of more local resource-tracking movements, such as altitudinal migrations. It is feasible that there is a continuum between seasonal local movements and long-distance migration that so far is not very well understood. Long-distance migration is observed at the northern extreme of the distribution of phyllostomid bats, where the relatively large species *Leptonycteris yerbabuenae*, *L. nivalis*, and *C. mexicana* track seasonal flowering peaks (e.g., of *Agave* spp. or columnar cacti such as *Carnegiea gigantea*) in the deserts between Mexico and the United States (Moreno-Valdez et al. 2004; Penalba et al. 2006; Scott 2004). The well-studied migrations of *L. yerbabuenae* in Mexico provide an especially interesting case, and there is evidence of two distinct migration patterns with different reproductive timing on the Mexican mainland (Fleming and Nassar 2002). While about half of the females migrate northwards during early spring and give birth in the

Sonoran Desert, the remaining half remain in central and southern Mexico and give birth in December/January (Fleming et al. 1993; Stoner et al. 2003). These different behavioral modes do not represent genetically distinct groups but probably reflect individual decisions by females concerning where to roost (Newton 2002). Males largely remain relatively sedentary and likely provide genetic connectivity between populations of more mobile females. Body size is probably a key trait in migration, as larger species are capable of storing the energy necessary for buffering long-distance flight (Winter and von Helversen 2001), while small species are probably restricted to more local movements, for example, altitudinal migration along mountain slopes.

Patterns of resource partitioning by nectar bats and the underlying plant traits that shape their guilds are still relatively unexplored. Territoriality is only rarely observed among nectar-feeding bats. When nectar bats gather at plants with high resource density, such as *Mucuna holtonii* (Fabaceae) or at *Quararibea cordata* (Malvaceae), animals are occasionally observed chasing each other from the flowers, and the typical distress calls of *Glossophaga* spp. or *Lonchophylla robusta* can be heard during these occasions. Similar observations can be made at artificial feeders, which offer an extremely rich and long-lasting resource. However, this behavior seems to be restricted to especially profitable resources, in marked contrast to the frequent occurrence of territoriality in the ecologically similar hummingbirds (Fleming et al. 2005). This may reflect differences in the sensory systems used by bats, in that echolocation functions over a much shorter distance than vision (Kalko 2004) and may not allow bats to detect competitors entering a territory in time to prevent consumption of resources. Defense of a foraging territory therefore may only be feasible at plants offering very high resource density—that is, a substantial amount of food concentrated in a very small area that can be easily defended. Correspondingly, field observations of resource defense by glossophagine bats come from plant species offering many flowers at large inflorescences, such as *Agave desmettiana* (Agavaceae) (Lemke 1984) or *Calyptrogyne ghiesbreghtiana* (Arecaceae). At the latter species, *G. commissarisi* defends access not only against conspecifics but also against small *Artibeus* spp. and katydid insects (Orthoptera: Ensifera) (Tschapka 2003). In most situations, however, community structuring by competition likely reflects exploitative rather than interference competition. This may have selected for small body size in glossophagines since smaller species have lower energy requirements (Winter and von Helversen 2001). Additional benefits of being small include increased maneuverability, which facilitates access to flowers. In turn, smaller size and lower energetic requirements may have facilitated the evolution of bat-adapted flowers in the New World. In line with this, trees provide the majority of flowers in the Old World, while a diversity of growth forms (vines, herbs, epiphytes, and trees) provide bat-adapted flowers in the New World (Fleming et al. 2005).

There is evidence that bats make foraging choices based on the energy density of floral resources, in terms of nectar volume, sugar concentration, number of flowers, and density of plants. In a Costa Rican lowland guild of four nectar bat species, *Lichonycteris obscura* and *L. robusta* visited the study site only during the main flowering season and favored plants with high energy density that permitted very efficient foraging (Tschapka 2004). Among the permanent residents at that study site, the common and less-specialized *G. commissarisi* favored plants offering high feeding efficiency for part of the year and then switched to feeding on abundant fruit when these were no longer available. The morphologically more specialized nectar-feeder *H. underwoodi* visited flowers year-round, including those with lower nectar density that required a higher flight effort per night.

Differences in wing characteristics support the idea

of resource partitioning based on foraging efficiency. Compared to differences between bat families (Findley et al. 1972), the differences in wing proportions among nectar-feeding phyllostomids are subtle, yet they may reflect adaptations to different spatial foraging strategies. One can quantify wing proportions by calculating the aspect ratio index and wing tip index, indices that reflect basic flight characteristics of bats (Findley et al. 1972). The aspect ratio index assesses relative antero-posterior width of the wing, with high values indicating narrow wings, which allow for rapid flight. The wing tip index measures the length of the wing tip relative to the forearm; relatively long tips also aid with high flight speeds and are found in hovering species. Data on these wing proportions compiled for nine nectar bat species from Costa Rica and Mexico reveal some interesting patterns (fig. 16.2). Within the La Selva guild, the specialized *H. underwoodi* has wings with both a higher aspect ratio index and higher wing tip index than the generalized *G. commissarisi*, which may be an adaptation to their efficient visitation to the more scattered low nectar density plants. Among all species in fig. 16.2, the highest aspect ratio indexes were found in *Anoura geoffroyi* and *Anoura cultrata*. Species of *Anoura* are characteristic inhabitants of montane forests. Perhaps the high aspect ratio index, indicating an adaptation to fast long-distance flight, in combination with their relatively large body size allows these species to occupy more flexible niches in montane habitats. While based at a permanent roosting cave, they may quickly reach

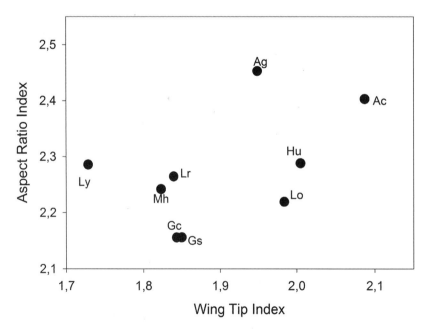

Figure 16.2. Wing proportions of nectar-feeding bats. *Anoura geoffroyi* and *A. cultrata* show a higher aspect ratio index than all other species. Species abbreviations and sample sizes: Ac: *A. cultrata* (4), Ag: *Anoura geoffroyi* (22), Gc: *Glossophaga commissarisi* (276), Gs: *G. soricina* (175), Hu: *Hylonycteris underwoodi* (41), Lo: *Lichonycteris obscura* (27), Lr: *Lonchophylla robusta* (71), Ly: *Leptonycteris yerbabuenae* (124), Mh: *Musonycteris harrisoni* (46).

foraging habitats in different altitudes with different floral resources, allowing these bats to track temporal shifts in local floral resource availability along mountain slopes.

Habitats with low nectar density will be unable to support large populations of nectar-feeding bats, particularly for those species that only rarely use alternative resource classes (fruits and/or insects). Small population size may ultimately also restrict mating behavior, in that species occurring at low densities will have fewer mating opportunities (see chap. 13, this vol.). The flexible diet of *Glossophaga* spp. allows the occurrence of large populations with roosts containing up to hundreds of individuals. In contrast, the largest colony reported of the more nectar-dependent *H. underwoodi* contained only eight individuals (LaVal and Fitch 1977), so mating opportunities should be more limited in this species. Although *G. commissarisi* and *H. underwoodi* are similar in size (body mass 8.8 g vs. 7.6 g, respectively; Tschapka 2004), maximum testis size differs by a factor of two (fig. 16.3). Testis size is a morphological correlate for the occurrence of sperm competition (Wilkinson and McCracken 2003; chap. 13, this vol.). The comparatively large testes of *G. commissarisi* suggest that sperm competition plays a more important role for it than for *H. underwoodi*. A female *Glossophaga* may easily encounter and mate with several males, who therefore compete with each other for paternity, while multiple matings should be much less frequent in the uncommon *H. underwoodi*. The small

testes of *Hylonycteris* and several other highly specialized and rare species, such as *M. harrisoni* and *L. obscura*, may therefore be an anatomical representation of a mating system adapted to low population densities in nectar-poor habitats.

Species of *Leptonycteris* (especially *L. curasoae* and *L. yerbabuenae*) are the largest glossophagines and differ from all other nectar-feeding phyllostomids by living in colonies containing tens of thousands of individuals (Fleming and Nassaar 2002). When feeding on the flowers of columnar cacti, these strong-flying bats commute long distances from their day roosts to their foraging areas (e.g., up to 90 km one-way; Y. Yossel, pers. comm.). Their large size (up to ca. 30 g) and high aspect ratio wings result in efficient long-distance commute flights (Sahley et al. 1993).

Bat-Pollinated Flowers

Despite the recent appearance of nectar-feeding phyllostomid bats (ca. 20 mya; Baker et al. 2012) relative to other pollinating animals, a substantial number of tropical and desert plants have adapted to bat pollination in the New World, including more than 500 species from 67 families (Fleming et al. 2009). Flowers of these plants typically share a suite of traits, termed the chiropterophilous pollination syndrome, that adapt them to pollination by bats. In many cases, chiropterophily has evolved from ornithophilous ancestors, specifically pollination by hummingbirds (Fleming et al. 2009;

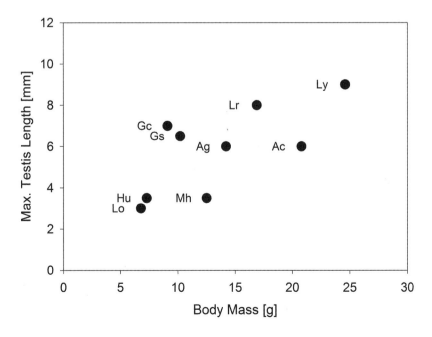

Figure 16.3. Testis size in nectar-feeding bats. Maximum testes length is plotted to account for the seasonal changes in testicle size. *Musonycteris harrisoni, Hylonycteris underwoodi,* and *Lichonycteris obscura* show remarkably small testicles for these species' overall size, suggesting a low level of sperm competition. Species abbreviations as in figure 16.2 (with sample sizes in parentheses): Ag (9), Ac (2), Gc (133), Gs (32), Hu (18), Lo (9), Lr (15), Ly (105), Mh (22).

Knox et al. 2008; Martén-Rodríguez et al. 2009; Tripp and Manos 2008).

Botanists have long recognized that different types of pollinators lead to convergent evolution on different suites of traits, and began to formalize descriptions of these pollination syndromes in the middle of the twentieth century (Baker 1961; van der Pijl 1961; Vogel 1954, 1969). For chiropterophily, these traits traditionally include (1) nocturnal anthesis, (2) whitish or drab colors, (3) large and sturdy flowers, (4) large amounts of nectar and pollen, (5) "unfresh" smell, and (6) well-exposed flowers positioned outside of the foliage (van der Pijl 1961; Vogel 1969). Since these initial descriptions, our understanding of the variety and evolutionary significance of these traits has progressed as bat pollination has been documented in more plants (Dobat and Peikert-Holle 1985; Fleming et al. 2009) and as the interaction between these mutualists has been studied in more detail.

The majority of chiropterophilous traits can be readily understood as outcomes of selective pressures imposed by relatively large nocturnal animals; these include nocturnal anthesis, large flower size, and the production of copious amounts of nectar. Given their exceptionally high metabolic rates, phyllostomid nectar bats require large quantities of nectar to satisfy their nightly energy needs. However, the copious nectar that bat flowers produce tends to be dilute compared to that of other flowers, typically with less than 20% sugar content (Perret et al. 2001; Roces et al. 1993; Sazima et al. 1999; von Helversen 1993). This is initially surprising, given that bats provided with options of different concentrations prefer higher sugar levels, up to a threshold of 60% (Roces et al. 1993; Rodríguez-Peña et al. 2007). This apparent discrepancy resolves when one considers that flowers have access to a set amount of sugar to dispense and thus can either increase quantity (more water, lower concentration) or quality (less water, higher concentration). Bats prefer both the greatest quantity and the highest quality per flower, but their ability to distinguish differences in either decreases at higher amounts. An elegant combination of experiments and computer simulations (Nachev et al. 2017) demonstrated that this leads to a preferred balance of quantity and quality involving moderate amounts of dilute nectar, close in concentration levels to that found in nature.

Along with relatively low overall sugar concentrations, bat-adapted flowers tend to have higher levels of hexoses (fructose and glucose) than sucroses in their nectar, in contrast to the sucrose-dominated nectar of insect- and bird-adapted flowers (Baker and Baker 1990; Baker et al. 1998; Rodríguez-Peña et al. 2016). However, bats actually prefer sucrose over hexose when given the choice (Aguirre et al. 2002; Herrera 1999) or fail to show any preference (Rodríguez-Peña et al. 2007). Thus, bats do not seem to select for hexose, and the evolutionary significance of high hexose levels in bat flowers remains a mystery.

Due to the relatively large size of phyllostomids and their rough handling of flowers during visits (compared to birds and insects), the petals and pedicels/peduncles of chiropterophilous flowers tend to be particularly sturdy. While hummingbirds and insects typically do not displace flowers during visits, bats either hover throughout visits, displacing them as they quickly enter and exit corollas (visits usually last only 0.3–0.6 s), or they cling briefly, hooking onto petals with their thumb claws and pulling the flowers down under their weight (Voigt 2004). Aside from sturdiness, the fit between bat and flower is another important aspect of floral size and shape. Two main morphologies can be found among chiropterophilous flowers: a "shaving brush" shape, with open flowers/inflorescences and multiple projecting stamens, and a "cup" or "bell" shape with petals forming a relatively wide tube that fits around the head of the bat. The former is more common among Old World flowers pollinated by pteropodids (Fleming and Muchhala 2008), including species of *Sonneratia* (Nor Zalipah et al. 2016) and *Syzygium* (Law 2001); they can also be seen in phyllostomid-pollinated New World species such as *Pseudobombax ellipticum* (Dobat and Peikert-Holle 1985). Tubular flowers are much more common in the New World (see plate 7) and range from the long tubes of *Pitcairnia* (Muchhala and Jarrín-V. 2002) and *Centropogon nigricans* (Muchhala 2006a) to the bell shapes of *Werauhia gladioliflora* (Tschapka and von Helversen 2007) and *Burmeistera* (Muchhala 2006b) to the conical shapes of cacti such as *Weberocereus tunilla* (Tschapka et al. 1999) and *Pachycereus* spp. (Fleming et al. 2001). For these tubular flowers, the width of the opening is particularly important to properly guide bats' heads and thus maximize contact with the flower's reproductive parts (Muchhala 2007). Chiropterophilous flowers tend to be wider than ornithophilous flowers, which are narrower to fit birds' bills (Castellanos et al. 2004) and sphingophilous flowers, which are particularly narrow to fit the hawkmoths' proboscos (Johnson et al. 2016). There are also instances of bat-adapted flowers that cannot be easily classified as either shaving brush or tubular, such as the papilionaceous flowers of *Mucuna* (von Helversen and von Helversen 2003) or the candelabra-like inflorescences of *Marcgravia* (Tschapka et al. 2006). For additional photographs and illustrations of the various shapes of bat-adapted flowers, please see Dobat and Peikert-Holle (1985).

Along with aspects of floral morphology to withstand and effectively transfer pollen during bat visits, selective pressures on chiropterophilous flowers will also favor traits that maximize attraction of their bat pollinators. As discussed above, vision, olfaction, and echolocation likely all play roles in nectar bat foraging. Of these, vision is arguably the least important, given the low light availability at night. While hummingbird and insect flowers are marked by a diversity of vibrant colors, bat flowers tend to be unsaturated (dull colored), reflecting the fact that bats are presumed to be color blind (Winter et al. 2003). In the Old World, despite low saturation, the flowers adapted to pteropodid fruit bats often display high brightness (e.g., white or cream coloration; Baker 1961), which likely reflects a heavier reliance on vision among the nonecholocating pteropodids (Liu et al. 2015; von Helversen 1993). In the New World, flowers adapted to phyllostomid nectar bats typically display both low saturation and low brightness, and shades of green or purple are the most common (although there are exceptions such as the bright cactus flowers of North American deserts). This marks a strong contrast with nocturnal flowers adapted to moth pollination, which are typically bright white (Grant 1983). In fact, the darker shades of some Neotropical bat flowers have been hypothesized to reflect an evolutionary tactic to "hide" the flowers from moths or other unwanted pollinators (von Helversen 1993). Experimental studies have shown that hawkmoths prefer white flowers under dim conditions, likely due to their high contrast and visibility (Goyret et al. 2008; Goyret and Yuan 2015). Given that hawkmoths typically do not contact stamens/stigmas during visits to bat-adapted flowers (Ibarra-Cerdeña et al. 2005; Machado et al. 1998), selection may favor darker colors to deter hawkmoth visits and conserve nectar for bats.

Chemical attractants are universally used for pollinator attraction, and bat-pollinated flowers are no exception. The smell of bat-pollinated flowers has been described as unpleasant, for example, garlicky, musky, or cadaver-like (Dobat and Peikert-Holle 1985). Interestingly, many Neotropical species share the presence of sulfurous compounds, which are only rarely found among floral attractants (Bestmann et al. 1997; Knudsen and Tollsten 1995). Experiments in a captive colony with naive bats showed that the compound dimethyl disulphide is an innate attractant for the nectar-feeding bat *G. soricina* (von Helversen et al. 2000). This substance is found in a number of Neotropical bat-pollinated flowers, but Old World bat plants seem not to use it at all (Pettersson et al. 2004). In fact, flowers of *Ceiba pentandra* in the New World contain dimethyl disulphide while those in the Old World do not (Pettersson et al. 2004; von Helversen et al. 2000). Supporting this, flight cage experiments found that the Old World nectar-feeding bat *Eonycteris spelaea* is not attracted to it (Carter and Stewart 2015).

Aside from scent and vision, phyllostomid bats also rely on echolocation while foraging. As in almost all bat families, echolocation is critical to their orientation at night (Gonzalez-Terrazas, Koblitz et al. 2016; Gonzalez-Terrazas, Martel et al. 2016). It has been recently discovered that some chiropterophilous plants have evolved to exploit this sensory system, providing highly conspicuous echoacoustic signals to guide bats. Von Helversen and von Helversen (1999) demonstrated that the concave upper petals (vexillum) of *M. holtonii* flowers function as acoustic "mirrors" for *Glossophaga*, concentrating and strongly reflecting their echolocation calls and thus signaling the presence of ripe flowers. Newly opened flowers offer the greatest reward, yet are hard to exploit because bats must land in a certain way to trigger the explosive pollination mechanism. Vexillae are common to all Fabaceae flowers, and in this case selection by nectar bats appears to have optimized it as a specialized echo reflector, maximizing successful visitation and thus pollen transfer (von Helversen and von Helversen 1999, 2003). *Marcgravia evenia* exhibit a similar phenomenon: specialized dish-shaped leaves positioned above their inflorescences provide consistently strong and invariant echoes over wide angles of incidence (Simon et al. 2011). Flight cage experiments have demonstrated that the presence of these leaves greatly reduce search times for flowers (Simon et al. 2011). The ubiquity of such echoacoustic "beacons" among bat-adapted flowers is still unclear, although it has been suggested that the bell-shaped corollas of many flowers may function in a similar fashion and that the waxy petals common to many bat flowers also serve to increase echo reflectance (Holderied and von Helversen 2006; von Helversen et al. 2003). Even in the absence of explicit beacons, chiropterophilous flowers can facilitate their detection by echolocating animals by reducing the presence of background obstructions that would otherwise lead to acoustic "clutter echoes." The well-exposed nature of chiropterophilous flowers noted early on in descriptions of pollination syndromes (*sensu* van der Pijl [1961, 51]: "a peculiar position outside of the foliage") may represent an adaptation to do this (Winter and von Helversen 2001). Different bat-adapted flowers accomplish this exposure differently, whether through flagelliflory (hanging inflorescences), cauliflory (inflorescences on trunks), or long pedicels (floral stems). Across the primarily bat-pollinated genus *Burmeistera*, the only species that does not possess long pedicels, *B. ceratocarpa*, instead has very small terminal leaves near its flowers, which likely provides an alternate solution to increasing floral exposure by minimizing clutter echoes from these leaves (Muchhala 2006b). Experimental work supports the idea that reducing background clutter aids in detection: in the absence of a complex background of leaves, bats readily located artificial flowers with or without odor cues, while with the leaf background they relied significantly more on odor cues (Muchhala and Serrano 2015). An alternate approach to reducing clutter echoes, rather than eliminating background plant structures, would be to provide background plant structures that are particularly effective at absorbing sound. In a possible example of this, flowers of the columnar cactus *Espostoa* emerge from an apical region that is covered by a cephalium, an area of dense, wooly hairs. These hairs may provide an acoustically nonreflective backdrop, thus enhancing the conspicuousness of echoes from emerging flowers (Schöner et al. 2016; von Helversen et al. 2003). The bat's sensory system may therefore exert selection on plants not only for ultrasound reflecting but also for nonreflecting surfaces.

Different pollinator types can also exert differing selection pressures on pollen, in terms of the amount produced per flower, its morphology, and its presentation schedule, that is, the manner in which it is made available to pollinators over the life of a flower (Castel-lanos et al. 2006; Thomson et al. 2000). An extensive comparison of pollen morphology of flowers adapted to bats versus other pollinators, however, found no consistent differences in shape or exine ornamentation, although bat-adapted pollen tended to be larger by around 10% (Stroo 2000). This likely represents a by-product of the larger size of bat flowers, rather than a specific adaptation to bats, since a strong correlation between style length and pollen size was also detected. Although morphological differences are not great, chiropterophilous flowers tend to produce much larger quantities of pollen (Faegri and van der Pijl 1979; Skog 1976). For example, among species in an Ecuadorian cloud forest, bat flowers were found to produce sevenfold more pollen on average than hummingbird flowers (Muchhala and Thomson 2010). Flight cage experiments comparing pollen transfer by bats and hummingbirds found a strong positive linear relationship between the amount of pollen bats picked up from one flower and subsequently delivered to a second flower. However, the same relationship for hummingbirds leveled off quickly; that is, there was an upper limit to how much pollen an individual hummingbird could carry (Muchhala and Thomson 2010). This may be due to behavioral differences or because fur can be loaded with pollen to a greater extent than can feathers. These results suggest that a shift to bat pollination will favor an evolutionary increase in pollen production, as flowers that produce more pollen will sire more seeds, while there is an upper limit to the profitability of increased investment in pollen production among hummingbird flowers. Bats may also select for greater pollen production because of the importance of pollen in their diet, although as mentioned previously, they have only rarely been documented to consume pollen directly from flowers during visits.

In the same way that phyllostomid frugivores, insectivores, and nectarivores do not always exclusively feed from their preferred resource, flowers often do not strictly conform to their pollination syndromes (Ollerton et al. 2009; Waser et al. 1996). Flowers are often visited and pollinated by taxa other than what would be predicted by their floral traits. Chiropterophily is actually one of the best-defined pollination syndromes in terms of the traits that characterize it and its predictive power (Rosas-Guerrero et al. 2014), likely due to the large differences in the size, behavior, and sensory systems of bats relative to more typical insect and hummingbird pollinators. The traits that adapt a flower to bats probably impose heavy trade-offs, decreasing the effectiveness of other pollinator types, as has been shown for floral width (Muchhala 2007). Even so, there are several well-documented cases of bat-visited flowers also relying on other pollinator types. For instance, bats, bees, and hawkmoths contribute to pollination of *Pachira aquatica* (Hernández-Montero and Sosa 2015), and both bats and hummingbirds can effectively pollinate *Siphocampylus sulfureus* (Sazima et al. 1994), *Aphelandra acanthus* (Muchhala et al. 2009), various Antillean *Gesneria* and *Rhytidophyllum* (Martén-Rodríguez et al. 2009), and several species of columnar cacti (Dar et al. 2006; Fleming et al. 2001).

For their part, nectarivorous phyllostomids will also stray on occasion from flowers that conform to their pollination syndrome. There are records of bats visiting insect- or wind-pollinated flowers (Hernandez-Montero and Sosa 2015; Sperr et al. 2011; Tschapka 2004), which they probably do not pollinate effectively and are thus essentially parasitizing floral resources adapted to other pollinators. It is possible that the high energy turnover and associated food requirements of nectar-feeding bats selected for a high degree of inquisitiveness and opportunism in their foraging behavior. Further evidence of their inquisitiveness can be seen in their rapid acceptance of hummingbird feeders across the Neotropics, structures that bear neither visual nor olfactory resemblance to the chiropterophilous syndrome (Tschapka 2003).

Within chiropterophily, there exists very little further specialization in terms of specific bat-flower interactions. All nectar-feeding phyllostomids in a particular habitat tend to visit all bat-adapted flowers available to them; although there often are distinct differences in intensity of use, typically there is almost complete overlap in terms of the floral species in their diets (Muchhala and Jarrín-V. 2002; Tschapka 2004). This is true even for *M. harrisonii*. Although it has the longest jaw of any nectar bat, no flowers are known to be adapted exclusively to its pollination (Tschapka et al. 2008). The only known exceptions to this rule are two species in the diet of *A. fistulata*. *C. nigricans* (Muchhala 2006a; Muchhala and Thomson 2009) and *Marcgravia williamsii* (Muchhala, unpublished data) have flower/nectary depths that prevent other bats from accessing their nectar

and are exclusively pollinated by *A. fistulata*. The fact that bat flowers do not typically specialize on different phyllostomid species comes with a fitness cost: bats regularly carry the pollen of multiple flowers on their fur (Muchhala and Jarrín-V. 2002; Tschapka 2004), losing pollen to foreign flowers and clogging stigmas with foreign pollen (Muchhala and Thomson 2012). In response to this cost, different bat flowers in a given habitat or clade will often place their pollen on different parts of bat bodies (Howell 1977; von Helversen 1995). For species with less precise pollen placement, this partitioning is relatively coarse, as in species of *Marcgravia* (Tschapka et al. 2006), which will use either the ventral or the dorsal surfaces of bat bodies and wings. *Burmeistera* flowers represent the opposite extreme because they pick up and deposit pollen from a precise spot on bats' heads, and multiple species can coexist with limited pollen mixing by varying this location from the end of the snout to between the shoulder blades (Muchhala 2008; Muchhala and Potts 2007).

Conclusions

Nectar-feeding bats are a large and diverse group of phyllostomid species that have enlisted plant partners in all major habitats of the Neotropics, from the deserts of Mexico and South America to Amazonian rainforests, up to the cloud forests of Central American cordilleras and the Andes. Nectarivory has evolved twice among phyllostomids, in the Glossophaginae and Lonchophyllinae. Species in these two subfamilies all possess elongated tongues and rostrums and the ability to hover, yet they differ in details of tongue morphology and feeding behavior. While the nectar of the coevolved flowers constitutes their main resource, for many species insects and fruits play an important role seasonally or even year-round. It is likely that the high energetic requirements for these animals selected for an extreme dietary flexibility and the ability to use many accessible resources. For plants, this flexibility results in a continuum ranging from coevolved mutualisms with highly efficient pollination (e.g., *M. neuranthum*, *C. gigantea*, *Crescentia alata*, *C. nigricans*) to parasitism where the bats are mere consumers of floral resources without providing benefits for the plants (e.g., insect- or wind- pollinated plants). This wide range of interactions offers interesting possibilities for studying evolutionary trade-offs because fitness consequences of bat-plant interactions are relatively easy quantified. Energetic benefits for bats can be estimated in kilojoules content of the nectar, while benefits of plants can be estimated by percentage of seed set.

While we have greatly improved our understanding of nectar bat biology over the past several decades, particularly in terms of physiology and behavior, much remains to be explored. We still do not fully understand nectar bat foraging behavior, in terms of what cues they rely on to find flowers, the importance of these cues when locating and approaching flowers, and typical movement patterns throughout nights and years. Some particularly interesting questions that we would like to highlight include the following. (1) Do nectar-feeding bats perform traplining, moving in predictable circuits each night? (2) Do they migrate, either locally or regionally, through the year? (3) Do they perceive colors at dusk/dawn, and what is the role of ultraviolet vision in movement and foraging? (4) To what extent do they actively hunt insects? (4) How do lonchophylline tongues extract nectar? (5) How do communities of nectar bats partition resources, and to what extent do they compete, either directly (territoriality/aggression) or indirectly (exploitative competition)? (6) Why are bat flowers hexose rich?

Finally, relatively little work has been done on conservation of nectar-feeding bats. As for most phyllostomids, nectar-feeding bats suffer from habitat loss and habitat fragmentation, which changes plant species composition and therefore also affects their native food resources. However, in contrast to, for example, most of the gleaning animalivores (chap. 14, this vol.), many nectar-feeding bats can also persist in largely anthropogenically transformed habitats. They roost in abandoned houses, sewer canals, and shacks and may use some cultivated or native species that thrive in disturbed areas. For instance, *Glossophaga* spp. are frequently observed in banana (*Musa* spp.) plantations. Although the cultivated forms of banana do not require pollination for fruit production, the inflorescences offer rich, year-round nectar sources that are readily accepted by bats. The long-term consequences for these bats of such a year-round uniform diet remain largely unexplored, as do the potential impacts of agrochemicals used in the plantations.

The potential impacts of global warming on nectar-feeding phyllostomids are also poorly understood. For instance, *L. yerbabuenae* colonies use the flowering and fruiting season of several columnar cactus species in the Sonoran Desert to be able to establish their maternity colonies there. The impacts of warming on these bats in the already extreme conditions of these deserts merits further study, especially given that this is a particularly sensitive period of their annual cycle. Finally, besides offering many intellectually stimulating scientific questions as well as urgent conservation challenges, the interaction between flower bats and bat-adapted flowers is also a particularly charismatic one, a fact which can and should be exploited to find more human friends for bats!

Acknowledgments

We would like to thank Theodore H. Fleming for inviting our contribution and for his careful and thorough editing of it. We are also grateful to two anonymous reviewers for their helpful suggestions. NM's research on nectar-feeding bats has been supported over the years by the National Geographic Society, Bat Conservation International, Fulbright Scholar Program, and National Science Foundation. MT's research has been supported by the German Academic Exchange Service (DAAD) and the German Research Foundation (DFG).

References

Aguirre, L. F., A. Herrel, R. van Damme, and E. Matthysen. 2002. Ecomorphological analysis of trophic niche partitioning in a tropical savannah bat community. Proceedings of the Royal Society of London Series B–Biological Sciences, 269:1271–1278.

Alvarez, J., M. R. Willig, J. K. Jones, and W. D. Webster. 1991. *Glossophaga soricina*. Mammalian Species Archive, 379:1–7.

Álvarez, T., and L. González Quintero. 1970. Analisis polinico del contenido gastrico de murcielagos Glossophaginae de Mexico. Anales de la Escuela Nacional de Ciencias Biologicas, 18:137–165.

Arkins, A., A. Winnington, S. Anderson, and M. Clout. 1999. Diet and nectarivorous foraging behaviour of the short-tailed bat (*Mystacina tuberculata*). Journal of Zoology, 247:183–187.

Ayala-Berdon, J., R. Galicia, C. Flores-Ortiz, R. A. Medellín, and J. E. Schondube. 2013. Digestive capacities allow the Mexican long-nosed bat (*Leptonycteris nivalis*) to live in cold environments. Comparative Biochemistry and Physiology Part A: Molecular and Integrative Physiology, 164:622–628.

Ayala-Berdon, J., and J. E. Schondube. 2011. A physiological perspective on nectar-feeding adaptation in phyllostomid bats. Physiological and Biochemical Zoology, 84:458–466.

Baker, H. G. 1961. The adaptation of flowering plants to nocturnal and crepuscular pollinators. Quarterly Review of Biology, 36:64–73.

Baker, H. G., and I. Baker. 1990. The predictive value of nectar chemistry to the recognition of pollinator types. Israel Journal of Botany, 39:157–166.

Baker, H. G., I. Baker, and S. A. Hodges. 1998. Sugar composition of nectars and fruits consumed by birds and bats in the tropics and subtropics. Biotropica, 30:559–586.

Baker, R. J., O. R. Bininda-Emonds, H. Mantilla-Meluk, C. A. Porter, and R. Van Den Bussche. 2012. Molecular timescale of diversification of feeding strategy and morphology in New World leaf-nosed bats (Phyllostomidae): a phylogenetic perspective. Pp. 385–409 in: Evolutionary History of Bats: Fossils, Molecules and Morphology (G. F. Gunnell and N. B. Simmons, eds.). Cambridge University Press, Cambridge.

Baker, R. J., R. L. Honeycutt, M. L. Arnold, V. M. Sarich, and H. H. Genoways. 1981. Electrophoretic and immunological studies on the relationship of the Brachyphyllinae and the Glossophaginae. Journal of Mammalogy, 62:665–672.

Baker, R. J., S. Solari, A. Cirranello, N. B. Simmons. 2016. Higher level classification of phyllostomid bats with a summary of DNA synapomorphies. Acta Chiropterologica, 18:1–38.

Barros, M. A., A. M. Rui, and M. E. Fabian. 2013. Seasonal variation in the diet of the bat Anoura caudifer (Phyllostomidae: Glossophaginae) at the southern limit of its geographic range. Acta Chiropterologica, 15:77–84.

Bestmann, H. J., L. Winkler, and O. von Helversen. 1997. Headspace analysis of volatile flower scent constituents of bat-pollinated plants. Phytochemistry, 46:1169–1172.

Bhatnagar, K. P., and F. C. Kallen. 1974. Cribriform plate of ethmoid, olfactory bulb and olfactory acuity in forty species of bats. Journal of Morphology, 142:71–89.

Biedinger, N., and W. Barthlott. 1993. Untersuchungen zur Ultraviolettreflexion von Angiospermenblüten. I. Monocotyledoeae. Tropische und Subtropische Pflanzenwelt, 86:1–122.

Bolzan, D. P., L. M. Pessôa, A. L. Peracchi, and R. E. Strauss. 2015. Allometric patterns and evolution in Neotropical nectar-feeding bats (Chiroptera, Phyllostomidae). Acta Chiropterologica, 17:59–73.

Brinkløv, S., E. K. V. Kalko, and A. Surlykke. 2009. Intense echolocation calls from two "whispering" bats, Artibeus jamaicensis and Macrophyllum macrophyllum (Phyllostomidae). Journal of Experimental Biology, 212:11–20.

Carstens, B. C., B. L. Lundrigan, and P. Myers. 2002. A phylogeny of the Neotropical nectar-feeding bats (Chiroptera: Phyllostomidae) based on morphological and molecular data. Journal of Mammalian Evolution, 9:23–53.

Carter, G. G., J. M. Ratcliffe, and B. G. Galef. 2010. Flower bats (Glossophaga soricina) and fruit bats (Carollia perspicillata) rely on spatial cues over shapes and scents when relocating food. PLOS ONE, 5:e10808.

Carter, G. G., and D. K. Riskin. 2006. Mystacina tuberculata. Mammalian Species, 790:1–8.

Carter, G. G., and A. B. Stewart. 2015. The floral bat lure dimethyl disulphide does not attract the palaeotropical dawn bat. Journal of Pollination Ecology, 17:129–131.

Casotti, G., L. G. Herrera, E. J. Braun, M. A. Carlos, and J. J. Flores. 2006. Relationships between renal morphology and

diets in 26 species of New World bats (suborder Microchiroptera). Zoology, 109:196–207.

Castellanos, M. C., P. Wilson, S. J. Keller, A. D. Wolfe, and J. D. Thomson. 2006. Anther evolution: pollen presentation strategies when pollinators differ. American Naturalist, 167:288–296.

Castellanos, M. C., P. Wilson, and J. D. Thomson. 2004. "Anti-bee" and "pro-bird" changes during the evolution of hummingbird pollination in Penstemon flowers. Journal of Evolutionary Biology, 17:876–885.

Chase, J. 1983. Differential responses to visual and acoustic cues during escape in the bat Anoura geoffroyi: cue preferences and behaviour. Animal Behaviour, 31:526–531.

Cirranello, A., N. B. Simmons, S. Solari, and R. J. Baker. 2016. Morphological diagnoses of higher-level phyllostomid taxa (Chiroptera: Phyllostomidae). Acta Chiropterologica, 18:39–71.

Clare, E. L., H. R. Goerlitz, V. A. Drapeau, M. W. Holderied, A. M. Adams, J. Nagel, E. R. Dumont, P. D. Hebert, and M. Brock Fenton. 2014. Trophic niche flexibility in Glossophaga soricina: how a nectar seeker sneaks an insect snack. Functional Ecology, 28:632–641.

Dar, S., M. D. Arizmendi, and A. Valiente-Banuet. 2006. Diurnal and nocturnal pollination of Marginatocereus marginatus (Pachycereeae : Cactaceae) in Central Mexico. Annals of Botany, 97:423–427.

Datzmann, T., O. von Helversen, and F. Mayer. 2010. Evolution of nectarivory in phyllostomid bats (Phyllostomidae Gray, 1825, Chiroptera: Mammalia). BMC Evolutionary Biology, 10:165.

Denzinger, A., E. K. Kalko, M. Tschapka, A. D. Grinnell, and H.-U. Schnitzler. 2016. Guild structure and niche differentiation in echolocating bats. Pp. 141–166 in: Bat Bioacoustics (M. B. Fenton, A. D. Grinnell, A. N. Popper, and R. R. Fay, eds) Springer, New York.

Dobat, K., and T. Peikert-Holle. 1985. Blüten und Fledermäuse—Bestäubung durch Fledermäuse und Flughunde (Chiropterophilie). Verlag Waldemar Kramer, Frankfurt am Main.

Faegri, K., and L. van der Pijl. 1979. The Principles of Pollination Ecology. 3rd rev. ed. Pergamon Press, Oxford.

Fenster, C. B., W. S. Armbruster, P. Wilson, M. R. Dudash, and J. D. Thomson. 2004. Pollination syndromes and floral specialization. Annual Review of Ecology, Evolution, and Systematics, 35:375–403.

Ferrarezzi, H., and E. d. A. Gimenez. 1996. Systematic patterns and the evolution of feeding habits in Chiroptera (Archonta: Mammalia). Journal of Comparative Biology, 1:75–94.

Findley, J. S., E. H. Studier, and D. E. Wilson. 1972. Morphologic properties of bat wings. Journal of Mammalogy, 53:429–444.

Fleming, T. H. 1982. Foraging strategies of plant-visiting bats. Pp. 287–325 in: Ecology of Bats (T. H. Kunz, ed.). Plenum Press, New York.

Fleming, T. H. 2002. The pollination biology of Sonoran Desert columnar cacti. Pp. 207–224 in: Columnar Cacti and Their Mutualists: Evolution, Ecology, and Conservation (T. H. Fleming and A. Valiente-Banuet, eds.). University of Arizona Press, Tucson.

Fleming T. H., C. Geiselman, and W. J. Kress. 2009. The evolution of bat pollination: a phylogenetic perspective. Annals of Botany, 104:1017–1043.

Fleming, T. H., and W. J. Kress. 2013. The Ornaments of Life: Coevolution and Conservation in the Tropics. University of Chicago Press, Chicago.

Fleming, T. H., and N. Muchhala. 2008. Nectar-feeding bird and bat niches in two worlds: pantropical comparisons of vertebrate pollination systems. Journal of Biogeography, 35: 764–780.

Fleming, T. H., N. Muchhala, and J. F. Ornelas. 2005. New World nectar-feeding vertebrates: community patterns and processes. Pages 161–184 in: Contribuciones Mastozoologicas en Homenaje a Bernardo Villa (V. Sánchez-Cordero and R. A. Medellín, eds.). Instituto de Biologia y Instituto de Ecologia, Universidad Nacional Autonoma de Mexico, Mexico City.

Fleming, T. H., and J. M. Nassar. 2002. The population biology of the lesser long-nosed bat Leptonycteris curasoae in Mexico and northern South America. Pp. 283–305 in: Columnar Cacti and Their Mutualists: Evolution, Ecology, and Conservation (T. H. Fleming and A. Valiente-Banuet, eds.). University of Arizona Press, Tucson.

Fleming, T. H., R. A. Nuñez, and L. D. S. L. Sternberg. 1993. Seasonal changes in the diets of migrant and non-migrant nectarivorous bats as revealed by carbon stable isotope analysis. Oecologia, 94:72–75.

Fleming, T. H., C. T. Sahley, J. N. Holland, J. D. Nason, and J. L. Hamrick. 2001. Sonoran Desert columnar cacti and the evolution of generalized pollination systems. Ecological Monographs, 71:511–530.

Freeman, P. W. 1995. Nectarivorous feeding mechanisms in bats. Biological Journal of the Linnean Society, 56:439–463.

Frick, W. F., P. A. Heady III, and J. P. Hayes. 2009. Facultative nectar-feeding behavior in a gleaning insectivorous bat (Antrozous pallidus). Journal of Mammalogy, 90:1157–1164.

Frick, W. F., R. D. Price, P. A. Heady III, and K. M. Kay. 2013. Insectivorous bat pollinates columnar cactus more effectively per visit than specialized nectar bat. American Naturalist, 181:137–144.

Garibaldi, L. A., N. Muchhala, I. Motzke, L. Bravo-Monroy, R. Olschewski, and A.-M. Klein. 2012. Services from plant-pollinator interactions in the Neotropics. Pp. 119–139 in: Ecosystem Services from Agriculture and Agroforestry: Measurement and Payment (B. Rapidel, F. DeClerck, J. F. Le Coq, and J. Beer, eds.) Earthscan, London.

Gentry, A. H. 1974. Coevolutionary patterns in Central American Bignoniaceae. Annals of the Missouri Botanical Garden, 61:728–759.

Giannini, N. P., and E. K. Kalko. 2004. Trophic structure in a large assemblage of phyllostomid bats in Panama. Oikos, 105:209–220.

Godinez-Alvarez, H., and A. Valiente-Banuet. 2000. Fruit-feeding behavior of the bats Leptonycteris curasoae and Choeronycteris mexicana in flight cage experiments: consequences for dispersal of columnar cactus seeds. Biotropica, 32:552–556.

Gonzalez-Terrazas, T. P., J. C. Koblitz, T. H. Fleming, R. A. Medellín, E. K. Kalko, H.-U. Schnitzler, and M. Tschapka.

2016. How nectar-feeding bats localize their food: echolocation behavior of Leptonycteris yerbabuenae approaching cactus flowers. PLOS ONE, 11:e0163492.

Gonzalez-Terrazas, T. P., C. Martel, P. Milet-Pinheiro, M. Ayasse, E. K. V. Kalko, and M. Tschapka. 2016. Finding flowers in the dark: nectar-feeding bats integrate olfaction and echolocation while foraging for nectar. Royal Society Open Science, 3:160199.

Gonzalez-Terrazas T. P., R. A. Medellín, M. Knörnschild, and M. Tschapka. 2012. Morphological specialization influences nectar extraction efficiency of sympatric nectar-feeding bats. Journal of Experimental Biology, 215:3989–3996.

Goyret J., M. Pfaff, R. A. Raguso, and A. Kelber. 2008. Why do Manduca sexta feed from white flowers? Innate and learnt colour preferences in a hawkmoth. Naturwissenschaften, 95:569–576.

Goyret, J., and M. L. Yuan. 2015. Influence of ambient illumination on the use of olfactory and visual signals by a nocturnal hawkmoth during close-range foraging. Integrative and Comparative Biology, 55:486–494.

Grant, V. 1983. The systematic and geographical distribution of hawkmoth flowers in the temperate North American flora. Botanical Gazette, 144:439–449.

Gregorin, R., and A. D. Ditchfield. 2005. New genus and species of nectar-feeding bat in the tribe Lonchophyllini (Phyllostomidae: Glossophaginae) from northeastern Brazil. Journal of Mammalogy, 86:403–414.

Griffiths, T. A. 1982. Systematics of the New World nectar-feeding bats (Mammalia, Phyllostomidae), based on the morphology of the hyoid and lingual regions. American Museum Novitates, 2742:1–45.

Griffiths, T. A. 1985. Molar cusp patterns in the bat genus Brachyphylla: some functional and systematic observations. Journal of Mammalogy, 66:544–549.

Haiduk, M. W., and R. J. Baker. 1982. Cladistical analysis of G-banded chromosomes of nectar feeding bats (Glossophaginae: Phyllostomidae). Systematic Biology, 31:252–265.

Håkansson, J., A. Hedenström, Y. Winter, and L. C. Johansson. 2015. The wake of hovering flight in bats. Journal of the Royal Society Interface, 12:20150357.

Harper, C. J., S. M. Swartz, and E. L. Brainerd. 2013. Specialized bat tongue is a hemodynamic nectar mop. Proceedings of the National Academy of Sciences, USA, 110:8852–8857.

Heithaus, E. R., T. H. Fleming, and P. A. Opler. 1975. Foraging patterns and resource utilization in seven species of bats in a seasonal tropical forest. Ecology, 56:841–854.

Hernández-Montero, J. R., and V. J. Sosa. 2015. Reproductive biology of Pachira aquatica Aubl. (Malvaceae: Bombacoideae): a tropical tree pollinated by bats, sphingid moths and honey bees. Plant Species Biology, 31:25–134.

Herrera, L. G. 1999. Preferences for different sugars in Neotropical nectarivorous and frugivorous bats. Journal of Mammalogy, 80:683–688.

Herrera, L. G., and C. Martínez del Río. 1998. Pollen digestion by New World bats: effects of processing time and feeding habits. Ecology, 79:2828–2838.

Hobson, K. A., M. M. Leticia, R. P. Nicte, M. C. Germán, and S.-C. Víctor. 2001. Sources of protein in two species of phytophagous bats in a seasonal dry forest: evidence from stable-isotope analysis. Journal of Mammalogy, 82:352–361.

Holderied, M. W., and O. von Helversen. 2006. "Binaural echo disparity" as a potential indicator of object orientation and cue for object recognition in echolocating nectar-feeding bats. Journal of Experimental Biology, 209:3457–3468.

Horner, M. A., T. H. Fleming, and C. T. Sahley. 1998. Foraging behaviour and energetics of a nectar-feeding bat, *Leptonycteris curasoae* (Chiroptera : Phyllostomidae). Journal of Zoology, 244:575–586.

Howell, D. 1974a. Bats and pollen: physiological aspects of the syndrome of chiropterophily. Comparative Biochemistry and Physiology Part A: Physiology, 48:263–276.

Howell, D. J. 1974b. Acoustic behavior and feeding in glossophagine bats. Journal of Mammalogy, 55:293–308.

Howell, D. J. 1977. Time sharing and body partitioning in bat-plant pollination systems. Nature, 270:509–510.

Howell, D., and N. Hodgkin. 1976. Feeding adaptations in the hairs and tongues of nectar-feeding bats. Journal of Morphology, 148:329–336.

Ibarra-Cerdeña, C. N., L. I. Iñiguez Dávalos, and V. Sánchez-Cordero. 2005. Pollination ecology of *Stenocereus queretaroensis* (Cactaceae), a chiropterophilous columnar cactus, in a tropical dry forest of Mexico. American Journal of Botany, 92:503–509.

Johnson, S. D., M. Moré, F. W. Amorim, W. A. Haber, G. W. Frankie, D. A. Stanley, A. A. Coccuci, and R. A. Raguso. 2016. The long and the short of it: a global analysis of hawkmoth pollination niches and interaction networks. Functional Ecology, 31:101 115.

Kalko, E. K. V. 2004. Neotropical leaf-nosed bats (Phyllostomidae): "Whispering" bats or candidates for acoustic survey? Pp. 63–69 *in*: Bat Echolocation Research: Tools, Techniques and Analysis (M. Brigham, E. K. V. Kalko, G. Jones, S. Parsons, and H. J. G. A. Limpens, eds.). Bat Conservation International, Austin, TX.

Kays, R., M. E. Rodríguez, L. M. Valencia, R. Horan, A. R. Smith, and C. Ziegler. 2012. Animal visitation and pollination of flowering balsa trees (*Ochroma pyramidale*) in Panama. Mesoamericana, 16:56–70.

Kelm, D. H., R. Simon, D. Kuhlow, C. C. Voigt, and M. Ristow. 2011. High activity enables life on a high-sugar diet: blood glucose regulation in nectar-feeding bats. Proceedings of the Royal Society of London B–Biological Sciences, 278:3490–3496.

Kelm, D. H., and O. von Helversen. 2007. How to budget metabolic energy: torpor in a small Neotropical mammal. Journal of Comparative Physiology B, 177:667–677.

Kier, W. M., and K. K. Smith. 1985. Tongues, tentacles and trunks: the biomechanics of movement in muscular-hydrostats. Zoological Journal of the Linnean Society, 83:307–324.

Knox, E. B., A. M. Muasya, and N. Muchhala. 2008. The predominantly South American clade of Lobeliaceae. Systematic Botany, 33:462–468.

Knudsen, J. T., and L. Tollsten. 1995. Floral scent in bat-pollinated plants—a case of convergent evolution. Botanical Journal of the Linnean Society, 119:45–57.

LaVal, R. K., and H. S. Fitch. 1977. Structure, movements and reproduction in three Costa Rican bat communities. Occasional Papers, Museum of Natural History, University of Kansas, 69:1–27.

Law, B. S. 2001. The diet of the common blossom bat (*Syconycteris australis*) in upland tropical rainforest and the importance of riparian areas. Wildlife Research, 28:619–626.

Lemke, T. O. 1984. Foraging ecology of the long-nosed bat, *Glossophaga soricina*, with respect to resource availability. Ecology, 65:538–548.

Lenoble, A., B. Angin, J.-B. Huchet, and A. Royer. 2014. Seasonal insectivory of the Antillean fruit-eating bat (*Brachyphylla cavernarum*). Caribbean Journal of Science, 48:127–131.

Liu, H.-Q., J.-K. Wei, B. Li, M.-S. Wang, R.-Q. Wu, J. D. Rizak, L. Zhong, L. Wang, F.-Q. Xu, and Y.-Y. Shen. 2015. Divergence of dim-light vision among bats (Order: Chiroptera) as estimated by molecular and electrophysiological methods. Scientific Reports, 5:11531.

Machado, I. C. S., I. Sazima, and M. Sazima. 1998. Bat pollination of the terrestrial herb *Irlbachia alata* (Gentianaceae) in northeastern Brazil. Plant Systematics and Evolution, 209:231–237.

MacSwiney G., M. C., B. Bolívar-Cimé, F. M. Clarke, and P. A. Racey. 2012. Transient yellow colouration of the bat *Artibeus jamaicensis* coincides with pollen consumption. Mammalian Biology, 77:221–223.

Mancina, C., and F. Balseiro. 2005. Pollen digestion by nectarivorous and frugivorous Antillean bats. Mammalian Biology–Zeitschrift für Säugetierkunde, 70:282–290.

Martén-Rodríguez, S., A. Almarales-Castro, and C. B. Fenster. 2009. Evaluation of pollination syndromes in Antillean Gesneriaceae: evidence for bat, hummingbird and generalized flowers. Journal of Ecology, 97:348–359.

Mora, E. C., and S. Macías. 2007. Echolocation calls of Poey's flower bat (*Phyllonycteris poeyi*) unlike those of other phyllostomids. Naturwissenschaften, 94:380–383.

Moreno-Valdez, A., R. L. Honeycutt, and W. E. Grant. 2004. Colony dynamics of *Leptonycteris nivalis* (Mexican long-nosed bat) related to flowering agave in northern Mexico. Journal of Mammalogy, 85:453–459.

Muchhala, N. 2006a. Nectar bat stows huge tongue in its rib cage. Nature, 444:701–702.

Muchhala, N. 2006b. The pollination biology of *Burmeistera* (Campanulaceae): specialization and syndromes. American Journal of Botany, 93:1081–1089.

Muchhala, N. 2007. Adaptive trade-off in floral morphology mediates specialization for flowers pollinated by bats and hummingbirds. American Naturalist, 169:494–504.

Muchhala, N. 2008. Functional significance of extensive interspecific variation in *Burmeistera* floral morphology: evidence from nectar bat captures in Ecuador. Biotropica, 40:332–337.

Muchhala, N., A. Caiza, J. C. Vizuete, and J. D. Thomson. 2009. A generalized pollination system in the tropics: bats, birds, and *Aphelandra acanthus*. Annals of Botany, 103:1481–1487.

Muchhala, N., and P. Jarrín-V. 2002. Flower visitation by bats in cloud forests of western Ecuador. Biotropica, 34:387–395.

Muchhala, N., P. V. Mena, and L. V. Albuja. 2005. A new species of *Anoura* (Chiroptera : Phyllostomidae) from the Ecuadorian Andes. Journal of Mammalogy, 86:457–461.

Muchhala, N., and M. D. Potts. 2007. Character displacement among bat-pollinated flowers of the genus *Burmeistera*: analysis of mechanism, process and pattern. Proceedings of the Royal Society B, 274:2731–2737.

Muchhala, N., and D. Serrano. 2015. The complexity of background clutter affects nectar bat use of flower odor and shape cues. PLOS ONE, 10:e0136657.

Muchhala, N., and J. D. Thomson. 2009. Going to great lengths: selection for long corolla tubes in an extremely specialized bat-flower mutualism. Proceedings of the Royal Society B, 276:2147–2152.

Muchhala, N., and J. D. Thomson. 2010. Fur versus feathers: pollen delivery by bats and hummingbirds, and consequences for pollen production. American Naturalist, 175:717–726.

Muchhala, N., and J. D. Thomson. 2012. Interspecific competition in pollination systems: costs to male fitness via pollen misplacement. Functional Ecology, 26:476–482.

Müller, B., M. Glösmann, L. Peichl, G. C. Knop, C. Hagemann, and J. Ammermüller. 2009. Bat eyes have ultraviolet-sensitive cone photoreceptors. PLOS ONE, 4:e6390.

Muñoz-Romo, M., J. F. Burgos, and T. H. Kunz. 2011. Smearing behaviour of male *Leptonycteris curasoae* (Chiroptera) and female responses to the odour of dorsal patches. Behaviour, 148:461–483.

Muñoz-Romo, M., L. T. Nielsen, J. M. Nassar, and T. H. Kunz. 2012. Chemical composition of the substances from dorsal patches of males of the Curacaoan long-nosed bat, *Leptonycteris curasoae* (Phyllostomidae: Glossophaginae). Acta Chiropterologica, 14:213–224.

Murray, K. L., and T. H. Fleming. 2008. Social structure and mating system of the buffy flower bat, *Erophylla sezekorni* (Chiroptera, Phyllostomidae). Journal of Mammalogy, 89:1391–1400.

Nachev, V., K. P. Stich, C. Winter, A. Bond, A. Kamil, and Y. Winter. 2017. Cognition-mediated evolution of low-quality floral nectars. Science, 355:75–78.

Nassar, J., N. Ramirez, and O. Linares. 1997. Comparative pollination biology of Venezuelan columnar cacti and the role of nectar-feeding bats in their sexual reproduction. American Journal of Botany, 84:918–927.

Newton, L. R. 2002. Mobility, reproduction, and population genetic structure of two phyllostomid bats. MS thesis, University of Miami, Coral Gables, FL.

Nogueira, M. R., A. L. Peracchi, and L. R. Monteiro. 2009. Morphological correlates of bite force and diet in the skull and mandible of phyllostomid bats. Functional Ecology, 23:715–723.

Norberg, U. M., T. H. Kunz, J. F. Steffensen, Y. Winter, and O. von Helversen. 1993. The cost of hovering and forward flight in a nectar-feeding bat, *Glossophaga soricina*, estimated from aerodynamic theory. Journal of Experimental Biology, 182:207–227.

Norberg, U. M. L., and Y. Winter. 2006. Wing beat kinematics of a nectar-feeding bat, *Glossophaga soricina*, flying at different flight speeds and Strouhal numbers. Journal of Experimental Biology, 209:3887–3897.

Nor Zalipah, M., S. Anuar, M. Sah, and G. Jones. 2016. The potential significance of nectar-feeding bats as pollinators in mangrove habitats of Peninsular Malaysia. Biotropica, 8:425–428.

Ollerton, J., R. Alarcón, N. M. Waser, M. V. Price, S. Watts, L. Cranmer, A. Hingston, C. I. Peter, and J. Rotenberry. 2009. A global test of the pollination syndrome hypothesis. Annals of Botany, 103:1471–1480.

Penalba, M. C., F. Molina-Freaner, and L. L. Rodriguez. 2006. Resource availability, population dynamics and diet of the nectar-feeding bat *Leptonycteris curasoae* in Guaymas, Sonora, Mexico. Biodiversity and Conservation, 15:3017–3034.

Perret, M., A. Chautems, R. Spichiger, M. Peixoto, and V. Savolainen. 2001. Nectar sugar composition in relation to pollination syndromes in Sinningieae (Gesneriaceae). Annals of Botany, 87:267–273.

Petit, S. 1997. The diet and reproductive schedules of *Leptonycteris curasoae curasoae* and *Glossophaga longirostris elongata* (Chiroptera: Glossophaginae) on Curaçao. Biotropica, 29:214–223.

Pettersson, S., F. Ervik, and J. T. Knudsen. 2004. Floral scent of bat-pollinated species: West Africa vs. the New World. Biological Journal of the Linnean Society, 82:161–168.

Ramírez, N., L. G. Herrera, and L. Miro. 2005. Physiological constraint to food ingestion in a new world nectarivorous bat. Physiological and Biochemical Zoology, 78:1032–1038.

Rex, K., B. I. Czaczkes, R. Michener, T. H. Kunz, and C. C. Voigt. 2010. Specialization and omnivory in diverse mammalian assemblages. Ecoscience, 17:37–46.

Rincón-Vargas, F., K. E. Stoner, R. M. Vigueras-Villaseñor, J. M. Nassar, Ó. M. Chaves, and R. Hudson. 2013. Internal and external indicators of male reproduction in the lesser long-nosed bat *Leptonycteris yerbabuenae*. Journal of Mammalogy, 94:488–496.

Roces, F., Y. Winter, and O. von Helversen. 1993. Nectar concentration preference and water balance in a flower visiting bat, *Glossophaga soricina antillarum*. Pp. 159–165 *in*: Animal-Plant Interactions in Tropical Environments (W. Barthlott, C. M. Naumann, K. Schmidt-Loske, and K. L. Schuchmann, eds.). Museum Koenig, Bonn.

Rodríguez-Peña, N., K. E. Stoner, C. M. Flores-Ortiz, J. Ayala-Berdón, M. A. Munguía-Rosas, V. Sánchez-Cordero, and J. E. Schondube. 2016. Factors affecting nectar sugar composition in chiropterophilic plants. Revista Mexicana de Biodiversidad, 87:465–473.

Rodríguez-Peña, N., K. Stoner, J. Schondube, J. Ayala-Berdón, C. Flores-Ortiz, and C. Martínez del Río. 2007. Effects of sugar composition and concentration on food selection by Saussure's long-nosed bat (*Leptonycteris curasoae*) and the long-tongued bat (*Glossophaga soricina*). Journal of Mammalogy, 88:1466–1474.

Rojas, D., A. Vale, V. Ferrero, and L. Navarro. 2011. When did plants become important to leaf-nosed bats? Diversification of feeding habits in the family Phyllostomidae. Molecular Ecology, 20:2217–2228.

Rojas, D., O. M. Warsi, and L. M. Dávalos. 2016. Bats (Chiroptera: Noctilionoidea) challenge a recent origin of extant Neotropical diversity. Systematic Biology, 65:432–448.

Root, R. 1967. The niche exploitation pattern of the blue-gray gnatcatcher. Ecological Monographs, 37:317–350.

Rosas-Guerrero, V., R. Aguilar, S. Martén-Rodríguez, L. Ashworth, M. Lopezaraiza-Mikel, J. M. Bastida, and M. Quesada. 2014. A quantitative review of pollination syndromes: do floral traits predict effective pollinators? Ecology Letters, 17:388–400.

Sahley, C. T., M. A. Horner, and T. H. Fleming. 1993. Flight speeds and mechanical power outputs of the nectar-feeding bat, *Leptonycteris curasoae* (Phyllostomidae: Glossophaginae). Journal of Mammalogy, 74:594–600.

Sánchez, R., and R. A. Medellín. 2007. Food habits of the threatened bat *Leptonycteris nivalis* (Chiroptera: Phyllostomidae) in a mating roost in Mexico. Journal of Natural History, 41:1753–1764.

Santana, S. E. 2015. Quantifying the effect of gape and morphology on bite force: biomechanical modelling and in vivo measurements in bats. Functional Ecology, 30:557–565.

Santana, S. E., I. R. Grosse, and E. R. Dumont. 2012. Dietary hardness, loading behavior, and the evolution of skull form in bats. Evolution, 66:2587–2598.

Sazima, I. 1976. Observations on the feeding habits of phyllostomatid bats (*Carollia*, *Anoura*, and *Vampyrops*) in southeastern Brazil. Journal of Mammalogy, 57:381–382.

Sazima, M., S. Buzato, and I. Sazima. 1999. Bat-pollinated flower assemblages and bat visitors at two Atlantic Forest sites in Brazil. Annals of Botany, 83:705–712.

Sazima, M., I. Sazima, and S. Buzato. 1994. Nectar by day and night—*Siphocampylus sulfureus* (Lobeliaceae) pollinated by hummingbirds and bats. Plant Systematics and Evolution, 191:237–246.

Schöner, M. G., R. Simon, and C. R. Schöner. 2016. Acoustic communication in plant-animal interactions. Current Opinion in Plant Biology, 32:88–95.

Scott, P. 2004. Timing of *Agave palmeri* flowering and nectar-feeding bat visitation in the Peloncillos and Chiricahua Mountains. Southwestern Naturalist, 49:425–434.

Silva Taboada, G. S., and R. H. Pine. 1969. Morphological and behavioral evidence for the relationship between the bat genus *Brachyphylla* and the Phyllonycterinae. Biotropica, 1:10–19.

Simon, R., M. W. Holderied, C. U. Koch, and O. von Helversen. 2011. Floral acoustics: conspicuous echoes of a dish-shaped leaf attract bat pollinators. Science, 333:631–633.

Simon R., M. Knörnschild, M. Tschapka, A. Schneider, N. Passauer, E. K. V. Kalko, and O. von Helversen. 2014. Biosonar resolving power: echo-acoustic perception of surface structures in the submillimeter range. Frontiers in Physiology, 5:1–9.

Skog, L. E. 1976. A study of the tribe Gesnerieae, with a revision of *Gesneria* (Gesneriaceae: Gesnerioideae). Smithsonian Contributions to Botany, 29:1–182.

Slaughter, B. H. 1970. Evolutionary trends of chiropteran dentitions. Pp. 51–83 *in*: About Bats (B. H. Slaughter and D. W. Walton, eds.). Southern Methodist University Press, Dallas.

Smith, J. D., and C. S. Hood. 1984. Genealogy of the New World nectar-feeding bats reexamined: a reply to Griffiths. Systematic Zoology, 33:435–460.

Sosa, M., and P. J. Soriano. 1993. Solapamiento de dieta entre *Leptonycteris curasoae* y *Glossophaga longirostris* (Mammalia: Chiroptera). Revista de Biologia Tropical, 41:529–532.

Soto-Centeno, J. A., and A. Kurta. 2006. Diet of two nectar-ivorous bats, *Erophylla sezekorni* and *Monophyllus redmani* (Phyllostomidae), on Puerto Rico. Journal of Mammalogy, 87:19–26.

Sperr, E. B., L. A. Caballero-Martínez, R. A. Medellín, and M. Tschapka. 2011. Seasonal changes in species composition, resource use and reproductive patterns within a guild of nectar-feeding bats in a west Mexican dry forest. Journal of Tropical Ecology, 27:133–145.

Stephan, H., and P. Pirlot. 1970. Volumetric comparisons of brain structures in bat. Journal of Zoological Systematics and Evolutionary Research, 8:200–236.

Stoner, K. E., K. A. Salazar, R. C. Fernandez, and M. Quesada. 2003. Population dynamics, reproduction, and diet of the lesser long-nosed bat (*Leptonycteris curasoae*) in Jalisco, Mexico: implications for conservation. Biodiversity and Conservation, 12:357–373.

Stroo, A. 2000. Pollen morphological evolution in bat pollinated plants. Plant Systematics and Evolution, 222:225–242.

Swanepoel, P., and H. H. Genoways. 1983a. *Brachyphylla cavernarum*. Mammalian Species, 205:1–6.

Swanepoel, P., and H. H. Genoways. 1983b. *Brachyphylla nana*. Mammalian Species, 206:1–3.

Tellez, G., and J. Ortega. 1999. *Musonycteris harrisoni*. Mammalian Species Archive, 622:1–3.

Thiele, J., and Y. Winter. 2005. Hierarchical strategy for relocating food targets in flower bats: spatial memory versus cue-directed search. Animal Behaviour, 69:315–327.

Thomson, J. D., P. Wilson, M. Valenzuela, and M. Malzone. 2000. Pollen presentation and pollination syndromes, with special reference to *Penstemon*. Plant Species Biology, 15:11–29.

Toelch, U., K. P. Stich, C. L. Gass, and Y. Winter. 2008. Effect of local spatial cues in small-scale orientation of flower bats. Animal Behaviour, 75:913–920.

Tripp, E. A., and P. S. Manos. 2008. Is floral specialization an evolutionary dead-end? Pollination system transitions in *Ruellia* (Acanthaceae). Evolution, 62:1712–1737.

Tschapka, M. 2003. Pollination of the understory palm *Calyptrogyne ghiesbreghtiana* by hovering and perching bats. Biological Journal of the Linnean Society, 80:281–288.

Tschapka, M. 2004. Energy density patterns of nectar resources permit coexistence within a guild of Neotropical flower-visiting bats. Journal of Zoology, 263:7–21.

Tschapka, M., and S. Dressler. 2002. Chiropterophily: on bat-flowers and flower-bats. Curtis's Botanical Magazine, 19:114–125.

Tschapka, M., S. Dressler, and O. von Helversen. 2006. Bat visits to *Marcgravia pittieri* and notes on the inflorescence diversity within the genus *Marcgravia* (Marcgraviaceae). Flora, 201:383–388.

Tschapka, M., T. P. Gonzalez-Terrazas, and M. Knörnschild. 2015. Nectar uptake in bats using a pumping-tongue mechanism. Science Advances, 1:e1500525.

Tschapka, M., E. B. Sperr, L. A. Caballero-Martínez, and R. A. Medellín. 2008. Diet and cranial morphology of *Musonycteris harrisoni*, a highly specialized nectar-feeding bat in western Mexico. Journal of Mammalogy, 89:924–932.

Tschapka, M., and O. von Helversen. 2007. Phenology, nectar

production and visitation behaviour of bats on the flowers of the bromeliad *Werauhia gladioliflora* in a Costa Rican lowland rain forest. Journal of Tropical Ecology, 23:385–395.

Tschapka, M., O. von Helversen, and W. Barthlott. 1999. Bat pollination of *Weberocereus tunilla*, an epiphytic rain forest cactus with functional flagelliflory. Plant Biology, 1:554–559.

van der Pijl, L. 1961. Ecological aspects of flower evolution. II. Zoophilous flower classes. Evolution, 15:44–59.

Vogel, S. 1954. Blütenbiologische typen als elemente der Sippengliederung: G. Fischer.

Vogel, S. 1969. Chiropterophilie in der neotropischen flora, neue mitteilungen II. Flora, 158:185–222.

Voigt, C. C. 2004. The power requirements (Glossophaginae : Phyllostomidae) in nectar-feeding bats for clinging to flowers. Journal of Comparative Physiology B: Biochemical Systemic and Environmental Physiology, 174:541–548.

Voigt, C. C., and J. Speakman. 2007. Nectar-feeding bats fuel their high metabolism directly with exogenous carbohydrates. Functional Ecology, 21:913–921.

von Helversen, D. 2004. Object classification by echolocation in nectar feeding bats: size-independent generalization of shape. Journal of Comparative Physiology A, 190:515–521.

von Helversen, D., M. W. Holderied, and O. von Helversen. 2003. Echoes of bat-pollinated bell-shaped flowers: conspicuous for nectar-feeding bats? Journal of Experimental Biology, 206:1025–1034.

von Helversen, D., and O. von Helversen. 1999. Acoustic guide in bat-pollinated flower. Nature, 398:759–760.

von Helversen, D., and O. von Helversen. 2003. Object recognition by echolocation: a nectar-feeding bat exploiting the flowers of a rain forest vine. Journal of Comparative Physiology A, 189:327–336.

von Helversen, O. 1993. Adaptations of flowers to the pollination by glossophagine bats. Pp. 167–174 *in*: Animal-Plant Interactions in Tropical Environments (W. Barthlott, C. M. Naumann, K. Schmidt-Loske, and K. L. Schuchmann, eds.). Museum Koenig, Bonn.

von Helversen, O. 1995. Blumenfledermäuse und Fledermausblumen–Wechselbeziehungen zwischen Blüte und Bestäuber und energetische Grenzbedingungen. Rundgespräche der Kommission für Okologie, 10:217–229.

von Helversen, O., L. Winkler, and H. G. Bestmann. 2000. Sulphur containing "perfumes" attract flower-visiting bats. Journal of Comparative Physiology A, 186:143–153.

von Helversen, O., and Y. Winter. 2003. Glossophagine bats and their flowers: costs and benefits for plants and pollinators. Pp. 346–397 *in*: Bat Ecology (T. H. Kunz and M. B. Fenton, eds.). University of Chicago Press, Chicago.

Warrick, D. R., B. W. Tobalske, and D. R. Powers. 2005. Aerodynamics of the hovering hummingbird. Nature, 435:1094.

Waser, N. M., L. Chittka, M. V. Price, N. M. Williams, and J. Ollerton. 1996. Generalization in pollination systems, and why it matters. Ecology, 77:1043–1060.

Wetterer, A. L., M. V. Rockman, and N. B. Simmons. 2000. Phylogeny of phyllostomid bats (Mammalia : Chiroptera): data from diverse morphological systems, sex chromosomes, and restriction sites. Bulletin of the American Museum of Natural History, 248:4–200.

Wilkinson, G. S., and G. F. McCracken. 2003. Bats and balls: sexual selection and sperm competition in the Chiroptera. Pp. 128–155 *in*: Bat Ecology (T. H. Kunz and M. B. Fenton, eds.). University of Chicago Press, Chicago.

Willig, M. R., G. R. Camilo, and S. J. Noble. 1993. Dietary overlap in frugivorous and insectivorous bats from edaphic Cerrado habitats of Brazil. Journal of Mammalogy, 74:117–128.

Winter Y., J. Lopez, and O. von Helversen. 2003. Ultraviolet vision in a bat. Nature, 425:612–614.

Winter, Y., and K. P. Stich. 2005. Foraging in a complex naturalistic environment: capacity of spatial working memory in flower bats. Journal of Experimental Biology, 208:539–548.

Winter, Y., and O. von Helversen. 2001. Bats as pollinators: foraging energetics and floral adaptations. Pp. 148–170 *in*: Cognitive Ecology of Pollination: Animal Behavior and Floral Evolution (L. Chittka and J. D. Thomson, eds.). Cambridge University Press, Cambridge.

Winter, Y., and O. von Helversen. 2003. Operational tongue length in phyllostomid nectar-feeding bats. Journal of Mammalogy, 84:886–896.

Zortéa, M. 2003. Reproductive patterns and feeding habits of three nectarivorous bats (Phyllostomidae: Glossophaginae) from the Brazilian Cerrado. Brazilian Journal of Biology, 63:159–168.

17

The Frugivores

Evolution, Functional Traits, and Their Role
in Seed Dispersal

Romeo A. Saldaña-Vázquez
and Theodore H. Fleming

Introduction

Most bats are insectivorous, but two families—the Old World Pteropodidae (with about 186 species) and New World Phyllostomidae (with about 216 species)—contain species that consume fruits and disperse the seeds of hundreds of species of tropical and subtropical plants. These two families are only distantly related and differ in their evolutionary ages and in many aspects of their morphology, sensory biology, and foraging ecology. Despite these differences, frugivorous bats of both families, along with fruit-eating birds, primates, and a variety of other mammals, play important functional roles in the maintenance and dynamics of tropical and subtropical terrestrial ecosystems worldwide (Fleming and Kress 2013).

Although some species of primarily insectivorous or carnivorous phyllostomids occasionally eat fruit (chap. 14, this vol.), this dietary specialization is concentrated mostly in two morphologically derived subfamilies, Carolliinae and Stenodermatinae, whose evolutionary ages are approximately 18–20 million years (Ma). Carolliines are much less species rich than stenodermatines (chap. 4, this vol.) and forage for fruit mainly in forest understories. Stenodermatines, in contrast, are the most species-rich phyllostomid clade and have diversified extensively in the last 10 Ma; they forage for fruit mainly in forest canopies (Dumont et al. 2012; Rojas et al. 2012, 2016; Saldaña-Vázquez et al. 2013).

Reflecting their relatively long evolutionary history of interacting with frugivorous bats, tropical fruits whose seeds are dispersed by bats (hereafter "bat fruits") have a series of morphological and nutritional characteristics that distinguish them from fruits dispersed by other kinds of vertebrate frugivores (e.g., birds and monkeys). Based on his knowledge of Old World pteropodid bats and their food plants, van der Pijl (1982) characterized bat fruits as drab in color sometimes musky smelling, often large and juicy, and sometimes displayed outside plant canopies via flagellicarpy (fruits on long pedicels) or caulicarpy (fruits directly on branches or trunks). Pteropodid fruits tend to be produced by large plants (e.g., canopy trees or megaherbs such as bananas) rather than shrubs or epiphytes. Phyllostomid fruits do not necessarily conform to these features, however, because phyllostomid bats are smaller than most pteropodids and they can echolocate. Characteristics of fruits eaten by phyllostomids include that they are usually green in color, either berries or drupes, usually with small, easily swallowed seeds, and often presented away from leaves. They are also produced by shrubs and epiphytes in addition to canopy trees. Lobova et al. (2009) discuss the characteristics of phyllostomid fruits in detail.

Bat fruits can also be viewed from the perspective of angiosperm phylogeny. Are these fruits concentrated in a limited number of plant lineages or are they widely distributed throughout this phylogeny? Based on the phylogenetic hypothesis proposed by APG III (2009), Fleming and Kress (2013) recognized five major angiosperm lineages: basal angiosperms (with 8 orders and 28 families), monocots (12 and 83), basal eudicots (12 and 64), and two lineages of advanced eudicots (asterids [14 and 105] and rosids [17 and 125]). The proportion of *orders* containing bat fruits (pteropodids and phyllostomids combined) in the five major lineages are basal angiosperms, 0.25; monocots, 0.33; basal eudicots, 0.08; asterids, 0.36; and rosids, 0.41. The proportion of *families* containing bat fruits in these lineages are basal angiosperms, 0.07; monocots, 0.08; basal eudicots, 0.02; asterids, 0.07; and rosids, 0.10. Proportions of bat fruits in angiosperm orders and families are considerably lower than those of bird fruits (Fleming and Kress 2013, table 6.3). From these results we conclude that, except for basal eudicots, bat fruits occur widely throughout angiosperm phylogeny and that more tropical plants rely on dispersal by birds than by bats.

Muscarella and Fleming (2007) review the diets of frugivorous phyllostomid and pteropodid bats and note that these bats generally feed on fruits from different plant families. The top five families for phyllostomids, based on number of genera producing bat fruits, are Cactaceae, Arecaceae, Sapotaceae, Moraceae, and Myrtaceae; the top five pteropodid fruit families are Sapotaceae, Anacardiaceae, Meliaceae, Arecaceae, and Rubiaceae. The most "popular" fruit families based on the number of bat genera (phyllostomids) or species (pteropodids) feeding on them are Solanaceae, Moraceae, Myrtaceae, Piperaceae, and Clusiasceae (phyl-

lostomids), and Moraceae, Myrtaceae, Anacardiaceae, Musaceae, and Arecaceae (pteropodids). Differences between the diets of the two families largely reflect biogeography (e.g., Cactaceae is an endemic New World family whereas Musaceae is an Old World endemic) and the fact that phyllostomid bats feed both in forest understories and canopies whereas pteropodids feed mainly in canopies.

Functional Traits and Diet Preferences

Diets of animals are associated primarily with two variables that occur on different scales of biological organization. The first variable is the diversity and density of food resources—the template upon which foraging and food choices take place. The second variable is the ability of animals to consume and obtain nutrients from these resources (Jara-Servín et al. 2017; Jordano 2000; Schondube et al. 2001). As discussed throughout this book, phyllostomid bats are one of the most species-rich bat families, and this diversity is associated with an exceptional diversity of functional traits (Schondube et al. 2001). Functional traits are morphological, behavioral, physiological, and phenological characteristics that are expressed in the phenotype of an organism and that reflect the response of organisms to their abiotic and biotic environments (Díaz et al. 2013; Violle et al. 2007).

Frugivorous phyllostomids exhibit a great variety of functional traits, including body mass (the range is 5–72 g), skull form, bite force, food handling behavior, renal morphology, and enzymatic activity (Dumont 2003; Santana and Dumont 2009; Schondube et al. 2001). This variation in functional traits is accompanied by variation in their diet preferences. Species of *Carollia*, for example, exhibit a strong preference for understory *Piper* fruits (Piperaceae, a family of basal angiosperms). About 60% of their diet is based on these fruits (Saldaña-Vázquez et al. 2013). In contrast, members of the Stenodermatinae exhibit a dichotomy in their fruit preferences. About 55% of the diet of species of *Sturnira*, an early member of this subfamily, comes from understory *Solanum* fruits (Solanaceae, an advanced eudicot), whereas about 67% of the diet of species of *Artibeus* and its relatives is based on canopy fruits produced by species of *Ficus* (Moraceae) and *Cecropia* (Urticaceae), which are advanced eudicots (Saldaña-Vázquez et al. 2013).

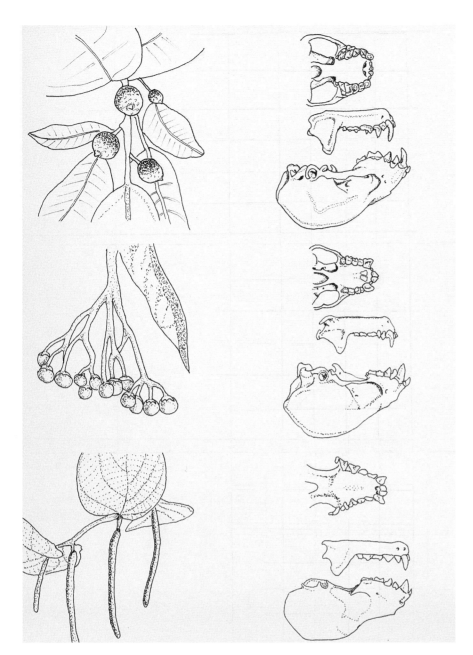

Figure 17.1. Skulls of three common lowland frugivores and their typical fruits. (*A*) *Carollia perspicillata* (*left*) and *Piper* species (*right*). (*B*) *Sturnira lilium* (*left*) and *Solanum* species (*right*). (*C*) *Artibeus jamaicensis* (*right*) and *Ficus* species (*right*). Sources of images are: Bats—*A. jamaicensis* and *S. lilium* redrawn from Simmons and Voss (1998); *C. perspicillata* redrawn from Cloutier and Thomas (1992). Fruits—redrawn from Lobova et al. (2009). Scale bar = 10 mm.

Not surprisingly, the functional traits of these bats usually match the physical characteristics of their major fruits. Most *Piper* fruits (actually infructescences or spadices) consumed by *Carollia* are small (1–4 g), green in color, and soft with a low pulp:seed ratio whereas those consumed by stenodermatines tend to be larger (1–30 g) and harder and come in a greater variety of colors (green, brown, yellow, orange, and dark violet), with a greater range of pulp:seed ratios and diaspore types (e.g., berry, spadix, and syconium; Hernández-Montero et al. 2011; Sánchez et al. 2012). Functional traits of *Carollia* species include relatively small size (12–25 g), relatively flat skulls with a narrow palate, low bite force,

and low wing loading compared with stenodermatinae bats, particularly the large species of *Artibeus* weighing up to 72 g (Dumont 2003; Santana and Dumont 2009; Vleut et al. 2015). In general, the skulls of canopy fig eaters differ from understory feeders by being taller and wider; their lower jaws have lower condyles and coronoid processes that are associated with strong masseter muscles for eating hard fruits (Nogueira et al. 2005). The skulls of *Carollia perspicillata*, *Sturnira lilium*, and *Artibeus jamaicensis* clearly illustrate salient differences in the trophic adaptations of these frugivores (fig. 17.1).

In their analysis of phyllostomid skulls based on the engineering principles of mechanical advantage

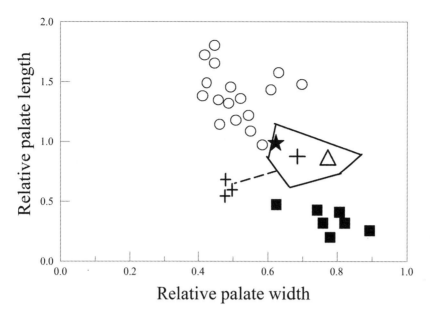

Figure 17.2. Ordination of the skulls of phyllostomid bats by relative palate width and length. The three cranial optima recognized by Dumont et al. (2014) include nectarivores (*circles*), generalists (*crosses*), frugivores (*triangles*), and short-faced bats (*squares*). The *polygon* contains many species of generalists and frugivores. The *star* indicates the location of *Carollia perspicillata*, a morphologically generalized frugivore. Redrawn from Dumont et al. (2014, fig. 2).

and von Mises stress, Dumont et al. (2014) found that at least three optima occur in the family. Mechanical advantage (MA) indicates the efficiency with which force is transferred from masticatory muscles through the skull to the food that is being eaten; it is strongly correlated with bite force. Von Mises stress (VMS) predicts the strength of systems (e.g., cortical bone) that fail in a ductile manner (i.e., the tendency to deform under tensile stress); high structural strength results in low von Mises stress. These three optima include: (1) low MA, high VMS found in nectarivores; (2) intermediate MA and VMS found in dietary generalists and many frugivores (e.g., the carolliines); and (3) high MA, low VMS found in short-faced stenodermatines (e.g., *Centurio senex*; Dumont et al. 2009) (fig. 17.2). Mechanical advantage has apparently more strongly influenced the evolution of these optima than VMS, and length of the palate relative to skull length is a good predictor of MA and VMS in these bats (fig. 17.1).

The food handling behavior of carolliines and stenodermatines also differs. The former bats are "gulpers" who quickly ingest the pulp and seeds of their relatively soft fruit, whereas the latter bats are "mashers" who slowly chew their more fibrous fruit and spit out spats of pulp and seeds while ingesting fruit juices (Bonaccorso and Gush 1987; Dumont 2003; Wagner et al. 2015). As discussed below, these two feeding modes likely have an important effect on the seed dispersal effectiveness of these groups of bats.

Functional traits that best explain the diversity and diet preferences of frugivorous phyllostomids include body mass, olfactory sensitivity, and digestive capacity (Fleming 1991; Kalko et al. 1996; Parolin et al. 2015; Saldaña-Vázquez et al. 2015). Body mass is positively related to bite force and wing loading in frugivorous bats (Santana and Dumont 2009; Vleut et al. 2015). These functional traits influence the ability of bats to forage in cluttered spaces and their ability to consume fruits that vary in hardness and fiber content. Frugivorous bats with high wing loading cannot forage easily in highly cluttered spaces where shrubs with small fruits are located (Caras and Korine 2009). Regardless of foraging zone, olfaction is one of the principal senses used by these bats to locate fruit (Korine and Kalko 2005; Thies et al. 1998). Experiments with fruits/oils of preferred bat fruits confirm the preference patterns proposed by Fleming (1986). Carollia species prefer *Piper* species more than other fruits, and *Artibeus* species prefer *Ficus* fruits more than others they are offered (Andrade et al. 2013; Parolin et al. 2015).

Finally, as the last step in the foraging-selection– assimilation of food process, the digestive capacity of these bats is positively correlated with their dietary diversity (Saldaña-Vázquez et al. 2015). Digestive capacity is the ability of animals to acquire nutrients independent of the quality of the food (Ayala-Berdon and Schondube 2011; Martínez del Río et al. 2001). This capacity is driven by the production of enzymes that break down nutrient molecules to make them available for absorption in the intestine. Frugivorous bats with a high digestive capacity include *A. jamaicensis* and *S. lilium*, which eat fruits of 108 and 50 plant genera, respectively; those with a lower digestive capacity include *Sturnira hondurensis* and *Carollia sowelli*, which eat fruits of 25 and 27 plant genera, respectively (Saldaña-Vázquez et al. 2015).

Up to this point, we have considered the intrinsic factors that influence the dietary preferences of frugivorous phyllostomids. As mentioned above, the diversity and density of food resources also influence the diets of these bats. For example, available evidence indicates that species of *Artibeus* decrease their consumption of *Ficus* and *Cecropia* with increasing elevation, whereas species of *Sturnira* increase their consumption of *Solanum* fruits with elevation and latitude (Giannini 1999; Saldaña-Vázquez et al. 2013). These changes reflect the different elevational and latitudinal ranges of these plant families (Fleming 1986). Therefore, the diversity of bat fruits within habitats is an important driver of fruit preferences and consumption. Species of *Carollia* represent an interesting case because the consumption of their core plant taxa (species of Piperaceae) is not related to elevation or latitude (Saldaña-Vázquez et al. 2013). This probably reflects the broad elevational range of these fruit as well as the steady-state fruiting phenology of many *Piper* species (Fleming 1986; Thies and Kalko 2004). Despite the progress we have made in research concerning factors that determine diet preferences and diversity in frugivorous phyllostomids, the integration of the intrinsic and extrinsic factors to explain how these bats structure their diet is still a challenge.

Like some phyllostomids, the functional trait most strongly related to diet preferences in pteropodid bats is body mass (Dumont 2003). Pteropodid bats weighing up to 1,500 g (e.g., species of *Acerodon* and *Pteropus*) often eat large and/or hard fruits (Dumont 2003). This pattern is similar to frugivorous phyllostomids, where large bats consume a high proportion of large fruits (Saldaña-Vázquez 2014). Like stenodermatines, most pteropodids feed as "mashers" rather than "gulpers." Interestingly, both pteropodids and phyllostomids consume *Ficus* fruits (Kalko et al. 1996), but New World *Ficus* differ from Old World species in the composition of their scents, reflecting differences in the olfactory sensitivities of phyllostomid and pteropodid bats (Hodgkison et al. 2013). In both frugivorous phyllostomids and pteropodids, intrinsic factors such as body

mass and extrinsic factors such as fruit mass and scent composition affect their diet choices.

Foraging for Fruits and Seed Dispersal

Unlike some pteropodid bats, which roost solitarily in canopy or subcanopy plants, most frugivorous phyllostomids are gregarious and roost in colonies of a few individuals to many thousands in a wide variety of natural and man-made structures (Kunz 1982; chap. 18, this vol.). They are thus refuging species (Hamilton and Watt 1970) and must commute some distance from their day roosts to their food resources every night. These distances range from a few hundred meters (e.g., harem males of *C. perspicillata* in Costa Rican tropical dry forest; Fleming 1988) to several kilometers (e.g., *A. jamaicensis* in Mexican tropical dry forest; Morrison 1978). Relatively short commute distances (1–2 km) occur in most carolliines, *A. watsoni*, *Rhinophylla pumilio*, *Stenoderma rufum*, and *S. lilium*, most of which are small and feed in forest understories, whereas longer commute distances (several kilometers) have been reported in larger canopy-feeding stenodermatines (e.g., *A. jamaicensis*, *A. lituratus*) and the phyllostomine *Phyllostomus hastatus* (e.g., Bonaccorso et al. 2006; Chaverri et al. 2007; Fleming et al. 1972; Gannon and Willig 1997; Henry and Kalko 2007; Mello et al 2008; Morrison 1978, 1980a; Wilkinson and Boughman 1998). In addition to commuting to and from their day roosts, frugivorous phyllostomids fly among their food plants and to and from night roosts where they actually consume fruits. Distances between food plants and night roosts range from a few meters to a kilometer or more (reviewed in Galindo-González 1998).

These foraging movements obviously have important seed dispersal consequences, which differ for large and small seeds. Phyllostomid bats do not ingest large seeds (e.g., those >3 cm in length). Instead, they carry single fruits containing these seeds to night roosts, where they consume the flesh surrounding them, usually dropping the seed beneath the roost. Dispersal distances of large seeds thus tend to be relatively short, that is, the distance from the fruiting plant to the night roost. The majority of fruits eaten by phyllostomid bats, however, are small and contain many small seeds (i.e., a few to hundreds) (Lobova et al. 2009). These fruits are also carried individually to night roosts where they are eaten, and seeds are ingested and ultimately defecated. Dispersal distances of these seeds depend on how long they are retained in the gut. Since most small seeds ingested by phyllostomids have retention times of ≤30 min (Bonaccorso and Gush 1987; Fleming 1988; Morrison 1980b), the movements made by bats during this period will determine the dispersal distances of these seeds. These distances can be highly variable. When bats sleep in their night roosts after consuming several fruits, ingested seeds will be defecated under the night roost and will be deposited relatively short distances from parent plants (like large seeds). If they are not defecated under night roosts, small seeds can be deposited in a variety of other places at different distances from parent plants. These include: (1) close to the parent plant when a bat returns to grab another fruit; (2) within the same feeding patch when it contains other fruiting plants; (3) between feeding patches; and (4) in new feeding patches. Distances (1) and (2) will generally result in short-distance dispersal distances (e.g., <ca. 100 m) whereas those of (3) and (4) will sometimes result in longer-distance dispersal distances (e.g., a kilometer or more).

As mentioned above, frugivorous phyllostomids eat fruit either as gulpers or as mashers, and these feeding modes might be expected to have different seed dispersal consequences. For example, gulpers excrete seeds that are mostly free of fruit pulp, whereas mashers spit out many seeds in a fibrous wad. Seeds that lack fruit pulp are significantly less likely to be killed by fungus than seeds surrounded by fruit pulp (Heer et al. 2010). As a result, gulpers are potentially more likely to disperse more viable seeds and are therefore potentially more effective seed dispersers than are mashers. Further research is needed to test this hypothesis.

A detailed study of the foraging behavior of *C. perspicillata*, a fruit gulper, in tropical dry forest in northwestern Costa Rica can illustrate these points (Fleming 1988). While this bat consumes at least 18 species of small-seeded fruit, including those of shrubs and trees, its diet in this habitat is centered on fruits produced by 5 species of *Piper* shrubs. At this site, most *Carollia* bats live in caves and hollow trees in colonies of a few dozen to several hundred individuals depending on season. Foraging behavior is related to social status, with harem males foraging close to the day roost and bachelor males and females foraging up to about 2 or 3 km from that roost. When feeding on most kinds of fruit,

bats take single fruits (or chunks of fruit) to night roosts located a few tens of meters from the parent plant to eat. In the wet season, bachelor males and females feed in two to four different feeding areas located up to about 2.5 km apart each night. In the dry season, they feed in up to eight different feeding areas located up to 3 km apart. In the wet season, Fleming (1981) estimated that these bats deposit about 90% of the seeds they ingest in their current feeding area and deposit most of the remaining 10% in another feeding area.

We can summarize the seed dispersal consequences of *C. perspicillata*'s foraging and feeding behavior in a plot of the probability of seed deposition as a function of distance from the parent plant using data on rates of seed defecation, rates of movement between feeding areas, and distances between feeding areas as reported by Fleming (1988) (fig. 17.3). This plot shows that, in both the wet and dry seasons, most seeds (up to 79%) are deposited <500 m from parent plants and that very few seeds are deposited >1.5 km from parent plants. Owing to the greater mobility of this bat during the dry season, seeds are likely to move somewhat greater distances, then, than in the wet season (fig. 17.3). Nonetheless, the seed dispersal curves in both seasons are highly leptokurtic, which is typical of most dispersal systems, including those involving phyllostomid bats.

Frugivorous phyllostomids are well-known for moving across open areas and between forest patches while foraging (e.g., Bernard and Fenton 2003; Henry et al. 2007; Loayza and Loiselle 2008; Medina et al. 2007; Sarmento et al. 2014) and often use riparian corridors when moving among resource patches (e.g., Galindo-González and Sosa 2003). They also routinely move between primary and nearby secondary forests in search of food (e.g., Fleming 1991; Voigt et al. 2012). These movements provide considerable mobility for the seeds they ingest and can result in seeds being dispersed to a variety of different sites and habitats (e.g., Arteaga et al. 2006; Henry et al. 2007).

What are the reproductive consequences to plants resulting from seed dispersal by phyllostomid bats? Does passage of seeds through bat guts have an effect on seed germination and do the foraging movements of phyllostomids result in seeds being deposited in good seed germination and seedling establishment sites? A meta-analysis conducted by Saldaña-Vázquez et al. (2019) indicates that seed passage through phyllostomid guts has neither a negative nor a positive effect on

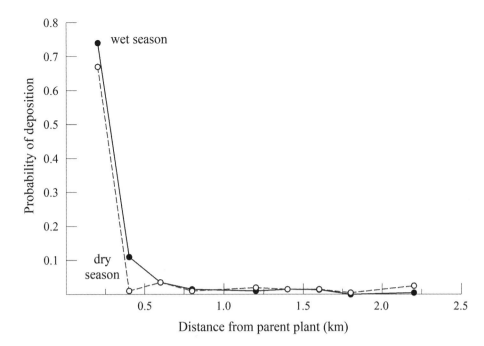

Figure 17.3. The seed dispersal curve of *Carollia perspicillata* in two different seasons in Costa Rican tropical dry forest based on data in Fleming (1988).

seed germination rates. What does matter for seedling establishment, however, is where seeds are deposited. For small seeds, being deposited in open areas away from overhanging vegetation (i.e., away from shaded night roosts, in forest gaps, and in open fields) can significantly increase the probability of seed germination and seedling establishment. The rapid colonization of abandoned pastures and agricultural fields by bat-dispersed plants in the genera *Cecropia*, *Piper*, *Solanum*, and *Vismia* clearly attests to the importance of phyllostomid bats as major agents of reforestation in the Neotropics (Muscarella and Fleming 2007).

In addition to differing in the quantitative details of their foraging movements, species of frugivorous phyllostomids also differ in the extent to which they undergo longer-distance seasonal movements. While none of these bats apparently undergo migrations involving hundreds of kilometers that occur in many other (nontropical) bats (Fleming and Eby 2003), various species do exhibit habitat shifts, undoubtedly in response to seasonal changes in resource availability in different habitats. For example, in the well-studied *C. perspicillata* living in tropical dry forest in northwestern Costa Rica, many females leave their lowland roosts in the dry season and presumably move tens of kilometers north to moister, more productive upland habitats (Fleming 1988). Similarly, seasonal changes in the abundance of frugivorous phyllostomids have been reported in tropical moist forests in Panama,

Brazil, Bolivia, and Argentina, probably in response to changes in resource availability among habitats (Aguirre et al. 2003; Bonaccorso 1979; Giannini 1999; Mello et al. 2009). Seasonal elevational movements occur in *S. lilium* in southwestern Mexico and *S. erythromos* in northern Argentina (Giannini 1999; L. Iñguez, pers. comm.). It is likely that altitudinal migrations also occur in other phyllostomids, but evidence for this is currently scarce (Esbérard et al. 2011; McGuire and Boyle 2013). Although seasonal habitat shifts might have seed dispersal consequences—specifically, they might occasionally result in seeds being dispersed in new habitats—we currently have no data on how often this actually occurs. Studies of the genetic structure of plant populations, including paternity analyses, are needed to document this (e.g., Gonzalez et al. 2017; Jordano et al. 2007; Mise et al. 2016).

Finally, it is important to note that not all frugivorous phyllostomids are legitimate seed dispersers. Two species of *Chiroderma* (*C. doriae* and *C. villosum*; Stenodermatinae) are actually seed predators (Nogueira and Peracchi 2003; Nogueira et al. 2005). Like other stenodermatines, these bats eat many fig fruits, but unlike other members of this subfamily, they do not spit out or defecate seeds. Instead, they thoroughly chew up seeds to obtain lipids and proteins. For example, Wagner et al. (2015) estimated that *C. villosum* obtains about 0.4 g of lipids from eating the seeds of a 7 g fig fruit. Nogueira et al. (2005) reported that the skulls and teeth of *Chiroderma* differ from those of other stenodermatines in the following features: height of the anterior zygomatic arch (increased to accommodate an enlarged masseter muscle); a longer masseter moment arm and reduced gape; one less lower molar; and an enlarged lower molar that is modified in conjunction with the second upper molar to serve as a "seed trap." Finally, unlike other fig-eating stenodermatines, these species of *Chiroderma* only eat the fruits from strangler figs and not those of free-standing figs (Wagner al. 2015). They apparently do this because the seeds of strangler figs are covered with a gelatinous coating, which probably makes it easier to separate them from fruit pulp; seeds of free-standing figs lack this coating.

Community Ecology of Seed Dispersal

As discussed above, frugivorous phyllostomid bats are important seed dispersers, and they play an important role in forest succession in Neotropical forests (Muscarella and Fleming 2007). However, the question remains, do all species play the same functional role in the seed dispersal loop (*sensu* Wang and Smith 2002)? A key concept for answering this question is ecological redundancy, which is defined as the the maintenance of ecosystem stability resulting from species (related or otherwise) that perform similar ecological functions (Morelli and Tryjanowski 2016; Walker 1992). By this definition, species maintain ecosystem stability via redundancy whenever they (1) play a similar functional role with other species in the community and (2) have a low degree of specialization in their interactions with other species, including their food species (Morelli and Tryjanowski 2016). In other words, high dietary overlap among species and broad diets can lead to functional redundancy within communities of mutualists such as frugivores and their food plants.

The degree of specialization of seed dispersal networks is negatively related to the degree of redundancy of interactions within a community (Blüthgen 2010). Since the degree of specialization in the interactions between bats and fruits is generally low (Hernández-Montero et al. 2015; Mello et al. 2011), there is potentially a high degree of redundancy in the interactions between bats and their fruit species. One factor that reduces this redundancy, however, is the modular or guild structure of frugivorous bat (and other frugivore)

communities. These communities often contain two groups of bats: the understory feeders (e.g., carolliines and *Sturnira* in the Stenodermatinae) and the canopy feeders (principally, stenodermatines plus *Phyllostomus* in the Phyllostominae) (Sánchez and Giannini 2018). Numerous studies have shown that dietary overlap between these two groups is relatively low (e.g., Giannini and Kalko 2004; Gorchov et al. 1995; Sarmento et al. 2014). To the extent that redundancy occurs in these communities, it is more likely to occur within rather than between modules.

Another factor influencing the degree of redundancy within modules of frugivorous bats is seed dispersal effectiveness. According to Schupp (1993), disperser effectiveness involves such things as number of seeds removed or ingested by particular frugivores and where those seeds are deposited relative to their germination requirements. As discussed above, frugivorous phyllostomids often defecate or drop large numbers of seeds under their feeding roosts. In the case of small, light-demanding seeds of early successional plants, these deposition sites are often inappropriate for successful germination and seedling establishment, especially whenever they occur in dark, heavily forested areas. Although disperser effectiveness has rarely been documented for frugivorous phyllostomids (e.g., Fleming 1988), bats that co-occur within modules are not likely to be equally effective seed dispersers because of differences in their roosting, foraging, and feeding behaviors. For example, Fleming (1991) studied the diets and local habitat distributions of four species of *Carollia* in two Costa Rican lowland forests. In both tropical dry and wet forests, the smallest *Carollia*s (*C. subrufa* in dry forest, *C. castanea* in wet forest) spent more time in secondary habitats and ate more *Piper* fruits than the largest species (*C. perspicillata* in both forests). Thus, a simple count of the number of species in a module of frugivores is unlikely to provide a realistic picture of the actual disperser redundancy within that module. Small species of *Carollia* are not necessarily redundant with large species. The same is likely to be true of canopy frugivores. This important topic deserves further attention.

If we ignore the complications of seed dispersal effectiveness, what happens to seed dispersal redundancy when communities differ in the number of their frugivores, either for biogeographic (e.g., along gradients of rainfall, elevation, or latitude) or anthropogenic reasons?

By definition, redundancy should decrease whenever frugivore diversity decreases. But does this put seed dispersal services provided by phyllostomids and other frugivores at greater risk of failing than in more diverse communities? While we currently lack a definitive answer to this question, available information indicates that a reduction of populations of some frugivorous bats results in higher specialization in bat-fruit ecological networks, which should reduce redundancy in seed dispersal services (Hernández-Montero et al. 2015). For example, the removal of understory plants in cloud forest to establish coffee plantations in the mountains of Veracruz, Mexico, resulted in a decrease in the abundance of *Sturnira* bats and an increase in the specialization of bat-fruit interactions. These changes could put at risk the resilience of the interaction network and seed dispersal provided by bats (Saldaña-Vázquez et al. 2010).

Anthropogenic land-use changes not only disrupt the structure of bat-fruit interaction networks and seed dispersal by bats via reductions in their populations, they can also rearrange their topology. Urbanization, for instance, promotes the creation of "new ecosystems" with new interactions for frugivorous bats. In a study conducted in Cuernavaca, Mexico, it was found that exotic plants supply about 40% of the diet of frugivorous bats such as *A. jamaicensis*, *Dermanura toltecus*, *Sturnira parvidens*, and *S. hondurensis* (Gurrusquieta-Navarro 2015). These kinds of anthropogenic changes thus have the potential to rearrange bat-fruit interactions with currently unknown consequences for seed dispersal.

Finally, climate change can also produce changes in the structure of bat-fruit interactions and seed dispersal by bats. These changes involve two principal factors: fruiting phenology and bat physiology. Phenology of plants and animals is driven by seasonal variation in climate and food resources, respectively (Durant et al. 2013; Stoner et al. 2003). Therefore, by affecting plant phenology, climate change could change the timing and availability of bat fruits and hence the structure of bat-fruit interactions. For example, in the dry season in montane forests of Argentina, bat-fruit production is reduced compared with the wet season, a pattern commonly found throughout the Neotropics. These changes in fruit availability result in changes in the composition of the diets and interaction structure of stenodermatine bats (Sánchez et al. 2012). Any further changes in the timing and duration of these seasons

will also have dietary and community structure consequences. Conversely, in the dry season in the tropical dry forest of Chamela, Mexico, the energy demands of the phytophagous bat *Glossophaga soricina* increase, resulting in its consumption of more plant resources, including some that contain low energy content, to meet their energy needs compared with the wet season (Ayala-Berdon et al. 2009). Secular changes in the climate of this habitat will undoubtedly continue to affect the energy budget and diet of this bat. Therefore, climate change could result in changes in the structure of bat-fruit interactions in this and other habitats.

In addition, by changing seasonal fruiting patterns and temperature, climate change could affect the migratory movements of some frugivorous bats. For example, *S. lilium* and *S. parvidens* become less abundant in the cold season of Brazil's Atlantic Forest and montane forests of Mexico, respectively, because they cannot tolerate cold temperatures (Mello et al. 2009; Saldaña-Vázquez, unpublished data). They probably migrate to warmer forests, which reduces diet overlap and competition among nonmigratory frugivorous bats such as *S. hondurensis* (in Mexico) that can tolerate cold temperatures. If climate change results in an increase in temperatures in these forests, we might expect that bat migrations will be less likely to occur or to occur less frequently, which could result in increased diet overlap and competition for food among bats in this community for much of the year. Evidence for this hypothesis is emerging from long-term monitoring in the mountains of Costa Rica. LaVal (2004) compared bat community structure in the Monteverde cloud forest at 1,350–1,500 m above sea level between 1973 and 1999. During this period, this locality experienced a change in land use that increased its local temperature. LaVal observed that captures of bats from lower elevations (e.g., the frugivore *S. lilium*, the nectarivore/frugivore *G. soricina*, and the vampire *Desmodus rotundus*) increased significantly in this period. The arrival of these new phytophagous bats could affect bat-bat interactions and bat-fruit interactions as well as seed dispersal patterns. However, the extent of this effect is currently unknown.

Conclusions and the Future

Frugivory is a feeding habit that has evolved independently in two families in the order Chiroptera.

Animals with this feeding habit obtain most of their energy and nutrients from fruit. In Neotropical phyllostomid bats, this feeding habit began about 20 Ma ago and occurs mostly in subfamilies Carolliinae and Stenodermatinae. The evolution of this feeding habit in phyllostomids has resulted in traits that allow them to detect, select, and acquire the nutrients from a wide variety of fruits. With its high species richness, the Stenodermatinae contains most of the variation in traits and diversity of fruit diets in Phyllostomidae. The structure of these bat-fruit interactions is not static. Certain biotic and abiotic changes can disrupt or rearrange them with unknown consequences for seed dispersal and seedling establishment. Despite decades of research on frugivorous bats, important information gaps still exist. For example, we lack basic information about the fruits consumed by many species, including rare stenodermatines such as species of *Chiroderma* (which are likely to be seed predators) and *Pygoderma* as well as species of *Sturnira*, including *S. tildae*, *S. magna*, and *S. luisi*. Without more information on the diets of frugivorous bats, we will not know their role in seed dispersal and will lack information needed to model the evolution of this feeding habit in the Phyllostomidae. We also lack basic information about the seed dispersal effectiveness of most frugivorous bats and the actual degree of redundancy that occurs within frugivore modules. Finally, we need more studies on the effects of anthropogenic disturbance on bat-fruit interactions and its effects on the seed dispersal provided by bats. Particularly important will be studies within an explicit landscape context, in order to find a balance between human land transformation and the conservation of bat-fruit interactions (Peters et al. 2016).

Acknowledgments

RAS-V thanks the Laboratory of Ecología de Paisajes Fragmentados of IIES (UNAM) and Laboratory of Análisis para la Conservación de la Biodiversidad of INIRENA (UMSNH) for allowing him to work on this research during a postdoctoral residence in both labs, and also CTIC-UNAM and SEP-PRODEP for their postdoctoral fellowships (CJIC/CTIC/0380/2015; 511-6/17-626). THF's work with frugivorous bats has been generously supported by the Smithsonian Institution, the University of Missouri–Saint Louis, the University of Miami, the U.S. National Science Foundation, and the Center for Field Research. We thank Marco Mello and two anonymous reviewers for their comments on a draft of this chapter.

References

Aguirre, L. F., A. Herrel, R. Van Damme, and E. Matthysen. 2003. The implications of food hardness for diet in bats. Functional Ecology, 17:201–212.

Andrade, T. Y., W. Thies, P. K. Rogeri, E. K. V. Kalko, and M. A. R. Mello. 2013. Hierarchical fruit selection by Neotropical leaf-nosed bats (Chiroptera: Phyllostomidae). Journal of Mammalogy, 94:1094–1101.

APG III. 2009. An update of the Angiosperm Phylogeny Group classification for the orders and families of flowering plants: APG III. Botanical Journal of the Linnean Society, 161: 105–121.

Arteaga, L. L., L. F. Aguirre, and M. I. Moya. 2006. Seed rain produced by bats and birds in forest islands in a Neotropical savanna. Biotropica, 38:718–724.

Ayala-Berdon, J., and J. E. Schondube. 2011. A physiological perspective on nectar-feeding adaptation in phyllostomid bats. Physiological and Biochemical Zoology, 84:458–466.

Ayala-Berdon, J., J. E. Schondube, and K. E. Stoner. 2009. Seasonal intake responses in the nectar-feeding bat Glossophaga soricina. Journal of Comparative Physiology B, 179:553–562.

Bernard, E., and M. B. Fenton. 2003. Bat mobility and roosts in a fragmented landscape in Central Amazonia, Brazil. Biotropica, 35:262–277.

Blüthgen, N. 2010. Why network analysis is often disconnected from community ecology: a critique and an ecologist's guide. Basic and Applied Ecology, 11:185–195.

Bonaccorso, F. J. 1979. Foraging and reproductive ecology in a Panamanian bat community. Bulletin of the Florida State Museum, Biological Sciences, 24:359–408.

Bonaccorso, F. J., and T. J. Gush. 1987. An experimental study of the feeding behaviour and foraging strategies of phyllostomid bats. Journal of Animal Ecology, 56:907–920.

Bonaccorso, F. J, J. R. Winkelmann, D. Shin, C. I. Agrawal, N. Aslami, C. Bonney, A. Hsu et al. 2006. Evidence for exploitative competition: comparative foraging behavior and roosting ecology of short-tailed fruit bats (Phyllostomidae). Biotropica, 39:249–256.

Caras, T., and C. Korine. 2009. Effect of vegetation density on the use of trails by bats in a secondary tropical rain forest. Journal of Tropical Ecology, 25:97–101.

Chaverri, G., O. E. Quiros, and T. H. Kunz. 2007. Ecological correlates of roost fidelity in the tent-making bat Artibeus watsoni. Ethology, 113:598–605.

Cloutier, D., and D. W. Thomas. 1992. Carollia perspicillata. Mammalian Species, 417:1–9.

Díaz, S., A. Purvis, J. H. C. Cornelissen, G. M. Mace, M. J. Donoghue, R. M. Ewers, P. Jordano, and W. D. Pearse. 2013. Functional traits, the phylogeny of function, and ecosystem service vulnerability. Ecology and Evolution, 3: 2958–2975.

Dumont, E. R. 2003. Bats and fruit: an ecomophological approach. Pp. 398–429 *in*: Bat Ecology (T. H. Kunz and M. B. Fenton, eds.). University of Chicago Press, Chicago.

Dumont, E. R., L. M. Dávalos, A. Goldberg, S. E. Santana, K. Rex, and C. C. Voigt. 2012. Morphological innovation, diversification and invasion of a new adaptive zone. Proceedings of the Royal Society B–Biological Sciences, 297:1797–1805.

Dumont, E. R., A. Herrel, R. A. Medellín, J. A. Vargas-Contreras, and S. E. Santana. 2009. Built to bite: cranial design and function in the wrinkle-faced bat. Journal of Zoology, 279: 329–337.

Dumont, E. R., K. Samadevam, I. Grosse, O. M. Warsi, B. Baird, and L. M. Dávalos. 2014. Selection for mechanical advantage underlies multiple cranial optima in New World leaf-nosed bats. Evolution, 68:1436–1449.

Durant, K. A., R. W. Hall, L. M. Cisneros, R. M. Hyland, and M. R. Willig. 2013. Reproductive phenologies of phyllostomid bats in Costa Rica. Journal of Mammalogy, 94:1438–1448.

Esbérard, C. E. L., I. Lima, P. H. Nobre, S. L. Althoff, T. Jordão-Nogueira, D. Dias, F. Carvalho, M. E. Fabián, M. L. Sekiama, and A. S. Sobrinho. 2011. Evidence of vertical migration in the Ipanema bat *Pygoderma bilabiatum* (Chiroptera: Phyllostomidae: Stenodermatinae). Zoologia, 28:717–724.

Fleming, T. H. 1981. Fecundity, fruiting pattern, and seed dispersal in *Piper amalago* (Piperaceae), a bat-dispersed tropical shrub. Oecologia, 51:42–46.

Fleming, T. H. 1986. Opportunism versus specialization: the evolution of feeding strategies in frugivorous bats. Pp. 105–118 *in*: Frugivores and Seed Dispersal (A. Estrada and T. H. Fleming, eds.). Dr. W. Junk Publishers, Dordrecht.

Fleming, T. H. 1988. The Short-Tailed Fruit Bat: A Study in Plant-Animal Interactions. University of Chicago Press, Chicago.

Fleming, T. H. 1991. The relationship between body size, diet, and habitat use in frugivorous bats, genus *Carollia* (Phyllostomidae). Journal of Mammalogy, 72:493–501.

Fleming, T. H., and P. Eby. 2003. Ecology of bat migration. Pp. 156–208 *in*: Bat Ecology (T. H. Kunz and M. B. Fenton. eds.). University of Chicago Press, Chicago.

Fleming, T. H., E. T. Hooper, and D. E. Wilson. 1972. Three Central American bat communities: structure, reproductive cycles, and movement patterns. Ecology, 53:555–569.

Fleming, T. H., and W. J. Kress. 2013. The Ornaments of Life: Coevolution and Conservation in the Tropics. University of Chicago Press, Chicago.

Galindo-González, J. 1998. Dispersion de semillas por murcielagos: su importancia en la conservacion y regeneracion del bosque tropical. Acta Zoologica Mexicana, 73:57–74.

Galindo-González, J., and V. J. Sosa. 2003. Frugivorous bats in isolated trees and riparian vegetation associated with human-made pastures in a fragmented tropical landscape. Southwestern Naturalist, 48:579–589.

Gannon, M. R., and M. R. Willig. 1997. The effect of lunar

illumination on movement and activity of the red fig-eating bat (*Stenoderma rufum*). Biotropica, 29:525–529.

Giannini, N. P. 1999. Selection of diet and elevation by sympatric species of *Sturnira* in an Andean rainforest. Journal of Mammalogy, 80:1186–1195.

Giannini, N. P., and E. K. V. Kalko. 2004. Trophic structure in a large assemblage of phyllostomid bats in Panama. Oikos, 105:209–220.

Gonzalez-Varo, J. P., C. S. Carvalho, J. M. Arroyo, and P. Jordano. 2017. Unravelling seed dispersal through fragmented landscapes: frugivore species operate unevenly as mobile links. Molecular Ecology, 26:4309–4321.

Gorchov, D. L., F. Cornejo, C. F. Ascorra, and M. Jaramillo. 1995. Dietary overlap between frugivorous birds and bats in the Peruvian Amazon. Oikos, 74:235–250.

Gurrusquieta-Navarro, M. C. 2015. Dieta de murciélagos frugívoros en la zona urbana de Cuernavaca, Morelos. Thesis, Universidad Autónoma del Estado de Morelos.

Hamilton, W. J., III, and K. E. F. Watt. 1970. Refuging. Annual Review of Ecology and Systematics, 1:263–286.

Heer, K., L. Albrecht, and E. K. V. Kalko. 2010. Effects of ingestion by Neotropical bats on germination parameters of native free-standing and strangler figs (*Ficus* sp., Moraceae). Oecologia, 163:425–435.

Henry, M., and E. K. V. Kalko. 2007. Foraging strategy and breeding constraints of *Rhinophylla pumilio* (Phyllostomidae) in the Amazon lowlands. Journal of Mammalogy, 88:81–93.

Henry, M., J. M. Pons, and J.-F. Cosson. 2007. Foraging behaviour of a frugivorous bat helps bridge landscape connectivity and ecological processes in a fragmented forest. Journal of Animal Ecology, 76:801–813.

Hernández-Montero, J. R., O. R. Rojas-Soto, and R. A. Saldaña-Vázquez. 2011. Consumo y dispersión de semillas de *Solanum schlechtendalianum* (Solanaceae) por el murciélago frugívoro *Sturnira ludovici* (Phyllostomidae). Chiroptera Neotropical, 17:1017–1021.

Hernández-Montero, J. R., R. A. Saldaña-Vázquez, J. R. Galindo-González, and V. J. Sosa. 2015. Bat-fruit interactions are more specialized in shaded-coffee plantations than in tropical mountain cloud forest fragments. PLOS ONE, 10:e0126084.

Hodgkison, R., M. Ayasse, C. Häberlein, S. Schulz, A. Zubaid, W. A. W. Mustapha, T. H. Kunz, and E. K. V. Kalko. 2013. Fruit bats and bat fruits: the evolution of fruit scent in relation to the foraging behaviour of bats in the New and Old World tropics. Functional Ecology, 27:1075–1084.

Korine, C., and E. K. V. Kalko. 2005. Fruit detection and discrimination by small fruit-eating bats (Phyllostomidae): echolocation call design and olfaction. Behavioral Ecology and Sociobiology, 59:12–23.

Jara-Servin, A. M., R. A. Saldaña-Vázquez, and J. E. Schondube. 2017. Nutrient availability predicts frugivorous bat abundance in an urban environment. Mammalia, 81:367–374.

Jordano, P. 2000. Fruits and frugivory. Pp. 125–166 *in*: Seeds: The Ecology of Regeneration in Plant Communities (M. Fenner, ed.). CABI Publishers, Wallingford.

Jordano, P., C. Garcia, J. A. Godoy, and J. L. Garcia-Castaño. 2007. Differential contribution of frugivores to complex seed dispersal patterns. Proceedings of the National Academy of Sciences, USA, 104:3278–3282.

Kalko, E. K. V., E. A. Herre, and C. O. Handley Jr. 1996. The relation of fig fruit syndromes to fruit-eating bats in the New and Old World tropics. Journal of Biogeography 23:565–576.

Kunz, T. H. 1982. Roosting ecology of bats. Pp. 1–55 in: Ecology of Bats (T. H. Kunz, ed.). Plenum Press, New York.

LaVal, R. 2004. Impact of global warming and locally changing climate on tropical cloud forest bats. Journal of Mammalogy, 85:237–244.

Loayza, A. P., and B. A. Loiselle. 2008. Preliminary information on the home range and movement patterns of Sturnira lilium (Phyllostomidae) in a naturally fragmented landscape in Bolivia. Biotropica, 40:630–635.

Lobova, T. A., C. K. Geiselman, and S. A Mori. 2009. Seed Dispersal by Bats in the Neotropics. New York Botanical Garden, Bronx.

Martínez del Río, C., J. E. Schondube, T. J. McWhorter, and L. G. Herrera. 2001. Intake responses in nectar feeding birds: digestive and metabolic causes, osmoregulatory consequences, and coevolutionary effects. American Zoologist, 41:902–915.

McGuire, L. P., and W. A. Boyle. 2013. Altitudinal migration in bats: evidence, patterns, and drivers. Biological Reviews, 88:767–786.

Medina, A., C. A. Harvey, D. S. Meerlo, S. Vilchez, and B. Hernandez. 2007. Bat diversity and movement in an agricultural landscape in Matiguas, Nicaragua. Biotropica, 39:120–128.

Mello, M. A. R., E. K. V. Kalko, and W. R. Silva. 2008. Diet and abundance of the bat Sturnira lilium (Chiroptera) in a Brazilian montane Atlantic Forest. Journal of Mammalogy, 89:485–492.

Mello, M. A. R., E. K. V. Kalko, and W. R. Silva. 2009. Ambient temperature is more important than food availability in explaining reproductive timing of the bat Sturnira lilium (Mammalia: Chiroptera) in a montane Atlantic Forest. Canadian Journal of Zoology, 87:239–245.

Mello, M. A. R., F. M. D. Marquitti, P. R. Guimarães, E. K. V. Kalko, P. Jordano, and M. A. M. De Aguiar. 2011. The missing part of seed dispersal networks: structure and robustness of bat-fruit interactions. PloS ONE, 6:e17395.

Mise, Y., K. Yamazaki, M. Soga, and S. Koike. 2016. Comparing methods of acquiring mammalian endozoochorous seed dispersal distance distributions. Ecological Research, 31:881–889.

Morelli, F., and P. Tryjanowski. 2016. The dark side of the "redundancy hypothesis" and ecosystem assessment. Ecological Complexity, 28:1–9.

Morrison, D. W. 1978. Influence of habitat on the foraging distances of the fruit bat, Artibeus jamaicensis. Journal of Mammalogy, 59:622–624.

Morrison, D. W. 1980a. Foraging and day-roosting dynamics of canopy fruit bats in Panama. Journal of Mammalogy, 61:20–29.

Morrison, D. W. 1980b. Efficiency of food utilization by fruit bats. Oecologia, 45:270–273.

Muscarella, R., and T. H. Fleming. 2007. The role of frugivorous bats in tropical forest succession. Biological Reviews, 82:573–590.

Nogueira, M. R., L. R. Monteiro, A. L. Peracchi, and A. F. B. de Araújo. 2005. Ecomorphological analysis of the masticatory apparatus in the seed-eating bats, genus Chiroderma (Chiroptera : Phyllostomidae). Journal of Zoology, 266:355–364.

Nogueira, M. R., and A. L. Peracchi. 2003. Fig-seed predation by two species of Chiroderma: discovery of a new feeding strategy in bats. Journal of Mammalogy, 84:225–233.

Parolin, L. C., S. B. Mikich, and G. V. Bianconi. 2015 Olfaction in the fruit-eating bats Artibeus lituratus and Carollia perspicillata: an experimental analysis. Anais da Academia Brasileira de Ciências, 87:2047–2053.

Peters, V. E., T. A. Carlo, M. Mello, R. A. Rice, D. W. Tallamy, S. A. Caudill, and T. H. Fleming. 2016. Using plant-animal interactions to inform tree selection in tree-based agroecosystems for enhanced biodiversity. Bioscience, 66:1046–1056.

Rojas, D., A. Vale, V. Ferrero, and L. Navarro. 2012. The role of frugivory in the diversification of bats in the Neotropics. Journal of Biogeography, 39:1948–1960.

Rojas, D., O. M. Warsi, and L. M. Dávalos. 2016. Bats (Chiroptera: Noctilionoidea) challenge a recent origin of extant Neotropical diversity. Systematic Biology, 65:432–448.

Saldaña-Vázquez, R. A. 2014. Intrinsic and extrinsic factors affecting dietary specialization in Neotropical frugivorous bats. Mammal Review, 44:215–224.

Saldaña-Vázquez, R. A., J. H. Castaño, J. Baldwin, and J. Pérez-Torres. 2019. Does seed ingestion by bats enhance germination? A new meta-analysis 15 years later. Mammal Review, 49 (3): 201–209.

Saldaña-Vázquez, R. A., E. Ruíz-Sánchez, L. Herrera-Alsina, and J. E. Schondube. 2015. Digestive capacity predicts diet diversity in Neotropical frugivorous bats. Journal of Animal Ecology, 84:1396–1404.

Saldaña-Vázquez, R. A., V. J. Sosa, J. R. Hernández-Montero, and F. López-Barrera. 2010. Abundance responses of frugivorous bats (Stenodermatinae) to coffee cultivation and selective logging practices in mountainous central Veracruz, Mexico. Biodiversity and Conservation, 19:2111–2124.

Saldaña-Vázquez, R. A., V. J. Sosa, L. I. Iñiguez Dávalos, and J. E. Schondube. 2013. The role of extrinsic and intrinsic factors in Neotropical fruit bat-plant interactions. Journal of Mammalogy, 94:632–639.

Sánchez, M. S., and N. P. Giannini. 2018. Trophic structure of frugivorous bats in the Neotropics: emergent patterns in evolutionary history. Mammal Review, 48:90–107.

Sánchez, M. S., N. P. Giannini, and R. M. Barquez. 2012. Bat frugivory in two subtropical rain forests of northern Argentina: testing hypotheses of fruit selection in the Neotropics. Mammalian Biology, 77:22–31.

Santana, S. E., and E. R. Dumont. 2009. Connecting behaviour and performance: the evolution of biting behaviour and bite performance in bats. Journal of Evolutionary Biology 22:2131–2145.

Sarmento, R., C. P. Alves-Costa, A. Ayub, and M. A. R. Mello. 2014. Partitioning of seed dispersal services between birds and bats in a fragment of the Brazilian Atlantic Forest. Zoologia, 31:245–255.

Schondube, J. E., G. Herrera, and C. Martínez del Río. 2001. Diet and the evolution of digestion and renal function in phyllostomid bats. Zoology—Analysis of Complex Systems, 104:59–73.

Schupp, E. W. 1993. Quantity, quality and the effectiveness of seed dispersal by animals. Pp. 15–29 *in*: Frugivory and Seed Dispersal: Ecological and Evolutionary Aspects (T. H. Fleming and A. Estrada, eds.). Kluwer Academic Publishers, Dordrecht.

Simmons, N. B., and R. S. Voss. 1998. The mammals of Paracou, French Guiana: a Neotropical lowland rainforest fauna. Part I: Bats. Bulletin of the American Museum of Natural History, no. 237. American Museum of Natural History, New York.

Stoner, K. E., K. A. O-Salazar, R. C. R. Fernandez, and M. Quesada. 2003 Population dynamics, reproduction, and diet of the lesser long-nosed bat (*Leptonycteris curasoae*) in Jalisco, Mexico: implications for conservation. Biodiversity and Conservation, 12:357–373.

Thies, W., and E. K. V. Kalko. 2004. Phenology of Neotropical pepper plants (Piperaceae) and their association with their main dispersers, two short-tailed fruit bats, *Carollia perspicillata* and *C. castanea* (Phyllostomidae). Oikos, 104:362–376.

Thies, W., E. K. V. Kalko, and H. Schnitzler. 1998. The roles of echolocation and olfaction in two Neotropical fruit-eating bats, *Carollia perspicillata* and *C. castanea*, feeding on *Piper*. Behavioral Ecology and Sociobiology, 42:397–409.

van der Pijl, L. 1982. Principles of Dispersal in Higher Plants. Springer-Verlag, Berlin.

Violle, C., M.-L. Navas, D. Vile, E. Kazakou, C. Fortunel, I. Hummel, and E. Garnier. 2007. Let the concept of trait be functional! Oikos, 116:882–892.

Vleut, I., J. Galindo-González, W. F. de Boer, S. I. Levy-Tacher, and L.-B. Vazquez. 2015. Niche differentiation and its relationship with food abundance and vegetation complexity in four frugivorous bat species in southern Mexico. Biotropica, 47:606–615.

Voigt, C. C., S. L. Voigt-Heucke, and A. S. Kretzschmar. 2012. Isotopic evidence for seed transfer from successional areas into forests by short-tailed fruit bats (*Carollia* spp.; Phyllostomidae). Journal of Tropical Ecology, 28:181–186.

Wagner, I., J. U. Ganzhorn, E. V. K. Kalko, and M. Tschapka. 2015. Cheating on the mutualistic contract: nutritional gain through seed predation in the frugivorous bat *Chiroderma villosum* (Phyllostomidae). Journal of Experimental Biology, 218:1016–1021.

Wang, B. C., and T. B. Smith. 2002. Closing the seed dispersal loop. Trends in Ecology and Evolution, 17:379–385.

Walker, B. H. 1992. Biodiversity and ecological redundancy. Conservation Biology, 6:18–23.

Wilkinson, G. S., and J. W. Boughman. 1998. Social calls coordinate foraging in greater spear-nosed bats. Animal Behaviour, 55:337–350.

Population and Community Ecology

18

Roosting Ecology

The Importance of Detailed Description

Armando Rodríguez-Durán

Introduction

Few mammals are as alien to humans as bats; their main sensory inputs, ultrasound and smell, differ substantially from those of our own visually oriented species. They navigate a three-dimensional medium that we are just beginning to dominate, and do it at hours when we normally rest. The study of bats is complicated by their secretiveness, combined with great vagility. When Nagel (1974) posed the question, "What is it like to be a bat?" he concluded that we would never know. But stubborn scientists do not like that kind of an answer. Great strides have been made in understanding bats, albeit not in the sense posed by Nagel.

Bats inhabit a variety of ready-made, natural sites, such as caves, rock crevices, hollow trees, exfoliating bark, and foliage, and they commonly occupy anthropogenic structures. In some cases, these mammals also create their own roosting environments, by modifying leaves, excavating termite nests, or even altering the microclimate and morphology of caves. Roosting under such different conditions influences the morphology, physiology, and behavior of bats in varied ways (Kunz 1982). Almost 40 years ago, Gaisler (1979) proposed a classification of roost types based on work by various researchers. Unfortunately, his system was not widely adopted, and even modern studies describing roosts still lack important details. One aim of this chapter is to stimulate the gathering of more detailed descriptions about the characteristics of roosts.

The number of reviews dealing extensively with roosting (e.g., Barclay and Kurta 2007; Bradbury 1977; Kunz 1982; Kunz and Lumsden 2003; Lewis 1995; Tuttle 1976) underscores the importance of this subject in understanding the biology of bats. It is not my intention here to repeat those thorough analyses but, rather, to illustrate the importance of understanding the roosting ecology of the Phyllostomidae (including related families within the Noctilionoidea) and to point out gaps in our knowledge and the steps needed to bridge them. Our knowledge of roosting by

311

tropical phyllostomids lags far behind that of bats from families common in temperate North America, Europe, New Zealand, or Australia (e.g., Brigham 2007; Kerth et al. 2001; Lumsden et al. 2002; O'Donnell and Sedgeley 1999). The number of phyllostomids for which we have basic information about roosts has increased from 81 (Tuttle 1976) to about 106 species in the last 40 years, about 49% of the species in the family as a whole (table 18.1). This somewhat low pace (less than one species/year) seems understandable when we consider the number of rare species in this group and the remote areas that some of them may inhabit. However, as the number of locally based researchers and collaborations increases (e.g., Aguirre 2007; LaVal and Rodríguez-Herrera 2002; Linares 1986; Rodríguez-Durán and Christenson 2012; Silva Taboada 1979), long-term observations should become more feasible.

The Significance of Roosts for Bats

The importance of understanding the roosting ecology of bats becomes evident when we see specializations such as pelage markings, which will make them less conspicuous while roosting in vegetation (Santana et al. 2011). Roosts can be ephemeral or permanent, scarce or abundant, or suitable for large congregations or small groups. The type of roost used by a bat species is also likely to influence its social interactions and organization. Many phyllostomids roosting in the foliage of trees appear to be polygynous and live in harems, a system in which one male defends and mates with multiple females (Wilkinson and McCracken 2003). But the mating systems of only a handful of phyllostomids have been well-studied (e.g., Chaverri and Kunz 2010; Kunz et al. 1998; McCracken and Wilkinson 2000; Murray and Fleming 2008). *Artibeus jamaicensis*, probably the best-studied species, has one of the broadest geographic distributions and uses a wide variety of roosts, including foliage, tree cavities, and caves (Ortega and Castro-Arellano 2001). This species exhibits intraspecific differences in size of the testes between populations in Panama and Mexico (Wilkinson and McCracken 2003). Although *A. jamaicensis* in Panama roosts mostly in hollow trees, to which females may show little fidelity (Morrison 1979), *A. jamaicensis* in Mexico form more stable harems in caves (Ortega and Arita 1999). Size of the testes is significantly larger in Panama than in Mexico, possibly as a result of varying

female fidelity to harems under these different roosting conditions (Wilkinson and McCracken 2003).

Two known exceptions to this pattern of polygyny, among the few species examined so far, are *Erophylla sezekorni* and *Macrotus californicus*. The phyllonycterine bat *E. sezekorni* roosts in large multimale-multifemale colonies. Murray and Fleming (2008) showed that the mating system of *E. sezekorni* has characteristics of a lek. Nevertheless, mating occurs in the cave where these bats roost, which violates the criterion that display sites do not contain resources required by females. Therefore, Murray and Fleming (2008) concluded that, although *E. sezekorni* demonstrates characteristics of a lek system, it is not a classical lek mating species. *Macrotus californicus*, another cave-roosting species, also shows some lek mating characteristics (Berry and Brown 1995). Many noctilionoids roost in large numbers in caves. Our lack of understanding about the relationship between roosts and mating behavior, as well as other aspects of their biology, offers some potentially interesting areas of research.

Availability of roosts may affect roost fidelity in bats, especially adult males, for which roosts could provide a defendable resource to attract females. For instance, Fleming (1988) found that young male *Carollia perspicillata* were more likely to remain in their natal cave than were young females. Similarly, male *Artibeus watsoni* exhibit greater roost fidelity than do females (Chaverri et al. 2007), a pattern observed in other tent-making species (Chaverri and Kunz 2010). *Artibeus jamaicensis*, a bat that uses a wide variety of roost types, appears less site faithful when roosting in foliage than in tree cavities (Morrison 1979) or caves (Ortega and Arita 1999). This could be a result of the more ephemeral nature of foliage roosts or of the greater abundance of potential foliage roosts, compared to cavities or caves. The few studies that have looked at roost availability seem to suggest that, by and large, potential roosting sites within a given geographical area are not limiting for most phyllostomids (Rodríguez-Durán 1998; Villalobos-Chaves 2016). However, identifying sites that have the "potential" to be a roost is problematic and, for many species, doing so must await a better understanding of the ecological, behavioral, and physiological functions of their roosts.

The type of roost used by a species may also influence interactions such as active information transfer (Wilkinson 1995), thus helping to shape the evolution

Table 18.1. Roosts reported for the bats of the superfamily Noctilionoidea

Species	Roost	Source
THYROPTERIDAE:		
Thyroptera tricolor	Leaves	11, 18
FURIPTERIDAE:		
Furipterus horrens	Caves, under fallen trees, tree cavity	11, 18
NOCTILIONIDAE:		
Noctilio albiventris	Tree cavity	18
N. leporinus	Tree cavity, caves, buildings, bell cavity	1, 5, 7, 10, 14, 18
MORMOOPIDAE:		
Mormoops blainvillei	Hot caves	5, 8, 12, 14, 15
M. megalophylla	Hot caves	1, 7, 9, 12
Pteronotus personatus	Hot caves	9, 12
P. parnellii complex	Hot caves	7, 8, 9, 12, 14, 15
P. macleayii	Hot caves	5
P. quadridens	Hot caves	2, 5, 8, 10, 14, 15
P. gymnonotus	Hot caves	12
P. davyi	Hot caves	1, 7, 9, 12
PHYLLOSTOMIDAE:		
MACROTINAE:		
Macrotus californicus	Caves, culverts	3
M. waterhousii	Caves, buildings	3, 5
MICRONYCTERINAE:		
Lampronycteris brachyotis	Tree cavity, caves, tunnels	1, 3, 4, 6, 7, 18
Micronycteris schmidtorum	Caves, tree cavity	3, 4, 7, 18
M. minuta	Tree cavity, cavern	3, 4, 7, 18
M. hirsuta	Tree cavity, under bridge	1, 3, 11, 18
M. brosseti	Tree cavity	11, 18
M. megalotis	Caves, tree cavity, buildings, culverts, under fallen tree, under bridge	1, 3, 4, 18
M. microtis	Tree cavity, culverts, caves	4, 11, 18
M. matses	Animal burrow	18
DESMODONTINAE:		
Diphylla ecaudata	Caves	3, 13
Desmodus rotundus	Caves, tree cavity, buildings, culverts, under bridge	1, 3, 13, 18
Diaemus youngi	Tree cavity, caverns	1, 3, 13
PHYLLOSTOMINAE:		
Mimon bennettii	Tree cavity	3, 11, 18
M. cozumelae	Caves, culverts	3
Vampyrum spectrum	Tree cavity, buildings	1, 3

(continued)

Table 18.1. Continued

Species	Roost	Source
PHYLLOSTOMINAE (cont.):		
Chrotopterus auritus	Caves	4, 13
Trachops cirrhosus	Tree cavity, culverts, caves, building	1, 4, 7, 11, 18
Macrophyllum macrophyllus	Culverts, caves, cavity in fallen tree	4, 7, 11, 13, 18
Tonatia bidens	Tree cavity	1, 3, 7
T. saurophila	Termite nest, caves, tree cavity	17
Lophostoma silvicolum	Tree cavity, termite nest	3, 14, 18
L. carrikeri	Tree cavity, termite nest	3, 18
L. brasiliense	Tree cavity, termite nest, buildings	4, 7, 18
Gardnerycteris crenulatum	Tree cavity, buildings	4, 7, 18
Phyllostomus discolor	Caves, tree cavity	1, 3, 4, 18
P. elongates	Tree cavity, culverts	3, 4, 11, 18
P. hastatus	Caves, tree cavity, buildings, leaves, termite nest	1, 3, 4, 7, 18
LONCHORHININAE:		
Lonchorhina orinocensis	Crevices	4, 7, 18
L. aurita	Caves, culverts	1, 4, 7, 13, 18
GLOSSOPHAGINAE:		
Anoura cultrata	Caves	4
A. caudifer	Caves, tree cavity, culverts	4, 7, 18
A. geoffroyi	Caves	1, 4, 7, 18
Hylonycteris underwoodi	Caves, culverts	3
Choeroniscus minor	Under fallen tree, tree cavity, caves	7, 11, 18
Choeronycteris mexicana	Caves, buildings	3
Musonycteris harrisoni	Foliage	3
Brachyphylla nana	Hot caves	5
B. cavernarum	Hot caves	10, 14
Erophylla bombifrons	Hot caves	10, 14
E. sezekorni	Hot caves	2, 5
Phyllonycteris poeyi	Hot caves	5
P. aphylla	Hot caves	5
Monophyllus redmani	Hot caves, bell cavity	2, 5, 10, 14
M. plethodon	Caves	3
Leptonycteris curasoae	Caves, buildings	4, 7
L. yerbabuenae	Caves, buildings	3
Glossophaga soricina	Caves, tree cavity, culverts, bridges, buildings	1, 4, 7, 11, 18
G. commissarisi	Caves	3
G. leachii	Caves, culverts, buildings	13
G. longirostris	Caves, tree cavity, buildings	1, 3, 4, 7
LONCHOPHYLLINAE:		
Hsunycteris thomasi	Tree cavity, under fallen tree, caves	3, 7, 11, 18
H. pattoni	Cavity in fallen tree	18

Table 18.1. Continued

Species	Roost	Source
LONCHOPHYLLINAE (*cont.*):		
Lonchophylla mordax	Caves	13
L. robusta	Caves	3, 7
Platalina genovensium	Caves	3
Lionycteris spurrelli	Caves, crevices	4, 18
GLYPHONYCTERINAE:		
Trinycteris nicefori	Tree cavity, tunnels	4, 7
Glyphonycteris daviesi	Tree cavity	18
G. sylvestris	Tree cavity, caves	4, 7, 18
CAROLLIINAE:		
Carollia castanea	Caves, tree cavity	3, 18
C. subrufa	Caves, tree cavity, culverts, crevices, buildings	3
C. brevicauda	Caves, culverts, tree cavity, buildings, bridges, leaves, under fallen tree	3, 4, 18
C. perspicillata	Caves, culverts, tree cavity, buildings, bridges, leaves	1, 3, 4, 11, 18
RHINOPHYLLINAE:		
Rhinophylla pumilio	Foliage, tents	11, 16, 18
STENODERMATINAE:		
Sturnira lilium	Tree cavity, caves, culverts, buildings	1, 3, 4
S. erythromos	Caves	17
S. tildae	Undercut earth bank	17
S. ludovici	Buildings	4, 13
Uroderma bilobatum	Tents, buildings	1, 11, 16, 18
U. magnirostrum	Tents, buildings	4, 16, 18
Vampyriscus nymphaea	Tents	16
Chiroderma salvini	Buildings	4
C. trinitatum	Caves, leaves	1, 3
C. villosum	Tree cavity, buildings	4
Mesophylla macconnelli	Tents, foliage	4, 11, 16, 18
Vampyressa pusilla	Tents, foliage, buildings	4, 16
V. thyone	Tents	16
Vampyrodes caraccioli	Leaves, foliage	1, 3
Platyrrhinus helleri	Tents, leaves, tree cavity, buildings, caves	1, 16, 18
P. dorsalis	Foliage, under fallen tree	18
P. infuscus	Caves, buildings	18
P. lineatus	Foliage	18
P. vittatus	Caves, under fallen tree	18
Ectophylla alba	Tents	16
Centurio senex	Foliage	1, 13
Sphaeronycteris toxophyllum	Caves	3

(continued)

Table 18.1. Continued

Species	Roost	Source
STENODERMATINAE (*cont.*):		
Ardops nichollsi	Foliage	3
Ariteus flavescens	Foliage	2, 8, 15
Stenoderma rufum	Foliage	10, 14
Phyllops falcatus	Foliage, buildings	5
Artibeus jamaicensis	Caves, tree cavity, foliage, leaves, tents, buildings, bell cavities	1, 2, 4, 5, 8, 10, 11, 14, 15, 16, 18
A. obscurus	Exfoliating bark, leaves, buildings	11, 18
A. planirostris	Foliage	18
A. intermedius	Caves, crevices	16
A. lituratus	Caves, tree cavity, leaves, foliage, crevices, buildings	1, 4, 18
A. hirsutus	Caves, buildings	3
A. inopinatus	Buildings	3
A. watsoni	Leaves, tents	16
A. aztecus	Caves	3
A. phaeotis	Caves, tents	16
A. anderseni	Tents	16
A. cinereus	Leaves, tents, crevices	1, 11, 16, 18
A. toltecus	Caves, buildings, leaves, tents	16
A. glaucus	Tents	16, 18
A. gnomus	Tents	11, 16, 18

Note: Based on (1) Goodwin and Greenhall 1961, (2) Goodwin 1970, (3) Tuttle 1976, (4) Handley 1976, (5) Silva Taboada 1979, (6) Medellín et al. 1983, (7) Linares 1986, (8) McFarlane 1986, (9) Aguilar-Morales and Ruiz-Castillo 1995, (10) Rodríguez-Durán 1998, (11) Simmons and Voss 1998, (12) Simmons and Conway 2001, (13) LaVal and Rodríguez-Herrera 2002, (14) Gannon et al) 2005, (15) Genoways et al. 2005, (16) Rodríguez-Herrera et al. 2007, (17) Aguirre 2007, and (18) Voss et al. 2016. Many of the species may be reported by multiple authors, in addition to the ones listed.

of sociality. The thyropterid bat *Thyroptera tricolor* shows anatomical specializations to roost in tubular, unfurling leaves. The ephemeral nature of unfurling leaves forces this bat to locate suitable roosts on a daily basis (Findley and Wilson 1974), posing a major challenge when trying to maintain a stable social group. *Thyroptera tricolor* advertises roosting locations to group mates by producing social calls (Chaverri and Gillam 2010; Chaverri et al. 2010), thus facilitating regrouping. Moreover, *T. tricolor* uses the hornlike shape of their roost to amplify their vocalizations and facilitate acoustic communication (Chaverri and Gillam 2013). Similar strategies of acoustic advertisement may be used by other species faced with the need to change roosts regularly, such as *A. watsoni* and *Ectophylla alba* (Gillam et al. 2013).

Parasitism is another important aspect of bat biology that is influenced by roosting preferences (Kunz 1982).

Many ectoparasites have been identified for phyllostomid bats (Webb and Loomis 1977); and some studies have shown that roosting habits in phyllostomids influence ectoparasitic densities and grooming behavior by the bats (Kurta et al. 2007; Patterson et al. 2008; ter Hofstede and Fenton 2005). By and large, bats with more permanent, enclosed roosts show greater parasite loads than do foliage-roosting species. However, our general lack of data about roosts used by bats introduces uncertainty in the understanding of the effect of roost type on parasitism (ter Hofstede and Fenton 2005). This assertion is further supported by a study of the interaction between streblid flies and bats in thermally structured caves (Morse et al. 2012), which concluded that roost specificity is an understudied but crucial factor in shaping parasite-host relationships.

As with other specialists, roost specialization in bats has been correlated with increased risk of extinction

(Sagot and Chaverri 2015). *Thyroptera tricolor* shows one of the most highly specialized morphologies for roosting (Wilson and Findley 1977), and when subjected to roost loss, this species increases mobility and decreases social cohesion, potentially increasing energy expenditure and exposure to predators and reducing cooperative interactions (Chaverri and Kunz 2011). Suitable roosting sites, together with adequate foraging grounds, are considered critical resources to avoid the extinction of *Leptonycteris yerbabuenae* (Fleming 1995). The recovery plan for this species also recognizes that interspecific interactions with *Myotis velifer* at the roost could be important for the conservation of *L. yerbabuenae*, but such interactions are poorly understood.

Roosts represent a major factor determining the distribution of bats (Humphrey 1975; Rodríguez-Durán and Kunz 2001; Voss et al. 2016). Many phyllostomids roost in trees (Rodríguez-Durán 2009; Silva Taboada 1979), and not surprisingly, deforestation affects their distribution (Fenton et al. 1992). The small number of phyllostomid species occurring today in the Nearctic region and Patagonian subregion suggests that it is difficult for members of this family to disperse beyond the limits of the tropics (Koopman 1976). Bell et al. (1986) showed that *Macrotus californicus*, the most northerly distributed phyllostomid, allocates 60% of its daily energy expenditure to resting metabolic costs while roosting. The capacity of this bat to invade the temperate zone year-round is due, to a large extent, to their use of geothermally heated caves as day roosts.

Detailed description of the thermal characteristics of phyllostomid roosts is often neglected, largely due to the traditional view that low temperature stress is limited to temperate zones. However, ambient temperatures typical of hibernacula used by temperate bats range from −10° to 21°C, while low temperatures during the cold season in the tropics vary from 6°C to about 18°C, thus overlapping the thermal zone in which temperate-zone bats resort to torpor to save energy (Geiser and Stawski 2011). Moreover, tropical bats have poorer insulation than temperate species of the same mass (McNab 1969) and should experience greater energetic stress at similar low temperatures. Although the capacity to enter torpor has been examined in few phyllostomids, it is evident that this family cannot be considered homeothermic (Geiser and Stawski 2011).

McNab (1969) suggested that phyllostomids' reduction of body temperature at mild ambient tem-

peratures was related to roosting behavior rather than to diet, and although few species have been examined, the thermal characteristics of phyllostomid roosts appear important for the conservation of energy even in the tropics. For instance, *Noctilio albiventris* selects its roost based on tree diameter and inner width of cavities (Aguirre et al. 2003) and can reduce energy expenditure by 47% through the formation of clusters of four or more individuals (Roverud and Chappell 1991). The mormoopid bats *Pteronotus quadridens* and *Mormoops blainvillei* and the phyllostomids *Monophyllus redmani* and *Erophylla bombifrons* live in hot caves in the West Indies (Rodríguez-Durán 1995, 2009; Silva Taboada 1979). The four species segregate within a cave based on microclimatic differences (Rodríguez-Durán and Soto-Centeno 2003), but all four species typically roost at temperatures over 26°C that allow energetic savings, compared to roosting at lower temperatures. When roosting at temperatures below 25°C, *M. blainvillei* appears to exhibits shallow daily torpor, whereas *M. redmani* forms clusters (Rodríguez-Durán 1995). These species are roost specialists, unable to survive or reproduce under any other roosting conditions. Thus, it appears that thermal characteristics of roosts are of utmost importance for the survival of noctilionoids and should not be neglected when studying their ecology.

A more complex roosting behavior results when the bats actually construct the roost. Given the energy expenditure associated with this activity, it is to be expected that some benefit must be derived from it. Although the benefit can come in different forms, such as reduced predation or greater mating success, the energetic benefits of two such roost types have been examined. *Lophostoma silvicolum* roosts in cavities that it excavates inside termite nests. Dechmann et al. (2004) concluded that by constructing their roosts in active termite nests, *L. silvicolum* increases its daily energy savings by around 5%, compared to roosting in tree cavities or dead termite nests. Among tent-roosting bats, Rodríguez-Herrera et al. (2016) found that tents of *E. alba* and *Uroderma bilobatum* had a significantly higher capacity to store heat, as compared to unmodified leaves, which likely resulted in energy savings in thermoregulation.

The advantages to bats of roosting in unmodified foliage remains to be examined. However, work on birds suggests that energetic benefits are derived from these roosts as well. Even though unmodified leaves are not as efficient in holding heat as tents built by bats (Rodríguez-Herrera et al. 2016), they are likely to affect air temperature, radiation balance, and serve as shelter from wind and rain. Walsberg (1986) found an important contribution of foliage in the reduction of heat loss due to forced convection for nocturnal roosts used by desert songbirds. Although he did not find a significant contribution of foliage to increases in ambient temperature within the roost, his results suggest that combined with clustering the effect could be significant. Protection from wind was found to provide a small but significant amount of energy to the vespertilionid bat *Lasiurus cinereus* roosting in temperate North America (Willis and Brigham 2005).

Many of the phyllostomids for which we have information appear to be roost generalists (table 18.1). However, given the rather anecdotal nature of some of this information, we cannot be certain about its accuracy. The fact that a bat is found in the foliage of a tree, in a crevice, or in a house does not necessarily mean that the structure represents a suitable roosting site, which is the type of roost where the species reproduces. Bats "caught by the day" away from their roost, may roost under conditions that are not conducive to long-term survival or reproduction (accidental roosts); therefore, it is important to collect adequate data about their roosting ecology. This lack of detailed knowledge about roosting behavior may preclude us, at this time, from establishing clear patterns based on phylogeny. Nevertheless, some general trends can be observed (table 18.1). For example, all phyllostomids using termite nests are phyllostomines, and most tent-roosting species are stenodermatines. Mainland and island mormoopids roost in hot caves, and all phyllostomids using this type of cave appear to be glossophagines. However, the glossophagines for which we have more detailed information are Antillean species (Rodríguez-Durán 1995; Silva Taboada 1979). Thus, it remains to be seen whether this is an adaptation to life on islands (Rodríguez-Durán 2009) or a pattern within the subfamily.

Types of Roosts

The lack of detailed information poses a difficulty when trying to develop a system of classification for roosts of noctilionoids. Voss et al. (2016, 24) developed a roosting-guild classification of Amazonian bats, based on nine roost types. In allocating bats to different categories, they stressed the need for "accurate information about roosting behavior of many more taxa." As mentioned above, Gaisler (1979) developed a scheme of classification of diurnal roosts of bats. The system was outlined in a matrix that considered both roost type and roosting behavior. Under this scheme, he classified bats as phytophilous, lithophilous, and anthropophilous. This system produced a matrix of 12 possible types of roost/roosting behaviors. For instance, a bat hanging freely inside a tree cavity would be a phytophilous-internal–free roosting individual. Gaisler discussed additional elements that could be used to enhance the classification but recognized the difficulty of applying a static system to a dynamic process, such as roosting. As we saw in the preceding section, individuals of one species may show different roosting behaviors under different situations. Therefore, in elaborating on previous schemes of classification of roost types, I will take into consideration only the physical characteristics of roosts, basing it on the three main structures selected, which I will denote as phytostructures, lithostructures, and anthropostructures. Phytostructures and lithostructures are easily defined, although within each category their characteristics vary widely. In contrast, anthropostructures pose greater difficulties of definition. Whereas a culvert or a bunker from World War II is certainly an anthropostructure, they share more characteristics with a cave than with a house. Spaces between cement blocks are not much different from crevices in rocks, and spaces in the walls of wooden houses not much different from tree cavities. In an admittedly imperfect criterion, I will consider only human-inhabited structures (following Voigt et al. 2016) as anthropostructures. Voigt et al. (2016) provide a detailed discussion of the complexities associated with synanthropic species.

At least 28% of the species considered in this chapter can be classified as roost generalists (table 18.1), which poses a problem for the development of any system of classification (e.g., Voss et al. 2016). To avoid the risk of counting accidental use of structures (see above), this estimate does not consider cases with only anecdotal reports of the use of an anthropostructure. At least 77 species (67%) use phytostructures, either tree cavities (35%) or foliage/leaves/trunk (32%), whereas 62% use lithostructures. As could be expected, these patterns vary by region, depending on the availability of roosts (Rodríguez-Durán 2009).

Most phyllostomids will be found using one of the roosts listed below (based mostly on Gaisler 1979; Rodríguez-Durán 2009; Rodríguez-Herrera et al. 2007; and Voss et al. 2016). This system is inevitably subjective and tentative, pending future research providing accurate information on roost use. However, I believe that it can serve as a useful guide to continue gathering this information.

A. Lithostructures
　1. Caves
　　a. Cool cave (frigidarium) with bell holes (solution cavities on cave ceiling)
　　b. Cool cave (frigidarium) without bell holes
　　c. Hot cave
　　　i. Tepidarium
　　　ii. Caldarium
　2. Culverts
　3. Undercut earth bank
　4. Rock crevices
　5. Animal burrow

Lithostructures are the second commonest type of roost for the families of bats that we are examining but are probably the best studied in the Neotropics. Lithostructures vary widely in characteristics such as temperature, humidity, illumination, ventilation, gas composition, and substrate (Brunet and Medellín 2001; Lundberg and McFarlane 2008; Rodríguez-Durán 2009). A detailed description of caves and their formation appears in Rodríguez-Durán (2009). For our purposes here, it's sufficient to define a cave as an underground cavity large enough to accommodate a human. Small caves are simple systems, although they may contain bell holes and crevices. The presence of these geological features is not insignificant. For instance, Lundberg and McFarlane (2008) found that rock temperatures in bell holes with bats were significantly higher than in empty bell holes. They go further to propose that shallow depressions selected by bats in cave ceilings become bell holes as a result of bat-mediated condensation corrosion.

Large caves, conversely, can be exceedingly complex systems. Several species of bats sharing a cave can all inhabit widely different roost types. Caves in the Neotropics can be classified as frigidarium (temperature below 25°C), tepidarium (temperature between 26°C and 28°C), and caldarium (temperature above 28°C)

(Rodríguez-Durán and Christenson 2012; modified after Rodríguez-Durán 2009). On top of these differences in temperature, some caused by the bats themselves (Rodríguez-Durán 2009), the morphological complexity of the cave can vary enormously. Brunet and Medellín (2001) found a relationship between cave complexity and number of species of bats. They defined several characteristics to quantify cave complexity: extent of passages and tunnels, density of bell holes, and spatial variation in temperature and humidity. The complexity of a cave can affect not only temperature (of the air and of the rock) and humidity but also the concentration of carbon dioxide (Rodríguez-Durán 2009), and possibly ammonia. Thus, in describing the roost of a lithophilous bat, it is not enough to indicate cave, cavern, or culvert; it is also important to describe the exact location of the bat within the lithostructure and the structure's characteristics.

B. Phytostructures
　1. Foliage
　2. Under modified leaves (tents)
　　a. Inverted boat
　　b. Paradox
　　c. Bifid
　　d. Boat/apical
　　e. Conical
　　f. Umbrella
　　g. Pinnate
　　h. Apical
　3. Under unmodified leaves
　4. Termite nests
　5. Cavity in standing tree
　6. Cavity in fallen tree
　7. Under fallen tree
　8. Exposed on standing tree

There is probably more variation among phytostructures than among lithostructures. Tents are amongst the best-studied phytostructures (Rodríguez-Herrera et al. 2007). Seventeen species of phyllostomid bats have been observed using tents (mostly Stenodermatinae), with eight different tent architectures on 77 species of plants (Rodríguez-Herrera et al. 2007). The thermal characteristics among these tents and with respect to unmodified leaves vary considerably (Rodríguez-Herrera et al. 2016), and these differences need to be measured to establish causal links with roost selection or investment in construction by the bats. Few species have been observed constructing tents, and although bats found in tents are often considered tent constructors, I submit that restraint should be exercised in making such assumptions. For instance, whereas *A. jamaicensis* can be found roosting in tents in Costa Rica, such behavior has not been observed in the West Indies faunal subregion, where other known tent builders are absent. Is it possible that bats learn to construct tents from other species? Does *A. jamaicensis* occupy tents that have been abandoned by other species? Or does *A. jamaicensis* parasitize other tent-constructing species by displacing them from their tents?

Tree diameter, canopy closure, tree height, elevation, and snag density are known to be important determinants in roost selection by cavity-roosting bats in northern North America (Fabianek et al. 2015). In Bolivia, after measuring various tree characteristics, Aguirre et al. (2003) found that *N. albiventris* selected roosts based on tree diameter and inner width of the cavities. As with other types of roosts, phytostructure microclimate is often ignored by researchers or measured when the bats are absent (Sedgeley 2001). Even in temperate North America, Boyles (2007) found that only five of 41 studies on bats roosting in tree cavities reported data on microclimate. He suggests some variables relatively easy to measure and potentially determinant of roost suitability: temperature throughout the year, water content of the tree or bark, and solar radiation. As evidenced by the work on bats roosting in tent and termite nests (see above), the importance of these variables should not be underestimated, given the tropical nature of phyllostomids.

    C. Anthropostructure
      1. Exposed under eave
      2. Exposed on ceiling
      3. In attic
      4. Inside wooden walls
      5. Inside cement block walls

Anthropostructures are probably the least studied of roosts for phyllostomids, perhaps because of the difficulties involved with intruding into people's homes or because most biologists are interested in studying bats in their "natural" habitat. The association of bats with anthropostructures in the Neotropics likely precedes the European conquest. Some molossid bats in the Neotropics are almost exclusively found occupying human dwellings (Silva Taboada 1979). The use of anthropostructures by bats can be quite complex, with bats probably moving around different parts of the house to take advantage of or avoid changes in temperature throughout the day (pers. obs.). The same detailed descriptions previously mentioned for litho- and phytostructures are needed for anthropostructures. Such descriptions may offer excellent opportunities to provide a better understanding of the roosting ecology of bats.

Conclusions

Some important developments have occurred within the last 40 years in our understanding of the roosting ecology of phyllostomids. Nevertheless, considerably more data on specific details of structural characteristics and microclimate are needed before we can establish causal links between them and their selection as roosts by bats. This is especially true with respect to phytostructures, in which most phyllostomids roost but for which, with the exception of tents, fewer advances in understanding have been made. As the Anthropocene advances, the importance of anthropostructures should not be underestimated.

The system and suggestions presented here are inevitably preliminary. Future research providing detailed data on morphological and thermal characteristics of roosts should result in a refinement of this classification scheme. But, in studying all types of roosts, it is also important to describe the behavior of the bats using them. The position and degree of bodily contact of the bat with the roost, as originally described by Gaisler (1979), should not be overlooked. The importance of these aspects of bats' daily life cannot be overstated, if we are to understand their ecology and try to work towards their conservation.

Acknowledgments

I extend special thanks to Ted Fleming for the invitation to write this chapter. The Universidad Interamericana de Puerto Rico, Bayamón Campus, provided time and space. Allen Kurta and two anonymous reviewers made comments on an earlier version of the manuscript, which greatly improved its clarity. I dedicate this chapter to Tom Kunz, mentor, visionary, and one of the pioneers in the study of the roosting ecology of bats.

References

Aguilar-Morales, S., and A. A. Ruiz-Castillo. 1995. Una comunidad de murciélagos en una cueva de calor como factor determinante en el sostenimiento de la diversidad animal cavernícola. Tesis conjunta de licenciatura, Universidad Nacional Autónoma de México, México, D.F.

Aguirre, L. F., ed. 2007. Historia natural, distribución y conservación de los murciélagos de Bolivia. Centro de Ecología Difusión Simón I.Patiño, Santa Cruz.

Aguirre, L. F., L. Lens, and E. Matthysen. 2003. Patterns of roost use by bats in a Neotropical savanna: implications for conservation. Biological Conservation, 111:435–443.

Barclay, R. M. R., and A. Kurta. 2007. Ecology and behavior of bats roosting in tree cavities and under bark. Pp. 17–60 *in*: Bats in Forests: Conservation and Management (M. J. Lacki, J. P. Hayes, and A. Kurta, eds.). Johns Hopkins University Press, Baltimore.

Bell, G. P., G. A. Bartholomew, and K. A. Nagy. 1986. The role of energetics, water economy, foraging behavior and geothermal refugia in the distribution of the bat, *Macrotus californicus*. Journal of Comparative Physiology, 156:441–450.

Berry, R. D., and P. E. Brown. 1995. Natural history and reproductive behavior of the California leaf-nosed bat (*Macrotus californicus*). Bat Research News, 36:49–50.

Boyles, J. G. 2007. Describing roosts used by forest bats: the importance of microclimate. Acta Chiropterologica, 9:297–303.

Bradbury, J. W. 1977. Social organization and communication. Pp. 2–73 *in*: Biology of Bats. Vol. 3. (W. A. Wimsatt, ed.). Academic Press, New York.

Brigham, R. M. 2007. Bats in forests: what we know and what we need to learn. Pp. 1–15 *in*: Bats in Forests: Conservation and Management (M. J. Lacki, J. P. Hayes, and A. Kurta, eds.). Johns Hopkins University Press, Baltimore.

Brunet, A. K., and R. A. Medellín. 2001. The species-area relationship in bat assemblages of tropical caves. Journal of Mammalogy, 82:1114–1122.

Chaverri, G., and E. H. Gillam. 2010. Cooperative signaling behavior of roost location in a leaf-roosting bat. Communicative and Integrative Biology, 3:1–4.

Chaverri, G., and E. H. Gillam. 2013. Sound amplification by means of a horn-like roosting structure in Spix's disc-winged bat. Proceedings of the Royal Society B, 280:2013.2362.

Chaverri, G., E. H. Gillam, and M. J. Vonhof. 2010. Social calls used by a leaf-roosting bat to signal location. Biology Letters, 6:441–444.

Chaverri, G., and T. H. Kunz. 2010. Ecological determinants of social systems: perspectives on the functional role of roosting ecology in the social behavior of tent-roosting bats. Advances in the Study of Behavior, 42:275–318.

Chaverri, G., and T. H. Kunz. 2011. Response of a specialist bat to the loss of a critical resource. PLOS ONE 6 (12): e28821. DOI: 10.1371/journal.pone.0028821.

Chaverri, G., O. E. Quirós, M. Gamba-Rios, and T. H. Kunz. 2007. Ecological correlates of roost fidelity in the tent-making bat *Artibeus watsoni*. Ethology, 113:598–605.

Dechmann, D. K. N., E. K. V. Kalko, and G. Kerth. 2004. Ecology of an exceptional roost: energetic benefits could explain why the bat *Lophostoma silvicolum* roosts in active termite nests. Evolutionary Ecology Research, 6:1037–1050.

Fabianek, F., M. A. Simard, and A. Desrochers. 2015. Exploring regional variation in roost selection by bats: evidence from meta-analysis. PLOS ONE 10 (9): e0139126. DOI: 10.1371/journal.pone.0139126.

Fenton, M. B., L. Acharya, D. Audet, M. B. C. Hickey, C. Merriman, M. K. Obrist, D. M. Syme, and B. Adkins. 1992. Phyllostomid bats (Chiroptera: Phyllostomidae) as indicators of habitat disruption in the Neotropics. Biotropica, 24:440–446.

Findley, J. S., and D. E. Wilson. 1974. Obsrvations on the Neotropical disk-winged bat, *Thyroptera tricolor*. Journal of Mammalogy, 55:562–571.

Fleming, T. H. 1988. The Short-Tailed Fruit Bat: A Study in Plant-Animal Interactions. University of Chicago Press, Chicago.

Fleming, T. H. 1995. Lesser Long-Nosed Bat Recovery Plan. U.S. Fish and Wildlife Service, Albuquerque, NM.

Gaisler, J. 1979. Ecology of bats. Pp. 281–342 *in*: Ecology of Small Mammals (D. M. Stoddard, ed.). Chapman and Hall, London.

Gannon, M. R., A. Kurta, A. Rodríguez-Durán, and M. R. Willig. 2005. Bats of Puerto Rico: An Island Focus and a Caribbean Perspective. Texas Tech University Press, Lubbock.

Geiser, F., and C. Stawski. 2011. Hibernation and torpor in tropical and subtropical bats in relation to energetics, extinction, and the evolution of endothermy. Integrative and Comparative Biology, 51:337–348.

Genoways, H. H., R. J. Baker, J. W. Bickham, and C. J. Phillips. 2005. Bats of Jamaica. Special Publications, Museum of Texas Tech University, no. 48. Museum of Texas Tech University, Lubbock.

Gillam, E. H., G. Chaverri, K. Montero, and M. Sagot. 2013. Social calls produced within and near the roost in two species of tent-making bats, *Dermanura watsoni* and *Ectophylla alba*. PLOS ONE, 8:e61731.

Goodwin, G. G., and A. M. Greenhall. 1961. A review of the bats of Trinidad and Tobago: descriptions, rabies infection, and ecology. Bulletin of the American Museum of Natural History, 122:187–302.

Goodwin, R. E. 1970. The ecology of Jamaican bats. Journal of Mammalogy, 51:571–579.

Handley, C. O., Jr. 1976. Mammals of the Smithsonian Venezuelan Project. Brigham Young University Science Bulletin, Biological Series, vol. 20, no. 5. Brigham Young University, Provo, UT..

Humphrey, S. R. 1975. Nursery roosts and the community diversity of Nearctic bats. Journal of Mammalogy, 56:321–346.

Kerth, G., K. Weissmann, and B. König. 2001. Day roost selection in female Bechstein's bats (*Myotis bechsteinii*): a field experiment to determine the influence of roost temperature. Oecologia, 126:1–9.

Koopman, K. F. 1976. Zoogeography. Pp. 39–48 in: Biology of Bats of the New World Family Phyllostomatidae. Pt. 1 (R. J. Baker, J. Knox Jones Jr., and D. C. Carter, eds). Special Publications, Museum, Texas Tech University, no. 10. Texas Tech Press, Lubbock..

Kunz, T. H. 1982. Roosting ecology of bats. Pp. 1–56 in: Ecology of Bats (T. H. Kunz, ed.). Plenum Publishing, New York.

Kunz, T. H., and L. F. Lumsden. 2003. Ecology of cavity and foliage roosting bats. Pp. 3–89 in: Bat Ecology (T. H. Kunz and M. B. Fenton, eds.). University of Chicago Press, Chicago.

Kunz. T. H., S. K. Robson, and K. A. Nagy. 1998. Economy of harem maintenance in the greater spear-nosed bat, *Phyllostomus hastatus*. Journal of Mammalogy, 79:631–642.

Kurta, A., J. O. Whitaker Jr., W. J. Wrenn, and J. A. Soto-Centeno. 2007. Ectoparasitic assemblages on mormoopid bats (Chiroptera: Mormoopidae) from Puerto Rico. Journal of Medical Entomology, 44:953–958.

LaVal, R. K., and B. Rodríguez-Herrera. 2002. Murciélagos de Costa Rica. Editorial INBio, San José, Costa Rica.

Lewis, S. E. 1995. Roost fidelity of bats: a review. Journal of Mammalogy, 76:481–496.

Linares, O. J. 1986. Murciélagos de Venezuela. Lagoven, Caracas.

Lumsden, L. F., A. F. Bennett, and J. E. Silinsa. 2002. Location of roosts of the lesser long-eared bat *Nyctophilus geofroyi* and Gould's wattled bat *Chalinolobus gouldii* in a fragmented landscape in south-eastern Australia. Biological Conservation, 106:237–249.

Lundberg, J., and D. A. McFarlane. 2008. Bats and bell holes: the microclimatic impact of bat roosting, using a case study from Runaway Bay Caves, Jamaica. Geomorphology, 106:78–85.

McCracken, G. F., and G. S. Wilkinson.2000. Bat mating systems. Pp. 321–362 in: Reproductive Biology of Bats (E. G. Crichton and P. H. Krutzch, eds.) Academic Press, San Diego, CA.

McFarlane, D. A. 1986. Cave bats in Jamaica. Oryx, 20:27–30.

McNab, B. K. 1969. The economics of temperature regulation in Neotropical bats. Comparative Biochemistry and Physiology, 31:227–268.

Medellín, R. D., D. Navarro, W. B. Davis, and V. J. Romero. 1983. Notes on the biology of *Micronycteris brachyotis* (Dobson) (Chiroptera), in southern Veracruz, México. Brenesia, 21:7–11.

Morrison, D. W. 1979. Apparent male defense of tree hollows in the fruit bat, *Artibeus jamaicensis*. Journal of Mammalogy, 60:11–15.

Morse, S. F., C. W. Dick, B. D. Patterson, and K. Dittmar. 2012. Some like it hot: evolution and ecology of novel endosymbionts in bat flies of cave-roosting bats (Hippoboscoidea, Nycterophiliinae). Applied Environmental Microbiology, 78:8639–8649.

Murray, K. L., and T. H. Fleming. 2008. Social structure and mating system of the buffy flower bat, *Erophylla sezekorni* (Chiroptera, Phyllostomidae). Journal of Mammalogy, 89: 1391–1400.

Nagel, T. 1974. What is it like to be a bat? Philosophical Review, 83:435–450.

O'Donnell, C. F. J., and J. A. Sedgeley. 1999. Use of roosts in the long-tailed bat, *Chalinolobus tuberculatus*, in temperate rainforest in New Zealand. Journal of Mammalogy, 80:913–923.

Ortega, J., and H. Arita. 1999. Structure and social dynamics of harem groups in *Artibeus jamaicensis* (Chiroptera: Phyllostomidae). Journal of Mammalogy, 80:1173–1185.

Ortega, J., and I. Castro-Arellano. 2001. *Artibeus jamaicensis*. Mammalian Species, 662:1–9.

Patterson, B. D., C. W. Dick, and K. Dittmar. 2008. Parasitism by bat flies (Diptera: Streblidae) on Neotropical bats: effects of host body size, distribution, and abundance. Parasitology Research, 103:1091–1100.

Rodríguez-Durán, A. 1995. Metabolic rates and thermal conductance in four species of Neotropical bats roosting in hot caves. Comparative Biochemistry and Physiology, 110A:347–355.

Rodríguez-Durán, A. 1998. Nonrandom aggregations and distribution of cave-dwelling bats in Puerto Rico. Journal of Mammalogy, 79:141–146.

Rodríguez-Durán, A. 2009. Bat assemblages in the West Indies: the role of caves. Pp. 265–280 in: Island Bats: Evolution, Ecology and Conservation (T. H. Fleming and P. A. Racey, eds.). University of Chicago Press, Chicago.

Rodríguez-Durán, A., and K. Christenson. 2012. Breviario sobre los murciélagos de Puerto Rico, La Española e Islas Vírgenes. Universidad Interamericana de Puerto Rico, Publicaciones Puertorriqueñas, San Juan, Puerto Rico.

Rodríguez-Durán, A., and T. H. Kunz. 2001. Biogeography of bats of the West Indies: an ecological perspective. Pp. 355–368 in: Biogeography of the West Indies (C. Woods and F. Sergile, eds.) CRC Press, Boca Raton, FL.

Rodríguez-Durán, A., and J. A. Soto-Centeno. 2003. Temperature selection by tropical bats roosting in caves. Journal of Thermal Biology, 28:465–468.

Rodríguez-Herrera, B., R. A. Medellín, and R. M. Tim. 2007. Murciélagos Neotropicales que acampan en hojas. Editorial INBio, Santo Domingo de Heredia, Costa Rica.

Rodríguez-Herrera, B., L. Víquez-R, E. Cordero-Schmidt, J. M Sandoval, and A. Rodríguez-Durán. 2016. Energetics of tent roosting bats: the case of *Ectophylla alba* and *Uroderma bilobatum* (Chiroptera: Phyllostomidae). Journal of Mammalogy, 97:246–252.

Roverud, R. C., and M. A. Chappell. 1991. Energetic and thermoregulatory aspects of clustering behavior in the Neotropical bat *Noctilio albiventris*. Physiological Zoology, 64:1527–1541.

Sagot, M., and G. Chaverri. 2015. Effects of roost specialization on extinction risk in bats. Conservation Biology, 29 (6): 1666–1673. DOI: 10.1111/cobi.12546.

Santana, S. E., T. O. Dial, T. P. Eiting, and M. E. Alfaro. 2011. Roosting ecology and the evolution of pelage markings in bats. PLOS ONE 6 (10):e25845. DOI: 10.1371/journal.pone .0025845.

Sedgeley, J. A. 2001. Quality of cavity microclimate as a factor influencing selection of maternity roosts by a tree-dwelling bat, *Chalinolobus tuberculatus*, in New Zealand. Journal of Applied Ecology, 38:425–438.

Silva Taboada, G. 1979. Los murciélagos de Cuba. Editorial Academia, Havana, Cuba.

Simmons, N. B., and T. M. Conway. 2001. Phylogenetic relationships of mormoopid bats (Chiroptera: Mormoopidae) based on morphological data. Bulletin of the American Museum of Natural History, 258:1–97.

Simmons, N. B., and R. S. Voss. 1998. The mammals of Paracou, French Guiana: a Neotropical lowland rainforest fauna. Pt. 1: Bats. Bulletin of the American Museum of Natural History, 237:1–219.

ter Hofstede, H. M., and M. B. Fenton. 2005. Relationships between roost preferences, ectoparasite density, and grooming behavior of Neotropical bats. Journal of Zoology, 266:333–340.

Tuttle, M. D. 1976. Collecting techniques. Pp. 71–88 *in*: Biology of Bats of the New World Family Phyllostomatidae. Pt. 1 (R. J. Baker, J. Knox Jones Jr., and D. C. Carter, eds.). Special Publications, Museum, Texas Tech University, no. 10. Texas Tech Press, Lubbock.

Villalobos-Chaves, D., J. Vargas Murillo, E. Rojas Valerio, B. W. Keeley, and B. Rodríguez-Herrera. 2016. Understory bats roosts, availability and occupation patterns in a Neotropical rainforest of Costa Rica. Revista de Biología Tropical, 64: 1333–1343.

Voigt, C. C., K. L. Phelps, L. F. Aguirre, M. C. Schoeman, J. Vanitharani, and A. Zubaid. 2016. Bats and buildings: the conservation of synanthropic bats. Pp. 427–462 *in*: Bats in the Anthropocene: Conservation of Bats in a Changing World (C. C. Voigt and T. Kingston, eds.). Springer, Cham, Switzerland.

Voss, R. S., D. W. Fleck, R. E. Strauss, P. M. Velazco, and M. B. Simmons. 2016. Roosting ecology of Amazonian bats: evidence for guild structure in hyperdiverse mammalian communities. American Museum Novitates, 3870:1–43.

Walsberg, G. E. 1986. Thermal consequences of roost-site selection: the relative importance of three modes of heat conservation. Auk, 103:1–7.

Webb, J. P., and R. B. Loomis.1977. Ectoparasites. Pp. 57–120 *in*: Biology of Bats of the New World Family Phyllostomatidae. Pt. 2 (R. J. Baker, J. Knox Jones Jr., and D. C. Carter, eds.). Special Publications, Museum, Texas Tech University, no. 13. Texas Tech Press, Lubbock.

Wilkinson, G. S. 1995. Information transfer in bats. Symposia of the Zoological Society of London, 67:345–360.

Wilkinson, G. S., and G. F. McCracken. 2003. Bats and balls: sexual selection and sperm competition in the Chiroptera. Pp. 128–148 *in*: Bat Ecology (T. H. Kunz and M. B. Fenton, eds.) University of Chicago Press, Chicago.

Willis, C. K. R., and R. M. Brigham. 2005. Physiological and ecological aspects of roost selection by reproductive female hoary bats (*Lasiurus cinereus*). Journal of Mammalogy, 86: 85–94.

Wilson, D. F., and J. S. Findley. 1977. *Thyroptera tricolor*. Mammalian Species, 71:1–3.

19

Population Biology

*Theodore H. Fleming
and Angela M. G. Martino*

Introduction

Studies of the population biology of animals have traditionally focused on topics such as population sizes and their seasonal and annual fluctuations, reproduction and survivorship, age and sex structure, dispersal patterns, and, relatively recently, genetic structure. For many kinds of animals, including birds and non-volant mammals, methods for studying these aspects of population biology are well developed. For small terrestrial mammals, for example, trapping grids have been widely used to determine population densities and structure whereas other techniques, including terrestrial transects, aerial surveys, and scat counts, have been used to assess population sizes of larger species. Bats are small mammals, but their population structure differs fundamentally from that of many non-volant small mammals in that they typically spend the daylight hours congregated in a roost from which they leave after sunset to forage. As a result, they tend to be highly clumped during the day and much more diffuse in their distribution at night. Their population ecology is thus analogous to that of colonial-nesting birds (e.g., Weimerskirch et al. 2008; Wooler et al. 1992). Because of this basic spacing pattern, studies of bat populations have traditionally been focused on their day roosts, and the term "bat population" usually refers to the size and structure of roost colonies or aggregations (but see below). In modern population ecology, the concept of a metapopulation—spatially separated populations connected by dispersal—has also become important as terrestrial landscapes become increasingly fragmented (Akçakaya et al. 2007; Hanski 1999). A series of bat colonies living in different areas (e.g., in different habitats or on true islands) but still connected by occasional dispersal thus represents a metapopulation. To our knowledge, no one has yet applied the metapopulation concept to phyllostomid populations (but see Muscarella et al. 2011). Thus, we will continue to use this roost-based tradition in discussing the population biology of phyllostomid bats in this chapter. Knowing the sizes of colonies and the number of colonies per unit area (typically per km²) allows researchers to estimate the population density of bats for comparison with other kinds of animals.

Estimating the size of bat colonies is often not easy, especially when they are large. Whereas video techniques, including thermal imaging (e.g., Ammerman et al. 2009; Best et al. 2015; Betke et al. 2008), are now recommended to accurately record the number of bats exiting a roost at night, this method is labor intensive and can be problematical when several bat species occupy a roost and depart simultaneously. In the pre-video era, visual counts conducted either inside a roost during the day or as bats leave a roost at night were a common way to estimate roost population sizes. Methods for estimating colony sizes are thoroughly discussed by Kunz et al. (2009) and O'Shea and Bogan (2003). Clement et al. (2015) discuss a method of estimating the population sizes of mobile animals such as bats using radiotelemetry and counts of unmarked individuals. Oyler et al. (2018) describe a genetic mark-recapture technique for estimating bat colony sizes. Finally, Meyer et al. (2010, 2015) discuss monitoring long-term population trends of bats using mist net and acoustic sampling methods away from roosts.

Colony size is just one parameter that we need to know to characterize bat populations. Important additional information includes the population's age and sex structure. While this information can sometimes be gained from visual inspection of colonies (e.g., when different age classes or sexes are spatially segregated within roosts), the most accurate method for estimating these parameters involves capturing individuals, either inside roosts with hand nets or outside roosts with mist nets or harp traps. Inspecting and handling captured bats provides data on age, sex, size (usually forearm length), body mass and condition, and reproductive status—important information for understanding a species' population ecology. Additionally, captured bats can be tagged with numbered arm bands, necklaces, or PIT tags (passive integrated transponders; Adams 2015; Frick et al. 2018) that provide valuable longitudinal information about the histories of particular individuals. Finally, tiny tissue samples or oral swabs can be obtained from captured bats to assess the genetic structure of populations (Rossiter 2009).

In this chapter, we will review what is known about the population biology of phyllostomid bats. Although it would be desirable to center this chapter around long-term population studies based on many tagged individuals, such studies are still distressingly uncommon for these bats (cf. Humphrey and Bonaccorso 1979). Only a handful of phyllostomids have been studied continuously for several years, whereas many such studies have been conducted on temperate-zone bats (e.g., Barlow et al. 2015; Frick et al. 2010; Lentini et al. 2015; O'Shea et al. 2011; Ransome 1989). As a result, only three species of phyllostomids (*Lonchophylla dekeyseri*, *Carollia perspicillata*, and *Artibeus jamaicensis*) were included in Lentini et al.'s (2015) analysis of survival rates in 44 species of microchiropteran bats. As Bernard et al. (2012) and Barlow et al. (2015) have noted for Brazilian and European bats, respectively, long-term studies are critically important for determining the conservation status of species. Despite the absence of many long-term studies, we do know a lot about the relative size and structure of populations of many species of phyllostomids and how their roost use changes seasonally. In addition, recent studies based on molecular genetics have begun to elucidate the genetic structure and dispersal history of populations of some species. Nonetheless, one of the major messages of this chapter is that we still have much to learn before we fully understand the population biology of most species in this diverse and ecologically important family.

We close this introduction by noting that focusing on bat roosts as "populations" has a historical and biogeographic bias. Studies of bat populations began in north temperate areas where bats often use caves or buildings as their day roosts and colony sizes can sometimes be large (Gaisler 1979). When relatively large numbers of bats roost together, it seems reasonable to consider the group to be a population. But what about species such as many phyllostomids that routinely roost in small numbers in hollow trees, foliage, or culverts as occurs in the Amazon Basin, where caves and rock shelters are absent (Voss et al. 2016)? It seems unreasonable in this situation to consider each roost colony to be a population. Instead, we suggest that collections of these colonies in a given area should constitute a spatially structured population (*sensu* Grimm et al. 2003). Estimating the sizes and other characteristics of these "diffuse" populations is likely to be much more difficult than estimating the sizes and characteristics of "concentrated" populations.

Relative Abundance of Phyllostomids

Because many phyllostomid bats are easy to catch in ground-level mist nets, they have been the subjects of many short-term studies and surveys. These studies clearly indicate these bats are very common in many Neotropical habitats. They also reveal that some phyllostomids are much more common than others (e.g., Fleming et al. 1972; Simmons and Voss 1998; Stevens et al. 2004 and included references). Relative abundances of these bats generally conform to the familiar reverse J-shaped curve, in which a few species are common and many species are uncommon. These results undoubtedly reflect the fact that the population densities (independent of colony sizes) of some species are much higher than those of many other species. These density differences clearly have a trophic, and hence a phylogenetic, basis. In general, plant-visiting phyllostomids, especially frugivores, are much more common than are nectarivores and animalivores in tropical forests (table 19.1). Table 19.1 summarizes the results of intensive mist net studies in three different habitats: tropical dry forest in northwestern Costa Rica, tropical moist forest in central Panama, and tropical wet forest in central French Guiana. Fruit-eating species in subfamilies Carolliinae and Stenodermatinae are by far the most common bats in each of those habitats. In tropical dry forest, Glossophaginae is the third most-common subfamily, whereas Phyllostominae with its relatively high species richness is third most-common in the other two habitats. It could be argued that the results from Santa Rosa and Barro Colorado are biased because the two most common species, *C. perspicillata* and *A. jamaicensis*, respectively, were the main subjects of study. No such bias exists at Paracou, however, and its results support the generalization that carolliines and stenodermatines are the numerically dominant bats among phyllostomids and presumably have the largest populations in many lowland mainland Neotropical habitats. Other studies supporting this conclusion include Estrada and Coates-Estrada (2001; lowland rainforest, Veracruz, Mexico); Pérez and Cortés (2009; montane forest, Colombia); Bernard and Fenton (2007; savannas and forest, Central Amazonia, Brazil);

Table 19.1. Data on the relative abundance of noctilionoid bats based on mist-net data from three intensively sampled habitats/sites

Habitat and Location	Family or Subfamily	Number of Species	Number of Captures and Recaptures (%)
Tropical dry forest, Santa Rosa, Costa Rica	Mormoopidae	1	42 (0.7)
	Micronycterinae	3	36 (0.6)
	Desmodontinae	2	96 (1.6)
	Phyllostominae	2	19 (0.3)
	Glossophaginae	2	739 (12.0)
	Glyphonycterinae	1	1 (0.0)
	Carolliinae	2	**3,139 (50.9)**
	Stenodermatinae	9	2,092 (33.9)
Tropical moist forest, Barro Colorado Island, Panama	Mormoopidae	1	86 (0.9)
	Micronycterinae	3	46 (0.5)
	Desmodontinae	1	12 (0.1)
	Phyllostominae	9	484 (5.3)
	Glossophaginae	2	36 (0.4)
	Lonchophyllinae	1	1 (0.0)
	Glyphonycterinae	2	112 (1.2)
	Carolliinae	3	583 (6.4)
	Stenodermatinae	13	**7,726 (85.0)**
Tropical wet forest, Paracou, French Guiana	Noctilionidae	2	4 (0.2)
	Mormoopidae	1	20 (0.8)
	Micronycterinae	7	59 (2.3)
	Desmodontinae	2	8 (0.3)
	Phyllostominae	15	426 (16.9)
	Glossophaginae	4	71 (2.8)
	Lonchophyllinae	1	55 (2.2)
	Glyphonycterinae	3	12 (0.5)
	Carolliinae	1	**1,144 (45.4)**
	Rhinophyllinae	1	128 (5.1)
	Stenodermatinae	15	594 (23.6)

Note: Numbers in **bold** indicate the most abundant clades at each site. Data come from Fleming (1988, app. 4, roost captures excluded), Handley et al. (1991, table 1.2), and Simmons and Voss (1998, table 69).

and Gorresen and Willig (2004; seasonal moist forest, Paraguay).

Just as abundances of different subfamilies are unequal, so are species' abundances within subfamilies at most sites. Thus, *C. perspicillata* tends to be much more common than its congeners in many habitats, and *A. jamaicensis* is more common than its congeners at two of the three sites listed in table 19.1; only *A. obscurus* is more common than *A. jamaicensis* at Paracou. Among nectar feeders, *Glossophaga soricina* is much more common than *Hylonycteris underwoodi* in dry forest (and at many other sites), whereas it and *Hsunycteris* (formerly *Lonchophylla*) *thomasi* are equally common in wet forest. Among the animalivorous subfamilies (i.e., Micronycterinae, Phyllostominae, and Glyphonycterinae), two species of *Micronycteris* and *Phyllostomus discolor* are most common (but still very uncommon) of six species at the dry forest site; *P. discolor* and *Glyphonycteris hirsuta* are most common at the moist forest site; and *P. elongatus*, *Tonatia saurophila*, and *Trachops cirrhosis* are most common at the wet forest site. Finally, among vampires, the mainly mammal-feeding *Desmodus rotundus* is usually much more common than the two bird-feeding species.

In summary, differences in the relative abundances of the different subfamilies undoubtedly reflect differences in resource densities. Fruit densities and biomass are much higher than those of nectar and leaf-associated insects and other animals in most Neotropical habitats (Fleming 1992; Fleming and Kress 2013). Prior to the evolution of the plant visitors, it is likely that animalivorous phyllostomids occurred at low densities and in small populations in most mainland Neotropical habitats. This began to change about 18–20 Ma with the evolution of dedicated frugivorous bats, and the numerical dominance of frugivores reached its peak with the evolution of stenodermatines, beginning about 10 Ma (chap. 2, this vol.).

Roost Use and Population Sizes

As described in detail in chap. 18, phyllostomid bats occupy many kinds of day roosts that vary widely in their temporal permanence, microclimate, and carrying capacity for bat colonies. These roosts range from relatively short-lived leaf tents that house a few bats to long-lived caves that sometimes house tens of thousands of bats. Overall, plant-based roosts (e.g., foliage, tree cavities, leaf tents, etc.) and caves and other so-called lithostructures (see chap. 18, this vol.) are used most frequently as day roosts by noctilionoid bats. In addition to resource densities, the types of roosts used by phyllostomids also play a major role in determining the density and distribution patterns of these bats (Villalobos et al. 2016).

In their analysis of the mating systems of noctilio-noid bats, Adams et al. (chap. 13, this vol.) summarized roost use of 154 species based on two ordinal variables — roost permanence and bat aggregation (colony) size within roosts (table 19.2). These data indicate that substantial variation exists in roost use and aggregation size in these bats. Mean roost permanence is high (i.e., at least 3.5 out of 4) in 9 of the 13 clades in table 19.2; most members of these clades use caves or similar long-lasting structures as day roosts. Caves are therefore likely to be the ancestral roost type in Noctilionoidea. In contrast, 2 animalivorous clades, Phyllostominae and Glyphonycterinae, often use tree cavities or other tree-based structures as roosts, and the highly frugivorous Stenodermatinae with its high species richness is exceptional in its frequent use of unmodified foliage and leaf tents as roosts (Kunz and Lumsden 2003).

Roost type has a very strong effect on colony or aggregation size because roost permanence and aggregation size are positively correlated (table 19.2; Spearman's r_s = 0.70, df = 11, P (two-tailed) = 0.0074). Bats living in less permanent roost types usually occur in small groups and, given the lack of temporal per-

Table 19.2. Data related to roost types and roost sizes of noctilionoid bats based on table S13.1 in chapter 13

Family or Subfamily	Number of Species	Mean Roost Permanence	Mean Aggregation Size
Noctilionidae	2	3.50	3.50
Mormoopidae	3	4.00	4.00
Macrotinae	2	4.00	3.00
Micronycterinae	7	3.50	1.17
Desmodontinae	3	3.67	2.33
Phyllostominae	19	3.33	1.35
Glossophaginae	26	3.74	2.28
Lonchophyllinae	9	3.96	2.00
Lonchorhininae	2	4.00	3.00
Glyphonycterinae	3	3.33	1.83
Carolliinae	5	3.50	1.83
Rhinophyllinae	3	2.00	2.00
Stenodermatinae	71	2.50	1.06

Note: Mean values are presented here. In table S13.1, roost permanence was scored on an ordinal scale as (1) foliage or roots, (2) tents, (3) hollow trees, logs, or excavated termite nests, and (4) caves, culverts, mines, or buildings. Roost size was also scored on an ordinal scale as (1) small or family groups less than 10, (2) groups containing 10–25 individuals, (3) small colonies of 26–100, (4) large colonies greater than 100.

manence of their roosts, are likely to change roosts regularly (e.g., Chaverri and Kunz 2006; Chaverri et al. 2007; Lewis 1995). Given the permanence of their roosts, cave dwellers might be expected to be less likely to change roosts regularly. But, as we will see, this is often not the case, perhaps for reproductive, social, or other reasons. In most habitats, the density of plant-based roosts is potentially far higher than that of cave roosts. As a result, from a landscape perspective, phyllostomid populations can be viewed as occurring in two basic spatial patterns: (1) a few widely scattered cave roosts often containing a few hundred to thousands of individuals (i.e., a "concentrated" population) and (2) many small colonies located in hollow trees or in foliage (i.e., a "diffuse" population). These configurations don't necessarily reflect the overall population size of a species, however. For example, despite living in small tree-based colonies (at least in many mainland habitats), *A. jamaicensis* is a very abundant bat, often more so than common cave-dwelling frugivores such as *C. perspicillata* in the same habitat.

Roost and Population Dynamics

Roost Dynamics

Roost occupancy by most bats—both temperate and tropical—is highly dynamic. A common pattern in temperate bats is the use of one set of roosts as winter hibernacula and another set as summer maternity or bachelor roosts. These two sets can be located a few dozen to hundreds or thousands of kilometers apart and often require bats to migrate substantial distances between them (Fleming and Eby 2003; Kunz 1982; Moussy et al. 2013). Likewise, although they do not hibernate, many phyllostomids also change roosts on a seasonal basis. Most of these changes probably involve short-distance moves (i.e., <50 km) between nearby habitats (e.g., in tropical dry forests in Mexico and Costa Rica in which the abundance of both nectarivorous and frugivorous bats changes seasonally; Stoner 2005; also see Ferreira et al. 2017; Mello et al. 2008; and Pereira et al. 2010 for other examples), but some involve elevational shifts (e.g., in nectarivorous and frugivorous bats in Mexico, Brazil, and Argentina; see, e.g., Esbérard et al. 2011; Giannini 1999; Herrera 1997; Mello et al. 2008), and a few involve longer-distance latitudinal migrations of hundreds of kilometers (Flem-

ing and Eby 2003; McGuire and Boyle 2013; Moussy et al. 2013). Overall, however, these movements tend to be shorter than those of many temperate-zone bats. And most of these movements occur in plant-visiting species and, as far as we know, not in animalivorous species or vampires.

While many of these roost changes are ultimately driven by changes in resource availability within habitats, they are also driven by reproductive activities (Klingbeil and Willig 2010). Regarding resource availability, we might expect animalivorous species that perhaps feed on less spatially patchy food than plant visitors to exhibit more stable roost use (and smaller, more stable home range sizes) than nectar and fruit eaters. As discussed in detail below, this scenario occurs in *Phyllostomus hastatus*. Another example comes from a comparison of the foraging ranges of two small bats living on islands in Gatun Lake, central Panama: *Micronycteris microtis* (an animalivore that roosts in the cavities of fallen trees) and *Artibeus watsoni* (a frugivore that roosts in leaf tents) (Albrecht et al. 2007; chap. 18, this vol.). Although sample sizes are small (three and five radio-tagged individuals in *M. microtis* and *A. watsoni*, respectively), home ranges of the frugivores averaged twice as large as those of the animalivores, and the former species often flew among islands to forage whereas the latter species foraged only on islands where they roosted.

To examine stability of roost use in more detail, we surveyed the noctilionoid roost use literature, looking for associations between roost type (permanent vs. ephemeral) and roost use (± year-round vs. seasonal). Our results are presented in table 19.3 (also see supplementary table S19.1). These data indicate that there is a clear association between roost type, as defined in that table, and seasonality of roost use. As expected, 75% of species using ephemeral roosts ($n = 16$), including many stenodermatids, switch roosts regularly or at least seasonally compared with only 17% of species using more permanent roost types ($n = 41$). These results highlight the fact that populations of some phyllostomids, especially many frugivorous phyllostomids, are very fluid regarding their fidelity to particular roosts, whereas others display high roost fidelity (Lewis 1995).

As discussed in the section "Case Studies" below, reproductive activities can also have a strong effect on seasonal patterns of roost use. Year-round use of the same roost occurs in only two of the four species

Table 19.3. Patterns of consistency of day roost use in noctilionoid bats

Family or Subfamily	Food Habits	Typical Roost Type[a]	Number of Species	Number of Species in Which Roost Use Is	
				Year-Round[b]	Seasonal[c]
Mormoopidae	Insects	Hard structure	7	5	2
Noctilionidae	Insects, fish	Hard structure	2	2	0
Macrotinae	Insects	Hard structure	2	0	2
Micronycterinae	Insects	Hard structure	1	1	0
Desmodontinae	Blood	Hard structure	2	1	1
Lonchorhininae	Mostly animals	Hard structure	1	1	0
Phyllostominae	Mostly animals	Hard structure	9	7	2
Glossophaginae	Mostly nectar and pollen	Hard structure	15	7	8
Lonchophyllinae	Mostly nectar and pollen	Hard structure	4	2	2
Carolliinae	Mostly fruit	Hard and/or ephemeral structures	2	2	0
Rhinophyllinae	Mostly fruit	Ephemeral structures	1	0	1
Stenodermatinae	Mostly fruit	Ephemeral structures	15	2	13
		Hard structures	4	3	1
		Both types	3	1	2

Note: Species were classified as either using the same roost year-round or only seasonally, as defined below. See supplementary table 19.S1 for names of species and sources.

[a] Hard structures are roosts that permit the development of more than one generation (*sensu* Sagot 2016), including hollow trees, fallen trees, tree holes, termite nests, caves, and also culverts and buildings. Ephemeral structures are roosts that do not permit the development of more than one generation (*sensu* Sagot 2016), including exfoliating bark, palm fronds, foliage, and so on.

[b] Year-round means that about 80% of a colony inhabits the roost for at least 10 mo per year.

[c] Seasonal in species using ephemeral roosts means that the colony and/or individuals change roost locations at least every 10 days, on average, or use the primary roost <70% of the time (*fide* Lewis 1995). Otherwise, it means roost use changes seasonally as a result of reproductive or migratory activities.

(*P. hastatus* and *A. jamaicensis* when roosting in caves) reviewed in that section. At the other extreme is the migratory nectar-feeder *Leptonycteris yerbabuenae* (and its congeners) that often uses geographically widely separated caves or mines for mating and maternity. The cave-dwelling frugivore *C. perspicillata* is intermediate between these two extremes. In tropical dry forest, males typically occupy the same roost year-round whereas many females switch roosts on a seasonal basis; female movements are correlated with seasonal reproductive cycles as well as with seasonal changes in resource availability (Fleming 1988).

Another factor that influences roost dynamics is sexual segregation during maternity or migration periods. This is common in many bats (Angell et al. 2013; McGuire and Boyle 2013; Senior et al. 2005), including some noctilionoids. We summarize available data in table 19.4 (also see supplementary table S19.2). Sexual segregation is present in three families or sub-

families and appears to be absent in five lineages. It occurs in mormoopids, *Macrotus*, and many species of glossophagines but appears to be absent in vampires, lonchophyllines, and many phyllostomines and stenodermatines. These results suggest that sexual roost segregation is likely to be ancestral in mormoopids and phyllostomids and is associated with highly gregarious species that inhabit long-lived roost structures.

Population Dynamics

Population dynamics are the result of four basic factors: reproduction, survivorship, immigration, and emigration. As discussed in detail in chapter 12, bats in general and phyllostomids specifically have "slow" life histories characterized by low reproductive rates and high annual survivorship rates (Barclay and Harder 2003). Whereas nearly all temperate-zone females are monestrous and produce a single young per year, female phyllostomids can be either monestrous or polyestrous, depending on clade, body size, geographic location, and trophic ecology. Plant-visiting species are more likely to be polyestrous than are animalivores (chap. 12, this vol.). Despite this reproductive dichotomy, phyllostomids clearly are low fecundity mammals that produce only one or two offspring per adult female per year.

We know much less about annual survival rates and longevity in phyllostomids than we do about their annual reproductive productivity. In captivity, some phyllostomids can live 10–29 years (summarized in chap. 12). The common vampire, *D. rotundus*, appears to hold the longevity record for this family (Weigl 2005). In the wild, however, maximum longevities and annual survival rates are likely to be far lower in most species. In two well-studied species (*C. perspicillata* and *A. jamaicensis*), for example, mortality rates early in life are high, and annual adult survivorship rates are about 65% or less (see below). As a result, life expectancy at birth in these bats averages just a few years, and rates of annual population turnover (i.e., the reciprocal of annual mortality rates) are likely to be high.

Dispersal, which is often defined as the movement of an animal from its place of birth to the place where it becomes resident, is also an important factor in population dynamics. More generally, it can be defined as movements of individuals, regardless of age, from one population (or colony in many bats) to another population. As in the case of survival rates, however, we still know little about its details in phyllostomids. In general, males are the dispersing sex in mammals (Greenwood 1980), but we don't yet know if this is true in these bats. Juvenile dispersal in phyllostomids is likely to be affected by a species' social system. In harem-based systems, for instance, whether juvenile males or females disperse from their natal roosts might depend on how easy it is for individuals to join (females) or acquire (males) harems (McCracken and Wilkinson 2000). In young males, this likely depends on the turnover rate of harem males. If it is high, then males might benefit from remaining in their natal roosts. Otherwise, they might benefit from dispersing. In contrast, females can probably join already-established groups of females (i.e., harems) much more easily than males can acquire harems. If this is true, then they should be able to do this in either their natal roost or some other roost. Regardless of whether dispersal is sex- or age-biased in phyllostomids, genetic studies discussed below in-

Table 19.4. Patterns of roost sexual segregation found in noctilionoids during the maternity period

Family or Subfamily	Food Habits	Number of Species	Sexual Segregation
Mormoopidae	Insects	7	Present
Noctilionidae	Insects, fish	1	Absent
Macrotinae	Insects	2	Sometimes present
Micronycterinae	Insects	4	Absent
Desmodontinae	Blood	2	Absent
Phyllostominae	Mostly animals	1	May be present
		1	Present
		8	Absent
Glossophaginae	Mostly nectar and pollen	1	May be present
		8	Present
		2	Absent
Lonchophyllinae	Mostly nectar and pollen	2	May be present
		2	Absent
Carolliinae	Mostly fruit	2	Sometimes present
Rhinophyllinae	Mostly fruit	1	No available data
Stenodermatinae	Mostly fruit	6	May be present
		3	Present
		12	Absent

Note: See supplementary table 19.S2 for names of species and sources.

dicate that dispersal rates among populations must be substantial because the extent of genetic subdivision in these bats is generally low.

Because reproduction is seasonal in most phyllostomids, the age structure of their populations is also seasonal, with peak numbers of newly weaned young (juveniles) following peaks in lactation. In bimodally polyestrous frugivores such as *C. perspicillata* and *A. jamaicensis*, peak numbers of newly volant juveniles occur at the ends of the dry and wet seasons in Costa Rica and Panama (Fleming 1988; Handley et al. 1991). Depending on the age at sexual maturity—usually one or two years in phyllostomids—populations will also contain substantial numbers of subadults (in a reproductive sense) year-round. The Mexican migratory nectar-feeder *L. yerbabuenae* clearly shows this in its northern maternity roosts, in which about 75% of the females are pregnant adults and about 25% are yearling (subadult) females (THF, unpubl. data). It is possible that these subadults migrate back to their natal roosts a year before their first pregnancy. Overall, the numerically dominant age classes year-round in phyllostomid populations are adults, especially in long-lived species such as *D. rotundus* and *P. hastatus*.

As a result of their "slow" life histories, phyllostomid populations generally have low growth potential and are unlikely to ever experience "explosive" population growth, as sometimes occurs in terrestrial mammals such as rodents with much faster life histories. In the absence of drastic habitat alterations, currently caused most often by human activities, many phyllostomid populations are thus likely to be at or near their habitats' carrying capacities as set by their food resources and roost availability. As has been widely discussed, their slow life histories makes phyllostomids (and all other bats) extremely vulnerable to extinction owing to a variety of negative effects caused primarily by human activities (e.g., Mickelburgh et al. 1992; 2002; Racey and Entwistle 2003; Voigt and Kingston 2016; chaps. 23 and 24, this vol.).

Hurricanes are natural disturbances that can provide us with insights into the population growth potential of certain phyllostomid bats. The effects of hurricanes on bats in the West Indies, especially Puerto Rico, have been particularly well-studied. Two major hurricanes, Hugo and Georges, brushed by or made direct landfall on that island in 1989 and 1998, respectively, and had strong effects on its plant communities and bird and bat populations, including three plant-visiting phyllostomids: *A. jamaicensis* (a frugivore), *Stenoderma rufum* (a frugivore), and *Monophyllus redmani* (a nectarivore). As described by Gannon and Willig (2009), populations of these species were monitored once or twice a year using mist nets set away from roosts in the Luquillo Experimental Forest beginning in 1987. Population responses to these storms are summarized in table 19.5. Populations of the two frugivores, which are both foliage roosters at this site, crashed after the storms because their fruit resources and roost sites were destroyed. That of *A. jamaicensis*, a polyestrous species, recovered faster, especially after Hurricane Hugo, than that of *S. rufum*, a monestrous species. Gannon and Willig (2009) suggested that the faster recovery by *A. jamaicensis* after Hugo resulted from a behavioral rather than a demographic response. That is, this strong-flying bat likely moved away from the zone of destruction in eastern Puerto Rico instead of suffering heavy mortality *in situ*. In contrast, it suffered heavy mortality after Georges, whose path of destruction covered much of the island. *Stenoderma rufum* on Puerto Rico was devastated by Hurricane Georges and apparently either disappeared or became very uncommon on the island. In contrast to the two frugivores, whose food supplies and roost sites were severely reduced by both hurricanes, populations of the nectarivore *M. redmani*, a cave dweller, did not crash after the hurricanes. Instead, in response to increased flower availability in the forest understory after the storms, its populations quickly returned to pre-storm levels after both hurricanes.

In summary, it took *A. jamaicensis* several years to recover its numbers via reproduction after Georges and took *S. rufum* even longer. Similar slow recoveries were

Table 19.5. Population responses of three species of Puerto Rican phyllostomid bats in the Luquillo Experimental Forest to two strong hurricanes, Hugo in 1989 and Georges in 1998 (Gannon and Willig 2009)

Species	Hurricane	Population response
Artibeus jamaicensis (a polyestrous frugivore and foliage rooster)	Hugo	Low, stable numbers prior to storm; population crashed after storm; numbers recovered 2 years after storm owing to a behavioral response (dispersal from intact populations)
	Georges	High numbers prior to storm; population crashed after storm; numbers took 4 years to recover via a demographic response after storm and continued to increase thereafter as early successional trees produced fruit
Stenoderm rufum (a monestrous frugivore and foliage rooster)	Hugo	Low, stable numbers prior to storm; population crashed and increased very slowly after storm
	Georges	Populations on the island apparently disappeared after the storm; numbers still very low 10 years post-Georges
Monophyllus redmani (a polyestrous nectarivore and cave rooster)	Hugo	No population crash after storm owing to a new pulse of flower and nectar resources
	Georges	No population crash after storm owing to a new pulse of flower and nectar resources

reported for frugivorous phyllostomids after Hurricane Hugo on Montserrat in the Lesser Antilles (Pedersen et al. 2009). Slow recoveries after strong Pacific cyclones have also been reported for foliage-roosting pteropodid bats (reviewed in Wiles and Brooke 2009), supporting the idea that plant-visiting bats with slow life cycles have low potential rates of population growth while their food supplies are recovering.

Population Genetics

In addition to their traditional ecological characteristics, populations have a genetic structure based on the genetic composition of their members. How much genetic variation occurs in phyllostomid populations and how is it distributed in space? These have been fundamental questions in population genetics for decades. Answers began to emerge with the development of electrophoretic techniques and the onset of allozyme analyses beginning in the 1960s and have continued with the development of advanced molecular genetics techniques such as the sequencing of specific genes (e.g., mitochondrial d-loop, cytochrome b, nuclear RAG2, and many others, including MHC genes) and nuclear microsatellites. The answers to these questions have an important bearing on the evolutionary potential of any species and also have important conservation implications. In general, low genetic variation, especially in highly subdivided populations, increases a species' risk of extinction. In contrast, species with highly subdivided populations have higher speciation potential than those lacking geographic subdivision. Here we will review what we currently know about phyllostomid population genetics.

In 1979, Donald Straney and coauthors published allozyme data from 14 species of phyllostomids to provide a preliminary overview of the genetic diversity that exists within populations of this family (Straney et al. 1979). We will use $\bar{H}$, or mean heterozygosity, to indicate the level of genetic diversity in some of these species. Heterozygosity indicates the proportion of alleles that are polymorphic within an individual. Straney et al.'s results indicated that, in six species, including two species of *Macrotus* plus *A. jamaicensis*, *C. perspicillata*, *A. geoffroyi*, and *G. soricina*, $\bar{H}$ ranged from 0.016 (in *A. geoffroyi*) to 0.080 (in *A. jamaicensis*) and averaged 0.035, a value that was substantially lower than $\bar{H} = 0.135$ in two species of *Myotis*. Recent Web of

Science and Google Scholar searches have failed to find any subsequent reviews of allozyme variation in phyllostomids, in contrast to many other organisms (e.g., flowering plants; Hamrick et al. 1992, 1993). Much more recently Marchesin et al. (2008) used random fragment length polymorphisms of one mitochondrial gene (12S/16S) and one nuclear gene (RAG2) to assess genetic variation in 23 New World bat species, including four phyllostomids (*A. lituratus*, *A. planirostris*, *C. perspicillata*, and *P. discolor*). Depending on the gene, these species exhibited less genetic diversity than species of Molossidae and Emballonuridae, a result that is similar to that of Straney et al.'s (1979) preliminary allozyme survey. Given that substantial genetic variation, as revealed by electrophoresis and other more recent molecular genetics techniques, exists in most populations of small mammals, however, it seems safe to conclude that species of phyllostomids with substantial population sizes are likely to harbor large amounts of genetic variation (for recent examples, see Arteaga et al. 2018; Del Real-Monroy and Ortega 2017; and Leiva-González et al. 2019). In their review of the structure of pteropodid bats in the Phillipines, for example, Heaney and Roberts (2009, 51) stated that "levels of genetic variation within all species are high, not low, and rather than showing evidence of an intrinsic vulnerability to extinction from natural causes, independent lineages of these bats have persisted in rather small areas for very long periods of time (often millions of years) in spite of frequent typhoons and volcanic eruptions." This is also likely to be true of many species of phyllostomids.

Our second question deals with how genetic variation is distributed geographically among populations of phyllostomids. Kevin Olival (2012) provided an overview of the genetic structure of bat populations based on data from 61 species (including nine phyllostomids) in 10 families. His data included mitochondrial and nuclear DNA markers. In his review, Olival sought to determine the extent to which various morphological, behavioral, and ecological variables predict degree of population subdivision as measured by Wright's F_{ST}, which indicates the probability that two alleles drawn at random from a population are identical by descent. Values of F_{ST} range from 0 (in a totally panmictic species with no genetic subdivision) to 1.0 (in a species whose populations are completely isolated from each other). Results of this review indicated that species with low values of F_{ST} were generally characterized by high aspect ratio wings (which explained about 36% of the variation in F_{ST} among all species), large body size, migratory behavior, and non-insectivorous diets. Overall, genetic subdivision was relatively low (the median value of F_{ST} was 0.100) across all species, and although it didn't differ significantly from that of other families, F_{ST} of phyllostomids was especially low (its median value was 0.030). Supporting Olival's results, Taylor et al. (2012) also reported a significant negative correlation between wing loading (a correlate of dispersal ability) and F_{ST} (determined using mitochondrial d-loop and cytochrome b gene sequences) in eight species of Afro-Malagasy molossids. Thus, relative mobility is likely to be an important factor influencing the population genetic structure of many species of bats.

Based on results of these studies, we can make the following predictions about the degree of genetic subdivision within phyllostomids. Low values of F_{ST} (i.e., low genetic subdivision) are likely to occur in species and clades characterized by wings with high aspect ratios and high wing loading, large body size, migratory behavior or otherwise high dispersability, seasonally polygynous breeding systems, a diet of nectar and/or fruit, and living on the mainland rather than on islands. Based on these characteristics, we expect the following clades (subfamilies) to contain species exhibiting low genetic subdivision: Desmodontinae, Glossophaginae, Carolliinae, and Stenodermatinae. The other clades (subfamilies)—Macrotinae, Micronyterinae, Phyllostominae (except for its large species such as *P. hastatus*), Glyphonycterinae, and Rhinophyllinae—should contain species exhibiting higher genetic subdivision.

Available data based on mitochondrial and microsatellite markers (table 19.6) are still quite limited for this family, especially for animalivorous clades, and do not allow us to rigorously test these predictions. No data are apparently available for four of the 10 phyllostomid subfamilies. Current estimates of F_{ST} range from 0.005 (surprisingly, in two island species, *B. cavernarum* and *A. nicholesi*) to 0.953 (in island populations of *Macrotus "waterhousi"*); the median value is about 0.06. In 12 of the 18 species in table 19.6, F_{ST} is low (≤ ~0.10). These species differ strongly in aspect ratio, size, and whether they occur on the mainland or on islands. As expected, two of the three species with very high values of F_{ST} occur on islands. Clearly, it is too early to tell whether any of our predictions for this family will be supported. Early in the study of the molecular genetics

of bat populations, Burland and Worthington Wilmer (2001) concluded, from relatively limited data, that bat populations generally exhibit low levels of subdivision between colonies and low relatedness within colonies. This conclusion is likely to stand for phyllostomids as more studies are conducted (e.g., see "Case Studies" below).

A variety of recent phylogeographic studies based on molecular genetics provide us with additional insights into genetic diversity and population structure of phyllostomids and their relatives. For example, in a broad geographic survey Ditchfield (2000) used 300–400 base pair sequences of the mitrochondrial cytochrome b gene to determine genetic variation and structure in 15 species of phyllostomids; his survey concentrated on four widespread lowland species (*A. lituratus*, *G. soricina*, *C. perspicillata*, and *S. lilium*). Genetic variation was expressed as haplotype diversity (H, as distinct from H̄), which indicates the probability that two individuals in a population have a *different* haplotype. Different haplotypes differ in their nucleotide sequences, and sequence differences among haplotypes within populations of phyllostomids typically range from about 1.5% to 2.5% (Martins et al. 2007). Ditchfield reported high haplotype diversity within populations of the 15 species, supporting the idea that phyllostomid populations harbor substantial amounts of genetic diversity (also see, e.g., Ripperger al. 2013). Regarding degree of genetic structuring, Ditchfield found little structuring, implying low values of F_{ST}, in *A. lituratus*, *C. perspicillata*, *S. lilium*, and most other species; *G. soricina*, in contrast, exhibited a greater degree of genetic structuring. Subsequent phylogeographic studies have reported similar results regarding haplotype diversity, but significant geographic structuring has been found in two species of *Carollia* (*C. perspicillata* and *C. brevicauda*) and in *D. rotundus* in South America (Martins et al. 2007; Pavan et al. 2011).

As discussed by Farneda et al. (2015), Meyer et al. (2008), and others, it is likely that habitat fragmentation, widespread throughout the tropics, will ultimately have negative demographic and genetic consequences for at least some phyllostomid bats, particularly gleaning animalivores. A microgeographic example of this effect comes from the results of Meyer et al.'s (2009) study of two frugivores, *C. perspicillata* and *Uroderma bilobatum*, in central Panama. These researchers used sequence variation in mitochondrial d-loop DNA to

Table 19.6. Summary of data on the genetic structure of phyllostomid bats based either on mitochondrial or nuclear markers

Subfamily	Species	IS or ML?	Migratory	Mass (g)	Aspect Ratio	Marker Type	Overall F_{ST}	Source
Macrotinae	Macrotus 'waterhousi"	IS	N	9.5	6.4	Mit	0.953	Fleming et al. 2009
Desmodontinae	Desmodus rotundus	ML	N	28.5	6.7	Mit	0.468	Martins et al. 2007; Wilkinson 1985
Phyllostominae	Phyllostomus hastatus	ML	N	107	7.6	Mit	0.031	McCracken and Bradbury 1981
	Trachops cirrhosus	ML	N	43.8	6.3	Nuc	0.0284	Halczok et al. 2018
Glossophaginae	Brachyphylla cavernarum	IS	N	...	...	Mit	0.005	Carstens et al. 2004
	Erophylla sezekorni and E. bombifrons	IS	N	16.3	6.1	Mit	0.566	Fleming et al. 2009
	Glossophaga longirostris	ML	N	...	...	Mit	0.725	Newton et al. 2003
	Leptonycteris yerbabuenae	ML	Y	19.5	5.9	Mit	0.109	Wilkinson and Fleming 1996
	L. yerbabuenae	Baja CA	Y	...	...	Nuc	0.012	Arteaga et al. 2018
	L. curasoae	ML	Y	...	...	Mit	0.167	Newton et al. 2003
Carolliinae	Carollia perspicillata	ML	Y	19.1	6.1	Mit	0.060	Meyer et al. 2009
	C. castanea	ML	N	15	...	Mit	0.008	Ripperger et al. 2014
Stenodermatinae	Artibeus jamaicensis	IS	N	47	6.4	Mit	0.013	Carstens et al. 2004
	A. jamaicensis	ML	N	...	...	Nuc	0.154	Del Real-Monroy and Ortega 2017
	A. lituratus	ML	N	60	6.1	Nuc	0.012	McCulloch et al. 2013
	A. watsoni	ML	N	12	...	Mit	0.050	Ripperger et al. 2013
	Uroderma bilobatum	ML	N	15.4	6.3	Mit	0.010	Meyer et al. 2009
	U. bilobatum	ML	N	...	...	Nuc	0.0055	Sagot et al. 2016
	Ardops nichollsi	IS	N	—	—	Mit	0.005	Carstens et al. 2004

Note: Genetic structure (F_{ST}) based on mitochondrial markers generally tends to be greater than that based on nuclear markers. Most of the data on mass and aspect ratio come from Norberg and Rayner (1987). Abbreviations: IS = islands; ML = mainland; N = no; Y = yes; Nuc = nuclear, Mit = mitochondrial.

* Based on genetic data (Fleming et al. 2009; Muscarella et al. 2011), this species is undoubtedly polyphyletic.

examine genetic diversity and structure of these species on 11 small islands in Gatun Lake and the Barro Colorado Island "mainland." The islands, which ranged from 2.5 to 50 ha in size and were 0.02–3.4 km from the mainland, were formed during the flooding of Gatun Lake in 1914. Based on a previous study of marked populations of these species and differences in their morphology, it was predicted that U. bilobatum, the more mobile species with higher aspect ratio wings, would exhibit less loss of genetic (haplotype) variation on the islands and less genetic structuring than the less mobile C. perspicillata. Results supported this prediction (and that of Olival [2012] for the effect of wing shape). Haplotype diversity was higher in U. bilobatum, with no difference between the islands (H = 0.92) and mainland (H = 0.96); its F_{ST} was 0.01. In contrast, hap-lotype diversity was significantly lower on the islands than the mainland in C. perspicillata (0.86 vs. 0.97), and its F_{ST} was 0.06.

Similar results regarding significant genetic structure in a Costa Rican fragmented agricultural landscape were reported for the understory-foraging frugivore Artibeus (Dermanura) watsoni by Ripperger et al. (2013) (table 19.6). In the same study system, however, Ripperger et al. (2014) found virtually no genetic structure in another small understory frugivore, Carollia castanea (table 19.6). McCulloch et al. (2013) also reported a lack of increased genetic structure as a result of widespread deforestation in Atlantic Forest populations of the large, canopy-foraging frugivore A. lituratus in Brazil and Paraguay. They found no difference in F_{ST}, which averaged 0.012 (table 19.6), between populations in forest fragments compared with continuous forest and concluded that large size, high mobility, and a relatively generalized diet are traits that likely result in high connectivity among populations of this species.

Clearly, much more work needs to be done in this important area before we fully understand what effects, if any, habitat fragmentation has on the genetic structure of populations of phyllostomid bats. Some species appear to be sensitive to this anthropogenic disturbance whereas others apparently are not. At this stage, it is difficult to generalize about what environmental and species' traits are best for predicting how different species will respond demographically and genetically to ever-increasing environmental disturbances. Nonetheless, it seems safe to say that mitigating the effects of these disturbances is important, not just for phyllostomid bats but for all of Earth's biodiversity (e.g., Farneda et al. 2015; Rocha et al. 2017).

Case Studies

As indicated at the beginning of this chapter, very few phyllostomids have been the subjects of long-term population studies. Here we will summarize what we know about the population biology of four relatively well-studied species that differ in a variety of ways (e.g., roosting ecology, trophic position, and social behavior). Vampires and their population ecology are discussed in chapter 15. At the very least, these species "snapshots" will highlight some of the important variation that occurs in the population biology of this family.

Phyllostomus hastatus

Weighing 80–100 g as adults, *P. hastatus* is the second largest phyllostomid bat. It exhibits significant sexual size dimorphism, with males weighing about 15% more than females. Locally uncommon to common in lowland tropical forests throughout most of the mainland Neotropics, it roosts primarily in caves but also in hollow trees, termite nests, and buildings. It is a truly omnivorous bat whose diet includes insects such as swarming termites and beetles, small vertebrates such as lizards, birds, rodents, and bats, nectar from large flowers (e.g., *Ochroma* and *Ceiba*), and fruit (e.g., *Cecropia, Ficus, Gurania, Mangifera*). Santos et al. (2003) provide a concise account of many aspects of the biology of this species.

The population biology of *P. hastatus* has been most thoroughly studied on Trinidad, beginning in the late 1960s (e.g., reviewed in McCracken and Bradbury 1981) where it lives in caves in colonies of several hundred individuals. Within colonies, this bat has a harem polygynous mating system based on female defense. Nearly all adult females live in "potholes" (i.e., solution or bell holes) in the ceilings of caves in harems of 15–25 individuals tended by a single adult male. Harems contain mixed age groups of unrelated females that are very stable in composition and that stay together for years, even with changes in harem males. Young females are not recruited into their natal harems but instead form new groups that break up as individuals join groups of older females. In these caves, non-harem (bachelor) males also live in temporally stable groups that sometimes contain young females; these groups contain up to 34 bats. Sex ratio in these colonies is strongly female biased (i.e., 3:1 in three intensively studied caves). McCracken and Bradbury (1981) suggest that this bias results from higher mortality rates in males.

The demography of *P. hastatus* includes a monestrous breeding system in which timing of births is highly synchronized within harems and colonies; timing is slightly different between colonies and is correlated with differences in rainfall (Porter and Wilkinson 2001). Females reach sexual maturity at two years of age, and nearly all adult females give birth to a single pup every year. Although they sometimes "babysit" babies while harem mates are feeding away from the roost (Wilkinson et al. 2016), females nurse only their own young and do not adopt babies that are not their own. Adult females have an annual survival rate of >90%, which is very high for phyllostomid bats (chap. 12, this vol.). Once they have gained control of a harem, males can hold them for several years and are constantly defending them against intruding males. As a result, harem males father 60–90% of the babies born in their harems, and a few harem males have the potential to father many offspring during their lives. Variation in reproductive success among adult males is thus vastly greater than that of adult females, which is characteristic of most polygynous species.

Foraging behavior of males and females strongly reflects their social status. Harem males typically forage within a kilometer of the day roost and are away from the roost for only about 1 h each night. Harem females have feeding areas that they use year-round and that are located 1–5 km from the roost; they average about 70 ha in area. Although they forage individually, harem mates tend to have adjacent and sometimes overlapping feeding areas; the feeding areas of different harems are located away from each other. Like females, bachelor males have stable feeding areas that are located away from those of harem females.

Finally, little genetic structure exists in the populations of *P. hastatus* on Trinidad. Within colonies, harems contain a random sample of genotypes, and different colonies also contain random samples of the overall genetic variation in the region. Occasional movement between colonies, mostly by females, and random recruitment within harems explains this (non)pattern.

In summary, *P. hastatus* is a long-lived polygynous bat that is notable because of the stability of its female groups, both within and away from its day roosts. This stability plus low adult mortality rates in females most likely also carry over to overall population stability. McCracken and Bradbury (1981) suggest that sharing of food information within harems, which sometimes occurs because the feeding areas of harem mates are close to each other, probably best explains female group stability. Their genetic analyses preclude kin selection as a cause of this stability.

Leptonycteris yerbabuenae

Except for its slightly larger congener *L. nivalis*, *L. yerbabuenae*, which weighs 20–27 g, and its sister species *L. curasoae* are the largest members of the Glossophaginae. Found mostly in dry and arid habitats from southern Arizona to Honduras, *L. yerbabuenae* is a nectar- and fruit-eating bat that specializes on food resources produced by columnar cacti and paniculate agaves. It (as well as its congeners) roosts primarily in caves and mines and is much more gregareous than other glossophagines. Colonies often contain tens of thousands and up to >100,000 individuals. Cole and Wilson (2006) and Fleming and Nassar (2002) provide summaries of many aspects of the biology of this species.

High mobility is a key population characteristic of this species. In many parts of its range, *L. yerbabuenae* is migratory, and as a result, many of its colonies fluctuate in size from a few hundred to tens of thousands of individuals on a seasonal basis. These numerical changes are associated with its migration and reproductive cycle.

The ultimate driver behind these changes likely is significant seasonal and geographic variation in resource levels. Like *P. hastatus*, this species is monestrous, and the timing of matings and births varies geographically. In coastal southwestern Mexico, for example, matings occur in October–December, and many females then migrate north to form maternity colonies in the Sonoran Desert where their birth season begins in mid-May. In central Mexico, matings occur in June–August, and some females migrate to southern Mexico to give birth in maternity colonies, beginning in December. In southern Baja California, matings occur in September–October, and females give birth in February–April, about midway between the birth periods in southern and northern Mexico (Frick et al. 2018).

Seasonal changes in the colony sizes and sex ratios of two well-studied roosts can illustrate the dynamic nature of populations of *L. yerbabuenae*. Ceballos et al. (1997) and Stoner et al. (2003) documented its annual population cycle in a sea cave on Isla Don Panchito a few kilometers off the coast of Jalisco in southwestern Mexico. Habitat in this region is tropical dry forest. The size of this colony in 1992–94 varied from a low of about 5,000, most of which were males, in March to a peak of about 75,000 with a sex ratio of nearly 1:1 in November. Matings took place there in October–December, after which large numbers of both males and females left the roost. Genetic studies indicate that many of these females migrate about 1,000 km north to form maternity colonies in the Sonoran Desert (Wilkinson and Fleming 1996). Stoner et al. (2003) reported somewhat different results in 1999–2000 by noting the presence of lactating females and juveniles in this roost in January–March. These females likely conceived their babies in the June–August mating period elsewhere in central Mexico.

A maternity roost in the Pinacate Biosphere Reserve in the Sonoran Desert near the U.S.-Mexico border houses the largest known population of *L. yerbabuenae*. Empty for much of the year, at least 100,000 females, most of whom are pregnant, occupy this large lava tube in April–August (Fleming and Nassar 2002). The resource base of flowers and fruit of columnar cacti around this roost is extremely low, and recent studies of geotagged *L. yerabuenae* indicate that some individuals fly over 90 km one-way to visit patches of flowering columnar cacti (Y. Yovel, pers. comm.).

Radio-tracking studies near another maternity roost

containing several thousand females and located on Isla Tiburon, off the coast of Sonora, Mexico, in the Gulf of California also confirm that this bat forages long distances from its day roost (Horner et al. 1998). At this site, some pregnant or lactating females fly over 100 km in total each night in order to commute to and forage in mainland feeding areas. Fleming et al. (2001) reported that the density of these bats in mainland feeding areas is low (about 1/ha or 100/km²).

Not all populations of *L. yerbabuenae* fluctuate as dramatically as those in western and northern Mexico. Galindo et al. (2004), for example, conducted a 2-year study of a colony living in a cave in tropical dry forest located in central Mexico. As in other areas, population size and adult sex ratio in this colony changed seasonally. This roost contained 22,000–27,000 *L. yerbabuenae*, with an equal sex ratio in February–July. Mating occurred in June–August, after which males departed, leaving behind up to 10,000 pregnant females, who gave birth to single pups beginning in December. About 8,000 of these males moved to a mine 8 km from the main colony. This study shows that some caves inhabited by this species can serve sequentially as both mating and maternity roosts. Galindo et al. (2004) reported that flowers and fruit were available year-round in this habitat, which helps explain why this population is more sedentary than ones in western Mexico. But throwing doubt on this explanation is the fact that, although bat-pollinated flowers and fruit are also available year-round near the sea cave in Jalisco, large numbers of both males and females leave that roost after the mating season (Stoner et al. 2003). Factors causing males and females to leave some caves in large numbers but not others deserve further study.

Annual surveys conducted by the Arizona Department of Game and Fish in southern Arizona indicate that many females and juveniles occupy post-maternity roosts in southern Arizona for a few weeks in the late summer and early fall before migrating back into Mexico (A. McIntire, pers. comm.). Most of these roosts contain hundreds to a few thousand bats and are located in upland habitats that contain paniculate agaves, whose flowers are its main food source at this time of the year (Ober et al. 2005).

Unlike other well-studied phyllostomids, the mating system of *L. yerbabuenae* and its congeners is not well-known yet. Given the high mobility and significant geographic separation of adult males and females of this species for much of the year, it is highly unlikely that these bats have a harem polygynous system. Mating colonies of these bats typically contain tens of thousands of bats, and males and females are often intermingled. Recent studies (e.g., Muñoz-Romo and Kunz 2009; Nassar et al. 2008; Rincón-Vargas et al. 2013), indicate that prior to the mating season some adult males in both *L. yerbabuenae* and *L. curasoae* develop a so-called dorsal patch in their interscapular region that contains a mixture of body fluids and is highly odoriferous. Most matings (e.g., 18 out of 21 in Muñoz-Romo and Kunz 2009) that have been videotaped have been between a female who had paired up with a male with a dorsal patch. Females tend to be attracted to the odors from these patches (Muñoz-Romo and Kunz 2009). Since a minority of adult males (e.g., 22%; Rincón-Vargas et al. 2013) in mating colonies have these seasonal dorsal patches, it is likely that male reproductive success in these species is highly skewed, as in other polygamous phyllostomids (see above). The question remains, however, about how many males females mate with and vice versa during the mating season. Do members of this genus have a truly promiscuous mating system unlike most other phyllostomids (McCracken and Wilkinson 2000)?

Genetic studies (e.g, Arteaga et al. 2018; Newton et al. 2003; Ramírez 2011; Wilkinson and Fleming 1996) indicate that genetic diversity is moderately high and that genetic subdivision is low in *L. yerbabuenae* and *L. curasoae* (table 19.6). Migration over long distances, occasional dispersal between reproductive demes (Newton 2002), and occupation of multiple caves (even on different Caribbean islands and the mainland in *L. curasoae*; Simal et al. [2015]) during the annual cycle assures that genetic admixture is extensive in populations of these species (Frick et al. 2018).

Carollia perspicillata

A medium-sized phyllostomid weighing 15–25 g, the short-tailed fruit bat is one of the most common mammals in the mainland Neotropical lowlands. Occurring in both wet and dry forests, it is particularly prevalent in disturbed habitats. It roosts in a wide variety of structures, including caves, hollow trees, culverts, and buildings. It is mostly frugivorous but also visits accessible flowers in the dry season and eats some insects. Its diet includes fruit from 40 plant families, but the bulk of its diet comes from understory fruits of *Piper* (Piperaceae) and *Solanum* (Solanaceae) (Lobova et al. 2009). Cloutier and Thomas (1992) and Fleming (1988) provide summaries of the biology of this species.

Carollia perspicillata has been intensively studied in tropical dry forest at Santa Rosa National Park in northwestern Costa Rica (Fleming 1988). Most aspects of its population biology, including its role as an important disperser of the seeds of pioneer plants, were investigated in this 12-year study (1974–85). Most individuals roost in caves and hollow trees at this site, and colonies living in three caves and one hollow tree were studied. Colony size and sex ratio in the caves changed seasonally, with peak numbers of 200 (in the Sendero cave) to over 400 (in the Cuajiniquil cave) in August–September during the wet season. Numbers were much lower in the dry season when many adult females left the Santa Rosa roosts and likely moved to moister, resource-rich habitats in the nearby uplands. Taking into account seasonal changes in their size, colonies in the Sendero and Red roosts were mostly stable between 1979 and 1985. In contrast, the Cuajiniquil colony decreased strongly in size between mid-1983 and mid-1984; some of its marked individuals moved to the other two colonies at that time. Additional evidence of relatively stable population numbers during 1974–85 comes from mist-netting data from a series of non-roost sites in which capture rates of *C. perspicillata* were relatively constant among years and averaged 0.53 individuals per net-hour. Based on mist-netting and radio-tracking data, Fleming (1988) estimated that its population density away from the roosts was about 0.093 bats per ha or about 9.3 bats per km^2, a rather low density for such a common bat. Bat densities in resource-rich patches, however, were much higher than this.

Like *P. hastatus* and *A. jamaicensis* (see below), *C. perspicillata* has a harem polygynous breeding system in which a few adult males defend groups of females. In the Sendero cave where behavioral observations took place (Williams 1986), harem males represented about 22% of the male population, and each of these males defended an average of 2.2 (range 1–18) females. Group size was positively correlated with male age as determined by tooth wear. Whereas many males defended their harem sites against intruding males both day and night and throughout the year, even in the absence of females during the dry season, females exhibited little stability in their associations with harem males and other females, in strong contrast to the situation in *P. hastatus*. They visited a number of harems before settling down with one male, suggesting that harem location was more important to females than the identity of any particular male. Harem males were thus defending a piece of prime real estate, probably based on safety from predators, which included two species of cave-dwelling snakes, rather than defending females. If this is true, then *C. perspicillata*'s mating system is likely to be resource-defense polygyny rather than female-defense polygyny as found in *P. hastatus*.

Like many plant-visiting phyllostomids (chap. 12, this vol.), *C. perspicillata* has a bimodal polyestrous breeding system, and once they are mature, usually as yearlings, most females give birth to a single pup twice a year. In northwestern Costa Rica, births occur in March–April in the dry season after a 4-month gestation period and again in August–September in the middle of the wet season. What is interesting about reproduction at this site is that, after they mate following their wet season birth, females leave the Santa Rosa colonies and give birth to their dry season babies somewhere else. After giving birth to this baby, they undergo another postpartum estrus and return to the Santa Rosa roosts, often to the same harem site that they had left the previous fall to give birth to their wet season babies. Thus, whereas females mate and give birth twice a year in different roosts, harem males (and also most bachelor males) remain in one roost throughout the year and thus participate in only one of the two mating seasons each year. A similar reproductive pattern and sexually based roost dynamics have been observed in *C. perspicillata* in tropical dry forest in southeastern Brazil (L. Monteiro, pers. comm.).

Fleming (1988) used reproductive data and mark-recapture data to construct a life and fecundity table for this species. These data indicated that this population's net reproductive rate, R_o, that is the number of daughters a female will produce during her lifetime, is close to 1.0, again signifying a stable population; that average life expectancy of a newborn (female) is about 2.6 years; and that both males and females die at a similar rate such that <5% of all bats live 8 years or more. Young bats are weaned at about 2 months of age, and young females join already-established harems rather than forming new same-age harems. Finally, young males are more likely to remain in their natal caves than are young females.

Like *P. hastatus*, the foraging areas of *C. perspicillata* away from the day roost reflect their social status. Harem males forage close to the roost and usually spend the time between feeding bouts back in the cave. Females and bachelor males forage up to 5 km away from the day roost, and, unlike females of *P. hastatus*, female harem mates do not forage near each other.

Finally, limited genetic data based on allozymes indicates that, as expected given the mobility of females in this species, little genetic subdivision occurs in its populations in western Costa Rica. It was not possible to estimate the degree of reproductive control males have over their harem females, but, again given the lack of social stability in this species, it is likely to be much lower than that of *P. hastatus* (see Fasel et al. 2017)

Artibeus jamaicensis

A relatively large bat weighing 30–50 g, *A. jamaicensis* is another very common frugivorous phyllostomid with a wide geographic distribution throughout the lowlands of tropical America and the West Indies. It often roosts in forest canopy foliage and hollow trees on the mainland and in caves on the Mexican Yucatán Peninsula and in the Caribbean. Like *C. perspicillata*, it opportunistically visits accessible flowers of canopy trees in the dry season but is basically frugivorous year-round. Although it is known to consume the fruits of species in at least 43 plant families (Lobova et al. 2009), most of its diet comes from fruit in the fig family (Moraceae). Whereas species of *Carollia* feed on fruits produced by understory plants, *A. jamaicensis* and many of its relatives feed on fruits produced by canopy plants. Handley et al. (1991) and Ortega and Castro-Arellano (2001) provide extensive summaries of the biology of this species.

Much of our knowledge about the population biology of this species comes from research conducted in tropical moist forest on Barro Colorado Island and the adjacent mainland in central Panama (Handley et al. 1991). Over a 6-year period (1975–80), this study tallied a total of nearly 18,000 captures and recaptures of marked *A. jamaicensis*. In this habitat, *A. jamaicensis* lives in scattered colonies either in canopy foliage (most males) or in slits in hollow trees located well above the forest floor (most females). Colonies in each of these roosts are small, averaging about one to three males or three to 14 females plus their young and a single harem male. Neither males nor females show long-term fidelity to particular roosts or particular groups of individuals. Because Leigh and Handley (1991) estimated that its total population size on Barro Colorado included about 1,500 adult females, 850 adult males, and 1,800 juveniles and subadults in 1980–81, several hundred colonies must have occurred on the island. They also estimated that the population density of this population was about 200 bats/km^2—a much higher density than its estimated density of 4.1/km^2 in tropical dry forest in northwestern Costa Rica (Fleming 1988). These differences in population density clearly reflect large differences in the density of *Ficus* trees in the two sites: nearly six *Ficus* trees/ha on Barro Colorado compared with an extremely low ($\ll$1/ha) density of *Ficus* trees at Santa Rosa (Fleming 1988; Morrison 1978a).

As described in detail in chapter 12 (this vol.), *A. jamaicensis* is bimodally polyestrous with most adult females giving birth in the dry season and again in the wet season. Leigh and Handley (1991) used the Barro Colorado Island mark-recapture database to construct a life and fecundity table for this species. Their results indicated that females produce about 1.2 surviving daughters per lifetime, which is probably close enough to an R$_o$ of 1.0, given the many assumptions behind their calculations, to conclude that the population was stable during that time period. Average female lifespan, based on this table, was about 1.6 years, and the annual survival rates of adult females were about 60–64%. Maximum lifespan of this species is probably about 10 years.

Like many other phyllostomids, *A. jamaicensis* has a harem polygynous social system based on resource (a safe roost site) defense. Although females exhibit longer fidelity to particular roost sites than foliage-roosting (non-harem) males, their fidelity to roosts and their female harem mates is much shorter than that of females of *P. hastatus* (but see below). After they are weaned, young females leave their natal roosts and join new roosts as subadults. As a result, harems of this species, like those of *P. hastatus* and *C. perspicillata*, contain mixed age groups. Also as in the other two species, harem males forage relatively close to the roost at night and spend a lot of time flying around the roost tree (Morrison and Morrison 1981). Harem females, in contrast, forage up to a kilometer or two away from the roost and away from each other; they do not forage in harem-specific areas (Morrison 1978a, 1978b).

Artibeus jamaicensis is also common throughout the Caribbean, where it often roosts in caves as well as in foliage. On Cuba and Puerto Rico, colonies living in caves are generally small and contain only a few hundred individuals rather than thousands (Silva Taboada 1979; THF, pers. obs.). As elsewhere, this species is harem polygynous, with harems living in cave potholes. Kunz et al. (1983) reported harems containing two to 14 adult females in a cave in Puerto Rico. Older and heavier males had larger harems than those of smaller and younger males. Attempts to radio-track this species on Puerto Rico were basically unsuccessful (Gannon et al. 2005). Bats that were radio-tagged in feeding areas did not return to those areas on subsequent nights, and their day roosts were never found. Gannon et al. (2005) concluded that these bats were foraging >10 km away from their roosts.

Ortega et al. (2003) used nuclear microsatellites to determine the genetic structure, relatedness, and degree of reproductive control by harem males in *A. jamaicensis* in two caves located 11 km apart on the Yucatán Peninsula, Mexico. Colonies in these caves contained 200–250 individuals. Unlike the situation in Panama, the harems they studied occurred in cave potholes and exhibited relatively high stability in both male and female membership over a 2-year period. Large harems (i.e, those with more than 14 females) were attended by two males: a dominant male and a subordinate male. Also in the caves were mixed-sex groups of bachelor (satellite) males and young females whose membership changed frequently (Ortega and Arita 1999). Genetic results indicated that very little subdivision occurred between the two caves ($F_{ST} = 0.008$); dominant males fathered about 69% of the babies in their harem; bachelor males fathered about 22% of the babies; and subordinate (young) males, which appeared to be the sons of dominant males, fathered about 9%. Neither harem females nor harem males were closely related, indicating the absence of genetic structure both within and between colonies.

Results of studies in Panama and the Yucatan (and probably the Caribbean) indicate that the population structure of *A. jamaicensis* varies among habitats. Whereas it roosts in foliage and hollow trees in many small colonies whose compositions are relatively fluid in forested areas, it roosts in larger colonies whose adult composition is more stable in karstic areas. Despite these differences in colony configurations, this bat is very common in areas where the densities of fig trees are high.

Conclusions

We have learned much about the characteristics of phyllostomid populations—their patterns of roost use, colony sizes, foraging behavior, genetic composition, and so on—since the 1970s when these kinds of studies began. But most of this knowledge has accumulated for only a few common species, especially for frugivores in the Carolliinae and Stenodermatinae. Except for *P. hastatus* (see above) and *Trachops cirrhosus* (Halczok et al. 2018), we know much less about the population biology of animalivorous species, which, because of their generally smaller population sizes and high dependence on tree roosts that occur more often in mature forests than in early successional forests (Voss et al. 2016), are more likely to be of conservation concern than are plant-visiting species.

Despite large gaps in our current knowledge about their population biology, it is clear that populations of these bats and their noctilionoid relatives occur in two basic roost or colony configurations: "concentrated" or "diffuse." Concentrated species are those that roost in caves (wherever available) and other long-lived structures that are often large enough to house large colonies, whereas diffuse species are those that often live in small colonies in relatively ephemeral, usually plant-based roosts. Although colony sizes in these two groups usually differ substantially, their overall population sizes do not necessarily differ. Thus, *A. jamaicensis* in central Panama roosts in many small colonies in canopy foliage or tree hollows but is an extremely common bat, owing to the high density of fig trees there. Regardless of roost type and colony size, plant-visiting phyllostomids, especially the frugivores, are much more common than nectarivores and animalivores in most lowland forests because of differences in the size of their resource bases as follows: biomass of fruit > biomass of insects > biomass of nectar and pollen.

Like most other bats, phyllostomids and their relatives have slow life cycles featuring relatively slow rates of sexual maturation, low fecundity, and relatively low annual adult mortality rates compared with similar-sized terrestrial mammals. This generalization, however, rests on relatively few data, and more demographic data for these bats would certainly enhance our understanding of their population ecology as well as their long-term stability. As in many areas of the biology of these bats, demographic data on phyllostomids other than frugivores is currently scarce. Nonetheless, given their slow life cycles, potential population growth rates of many phyllostomids is likely to be low, and it seems reasonable to suggest that, in the absence of anthropogenic disturbances, many species are at or near their habitat's carrying capacity as determined by food resources and roost availability. Since anthropogenic disturbances are widespread in the tropics, however, it remains to be seen how common population stability actually is in these bats.

Regardless of roost type, many phyllostomid populations are very fluid and undergo seasonal or short-term changes in roost use and habitat distributions. These changes occur in response to changes in resource distributions and reproductive activity and are probably more common in plant-visiting species than in animalivores and sanguinivores. Some of these changes involve altitudinal and latitudinal movements. In addition, seasonal sexual segregation often occurs in gregarious species that inhabit long-lived roosts. These movements obviously have important conservation implications because of the multiple habitat requirements, including ones that cross international borders, of many species.

Populations of some phyllostomid bats appear to harbor substantial levels of genetic diversity and often exhibit low levels of between-population genetic subdivision. This situation suggests that these bats have relatively large genetically effective population sizes and that dispersal rates among populations, even in some island-dwelling species, are likely to be high. But, as in other areas of phyllostomid population biology, we need more basic genetic data on species other than frugivores. Especially important is increased knowledge about how habitat destruction and fragmentation affect the population genetics of low-density species. As discussed in chapters 23 and 24, long-term monitoring of population sizes and their genetic composition is critical for determining management strategies that will minimize the loss of phyllostomid populations and species.

Acknowledgments

We thank Kathy Stoner for help in developing the conceptual framework for this chapter. THF thanks the following agencies for supporting his study of phyllostomid populations: the Smithsonian Institution, U.S. National Science Foundation, National Geographic Society, Center for Field Studies, Arizona Department of Game and Fish, the Ted Turner Endangered Species Fund as well as the University of Missouri–Saint Louis and University of Miami. AMGM thanks the following agencies for supporting her studies of phyllostomid populations: Fundacite Centro-Occidente and Universidad Nacional Experimental Francisco de Miranda. We thank Winifred Frick and Paul Racey for references and for commenting on this chapter.

References

Adams, E. R. 2015. Seasonal and nightly activity of Mexican long-nosed bats (*Leptonycteris nivalis*) in Big Bend National Park, Texas. MS thesis, Angelo State University, San Angelo, TX.

Akçakaya, H. R., G. Mills, and C. P. Doncaster. 2007. The role of metapopulations in conservation. Pp. 64–84 *in*: Key Topics in Conservation Biology (D. W. MacDonald and K. Service, eds.). Blackwell Publishing, Oxford.

Albrecht, L., C. F. J. Meyer, and E. K. V. Kalko. 2007. Differential mobility in two small phyllostomid bats, *Artibeus watsoni* and *Micronycteris microtis*, in a fragmented Neotropical landscape. Acta Theriologica, 52:141–149.

Ammerman, L. K., M. McDonough, N. I. Hristov, and T. H. Kunz. 2009. Census of the endangered Mexican long-nosed bat *Leptonycteris nivalis* in Texas, USA, using thermal imaging. Endangered Species Research, 8:87–92.

Angell, R. L., R. K. Butlin, and J. D. Altringham. 2013. Sexual segregation and flexible mating patterns in temperate bats. PLOS ONE, 8:e54194.

Arteaga, M. C., R. A. Medellín, P. A. Luna-Ortiz, P. A. Heady, and W. F. Frick. 2018. Genetic diversity distribution among seasonal colonies of a nectar-feeding bat (*Leptonycteris yerbabuenae*) in the Baja California Peninsula. Mammalian Biology, 92:78–85.

Barclay, R. M. R., and L. D. Harder. 2003. Life histories of bats: life in the slow lane. Pp. 209–253 *in*: Bat Ecology (T. H. Kunz and M. B. Fenton, eds.). University of Chicago Press, Chicago.

Barlow, K. E., P. A. Briggs, K. A. Haysom, A. M. Hutson, N. L. Lechiara, P. A. Racey, A. L. Walsh, and S. D. Langton. 2015. Citizen science reveals trends in bat populations: the National Bat Monitoring Programme in Great Britain. Biological Conservation, 182:14–26.

Bernard, E., L. M. S. Aguiar, D. Brito, A. P. Cruz-Neto, R. Gregorin, R. B. Machado, M. Oprea, A. P. Paglia, and V. C. Tavares. 2012. Uma análise de horizontes sobre a conservação de morcegos no Brasil. Pp. 19–35 *in*: Mamíferos do Brasil: Genética, Sistemática, Ecologia e Conservação. Vol. 2 (T. R. O. Freitas and E. M. Vieira, eds.). Sociedade Brasileira de Mastozoologia, Rio de Janeiro.

Bernard, E., and M. B. Fenton. 2007. Bats in a fragmented landscape: species composition, diversity and habitat interactions in savannas of Santarém, Central Amazonia, Brazil. Biological Conservation, 134:332–343.

Best, A., G. Diamond, J. Diamond, D. C. Buecher, R. Sidner, D. Cerasale, and J. Tress. 2015. Survey of an endangered bat roost at Coronado National Memorial, Arizona. Park Science, 32:49–56.

Betke, M., D. E. Hirsh, N. C. Makris, G. F. McCracken, M. Procopio, N. I. Hristov, S. Tang et al. 2008. Thermal imaging reveals significantly smaller Brazilian free-tailed bat colonies than previously estimated. Journal of Mammalogy, 89:18–24.

Burland, T. M., and J. Worthington Wilmer. 2001. Seeing in the dark: molecular approaches to the study of bat populations. Biological Reviews, 76:389–409.

Carstens, B. C., J. Sullivan, L. M. Dávalos, P. A. Larsen, and S. C. Pedersen. 2004. Exploring population genetic structure in three species of Lesser Antillean bats. Molecular Ecology, 13:2557–2566.

Ceballos, G., T. H. Fleming, C. Chavez, and J. Nassar. 1997. Population dynamics of *Leptonycteris curasoae* (Chiroptera: Phyllostomidae) in Jalisco, Mexico. Journal of Mammalogy, 78:1220–1230.

Chaverri, G., and T. H. Kunz. 2006. Roosting ecology of the tent-roosting bat *Artibeus watsoni* (Chiroptera: Phyllostomidae) in southwestern Costa Rica. Biotropica, 38:77–84.

Chaverri, G., O. E. Quiros, M. Gamba-Rios, and T. H. Kunz. 2007. Ecological correlates of roost fidelity in the tent-making bat *Artibeus watsoni*. Ethology, 113:598–605.

Clement, M. J., J. M. O'Keefe, and B. Walter. 2015. A method for estimating abundance of mobile populations using telemetry and counts of unmarked animals. Ecosphere, 6:1–13.

Cloutier, D., and D. W. Thomas. 1992. *Carollia perspicillata*. Mammalian Species, 417:1–9.

Cole, F. R., and D. E. Wilson. 2006. *Leptonycteris yerbabuenae*. Mammalian Species, 797:1–7.

Del Real-Monroy, M., and J. Ortega. 2017. Spatial distribution of microsatellite and MHC-DRB exon 2 gene variability in the Jamaican fruit bat (*Artibeus jamaicensis*) in Mexico. Mammalian Biology, 84:1–11.

Ditchfield, A. D. 2000. The comparative phylogeography of Neotropical mammals: patterns of intraspecific mitochondrial DNA variation among bats contrasted to nonvolant small mammals. Molecular Ecology, 9:1307–1318.

Esbérard, C. E., I. P. de Lima, P. H. Nobre, S. L. Althoff, T. Jordão-Nogueira, D. Dias, F. Carvalho, M. E. Fábián, M. L. Sekiama, and A. Stanke Sobrinho. 2011. Evidence of vertical migration in the Ipanema bat *Pygoderma bilabiatum* (Chiroptera: Phyllostomidae: Stenodermatinae). Zoologia, 28:717–724.

Estrada, A., and R. Coates-Estrada. 2001. Species composition and reproductive phenology of bats in a tropical landscape at Los Tuxtlas, Mexico. Journal of Tropical Ecology, 17:627–646.

Farneda, F. Z., R. Rocha, A. Lopez-Baucells, M. Groenenberg, I. Silva, J. M. Palmeirim, P. E. D. Bobrowiec, and C. F. J. Meyer. 2015. Trait-related responses to habitat fragmentation in Amazonian bats. Journal of Applied Ecology, 52:1381–1391.

Fasel, N. J., C. Wesseling, A. A. Fernandez, A. Vallat, G. Glauser, F. Helfenstein, and H. Richner. 2017. Alternative reproductive tactics, sperm mobility and oxidative stress in *Carollia*

perspicillata (Seba's short-tailed bat). Behavioral Ecology and Sociobiology, 71:11.

Ferreira, D. F., R. Rocha, A. Lopez-Baucells, F. Z. Farneda, J. M. B. Carreiras, J. M. Palmeirim, and C. F. J. Meyer. 2017. Season-modulated responses of Neotropical bats to forest fragmentation. Ecology and Evolution, 7:4059–4071.

Fleming, T. H. 1988. The Short-Tailed Fruit Bat: A Study in Plant-Animal Interactions. University of Chicago Press, Chicago.

Fleming, T. H. 1992. How do fruit- and nectar-feeding birds and mammals track their food resources? Pp. 355–391 *in*: Resource Distributions and Plant-Animal Interactions (M. D. Hunter, T. Ohgushi, and P. W. Price, eds.). Academic Press, Orlando, FL.

Fleming, T. H., and P. Eby. 2003. Ecology of bat migration. Pp. 156–208 *in*: Bat Ecology (T. H. Kunz and M. B. Fenton, eds.). University of Chicago Press, Chicago.

Fleming, T. H., E. T. Hooper, and D. E. Wilson. 1972. Three Central American bat communities: structure, reproductive cycles, and movement patterns. Ecology, 53:555–569.

Fleming, T. H., and W. J. Kress. 2013. The Ornaments of Life: Coevolution and Conservation in the Tropics. University of Chicago Press, Chicago.

Fleming, T. H., K. L. Murray, and B. C. Carstens. 2009. Phylo-geography and genetic structure of three evolutionary lineages of West Indian phyllostomid bats. Pp. 116–150 *in*: Island Bats: Evolution, Ecology, and Conservation (T. H. Fleming and P. A. Racey, eds). University of Chicago Press, Chicago.

Fleming, T. H., and J. M. Nassar. 2002. The population biology of the lesser long-nosed bat *Leptonycteris curasoae* in Mexico and northern South America. Pp. 283–305 *in*: Columnar Cacti and Their Mutualists: Evolution, Ecology, and Conservation (T. H. Fleming and A. Valiente-Banuet, eds.). University of Arizona Press, Tucson.

Fleming, T. H., C. T. Sahley, J. N. Holland, J. D. Nason, and J. L. Hamrick. 2001. Sonoran Desert columnar cacti and the evolution of generalized pollination systems. Ecological Monographs, 71:511–530.

Frick, W. E., P. A. Heady, A. D. Earl, M. C. Arteaga, P. Cortes-Calva, and R. A. Medellín. 2018. Seasonal ecology of a migratory nectar-feeding bat at the edge of its range. Journal of Mammalogy, 99:1072–1081.

Frick, W. F., J. F. Pollock, A. C. Hicks, K. E. Langwig, D. S. Reynolds, G. G. Turner, C. M. Butchkoski, and T. H. Kunz. 2010. An emerging disease causes regional population collapse of a common North American bat species. Science, 329:679–682.

Gaisler, J. 1979. Ecology of bats. Pp. 281–342 *in*: Ecology of Small Mammals (D. M. Stoddart, ed.). Chapman and Hall, London.

Galindo, C., A. Sánchez, R. H. Quijano, and L. G. Herrera. 2004. Population dynamics of a resident colony of *Leptonycteris curasoae* (Chiroptera: Phyllostomidae) in central Mexico. Biotropica, 36:382–391.

Gannon, M. R., A. Kurta, A. Rodríguez-Durán, and M. R. Willig. 2005. Bats of Puerto Rico. Texas Tech University Press, Lubbock, Texas.

Gannon, M. R., and M. R. Willig. 2009. Island in the storm: disturbance ecology of plant-visiting bats in the hurricane-prone island of Puerto Rico. Pp. 281–301 *in*: Island Bats: Evolution,

Ecology, and Conservation (T. H. Fleming and P. A. Racey, eds.). University of Chicago Press, Chicago.

Giannini, N. P. 1999. Selection of diet and elevation by sympatric species of *Sturnira* in an Andean rainforest. Journal of Mammalogy, 80:1186–1195.

Gorresen, P. M., and M. R. Willig. 2004. Landscape responses of bats to habitat fragmentation in Atlantic Forest of Paraguay. Journal of Mammalogy, 85:688–697.

Greenwood, P. J. 1980. Mating systems, philopatry and dispersal in birds and mammals. Animal Behaviour, 28:1140–1162.

Grimm, V., K. Reise, and M. Strasser. 2003. Marine metapopulations: a useful concept? Helgol Marine Research, 56: 222–228.

Halczok, T. K., S. D. Brandel, V. Flores, S. J. Puechmaille, M. Tschapka, R. A. Page, and G. Kerth. 2018. Male-biased dispersal and the potential impact of human-induced habitat modifications on the Neotropical bat *Trachops cirrhosus*. Ecology and Evolution, 8:6065–6080.

Hamrick, J. L., M. J. Godt, and S. L. Sherman-Broyles. 1992. Factors influencing levels of genetic diversity in woody plant species. New Forests, 6:95–124.

Hamrick, J. L., D. A. Murawski, and J. D. Nason. 1993. The influence of seed dispersal mechanisms on the genetic structure of tropical tree populations. Pp. 281–297 *in*: Frugivory and Seed Dispersal: Ecological and Evolutionary Aspects (T. H. Fleming and A. Estrada, eds). Kluwer Academic Publishers, Dordrecht.

Handley, C. O., Jr, D. E. Wilson, and A. L. Gardner. 1991. Demography and natural history of the common fruit bat, *Artibeus jamaicensis*, on Barro Colorado Island, Panama. Smithsonian Contributions to Zoology, 511:1–173.

Hanski, I. 1999. Metapopulation Ecology. Oxford University Press, Oxford.

Heaney, L. R., and T. E. Roberts. 2009. New perspectives on the long-term biogeographic dynamics and conservation of Philippine fruit bats. Pp. 17–58 *in*: Island Bats: Evolution, Ecology, and Conservation (T. H. Fleming and P. A. Racey, eds.). University of Chicago Press, Chicago.

Herrera, L. G. 1997. Evidence of altitudinal movements of *Leptonycteris curasoae* (Chiroptera: Phyllostomidae) in central Mexico. Revista Mexicana de Mastozoologia, 2:116–118.

Horner, M. A., T. H. Fleming, and C. T. Sahley. 1998. Foraging behaviour and energetics of a nectar-feeding bat, *Leptonycteris curasoae* (Chiroptera : Phyllostomidae). Journal of Zoology, 244:575–586.

Humphrey, S. R., and F. J. Bonaccorso. 1979. Population and community ecology. Pp. 409–441 *in*: Biology of Bats of the New World family Phyllostomatidae. Pt. 3 (R. J. Baker, J. K. Jones Jr., and D. C. Carter, eds.). Special Publications of the Museum, Texas Tech University, no. 16. Texas Tech Press, Lubbock.

Klingbeil, B. T., and M. R. Willig. 2010. Seasonal differences in population-, ensemble- and community-level responses of bats to landscape structure in Amazonia. Oikos, 119:1654–1664.

Kunz, T. H. 1982. Roosting ecology of bats. Pp. 1–55 *in*: Ecology of Bats (T. H. Kunz, ed.). Plenum Press, New York.

Kunz, T. H., P. V. August, and C. D. Burnett. 1983. Harem social organization in cave roosting *Artibeus jamaicensis* (Chiroptera: Phyllostomidae). Biotropica, 15:133–138.

Kunz, T. H., M. Betke, N. I. Hristov, and M. J. Vonhof. 2009. Methods for assessing colony size, population size, and relative abundance of bats. Pages 133–157 *in*: Ecological and Behavioral Methods for the Study of Bats (T. H. Kunz and S. Parsons, eds.). Johns Hopkins University Press, Baltimore.

Kunz, T. H., and L. F. Lumsden. 2003. Ecology of cavity and foliage-roosting bats. Pp. 3–89 *in*: Bat Ecology (T. H. Kunz and M. B. Fenton, eds.). University of Chicago Press, Chicago.

Leigh, E. G., Jr., and C. O. Handley Jr. 1991. Population estimates. Pp. 77–87 *in*: Demography and Natural History of the Common Fruit Bat, *Artibeus jamaicensis*, on Barro Colorado Island, Panama (C. O. Handley Jr., D. E. Wilson, and A. L. Gardner, eds.). Smithsonian Contributions to Zoology, no. 511. Smithsonian Institution Press, Washington, DC.

Leiva-González, E. M., D. Navarrete-Gutiérrez, L.Ruiz-Montoya, A. Santos-Moreno, C. Kraker-Castañeda, and M. García-Bautista. 2019. Analysis of the contribution of landscape attributes on the genetic diversity of *Artibeus jamaicensis* Leach, 1821. Mammal Research, 64:223–233.

Lentini, P. E., T. J. Bird, S. R. Griffiths, L. N. Godhino, and B. A. Wintle. 2015. A global synthesis of survival estimates for microbats. Biological Letters, 11:20150371.

Lewis, S. E. 1995. Roost fidelity of bats—a review. Journal of Mammalogy, 76:481–496.

Lobova, T. A., C. K. Geiselman, and S. Mori. 2009. Seed Dispersal by Bats in the Neotropics. New York Botanical Garden, Bronx.

Marchesin, S. R. C., M. R. Beguelini, K. C. Faria, P. R. L. Moreira, and E. Morielle-Versute. 2008. Assessing genetic variability in bat species of Emballonuridae, Phyllostomidae, Vespertilionidae and Molossidae families (Chiroptera) by RFLP-PCR. Genetics and Molecular Research, 7:1164–1178.

Martins, F. M., A. D. Ditchfield, D. Meyer, and J. S. Morgante. 2007. Mitochondrial DNA phylogeography reveals marked population structure in the common vampire bat, *Desmodus rotundus* (Phyllostomidae). Journal of Zoological Systematics and Evolutionary Research, 45:372–378.

McCracken, G. F., and J. W. Bradbury. 1981. Social organization and kinship in the polygynous bat *Phyllostomus hastatus*. Behavioral Ecology and Sociobiology, 8:11–34.

McCracken, G. F., and G. S. Wilkinson. 2000. Bat mating systems. Pp. 321–362 *in*: Reproductive Biology of Bats (E. G. Crichton and P. H. Krutzsch, eds.). Academic Press, San Diego.

McCulloch, E. S., J. S. Tello, A. Whitehead, C. M. J. Rolón-Mendoza, M. C. D. Maldonado-Rodríguez, and R. D. Stevens. 2013. Fragmentation of Atlantic Forest has not affected gene flow of a widespread seed-dispersing bat. Molecular Ecology, 22:4619–4633.

McGuire, L. P., and W. A. Boyle. 2013. Altitudinal migration in bats: evidence, patterns, and drivers. Biological Reviews, 88: 767–786.

Mello, M. A. R., E. K. V. Kalko, and W. R. Silva. 2008. Diet and abundance of the bat *Sturnira lilium* (Chiroptera) in a

Brazilian montane Atlantic Forest. Journal of Mammalogy, 89:485–492.

Meyer, C. F. J., L. M. S. Aguiar, L. F. Aguirre, J. Baumgarten, F. M. Clarke, J. F. Cosson, S. E. Villegas et al. 2010. Long-term monitoring of tropical bats for anthropogenic impact assessment: gauging the statistical power to detect population change. Biological Conservation, 143:2797–2807.

Meyer, C. F. J., L. M. S. Aguiar, L. F. Aguirre, J. Baumgarten, F. M. Clarke, J. F. Cosson, S. E. Villegas et al. 2015. Species undersampling in tropical bat surveys: effects on emerging biodiversity patterns. Journal of Animal Ecology, 84:113–123.

Meyer, C. F. J., J. Frund, W. P. Lizano, and E. K. V. Kalko. 2008. Ecological correlates of vulnerability to fragmentation in Neotropical bats. Journal of Applied Ecology, 45:381–391.

Meyer, C. F. J., E. K. V. Kalko, and G. Kerth. 2009. Small-scale fragmentation effects on local genetic diversity in two phyllostomid bats with different dispersal abilities in Panama. Biotropica, 41:95–102.

Mickleburgh, S. P., A. M. Hutson, and P. A. Racey, editors. 1992. Old World fruit bats, an action plan for their conservation. International Union for the Conservation of Nature and Natural Resources, Gland, Switzerland.

Mickleburgh, S. P., A. M. Hutson, and P. A. Racey. 2002. A review of the global conservation status of bats. Oryx, 36:18–34.

Morrison, D. W. 1978a. Foraging ecology and energetics of the frugivorous bat *Artibeus jamaicensis*. Ecology, 59:716–723.

Morrison, D. W. 1978b. Influence of habitat on the foraging distances of the fruit bat, *Artibeus jamaicensis*. Journal of Mammalogy, 59:622–624.

Morrison, D. W., and S. H. Morrison. 1981. Economics of harem maintenance by a Neotropical bat. Ecology, 62:864–866.

Moussy, C., D. J. Hosken, F. Mathews, G. C. Smith, J. N. Aegerter, and S. Bearhop. 2013. Migration and dispersal patterns of bats and their influence on genetic structure. Mammal Review, 43:183–195.

Muñoz-Romo, M., and T. H. Kunz. 2009. Dorsal patch and chemical signaling in males of the long-nosed bat, *Leptonycteris curasoae* (Chiroptera: Phyllostomidae). Journal of Mammalogy, 90:1139–1147.

Muscarella, R. A., K. L. Murray, D. Ortt, A. L. Russell, and T. H. Fleming. 2011. Exploring demographic, physical, and historical explanations for the genetic structure of two lineages of Greater Antillean bats. PLOS ONE, 6:e17704.

Nassar, J. M., M. V. Salazar, A. Quintero, K. E. Stoner, M. Gomez, A. Cabrera, and K. Jaffe. 2008. Seasonal sebaceous patch in the nectar-feeding bats *Leptonycteris curasoae* and *L. yerbabuenae* (Phyllostomidae : Glossophaginae): phenological, histological, and preliminary chemical characterization. Zoology, 111:363–376.

Newton, L. R. 2002. Mobility, reproduction, and population genetic structure of two phyllostomid bats. MS thesis, University of Miami, Coral Gables, FL.

Newton, L. R., J. Nassar, and T. H. Fleming. 2003. Genetic population structure and mobility of two nectar-feeding bats from Venezuelan deserts: inferences from mitochondrial DNA. Molecular Ecology, 12:3191–3198.

Norberg, U. M., and J. M. V. Rayner. 1987. Ecological morphology and flight in bats (Mammalia; Chiroptera): wing adaptations, flight performance, foraging strategy, and echolocation. Philosophical Transactions of the Royal Society of London B, 316:335–427.

Ober, H. K., R. J. Steidl, and V. M. Dalton. 2005. Resource and spatial-use patterns of an endangered vertebrate pollinator, the lesser long-nosed bat. Journal of Wildlife Management, 69:1615–1622.

Olival, K. J. 2012. Evolutionary and ecological correlates of population genetic structure in bats. Pp. 267–316 *in*: Evolutionary History of Bats (G. F. Gunnell and N. B. Simmons, eds.). Cambridge University Press, Cambridge.

Ortega, J., and H. T. Arita. 1999. Structure and social dynamics of harem groups in *Artibeus jamaicensis* (Chiroptera: Phyllostomidae). Journal of Mammalogy, 80:1173–1185.

Ortega, J., and I. Castro-Arellano. 2001. *Artibeus jamaicensis*. Mammalian Species, 662:1–9.

Ortega, J., J. E. Maldonado, G. S. Wilkinson, H. T. Arita, and R. C. Fleischer. 2003. Male dominance, paternity, and relatedness in the Jamaican fruit-eating bat (*Artibeus jamaicensis*). Molecular Ecology, 12:2409–2415.

O'Shea, T.J., and M. A. Bogan, eds. 2003. Monitoring Trends in Bat Populations of the United States and Territories: Problems and Prospects. Information and Technology Report, USGS/BRD/ITR-2003-0003. U.S. Geological Survey, Biological Resources Discipline, [Fort Collins, CO].

O'Shea, T. J., D. J. Neubaum, M. A. Neubaum, P. M. Cryan, L. E. Ellison, T. R. Stanley, C. E. Rupprecht, W. J. Pape, and R. A. Bowen. 2011. Bat ecology and public health surveillance for rabies in an urbanizing region of Colorado. Urban Ecosystems, 14:665–697.

Oyler-McCance, S. J., J. A. Fike, P. M. Lukacs, D. W. Sparks, T. J. O'Shea, and J. O. Whitaker. 2018. Genetic mark-recapture improves estimates of maternity colony size for Indiana bats. Journal of Fish and Wildlife Management, 9:25–35.

Pavan, A. C., F. Martins, F. R. Santos, A. Ditchfield, and R. A. F. Redondo. 2011. Patterns of diversification in two species of short-tailed bats (*Carollia* Gray, 1838): the effects of historical fragmentation of Brazilian rainforests. Biological Journal of the Linnean Society, 102:527–539.

Pedersen, S. C. G. G. Kwiencinski, P. A. Larsen, M. N. Morton, R. A. Adams, H. H. Genoways, and V. J. Swier. 2009. Bats of Montserrat: population fluctuation and response to hurricanes and volcanoes, 1978–2005. Pp. 302–340 *in*: Island Bats: Evolution, Ecology, and Conservation (T. H. Fleming and P. A. Racey, eds.). University of Chicago Press, Chicago.

Pereira, M. J. R., J. T. Marques, and J. M. Palmeirim. 2010. Ecological responses of frugivorous bats to seasonal fluctuation in fruit availability in Amazonian forests. Biotropica, 42:680–687.

Pérez-Torres, J., and N. Cortés-Delgado. 2009. Murciélagos de la Reserva Natural La Montaña del Ocaso (Quindío, Colombia). Chiroptera Neotropical, 15:456–460.

Porter, T. A., and G. S. Wilkinson. 2001. Birth synchrony in greater spear-nosed bats (*Phyllostomus hastatus*). Journal of Zoology, 253:383–390.

Racey, P. A., and A. C. Entwistle. 2003. Conservation ecology of bats. Pp. 680–743 *in*: Bat Ecology (T. H. Kunz and M. B. Fenton, eds). University of Chicago Press, Chicago.

Ramírez, J. 2011. Population genetic structure of the lesser long-nosed bat (*Leptonycteris yerbabuenae*) in Arizona and Mexico. MS thesis, University of Arizona, Tucson.

Ransome, R. D. 1989. Population changes of greater horseshoe bats studied near Bristol over the past 26 years. Biological Journal of the Linnean Society, 38:71–82.

Rincón-Vargas, F., K. E. Stoner, R. M. Vigueras-Villasenor, J. M. Nassar, O. M. Chaves, and R. Hudson. 2013. Internal and external indicators of male reproduction in the lesser long-nosed bat *Leptonycteris yerbabuenae*. Journal of Mammalogy, 94:488–496.

Ripperger, S. P., M. Tschapka, E. K. V. Kalko, B. Rodriguez-Herrera, and F. Mayer. 2013. Life in a mosaic landscape: anthropogenic habitat fragmentation affects genetic population structure in a frugivorous bat species. Conservation Genetics, 14:925–934.

Ripperger, S. P., M. Tschapka, E. K. V. Kalko, B. Rodriguez-Herrera, and F. Mayer. 2014. Resisting habitat fragmentation: high genetic connectivity among populations of the frugivorous bat *Carollia castanea* in an agricultural landscape. Agriculture Ecosystems and Environment, 185:9–15.

Rocha, R., A. Lopez-Baucells, F. Z. Farneda, M. Groenenberg, P. E. D. Bobrowiec, M. Cabeza, J. M. Palmeirim, and C. F. J. Meyer. 2017. Consequences of a large-scale fragmentation experiment for Neotropical bats: disentangling the relative importance of local and landscape-scale effects. Landscape Ecology, 32:31–45.

Rossiter, S. J. 2009. Parentage and kinship analysis in bats. Pp. 695–713 *in*: Ecological and Behavioral Methods for the Study of Bats (T. H. Kunz and S. Parsons, eds.). Johns Hopkins University Press, Baltimore.

Sagot, M. 2016. Effects of range, habitat and roosting ecology in patterns of group association in bats. Pp 247–259 *in*: Sociality in Bats (J. Ortega, ed.). Springer International Publishing, Cham, Switzerland. DOI 10.1007/978-3-319-38953-0_12.

Sagot, M., C. D. Phillips, R. J. Baker, and R. D. Stevens. 2016. Human-modified habitats change patterns of population genetic structure and group relatedness in Peter's tent-roosting bats. Ecology and Evolution, 6:6050–6063.

Santos, M., L. F. Aguirre, L. B. Vázquez, and J. Ortega. 2003. *Phyllostomus hastatus*. Mammalian Species, 722:1–6.

Senior, P., R. K. Butlin, and J. D. Altringham. 2005. Sex and segregation in temperate bats. Proceedings of the Royal Society B, 272:2467–2473.

Silva Taboada, G. 1979. Los Murciélagos de Cuba. Editora de la Academia de Ciencias de Cuba, Havana.

Simal, F., C. de Lannoy, L. Garcia-Smith, O. Doest, J. A. de Freitas, F. Franken, I. Zaandam et al. 2015. Island-island and island-mainland movements of the Curacaoan long-nosed bat, *Leptonycteris curasoae*. Journal of Mammalogy, 96:579–590.

Simmons, N. B., and R. S. Voss. 1998. The mammals of Paracou, French Guiana: a Neotropical lowland rainforest fauna. Pt. 1: Bats. Bulletin of the American Museum of Natural History, no. 237. American Museum of Natural History, New York.

Stevens, R. D., M. R. Willig, and I. G. De Fox. 2004. Comparative community ecology of bats from eastern Paraguay: taxonomic, ecological, and biogeographic perspectives. Journal of Mammalogy, 85:698–707.

Stoner, K. E. 2005. Phyllostomid bat community structure and abundance in two contrasting tropical dry forests. Biotropica, 37:591–599.

Stoner, K. E., K. A. O. Salazar, R. C. R. Fernandez, and M. Quesada. 2003. Population dynamics, reproduction, and diet of the lesser long-nosed bat (*Leptonycteris curasoae*) in Jalisco, Mexico: implications for conservation. Biodiversity and Conservation, 12:357–373.

Straney, D. O., M. H. Smith, I. F. Greenbaum, and R. J. Baker. 1979. Biochemical genetics. Pp. 157–176 *in*: Biology of Bats of the New World Family Phyllostomatidae. Pt. 3 (R. J. Baker, J. K. Jones, and D. C. Carter, eds.). Special Publications of the Museum, Texas Tech University, no. 16. Texas Tech Press, Lubbock.

Taylor, P. J., S. M. Goodman, M. C. Schoeman, F. H. Ratrimomanarivo, and J. M. Lamb. 2012. Wing loading correlates negatively with genetic structuring of eight Afro-Malagasy bat species (Molossidae). Acta Chiropterologica, 14:53–62.

Villalobos-Chaves, D., J. Vargas Murillo, E. Rojas Valerio, B. W. Keeley, and B. Rodríguez-Herrera. 2016. Understory bat roosts, availability and occupation patterns in a Neotropical rainforest of Costa Rica. Revista Biología Tropical, 64: 1333–1343.

Voigt, C. C., and T. Kingston, eds. 2016. Bats in the Anthropocene: Conservation of Bats in a Changing World. Springer Open, Heidelberg.

Voss, R. S., D. W. Fleck, R. E. Strauss, P. M. Velazco, and N. B. Simmons. 2016. Roosting ecology of Amazonian bats: evidence for guild structure in hyperdiverse mammalian communities. American Museum Novitates, 3870:1–43.

Weigl, R. 2005. Longevity of mammals in captivity: from the living collections of the world. Kleine Senckenberg-Reihe, no. 48. Schweizerbart, Stuttgart .

Weimerskirch, H., J. C. Stahl, and P. Jouventin. 2008. The breeding biology and population dynamics of King Penguins *Aptenodytes patagonica* on the Crozet Islands. Ibis, 134: 107–117.

Wiles, G. J., and A. P. Brooke. 2009. Conservation threats to bats in the Pacific islands and insular Southeast Asia. Pp. 405–459 *in*: Island Bats: Evolution, Ecology, and Conservation (T. H. Fleming and P. A. Racey, eds.). University of Chicago Press, Chicago.

Wilkinson, G. S. 1985. The social organization of the common vampire bat. II. Mating system, genetic structure, and relatedness. Behavioral Ecology and Sociobiology, 17:123–134.

Wilkinson, G. S., G.G. Carter, K. M. Bohn and D. M. Adams. 2016. Non-kin cooperation in bats. Philosophical Transactions of the Royal Society B, 371:20150095.

Wilkinson, G. S., and T. H. Fleming. 1996. Migration and evolution of lesser long-nosed bats *Leptonycteris curasoae*, inferred from mitochondrial DNA. Molecular Ecology, 5:329–339.

Williams, C. F. 1986. Social organization of the bat *Carollia perspicillata* (Chiroptera, Phyllostomidae). Ethology, 71: 265–282.

Wooler, R. D., J. S. Bradley, and J. P. Croxall. 1992. Long-term population studies of seabirds. Trends in Ecology and Evolution, 7:111–114.

20

Community Ecology

Richard D. Stevens
and Sergio Estrada-Villegas

Introduction

As many of the chapters in this volume assert, Phyllostomidae is an extremely species-rich taxon of remarkable ecological, phylogenetic, and morphological diversity that is the product of millions of years of evolution. Indeed, Neotropical communities dominated by phylostomids are arguably the most trophically diverse mammalian communities on earth (Patterson et al. 2003). Much of the rich history of studying phyllostomids has focused on their community ecology. Because of their diversity, communities that are dominated by phyllostomids represent ideal model systems for trying to tease apart the mechanistic basis to community ecology in general and that of Neotropical bat communities in particular. Nonetheless, it is the remarkable high diversity that distinguishes phyllostomid communities that also presents a substantive challenge when trying to mechanistically explain their structure.

Two main avenues have traditionally been explored when studying phyllostomid bat communities: (1) better characterization and minimization of sampling biases and (2) biological processes that determine their structure. Neotropical bat communities are notoriously difficult to sample in an adequate and unbiased fashion, and much effort and care needs to be employed to minimize these limitations. Indeed, volumes have been written to this effect. Comprehensive references to these can be found in Bergallo et al. (2003), Kalko (1998), Kingston (2009), Meyer et al. (2011, 2015), Moreno and Halffter (2000), and Stevens (2013). Similarly, much effort has been employed to understand why phyllostomid communities are structured the way they are. Since the first synthesis of phyllostomid community ecology (Humphry and Bonaccorso 1979), hundreds of manuscripts have been published describing and comparing which biological processes structure phyllostomid communities. Our aim is to summarize these studies and to place this large body of literature into a new conceptual framework that hopefully will better organize these findings as well as point out areas that demand future attention.

Initial Caveat

Phyllostomidae is a Neotropical family of bats whose geographic distribution ranges from the middle of Argentina to the southern United States (Willig and Selcer 1989). Within this geographic range, eight other families of bats are also distributed, but phyllostomids dominate community structure in terms of species richness and abundance of individuals (Stevens 2004). To this end, the vast majority of studies focusing on Neotropical bat communities have not focused exclusively on Phyllostomidae but on the entire assemblage of bats. Because bats from these other families typically make up only a minor contribution to these communities, and are likely rarely sampled with mist nets, we feel that we can use studies that focused on all bats to guide our understanding of phyllostomid community structure.

Foundations of Mechanistic Study of Phyllostomid Community Organization

Examination of the structure of phyllostomid bat communities began as detailed descriptions of bat assemblages became available in the Neotropics decades ago (Bonaccorso 1975; Fleming et al. 1972; LaVal and Fitch 1977; Reis 1984; Thomas 1972; Willig 1983). These studies provided the first patterns of community structure and concluded that phyllostomid bat communities were distinct not only in their relatively high species richness when compared to bat communities in other parts of the world (Findley and Wilson 1983) but also in their impressively high trophic diversity (Bonaccorso 1975). Indeed, Phyllostomidae has the greatest trophic and morphological diversity when compared to any other family-level group within Mammalia (Patterson et al. 2003).

From the late 1960s to the 1990s, a broad and robust conceptual foundation of community ecology was formed, and it offered a substantial springboard from which to improve our understanding of the mechanistic basis to the structure of phyllostomid bat communities (fig. 20.1). Much of this foundation, and many of the studies conducted during that time, addressed various aspects of niche theory. Indeed, addressing ideas introduced by Hutchinson (1959) on size ratios and those put forward by MacArthur and Levins (1967) on limiting similarity formed much of the basis to these efforts.

For example, Handley (1967) and then Bonaccorso (1975) first suggested vertical partitioning of different layers of the forest canopy in tropical phyllostomid bat communities (fig. 20.1A), and while certainly not ubiquitous across all tropical bat communities, vertical partitioning is a common pattern describing community structure (Kalko and Handley 2001; Pereira et al. 2010).

A number of early studies also used niche matrices (fig. 20.1B) to describe the extent of organization of communities given an absence of details on diet or even distribution of rare species within communities (Fleming 1986; LaVal and Fitch 1977; Willig 1986). Such matrices were typically arranged such that different guilds spanned one dimension and body size spanned the other. The aim was to examine regularities in the distribution of body sizes within guilds. According to idealized Hutchinsonian size ratios (Hutchinson 1959) and MacArthurian limiting similarities (MacArthur and Levins 1967), there should be a uniform distribution of species across niche cells, if not just one species per cell. Such expectations were not found for phyllostomids (Fleming 1986; Willig 1986). Oftentimes in the same community there were numerous empty cells as well as cells that were occupied by numerous species. Such patterns may result from lack of competition, inadequate sampling of communities, uneven food-size distributions, or random assembly from a regional fauna that also exhibits an uneven distribution of body sizes (Willig 1986). Moreover, many phyllostomids are, to a varying degree, euryphagic (i.e., consuming a wide variety of food types). For example, *Chrotopterus auritus* consumes not only arthropods but also fruit and pollen (Munin et al. 2012), and *Sturnira lilium*, while foraging for mostly fruit, consumes appreciable amounts of insects, especially during certain portions of the year (Herrera et al. 2002). An equally important outcome of these studies was the formalization of ideas as to how a number of phenomena not related to interspecific interactions could modify patterns predicted by theory. Bats are highly mobile, and rescue effects may be common (Willig and Moulton 1989), thereby allowing competitors to coexist. Competitive interactions may take place, but unrecognized patterns, such as low morphological dispersion within and between communities, may also manifest (Stevens and Willig 1999). Bats may not even compete, or interactions may not be sufficiently strong

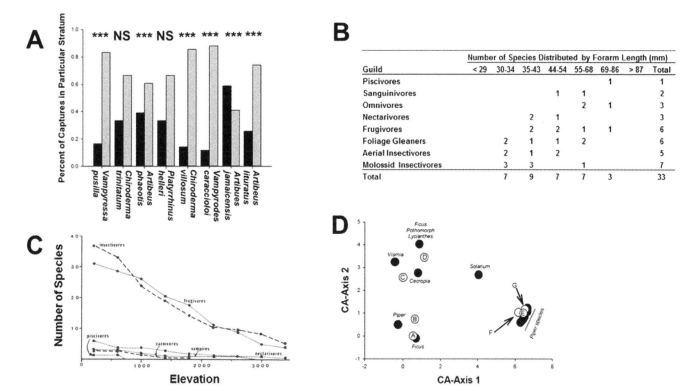

Figure 20.1. Examples of insights stemming from classical studies of phyllostomid communities. (A) In a number of phyllostomid communities, species are active in different portions of forest strata. Black bars represent the percent of individuals per species captured at ground level, whereas *gray* represents number of individuals captured in the forest canopy. *Asterisks* represent a significant difference, whereas NS indicates no significant difference. Figure produced from the data in Bonaccorso (1975). (B) Species in Neotropical bat communities do not follow the expectation of an even distribution of body sizes across cells of a niche matrix. Table was modified from Willig (1986). (C) Elevational gradients in bat diversity in Peru (redrawn from Graham [1983]). (D) Relationships between frugivorous phyllostomids and the fruit they consume. *Black dots* are fruit taxa, *white dots* are bat taxa. Strong differences between large stenodermatines, small stenodermatines, and carolliines in terms of dietary affinities exist in Peruvian communities. Data used to construct this figure come from Palmeirim et al.(1989).

enough to generate community-wide patterns in morphology or body size (Willig and Moulton 1989). The overarching conclusion of this research was that little evidence exists to suggest that competition causes patterns of body size in phyllostomid communities consistent with the expectations of niche matrices.

Niche matrices are not the only means of evaluating structure based on morphology or body size. A better way to examine structure may be to determine the degree of dispersion of phenotypes among species by comparing several morphological features simultaneously. For example, if morphological similarity is related to ecological similarity and competition is an important driver of community structure, then there should be a regular distribution of species in multivariate morphological space (Willig and Moulton 1989). The seminal paper comparing morphological structure of real communities to those generated from a null model found that such patterns were no different from random

expectations and that patterns consistent with effects of competitive interactions did not characterize the morphological structure of bat communities, at least in Northeast Brazil (Willig and Moulton 1989). Subsequent studies confirmed these findings for a larger number of Neotropical bat communities both at local and larger spatial scales (Arita 1996, 1997; Moreno et al. 2006; Stevens and Willig 1999).

Some groups of phyllostomids are better than others for examining effects of competitive interactions. For example, most frugivorous phyllostomids consume many fruits with small seeds that pass quickly through the digestive system (Palmeirim et al. 1989). This greatly facilitates direct comparisons of diets and the extent of overlaps and potential competition, at least for this portion of the diet (fig. 20.1D). Palmeirim et al. (1989) were perhaps the first to attempt to examine community-wide dietary overlap and included both birds and bats. Bats and birds tended to exhibit quite distinct diets, a pattern that has been repeated across studies (Fleming and Muchhala 2008; Gorchov et al. 1995). Moreover, bats exhibited lower levels of overlap than birds, again suggesting that competitive interactions do not drive community organization.

Despite the inability of competition theory to explain the structure of phyllostomid communities, other foundational approaches examined how environmental characteristics could determine why a set of species coexisted. In the Peruvian Andes, Graham (1983) demonstrated strong elevation gradients whereby

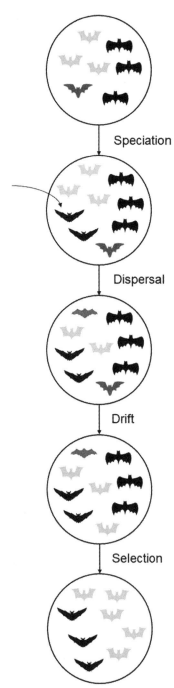

Figure 20.2. Schematic model of how the four main higher-level fundamental processes affecting community structure may increase or decrease richness and abundance. Each type of silhouette represents a variant or species. Speciation adds new variants to a community via diversification. Dispersal allows new variants and more individuals of preexisting variants to invade the community. Drift causes random changes in the abundances of species within a community, sometimes eliminating rare species. Selection can either enhance or diminish the abundance of particular species based on their replication rates and their morphological or physiological traits. The order of the higher-level fundamental processes is arbitrary; all processes can take place simultaneously. Modified from Vellend 2016.

species richness decreased with increased elevation (fig. 20.1C). Related to this elevational gradient were decreases in temperature, foliage height diversity, and food resources at higher elevations. The concomitant decrease of these three variables with species richness makes it difficult to disentangle which were the driving variables. Moreover, many species likely resulted from diversification in the lowlands and were bound to have low-elevation distributions (Graham 1983). Thus, this particular elevation gradient probably results from a complex mix of historical effects that determine the contemporary niches of species and from present-day environmental gradients that determine how climate and resources change with elevation (Patterson et al. 1996).

Another niche characteristic that has been highly influential on structure of phyllostomid communities is sensitivity to anthropogenic habitat modification and fragmentation. Efficacy of phyllostomid bats, in particular those in the subfamily Phyllostominae, as indicators of anthropogenic habitat modification has been well-established (Fenton et al. 1992). Pioneering and comprehensive work on community-wide effects of habitat fragmentation was conducted in the Las Tuxtlas region of southern Mexico (Estrada et al. 1993). In Los Tuxtlas, as well in many other localities in the Neotropics, bats respond strongly to habitat modification on a species-by-species basis as well as at the community level. Studies on habitat fragmentation also evidenced that size of fragments and isolation among them are important determinants of phyllostomid community diversity and evenness (Cosson et al. 1999).

Much of the seminal work on the mechanistic basis to phyllostomid community organization focused almost exclusively on niche characteristics (Fleming 1986; Willig 1986). Subsequent work on phyllostomid communities has been voluminous and has matured to address a wide number of structuring mechanisms. Indeed, we found more than 230 manuscripts that dealt with the mechanistic basis of phyllostomid community organization (supplementary table S20.1). To amass this body of literature, we began with a Web of Knowledge search on 29 January 2016 using the keywords "bat" and "community." Upon identifying an appropriate publication, we also searched its reference section and looked for other publications that could be appropriate. We also searched in our personal libraries for studies on community ecology of Phyllostomidae, looked for references therein, and finally searched the World Wide Web for unpublished undergraduate and graduate theses. To do so, we typed the name of the countries of Central America, South America, and the Caribbean with the keywords "murciélago" and "comunidades," "ensambles," or "ensamblajes." Maria Joao Pereira kindly provided studies published in Portuguese. We are aware that some studies could have slipped past our search, but we are confident that we reviewed the majority of the literature, and this literature represents the trends of community ecology of Phyllostomidae.

Organizing such a multifaceted literature can be complicated. Nonetheless, a recent conceptual framework based on four higher-level fundamental processes can furnish many of the mechanisms on phyllostomid communities that have been addressed to date (Vellend 2016). This conceptual framework can also point out areas that lack substantive and highly needed research (Vellend 2016). Therefore, we used this new framework to organize this literature.

New Conceptual Framework

Recently Mark Vellend and colleagues (Vellend and Gerber 2005; Vellend and Orrock 2009; Vellend 2016) constructed a conceptual framework for community ecology that organizes all potential mechanisms that act on communities as stemming from a combination of four higher-level fundamental processes: (1) selection, (2) speciation, (3) drift, and (4) dispersal (fig. 20.2). All contemporary and historical models of community organization incorporate at least one or more of these four processes. The field of community ecology in general, but also the study of phyllostomid bat communities in particular, has traditionally focused on selection (fig. 20.3, table S20.1). In the context of the framework, selection is simply the differential success of different species. It is analogous to natural selection on individuals that leads to a particular population-level characteristic (i.e., relative abundance). Communities are composed of variants (species), and these variants reproduce or replicate (i.e., change their abundances) at different rates. In this case, the characteristic under selection is simply the species identity. Those species that are best suited for a particular environment will exhibit the highest nonzero abundance (high replication rates) in the community whereas others that are less suited

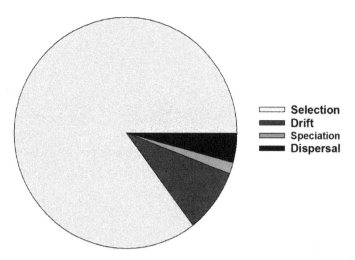

Figure 20.3. Distribution of effort of studies addressing the four main higher-level fundamental processes affecting community structure. Selection has by far been studied the most whereas speciation and dispersal the least.

will have lower nonzero abundance (low replication rates). If a particular species is ill-suited in a particular community, it will exhibit zero abundance and will not occur at that site in the absence of dispersal. Most of traditional community ecology falls under the purview of selection. For example, species may be differentially adapted to different environmental regimes, or biotic interactions may determine their relative success in different communities, and thus they will have different abundances across communities (Vellend 2016).

Speciation, drift and dispersal are less-studied processes in community ecology but have very important effects on community structure (Vellend 2016). Speciation is the addition of species to the regional species pool via diversification. Many questions in community ecology address phenomena that occur within local communities and are of relatively rapid temporal duration. Thus, often times understanding how species in particular communities came about does little to inform us about how they coexist after the diversification event. Nonetheless, when comparisons are made across communities with different species pools, then assessing differential speciation becomes important for interpreting differences in terms of species richness and composition at the regional scale. Differential diversification becomes central in macroevolutionary, macroecological, or biogeographical comparisons because differences in species richness and community composition between sites depend on rates at which species pools are formed (Ricklefs 1987).

Drift (aka ecological drift) was described by Hubbell (2001) as cumulative small changes in species composition related to stochasticity in birth, death, and reproduction that contribute to the abundances of each species in a community. When the number of individuals in a community is finite, stochastic generational variation in birth, death, and reproduction across all species will cause stochastic variation in species composition (Hubbell 2001). As with genetic drift, the fewer the number of individuals within a community, the larger the stochastic fluctuations in species composition. In contrast, communities composed of relatively greater numbers of individuals will vary less and changes in species composition due to drift will be smaller. For communities composed of the same number of individuals, those with more species will exhibit greater amounts of drift because, on average, each species has lower abundance. Moreover, the probability of any species dominating the composition of the community is directly related to its initial abundance and the total number of individuals in the community. Drift by itself is of primary importance when the total abundance in communities is small. Nonetheless, this process interacts in important ways with other processes (e.g., community isolation) and can have important impacts on community organization (Vellend 2016).

Finally, dispersal can also have important impacts on community structure. Dispersal is the interchange of individuals among communities and serves as an important rescue effect (Brown and Kodric-Brown 1977) that prevents local extinction of species in the recipient community. In many source-sink systems, dispersal is the only means for some species to prevent local extinction in sink communities (Shmida and Wilson 1985). In addition, dispersal serves to homogenize variation in species composition at the regional level. When dispersal is intense/pervasive/widespread in the absence of the other three processes, species composition of all communities within a regional metacommunity will eventually become identical (Vellend 2016).

Relating Existing Approaches to Higher-Level Processes of Selection, Speciation, Drift, and Dispersal on Phyllostomid Communities

Focus on higher-level processes and their effects on phyllostomid community structure has not been even (fig. 20.3, table S20.1). Many of the research themes of individual studies address lower-level processes associated with selection (fig. 20.4). This is primarily due to the important role that competition theory has played in the development of community ecology in general (Hubbell 2001). Nonetheless, more recently, other research themes—especially those addressing the higher-level processes of speciation, drift, and dispersal—have increased in frequency.

Selection

Most studies of phyllostomid community ecology can be grouped under the higher-level process of selection because most factors that modify the ranking of species abundances can be considered as factors that differentially select among species (Vellend 2016). As stated earlier, much of the historical development of the study of phyllostomid communities demonstrated that predictions from competition theory do not provide, in general, accurate descriptions of community patterns. However, the study of factors that affect the relative abundance of phyllostomid species beyond competition theory has also been quite prolific (see table S20.1). At a local scale, important subsets of studies have been devoted to determining the effect of food type, habitat type, vertical vegetation complexity, and climatic variation on community structure. Another important subset has addressed the effect of elevation and latitude on communities. However, much research on phyllostomid community ecology, especially in the last decades, has been devoted to quantifying how anthropogenic habitat modifications have affected species distribution and abundance. Here we summarize an important part of this literature, going beyond the classic studies cited above, and provide an overview on how these factors might be selective forces that shape community structure.

Food, Habitat, Vertical Stratification, and Climatic Variation

Food type and food availability can have profound consequences on whether species have high or low abundance within communities (Kalko 1998). Although early analyses were unable to show that niche differences between species were structured by competition, food type and food availability exert a strong effect on species abundances across phyllosto-

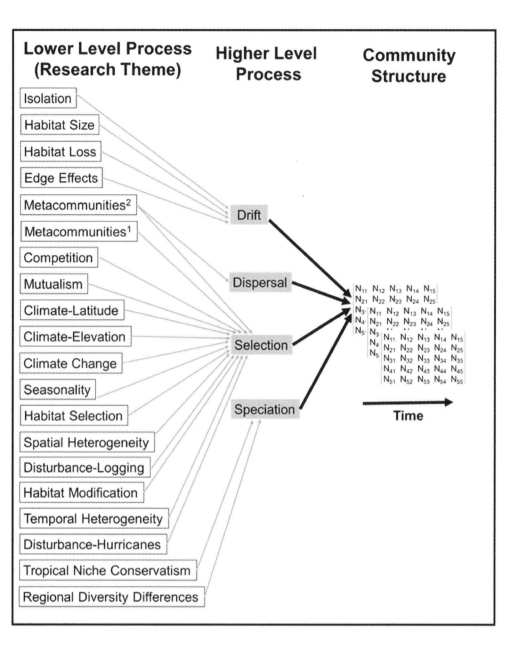

Figure 20.4. Popular themes (lower-level processes) pursued by phyllostomid ecologists when investigating the mechanistic basis to phyllostomid community ecology. Community structure is represented by spatial variation in species composition and is characterized by a species-by-site matrix of either presence/absence or abundance data. Different matrices represent different points in time. Spatial and temporal variation in species composition is driven by higher-level processes of selection, speciation, drift, and dispersal. These higher-level processes sum a number of different lower-level processes that often are the focus of particular research themes. As the figure demonstrates, much of the focus of phyllostomid ecologists has been on selection, with little focus on the remaining higher-level processes (see fig. 20.3). *Metacommunities[1]* refers to elements of metacommunity structure, *Metacommunities[2]* refers to variance decomposition.

mid communities (Fleming 1986; Saldaña-Vázquez 2014). Different groups of phyllostomids show clear preferences for particular types of food (i.e., core food taxa) despite dietary overlap when preferred resources become scarce (Saldaña-Vázquez 2014; Sánchez et al. 2012). For example, there is strong correlative and experimental support showing that species from the genus *Carollia* prefer fruits from the genus *Piper*, species from the genus *Sturnira* prefer fruits from the genus *Solanum*, and species from the tribe Ectophylline prefer fruits from the genus *Ficus* and *Cecropia* (Heithaus et al. 1975). These strong dietary associations are not only a contemporary phenomenon, they also show a strong evolutionary basis (Giannini and Kalko 2004). More-

over, these relationships are reflected in a strong positive relationship between food size and body size for frugivorous phyllostomids (Heithaus et al. 1975; Kalko, Herre et al. 1996), and a strong pattern from insectivory to carnivory with increased body size in animalivorous phyllostomids (Giannini and Kalko 2004; chap. 14, this vol.). Therefore, it comes as no surprise that food preferences are prevalent in phyllostomid communities in lowland forests (Giannini and Kalko 2004), mountain and cloud forests (Estrada-Villegas et al. 2010), dry forests (Novoa et al. 2011), savannas and Cerrados (Willig et al. 1993), subtropical forests (Sánchez et al. 2012), and likely in all habitats across the geographic distribution of Phyllostomidae.

Food abundance is strongly related to variation in the abundance of bat species among communities through space or time. For example, Kalko, Handley et al. (1996) commented that the relative abundance of *Artibeus* and other fig-eating bats increased substantially where the abundance of *Ficus* was high across Barro Colorado Island, Panama. Conversely, areas with low fig abundance were related to low relative bat abundance (Kalko, Handley et al. 1996). This same pattern was observed by Aguirre et al. (2003) in Bolivia; food availability influenced species turnover across the landscape not only for fig-eating bats but also for gleaning insectivores. Likewise, Tschapka (2004) related relative abundance of glossophagines and the abundance of their food source in La Selva, Costa Rica, and found that relative abundance was correlated with fluctuations in energy availability (i.e., food availability) across time. At a large spatial scale, Saldaña-Vázquez et al. (2013) showed that plant availability between eco-regions across the Americas affects community composition due to strong dietary preferences. This study concluded that the availability of *Solanum* and *Ficus* with respect to other fruits consumed by bats in each eco-region explained the presence and, potentially, the abundance of *Sturnira* and *Artibeus*, respectively. In sum, food type and food availability are crucial factors that structure phyllostomid communities because many species have strong preferences, and the type of food available across the landscape will ultimately affect differential nonzero abundances among species.

The distribution and abundance of vegetation inside forests is a differential selective pressure because it can determine which species are able to maneuver and navigate (Schnitzler and Kalko 2001). Although niche theory could be invoked to explain vertical stratification in a number of phyllostomid communities (e.g., Lim and Engstrom 2001), it seems more productive to first determine the selective pressure that vegetation clutter exerts on community structure before adducing niche differentiation via limiting similarity. A number of studies have demonstrated vertical stratification, as described by the pioneer studies of Handley (1967) and Bonaccorso (1975). It is usually assumed that vegetation clutter contributes to vertical stratification. For example, Weise (2007) showed that gleaning insectivores were predominantly found in lower strata while larger frugivores were usually found in the upper strata in Panama. The same pattern was found for both guilds

in two other locations in Brazil (Bernard 2001; Pereira et al. 2010). Inter-habitat comparisons in Amazonia have also confirmed vertical stratification; Pereira et al. (2010) demonstrated that community composition between low strata and high strata is significantly different in terra firme, várzea, and igapó. Nevertheless, exceptions exist whereby phyllostomids do not stratify vertically (Scultori et al. 2008; Weber et al. 2011), and the absence of such a pattern is exactly the kind of result that would contribute to the "file drawer" effect (i.e., failure to publish nonsignificant results, Rosenthal 1979). Moreover, some species seem to be idiosyncratic with regards to vertical stratification when compared among locations (Kalko and Handley 2001; Weber et al. 2011; Weise 2007). The presence of these idiosyncrasies led Voigt (2010) and Rex et al. (2011) to question the long-standing idea of stratification of foraging activity among frugivores. They tested the hypothesis that each species uses a particular fraction of the vertical profile of the forest in Costa Rica and Ecuador by comparing stable carbon isotopes from bats and from available fruits and insects along the vertical profile of the forest (the $^{13}C/^{12}C$ ratio of plant and insect biomass decreases from the canopy to the understory). Indeed, they concluded that there is no correlation between where species are captured and the source of the carbon they consume. Therefore, bats should not be grouped into guilds that putatively forage at a particular stratum, and many bats, both large and small, use the understory just for commuting purposes and not for foraging (Rex et al. 2011). In sum, from a multisensorial stand point, vegetation clutter is likely an important selective factor reflecting vertical stratification based on aerodynamic responses to clutter, rather than as responses to differences in food availability along the forest profile.

Climatic variation is a strong selective force that structures phyllostomid communities. In its most basic form, spatial distribution of environmental conditions drives spatial variation of species richness and community composition across phyllostomid communities. For example, diversity of phyllostomid communities exhibits a strong latitudinal gradient (Willig et al. 2009) that is likely related to either contemporary or historical effects of climate. The current evidence suggests that phyllostomids are of tropical origin (Rojas et al. 2016) and have subsequently diversified from tropical environs to subtropical and even temperate latitudes. But despite the wide latitudinal range of the family, phyllostomids are strict thermoregulators (i.e., do not use long-term hibernation or short-term torpor), and this factor limits their geographical distribution across the latitudinal gradient (McNab 1969). Indeed, temperature seasonality is an important influence on species composition of phyllostomid communities and is likely an important driver of structure, especially toward the limit of the geographic range of the family (Stevens 2013). Temperature not only selects species ranges, it also has a strong selective impact on richness and composition (Stevens and Amarilla-Stevens 2012). Phylogenetic diversity is also strongly affected by temperature seasonality; in the Atlantic Forest of Brazil and in Paraguay, higher seasonality was correlated with phylogenetic clustering, and lower seasonality was correlated with phylogenetic overdispersion (Stevens and Gavilanez 2015).

Seasonality of precipitation may be a selective force as important as seasonality of temperature (Pereira et al. 2009; Sánchez-Palomino et al. 1993). Much of the Neotropics experiences distinct wet and dry seasons, with some dry seasons being protracted, and certain phyllostomid species respond to these oscillations in rainfall by regional migrations to other environs with more resources (Esbérard et al. 2011; Fleming 1988) or by simple changes in local abundances between seasons (Mello 2009). Rainfall not only affects resource availability for both frugivores and nectarivores (Moya et al. 2008; Weise 2007), it also affects local habitat availability. For example, Pereira et al. (2009) showed that seasonal flooding affected both várzea and terra firme forests in Amazonia; there was a marked difference in composition between habitats due to seasonality, in part due to the loss of roosting sites or foraging grounds for understory bats due to the flooding of the várzea. Additionally, seasonality, coupled with habitat area and configuration, had a differential effect on guild structure in a fragmented system in the Peruvian Amazon (Klingbeil and Willig 2010). Despite the fact that some studies have failed to detect a strong effect of seasonality on community structure, probably due to comparisons between sites that differ strongly in forest composition (Falcão et al. 2014) or fragment size (Ortêncio-Filho et al. 2014), precipitation commonly exerts a selective pressure on community structure. In fact, because community composition of phyllostomids responds to both temperature and precipitation (Ruggiero and Kitzberger 2004), seasonality may be

inescapable across the entire geographic range of Phyllostomidae, and it may be a ubiquitous climatic factor that affects the structure of their communities.

Elevation

Another strong environmental selective force on phyllostomid communities is elevation. There are a number of examples of niche-related responses of phyllostomids to different environmental conditions that are correlated with elevation (Graham 1983; Patterson et al. 1996; Sánchez-Cordero 2001), but the relationship between elevation and structure has its nuances (Cisneros et al. 2014; McCain 2007a, 2007b). For example, richness declines with elevation, but this response is inconsistent and not the only form of variation. In fact, mid-elevational peaks in richness occur with equal frequency along elevation gradients, especially when the effect of area across elevation is taken into account (McCain 2007b). However, changes in species richness are consistent when the underlying environmental gradients associated with elevation (e.g., climate) are taken into consideration (McCain 2007a). There are two interesting and contrasting patterns of species richness along elevation gradients. In the first, phyllostomid richness and abundance peaks at low elevation when temperature and water availability are high at lower elevation (McCain 2007a). In the second, species richness peaks at mid-elevation when temperature is high but water availability is low at lower elevation (McCain 2007a). Moreover, taxonomic, phylogenetic, and functional diversity covary positively along an elevational gradient in Peru, but different dimensions of biodiversity neither peak at mid-elevation nor follow the linear decline exhibited by species richness (Cisneros et al. 2014; Graham 1983). Similar conclusions with regards to taxonomic and functional diversity were found for gradients in Colombia and Venezuela but using number of guilds as the basis of functional diversity (Bejarano-Bonilla et al. 2007; Soriano 2000). Furthermore, other aspects of community structure are also affected by elevation. Some metacommunities exhibit a nested distribution, in which higher-elevation communities represent subsets of lower-elevation communities (Patterson et al. 1996; Presley et al. 2012) while others exhibit a clumped (i.e., Clementian) distribution in which nonoverlapping sets of species are found along the elevation gradient (Ríos Blanco 2014). The former pattern might stem from the ability of fewer species to thermoregulate as elevation increases; the latter might stem from a combination of thermoregulation ability and the consequence of anthropogenic habitat modification on phyllostomid community structure (Presley et al. 2012; Ríos Blanco 2014).

Anthropogenic Habitat Modification

Disturbance due to anthropogenic habitat modification is also a very important selective force that has significant impacts on phyllostomid communities. After the initial descriptions of responses of phyllostomid bats to deforestation (Fenton et al. 1992; Brosset et al. 1996), many studies have assessed effects of anthropogenic habitat modification on the structure of phyllostomid communities. In fact, the majority of studies involving the higher-level process of selection, and for that matter speciation, drift, and dispersal, have focused on describing and predicting changes in community structure after anthropogenic habitat modifications. Undeniably, assessing how habitat modifications affect phyllostomid communities has greatly stimulated the study of bat ecology in the last decades. The most general finding across studies is that species composition changes with anthropogenic habitat modification, but responses are complex (Meyer et al. 2016). For example, while some species increase in abundance, others decrease, and these changes depend on the kind of habitat modification, type of matrix, geographic location and species identity (Meyer et al. 2016). Moreover, responses appear to be scale dependent, spatially heterogeneous, and temporally dynamic (Moreno and Halffter 2001; Gorresen and Willig 2004).

Effects of anthropogenic habitat modification are multifaceted. It is paramount to distinguish short-term effects of cutting down forest (i.e., response to logging) from long-term effects of modifying a primary forest into that of secondary forest (i.e., habitat modification), from the effects of altering habitat heterogeneity and structure of the landscape (i.e., fragmentation). These three factors affect species composition but represent different forms of selection that may drive changes in relative success of species. The most common bats in a community are often relatively unaffected by logging in terms of their abundance (Castro-Arellano et al. 2007). Nonetheless, of the rarer species, animalivorous bats tend to decrease while the common frugivorous and nectarivorous species tend to increase in abundance after logging (Clarke et al. 2005). As a result, logging can result in increased diversity and decreased dominance of bat assemblages (Castro-Arellano et al. 2007), at least in the short term. Effects of logging are often on relatively small spatial scales and may not necessarily manifest as changes in abundance or reductions in species richness at the landscape level. Finally, logging also can affect the activity patterns of species. For example, in lowland tropical forest of Brazil, activity patterns of frugivores overlap significantly in primary forest, whereas overlap is random in reduced-impact logging sites. These differences may be due to increased predation risk and reduced feeding areas in reduced-impact logging sites compared with primary forest (Castro-Arellano et al. 2009). Activity patterns of aerial insectivores, gleaning animalivores, and nectarivores exhibited no change between logged and unlogged areas, at least in this study.

Long-term effects of habitat modification on phyllostomids are similar. Bats tend to exhibit higher species richness in old-growth forest compared to modified habitats (Estrada-Villegas and Ramírez 2013; Pérez-Torres and Ahumada P. 2004). Frugivorous, and in some cases nectarivorous, bats exhibit higher abundance in modified environments (Delaval and Charles-Dominique 2006), regardless of whether modified habitats are secondary forests or coffee, cacao, or oil palm plantations (Faria et al. 2006; Harvey and González Villalobos 2007). Animal and insect gleaners, however, show lower abundances in modified environments (Fenton et al. 1992). The type and permeability of edges between modified habitats and contiguous forests can also act as a selection force on species abundances (Cortés-Delgado and Pérez-Torres 2011). Overall, the trend in agricultural landscapes is that, as cultivation intensifies, forest cover and food resources decrease, phyllostomid richness decreases, and the bat community is dominated by a subset of frugivores (Pérez-Torres et al. 2009; Reis et al. 2006).

The trend in community structure across successional forests mirrors the trend through agricultural intensification; bat richness increases and composition diversifies as pastures regain forest cover, and as forests age and trees gain stature (MacSwiney G. et al. 2007). For example, de la Peña-Cuéllar et al. (2012) showed that species richness increased while community dissimilarity decreased as stands aged in a secondary forest in Mexico. Similarly, Castro-Luna et al. (2007), in a lowland forest in Mexico, found that composition shifted from almost an exclusively frugivore community in secondary vegetation to a community composed of both animalivores and frugivores in old-growth forests. Likewise, habitat modification can have a significant effect on activity patterns. For example, Montaño-Centellas et al. (2015) found that, in the most disturbed habitats in the Andes of Bolivia, *Carollia perspicillata* and *Sturnira lilium* are segregated temporally, but this segregation erodes in more forested habitats. Finally, Avila-Cabadilla et al. (2014) showed that the effect of succession is modified by seasonality; in three dry forests across the Neotropics, they found that frugivores were more abundant in early successional stages in dry-season dry forests, whereas nectarivores were more abundant in the same stage of succession but during the rainy season in Mexico and the dry season in Brazil. Indeed, the consequences of landscape modification may be exacerbated by seasonality (Klingbeil and Willig 2010). Seasonality can cause modified landscapes to have greater seasonal fluctuations in species composition than do their continuous forest counterparts (García-García and Santos-Moreno 2014). Seasonality can even switch the guild or which dimension of biodiversity is most affected by habitat modification (Cisneros et al. 2015).

The complexity and variety of responses to man-made habitat modifications is also evidenced by the lack of congruence between the aforementioned studies and others that have demonstrated weak or no differences in species richness or community composition between modified areas and contiguous forests (Castro-Luna et al. 2007; Estrada and Coates-Estrada 2002). For example, in a meta-analysis by García-Morales et al. (2013), the difference in frequency of Phyllostomidae between well-preserved forests and human-use areas demonstrated that habitat modification had a positive effect on phyllostomid abundance. However, this result is driven by the large number of frugivores and nectarivores that benefit from the increased abundance of resources in modified habitats (García-Morales et al. 2013). When the analysis was broken down by guilds, the occurrence of animal gleaners was lower in human-use areas. In sum, all studies that have assessed direction and magnitude of response to anthropogenic habitat modification have demonstrated that species abundances change in comparison to unmodified habitats. This change, by definition, is the selective effect that habitat modifications exerts on phyllostomid communities (Vellend 2016).

Man-made fragmentation and landscape-level effects of changes in heterogeneity also have significant impacts on phyllostomid communities. Indeed, configuration, quality, and the age of the matrix that surrounds fragments are important determinants of responses to anthropogenic modification (Meyer et al. 2016). For example, the relative importance of area and isolation appears to change relative to the contrast of the matrix. In relatively lower-contrast matrices such as in agricultural landscapes, fragment area is more important for predicting aspects of bat species composition, whereas in high-contrast matrices such as water, isolation is a better predictor of such characteristics (Mendenhall et al. 2014; Meyer and Kalko 2008). In landscapes where the effect of fragment area is weak, isolation or forest structure and diversity between fragments exert an important influence on richness (Estrada et al. 1993; Reis et al. 2006). Moreover, much of the variation among studies, over and beyond species-specific idiosyncratic responses to fragmentation, may be accounted for by differences in contrast between the matrix and habitat fragments, which in itself may form a strong selection gradient acting on community structure. For example, a matrix composed of agricultural monocultures may become a high-contrast/low-quality habitat that benefits some species but selects against others (Farneda et al. 2015). Even a matrix of secondary forest may have sufficient contrast so as to select for or against different species, so the matrix ends up providing a different set of selective forces relative to the original habitat (Mendenhall et al. 2014). It is important to point out that effects of a high-contrast matrix weaken in savannas, which are naturally fragmented systems (Bernard and Fenton 2002).

The differential response among guilds to fragmentation parallels the effects among guilds found in logging and habitat modification studies; frugivores increase in richness and abundance after fragmentation, while animalivores decrease in richness and abundance after fragmentation (Meyer and Kalko 2008; Schulze et al. 2000). Animalivorous bats are negatively affected by fragmentation because they are less mobile than other guilds, lower vagility reduces bat ability to move into or out of fragments (Meyer and Kalko 2008), and prey density may be lower inside fragments (Gorresen and Willig 2004). Moreover, animalivores' responses are also contingent on landscape configuration across spatial scales; Klingbeil and Willig (2009) showed that, as forest edges increased across spatial scales, the abundance of animalivores increased. They argued that higher edge density created low-contrast edges between fragments and the matrix, which allowed animalivores to forage and commute. In fact, species or guild responses to fragmentation are substantially influenced by spatial scale because there is no single scale at which either species or guilds consistently respond (Gorresen et al. 2005). Bats might be familiar with their landscape at multiple scales, which makes it necessary to assess how configuration of the landscape at multiple spatial scales selects species composition and guild structure. In conclusion, disturbance, in the form of logging, habitat modification, or fragmentation, proves to be a strong selective force that has significant impacts on phyllostomid communities.

Speciation

The best expression of differential speciation between regions is contemporary differences in species richness. Differences in species richness between two regions likely result from different histories characterizing the diversification of taxa. Nonetheless, differences between areas in terms of species richness can result from either differential speciation or differential extinction (Wiens and Donoghue 2004). In other words, species richness in regions with more species could be the product of higher rates of speciation and/or lower rates of extinction. Currently, it is quite challenging to determine the independent strength of these two processes. To this end, examination of historical effects on richness has typically focused on differences in "diversification" (difference between extinction and speciation). For the most part, differences between continents in terms of species richness are most likely the product of differential diversification. Findley and Wilson (1983) compared bat species richness among Nearctic, Neotropical, and African bat communities and demonstrated that African communities were substantially impoverished in terms of numbers of species. Similarly, Fleming et al. (1987) described a similar pattern for Neotropical, African, and southeast Asian bat communities. High species richness of Neotropical

bat communities likely is the result of greater rates of diversification in the New World.

Patterns of diversification of bats in the New World, particularly for endemic clades such as the Noctilionoidea, have not been constant through either space or time. Not only has the rate of diversification of Noctilionoidea likely decreased over the last 2.5 Ma but it has not been equal across North, Central, and South America (Rojas et al. 2016). Noctilionoidea, as well as most phyllostomid subfamilies and the most species-rich genera (e.g., *Artibeus*, *Platyrrhinus*, *Sturnira*), have a South American origin. However, Central and North America were also important centers of diversification for the family (Rojas et al. 2016). Such differences in patterns of diversification probably gave rise to the geographic differences in structure of species pools from which communities have been assembled.

Another manifestation of differential diversification across regions of the New World can be identified by examining the phylogenetic characteristics of species within those regions. For example, short branch lengths on a phylogeny result from less sequence divergence between a species and their most recent common ancestor, and taxa with shorter branch lengths can be considered to have diverged more recently than those with long branch lengths (Nei and Kumar 2000). Comparing different metrics of these characteristics (i.e., mean or variance of branch lengths) for taxa between regions can shed light on the diversification process. In this regard, phyllostomid communities from more temperate regions exhibit smaller mean branch lengths, suggesting that they are on average younger than their equatorial counterparts (Stevens 2013). Moreover, these same taxa are the product of more evolution, as measured by the total branch length separating a species from the root of the phyllostomid tree (Stevens 2013). These observations suggest that differential diversification has been an important process structuring phyllostomid bat communities.

Drift

Drift results from cumulative random changes in species composition within communities, and it can be hard to distinguish from unexplained variation in species composition through space or time. Typically, unexplained variation is attributed to effects of unmeasured explanatory variables, and this might be why drift is hard to detect or why tools to detect drift are not prevalent in ecological research. However, drift is a particularly strong structuring process when some or all species' abundances are low; stochasticity of births and deaths have a stronger proportional effect in smaller communities (i.e., fewer total number of individuals) (Ruokolainen et al. 2009). Moreover, drift is also considered to be the preponderant driver of change in communities where all species are "ecological equivalents"; differences in abundance are not due to dissimilarities among morphological or physiological traits but are due to small and stochastic changes in births and deaths (Hubbell 2001). Ecological equivalency may not apply to phyllostomid communities because morphological traits are sufficiently different between species (Moreno et al. 2006). Moreover, it is unclear to what degree assumptions of strict ecological equivalence are necessary and when this can be substituted with "roughly similar" (Vellend 2016). To our knowledge no study has yet directly assessed the effect of drift combined with ecological equivalency on phyllostomid communities.

For phyllostomid communities, there have been few studies that argue for or against ecological drift as an important structuring mechanism. These studies typically have shown that small habitat size creates conditions for drift via reduction of community size. Reductions in habitat size can be a precursor of drift because smaller habitats possess fewer resources, which can only sustain small communities of species that face higher risks of extinction (Hubbell 2001). It comes as no surprise that most studies addressing the effect of habitat size come from those assessing the tenets of island biogeography in archipelagos or in fragmented systems, either man-made or natural (i.e., caves). The general finding for archipelagos is that habitat size, usually measured by island area, has a strong and positive effect on species richness, composition, and nestedness (Dávalos and Russell 2012; Willig et al. 2010). Fleming (1982) assessed patterns of species-area relationships in Caribbean birds and bats and concluded that higher species richness is correlated with larger island size for most families. He adduced that area per se explained the trend because species absent from islands could disperse onto them but were not able to sustain nonzero abundances given that resource availability decreases with decreasing island size. Interestingly, he

found that stenodermatines and glossophagines did not follow the trend. Further analysis confirmed Fleming's conclusion; a reduction in area, but not in habitat diversity or disturbance, explained the decrease in species richness (Ricklefs and Lovette 1999; Willig et al. 2010). The most plausible explanation was that species with low population densities on smaller islands face more drift and hence higher extinction rates (Ricklefs and Lovette 1999; Willig et al. 2010).

Contrary to studies on archipelagos, communities on land-bridge fragmented systems do not seem to be greatly affected by reduction in habitat size. In general, richness, abundance, and nestedness on small islands were not statistically different from mainland control sites (Meyer and Kalko 2008; Pons and Cosson 2002). Habitat size only had limited explanatory power on community composition when analyses were performed at small spatial scales (landscapes of 0.5 km radius) (Meyer and Kalko 2008). Two factors may explain the disparate effects of habitat size between land-bridge systems and archipelagos. First, newly formed islands on land-bridge systems may not have reached equilibrium, so other factors may exert a stronger effect on their communities (see the effect of isolation below). In comparison, the Caribbean archipelago, formed between the Eocene and Miocene (James 2005), is probably close to equilibrium. Second, fragments surrounded by water may suffer from a small-island effect, where episodic disturbances or habitat characteristics are more likely to control species richness and community composition than is size alone (Lomolino and Weiser 2001).

The effect of habitat size on community structure in fragmented terrestrial systems is more complex. Although species inventories have shown that smaller fragments harbor impoverished communities (Reis et al. 2003), and that forest edges are even poorer than fragments (da Silva et al. 2013), species-area patterns are more complex in countryside fragmented systems than land-bridge systems or archipelagos. For example, Montiel et al. (2006) were only able to detect an effect of habitat size on diversity and abundance in a dry forest in Mexico when seasonality was factored into the analysis. Avila-Cabadilla et al. (2012), also in a dry forest in Mexico, demonstrated that the abundance of nectar-feeding phyllostomids decreased with increasing area of dry forests, whereas the abundance of fruit-eating bats increased significantly with area of riparian forests. The authors argued that nectarivorous bats did not exhibit the expected positive species-area relationship because these bats are feeding generalists, while frugivorous bats decreased in abundance because habitat loss reduced feeding areas. Trait-based comparisons have also shown that habitat loss has had differential effects on the abundance of different guilds (Farneda et al. 2015). While less mobile and larger animalivorous bats decreased in abundance in altered habitats (a proxy for smaller habitats in a low-contrast matrix), phytophagous bats with intermediate mobility and high relative wing loading increased in abundance in smaller habitats (Farneda et al. 2015). Lastly, in a comparison between a land-bridge and a countryside fragmented system, Mendenhall et al. (2014) concluded that habitat size was a good predictor of species richness on islands but not on fragments in an agricultural matrix. However, community abundance decreased as habitat size decreased in forest fragments, but the same pattern was not observed across islands in the land-bridge system (Mendenhall et al. 2014). In sum, the effect of habitat size on communities in anthropogenically modified fragmented systems is modulated by seasonality, trophic level, and species traits.

Finally, and contrary to studies in anthropogenic fragmented systems, communities within caves show a clear positive species-area relationship. Caves are naturally fragmented systems that provide essential roosting requirements for many phyllostomids (see chap. 18, this vol.). Larger caves harbor richer bat assemblages (Brunet and Medellín 2001; Cardona et al. 2013). The effect of habitat size on richness is enhanced by the fact that larger caves offer a wider range of microclimatic conditions (e.g., humidity and temperature) and more types of chambers where species can roost (Brunet and Medellín 2001). Despite the clear positive species-area relationship, the effect of drift on small communities in caves is yet to be determined.

Dispersal

Phyllostomid bats are often cited as having substantial dispersal abilities, which contribute importantly to community structure (Stevens and Willig 1999; Willig and Moulton 1989). Nonetheless, because bats are mobile and move around in the dark, actual estimates of dispersal are difficult to ascertain. Phyllostomid bats can travel large distances, up to 120 km or more (Esbérard et al. 2017). Additionally, phyllostomid bats tend to have large home ranges, although they differ greatly in extent (Aguiar et al. 2014; Gannon and Willig 1997). Lastly, phyllostomids tend to exhibit patterns of genetic structure indicative of integration of distant sites by relatively high rates of dispersal (McCulloch et al. 2013; Oprea 2013). All of these patterns suggest that dispersal in these bats may be a much more important factor to the structure of communities than previously realized.

Isolation decreases the probability of dispersal to distant habitats and diminishes the probability of species to be rescued from extinction. Accordingly, isolation can be used as a proxy for dispersal. In this regard, the way that isolation affects diversity patterns has often been analyzed either in the context of island biogeography on archipelagos or in fragmented systems with different types of matrices. Studies of island biogeography in the Caribbean and off the coast of Brazil have determined that isolation (i.e., distance to mainland) has had a strong effect on community composition and richness (Pathek et al. 2014; Presley and Willig 2008). Species colonization depends on random dispersal events, which are inversely proportional to isolation. On land-bridge systems, which in some respects emulate ocean archipelagos, communities on far islands were less rich and more structurally simplified than mainland control sites (Meyer and Kalko 2008; Pons and Cosson 2002). Across land-bridge islands of Lake Gatun, Panama, isolation strongly reduced species abundances, especially affecting gleaning animalivores. Isolation can be so strong that even the most abundant species, with high recapture rates and high dispersal capabilities, avoid long-distance flights to far islands (Meyer and Kalko 2008). Interestingly, however, species co-occurrence patterns were random for phytophagous bats even after accounting for the effect of isolation. Such random co-occurrence was explained by high dispersal capabilities (i.e., high wing aspect ratios) (Meyer and Kalko 2008). As in ocean archipelagos, communities on isolated islands surrounded by an inhospitable matrix (water) may be dominated by a limited subset of species while the other species have small populations that may fluctuate stochastically. Examples from anthropogenically fragmented systems have also shown that isolation plays an important role in shaping communities, but the amount and type of habitat (e.g., early secondary forest or heterogeneous orchards) surrounding fragments may buffer the effect of isolation (Mendenhall et al. 2014; Montiel et al. 2006). In scenarios where the matrix provides rescue effects or has novel bat communities altogether, the role of dispersal may not be as important as in archipelagos or land-bridge systems where isolation is greater.

In communities where dispersal exerts a strong effect on community structure, the supply of individuals among patches may override the effect of selection (i.e., species sorting) across various habitats. Without dispersal, local extinction of inferior competitors is expected. Within the metacommunity paradigm (Leibold et al. 2004), mass effects incorporate both dispersal and differential performance of species between patches; patches offer different resources and conditions, with some patches becoming sources of species and others becoming sinks, but dispersal regulates flow of species among patches (Leibold et al. 2004). Accordingly, the distribution of species among communities within metacommunities can be compared, and different patterns of distributions can be used to infer the importance of different ecological or biogeographical processes (Presley et al. 2010). Where there is high dispersal, species should have wide distributions but when dispersal is limited, strong species sorting should characterize distributions across metacommunities.

Only two studies have evaluated directly or indirectly the importance of dispersal on phyllostomid community structure within a metacommunity context. The first study to do so was Stevens et al. (2007), who showed that the mass effects paradigm best explained variation in species composition among 26 bat communities distributed across Paraguay. A lower-level process (habitat selection in Vellend's terminology) and dispersal were important determinants of metacommunity structure at the regional scale, which is understandable given the tight association of bats with particular ecosystems (Lopez-Gonzalez 2004) or resources (Giannini and Kalko 2004) and given their capacity to travel between patches (Montiel et al. 2006) or to migrate (Cryan et al. 2004).

The second study, by Presley and Willig (2008), showed that patterns of distribution of phyllostomid bats across Caribbean islands could be explained by the relative proximity of islands to different sources on the mainland, which directly determines the probability of successful dispersal in island biogeography (Presley and Willig 2008). For example, all the Bahamian islands have very similar bat communities because they are close to their source (subtropical North America), whereas the communities in the Lesser Antilles, which are farther away from their sources (i.e., lower probability of dispersal), show higher turnover rates from island to island. Bat communities in the Greater Antilles also seem to be shaped by dispersal; the majority of species originated in the larger islands but idiosyncratic dispersal likely has produced nested distributions across the smaller islands.

Future Prospectus

While a focus on selection has illuminated a number of lower-level processes that do not ubiquitously account for structure of phyllostomid communities, such a focus has not generated consensus on processes that do. Even some processes that fall under selection are in need of detailed study.

Selection

The effects of predation on bat community structure is poorly understood and conspicuously absent from figure 20.1 and table S20.1. Indeed, the traditional adage regarding effects of predation on bats is that predation happens, but probably no bat population is controlled by this process. Nonetheless, some recent findings bring this into question and suggest that bats succumbing to predation may be much more common than traditionally thought (Roulin and Criste 2013). Moreover, simple abstractions of predator-prey interactions are pathogen-prey interactions. Bat-pathogen interactions have come to the forefront because of the rapid and powerful effects of the spread of white-nose syndrome on vespertilionid bats of North America, having drastically altered the local and global abundance of a number of bat species and has led to their local extinction of some species (Frick et al. 2015). Effects of these pathogens undoubtedly change community structure (Jachowski et al. 2014). As a group, predator-prey, parasite-prey, and pathogen-prey interactions need be better appreciated as selective forces that affect phyllostomid communities and are worthy of increased focus in the future.

Perhaps more pragmatically and ethically, future research focus should strive for more comprehensive understanding of anthropogenic effects on bat communities as a selective force on community structure, in particular with respect to land-use change, habitat fragmentation, and global climate change. Much has been reported regarding the responses of Neotropical bats to land-use change, but this body of research has provided mixed results. Phyllostomids do often respond to land-use change, but responses are idiosyncratic and dependent on feeding and foraging guild (Gorresen and Willig 2004). Phyllostomines offer the most opportunity as bioindicators because they decrease in abundance and species richness fairly rapidly with land-use change (Fenton et al. 1992; García-Morales et al. 2013; Meyer and Kalko 2008). Nonetheless, other guilds either show no response or show an increase in abundance and diversity in response to human modification (García-Morales et al. 2013). A sufficient number of studies are now available so as to perform more detailed meta-analyses, some of which have provided insights into generalities regarding the effects of anthropogenic modification on bat communities (García-Morales et al. 2013; Mendenhall et al. 2014).

The selective force imposed by fragmentation is subtly different from other anthropogenic habitat modifications. Fragmentation describes the process of habitat loss whereby large continuous habitat is divided into a larger number of smaller fragments of lower individual and total area (Fahrig 2003). Along with loss of total habitat, effects of fragmentation include isolation of fragments, changes in configuration and composition of the landscape, and how that translates into modification of original biodiversity (Fahrig 2003). Thus, these different effects of fragmentation are expected to select different species in different ways, which will ultimately affect community structure. In this regard, fragment-focused and landscape-focused approaches can differentially address these different effects of fragmentation. Fragment-focused approaches are ideal for examining effects of changes in size of fragments on biodiversity. Nonetheless, many effects of fragmentation are themselves landscape characteristics such as average degree of isolation, configuration and geometry of fragments, or changes in the composition of fragments and the matrix (Fahrig 2003). Much work has been fragment focused with relatively less being landscape focused. Landscape-focused research on phyllostomids has demonstrated that bats respond primarily to losses in amount of more desirable habitat (Gorresen and Willig 2004; Muyleart et al 2016; but see Klingbeil and Willig

2009). Indeed, because many landscape compositional and configurational indices are correlated with reductions in amount of habitat, there is substantial controversy as to whether the greatest selective force on biodiversity is from habitat loss per se (habitat amount hypothesis) or from changes in composition or configuration (Fahrig 2013). In this regard, managing amount of desirable habitat is likely much more feasible than managing a particular configuration or composition of fragments. Therefore, testing the habitat amount hypothesis, which makes simple and testable predictions, can expedite decisions when trying to determine the primacy of amount of habitat over more complicated hypotheses involving composition and configuration (Fahrig 2013). More direct tests of the habitat amount hypothesis on phyllostomid communities need to be conducted.

Assessing thresholds at which communities precipitously lose species (local extinction) due to selection may be a fruitful avenue for management and conservation of landscapes. Relationships between population size or species richness within fragments and habitat loss are not linear but are precipitously nonlinear (Andren 1994; Fahrig 2003). An extinction threshold is simply the lowest amount of suitable habitat in a landscape that will sustain either a population or the same species richness as a similar but unfragmented landscape. Andren (1994) estimated the fragmentation threshold for birds and terrestrial mammals from 35 different studies to be between 20% and 30% of remaining suitable habitat. Such hypotheses have only been recently addressed for phyllostomid bats. Muyleart et al. (2016) found the extinction threshold in Cerrado habitats of Brazil to be 47% of the remaining native habitat. Although this result suggests that phyllostomids may be fairly sensitive to reductions of native habitat and fragmentation, this threshold might not generalize to other ecosystems throughout the Neotropics; other systems may be differentially sensitive to reductions in native habitat and exhibit different thresholds. Differences among landscapes in terms of extinction thresholds can illuminate selection gradients regarding fragmentation. Identification of a gradient of thresholds that is related to characteristics of fragmented landscapes or traits of species would go a long way to provide a more comprehensive understanding of the effects of this aspect of anthropogenic habitat modification on phyllostomid community structure. Moreover, most landscape-focused studies of effects of fragmentation on phyllostomids have occurred in extensively modified regions that are likely below this 47% threshold. Indeed, future studies should examine extinction thresholds in less modified systems to determine how selective different fragmented habitats are and then verify the generality of this threshold across systems with varying levels of fragmentation.

Climate change is a natural process, a very strong selective force, and an integral part of the geological history of Earth. Nonetheless, the rate of change over the last several decades has been astonishingly fast and higher than changes that have occurred over any other geological time (Parry et al. 2008). Indeed, bats respond to climate change in a wide variety of ways (LaVal 2004; Sherwin et al. 2012), including changes in geographic distribution, local relative abundance, reproductive phenology, and foraging and roosting habits. Many species have demonstrated more polar and more upward range expansions with the onset of human-induced climate change (LaVal 2004; Scheel et al. 1996). As with responses to habitat modification and fragmentation, responses are species specific, rendering community-level predictions difficult. Nonetheless, some patterns exist, at least for non-phyllostomid bats; biogeographic affinities determine responses of species to climate change because these affinities reflect ecological characteristics and environmental limits of species (Rebelo et al. 2010). Accordingly, it might be expected that phyllostomids occurring in more subtropical communities or those occurring at mid-elevations may change their distribution more than those in tropical lowland communities because temperatures will likely change the most in these geographic areas (Larsen et al. 2011). We have only begun to scratch the surface in terms of how future climate will affect bats in general, and phyllostomids in particular. Better understanding of such effects should be a research priority. Moreover, how biotic interactions are modulated by climate change is one of the big unknowns regarding our understanding of future distributions of bats and the diversity of the communities that they form. There likely is a definite climate change–species interaction synergy that should be considered in future scenarios. For example, climate change in Costa Rica has allowed a superior competitor, *Sturnira lilium*, to move up in its elevational distribution and negatively impact the abundance of *Sturnira ludovici* (LaVal 2004). While the "ghost of competition past" (Connell 1980) may have provided stability in the structure of at least some bat communities in the past, future dynamics may provide many opportunities for new interaction among species.

Dispersal

Focus on the higher-level process of dispersal is increasing, primarily as a result of the application of metacommunity approaches. If bats do form metacommunities, then we can no longer rely on single-community or even single spatial-scale analyses to try to understand structure. First, dispersal is likely important for homogenization of community structure across heterogeneous regions. Perhaps more importantly, however, dispersal interacts with the other three higher-level processes (Vellend 2016), potentially counteracting their complete manifestation and potentially creating unstable states that may be responsible for much of the seemingly stochastic nature of phyllostomid community structure. Quantifying effects of dispersal is methodologically challenging. Moreover, current estimation of dispersal effects via spatial statistics (Stevens et al. 2007) is likely very coarse and treats all species equally. Unfortunately, quantifying the amount of dispersal that unites two communities and thus how important this mechanism is in real time cannot be directly measured yet. Nevertheless, there are different techniques that can serve as proxies to assess different aspects of dispersal. Technology is constantly improving and smaller telemetry transmitters with longer battery life (Sapir et al. 2014) will provide more comprehensive information on home range, movements, and residence times at the local scale. Measures of dispersal can also be determined based on a few genetic markers (McCulloch et al. 2013; Oprea 2013), and modern, entire genome approaches may be able to provide even more detailed measures of dispersal at a larger spatial scale. Moreover, other population genetics techniques such as genotype-based individual assignments (Pritchard et al. 2000) can identify individuals to a source population as well as assess the relative contributions of mass effects and auto-recruitment on the structure of communities. Indeed, this represents an important avenue to better understand effects of dispersal on the structure of phyllostomid bat communities.

Other aspects of dispersal, such as interspecific variation and how differences contribute to metacommunity structure, should be explored further. Indeed, species likely do not contribute equally to the mass effects that have been reported for phyllostomid bat communities (Stevens et al. 2007), and understanding the life-history, trophic, functional, and phylogenetic determinants of this variation would be quite illuminating. Important trade-offs between dispersal and competitive abilities may facilitate coexistence for a number of species (i.e., patch dynamics; Coomes and Grubb 2003; Meynard and Quinn 2008) and may very well do so for bats. Nonetheless without species-specific estimates of dispersal such a possibility cannot be evaluated.

Currently, two different ways of analyzing metacommunities are available, namely, elements of metacommunity structure (Presley et al. 2010) and equally popular variation decomposition (Stevens et al. 2007). Both of these approaches make an important step forward in that they examine local and regional determinants of community structure. Elements of metacommunity structure identifies the latent gradient that accounts for the most variation in species composition across a metacommunity, whereas variance decomposition examines the relative contributions of environmental sorting and dispersal to variation in community structure across a metacommunity. Both approaches can be improved, however. Directly integrating dispersal into elements of metacommunity structure would be fruitful. The primary process underlying the metacommunity paradigm is dispersal among communities at the regional scale. Nonetheless, elements of metacommunity structure is silent to dispersal effects. While variance decomposition directly incorporates the concept of dispersal, use of spatial statistics based solely on distances among sites may not fully characterize dispersal of organisms across a landscape and says nothing about species-specific contributions to the dispersal effect. Indeed, better integration and measurement of dispersal among sites will greatly enhance our understanding of phyllostomid community ecology. Elements of metacommunity structure and variance decomposition have yet to be united, or even compared and contrasted based on the same data set. Such comparisons would greatly facilitate better understanding of the methods and patterns of metacommunity ecology.

The role of dispersal likely interacts with other higher-level processes, which increases the complexity of metacommunity structure but provides interesting challenges to understanding such complexity. Given the high functional diversity of Phyllostomidae, its wide evolutionary diversification, and its high abundance in many Neotropical systems, this family provides a powerful springboard from which to expand on recent developments in metacommunity ecology. For example, phyllostomids may serve to integrate community ecology with ecosystem ecology via spatial dynamics. Not only does the rich trophic diversity of phyllostomids translate into high functional diversity, but this family is also involved in a variety of ecosystem processes such as secondary production, immobilization, and mineralization of nutrients. Theoretically, at least some of the variation among ecosystems within a region is the product of spatial flow of nutrients and energy stemming from dispersal (Massol et al. 2011). For example, differences among sites in productivities and rates of population growth could generate spatial flows of nutrients that are influenced by concomitant differences in community structure. Such an effect could create mass effect–mediated spatial flows of nutrients and energy across landscapes. By examining both sides of the coin of species distribution and abundance (communities on one side and ecosystems on the other), a much richer understanding of ecosystem and community ecology may result in a general and better understanding of the manifold effects of dispersal on the ecology of organisms.

Speciation

Understanding how ecological speciation may arise and how this higher-level regional process may play an important role in local community dynamics will provide many insights. Ecological speciation highlights cladogenesis under ongoing gene flow that occurs at small spatial and temporal scales (Ai et al. 2013). Such speciation is often the result of adaptation to local conditions (local communities) in heterogeneous regions. Niche-based dynamics and adaptation can drive ecological speciation that produces incipient species in source populations (Weins and Donoghue 2004). It is possible that sister clades within subfamilies of Phyllostomidae might have experienced recent events of ecological speciation or incipient speciation events driven by local adaptation. Moreover, dispersal via a mass effect allows the persistence of species in communities where they are maladapted and as such can enhance the persistence of incipient species for longer periods of time (Ai et al. 2013). Phyllostomids are indeed highly mobile (Meyer and Kalko 2008; Stevens et al. 2007; Willig and Moulton 1989). High dispersal rates among local communities can scale up ecological dynamics thereby making the effects of dispersal both a local and a regional process. Such effects can modify the balance between local and regional processes as well as the ecological and evolutionary processes that underlie them (Hubert et al. 2015). Future studies should better explore ecological speciation at the regional scale, how ecological speciation is modified by dispersal, and how the joint effect of ecological speciation and dispersal affects the distribution and abundance of species within metacommunities.

Long-Term Perspectives

Higher-level processes, as well as a number of lower-level processes, require long time series of data to be adequately detected. For example, the random fluctuations and ultimate local extinction that result from the higher-level process of drift are likely best characterized by long-term data sets (Vellend 2016). Similarly, it has been estimated that an approximately 20-year time span may be required to detect even directional changes in abundance of bats in many tropical systems (Meyer et al. 2010). Moreover, detection of species interactions though changes in abundance and responses to global climate change often require long-term data (LaVal 2004). Indeed, while a number of sites exist where long-term sampling is underway (see those in Meyer et al. 2010), long-term insights are conspicuously few regarding phyllostomid community ecology (also see chap. 19, this vol.). Logistically, long-term data sets are difficult to produce because most funding is for projects of short duration and it is difficult to staff projects with the same people for decades. Future long-term studies should strive to be systematic through both space and time. In other words, a number of plots of the same size should be sampled regularly through time. Ideally, efforts at numerous sites should be integrated not only to collect long-term data but to do it in the same way across sites. Such a movement is underway and is producing long-term insights (Meyer et al. 2010; 2015). Long-term perspectives promise much improved clarity on phyllostomid community structure.

Lastly, to date, in terms of phyllostomid community ecology, focus has primarily been given to independent

effects of lower-level processes (e.g., mutualism independent of elevation), with much less focus given to the independent effects of the four higher-level processes of selection, speciation, dispersal, and drift. Nonetheless, these processes undoubtedly interact. For example, from a theoretical standpoint, dispersal can counteract selection for certain environmental conditions (Vellend 2016). Additionally, dispersal and its associated gene flow can slow or prevent the speciation process (Nei and Kumar 2000), whereas drift or selection can expedite the speciation process (Nei and Kumar 2000). Nonetheless, these are theoretical predictions and whether these interactions occur within phyllostomid communities remains to be explored. This survey has demonstrated that all four higher-level processes act on phyllostomid community structure, and future research should strive to more thoroughly explore effects of dispersal, drift, and speciation on phyllostomid community structure to better ascertain their relative importance. Moreover, we should move beyond simply documenting the relative importance of single low-level processes and instead move to simultaneous comparisons of several low-level processes, or even determine the interactions between higher-level processes and how such interactions give rise to variation in phyllostomid communities through space and time.

Conclusions

The study of structure of phyllostomid bat communities has a rich history that parallels the history of community ecology as a whole. Classical approaches focused on deterministic structure resulting from biotic interactions or environmental heterogeneity. Recently, the mechanistic foundations of community structure have been distilled to four higher-level fundamental processes: selection, speciation, drift, and dispersal. All contemporary models of community structure can be cast as a combination of the effects of these four processes. The extensive research on determinants of phyllostomid community structure allows the application of this new theoretical framework to improve our understanding phyllostomid community structure and to simplify venues for future studies. The majority of studies on community ecology of phyllostomids have focused on processes related to selection, in particular with regards to responses to anthropogenic habitat modification or other environmental characteristics such as food availably, elevation, latitude, and seasonality. However, the paucity of studies focusing on speciation, drift, and dispersal is striking. We propose a roadmap for future investigations that can bridge the gap between the effects of selection and the other three higher-level processes. Focus on selection has not generated a unified understanding as to the selective forces that control the structure of phyllostomid bat communities, especially when it comes to the forces imposed by fragmentation versus those imposed by land-use change. Moreover, speciation rates need to be integrated into phyllostomid community structure in order to understand regional differences in species richness, abundances, and trait distributions. We strongly advocate for long-term studies to distinguish effects of drift from other forms of stochasticity, and a better grasp of dispersal is needed to determine not only how it homogenizes beta diversity but also how dispersal interacts with other higher-level processes. Studying how selection, speciation, drift, and dispersal interact would greatly enhance our understanding of how phyllostomid communities are organized and structured.

Acknowledgments

We would like to thank the editors, in particular Marco Mello and Ted Fleming, for shepherding this manuscript through the review process. Maria Joao Pereira kindly provided studies published in Portuguese for our literature review. We appreciated the valuable feedback from two anonymous reviewers. Writing of this manuscript was supported by College of Agricultural Sciences and Natural Resources of Texas Tech University, the Office of the Vice President for Research of Texas Tech University, and the Department of Biological Sciences of Marquette University.

References

Aguiar, L. M. S., E. Bernard, and R. B. Machado. 2014. Habitat use and movements of *Glossophaga soricina* and *Lonchophylla dekeyseri* (Chiroptera: Phyllostomidae). Zoologia, 31:223–229.

Aguirre, L. F., L. Lens, R. van Damme, and E. Matthysen. 2003. Consistency and variation in the bat assemblages inhabiting two forest islands within a Neotropical savanna in Bolivia. Journal of Tropical Ecology, 19:367–374.

Ai, S., D. Gravel, C. Chu, and G. Wang. 2013. Spatial structures of the environment and of dispersal impact species distributions in competitive metacommunities. PLOS ONE, 8:e68927.

Andren, H. 1994. Effects of habitat fragmentation on birds and mammals in landscapes with different proportion of suitable habitat. Oikos, 71:355–366.

Arita, H. T. 1996. The conservation of cave-roosting bats in Yucatan, Mexico. Biological Conservation, 76:177–185.

Arita, H. T. 1997. Species composition and morphological structure of the bat fauna of Yucatan, Mexico. Journal of Animal Ecology, 66:83–97.

Avila-Cabadilla, L. D., G. A. Sanchez-Azofeifa, K. E. Stoner, M. Y. Alvarez-Añorve, M. Quesada, and C. A. Portillo-Quintero. 2012. Local and landscape factors determining occurrence of phyllostomid bats in tropical secondary forests. PLOS ONE, 7:e35228.

Avila-Cabadilla, L. D., K. E. Stoner, J. M. Nassar, M. M. Espírito-Santo, M. Y. Alvarez-Añorve, C. I. Aranguren, M. Henry, J. A. González-Carcacía, L. A. Dolabela Falcão, and G. A. Sanchez-Azofeifa. 2014. Phyllostomid bat occurrence in successional stages of Neotropical dry forests. PLOS ONE, 9:e84572.

Bejarano-Bonilla, D. A., A. Yate-Rivas, and M. H. Bernal-Bautista. 2007. Diversidad y distribucion de la fauna quiroptera en un transecto altitudinal en el departamento del Tolima, Colombia. Caldasia, 29:297–308.

Bergallo, H. G., E. E. L. Esbérard, M. A. R. Mello, V. Lins, R. Mangolin, G. G. S. Melo, and M. Baptista. 2003. Bat species richness in Atlantic Forest: what is the minimum sampling effort? Biotropica, 35:278–288.

Bernard, E. 2001. Vertical stratification of bat communities in primary forests of Central Amazon, Brazil. Journal of Tropical Ecology, 17:115–126.

Bernard, E., and M. B. Fenton. 2002. Species diversity of bats (Mammalia: Chiroptera) in forest fragments, primary forests, and savannas in central Amazonia Brazil. Canadian Journal of Zoology, 80:1124–1140.

Bonaccorso, F. J. 1975. Foraging and reproductive ecology in a community of bats in Panama. PhD diss., University of Florida.

Brosset, A., P. Charles-Dominique, A. Cockle, J.-F. Cosson, and D. Masson. 1996. Bat communities and deforestation in French Guiana. Canadian Journal of Zoology, 74:1974–1982.

Brown, J. H., and A. Kodric-Brown. 1977. Turnover rates in insular biogeography: effect of immigration on extinction. Ecology, 58:445–449.

Brunet, A. K., and R. A. Medellín. 2001. The species-area relationship in bat assemblages of tropical caves. Journal of Mammalogy, 82:1114–1122.

Cardona, D. M., J. Pérez-Torres, and L. S. Castillo. 2013. Patrón Anidado de Distribución de Murciélagos en un Conjunto de Cuevas del Enclave Seco del Chicamocha (Santander-Colombia). MSc, Universidad Internacional Menéndez Pelayo, Ecuador.

Castro-Arellano, I., S. J. Presley, L. N. Saldanha, M. R. Willig, and J. M. Wunderle Jr. 2007. Effects of reduced impact logging on bat biodiversity in terra firme forest of lowland Amazonia. Biological Conservation, 138:269–285.

Castro-Arellano, I., S. J. Presley, M. R. Willig, J. M. Wunderle, and L. N. Saldanha. 2009. Reduced-impact logging and temporal activity of understorey bats in lowland Amazonia. Biological Conservation, 142:2131–2139.

Castro-Luna, A. A., V. J. Sosa, and G. Castillo-Campos. 2007. Quantifying phyllostomid bats at different taxonomic levels as ecological indicators in a disturbed tropical forest. Acta Chiropterologica, 9:219–228.

Cisneros, L. M., K. R. Burgio, L. M. Dreiss, B. T. Klingbeil, B. D. Patterson, S. J. Presley, and M. R. Willig. 2014. Multiple dimensions of bat biodiversity along an extensive tropical elevational gradient. Journal of Animal Ecology, 83:1124–1136.

Cisneros, L. M., M. E. Fagan, and M. R. Willig. 2015. Effects of human-modified landscapes on taxonomic, functional and phylogenetic dimensions of bat biodiversity. Diversity and Distributions, 21:523–533.

Clarke, F. M., D. V. Pio, and P. A. Racey. 2005. A comparison of logging systems and bat diversity in the Neotropics. Conservation Biology, 19:1194–1204.

Connell, J. H. 1980. Diversity and the coevolution of competitors, or the ghost of competition past. Oikos, 35:131–138.

Coomes, D. A., and P. J. Grubb. 2003. Colonization, tolerance, competition and seed-size variation within functional groups. Trends in Ecology and Evolution, 18:283–291.

Cortés-Delgado, N., and J. Pérez-Torres. 2011. Habitat edge context and the distribution of phyllostomid bats in the Andean forest and anthropogenic matrix in the Central Andes of Colombia. Biodiversity and Conservation, 20:987–999.

Cosson, J.-F., J.-M. Pons, and D. Masson. 1999. Effects of forest fragmentation on frugivorous and nectarivorous bats in French Guiana. Journal of Tropical Ecology, 15:515–534.

Cryan, P. M., M. A. Bogan, R. O. Rye, G. P. Landis, and C. L. Kester. 2004. Stable hydrogen isotope analysis of bat hair as evidence for seasonal molt and long-distance migration. Journal of Mammalogy, 85:995–1001.

da Silva, J. R. R., H. O. Filho, and T. E. Lacher. 2013. Species richness and edge effects on bat communities from Perobas Biological Reserve, Paraná, southern Brazil. Studies on Neotropical Fauna and Environment, 48:135–141.

Dávalos, L. M., and A. L. Russell. 2012. Deglaciation explains bat extinction in the Caribbean. Ecology and Evolution 2:3045–3051.

de la Peña-Cuéllar, E., K. E. Stoner, L. D. Avila-Cabadilla, M. Martínez-Ramos, and A. Estrada. 2012. Phyllostomid bat assemblages in different successional stages of tropical rain forest in Chiapas, Mexico. Biodiversity and Conservation, 21:1381–1397.

Delaval, M., and P. Charles-Dominique. 2006. Edge effects on frugivorous and nectarivorous bat communities in a Neotropical primary forest in French Guiana. Revue d'Ecologie 61:343–352.

Esbérard, C. E. L., I. P. de Lima, P. H. Nobre, S. L. Althoff, T. Jordao-Nogueira, D. Dias, F. Carvlaho, M. E. Fabian, M. L. Sekiama and A. S. Sobrinho. 2011. Evidence of vertical migration in the Ipanema bat *Pygoderma bilabiatum* (Chiroptera: Phyllostomidae: Stenodermatinae). Zoologia, 28:717–724.

Esbérard, C. E. L., M. S. M. Godoy, L. Renovato and W. D. Carvalho. 2017. Novel long-distance movements by Neotropical bats (Mammalia: Chiroptera: Phyllostomidae) evidenced by recaptures in southeastern Brazil. Studies in Neotropical Fauna and Environment 52:75–80.

Estrada, A., and R. Coates-Estrada. 2002. Bats in continuous forest, forest fragments and in an agricultural mosaic habitat-island at Los Tuxtlas, Mexico. Biological Conservation, 103:237–245.

Estrada, A., R. Coates-Estrada, and D. Meritt Jr. 1993. Bat species richness and abundance in tropical rain forest fragments and in agricultural habitats at Los Tuxtlas, Mexico. Ecography, 16:309–318.

Estrada-Villegas, S., J. Pérez-Torres, and P. R. Stevenson. 2010. Ensamblaje de murciélagos en un bosque subandino colombiano y análisis sobre la dieta de algunas especies. Mastozoología Neotropical, 17:31–41.

Estrada-Villegas, S., and B. Ramírez. 2013. Bats of Casanare, Colombia. Chiroptera Neotropical, 19:1–13.

Fahrig, L. 2003. Effects of habitat fragmentation on biodiversity. Annual Review of Ecology, Evolution, and Systematics, 34:487–515.

Fahrig, L. 2013. Rethinking patch size and isolation effects: the habitat amount hypothesis. Journal of Biogeography, 40:1649–1663.

Falcão, L. A. D., M. M. do Espírito-Santo, L. O. Leite, R. N. S. L. Garro, L. D. Avila-Cabadilla, and K. E. Stoner. 2014. Spatio-temporal variation in phyllostomid bat assemblages over a successional gradient in a tropical dry forest in southeastern Brazil. Journal of Tropical Ecology, 30:123–132.

Faria, D., R. R. Laps, J. Baumgarten, and M. Cetra. 2006. Bat and bird assemblages from forests and shade cacao plantations in two contrasting landscapes in the Atlantic Forest of southern Bahia, Brazil. Biodiversity and Conservation, 15:587–612.

Farneda, F. Z., R. Rocha, A. López-Baucells, M. Groenenberg, I. Silva, J. M. Palmeirim, P. E. D. Bobrowiec, and C. F. J. Meyer. 2015. Trait-related responses to habitat fragmentation in Amazonian bats. Journal of Applied Ecology, 52 (5): 1381–1391. https://doi.org/10.1111/1365-2664.12490.

Fenton, M. B., L. Acharya, D. Audet, M. B. C. Hickey, C. B. Merriman, M. K. Obrist, D. M. Syme, and B. Adkins. 1992. Phyllostomid bats (Chiroptera: Phyllosomatidae) as indicators of habitat disruption in the Neotropics. Biotropica, 24:440–446.

Findley, J. S., and D. E. Wilson. 1983. Are bats rare in tropical Africa? Biotropica, 15:299–303.

Fleming, T. H. 1982. Parallel trends in the species diversity of West Indian birds and bats. Oecologia, 53:56–60.

Fleming, T. H. 1986. The structure of Neotropical bat communities: a preliminary analysis. Revista Chilena de Historia Natural, 59:135–150.

Fleming, T. H. 1988. The Short-Tailed Fruit Bat: A Study in Plant-Animal Interactions. University of Chicago Press, Chicago.

Fleming, T. H., R. Breitwisch, and G. H. Whitesides. 1987. Patterns of tropical vertebrate frugivore diversity. Annual Review of Ecology and Systematics, 18:91–109.

Fleming, T. H., E. T. Hooper, and D. E. Wilson. 1972. Three Central American bat communities: structure, reproductive cycles and movement patterns. Ecology, 53:555–569.

Fleming, T. H., and N. Muchhala. 2008. Nectar-feeding bird and bat niches in two worlds: pantropical comparisons of vertebrate pollination systems. Journal of Biogeography, 35:764–780.

Frick, W. F., S. J. Puechmaille, J. R. Hoyt, B. A. Nickel, K. E. Langwig, J. T. Foster, K. E. Barlow et al. 2015. Disease alters macroecological patterns of North American bats. Global Ecology and Biogeography, 24:741–749.

Gannon, M. R., and M. R. Willig. 1997. The effect of lunar illumination on movement and activity of the red fig-eating bat (Stenoderma rufum). Biotropica, 29:525–529.

García-García, J. L., and A. Santos-Moreno. 2014. Variación estacional en la diversidad y composición de ensambles de murciélagos filostómidos en bosques continuos y fragmentados en Los Chimalapas, Oaxaca, México. Revista Mexicana de Biodiversidad, 85:228–241.

García-Morales, R., E. I. Badano, and C. E. Moreno. 2013. Response of Neotropical bat assemblages to human land use. Conservation Biology, 27:1096–1106.

García-Morales, R., L. Chapa-Vargas, E. Badano, J. Galindo-González, and K. Monzalvo-Santos. 2014. Evaluating phyllostomid bat conservation potential of three forest types in the northern Neotropics of Eastern Mexico. Community Ecology, 15:158–168.

Giannini, N. P., and E. K. V. Kalko. 2004. Trophic structure in a large assemblage of phyllostomid bats in Panama. Oikos, 105:209–220.

Gorchov, D. L., F. Cornejo, C. F. Ascorra, and M. Jaramillo. 1995. Dietary overlap between frugivorous birds and bats in the Peruvian Amazon. Oikos, 74:235–250.

Gorresen, P. M., and M. R. Willig. 2004. Landscape responses of bats to habitat fragmentation in Atlantic Forest of Paraguay. Journal of Mammalogy, 85:688–697.

Gorresen, P. M., M. R. Willig, and R. E. Strauss. 2005. Multivariate analysis of scale-dependent associations between bats and landscape structure. Ecological Applications, 15:2126–2136.

Graham, G. L. 1983. Changes in bat species diversity along an elevational gradient up the Peruvian Andes. Journal of Mammalogy, 64:559–571.

Handley, C. O. 1967. Bats of the canopy of an Amazonian forest. Pages 211–215 in: Atas do Simposio sobre a biota Amazonica. Conselho Nacional de Pesquisas Rio de Janeiro.

Harvey, C. A., and J. A. González Villalobos. 2007. Agroforestry systems conserve species-rich but modified assemblages of tropical birds and bats. Biodiversity and Conservation, 16:2257–2292.

Heithaus, E. R., T. H. Fleming, and P. A. Opler. 1975. Foraging patterns and resource utilization in seven species of bats in a seasonal tropical forest. Ecology, 56:841–854.

Herrera, L. G., E. Gutierrez, K. A. Hobson, B. Altube, W. G. Diaz, and V. Sánchez-Cordero. 2002. Sources of assimilated protein in five species of New World frugivorous bats. Oecologia, 133:280–287.

Hubbell, S. P. 2001. The Unified Neutral Theory of Biodiversity and Biogeography. Princeton University Press, Princeton, NJ.

Hubert, N., V. Calcagno, R. S. Etienne, and N. Moquet. 2015. Metacommunity speciation models and their implications for diversification theory. Ecology Letters, 18:864–881.

Humphrey, S. R., and F. J. Bonaccorso. 1979. Population and community ecology. Pages 409–441 in: Biology of Bats of the New World Family Phyllostomatidae. Pt. 3 (R. J. Baker, J. K. Jones Jr., and D. C. Carter, eds.). Special Publications of the Museum of Texas Tech University, no. 16. Texas Tech Press, Lubbock.

Hutchinson, G. E. 1959. Homage to Santa Rosalia or why are there so many kinds of animals? American Naturalist, 93:145–159.

Jachowski, D. S., C. A. Dobony, L. S. Coleman, W. M. Ford, E. R. Britzke, and J. L Rodrighe. 2014. Disease and community structure: white-nose syndrome alters spatial and temporal niche partitioning in sympatric bat species. Diversity and Distributions, 20:1002–1015.

James, K. H. 2005. A simple synthesis of Caribbean geology. Caribbean Journal of Earth Science, 39:69–82.

Kalko, E. K. V. 1998. Organisation and diversity of tropical bat communities through space and time. Zoology, 101:281–297.

Kalko, E. K. V., and C. O. Handley Jr. 2001. Neotropical bats in the canopy: diversity, community structure, and implications for conservation. Plant Ecology, 153:319–333.

Kalko, E. K. V., C. O. Handley Jr., and D. Handley. 1996. Organization, diversity and long-term dynamics of a Neotropical bat community. Pages 503–553 in: Long-Term Studies in Vertebrate Communities (M. L. Cody and J. Smallwood, eds.). Academic Press, Los Angeles.

Kalko, E. K. V., E. A. Herre, and C. O. Handley Jr. 1996. Relation of fig fruit characteristics to fruit-eating bats in the New and Old World tropics. Journal of Biogeography, 23:565–576.

Kingston, T. 2009. Analysis of species diversity of bat assemblages. Pages 195–215 in: Ecological and Behavioral Methods for the Study of Bats (T. H. Kunz and S. Parsons, eds.). Johns Hopkins University Press, Baltimore.

Klingbeil, B. T., and M. R. Willig. 2009. Guild-specific responses of bats to landscape composition and configuration in fragmented Amazonian rainforest. Journal of Applied Ecology, 46:203–213.

Klingbeil, B. T., and M. R. Willig. 2010. Seasonal differences in population-, ensemble- and community-level responses of bats to landscape structure in Amazonia. Oikos, 119:1654–1664.

Larsen, T. H., G. Brehm, H. Navarrete, P. Franco, H. Gomez, J. L. Mena, V. Morales, J. Argollo, L. Blacutt, and V. Canhos. 2011. Range shifts and extinctions driven by climate change in the tropical Andes: synthesis and directions. Pages 47–67 in: Climate Change and Biodiversity in the Tropical Andes (S. K. Herzog, R. Martínez, P. M. Jørgensen, and T. Holm, eds.). Inter-American Institute for Global Change Research (IAI) and Scientific Committee on Problems of the Environment (SCOPE), [São José dos Campos, São Paulo, Brazil].

LaVal, R. K. 2004. Impact of global warming and locally changing climate on tropical cloud forest bats. Journal of Mammalogy, 85:237–244.

LaVal, R. K., and H. S. Fitch. 1977. Structure, movements and reproduction in three Costa Rican bat communities. Occasional Papers, Museum of Natural History, University of Kansas, 69:1–28.

Leibold, M. A., M. Holyoak, N. Mouquet, P. Amarasekare, J. M. Chase, M. F. Hoopes, R. D. Holt et al. 2004. The metacommunity concept: a framework for multi-scale community ecology. Ecology Letters, 7:601–613.

Lim, B. K., and M. D. Engstrom. 2001. Bat community structure at Iwokrama Forest, Guyana. Journal of Tropical Ecology, 17:647–665.

Lomolino, M. V., and M. D. Weiser. 2001. Towards a more general species-area relationship: diversity on all islands, great and small. Journal of Biogeography, 28:431–445.

Lopez-Gonzalez, C. 2004. Ecological zoogeography of the bats of Paraguay. Journal of Biogeography, 31:33–45.

MacArthur, R., and R. Levins. 1967. The limiting similarity, convergence and divergence of coexisting species. American Naturalist, 101:377–385.

MacSwiney G., M. C., P. Vilchis L, F. M. Clarke, and P. A. Racey. 2007. The importance of cenotes in conserving bat assemblages in the Yucatan, Mexico. Biological Conservation, 136:499–509.

Massol, F., D. Gravel, N. Mouquet, M. W. Cadotte, T. Fukami, and M. A. Leibold. 2011. Linking community and ecosystem dynamics through spatial ecology. Ecology Letters, 14: 313–323.

McCain, C. M. 2007a. Could temperature and water availability drive elevational species richness patterns? A global case study for bats. Global Ecology and Biogeography, 16:1–13.

McCain, C. M. 2007b. Area and mammalian elevational diversity. Ecology, 88:76–86.

McCulloch, E. S., A. Whitehead, J. S. Tello, C. M. J. Rolón-Mendoza, M. Maldonado, and R. D. Stevens. 2013. Fragmentation of Atlantic Forest does not yet prevent gene flow of an important seed-dispersing bat. Molecular Ecology, 22:4619–4633.

McNab, B. K. 1969. The economics of temperature regulation in Neotropical bats. Comparative Biochemistry and Physiology, 31:227–268.

Mello, M. A. R. 2009. Temporal variation in the organization of a Neotropical assemblage of leaf-nosed bats (Chiroptera: Phyllostomidae). Acta Oecologica, 35:280–286.

Mendenhall, C. D., D. S. Karp, C. F. J. Meyer, E. A. Hadly, and G. C. Daily. 2014. Predicting biodiversity change and averting collapse in agricultural landscapes. Nature, 509:213–217.

Meyer, C. F. J., L. Aguiar, L. F. Aguirre, J. Baumgarten, F. M. Clarke, J.-F. Cosson, S. Estrada-Villegas et al. 2011. Accounting for detectability improves estimates of species richness in tropical bat surveys. Journal of Applied Ecology, 48:777–787.

Meyer, C. F. J., L. M. S. Aguiar, L. F. Aguirre, J. Baumgarten, F. M. Clarke, J. F. Cosson, S. E. Villegas et al. 2010. Long-term monitoring of tropical bats for anthropogenic impact assessment: gauging the statistical power to detect population change. Biological Conservation, 143:2797–2807.

Meyer, C. F. J., L. Aguiar, L. F. Aguirre, J. Baumgarten, F. M. Clarke, J.-F. Cosson, S. Estrada-Villegas et al. 2015. Species undersampling in tropical bat surveys: effects of emerging biodiversity patterns. Journal of Animal Ecology, 84:113–123.

Meyer, C. F. J., and E. K. V. Kalko. 2008. Assemblage-level responses of phyllostomid bats to tropical forest fragmentation: land-bridge islands as a model system. Journal of Biogeography, 35:1711–1726.

Meyer, C. F. J., M. J. Struebig, and M. R. Willig. 2016. Responses of tropical bats to habitat fragmentation, logging, and deforestation. Pages 63–103 in: Bats in the Anthropocene: Conservation of Bats in a Changing World (C. C. Voigt and

T. Kingston, eds.). Springer International Publishing, Cham, Switzerland.

Meynard, C. N., and J. F. Quinn. 2008. Bird metacommunities in temperate South American forest: vegetation structure, area, and climate effects. Ecology, 89:981–990.

Montaño-Centellas, F., M. I. Moya, L. F. Aguirre, R. Galeón, O. Palabral, R. Hurtado, I. Galarza, and J. Tordoya. 2015. Community and species-level responses of phyllostomid bats to a disturbance gradient in the tropical Andes. Acta Oecologica, 62:10–17.

Montiel, S., A. Estrada, and P. León. 2006. Bat assemblages in a naturally fragmented ecosystem in the Yucatan Peninsula, Mexico: species richness, diversity and spatio-temporal dynamics. Journal of Tropical Ecology, 22:267–276.

Moreno, C., H. Arita, and L. Solis. 2006. Morphological assembly mechanisms in Neotropical bat assemblages and ensembles within a landscape. Oecologia, 149:133–140.

Moreno, C. E., and Halffter, G. 2000. Assessing the completeness of bat biodiversity inventories using species accumulation curves. Journal of Applied Ecology, 37:149–158.

Moreno, C. E., and G. Halffter. 2001. Spatial and temporal analysis of α, β and γ diversities of bats in a fragmented landscape. Biodiversity and Conservation, 10:367–382.

Moya, M. I., F. Montaño-Centellas, L. F. Aguirre, J. Tordoya, J. Martínez, and M. I. Galarza. 2008. Variación temporal de la quiropterofauna en un bosque de Yungas en Bolivia. Mastozoología Neotropical, 15:349–357.

Munin, R. L., E. Fischer, and F. Goncalves. 2012. Food habits and dietary overlap in a phyllostomid bat assemblage in the Pantanal of Brazil. Acta Chiropterologica, 14:195–204.

Muylaert, R. L., R. D. Stevens, and M. C. Ribeiro. 2016. Threshold effect of habitat loss on bat richness in Cerrado-forest landscapes. Ecological Applications, 26:1854–1867.

Nei, M., and S. Kumar. 2000. Molecular Evolution and Phylogenetics. Oxford University Press, Oxford.

Novoa, S., R. Cadenillas, and V. Pacheco. 2011. Dispersión de semillas por murciélagos frugívoros en bosques del parque nacional Cerros de Amotape, Tumbes, Perú. Mastozoología Neotropical, 18:81–93.

Oprea, M. 2013. Variabilidade e estructura genetica espacial em *Glossophaga soricina* com ocorrencia no Cerrado. Thesis, Universidade Federal de Goias, Goias, Brazil.

Ortêncio-Filho, H., T. E. Lacher, and L. C. Rodrigues. 2014. Seasonal patterns in community composition of bats in forest fragments of the Alto Rio Paraná, southern Brazil. Studies on Neotropical Fauna and Environment, 49:169–179.

Palmeirim, J. M., D. L. Gorchov, and S. Stoleson. 1989. Trophic structure of a Neotropical frugivore community: is there competition between birds and bats? Oecologia, 79:403–411.

Parry, M., O. Canziani, J. Palutikof, P. van der Linden, and C. Hanson, eds. 2008. Climate Change 2007: Impacts, Adaptation and Vulnerability. Working Group II Contribution to the Fourth Assessment Report of the Intergovernmental Panel on Climate Change. Cambridge University Press, Cambridge.

Pathek, D. B., G. L. Melo, J. Sponchiado, and N. C. Cáceres. 2014. Distance from the mainland is a selective pressure for Phyllostomidae bats: the case of Maracá-Jipioca Island on the northern coast of Brazil. Mammalia, 78:487–495.

Patterson, B. D., V. Pacheco, and S. Solari. 1996. Distribution of bats along an elevational gradient in the Andes of Peru. Journal of Zoology, 240:637–658.

Patterson, B. D., M. R. Willig, and R. D. Stevens. 2003. Trophic strategies, niche partitioning, and patterns of ecological organization. Pages 536–557 in: Bat Ecology (T. H. Kunz and B. Fenton, eds.). University of Chicago Press, Chicago.

Pereira, M. J. R., J. T. Marques, and J. M. Palmeirim. 2010. Vertical stratification of bat assemblages in flooded and unflooded Amazonian forests. Current Zoology, 56:469–478.

Pereira, M. J. R., J. T. Marques, J. Santana, C. D. Santos, J. Valsecchi, H. L. De Queiroz, P. Beja, and J. M. Palmeirim. 2009. Structuring of Amazonian bat assemblages: the roles of flooding patterns and floodwater nutrient load. Journal of Animal Ecology,78:1163–1171.

Pérez-Torres, J., and J. A. Ahumada P. 2004. Murcielagos en bosques alto-Andinos, fragmentados, y continuous, en el sector occidental de la sabana de Bogota (Colombia). Universitas Scientiarum, 9:14.

Pérez-Torres, J., C. Sánchez-Lalinde, and N. Cortés-Delgado. 2009. Murciélagos asociados a sistemas naturales y transformados en la ecorregión Eje Cafetero. Pages 157–167 in: Valoración de la Biodiversidad en la Ecorregión del Eje Cafetero (J. M. Rodríguez, J. C. Camargo, A. M. Niño, J. Pineda, L. M. Arias, M. A. Echeverry, and C. L. Miranda, eds.). Centro de Investigaciones y Estudios en Biodiversidad y Recursos Genéticos, Pereira, Colombia.

Pons, J.-M., and J.-F. Cosson. 2002. Use of forest fragments by animalivorous bats in French Guiana. Revue D Ecologie-Ta Terre et la Vie, 57:117–130.

Presley, S. J., L. M. Cisneros, B. D. Patterson, and M. R. Willig. 2012. Vertebrate metacommunity structure along an extensive elevational gradient in the tropics: a comparison of bats, rodents and birds. Global Ecology and Biogeography, 21:968–976.

Presley, S. J., C. L. Higgins, and M. R. Willig. 2010. A comprehensive framework for the evaluation of metacommunity structure. Oikos, 119:908–917.

Presley, S. J., and M. R. Willig. 2008. Composition and structure of Caribbean bat (Chiroptera) assemblages: effects of inter-island distance, area, elevation and hurricane-induced disturbance. Global Ecology and Biogeography, 17:747–757.

Pritchard J. K., M. Stephens, and P. Donnelly. 2000. Inference of population structure using mutlilocus genotype data. Genetics, 155:945–959.

Rebelo, H., P. Tarroso, and G. Jones. 2010. Predicted impact of climate change on European bats in relation to their biogeographic patterns. Global Change Biology, 16:561–576.

Reis, N. R. 1984. Estructura de comunidade de morcegos na regiao de Manaus, Amazonas. Revista Braileira de Biologia, 44:247–254.

Reis, N. R. d., M. L. d. S. Barbieri, I. P. d. Lima, and A. L. Peracchi. 2003. O que é melhor para manter a riqueza de espécies de morcegos (Mammalia, Chiroptera): um fragmento florestal grande ou vários fragmentos de pequeno tamanho? Revista Brasileira de Zoologia, 20:225–230.

Reis, N. R., A. L. Peracchi, I. P. d. Lima, and W. A. Pedro. 2006. Riqueza de espécies de morcegos (Mammalia, Chiroptera) em dois diferentes habitats, na região centro-sul do Paraná, sul do Brasil. Revista Brasileira de Zoologia, 23:813–816.

Rex, K., R. Michener, T. H. Kunz, and C. C. Voigt. 2011. Vertical stratification of Neotropical leaf-nosed bats (Chiroptera: Phyllostomidae) revealed by stable carbon isotopes. Journal of Tropical Ecology, 27:211–222.

Ricklefs, R. E. 1987. Community diversity: relative roles of local and regional processes. Science 235:167–171.

Ricklefs, R. E., and I. J. Lovette. 1999. The roles of island area per se and habitat diversity in the species-area relationships of four Lesser Antillean faunal groups. Journal of Animal Ecology, 68:1142–1160.

Ríos Blanco, M. C. 2014. Estructura del metaensamblaje de murciélagos en un paisaje antropogénico: (ecorregión eje cafetero-Colombia). Pontificia Universidad Javeriana, Bogotá, Colombia.

Rojas, D., O. M. Warsi, and L. M. Dávalos. 2016. Bats (Chiroptera: Noctilionidae) challenge a recent origin of extant Neotropical diversity. Systematic Biology, 65:432–448.

Rosenthal, R. 1979. The file drawer problem and tolerance for null results. Psychological Bulletin, 86 (3): 638–641.

Roulin, A., and P. Christe. 2013. Geographic and temporal variation in the consumption of bats by European barn owls. Bird Study, 60:561–569

Ruggiero, A., and T. Kitzberger. 2004. Environmental correlates of mammal species richness in South America: effects of spatial structure, taxonomy and geographic range. Ecography, 27:401–417.

Ruokolainen, L., E. Ranta, V. Kaitala, and M. S. Fowler. 2009. When can we distinguish between neutral and non-neutral processes in community dynamics under ecological drift? Ecology Letters, 12:909–919.

Saldaña-Vázquez, R. A. 2014. Intrinsic and extrinsic factors affecting dietary specialization in Neotropical frugivorous bats. Mammal Review, 44:215–224.

Saldaña-Vázquez, R. A., V. J. Sosa, L. I. Iñiguez Dávalos, and J. E. Schondube. 2013. The role of extrinsic and intrinsic factors in Neotropical fruit bat-plant interactions. Journal of Mammalogy, 94:632–639.

Sánchez, M. S., N. P. Giannini, and R. M. Barquez. 2012. Bat frugivory in two subtropical rain forests of northern Argentina: testing hypotheses of fruit selection in the Neotropics. Mammalian Biology—Zeitschrift für Säugetierkunde, 77:22–31.

Sánchez-Cordero, V. 2001. Elevation gradients of diversity for rodents and bats in Oaxaca, Mexico. Global Ecology and Biogeography Letters, 10:63–76.

Sánchez-Palomino, P., P. Rivas-Pava, and A. Cadena. 1993. Composición, abundancia y riqueza de especies de la comunidad de murciélagos en bosques de galería en la serranía de La Macarena (Meta-Colombia). Caldasia, 17:301–312.

Sapir, N., N. Horvitz, D. K. N. Dechmann, J. Fahr, and M. Wikelski. 2014. Commuting fruit bats beneficially modulate their flight in relation to wind. Proceedings of the Royal Society B–Biological Sciences, 281:20140018.

Scheel, D., T. L. S. Vincent, and G. N. Cameron. 1996. Global warming and the species richness of bats in Texas. Conservation Biology 10:452–464.

Schnitzler, H. U., and E. K. V. Kalko. 2001. Echolocation by insect-eating bats. Bioscience 51:557–569.

Schulze, M. D., N. E. Seavy, and D. F. Whitacre. 2000. A comparison of the phyllostomid bat assemblages in undisturbed neotropial forest and in forest fragments of a slash-and-burn farming mosaic in Petén, Guatemala. Biotropica, 32:174–184.

Scultori, C., S. Von Matter, and A. L. Peracchi. 2008. Métodos de amostragem de morcegos em sub-dossel e dossel florestal, com ênfase em redes de neblina. Pages 17–32 in: Ecologia de Morcegos (N. R. Reis, A. L. Peracchi, and G. A. S. D. dos Santos, eds.). Technical Books Editora, Londrina.

Sherwin, H. A., W. I. Montgomery, and M. G. Lundy. 2012. The impact and implications of climate change for bats. Mammal Review, 43:171–182.

Shmida, A., and M. V. Wilson. 1985. Biological determinants of species diversity. Journal of Biogeography, 12:1–20.

Soriano, P. 2000. Functional structure of bat communities in tropical rainforest and Andean cloud forest. Ecotropicos, 13:1–20.

Stevens, R. D. 2004. Untangling latitudinal richness gradients at higher taxonomic levels: familial perspectives on the diversity of New World bat communities. Journal of Biogeography, 31:665–674.

Stevens, R. D. 2013. Gradients of bat diversity in Atlantic Forest of South America: environmental seasonality, sampling effort and spatial autocorrelation. Biotropica, 45:764–770.

Stevens, R. D., and H. N. Amarilla-Stevens. 2012. Seasonal environments, episodic density compensation and dynamics of structure of chiropteran frugivore guilds in Paraguayan Atlantic Forest. Biodiversity and Conservation, 21:267–279.

Stevens, R. D., and M. M. Gavilanez. 2015. Dimensionality of community structure: phylogenetic, morphological and functional perspectives along biodiversity and environmental gradients. Ecography, 38:861–875.

Stevens, R. D., C. López-Gonzales, and S. J. Presley. 2007. Geographical ecology of Paraguayan bats: spatial integration and metacommunity structure of interacting assemblages. Journal of Animal Ecology, 76:1086–1093.

Stevens, R. D., and M. R. Willig. 1999. Size assortment in New World bat communities. Journal of Mammalogy, 80:644–658.

Stevens, R. D., and Willig, M. R. 2000. Density compensation in New World bat communities. Oikos, 89:367–377.

Thomas, M. E. 1972. Preliminary study of the annual breeding patterns and population fluctuations of bats in three ecologically distinct habitats in southwestern Colombia. PhD diss., Tulane University, New Orleans.

Tschapka, M. 2004. Energy density patterns of nectar resources permit coexistence within a guild of Neotropical flower-visiting bats. Journal of Zoology, 263:7–21.

Velend, M. 2016. The Theory of Ecological Communities. Princeton University Press, Princeton, NJ.

Velend, M., and Gerber, M. A. 2005. Connections between species diversity and genetic diversity. Ecology Letters, 8:767–781.

Velend, M., and Orrock, J. L. 2009. Genetic and ecological models of diversity: lessons across disciplines. Pages 439–461 in: The Theory of Island Biogeography Revisted. (J. B. Losos and R. E. Ricklefs, eds.). Princeton University Press, Princeton, NJ.

Voigt, C. C. 2010. Insights into strata use of forest animals using the "canopy effect." Biotropica, 42:634–637.

Weber, M. d. M., J. L. Steindorff de Arruda, B. Oliveira Azambuja, V. L. Camilotti, and N. C. Cáceres. 2011. Resources partitioning in a fruit bat community of the southern Atlantic Forest, Brazil. Mammalia, 75:217–225.

Weise, C. D. 2007. Community structure, vertical stratification and seasonal patterns of Neotropical bats. PhD diss., University of New Mexico, Albuquerque.

Wiens, J. J., and M. J. Donoghue. 2004. Historical biogeography, ecology and species richness. Trends in Ecology and Evolution, 19:639–644.

Willig, M. R. 1983. Composition, microgeographic variation, and sexual dimorphism in Caatingas and Cerrado bat communities from northeast Brazil. Bulletin of Carnegie Museum of Natural History, 23:1–131.

Willig M. R. 1986. Bat community structure in South America: a tenacious chimera. Revista Chilena de Historia Natural, 59:151–168.

Willig, M. R., G. R. Camilo, and S. J. Noble. 1993. Dietary overlap in frugivorous and insectivorous bats from edaphic Cerrado habitats of Brazil. Journal of Mammalogy, 74:117–128.

Willig, M. R., S. K. Lyons, and R. D. Stevens. 2009. Spatial methods for the macroecological study of bats. Pages 216–245 in: Ecological and Behavioral Methods for the Study of Bat (T. H. Kunz and S. Parsons, eds.). Johns Hopkins University Press, Baltimore.

Willig, M. R., and M. P. Moulton. 1989. The role of stochastic and deterministic processes in structuring Neotropical bat communities. Journal of Mammalogy, 70:323–329.

Willig, M. R., S. J. Presley, C. P. Bloch, and H. H. Genoways. 2010. Macroecology of Caribbean bats: effects of area, elevation, latitude, and hurricane-induced disturbance. Pages 216–264 in: Island Bats: Evolution, Ecology, and Conservation (T. H. Fleming and P. A. Racey, eds.). University of Chicago Press, Chicago.

Willig, M. R., and K. W. Selcer. 1989. Bat species density gradients in the New World: a statistical assessment. Journal of Biogeography, 16:189–195.

21

Network Science as a Framework for Bat Studies

Marco A. R. Mello
and Renata L. Muylaert*

From Natural History to Complexity

Ecology has always had a strong focus on description since its early days but has slowly begun to diversify its scope (McIntosh 1987). Now it aims at conceptual synthesis and transdisciplinary research (Sutherland et al. 2013). In our review, we show that the study of phyllostomid bats in the twenty-first century has gone far beyond the original scope observed in the 1970s, when the seminal monographs edited by Robert Baker and colleagues were published. Traditionally, most bat books focused on taxonomy and systematics (Nowak 1994), and some on natural history and geographic distribution (Gardner 2008). Nevertheless, a number of pioneering books have started dealing with bat population and community ecology since the 1980s (Findley 1993; Fleming 1988; Kunz 1982). Our chapter, as well as others in this book, follows the same path and tries to point to new directions in bat research with a focus on ecology.

Among several other frameworks that have been imported to the study of phyllostomids in the past decades (e.g., chemical ecology, telemetry, bioinformatics, genomics, neuroscience, and sensory physiology), network science (*sensu* Barabasi 2016) has helped advance considerably the fields of species interactions and community ecology in ours and many other disciplines (Poisot et al. 2016). Ultimately, network science helps understand the interactions between bats among themselves and with other organisms in the light of complex systems, an important frontier of contemporary science (Ings et al. 2009). And it also helps answer fundamental ecological questions, for instance, those related to how indirect and weak interactions may play an important role in community structuring (Sutherland et al. 2013). It also helps reduce ecological shortfalls, mainly Eltonian, by allowing us to link organismal traits to system structure (Hortal et al. 2015).

---

* Both authors contributed equally to this chapter.

The analysis of complex systems modeled as networks started almost 300 years ago in mathematics (Euler 1741). Nevertheless, only since the mid-twentieth century has it spread to all fields of knowledge (Barabasi 2016). Network science differs considerably from other analytical approaches because it allows assessing not only the elements of a set of interest but also especially the relationships between them. This is an alternative to mean field theory, which was exported from physics to several other disciplines long ago and has dominated our way of doing science in the past centuries. Instead of focusing on the properties of a hypothetical mean element that should represent all the elements of a given set, by modeling the set as a system (i.e., elements plus relationships), it is possible to assess its nontrivial emergent properties. Imagine, for instance, the difference between graphite and diamond, two substances made of the same chemical element (carbon), but that differ markedly in their electrical conductivity, transparency, and hardness. By looking only at a hypothetical mean carbon atom, it would be impossible to grasp the complexity of the alternate systems formed by this element.

The same holds true for the ecological systems formed by phyllostomid bats and other organisms. Using network science as a theoretical framework, a bat researcher can work with all variability observed within and between the populations of a given model species. Why, for instance, draw conclusions about dietary specialization based on mean proportions of fruits and other food items in the menu of a particular phyllostomid species, if you can consider its diet as an emergent property of the collective behavior of different individuals? The agenda behind this approach is to discover nontrivial emergent properties, such as those observed in carbon-based substances. Therefore, network science can investigate multiple phenomena and relationships at varying spatial and temporal scales by modeling individuals, functional groups, populations, and communities as nodes and the relationships between them as links. Emergent properties arise from the assembly of those networks and can then be detected using sophisticated analytical methods.

This quest for emergent properties is not new. The first use of a network approach in ecology dates back to the late nineteenth century (Camerano 1880). In ecology, during most of the twentieth century, a network approach was used mainly to study food webs. However, early in the first decade of the twenty-first century, there was a shift in interest to mutualistic networks (Ings and Hawes 2018). Ecology has benefited much from this approach, which serves as a cognitive map for answering questions related to ecological systems at all levels of biological complexity and spatial scales (Dormann et al. 2017; Poisot et al. 2016). Network science provides useful analytical and theoretical frameworks that have also helped bat research to advance. Therefore, it is important to become acquainted with network dialect to better understand the literature (table 21.1).

To our knowledge, the first two bat studies that used a network approach were published only in the middle of the first decade of this century and focused on sociality (Ortega and Maldonado 2006; Rhodes et al. 2006). Since then, many other studies that combine bat biology and network science have been published, focusing on a variety of topics. For example, we have seen this approach in studies on metacommunity structure in bat assemblages (Meyer and Kalko 2008) and viral infection in bats (Luis et al. 2015).

In this chapter, we review how network science has been used to disentangle bats from the "web of life" (*sensu* Bascompte 2009; Darwin 1859; Lewinsohn and Cagnolo 2012), in addition to the main issues it is helping to solve and how its application varies among studies that include phyllostomids and other groups. It is not our goal, here, to describe in detail the findings of each study covered by our survey or to discuss the natural history of the interactions between bats and other organisms. Still, the reader who wants to look into those details will find them in in the online supplementary material for this chapter and also in other chapters of this book. For instance, chapters 16 and 17 also report on bat studies that have made important discoveries about pollination and seed dispersal using a network approach. In general, network science applied to bat research is still young but very diverse in terms of the topics covered. We aim to document the multiple avenues for research that have been opened by those pioneering studies.

Our Review Methods

Here we present a full narrative literature review (*sensu* Pautasso 2013) of the current research on bats (mainly phyllostomids) using network theory. We chose this

Table 21.1. A small dictionary of network science

Term	Definition and Usage
Adjacency matrix	A matrix that defines which *vertices* in a *graph* are connected to one another by an *edge*.
Average path length	A metric of *connectivity*. Considering all small *paths* between all pairs of *nodes* in a *network*, the average path length is the arithmetic mean of the length of those small paths. For example, "six degrees of separation" is a concept related to how many social relationships, on average, separate any two persons in the world.
Betweenness	A metric of *centrality*. The proportion of *small paths* in the *network* in which the *node* is present. For example, a bat species that feeds on flower species of different *modules* in a network is expected to have high betweenness, as it is a bridge between regions of the network.
Binary	A *link* whose value is either 0 or 1. A *network* based on presence or absence of links is called binary. For example, a bat species visiting or not visiting a plant species in a pollination network.
Bipartite	A *network* that is divided into two *sets* of *nodes*, in which *links* may exist only between nodes of different sets. For example, a network formed between nectarivorous bat species and the plant species visited by them: bats may visit plants but not other bats. Synonym: *two-mode network*.
Centrality	The relative importance of a *node* to the *topology* of its *network*. There are many different concepts of centrality focused on different aspects of a node's importance. For example, the *degree* of a node is a kind of centrality.
Closeness	A metric of *centrality*. The average number of *small paths* that separates a given *node* to any other nodes of the same *network*. For example, a bat species that visits a set of flower species, which are visited by many other bats in a network, is expected to have high closeness. In other words, the bat species has a very common niche.
Combined	A *network* with two or more predominant *topologies*, usually hierarchical to one another. For instance, a network that has a *modular* topology and whose modules that are internally *nested*.
Community	In network science, a *community* is a cohesive subgroup of *nodes* or *links* in the *network*, which are more strongly related to one another than to other nodes or links of the same network. Communities are also called *modules*.
Complementary specialization	A *connectivity* metric. It assesses how much the *nodes* in a *network* establish unique *links*. For example, in a bat-plant network, if each bat species visits a different *set* of plant species, and niche overlap is close to zero, then the network scores high complementary specialization; in other words, the bat species play complementary functional roles.
Complex system	A *system* whose properties cannot be inferred only from its *elements* or *relationships* is called a complex system. The properties of a complex system emerge from its assembly, and so they are called emergent properties. For example, a *network* may be *nested*, but a *node* may not.
Connectance	A metric of *connectivity*. The proportion of *links* observed in a *network* in relation to the number of potential links it could maximally have. For example, interaction types with higher specificity, such as parasitism, are expected to form networks with lower connectance than interaction types with lower specificity, such as seed dispersal.
Connectivity	The number and distribution of *links* in a *network* or *edges* in a *graph*. For example, if two networks have the same *size*, but one has fewer links, it has lower connectivity. Another example: if two networks have the same size and *degree*, but one is *nested* and the other is *modular*, they have different connectivity.
Degree	A metric of *centrality*. The number of links that a *node* has in a *network*, or the number of *vertices* in a *graph*. For example, a bat species that visits several flower species in a network has higher degree than a bat species that visits few flower species.
Drawing	The visual representation of a *network* or *graph*. There are several methods for drawing networks and graphs, based on algorithms focused on highlighting different properties of the *system*. For instance, *bipartite* algorithms emphasize the different sets of nodes in the network, while energy-minimization algorithms emphasize the *centrality* of different *nodes* and the *modularity* of the network, and circular algorithms try to arrange the nodes in a circle or sphere.
Edge	Used in *graph* theory. A *relationship* between *vertices* of a graph.
Element	An entity that belongs to a *set* or a *system*.
Emergent property	A property that emerges from the assembly of a complex *network*. Emergent properties may be weak, when they exist by definition, or strong, when they emerge from the *system*. For example, if a bat-plant network is *nested*, *nestedness* is one of its weak emergent properties, while *robustness* to extinctions may be one of its strong emergent properties.
Foodweb	A *network* whose *links* represent trophic interactions ("who eats whom?"). For example, a network formed by bats and their predators, or bats and their prey.
Graph	Used in *graph* theory. A *set* of *vertices* (*elements*) and the *edges* (*relationships*) between those vertices. Graphs are abstract systems studied in pure mathematics. When a graph is studied in applied mathematics and represents a real-word system, it is called a *network*.
Layer	A set of *links* of one type in a *multilayer network*. For example, in a bat-plant network that contains two types of interactions, let us say frugivory and nectarivory, the frugivory links form one layer of the network.

(continued)

Table 21.1. Continued

Term	Definition and Usage
Link	Used in *network* theory. A *relationship* between nodes of a network. For example, an interaction of pollination between a bat species and a plant species.
Modular	A *network* in which *modularity* predominates as a *topology*. For example, a pollination network in which different plant families form separate subgroups visited by different bat species.
Modularity	The *topology* of a *network* in terms of how many cohesive subgroups (i.e., *modules* or *communities*) it contains and how much *connectivity* there is between those modules.
Module	A subgroup of *nodes* in a *network* that are more densely (in a *binary* network) or strongly (in a *weighted* network) connected to one another than to other nodes of the same network. It may also refer to a subgroup of links in a network that share a very similar subgroup of vertices. For example, in a bat-plant network, a subgroup of bat species that feed on the same plant species.
Multilayer	A *network*, in which two *nodes* may be connected to one another by two or more types of *links*. For example, a network formed by bats and the plants that they visit to feed on fruits (link type 1) or nectar (link type 2).
Multipartite	A *network* that is divided into three or more *sets* of nodes, in which links may exist only between nodes of different sets. For example, a network formed by nectarivorous bats, the plant species visited by them, and the ectoparasites of those bats. Tripartite *foodwebs* are also called tritrophic networks.
Nested	A *network* in which *nestedness* predominates as a *topology*. For example, a pollination network, in which the plant species visited by a bat species with lower *degree* are also visited by a bat species with higher degree.
Nestedness	The *topology* of a *network* in terms of how much the *links* of *nodes* with lower degree represent a subset of the links of nodes with higher degree.
Network	Used in *network* theory. A *set* of nodes (*elements*) and the *links* (*relationships*) between those nodes. For example, in biology, real-word systems, such as ensembles formed by plants and pollinators, are modeled as networks.
Node	Used in *network* theory. An *element* of a *network*. For example, a bat species in a pollination network at the community level.
One-mode	See *unipartite*.
Path	A *set* of *links* in a *network* that connect two *nodes*. A *small path* is a geodesic, i.e., the smallest set of links between two nodes.
Projection	A *subnetwork* formed by one set of *nodes* from a *multipartite network*. For example, in a bat-plant network, if one builds a subnetwork with bat species only, in which two bat species are connected to one another by a *link* when they share at least one plant species, this is a *unipartite* projection.
Relationship	A *connection* between *elements* in a *system*, *network*, or *graph*.
Robust	A *network* that has the *emergent property* of *robustness*.
Robustness	An *emergent property* of some *networks* that suffer little change in *topology* when a given proportion of its *nodes* or *links* are removed. For example, a bat-plant network is robust when it can lose several flower species without changing from *nestedness* into another topology.
Set	A group of *elements* in a *system*, *network*, or *graph*.
Size	The number of *nodes* in a *network* or the number of *edges* in a *graph*. For example, the sum of bat and plant species in a pollination network.
Subnetwork	A subset of the *nodes* and *links* of a *network*. For example, in a bat-plant network that includes different plant families, if one builds a network with only the Piperaceae and the bat species that visit those plants, this is a *subnetwork*.
System	A *set* of *elements* and the *relationships* between those *elements*. *Networks* and *graphs* are kinds of systems.
Topology	The structure of a *network* or *graph* in terms of *degree*, *size*, and *connectivity*. For example, *nestedness* and *modularity* are two kinds of topology.
Two-mode	See *bipartite* network.
Unipartite	A *network* that that contains only one *set* of *nodes*, in which *links* may exist between any nodes. For example, a network of social relationships between individuals within a bat population. Synonym: *one-mode network*.
Vertex	Used in *graph* theory. An *element* of a graph.
Weighted	A *link* to which some weight is attributed. A *network* based not only on presence or absence of links but also on the weight of those links is called weighted. For example, the frequency of visits observed in the field of a given bat species to a given plant species may be used to attribute a weight to the link between them.

Note: In some cases, the same entities are represented by different terms in network theory and graph theory. For further vocabulary, explanations, and mathematical definitions, consult the specialized literature (e.g., Barabasi 2016; Mello et al. 2016; Pocock et al. 2016). Terms explained circularly in this dictionary are written in *italics*.

review method in order to minimize personal biases in the search and choice of studies. Our goal was to avoid overlooking studies that are not directly related to our own fields of expertise. In addition, considering that the first studies with a network approach focused on bats were published only in 2006, both the amount of published research and the amount of literature reviews in this field are small. Therefore, following the framework of Pautasso (2013), we established as our main goal in this review to identify the primary research questions pursued. In the online supplementary material, we extracted qualitative information about the studies included in this review, such as central findings, caveats, and suggestions for future research.

First, we surveyed the literature on Google Scholar (https://scholar.google.com) using the following keyword combinations: combination 1—"Ecology AND bat AND Chiroptera AND Phyllostomidae AND networks AND graphs"; combination 2—"Ecology AND bat AND Chiroptera AND networks AND graphs." Our initial target was studies that focused only on the family Phyllostomidae (keyword combination 1), but we quickly noticed that some interesting studies on bat networks also used other families as models (e.g., Fortuna et al. 2009), so we decided to expand the scope of our review. We restricted our survey to studies published up to June 2016 in order to cover the first 10 years of bat research with a network approach. This survey led to 128 results. Second, to increase the range of search and avoid biases, we also checked for the references cited

by those studies, looking for the same keywords used in the survey. Third, we performed the same survey one more time, excluding the keyword Phyllostomidae (keyword combination 2), which resulted in a list of 831 studies, but no further studies were added to our review.

We restricted our review to studies that used analytical methods directly derived from network science. Thus, we did not consider studies that are based on similar analyses but that come from other theoretical frameworks, such as multivariate statistics and niche overlap analysis. After finishing our survey, we included one more recently published study (Zarazúa-Carbajal et al. 2016).

All figures and analyses were made in R 3.6.0 (R Core Team 2019) using the packages *bipartite* (Dormann et al. 2009), *igraph* (Csárdi and Nepusz 2006), and *ggplot2* (Wickham 2009).

General Trends

Among the 128 search results mentioned above, only 29 studies linked bats and network science directly (supplemental table S21.1). Twenty-three of them used phyllostomid bats as a model. We classified them according to main topic: (*A*) host-pathogen interactions, (*B*) metacommunities, (*C*) mutualistic interactions, (*D*) social interactions, and (*E*) use of space (fig. 21.1). The number of studies varied with main topic, and, except for use of space, phyllostomids were used as

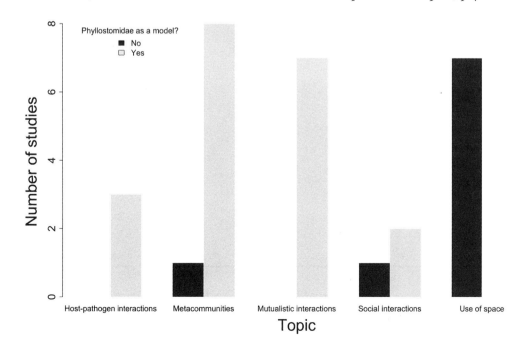

Figure 21.1. Number of studies by topic included in our review. Most studies ($N = 9/29$) focused on bat metacommunities.

models to assess all topics. Mutualistic interactions and metacommunities were the most common topics.

As explained in table 21.1, a graph is an abstract mathematical system, while a network represents a real-world system. A network is composed of nodes (elements of the system) and links (relationships between elements), and the biological meaning of the links defines the type of network.

In the 29 studies we analyzed, the biological meaning of nodes and links varied considerably. For example, some studies were interested in ecological systems at the community level, and their nodes represent bat species, while other studies focused on bat populations, and their nodes represent individual bats. Depending on the topic being investigated, links in those studies represented interactions of frugivory or social relationships.

The studies considered in our review differed from one another not only in terms of what the links in the network mean (social relationships, frugivory, parasitism, etc.) but also in terms of network type (one mode = 10 studies, two mode = 14, both = 5) and data type (binary = 17, weighted = 9, both = 3). Despite its late debut in bat research, the number of studies with a network approach has been increasing, and so more studies are to be expected with time (fig. 21.2). Several software packages are available for network analysis and drawing (see comprehensive lists in Barabasi 2016; Mello et al. 2016; Pocock et al. 2016). We found that at least 12 programs were used in the studies included in our review (fig. 21.3).

Our survey clearly shows that, despite the variety of programs available and used, R (R Core Team 2019) is a hub among bat studies. The studies considered in our review used between one and five programs for network analysis, and each program was used by at least three studies. These results suggest that bat researchers still have many possibilities regarding software and metrics to explore. Nestedness (see table 21.1) was by far the most popular concept used in those studies (T and NODF metrics, $n = 15$), followed by modularity (Newman, Guimerà, Barber, and Louvain metrics, $n = 5$), and specialization (H_2' metric, $n = 4$).

Some drivers of ecological interactions need to be more broadly assessed with network science. For instance, functional diversity has not been directly investigated with a network approach in bat studies, although some papers used species traits to assess antagonisms (Zarazúa-Carbajal et al. 2016) and mutualisms (Mello et al. 2015). Biodiversity has been studied mainly from a taxonomic perspective (e.g., Presley et al. 2012), but in network studies, it has also been viewed from a functional perspective. However, we found no network studies on bats that used phylogeny explicitly.

Although network science has provided many new tools for analyzing systems formed by bats and other organisms, there is some controversy regarding the assumptions and properties of network indexes, especially in the analysis of social relationships (Johnson et al. 2013) and interspecific interactions (Blüthgen 2010). The main concerns are related to the dependence of some metrics on sampling sufficiency (Dormann et al. 2009; Jordano 2016), network size (Fründ et al. 2016), and the ecological interpretation of network metrics that have been used as proxies for ecological concepts (Mello et al. 2015). Considering those biases, comparing topology between networks is even harder than describing each network separately, so caution is advised.

Metacommunities

Systems formed by local communities (sets of populations containing organisms of different species) that are interconnected to one another by processes of dispersal, that is, metacommunities, have been studied in ecology for decades. Already in the seminal book that

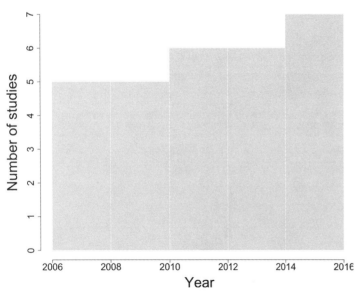

Figure 21.2. Growth in the number of studies on bats that used network science as a framework from 2006 to 2016.

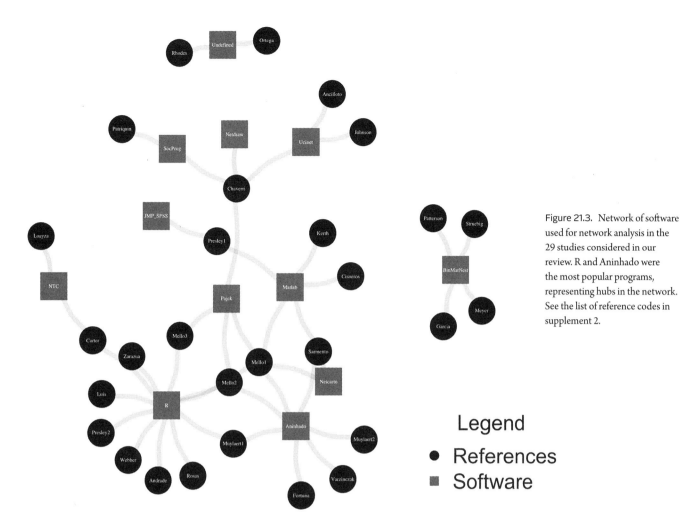

Figure 21.3. Network of software used for network analysis in the 29 studies considered in our review. R and Aninhado were the most popular programs, representing hubs in the network. See the list of reference codes in supplement 2.

Legend

● References
■ Software

formalized the analytical framework of metacommunities (*The Unified Neutral Theory of Biodiversity and Biogreography* [Hubbell 2001]), bats have been used as models (see the data from Barro Colorado Island, Panama).

Due to their high mobility, which stands out among mammals, bats are assumed to have high dispersal capacity, which would lead to local communities having structures similar to their respective regional metacommunities (Meyer and Kalko 2008; Presley and Willig 2010). In fact, from a philosophical perspective, metacommunities are usually considered to be communities in which dispersal is limited and spatially heterogeneous (*sensu* Vellend 2016). In the case of bats, however, this limitation seems to be weaker than in other groups. The network approach applied to metacommunities has helped with understanding complex cases such as those in which bats are involved, as it expanded the analytical tool set of the neutral theory. Its main contribution was providing new metrics to calculate

topology in order to differentiate between nestedness, modularity, and turnover in species occurrences.

A recent framework (Presley et al. 2010) suggests the existence of different patterns in metacommunities along a gradient involving those three main topological patterns. There is indeed evidence of nestedness in bat communities on islands (Meyer and Kalko 2008). Bat communities in anthropogenically fragmented landscapes also seem to be nested, despite evidence of a fragmentation threshold (Muylaert et al. 2016). Nestedness in bat communities that live in fragmented landscapes seems to be related, again, to the high mobility of those animals, which can cross several types of barriers and use several types of habitats (Heer et al. 2015).

Their high mobility also gives them a wider variety of habitat patch choices in heterogeneous landscapes, therefore increasing the importance of species-area relationships and neutral processes in the assembly of metacommunities. In fact, in bats and in other groups the balance between niche and neutral processes as drivers of community structure seems to be a continuum (Gravel et al. 2006) that varies with scale (Pinheiro et al. 2016) and depends on the balance between selection, drift, dispersal, and speciation (Vellend 2016).

Social Relationships

Sociality is a very popular topic assessed with a network approach in bat studies. It can be divided into two broad themes: social relationships and use of space. For a broad review on sociality in bats, see Johnson et al. (2013). Social networks are a textbook example of how a systems perspective helps unveil emergent properties that could not be observed with a mean field approach (Christakis and Fowler 2009). For instance, social graphs of bats helped advance our understanding of how they organize their populations into social groups (Johnson et al. 2013) and also how those groups are subjected to fission-fusion dynamics, which ultimately influence the transmission of diseases and the conservation of threatened populations (Fortuna et al. 2009).

The analysis of bat social structure has developed considerably in the past years, especially thanks to new tools such as miniature transmitters that allow tracking their behavior in detail (Fleischmann and Kerth 2014). Although we now have a good variety of methods for acquiring data to build bat social networks, there is still much to be done, especially in terms of incorporating new technologies for observation of bat behavior *in situ*. The small size of phyllostomid bats and their nocturnal habits make direct observation and tagging hard to accomplish (O'Mara et al. 2014).

A problem with studies on social relationships is the lack of broad sampling. For instance, rarely were more than ten groups of bats analyzed in the studies we found, which often leads to an increase in errors and can limit statistical inference. We need to improve our sampling designs since there are many open avenues for research in the field of bat social networks (Ancillotto et al. 2012).

One of the main insights provided by network science in the field of sociality is the fission-fusion structure observed in many bat societies; in many cases bat populations are composed of spatially separate colonies bound to one another by gene flow (as in Fortuna et al. 2009; see chap. 19, this vol.). The potential for using networks to study bat sociality is huge and has only begun to be explored. Social network analysis is one of the largest fields in network science with many tools available.

Use of Space

There are many possibilities for understanding the spatial ecology of bats using network science. Thanks to the new paradigm of movement ecology, we now have a better comprehension of the processes that shape movement and use of space patterns. Another key development in the past decades was methodological: GPS tracking of medium- and small-sized bats (O'Mara et al. 2014). The large data sets generated by modern tracking methods are now being analyzed mainly using a network approach. Spatial networks lead to important developments in this discipline, especially related to understanding flight behavior, foraging routes, territories, and migration in bats.

Network metrics are now considered good proxies for landscape connectivity, thanks to the development of ecologically scaled landscape metrics (Minor and Urban 2008). As our understanding of the dispersal capacity of bats improves, we can calibrate models that consider bat movements, colonization of new areas (in the field of restoration ecology), and population viability (Niebuhr et al. 2015).

Tracking fine-scale movement also helps us understand how bats are influenced by the surrounding landscape and how bats can change landscapes in time via seed or pathogen dispersal. In this sense, field studies on bat movements are needed, as well as individual-based modeling in complex landscapes (Jeltsch et al. 2013). Spatial network analysis is helping us reveal the role of bats as "mobile links," including resource, genetic, and process links (Jeltsch et al. 2013; Kremen et al. 2007).

One very interesting finding made using network science was that the fission-fusion structure of bat societies might affect viral outbreaks and the conservation of endangered species (Fortuna et al. 2009). In that study, a modular structure in spatial networks of insectivorous bats slowed down the potential spread of diseases and the exchange of information through the network. This approach could be also used for phyllostomids, since their roosting and social network structure may be key to understanding their role in pathogen spillover and infection dynamics.

Mutualistic Interactions

Traditionally applied to the study of food webs in ecology, network science blossomed in the field of mutualisms in the early part of the current century (Ings and Hawes 2018). Traditionally, mutualistic interactions between bats and plants were assessed mainly through studies that resulted in diet inventories or behavioral description of bat behavior towards fruits and flowers. In some cases, experimental approaches were used to test whether the net effect of a given bat species on a given plant species was mainly positive, negative, or neutral. Network science has allowed connecting the dots between fragmented naturalistic information (Poisot et al. 2016) in order to understand the emergent properties of bat-plant systems. Putting the pieces together in the puzzles of seed dispersal and pollination is crucial because the functional role played by a species, as well as that species' evolutionary influence on other species, is the result of its direct and indirect paths of interaction (Mello et al. 2015).

In the context of mutualistic networks, evidence gathered using bats as models is helping solve the dilemma of prevalent topologies among interaction networks (Lewinsohn et al. 2006). The current paradigm says that mutualisms tend to form nested networks, while antagonisms lead to modular networks (Bascompte et al. 2003; Cohen and Havlin 2010; Montoya et al. 2015). Nevertheless, recent studies, including some that used phyllostomids as models, suggest that those structural patterns might occur together as compound topologies (Mello et al. 2011; Pinheiro et al. 2016).

The archetype of compound topologies helps us understand how redundant or complementary the ecosystem functions and services delivered by bats are in comparison to other animals that participate in the same mutualistic networks. For instance, bats and birds seem to form separate, internally nested modules within mixed-taxon frugivory and seed dispersal networks with a compound topology (Mello et al. 2011; Sarmento et al. 2014) (fig. 21.4). This means that bats interact with different subsets of plants than birds and other animals (a similar pattern was observed by Donatti et al. 2011), so their functional roles cannot be easily replaced in case of local extinctions (also see chap. 17, this vol.).

So far, the only study that has applied the keystone-species concept to a mutualistic system formed by bats and plants suggested that dietary specialization is key to understanding why some bat species are more important than others for maintaining vital ecosystem functions, such as seed dispersal (Mello et al. 2015). Another study using the keystone-species framework aimed at suggesting how the knowledge of plant-animal interactions, including bat-plant networks, might be applied to tree species selection in agroecosystems (Peters et al. 2016).

Another trend that stands out in our sample is that, despite the fact that mutualisms in general are known to have been largely neglected in the ecological literature until recently (Bronstein 2015), it seems that bat-plant mutualistic interactions have been studied with a network approach more frequently than other types of ecological interactions. For instance, nearly a quarter of our selected studies (seven out of 29) focused on frugivory, seed dispersal, nectarivory, or pollination. Studies on bat-plant mutualisms range from the community (e.g., Mello et al. 2011) to individual levels of biological complexity (e.g., Muylaert et al. 2014). It is interesting to note that potential mutualisms comprised more than twice the number of studies on potential antagonisms, such as parasite infestation or viral infection. And mutualisms continue to receive much more attention from bat researchers who use a network approach, as seen in the examples reported in a recent book on ecological networks in the tropics (Dáttilo and Rico-Gray 2018).

Finally, in studies of ecological networks, we need to also consider that interaction events are usually classified on the basis of indirect evidence (Johnson et al. 2013). For instance, interactions of simple nectarivory and true pollination are sometimes hard to tell apart (King et al. 2013), as well as interactions of frugivory and seed dispersal (Schupp et al. 2010). Therefore, studies on interaction networks formed by bats and other organisms need to account for Darwin's "tangled bank"—in other words, the multilayered nature of ecological and social associations, considering positive, negative, and neutral outcomes for both sides involved (as in Genrich et al. 2017). With the exponential advancement of computing power, data visualization and processing are also developing quickly, so that it is now possible to decode systems of very high complexity (Lima 2011). The new framework provided by multilayered networks is very promising for untangling complex interactions (Boccaletti et al. 2014; Pilosof et al. 2017).

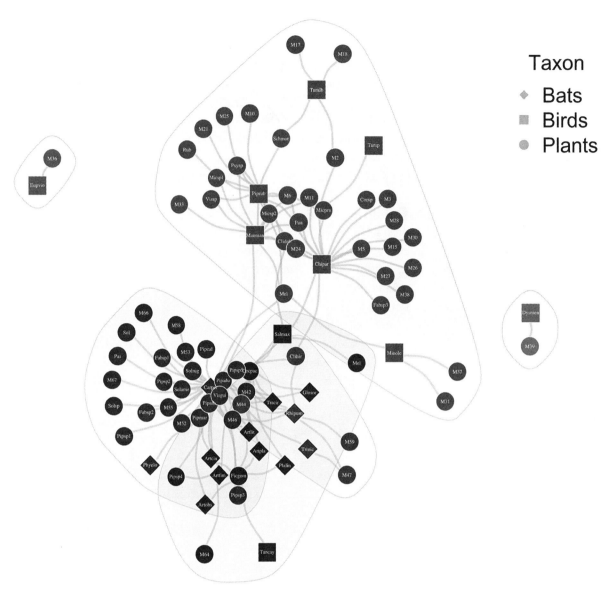

Figure 21.4. A network with bat (*diamonds*), bird (*squares*), and plant species (*circles*), showing a combined topology: modules that are internally nested. The *lines* represent interactions of frugivory and seed dispersal. The *shades of gray* in the symbols, lines, and surrounding clouds represent the modules found in the network using the Louvain weighted algorithm (resolution = 1.0, Blondel et al. 2008). Figure redrawn using the same data analyzed by Sarmento et al. (2014). The list of bat, bird, and plant species represented in the network is provided in supplement 3; codes were made using the first three letters of the respective genus and epithet.

Host-Pathogen Interactions

Disease emergence, as well as the part that bats play in it, is a broad and polemic topic—one that needs to be carefully addressed and openly discussed. Pathogen spillover can be harmful both for bat populations (e.g., in the case of the white-nose syndrome) and human populations (e.g., in the case of rabies) that rely on bats for a variety of ecosystem services (Kunz et al. 2011). Information about bats as reservoirs for viruses and other pathogens relies on a set of studies that contain considerable evidence pointing to bats as pathogen hosts (Hayman 2016a)—and this body of evidence keeps growing (Goés et al. 2016; Jones et al. 2008; Kretzschmar and Morris 1996; Lutz et al. 2016; Young and Olival 2016).

Network science can be helpful for understanding host-pathogen interactions. It may help track pathogen diffusion, potential spillover events, levels of interdependence between pathogen and hosts, and many other aspects of disease. It was first used as a framework for understanding infection and disease over two decades ago (Kretzschmar and Morris 1996). Years later, networks formed by bats and parasitic flies have been studied and compared to other network types (Patterson et al. 2009).

Recent viral outbreaks related to bats have led to innovative studies using network science. Bats potentially carry Ebola virus (Sun 2019), coronaviruses (Goés et al. 2016), and hantavirus (Sabino-Santos et al. 2015). This is another open avenue for research because other bat infections may be potentially discovered (Young and Olival 2016). A recent global network study highlighted the importance of bats in spreading and coevolving with viruses for millions of years (Luis et al. 2015). This study highlighted the need for understanding the role of long- and medium-scale migration on viral spread across species, since migration is an important driver of pathogen sharing.

Recent efforts have also been made to consider parallel phylogenies for understanding pathogen-host interactions and viral diversity. After an Ebola outbreak a few years ago, and together with the emergence of other viruses potentially carried by bats, such as the severe acute respiratory syndrome–like (SARS-like) virus and Middle East respiratory syndrome (more commonly known as MERS), attention has turned to viral diversity in bats (Daszak et al. 2000). For instance, viral diversity for many bat species has been assessed in viral discovery hotspots such as the Brazilian Atlantic Forest (Goés et al. 2016). Moreover, new findings on viral infection have been made, such as the discovery of hantavirus antibodies among phyllostomids in the Americas (Melo et al. 2015). In the case of phyllostomids, more attention should be given to studying bat-virus networks, as there are two hotspots of risk for the emergence of new diseases, mainly in Central America and southeastern Brazil (Jones et al. 2008).

Regarding disease spread caused by the Ebola virus, some studies show that deforestation and hunting are linked to higher disease risk, which supports the idea that biological conservation is not detached from disease prevention (Hayman 2016b, 2019; Rulli et al. 2017). Some recent findings provide concrete evidence of a link between the Ebola virus and bats. Despite the fact that much of the Ebola virus transmission cycle is still puzzling (Olival and Hayman 2014), we cannot remove bats from this puzzle and ignore the evidence that points to pteropodids as reservoirs (Han et al. 2016; Leroy et al. 2005). Several years ago, Leroy et al. (2005) detected Ebola virus RNA in bats, and in January 2019 the Ecohealth Alliance, in partnership with

the government of Liberia, reported finding Ebola virus RNA sequences in *Miniopterus inflatus* (Sun 2019). Evidence from viral sequences extracted from pteropodids such as *Epomops franqueti*, *Hypsignathus monstrosus*, and *Myonycteris torquota* and phylogenetic analysis suggest that fruit bats have played a role in the evolution of Zaire Ebola virus (Leendertz et al. 2016). Other studies have detected anti-Ebola virus antibodies in apparently healthy bats (Hayman et al. 2010; Pourrut et al. 2009).

These results suggest that network science can help describe and forecast complex interactions and processes worldwide, such as the white-nose syndrome dynamics and viral epidemics, helping to inform bat conservation and public health initiatives.

Shortfalls

Although we have made great progress in understanding the world's biodiversity in the past decades, there is still a lot of work to do. There are seven main shortfalls in our knowledge of biodiversity (Hortal et al. 2015), and network science applied to bat research can help solve at least two of them.

The first is the Eltonian shortfall, which is related to the lack of knowledge of species interactions and the effects of these interactions on individuals. We are far from understanding how these interspecific interactions can influence the survival and fitness of interacting species. This shortfall is closely related to the framework of keystone species, which are very hard to define and identify in natural and agricultural systems (Cottee-Jones and Whittaker 2012). Thanks to the tools and framework provided by network science, however, some important developments have been achieved in this area (Benedek et al. 2007). Keystones may now be identified using centrality metrics, which help describe the relative contribution of an element to the structure and dynamics of its system (Cirtwill et al. 2018; Mello 2019).

Finally, the Raunkiaeran shortfall is related to the lack of knowledge of species traits and ecological functions. Functional traits are usually measured at the individual level but need not be restricted to it (Hortal et al. 2015). Some studies have been trying to understand how species traits affect the structure of interaction networks, but our knowledge in this area is still in its infancy (Dehling et al. 2016; Mello et al. 2015). So far, network science has helped us find out that ecological specialization is key to understanding ecological functions in interaction networks (Aizen et al. 2012; Stang et al. 2006).

Conclusions

The main message from our review is that network science as a framework for studying bats is still in its infancy, but it has already helped advance knowledge in many different disciplines.

From 2006 to 2016, only 29 studies had used this approach. Those pioneering studies point to exciting avenues of research, which could lead to important advances in bat research. One of those avenues is mutualistic interactions. We now have a lot of data on the role of bats as consumers of plant parts and their role in plant reproduction (Fleming and Kress 2013). Network science provides us with tools to put all that information together in models that consider not only sets of species but also systems of species that are bound together by different kinds of associations. Future studies might, for instance, investigate the relative importance of different plant species and genera in maintaining bat populations and communities. With a network approach, it is also possible to use this kind of knowledge to inform ecological restoration initiatives, by suggesting which tree and shrub species are most important in seed dispersal and pollination networks—ones that might be used to accelerate forest and savanna regeneration (as in Howe 2016; Kaiser-Bunbury and Blüthgen 2015; Peters et al. 2016).

Likewise, the study of host-pathogen interactions—for instance, those focused on Ebola viruses and coronaviruses—might also benefit from a network approach. Disease is traditionally studied from a clinical perspective, but studying the ecological systems formed by pathogens, vectors, and hosts might help to understand and control outbreaks (Pinheiro et al. 2016; Svensson-Coelho et al. 2014). The ecology of bats in the context of disease (Luis et al. 2015) and parasitism (Zarazúa-Carbajal et al. 2016) is starting to be modeled using network science. However, much remains to be done, as bats harbor a huge diversity of viruses, bacteria, and fungi. We still need to understand, for instance, how some of those pathogens may be transmitted from bats to humans and vice versa.

Great potential also exists to develop further the spatial and movement ecology of bats by using a net-work approach. Tracking devices are getting smaller every year, and GPS telemetry is starting to be used more frequently to assess bat home ranges, commuting behavior, orientation, and migration (Cvikel et al. 2014; Holland et al. 2006). The Icarus initiative hosted by the Max Planck Institute of Animal Behavior in Germany (https://www.icarus.mpg.de/en) is a disruptive innovation in the study of animal movements, and prominent bat researchers are directly involved in it. In a few years, our understanding of bat dispersal, foraging, commuting, and migration might change dramatically thanks to Icarus. In addition, a new generation of digital sensors, developed specially by bat researchers from the Museum of Natural History in Berlin (https://www.museumfuernaturkunde.berlin/de/taxonomy/term/234/frieder.mayer), Germany, is now making it possible to record individual to individual interactions between bats in flight, which will allow us to build much more realistic social networks and to understand the role of dispersal and drift in shaping bat societies. Furthermore, the theoretical framework of social network analysis as applied to wildlife is growing and starting to develop more specialized approaches (Farine 2017; Farine and Whitehead 2015).

In conclusion, the time is ripe for innovative bat research, and network science is one of our most valuable tools for the task of disentangling bats from the web of life. So far, bat studies, as well as studies using other organisms as models, have focused their attention on trivial emergent properties of complex networks, such as nestedness and modularity. In the future, we should also investigate nontrivial emergent properties in all kinds of networks formed by bats. This next step might help us, for instance, to understand how natural and anthropic changes in network structure affect dynamical processes, ultimately having impacts on vital ecosystem services such as pollination, seed dispersal, pest suppression, social information transfer, and pathogen transmission. Those ecosystem services could be assessed as the nontrivial emergent properties of bat networks, like high hardness emerges from the tridimensional crystalline assembly of carbon atoms in a diamond.

Acknowledgments

We thank our colleagues from several institutions all over the world, who have published inspiring papers on bats using network science as a framework. Special thanks go to Elisabeth K. V. Kalko, who helped us interpret the biological meaning of network metrics, and to Pedro Jordano and Paulo Guimarães Jr., who opened the doors of the amazing network world to us. Pedro Jordano and Christoph Meyer reviewed the first version of our chapter and helped us improve it considerably as did Theodore Fleming and Liliana Dávalos. We are also grateful to Carsten Dormann, who created *bipartite* for R and made network analysis much easier. Katherine Ognyanova gave us invaluable tips on how to draw networks in R, and we strongly recommend her online tutorial (http://kateto.net/network-visualization). MARM was funded by the Alexander von Humboldt Foundation (AvH, # 3.4–8151/15037), Brazilian Council for Scientific and Technological Development (CNPq, #302700/2016–1, and scholarships given to students), Brazilian Coordination for the Improvement of Higher Education Personnel (CAPES, scholarships given to students), and Dean of Research of the University of São Paulo (PRP-USP, 18.1.660.41.7). RLM received PhD scholarships from the São Paulo Research Foundation (FAPESP, #2015-17739-4, #2017-21816-0).

Online Supplementary Material

All supplements are provided online as a single XLSX file.

Supplement 1. Information extracted from the 29 papers included in our review. We made our survey on Google Scholar and restricted it to studies published from 2006 to 2016.

Supplement 2. References and short names of the studies included in our review and represented in figure 21.3.

Supplement 3. List of bat, bird, and plant species represented in figure 21.4.

References

Aizen, M. A., M. Sabatino, and J. M. Tylianakis. 2012. Specialization and rarity predict nonrandom loss of interactions from mutualist networks. Science 335 (6075): 1486–1489. https://doi.org/10.1126/science.1215320.

Ancillotto, L., M. T. Serangeli, and D. Russo. 2012. Spatial proximity between newborns influences the development of social relationships in bats. Ethology 118 (4): 331–340. https://doi.org/10.1111/j.1439-0310.2011.02016.x.

Barabasi, A.-L. 2016. Network Science. Cambridge: Cambridge University Press. http://barabasi.com/networksciencebook/.

Bascompte, J. 2009. Disentangling the web of life. Science 325 (5939): 416–419. https://doi.org/10.1126/science.1170749.

Bascompte, J., P. Jordano, C. J. Melian, and J. M. Olesen. 2003. The nested assembly of plant-animal mutualistic networks. Proceedings of the National Academy of Sciences, USA 100 (16): 9383–9387. https://doi.org/DOI 10.1073/pnas.1633576100.

Benedek, Z., F. Jordán, and A. Báldi. 2007. Topological keystone species complexes in ecological interaction networks. Community Ecology 8 (1): 1–7. https://doi.org/10.1556/ComEc.8.2007.1.1.

Blondel, V. D., J.-L. Guillaume, R. Lambiotte, and E. Lefebvre. 2008. Fast unfolding of communities in large networks. Journal of Statistical Mechanics: Theory and Experiment 2008 (10): P10008. https://doi.org/10.1088/1742-5468/2008/10/P10008.

Blüthgen, N. 2010. Why network analysis is often disconnected from community ecology: a critique and an ecologist's guide. Basic and Applied Ecology 11 (3): 185–195. https://doi.org/10.1016/j.baae.2010.01.001.

Boccaletti, S., G. Bianconi, R. Criado, C. I. del Genio, J. Gómez-Gardeñes, M. Romance, I. Sendiña-Nadal, Z. Wang, and M. Zanin. 2014. The structure and dynamics of multilayer networks. Physics Reports 544 (1): 1–122. https://doi.org/10.1016/j.physrep.2014.07.001.

Bronstein, J. L., ed. 2015. Mutualism. Oxford: Oxford University Press.

Camerano, L. 1880. On the equilibrium of living beings by means of reciprocal destruction. Atti Della Reale Accademia Delle Scienze Di Torino 15 (8): 393–414.

Christakis, N. A., and J. H. Fowler. 2009. Connected: The Surprising Power of Our Social Networks and How They Shape Our Lives. New York: Little, Brown and Company.

Cirtwill, A. R., G. V. D. Riva, M. P. Gaiarsa, M. D. Bimler, E. F. Cagua, C. Coux, and D. M. Dehling. 2018. A review of species role concepts in food webs. Food Webs 16 (September): e00093. https://doi.org/10.1016/j.fooweb.2018.e00093.

Cohen, R., and S. Havlin. 2010. Complex Networks: Structure, Robustness, and Function. Cambridge: Cambridge University Press.

Cottee-Jones, H. E. W., and R. J. Whittaker. 2012. Perspective: the keystone species concept: a critical appraisal. Frontiers of Biogeography 4 (3): 117–127. http://www.escholarship.org/uc/item/15d2b65t.

Csárdi, G., and T. Nepusz. 2006. The Igraph software package for complex network research. InterJournal Complex Systems 1695:1695. http://igraph.org.

Cvikel, N., E. Levin, E. Hurme, I. Borissov, A. Boonman, E. Amichai, and Y. Yovel. 2014. On-board recordings reveal no jamming avoidance in wild bats. Proceedings of the Royal Society B–Biological Sciences 282 (1798): 20142274. https://doi.org/10.1098/rspb.2014.2274.

Darwin, C. R. 1859. The Origin of Species by Means of Natural Selection or the Preservation of Favoured Races in the Struggle for Life. London: Bantam Classics.

Daszak, P., A. A. Cunningham, and A. D. Hyatt. 2000. Emerging infectious diseases of wildlife—threats to biodiversity and human health. Science 287 (January): 443–449. https://doi.org/10.1126/science.287.5452.443.

Dáttilo, W., and V. Rico-Gray. 2018. Ecological Networks in the Tropics. Cham, Switzerland: Springer International Publishing. https://doi.org/10.1007/978-3-319-68228-0.

Dehling, D. M., P. Jordano, H. M. Schaefer, K. Böhning-Gaese, and M. Schleuning. 2016. Morphology predicts species' functional roles and their degree of specialization in plant-frugivore interactions. Proceedings of the Royal Society B–Biological Sciences 283 (1823): 20152444. https://doi.org/10.1098/rspb.2015.2444.

Donatti, C. I., P. R. Guimarães, M. Galetti, M. A. Pizo, F. M. D. Marquitti, and R. Dirzo. 2011. Analysis of a hyper-diverse seed dispersal network: modularity and underlying mechanisms. Ecology Letters 14 (8): 773–781. https://doi.org/10.1111/j.1461-0248.2011.01639.x.

Dormann, C. F., J. Frund, N. Blüthgen, and B. Gruber. 2009. Indices, graphs and null models: analyzing bipartite ecological networks. Open Ecology Journal 2 (1): 7–24. https://doi.org/10.2174/1874213000902010007.

Dormann, C. F., J. Fründ, and H. M. Schaefer. 2017. Identifying causes of patterns in ecological networks: opportunities and limitations. Annual Review of Ecology, Evolution, and Systematics 48 (1): 559–584. https://doi.org/10.1146/annurev-ecolsys-110316-022928.

Euler, L. 1741. Solutio problematis ad geometriam situs pertinentis. Commentarii Academiae Scientiarum Petropolitanae 8:128–140. http://www.math.dartmouth.edu/~euler/pages/E053.html.

Farine, D. R. 2017. A guide to null models for animal social network analysis. Methods in Ecology and Evolution 8 (10): 1309–1320. https://doi.org/10.1111/2041-210X.12772.

Farine, D. R., and H. Whitehead. 2015. Constructing, conducting and interpreting animal social network analysis. Journal of Animal Ecology 84 (5): 1144–1163. https://doi.org/10.1111/1365-2656.12418.

Findley, J. S. 1993. Bats: A Community Perspective. New York: Cambridge University Press.

Fleischmann, D., and G. Kerth. 2014. Roosting behavior and group decision making in 2 syntopic bat species with fission-fusion societies. Behavioral Ecology 25 (5): 1240–1247. https://doi.org/10.1093/beheco/aru117.

Fleming, T. H. 1988. The Short-Tailed Fruit Bat: A Study in Plant-Animal Interactions. Chicago: University of Chicago Press.

Fleming, T. H., and W. J. Kress. 2013. The Ornaments of Life: Coevolution and Conservation in the Tropics. Chicago: University of Chicago Press. http://books.google.com/books?id=zW4zAAAAQBAJ&pgis=1.

Fortuna, M. A., A. G. Popa-Lisseanu, C. Ibanez, and J. Bascompte. 2009. The roosting spatial network of a bird-predator bat. Ecology 90 (4): 934–944. https://doi.org/doi:10.1890/08-0174.1.

Fründ, J., K. S. McCann, and N. M. Williams. 2016. Sampling bias is a challenge for quantifying specialization and network structure: lessons from a quantitative niche model. Oikos 125 (4): 502–513. https://doi.org/10.1111/oik.02256.

Gardner, A. L., ed. 2008. Mammals of South America. Vol. 1. Chicago: University of Chicago Press. https://doi.org/10.7208/chicago/9780226282428.001.0001.

Genrich, C. M., M. A. R. Mello, F. A. O. Silveira, J. L. Bronstein, and A. P. Paglia. 2017. Duality of interaction outcomes in a plant-frugivore multilayer network. Oikos 126 (3): 361–368. https://doi.org/10.1111/oik.03825.

Goés, L. G. B., A. C. de Almeida Campos, C. de Carvalho, G. Ambar, L. H. Queiroz, A. P. Cruz-Neto, M. Munir, and E. L. Durigon. 2016. Genetic diversity of bats coronaviruses in the Atlantic Forest hotspot biome, Brazil. Infection, Genetics and Evolution 44 (November): 510–513. https://doi.org/10.1016/j.meegid.2016.07.034.

Gravel, D., C. D. Canham, M. Beaudet, and C. Messier. 2006. Reconciling niche and neutrality: the continuum hypothesis. Ecology Letters 9 (4): 399–409. https://doi.org/10.1111/j.1461-0248.2006.00884.x.

Han, B. A., J. P. Schmidt, L. W. Alexander, S. E. Bowden, D. T. S. Hayman, and J. M. Drake. 2016. Undiscovered bat hosts of filoviruses. PLOS Neglected Tropical Diseases 10 (7): 1–10. https://doi.org/10.1371/journal.pntd.0004815.

Hayman, D. T. S. 2016a. Bats as viral reservoirs. Annual Review of Virology 3 (1): 77–99.

Hayman, D. T. S. 2016b. Conservation as vaccination. EMBO Reports 17 (3): 286–291.

Hayman, D. T. S. 2019. African primates: likely victims, not reservoirs, of Ebolaviruses. Journal of Infectious Diseases, 220 (10): 1547–1550. https://doi.org/10.1093/infdis/jiz007.

Hayman, D. T. S., P. Emmerich, M. Yu, L.-F. Wang, R. Suu-Ire, A. R. Fooks, A. A. Cunningham, and J. L. N. Wood. 2010. Long-term survival of an urban fruit bat seropositive for Ebola and Lagos bat viruses. PLOS ONE 5 (8): e11978.

Heer, K., M. Helbig-Bonitz, R. G. Fernandes, M. A. R. Mello, and E. K. V. Kalko. 2015. Effects of land use on bat diversity in a complex plantation-forest landscape in northeastern Brazil. Journal of Mammalogy 96 (4): 720–731. https://doi.org/10.1093/jmammal/gyv068.

Holland, R. A., K. Thorup, M. J. Vonhof, W. W. Cochran, and M. Wikelski. 2006. Bat orientation using earth's magnetic field. Nature 444 (7120): 702. https://doi.org/10.1038/444702a.

Hortal, J., F. de Bello, J. A. F. Diniz-Filho, T. M. Lewinsohn, J. M. Lobo, and R. J. Ladle. 2015. Seven shortfalls that beset large-scale knowledge of biodiversity. Annual Review of Ecology, Evolution, and Systematics 46 (1): 523–549. https://doi.org/10.1146/annurev-ecolsys-112414-054400.

Howe, H. F. 2016. Making dispersal syndromes and networks useful in tropical conservation and restoration. Global Ecology and Conservation 6 (April): 152–178. https://doi.org/10.1016/j.gecco.2016.03.002.

Hubbell, S. P. 2001. The Unified Neutral Theory of Biodiversity and Biogeography. Princeton, NJ: Princeton University Press. http://books.google.com/books/about/The_unified_neutral_theory_of_biodiversi.html?id=EIQpFBu84NoC.

Ings, T. C., and J. E. Hawes. 2018. The history of ecological networks. In Ecological Networks in the Tropics, edited by W. Dáttilo and V. Rico-Gray, 15–28. Cham, Switzerland: Springer International Publishing. https://doi.org/10.1007/978-3-319-68228-0_2.

Ings, T. C., J. M. Montoya, J. Bascompte, N. Blüthgen, L. Brown, C. F. Dormann, F. Edwards et al. 2009. Ecological networks—beyond food webs. Journal of Animal Ecology 78 (1): 253–269. https://doi.org/10.1111/J.1365-2656.2008.01460.X.

Jeltsch, F., D. Bonte, G. Pe'er, B. Reineking, P. Leimgruber, N. Balkenhol, B. Schröder et al. 2013. Integrating movement ecology with biodiversity research—exploring new avenues to address spatiotemporal biodiversity dynamics. Movement Ecology 1 (1): 6. https://doi.org/10.1186/2051-3933-1-6.

Johnson, J. S., J. N. Kropczynski, and M. J. Lacki. 2013. Social network analysis and the study of sociality in bats. Acta Chiropterologica 15 (1): 1–17. https://doi.org/10.3161/150811013X667821.

Jones, K. E., N. G. Patel, M. A .Levy, A. Storeygard, D. Balk, J. L. Gittleman, and P. Daszak. 2008. Global trends in emerging infectious diseases. Nature Letters 451 (February): 990–994. https://doi.org/10.1038/nature06536.

Jordano, P. 2016. Sampling networks of ecological interactions. Functional Ecology 30 (12): 1883–1893. https://doi.org/10.1111/1365-2435.12763.

Kaiser-Bunbury, C. N., and N. Blüthgen. 2015. Integrating network ecology with applied conservation: a synthesis and guide to implementation. AoB PLANTS 7 (0): plv076. https://doi.org/10.1093/aobpla/plv076.

King, C., G. Ballantyne, and P. G. Willmer. 2013. Why flower visitation is a poor proxy for pollination: measuring single-visit pollen deposition, with implications for pollination networks and conservation. Methods in Ecology and Evolution 4 (9): 811–818. https://doi.org/10.1111/2041-210X.12074.

Kremen, C., N. M. Williams, M. A. Aizen, B. Gemmill-Herren, G. LeBuhn, R. Minckley, L. Packer et al. 2007. Pollination and other ecosystem services produced by mobile organisms: a conceptual framework for the effects of land-use change. Ecology Letters 10 (4): 299–314. https://doi.org/10.1111/j.1461-0248.2007.01018.x.

Kretzschmar, M., and M. Morris. 1996. Measures of concurrency in networks and the spread of infectious disease. Mathematical Biosciences 133 (2): 165–195. https://doi.org/10.1016/0025-5564(95)00093-3.

Kunz, T. H. 1982. Ecology of Bats. New York: Plenun Press.

Kunz, T. H., E. Braun de Torrez, D. Bauer, T. Lobova, and T. H. Fleming. 2011. Ecosystem services provided by bats. Annals of the New York Academy of Sciences 1223 (1): 1–38. https://doi.org/10.1111/j.1749-6632.2011.06004.x.

Leendertz, S. A. J., J. F. Gogarten, A. Düx, S. Calvignac-Spencer, and F. H. Leendertz. 2016. Assessing the evidence supporting fruit bats as the primary reservoirs for Ebola viruses. EcoHealth 13 (1): 18–25. https://doi.org/10.1007/s10393-015-1053-0.

Leroy, E. M., B. Kumulungui, X. Pourrut, P. Rouquet, A. Hassanin, P. Yaba, A. Délicat, J. T. Paweska, J. P. Gonzalez, and R. Swanepoel. 2005. Fruit bats as reservoirs of Ebola virus. Nature 438 (7068): 575–576. https://doi.org/10.1038/438575a.

Lewinsohn, T. M., and L. Cagnolo. 2012. Keystones in a tangled bank. Science 335 (6075): 1449–1451. https://doi.org/10.1126/science.1220138.

Lewinsohn, T. M., P. I. Prado, P. Jordano, J. Bascompte, and J. M. Olesen. 2006. Structure in plant-animal interaction assemblages. Oikos 113 (1): 174–184. https://doi.org/10.1111/j.0030-1299.2006.14583.x.

Lima, M. 2011. Summary for policymakers. In Climate Change 2013—the Physical Science Basis, edited by Intergovernmental Panel on Climate Change, 1–30. Cambridge: Cambridge University Press. https://doi.org/10.1017/CBO9781107415324.004.

Luis, A. D., T. J. O'Shea, D. T. S. Hayman, J. L. N. Wood, A. A. Cunningham, A. T. Gilbert, J. N. Mills, and C. T. Webb. 2015. Network analysis of host-virus communities in bats and rodents reveals determinants of cross-species transmission. Ecology Letters 18 (11): 1153–1162. https://doi.org/10.1111/ele.12491.

Lutz, H. L, B. D. Patterson, J. C. Kerbis Peterhans, W. T. Stanley, P. W. Webala, T. P. Gnoske, S. J. Hackett, and M. J. Stanhope. 2016. Diverse sampling of East African haemosporidians reveals chiropteran origin of malaria parasites in primates and rodents. Molecular Phylogenetics and Evolution 99 (March): 7–15. https://doi.org/10.1016/j.ympev.2016.03.004.

McIntosh, R. P. 1987. The Background of Ecology: Concept and Theory. Cambridge: Cambridge University Press.

Mello, M. A. R. 2019. Keystone species. In Oxford Bibliographies in Ecology, edited by David Gibson. New York: Oxford University Press. https://doi.org/10.1093/OBO/9780199830060-0213.

Mello, M. A. R., A. R. Francisco, L. F. Costa, W. D. Kissling, Ç. H. Şekercioğlu, F. M. D. Marquitti, and E. K. V. Kalko. 2015. Keystone species in seed dispersal networks are mainly determined by dietary specialization. Oikos 124 (8): 1031–1039. https://doi.org/10.1111/oik.01613.

Mello, M. A. R., F. M. D. Marquitti, P. R. Guimarães, E. K. V. Kalko, P. Jordano, and M. A. M. Aguiar. 2011. The modularity of seed dispersal: differences in structure and robustness between bat- and bird-fruit networks. Oecologia 167 (1): 131–140. https://doi.org/10.1007/s00442-011-1984-2.

Mello, M. A. R., R. L. Muylaert, R. B. P. Pinheiro, and G. M. F. Félix. 2016. Guia Para Análise de Redes Ecológicas. Belo Horizonte: Os autores. https://doi.org/10.13140/RG.2.2.27886.41280.

Melo, M. N., F. G. M. Maia, T. M. Vieira, C. B. Jonsson, C. B. Goodin, C. B. Gonçalves, P. D. Barroso et al. 2015. Evidence of Hantavirus infection among bats in Brazil. American Journal of Tropical Medicine and Hygiene 93 (2): 404–406. https://doi.org/10.4269/ajtmh.15-0032.

Meyer, C. F. J., and E. K. V. Kalko. 2008. Bat assemblages on neotropical land-bridge islands: nested subsets and null model analyses of species co-occurrence patterns. Diversity and Distributions 14 (4): 644–654. http://dx.doi.org/10.1111/j.1472-4642.2007.00462.x.

Minor, E. S., and D. L. Urban. 2008. A graph-theory framework for evaluating landscape connectivity and conservation planning. Conservation Biology 22 (2): 297–307. https://doi.org/10.1111/j.1523-1739.2007.00871.x.

Montoya, D., M. L. Yallop, and J. Memmott. 2015. Functional Group Diversity Increases with Modularity in Complex Food Webs. Nature Communications 6 (June): 7379. https://doi.org/10.1038/ncomms8379.

Muylaert, R. L., D. M. S. Matos, and M. A. R. Mello. 2014. Interindividual variations in fruit preferences of the yellow-shouldered bat *Sturnira lilium* (Chiroptera: Phyllostomidae) in a cafeteria experiment. Mammalia 78 (1): 93–101. https://doi.org/10.1515/mammalia-2012-0103.

Muylaert, R. L., R. D. Stevens, and M. C. Ribeiro. 2016. Threshold effect of habitat loss on bat richness in cerrado-forest landscapes. Ecological Applications 26 (6): 1854–1867. https://doi.org/10.1890/15-1757.1.

Niebuhr, B. B. S., M. E. Wosniack, M. C. Santos, E. P. Raposo, G. M. Viswanathan, M. G. E. da Luz, and M. R. Pie. 2015. Survival in patchy landscapes: the interplay between dispersal, habitat loss and fragmentation. Scientific Reports 5:11898. https://doi.org/10:1038/srep11898.

Nowak, R M. 1994. Walker's Bats of the World. London: Johns Hopkins University Press.

Olival, K., and D. Hayman. 2014. Filoviruses in bats: current knowledge and future directions. Viruses 6 (4): 1759–1788. https://doi.org/10.3390/v6041759.

O'Mara, M. T., M. Wikelski, and D. K. N. Dechmann. 2014. 50 years of bat tracking: device attachment and future directions. Methods in Ecology and Evolution 5 (4): 311–319. https://doi.org/10.1111/2041-210X.12172.

Ortega, J., and J. E. Maldonado. 2006. Female interactions in harem groups of the Jamaican fruit-eating bat, *Artibeus jamaicensis* (Chiroptera: Phyllostomidae). Acta Chiropterologica 8 (2): 485–495. https://doi.org/10.3161/1733-5329(2006)8[485:FIIHGO]2.0.CO;2.

Patterson, B. D., C. W. Dick, and K. Dittmar. 2009. Nested distributions of bat flies (Diptera: Streblidae) on neotropical bats: artifact and specificity in host-parasite studies. Ecography 32 (3): 481–487. https://doi.org/10.1111/j.1600-0587.2008.05727.x.

Pautasso, M. 2013. Ten simple rules for writing a literature review. PLOS Computational Biology 9 (7): e1003149. https://doi.org/10.1371/journal.pcbi.1003149.

Peters, V. E., T. A. Carlo, M. A. R. Mello, R. A. Rice, D. W. Tallamy, S. A. Caudill, and T. H. Fleming. 2016. Using plant-animal interactions to inform tree selection in tree-based agroecosystems for enhanced biodiversity. BioScience 66 (12): 1046–1056. https://doi.org/10.1093/biosci/biw140.

Pilosof, S., M. A. Porter, M. Pascual, and S. Kéfi. 2017. The multilayer nature of ecological networks. Nature Ecology and Evolution 1 (4): 0101. https://doi.org/10.1038/s41559-017-0101.

Pinheiro, R. B. P., G. M. F. Félix, A. V. Chaves, G. A. Lacorte, F. R. Santos, É. M. Braga, and M. A. R. Mello. 2016. Trade-offs and resource breadth processes as drivers of performance and specificity in a host-parasite system: a new integrative

hypothesis. International Journal for Parasitology 46 (2): 115–121. https://doi.org/10.1016/j.ijpara.2015.10.002.

Pocock, M. J. O., D. M. Evans, C. Fontaine, M. Harvey, R. Julliard, Ó. McLaughlin, J. Silvertown, A. Tamaddoni-Nezhad, P. C. L. White, and D. A. Bohan. 2016. The visualisation of ecological networks, and their use as a tool for engagement, advocacy and management. In Advances in Ecological Research, edited by Guy Woodward and David A. Bohan, 41–85. Cambridge: Academic Press. https://doi.org/10.1016/bs.aecr.2015.10.006.

Poisot, T., D. B. Stouffer, and S. Kéfi. 2016. Describe, understand and predict: why do we need networks in ecology? Functional Ecology 30 (12): 1878–1882. https://doi.org/10.1111/1365-2435.12799.

Pourrut, X., M. Souris, J. S. Towner, P. E. Rollin, S. T. Nichol, J.-P. Gonzalez, and E. Leroy. 2009. Large serological survey showing cocirculation of Ebola and Marburg viruses in Gabonese bat populations , and a high seroprevalence of both viruses in *Rousettus aegyptiacus*. BMC Infectious Diseases 10:1–10. https://doi.org/10.1186/1471-2334-9-159.

Presley, S. J., L. M. Cisneros, B. D. Patterson, and M. R. Willig. 2012. Vertebrate metacommunity structure along an extensive elevational gradient in the tropics: a comparison of bats, rodents and birds. Global Ecology and Biogeography 21 (10): 968–976. https://doi.org/10.1111/j.1466-8238.2011.00738.x.

Presley, S. J., C. L. Higgins, and M. R. Willig. 2010. A comprehensive framework for the evaluation of metacommunity structure. Oikos 119 (6): 908–917. https://doi.org/10.1111/j.1600-0706.2010.18544.x.

Presley, S. J., and M. R. Willig. 2010. Bat metacommunity structure on Caribbean Islands and the role of endemics. Global Ecology and Biogeography 19 (2): 185–199. https://doi.org/10.1111/j.1466-8238.2009.00505.x.

R Core Team. 2019. R: A Language and Environment for Statistical Computing. Vienna: R Foundation for Statistical Computing. https://www.r-project.org/.

Rhodes, M., G. W. Wardell-Johnson, M. P. Rhodes, and B. Raymond. 2006. Applying network analysis to the conservation of habitat trees in urban environments: a case study from Brisbane, Australia. Conservation Biology 20 (3): 861–870. https://doi.org/10.1111/j.1523-1739.2006.00415.x.

Rulli, M. C., M. Santini, D. T. S .Hayman, and P. D'Odorico. 2017. The nexus between forest fragmentation in Africa and Ebola virus disease outbreaks. Scientific Reports 7:41613.

Sabino-Santos, G., F. G. M. Maia, T. M. Vieira, R. L. Muylaert, S. M. Lima, C. B. Goncalves, P. D. Barroso et al. 2015. Evidence of Hantavirus infection among bats in Brazil. American Journal of Tropical Medicine and Hygiene, 93:404–406.

Sarmento, R., C. P. Alves-Costa, A. Ayub, and M. A. R. Mello. 2014. Partitioning of seed dispersal services between birds and bats in a fragment of the Brazilian Atlantic Forest. Zoologia 31 (3): 245–255. https://doi.org/10.1590/S1984-46702014000300006.

Schupp, E. W., P. Jordano, and J. M. Gómez. 2010. Seed dispersal effectiveness revisited: a conceptual review. New Phytologist

188 (2): 333–353. https://doi.org/10.1111/j.1469-8137
.2010.03402.x.

Stang, M., P. G. L. Klinkhamer, and E. van der Meijden. 2006.
Size constraints and flower abundance determine the number
of interactions in a plant-flower visitor web. Oikos 112 (1):
111–121.

Sun, L. H. 2019. Scientists find deadly Ebola virus for first time in
West African bat. Washington Post, January 24. https://www
.washingtonpost.com/health/2019/01/24/scientists-find
-deadly-ebola-virus-first-time-west-african-bat/?noredirect
=on&utm_term=.1ba24c0ce8f0.

Sutherland, W. J., R. P. Freckleton, H. C. J. Godfray, S. R.
Beissinger, T. Benton, D. D. Cameron, Y. Carmel et al. 2013.
Identification of 100 fundamental ecological questions.
Journal of Ecology 101 (1): 58–67. https://doi.org/10.1111
/1365-2745.12025.

Svensson-Coelho, M., V. A. Ellis, B. A. Loiselle, J. G. Blake, and
R. E. Ricklefs. 2014. Reciprocal specialization in multihost
malaria parasite communities of birds: a temperate-tropical

comparison. American Naturalist 184 (5): 624–635. https://
doi.org/10.1086/678126.

Vellend, M. 2016. The Theory of Ecological Communities
(MPB-57). Princeton, NJ: Princeton University Press.
https://books.google.com.br/books/about/The
_Theory_of_Ecological_Communities_MPB.html?id=
2Yn8CwAAQBAJ&redir_esc=y.

Wickham, H. 2009. Ggplot2: Elegant Graphics for Data Analysis.
New York: Springer New York. https://doi.org/10.1007/978
-0-387-98141-3.

Young, C. C. W., and K. J. Olival. 2016. Optimizing viral
discovery in bats. PLOS ONE 11 (2): 1–18. https://doi.org
/10.1371/journal.pone.0149237.

Zarazúa-Carbajal, M., R. A. Saldaña-Vázquez, C. A. Sandoval-
Ruiz, K. E. Stoner, J. Benitez-Malvido, G. D. Maganga, E.-M.
Leroy, D. Fontenille, D. Ayala, and C. Paupy. 2016. The specificity of host-bat fly interaction networks across vegetation
and seasonal variation. Parasitology Research 115 (10): 1–8.
https://doi.org/10.1007/s00436-016-5176-1.

22

Contemporary Biogeography

Richard D. Stevens,
Marcelo M. Weber,
and Fabricio Villalobos

Introduction

The contemporary Phyllostomidae is mostly a Neotropical clade whose geographic distribution ranges from the southwestern United States to Uruguay and northern Argentina. In addition to its continental distribution, species of this family commonly occur in, if not dominate, islands of the Caribbean Sea and most continental islands of North, Central, and South America. Elevational range of this family spans from sea level to approximately 3,600 m (Mena et al. 2011). Within these limits, phyllostomids form strong contemporary gradients of biodiversity. When all mammals are considered as a group, phyllostomids are often the main drivers of gradients of diversity (Kaufman 1995). Indeed, bats in general and phyllostomids in particular have served as a model system to better understand gradients of diversity, and much of what has been learned has come from a Neotropical perspective. Neotropical bat assemblages represent a varied mix of nine different families but as a rule, species richness and abundance of Phyllostomidae dominate. Typically, biogeographic investigation does not focus on Phyllostomidae or even on all Noctilionoidea but, rather, on all Neotropical bats. Nonetheless, Phyllostomidae dominates the signal of biogeographic pattern (Stevens 2004), and for this reason, we consider all studies where Phyllostomidae dominates in our discussion of contemporary biogeography.

Traditional Approaches: Secondary Gradients

Traditional approaches to examining variation in biodiversity have focused on secondary gradients or those that are purely spatial but correlated with primary environmental characteristics. Secondary gradients such as latitude, elevation, and area can serve as good surrogates for primary environmental characteristics, such as temperature, precipitation, productivity, or habitat diversity, to name only a few (e.g., Patten 2004; Willig and Selcer 1989; Willig et al. 2003). Indeed, in the absence of primary environmental data, examination of biodiversity along these secondary gradients is an informative place to begin to characterize patterns and search for mechanisms.

Species-Area Curves

The species-area relationship is perhaps the most fundamental biogeographic pattern (Rosenzweig 1995). Indeed, the positive relationship between the area of a sample and number of species in that sample is ubiquitous for both plants and animals. Examination of species-area relationships in phyllostomid-dominated systems has typically been from two different perspectives. The first is in the context of anthropogenically modified fragmentation in which the area of fragments or the amount of forested area within spatial landscapes is related to species richness (Klingbeil and Willig 2009; Meyer and Kalko 2008; Muylaert et al. 2016). Indeed, species-area relationships characterize effects of habitat fragmentation on bat diversity. Nonetheless, such studies are primarily local in scope and not biogeographic and will not be considered here.

The second perspective on bat species-area relationships considers natural settings, mainly taking advantage of island systems such as that of the Caribbean Sea (Baker and Genoways 1978; Rodriguez-Duran and Kunz 2001; Willig et al. 2009). All of these studies have found strong and significant patterns between species richness and area of islands. Indeed, such patterns characterize each of the three island chains that compose the Caribbean: Bahamas, Greater Antilles, and Lesser Antilles (Willig et al. 2009).

While the species-area relationship is well-characterized, the mechanistic basis to this pattern is not as well understood. Numerous island characteristics such as elevation or topographic, geologic, or habitat diversity are all correlated with island size, making it difficult to distinguish if area per se or these other characteristics are the true drivers of species-area relationships (Simberloff 1976). Ricklefs and Lovette (1999) used a multiple regression approach to account for the confounded nature of area, habitat diversity, and elevation whereby the correlated and unique effects of each were determined. For bats, only size of the island accounted for significant unique variation in bat species richness, indicating that area per se best explained the species-area relationship. Morand (2000) reanalyzed the data of Ricklefs and Lovette (1999) but also included inter-island distances as a measure of isolation. He found, for bats, that when distances were incorporated into analyses, the unique and significant positive relationship between area and species richness disappeared and that island distance was the most important contributor to species richness. Willig et al. (2009) analyzed a data set including five more islands from the Lesser Antilles, as well as distributional updates for the 18 original ones, and found support for the Ricklefs and Lovette (1999) result instead of those of Morand (2000).

The Caribbean is a region characterized by an active disturbance regime from frequent hurricane activity. Moreover, such activity has likely had an impact on the flora and fauna of the Caribbean for millennia. Willig et al. (2009) expanded prior work on species-area relationships in the Caribbean but also examined effects of the hurricane regime experienced on each island. Moreover, they examined the entire Caribbean, as well as conducting analyses for the Bahamas and the Greater and Lesser Antilles separately. Again, island area accounted for the most unique variation in species richness, although the effect varied based on island group. Interestingly, hurricane effects did not significantly contribute to variation in species richness.

In a novel application, Dávalos and Russell (2012) reconstructed species-area relationships for the Bahamas and the Greater and Lesser Antilles for the Last Glacial Maximum (LGM) and compared species loss to area loss due to climate change. Slopes for current and LGM species-area relationships were similar for the Bahamas and the Greater Antilles but significantly steeper for contemporary data in the Lesser Antilles. Perhaps more importantly, across the Bahamas and Greater Antilles, postglacial species loss was explained by inferred loss of land due to sea-level rise. These results have strong implications for not only the area per se hypothesis but also scenarios of future climate change.

Elevation Gradients

In many ways, elevational contexts are ideal for the study of biodiversity gradients because they exhibit sharp and short clines of environmental variation that are often of similar change in magnitude as those along spatially more extensive gradients such as latitude. Moreover, while spatially proximal elevation gradients are related due to shared species pools, no two elevation gradients are the same and they therefore likely provide more replication than do latitudinal gradients (McCain 2007a).

Four widespread patterns characterize changes in species richness in general with elevation: (1) a decrease with increased elevation, (2) a low-elevation plateau then a decrease at higher elevation, (3) a low-elevation plateau with a mid-elevation peak, and (4) a mid-elevation peak (McCain and Grytnes 2010). The first comprehensive descriptions of elevational gradients in bat faunas dominated by phyllostomids were both in Peru and demonstrated decreases in species richness with elevation (Graham 1983, 1990; Patterson et al. 1996, 1998). Moreover, species diversity (i.e., Hill's index, see Magurran 2004) decreased and evenness (i.e., Hill's ratio) increased with increased elevation (Graham 1983) and different guilds exhibited different elevational extents (Patterson et al. 1996) as well as different rates of change with elevation (Graham 1983).

Elevation gradients in the Andes of Peru appear to be strong, linear, and general. Nonetheless, elevation gradients in other parts of the distribution of Phyllostomidae exhibit significant variation, in particular a variety of mid-elevation peaks and decreases with elevation (McCain 2007a). For example, mid-elevation peaks have been demonstrated in Ecuador (Carrera-E 2003) as well as the Sierra Mazateca (Sánchez-Cordero 2001) and Jalisco (Iñiguez Dávalos 1993), Mexico. In contrast, decreases in species richness with elevation have been reported from Colombia (Muñoz Arango 1990), Venezuela (Handley 1976), and Mexico (Sánchez-Cordero 2001). Indeed, such variation may at first appear to frustrate generalization of elevational effects. Nonetheless, such variation may actually be advantageous, for it provides a variety of situations from which more powerful tests of mechanisms can be applied; if a mechanism is general, it should account for the entire variety of elevational patterns (McCain and Grytnes 2010).

Proposed explanatory mechanisms for elevation gradients are many but fall into three main groups: spatial, historical, and climatic. Spatial mechanisms involve either species-area relationship (Rosenzweig 1995) or mid-domain effects (Colwell and Lees 2000; Willig and Lyons 1998). There is mixed support for species-area relationships along elevation gradients with studies exhibiting no, positive, or negative relationships with area. There is even less support for the idea that Neotropical bats exhibit mid-domain effects along elevation gradients (McCain 2007a).

Historical mechanisms involve differential extinction, speciation, and dispersal along elevation gradients. To characterize such phenomena, a number of data-intensive tools are required such as a fully resolved species-level phylogeny and sophisticated computational methods (Hernandez et al. 2013; Pennel and Harmon 2013). Such tools have not been readily available until recently, and this has stymied research on historical hypotheses that explain phyllostomid biodiversity gradients. A prominent and well-conceived historical mechanism is that of niche conservatism. Niche conservatism assumes that species within a clade have a strong tendency to retain the environmental requirements of their last common ancestor and that movements of species outside the ancestral range of environmental conditions are relatively uncommon (Wiens et al. 2007). As such, during diversification species accumulate more in areas with environmental conditions like those of their ancestors and to a lesser degree in areas that are very different. As Graham (1983) pointed out for South America, the Amazon Basin is much older than the uplift of the Andes. Moreover, most if not all South American bat families were present before the Andes uplift (Graham 1983). If phyllostomid bats originated at low elevations and requirements for these climatic conditions are disproportionately conserved in the contemporary fauna, then this might explain the strong, monotonic elevational decreases in species richness that appear to characterize Andean elevation gradients. Although other mechanisms also predict a monotonic negative pattern, the niche conservatism hypothesis is the only one that makes predictions regarding species richness and phylogenetic characteristics of species that can be used to tease out its potential primacy. Modern species-level phylogenies are currently available (e.g., Bininda-Emonds et al. 2007; Fritz et al. 2009; Rojas et al. 2016) and a test of this hypothesis appears to be low-hanging fruit for increasing our understanding of the mechanistic basis of elevation gradients of species richness in phyllostomid bats.

Climatic hypotheses, in contrast, have received much more attention. McCain (2007a) examined 12 elevational gradients of bats distributed globally and found a mix of mid-elevation peaks and decreases in species richness with increasing elevation. McCain (2007b) was able to relate differences among gradients, in particular whether there was a decrease or mid-elevational peak, to a combination of temperature and precipitation. Most but not all tropical mountains have wet bases whereas most temperate mountains examined have arid bases. Therefore, productivity in tropical areas is primarily based on temperature because water is readily available along the entire gradient whereas it is based on both temperature and precipitation in temperate areas. Moreover, she proposed that temperature and precipitation express linear decreasing and mid-elevation peaks, respectively, along elevation gradients. In addition, temperate gradients are longer (i.e., span longer gradients of temperature and precipitation) than are tropical gradients, and as a result temperate gradients span the mid-elevational peak and decline, whereas tropical gradients only span the portion of the gradient where temperature and precipitation decline together (fig. 22.1).

As with latitudinal gradients, examination of variation in multiple dimensions of biodiversity along elevation gradients has recently increased. For an extensive elevation gradient in Manu National Park, Peru, Cisneros et al. (2014) examined concerted variation in species richness as well as functional and phylogenetic dispersion of species. Along this elevation gradient, species richness was not significantly correlated with phylogenetic dispersion but was significantly correlated with four of seven measures of functional dispersion. Phylogenetic and functional dispersion were significantly correlated with each other. However, species richness but not phylogenetic or functional dispersion exhibited a significant elevational gradient, with species richness deviating from phylogenetic and functional dispersion at high elevations. Such a result suggests that

more than one mechanism may determine variation in biodiversity along elevation gradients.

Latitudinal Gradients

Latitudinal gradients of species richness of New World bats are now well-documented, despite methodological differences among studies (Fleming 1973; Willig and Sandlin 1991; Willig and Selcer 1989). All bat families exhibit richness patterns related to latitude but the Phyllostomidae presents the strongest gradient (Stevens 2004). Moreover, complementary latitudinal gradients exist across multiple dimensions of biodiversity including taxonomic, functional, phenetic, and phylogenetic perspectives (plate 8, Stevens et al. 2003).

While the occurrence of latitudinal gradients in species richness is indisputable (Willig et al. 2003), the mechanistic basis for these patterns is still unclear. A number of proposed hypotheses can likely be eliminated. For example, the area hypothesis (Rosenzweig 1995) has been a long-standing explanation for latitudinal gradients. If the area hypothesis could account for gradients in species richness, then area should increase toward the equator and species richness should be positively related to area. However, Willig and Bloch (2006) demonstrated that there is no latitudinal gradient in area of biomes or geographic provinces and that there is no relationship between species richness and the area of these two kinds of units.

Mid-domain effects have also been commonly suggested to cause latitudinal gradients in diversity (Colwell

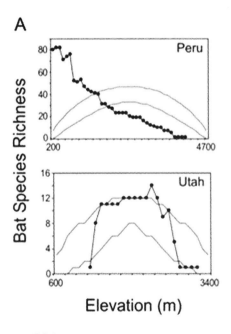

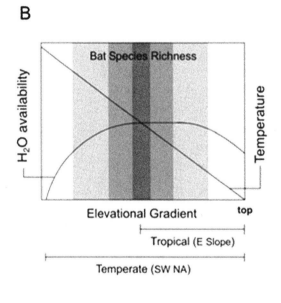

Figure 22.1. Elevational Gradients. (*A*) Two empirical examples: a linear decline in species richness with increasing elevation (*top*) and a mid-elevation peak in species richness (*bottom*). (*B*) Mechanistic model proposed by McCain (2007a). Precipitation exhibits a humped distribution regarding elevation, whereas temperature exhibits a linear decline. In terms of the environment, temperate gradients are longer than tropical gradients because they span more environmental variation. Accordingly, temperate species-richness gradients are hump shaped, whereas tropical ones are linear.

and Lees 2000; Willig and Lyons 1998). Accordingly, when positions of geographic distributions are randomly shuffled within a bounded domain, a peak in diversity should result in the middle. Since the latitudinal edges of the New World domain are more polar and the middle is around the equator, expectations of mid-domain effect (MDE) models are similar to the observed latitudinal gradient in species richness. The shape of the relationship between species richness and latitude are similar for MDE-produced and -observed richness gradients. Nonetheless, at high latitudes, the MDE consistently overpredicts and at low latitudes underpredicts species richness (Arita et al. 2014; Tello and Stevens 2012a; Willig and Lyons 1998). For Noctilionoidea, variables representing climate and energy account for significantly more variation in species richness than does the MDE (Tello and Stevens 2012a). Moreover, if the MDE is what drives contemporary gradients of biodiversity of phyllostomids, then such a model should recapitulate not only species-richness gradients but also gradients across other dimensions of biodiversity. It does not. Mid-domain effects produce gradients in functional, phenetic, and phylogenetic dimensions of biodiversity, but they are qualitatively and quantitatively different from observed patterns (plate 8, Stevens et al. 2013).

Other studies have contrasted observed gradients with those predicted by MDE for other characteristics of biodiversity (Arita et al. 2005; Stevens et al. 2013; Villalobos and Arita 2010). Similar to results found for different dimensions of biodiversity, MDE does not pre-

dict these other characteristics well. This suggests that continued investigation of causes behind unexplained deviations from MDE may illuminate additional ways that biodiversity gradients deviate from such a random expectation. For instance, Arita et al. (2005) used deviations from MDE to confirm the positive link between the species-richness gradient and Rapoport's rule (i.e., a positive relationship between latitude and species range sizes) in North American mammals, including phyllostomid bats. Later, Villalobos and Arita (2010) showed that MDE was unable to account for species-level patterns such as the degree of co-occurrence among species. This study focused exclusively on phyllostomid bats and used a novel framework based on the intrinsic relationship between species diversity and distribution (fig. 22.2). They found that phyllostomids have a higher level of co-occurrence than expected from MDE, which results from processes in addition to those that affect the size, shape, and location of geographic ranges.

Primary Gradients of Diversity: Climatic and Historical Processes

Over the last two decades, availability of global data sets on primary environmental characteristics such as temperature, precipitation (Hijmans et al. 2005), and productivity (UNEP 2014), as well as a wealth of species-level phylogenies, have revolutionized how biogeographers examine the mechanistic basis of broadscale biodiversity gradients. Explanations for

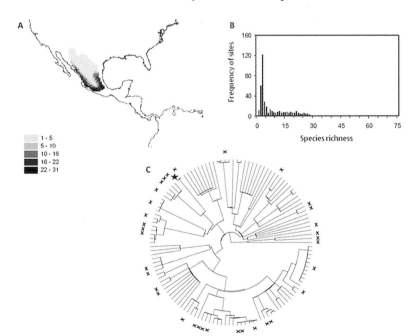

Figure 22.2. Spatial and phylogenetic structure of co-occurrence within a focal species' range: the Mexican long-nosed bat *Leptonycteris nivalis*. (*A*) Range map of *L. nivalis* depicting the spatial structure of co-occurrence (i.e., richness) and (*B*) the resulting species-richness frequency distribution across its occupied sites. Considering the phylogenetic relationships among phyllostomids (*C*), species co-occurring with *L. nivalis* (*star*) represent different phyllostomid lineages (*Xs*) resulting in an overdispersed phylogenetic field (see text).

variation in biodiversity along primary environmental gradients can be categorized into two groups: ecological and evolutionary hypotheses. Ecological hypotheses focus on biotic (e.g., interactions among species) and abiotic factors (e.g., environmental conditions) as determinants of species coexistence, whereas evolutionary hypotheses focus on the age (i.e., time since origin) and diversification of lineages (Brown 2014; Mittelbach et al. 2007). Both types of hypotheses have been tested on Phyllostomidae in particular (Stevens 2006, 2011; Villalobos et al. 2013, 2014) or including this family within higher taxa such as the superfamily Noctilionoidea (Rojas et al. 2016; Stevens et al. 2013) or on Chiroptera as a whole (Patten 2004; Rolland et al. 2014; Tello and Stevens 2010).

Ecological Hypotheses

Environmental correlates of bat diversity have been evaluated more recently, although less frequently than the sole effect of latitude (Patten 2004; Tello and Stevens 2010; Willig et al. 2003). For instance, Patten (2004) found that precipitation, topography, and temperature range were the best predictors of bat species richness, especially for phyllostomids, in North America. He also showed that these environmental factors varied in their importance depending on the bat family under study (e.g., Phyllostomidae vs. Vespertilionidae). More recently, Tello and Stevens (2010) combined these main environmental factors—namely, precipitation and temperature—along with elevation and primary productivity to test the effect of environmental energy, seasonality, and topographic heterogeneity in determining bat species richness in the New World. They showed that seasonality and energy act independently but complementarily in determining bat species richness, whereas heterogeneity has a smaller effect. However, they also demonstrated that the relative importance of these factors depends on the range size of the species being considered, with richness of large-ranged species being explained primarily by seasonality and energy, whereas richness of species with restricted distributions being best explained by heterogeneity. These findings highlight the need to consider different phylogenetic/taxonomic (e.g., families) and distributional scales (e.g., range size) in order to better understand the processes driving diversity gradients.

At a smaller but still extensive scale, species richness of phyllostomids is strongly related to climatic variables in the Atlantic Forest of South America, the tropical forest with the longest latitudinal extent (Stevens 2013). Species richness of phyllostomids was related most strongly to temperature seasonality. Moreover, patterns of phylogenetic dispersion exhibit similar patterns (Stevens and Gavilanez 2015). Phyllostomids do not use hibernation or torpor to endure bouts of low ambient temperature (McNab 1969). Thus, it is not surprising that temperature seasonality is an important determinant of bat biodiversity patterns and likely is a general phenomenon across the distribution of Phyllostomidae.

The relevance of environmental conditions to diversity gradients relies on the assumption of biological responses of species to these conditions (Currie et al. 2004). However, such richness-environment correlations are silent on the actual processes that change species numbers in an area (Ricklefs 2004). Moreover, these richness-environment relationships may simply represent spurious correlations between these variables, with no mechanistic link to particular biological processes (Tello and Stevens 2012b). This implies that a better null expectation for such richness-environment correlations is different than the traditional null (i.e., $r = 0$) and that this should be taken into account when inferring the effect of environment on species-richness gradients (Tello and Stevens 2012b).

Evolutionary Hypotheses

Tropical Niche Conservatism

Recent availability of appropriate phylogenetic data and more sophisticated quantitative tools has allowed more explicit tests of evolutionary hypotheses. These hypotheses consider time and history of lineages as a result of evolutionary processes—namely, speciation, extinction, and dispersal—as well as their interactions, which ultimately create, maintain, and modify geographical diversity gradients (Ricklefs 2004). One such hypothesis that has attracted much attention is the tropical niche conservatism hypothesis (TNC, Wiens and Donoghue 2004), which intends to explain latitudinal gradients of diversity. Tropical niche conservatism is subtly different from the more general niche conservatism hypothesis and posits that most clades have had a tropical origin and, by retaining their ancestral environmental preferences during diversification, have been slow to colonize temperate regions and thus have had more time to accumulate species in the tropics (Wiens and Donoghue 2004). These mechanisms, tropical origin and time for speciation (Stephens and Wiens 2003), provide a historical explanation for species-richness gradients with several testable predictions (Stevens 2006). Four predictions can be derived from center of origin and time-for-speciation effects on diversity gradients, from the center to the periphery of the family's geographical range: (1) species richness declines, (2) species are more derived phylogenetically, (3) species have younger ages, and (4) these ages and other measures of derivedness are less variable. Support is strong for these predictions, with phyllostomid bat richness decreasing away from the equator, exhibiting higher rates of sequence divergence and lower relative ages towards the periphery of the family's range (Stevens 2006).

Stevens (2011) aimed to distinguish between effects of TNC and time for speciation on bat diversity gradients, given that both hypotheses make mutually exclusive predictions based on environmental (TNC) and spatial (time for speciation) gradients. Time for speciation is a passive outcome of the diversification of a taxon, whereas TNC is a direct response to environmental gradients related to latitude, especially temperature seasonality. The mismatch between environmental and spatial gradients was then used to assess the relative importance of each hypothesis. Tropical niche conservatism has been more important than time for speciation in driving the phyllostomid diversity gradient, inferred from stronger relationships between environmental variables, especially temperature seasonality, and phylogenetic characteristics (e.g., species relative ages and sequence divergence) of phyllostomids across the latitudinal gradient (Stevens 2011). Other studies have found support for TNC in bats (Buckley et al. 2010; Villalobos et al. 2013). For instance, Buckley et al. (2010) suggested strong niche conservatism in the entire order Chiroptera but especially for New World families, as evidenced by a strong positive richness-environment relationship and significant phylogenetic signal in species tolerances (e.g., temperature).

Additional analyses suggest that TNC is a strong diversifier of groups of tropical origin. Using a species-level approach and focusing solely on phyllostomid bats, Villalobos et al. (2013) found support for TNC in driving diversity gradients. Their approach considered the phylogenetic structure of species co-occurrences

within species ranges ("phylogenetic fields"; fig. 22.2) and described a positive relationship between phylogenetic clustering (i.e., co-occurences of closely related species) and species richness within individual ranges as well as a pattern ranging from clustered to overdispersed (i.e., co-occurences of distantly related species) phylogenetic fields from low to higher latitudes as a result of shared environmental preference, in accordance with TNC. More interestingly, these findings were also supported at the level of phyllostomid subfamilies, suggesting that their species co-occurrence patterns respond to similar historical processes (e.g., niche conservatism) and that these processes are equally important among subfamilies regardless of their ecological differences (Villalobos et al. 2013).

The prevalence of niche conservatism and the particular predictions of TNC on diversity gradients continue to be intensely debated, both methodologically and conceptually (Losos 2008; Munkemüller et al. 2015; Pyron et al. 2015; Wiens 2008), including its ability to explain and its relevance for demonstrating diversity patterns of bats (Peixoto et al. 2017). Pereira and Palmeirim (2013), for example, rejected the predictions of TNC when analyzing New World bats as a group (Chiroptera) and separately for the most species-rich families (Phyllostomidae, Molossidae, and Vespertilionidae). They found neither a clear gradient of higher species derivedness towards the periphery of the geographic range of the order or constituent families (*contra* Stevens 2011) nor a greater spatial correlation between total and basal species richness than with derived species richness (*contra* Hawkins et al. 2012). A spatial correlation between total and basal species richness is expected under TNC, given the assumption of a tropical origin of clades. As a result, we expect a more important role for basal taxa compared to derived taxa in determining the latitudinal gradient. According to Pereira and Palmeirim (2013), the discrepancy between their findings and those of Stevens (2006), specifically regarding Phyllostomidae, may be due to differences in the distributional data used by the two studies, whereby Pereira and Palmeirim used range maps spanning the complete distribution of the family (37°N and 35°S), whereas Stevens (2006) used community data (i.e., groups of species that actually co-occur) but that did not reach the distributional limits of the family (21.1°N and 24.1°S). Therefore, no consensus has yet been reached regarding the validity of TNC for explaining diversity gradients for phyllostomids or for Chiroptera as a whole (but see Peixoto et al. 2017).

Diversification Rates

Another important evolutionary hypothesis to explain diversity gradients posits that the greater number of species in the tropics compared to the temperate zone results from higher diversification rates in the former than in the latter area (Mittelbach et al. 2007; Rohde 1992). The rate hypothesis has now been evaluated and supported as an explanation of the global latitudinal gradient in different taxa (flowering plants, Davies et al. 2004; birds, Cardillo et al. 2005; amphibians, Pyron and Wiens 2013; mammals, Rolland et al. 2014). Rolland et al. (2014) demonstrated that higher diversification rates in the tropics are due to both higher speciation and lower extinction rates in this zone compared to the temperate zones. Thus, contrary to TNC, which does not assume any differences in diversification rates between tropical and temperate regions, Rolland et al.'s (2014) findings suggest that more factors than TNC (i.e., time of origin and limited dispersal) may be necessary to explain the latitudinal diversity gradient and that differences in evolutionary processes of diversification also need to be considered when explaining this pattern. In the most recent phylogenetic analysis of phyllostomid bats (including the rest of the Noctilionoidea), Rojas et al. (2016) detected a single shift in speciation rate across the entire Noctilionoidea that corresponded to the evolution of Stenodermatinae. Interestingly, Shi and Rabosky (2015) found the same result when evaluating diversification rates across the entire Chiroptera. Both of these studies confirmed the relevance of Stenodermatinae, and thus Phyllostomidae, in producing the outstanding diversity of Noctilionoidea and Chiroptera as a whole. In a geographical context, the Rojas et al. (2016) analysis demonstrated that within-continent speciation and dispersal have been the most important processes in the diversification of Noctilionoidea. The former process is also consistent with diversification resulting from ecological differentiation, as proposed by Rojas et al. (2012), in which lineages diversified in the same region owing to dietary differences.

Recent studies have considered alternative approaches for investigating the role of evolutionary processes in creating contemporary bat diversity gradients. For example, Arita et al. (2014) introduced a higher-taxon approach for testing the traditional hypothesis on the origin and evolution of the New World bat fauna. The traditional explanation posits a single center of diversification for New World bat families (North America for Vespertilionidae and South America for Molossidae and five Noctilionoidea families). However, recent findings have questioned this traditional view and postulated the existence of two diversification centers (South and North America) for bat families (Czaplewski and Morgan 2012; Dávalos 2006, 2007). Based on the assumption that the current distribution of higher taxa should reflect the diversification processes occurring when these taxa diverged (Davies and Buckley 2011; Smith et al. 2012), Arita et al. (2014) evaluated the possibility that areas with high richness and endemism of genera coincide with old diversification centers. If the traditional hypothesis was correct, these areas should be found in North America for the Vespertilionidae and South America for the rest of the families and should coincide with high species-richness regions. Conversely, if there were two centers of diversification instead of one for South American families, the richness and endemism of bat genera should coincide with such centers and differ from regions of high species richness (Arita et al. 2014). Arita et al. (2014) found support for this latter hypothesis, particularly for noctilionoids. The existence of higher than expected (based on species richness) genus richness in North America as well as genera endemic to that region, mainly from Phyllostomidae (e.g., *Choeronycteris*, *Macrotus*, and *Musonycteris*), suggests that noctilionoids diversified in both North and South America, thus contradicting the traditional hypothesis (also, see chap. 6, this vol.). These results, and those of others (e.g., Pereira and Palmeirim 2013; Rolland et al. 2014; Stevens 2006), imply that a complex set of phenomena ranging from niche conservatism to niche evolution and differential diversification are driving the current diversity gradient in Phyllostomidae.

Methodological Limitations to the Study of Phyllostomid Biodiversity Gradients

Data

Perhaps all scientific enterprises are data deficient, either in terms of amount or quality, and the study of phyllostomid biogeography is no different. Amounts of data have increased drastically over the last decade due to accelerated generation by scientists and freely available compendia on distributions such as Nature Serve and the International Union for Conservation of Nature (IUCN) and on abundance such as the Atlantic project (Muylaert et al. 2017). Typically two distinct types of data are used to characterize spatial distributions of taxa in biogeographic studies, both of which have strengths and limitations: range maps and community data.

Often, data used to evaluate patterns of diversity come from overlapping geographic range maps of species and then determining species richness and composition within grid cells overlaid on these range maps. Range maps are often constructed by drawing a polygon around marginal records of a species distribution (e.g., extent-of-occurrence maps; see below). Advantages of such an approach are that comprehensive sets of range maps are available for many groups of species. Moreover, maps are easy to manipulate and score, given modern geographic information system (GIS) approaches, and such range maps provide robust estimates of the position and extent of species geographic ranges. Many studies have employed range maps, and so their results are reasonably comparable among taxa and geographic extents.

There are also limitations related to the use of range maps for the generation of biodiversity data. Many range maps are based on the collection of marginal data over the entire temporal span of the study of a particular species. This can often be a 100–250-year time period. Species distributions change, and the incorporation of all such data into the drawing of a geographic range likely leads to an overestimate. Perhaps an even greater issue is that geographic range maps imply that species are found everywhere within the polygon representing their geographic range; in most cases this is not realistic. As a result, overlapping species geographic range maps and enumerating species composition almost certainly overestimates the group of species that actually coexist in each grid cell and hence inflates the estimate of species richness.

A second common way of generating data to examine patterns of biodiversity is to compile a large number of sites where extensive sampling of the bat community has been conducted. These data have as an advantage that they enumerate lists of species that actually coexist at a particular place and time. Moreover, these lists often include data on numbers of captures of each species that can be used to estimate patterns of abundance.

They also form the basis for rarefication and species-richness estimators that can be used to provide potentially more precise estimates of magnitude and variation of species richness (Gotelli and Colwell 2001). Because most species in Neotropical bat communities are rare, it is difficult to detect many, especially those in the right tail of the rank-abundance distribution. Thus, the downside to such data is that they almost always represent underestimates of the species that coexist within a particular community.

It is still unclear whether the kind of data (local communities or range maps) influences biogeographic interpretation. For example, comparisons of Willig and Selcer (1989) with Stevens (2004), which used range map and community data, respectively, and Stevens and Willig (2002), which compared range map and community data for the same sites, indicate that the same patterns are qualitatively described by both types of data sets. Moreover, phylogenetic clustering was found to characterize the structure of bat communities using community data (Patrick and Stevens 2016), presence/absence in sampling grids over a similar region (Riedinger et al. 2013), or range maps (Villalobos et al. 2013). This again suggests that differences in kinds of data may not be the most important form of variation among studies. Nonetheless, some discrepancies do exist. For example, Pereira and Palmeirim (2013) used range maps to analyze the latitudinal diversity gradient in Chiroptera and found no support for the TNC hypothesis, whereas Stevens (2006), using species composition obtained from local communities, did. Indeed, there are other differences between these two studies besides the use of community versus range map data, and it is still unclear if differences in findings are related to the kind of data or to the scale (i.e., grain size) used in the different studies. Despite these differences, different kinds of data likely often provide similar answers, particularly when the patterns are strong. Analyses evaluating the sensitivity of results and how they are related to the kind of data used are still lacking and will be important for elucidating how precision and power are affected by different input data.

Species Discovery, Changing Taxonomy, and the Lack of Resolved Phylogenies

Understanding patterns of biodiversity relies heavily on systematics and taxonomy. The number of known species in a group is generally related to the number of active taxonomists of that group (Lewinsohn and Prado 2002). In 2005, about 160 species distributed in 57 genera were known for Phyllostomidae (Simmons 2005). Active systematic work over that last decade has changed those numbers, and both new species and new genera have been described. Many complex taxa, such as *Platyrrhinus* (e.g., Velazco and Gardner 2009; Velazco et al. 2010), *Sturnira* (e.g., Velazco and Patterson 2014), and *Lonchophylla* (e.g., Dias et al. 2013; Parlos et al. 2014), have been revised, providing major increases in our understanding of the biodiversity of Phyllostomidae.

Recently, deficient knowledge of species taxonomy and systematics has been identified as the Linnean shortfall (Bini et al. 2006; Hortal et al. 2015). This shortfall refers to the fact that most living species are not formally described or known to science. For leaf-nosed bats, this shortfall has decreased in recent years thanks to the work of many taxonomists. Interestingly though, most species described lately have not come from unexplored geographic regions as was common in the nineteenth and first half of the twentieth century. These newly described species have come primarily from taxonomic revisions based on previously collected specimens held in natural history museums (e.g., Dias et al. 2013; Velazco et al. 2010; also see chap. 4, this vol.). Indeed, taxonomic reviews will remain important for improving our knowledge of phyllostomid taxonomy.

Incorporation of newly described or elevated species into biogeographic and macroecological analyses of biodiversity tends to be slow. In particular, description of new species is often not promptly followed by updating species geographic range maps available from online data sets such as IUCN and Nature Serve. For instance, *Platyrrhinus helleri* was distributed from southern Mexico to central South America before the last taxonomic revision. After, the *P. helleri* complex was split into five species: *P. matapalensis* (Velazco 2005), *P. incarum*, *P. angustirostris*, *P. fusciventris*, and *P. helleri* (Velazco et al. 2010). However, range maps for these species are available only for *P. matapalensis*, which was described earlier than the other three new species in the IUCN database (http://www.iucnredlist.org/, accessed on 25 August 2016). As a result, these recently described species are excluded from studies (e.g., Weber et al. 2014) or, even worse, they may have been included in analyses without considering changing taxonomy. The effects of

such omissions of data on the description and our understanding of biodiversity gradients is still unclear and in need of assessment.

Another limitation is the lack of resolved phylogenetic relationships of species in general but particularly for those that have been recently described, including where these new taxa fit into preexisting phylogenies (e.g., Taddei and Lim 2010; Tavares et al. 2014). This situation has changed recently so that a description of a species is usually accompanied by a phylogeny of the group from which it belongs (e.g., Parlos et al. 2014; Velazco and Patterson 2008, 2013). Phylogenetic information is crucial for understanding patterns of phylogenetic diversity (Stevens 2006), species range dynamics (Weber et al. 2014), and community structure (Patrick and Stevens 2014). However, It is not known how the exclusion of recently described species from phylogenetic analyses may affect general biogeographic conclusions. Not including these species may cause loss of information whereas including species with changing taxonomy into analyses may provide misleading results, especially when a restricted set of species (e.g., species within a genus) is under examination. If species are missing from a phylogeny, this may affect tree topology, making both species relatedness and estimated rates of speciation less reliable (FitzJohn et al. 2009; Weber et al. 2014). Up until now there have been few analyses assessing the sensitivity of results to changes in taxonomy as well as how such variability affects the magnitude of effect sizes. For example, Patrick and Stevens (2014) analyzed how phylogenetic community structure metrics might be affected by changing phylogenies and found that results were robust to even substantial changes in phylogeny.

Novel Approaches and Improvements

Multiple Dimensions of Biodiversity

Geographic gradients of biodiversity have traditionally been evaluated by focusing on a single dimension—namely, taxonomic (alpha) diversity or simply species richness. Current efforts to understand diversity gradients now include the evaluation of numerous dimensions of biodiversity represented by taxonomic, phylogenetic (i.e., evolutionary differences), and functional (i.e., ecological variability) diversity (Stevens and Tello 2014). One goal of including these additional dimensions is to advance our understanding of the processes determining biodiversity gradients by developing stronger tests of potentially causal mechanisms (Cisneros et al. 2014; Stevens et al. 2013).

A number of studies have found complementary spatial variation among dimensions of biodiversity, particularly for the Phyllostomidae at regional (Cisneros et al. 2015, 2016; Stevens and Gavilanez 2015) and continental scales (Stevens and Tello 2014; Stevens et al. 2013). Seemingly unique dimensions such as taxonomic, phylogenetic, and functional diversity are correlated (Stevens and Tello 2014), and whenever more than one dimension is examined or compared, such correlations should be considered. Moreover, correlation among indices due to mathematical similarity or to shared responses to biological processes should also be taken into account. Taxonomic, phylogenetic, phenetic, and functional dimensions of biodiversity for noctilionoid bats are significantly correlated across geographic space (Stevens and Tello 2014). In fact, only two main axes of variation in biodiversity exist for Noctilionoidea in the New World. The first is related to a spatial gradient in general changes in the magnitude of biodiversity indices with an increase in the number of objects (i.e., species), whereas the second axis is related to spatial variation in functional diversity. Empirical dimensionality (i.e., degree of redundancy in variation of multiple dimensions of biodiversity) is different from that resulting from a null model, which suggests that the mathematical relationships among indices are not necessarily driving observed correlations and that some biological mechanism (e.g., niche conservatism, pattern of diversification) must be influencing correlations.

At a regional scale, Stevens and Gavilanez (2015) found that while phylogenetic, phenetic, and functional structure of phyllostomid bat communities in the Atlantic Forest of South America are spatially variable, they are substantially correlated among each other. In fact, the dimensionality of phyllostomid community structure is lower than the three dimensions examined. Phylogenetic diversity exhibited the most unique variation, perhaps because it captures more ecological variation than other dimensions. In addition, the phylogenetic dimension was significantly related to temperature seasonality across the Atlantic Forest, which may act as a filter allowing phylogenetic clustering in highly seasonal sites and phylogenetic overdispersion in low seasonality sites. Importantly, Stevens and Gavilanez

(2015) also demonstrated that, despite the low dimensionality of phyllostomid communities, it was different than expected from a regional sampling effect. This, in turn, suggests that different dimensions of biodiversity may be more important than others during the community assembly process, highlighting the relevance of considering different dimensions of biodiversity when studying bat community structure. It is also important to note that low dimensionality does not mean complete redundancy among diversity dimensions or that species richness is sufficient to characterize biodiversity. In fact, even small differences in the magnitude of spatial variation among dimensions of biodiversity can be quite informative and can serve as a means to test predictions derived from potential effects of mechanisms thought to be important in driving diversity gradients as well as for identifying particular areas of conservation interest. Nonetheless, different dimensions should not be considered orthogonal, and correlations should be an important consideration when evaluating multiple dimensions of biodiversity. Finally, comparative studies among different regions and taxa should be conducted in order to assess the generality of the results found for New World bats and the extent to which Phyllostomidae continues to be a driver of diversity patterns in higher clades (e.g., mammals).

Multiple Response Variables and Modeling Approaches

Given the large spatial and temporal scales of geographic diversity gradients, causal mechanisms have been commonly inferred from statistical analyses such as richness-environment correlations rather than by using controlled experiments (Currie et al. 2004). More recently, simulation models have been advocated to infer or even more directly test effects of particular biological mechanisms along geographic gradients (Gotelli et al. 2009). These simulation models generate expected patterns either by randomizing observed data while keeping some biological information, as in null models, or based on particular hypothesized mechanisms (e.g., range construction and temporal dynamics) such as stochastic or pattern-oriented modeling (Grimm et al. 2005; Villalobos and Rangel 2014). The MDE model explained above (see "Latitudinal Gradients") exemplifies the null model simulation approach, from which important insights into bat diversity gradients have been obtained (see Willig et al. 2003). Conversely, the application of more novel stochastic simulation and pattern-oriented approaches has only recently been considered for evaluating bat diversity gradients (Tello and Stevens 2012b).

One example is the study by Villalobos et al. (2014), which modeled species' range construction and placement under a stochastic simulation approach to understand the diversity and distribution of Phyllostomidae. These authors conducted spatially explicit simulations that included effects of climate and niche conservatism on species-range structures and locations. Focusing on composite response variables (i.e., co-diversity of sites and co-occurrence among species), Villalobos et al. (2014) were able to compare and rank different models in a more rigorous way than would have been possible by simply comparing richness gradients. They found that co-occurrence among phyllostomid species, which ultimately determines geographic variation in species richness, and co-diversity among sites are higher than expected, given the effect of climate or niche conservatism or their joint action. Thus, other processes besides those evaluated must be driving the high co-occurrence among phyllostomids. These processes could be related to evolutionary range dynamics (e.g., diversification, geographic speciation, or dispersal) of these bats throughout their history (Villalobos et al. 2014). This, in turn, highlights the need to consider response variables over and beyond that of species richness, such as species turnover (López-González et al. 2015) and phylogenetic/functional community structure (Cisneros et al. 2014; Stevens and Gavilanez 2015), to better understand mechanisms underlying diversity gradients.

Turnover of species across space (beta diversity) has also been considered an informative response variable in addition to species richness (alpha diversity) for developing stronger tests of causes of diversity gradients (Tuomisto 2010). For bats, beta diversity is an important component of spatial diversity gradients, at least in the New World, where most studies describing these patterns have been conducted (Rodríguez and Arita 2004; Stevens and Willig 2002; Willig and Sandlin 1991; but see Peixoto et al. 2014). Recently, however, there has been some debate on the relative importance of beta diversity for bat diversity gradients (López-González et al. 2015). On one hand, large-scale studies have shown the absence of a clear beta diversity gradi-

ent accompanying the alpha diversity gradient (Rodríguez and Arita 2004). In particular, the beta diversity pattern is constant across latitude given that both local (alpha) and regional (gamma) diversity covary along the gradient. On the other hand, regional and local-scale studies argue that a beta diversity gradient greatly contributes to the bat diversity gradient, particularly towards southern latitudes in North America (e.g., Mexico), where phyllostomids dominate bat communities (Stevens and Willig 2002; Willig and Sandlin 1991). For example, López-González et al. (2015), evaluated the diversity gradient of bats across Mexico and showed that environmental factors, mainly heterogeneity in vegetation and climate, determine species turnover. Such discrepancies concerning the importance of beta diversity may simply highlight the fact that bat diversity gradients may have different causal mechanisms at different scales, with environmental and biotic factors being more important at local scales, whereas regional and historical factors are most important at large scales (Villalobos et al. 2014). This is indeed what has been argued for alpha diversity (Fine 2015) and thus highlights the usefulness of considering multiple response variables to more fully understand the causes of diversity gradients.

Niche Models: Filling Gaps and Forecasting Changes

There are at least two ways of representing a species geographic range map. The first and most common in biogeographic analyses is based on extent of occurrence, a map drawn by specialists (e.g., IUCN and Nature Serve) indicating the area that lies within the outermost geographic limits of a species occurrence and that is often based on marginal records (Gaston and Fuller 2009). The second, much more resolved, approach is based on ecological niche modeling (ENM). Ecological niche modeling is the practice of using statistical and related methods with mapped biological and environmental data to model species distributions and other spatial variables of interest, such as abundance and species occupancy (Franklin 2010). Extent-of-occurrence maps are based on the knowledge of researchers of a focal species but tend to include areas where a species does not occur (i.e., commission error) (Franklin 2010). Furthermore, it is common to obtain occurrence data that falls outside an extent-of-occurrence map (e.g., Weber and Grelle 2012). Thus, although extent-of-occurrence maps provide a good approximation of a geographic range, they lack precision. Range maps estimated from an ecological niche model tend to provide more refined distributions than do extent-of-occurrence maps. Although they are more sophisticated, ecological niche models have many caveats that generate uncertainties since they are affected by the types of presence data (e.g., presence-only or presence-absence), environmental predictors, and/or algorithms. Studies have shown that the main source of variation among predictions derived from ecological niche models comes from the type of algorithm used to construct the map (e.g., Diniz-Filho et al. 2009). Therefore, when applying an ecological niche model to predict the distribution of a phyllostomid species, this uncertainty should be considered when trying to produce more reliable models. The ensemble approach is the most popular and efficient way of combining different predictions and minimizing uncertainty (Araújo and New 2006). This approach consists of generating several ecological niche models built with different algorithms that are assessed individually and then the selected individual models are combined providing a consensus model of a geographic distribution. There are many methods of creating a consensus model, which are evaluated in Marmion et al. (2009).

The most common goal of using ecological niche models is to predict the actual geographic range of a species (Franklin 2010), especially for those that are rare and/or endangered or that are only known from a few collecting sites (e.g., Ramoni-Perazzi et al. 2012; Teixeira et al. 2014; Weber et al. 2010). For species for which there is no estimate of its geographic distribution, ENM has proven to be useful. For instance, the nectar-feeding species *Lonchophylla peracchi* was described after a taxonomic revision of specimens of *L. bokermanni*, and it was suggested that it might occur in the Atlantic Forest in southeastern Brazil (Dias et al. 2013). However, this is an imprecise description of its distribution. An ecological niche model carried out for this species later identified areas of high environmental suitability, thereby providing a clearer estimate of its distribution, and indicated where undiscovered populations might occur (Teixeira et al. 2014). In fact, field sampling found a new population of *L. peracchi* in an area assigned as highly suitable for the species (Teixeira et al. 2013). Therefore, when carried out properly, ecological niche models may provide useful and realistic estimates of a geographic distribution.

Ecological niche models also provide insights into range size variation across evolutionary and ecological time. For example, Weber et al. (2014) modeled the unfilled potential range (i.e., the portion of the potential range that is not occupied by a species) of 49 phyllostomids and compared them to realized ranges. They demonstrated that old phyllostomid species have smaller unfilled potential ranges than do younger species. Therefore, old species have larger realized ranges because they filled a larger part of their potential range, whereas new species have larger unfilled potential ranges and smaller realized ranges.

One hot topic regarding ecological niche models is their use to estimate the potential effects of climate change on species distributions. Several studies have reported the negative effects of climate change on biodiversity (Hannah 2011). Rosa (2014) modeled the potential effects of climate change on 19 endemic bat species from the Amazon using both climatic and land-use variables, six algorithms, and an ensemble approach. He found that 14 species may lose more than 50% of their ranges by 2050 regardless of the scenario under consideration. This change will also affect the geographic pattern of species richness in this biome, increasing species richness towards Colombia, Ecuador, and the Guianas and decreasing richness in the Central Amazon, currently the richest portion of the biome.

Abundance Models

Abundance models describe variation measured either among species in an assemblage or across sites occupied by a single species. At the community level, generally a few species of phyllostomids dominate while most other species are rare (Alroy 2015; also see chap. 19, this vol.). The distribution of species abundances within a community is generally interpreted as a consequence of niche partitioning, in which a species that is able to consume many resources is abundant, whereas a species restricted to few resources is rare (Magurran 2004). Species abundance distributions may be fitted by statistical or biological models, such as the lognormal (Preston 1948), logarithmic series (Fisher et al. 1943), geometric series (May 1975), broken stick (MacArthur 1957), or the neutral model (Hubbell 2001). A review of these models and more can be found in Tokeshi (1990) and Magurran (2004). Despite having mechanistic explanations for the structure of communities, fitting abundance distribution models to phyllostomid bat communities is rare and deserves more attention from phyllostomid bat ecologists. For example, Alroy (2015) fitted theoretical abundance distributions for a very well sampled bat community in Mexico and found that the log-series distribution had the best fit, followed by the lognormal distribution. The ecological interpretation of this fit is that one or a few factors such as resource availability drive community structure (Magurran 2004). Schulze et al. (2000) observed that the broken stick distribution had the best fit and described better the rank-abundance patterns of phyllostomids in both continuous and fragmented forest in Guatemala, whereas the geometric series did not fit the rank-abundance distribution of bat species. Fit of the broken stick suggested that ecological resources were being divided more or less equally among bat species (May 1975) or that species had similar competitive abilities and were therefore partitioning niche space (Tokeshi 1993). Given the lack of studies on phyllostomid abundance distributions, no biogeographic trend has been proposed in this regard. Nonetheless, hypotheses on abundance distributions based on theory underlying latitudinal gradients of diversity can be formulated. For example, abundance distributions from species-rich communities are likely different than those from species poor communities. Nonetheless, are differences any greater than the effect of more individuals and thus simply a change in the veil line of Preston (1948)? Can these differences be used to determine if competition in species-rich communities is more intense than in species poor communities? Are environmental variables (e.g., productivity, temperature, and precipitation) related to the shape of species abundance distributions? These are relevant biogeographic questions that have not yet been addressed by bat ecologists, although there are data available for doing so. Both competition and environmental gradients (e.g., temperature, precipitation, and productivity) can influence patterns of species abundance in bat communities (Stevens and Amarilla-Stevens 2012; Stevens and Willig 2000). Thus, further investigation considering both species interactions and environmental variables are necessary to better understand patterns of abundance distributions of phyllostomid bats.

At the species level, abundance models are also rarely employed. Obtaining data on abundance of a species throughout its distribution is expensive and time consuming. One way to avoid these costs is to gather data on relative abundance from the literature (e.g., Stevens and Willig 2000; Weber and Grelle 2012) or to count the number of specimens deposited in museum collections (Pinto et al. 2013). Each method has limitations, such as spatial bias and undersampling (Rowe 2005), which should be considered by researchers before conducting analyses and making ecological interpretations. Once collected, such abundance data sets can be used to examine a number of macroecological questions. For example, for the nectar-feeding bat *Anoura caudifer* there is a strong and positive relationship between relative abundance and environmental suitability derived from ENM (Weber and Grelle 2012). However, for other phyllostomid species, this relationship may vary from no correlation between abundance and environmental suitability to species presenting high correlations (Weber et al. in prep., fig. 22.3). In fact, it has been shown recently that the relationship between species abundances and environmental suitability can be considered a general pattern (Weber et al. 2017). However, there is wide variation in patterns that have been found, which suggests that environmental suitability is not always linearly correlated with abundance and that this relationship may be better described by a triangular constraint envelope (Weber et al. 2017). Nonetheless, ecological niche models based on occurrence data may be applied as a reasonable proxy for species abundance data. This would allow description of patterns of species abundances at regional and/or continental scales and could lead to a mechanistic analysis of factors that may be generating these patterns.

Conclusions

Currently, there is an increasing amount of data and a quickly growing body of quantitative methods that allow an unprecedented level of rigor and detail when studying biogeographic patterns, especially for relatively well-known groups such as Neotropical bats. Focus on Neotropical bats, especially Phyllostomidae,

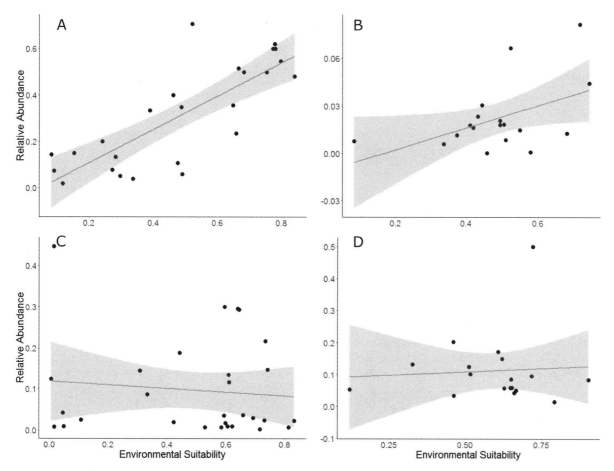

Figure 22.3. Panel showing the linear relationship between relative abundance and environmental suitability derived from ecological niche modeling built in Maxent version 3.3.3.k for four phyllostomid species: (A) *Anoura caudifer* ($r = 0.8$, $p < 0.001$, $n = 25$), (B) *Sturnira tildae* ($r = 0.48$, $p = 0.04$, $n = 18$), (C) *Platyrrhinus lineatus* ($r = -0.1$, $p = 0.59$, $n = 29$), and (D) *Tonatia saurophila* ($r = 0.07$, $p = 0.78$, $n = 18$). *Gray areas* indicate standard error. Sources: Weber and Grelle (2012) and Weber et al. (in prep.).

has provided many rich insights into contemporary biogeography of Earth's biota, in particular from perspectives of describing patterns and searching for mechanisms underlying broadscale gradients of biodiversity. Phyllostomid bats exhibit arguably some of the strongest biodiversity gradients in the world and are the main drivers of mammalian diversity gradients in the New World. We have advanced our understanding of contemporary gradients of diversity, including their main causes, comprising evolutionary mechanisms such as diversification and phylogenetic niche conservatism as well as ecological factors linked to climatic variation such as temperature seasonality. Despite these advances, several knowledge shortfalls (e.g., taxonomic and distributional) still exist, and overcoming them will require not only more sampling but also novel methods that integrate models (e.g., ecological niche models) and observational data. We hope that the approaches

and findings described in this chapter foster future studies linking different types of data, scales, and modeling techniques to identify the ultimate causes behind diversity gradients of bats in general and its main component in the New World, the family Phyllostomidae.

Acknowledgments

We would like to thank the editors, in particular Marco Mello and Ted Fleming, for shepherding this manuscript through the review process. We appreciated the valuable feedback from two anonymous reviewers. Writing of this manuscript was supported by the College of Agricultural Sciences and Natural Resources and Office of the Vice President of Research, Texas Tech University. This study was financed in part by the Coordenação de Aperfeiçoamento de Pessoal de Nível Superior—Brasil (CAPES)—Finance Code

001. MMW received a postdoctoral scholarship from PNPD/CAPES (Brazil). FV thanks Héctor Arita for introducing him to the study of bats and CONACyT for financial support.

References

Alroy, J. 2015. The shape of terrestrial abundance distributions. Science Advances, 1:e1500082. DOI: 10.1126/sciadv .1500082.

Araújo, M. B., and M. New. 2006. Ensemble forecasting of species distributions. Trends in Ecology and Evolution, 22:42–47.

Arita, H. T., P. Rodríguez, and E. Vázquez-Domínguez. 2005. Continental and regional ranges of North American mammals: Rapoport's rule in real and null worlds. Journal of Biogeography, 32:961–971.

Arita, H. T., J. Vargas-Barón, and F. Villalobos. 2014. Latitudinal gradients of genus richness and endemism and the diversification of New World bats. Ecography, 37:1024–1033.

Baker, R. J., and H. H. Genoways. 1978. Zoogeography of Antillean bats. Special Publication, Academy of Natural Sciences of Philadelphia, 13:53–97.

Bini, L. M., J. A. F. Diniz-Filho, T. F. L. V. B. Rangel, R. P. Bastos, and M. P. Pinto. 2006. Challenging Wallacean and Linnean shortfalls: knowledge gradients and conservation planning in a biodiversity hotspot. Diversity and Distributions, 12: 475–82.

Bininda-Emonds, O. R. P., M. Cardillo, K. E. Jones, R. D. E. MacPhee, R. M. D. Beck, R. Grenyer, S. A. Price, R. A. Vos, J. L. Gittleman, and A. Parvis. 2007. The delayed rise of present-day mammals. Nature, 446:507–512.

Brown, J. H. 2014. Why are there so many species in the tropics? Journal of Biogeography, 41:8–22.

Buckley, L. B., T. J. Davies, D. D. Ackerly, N. J. Kraft, S. P. Harrison, B. L. Anacker, H. V. Cornell et al. 2010. Phylogeny, niche conservatism and the latitudinal diversity gradient in mammals. Proceedings of the Royal Society of London B–Biological Sciences, 277:2131–2138.

Cardillo, M., C. D. L. Orme, and I. P. F. Owens. 2005. Testing for latitudinal bias in diversification rates: an example using New World birds. Ecology, 86:2278–2287.

Carrera-E., J. P. 2003. Distribución de murcielagos (Chiroptera) a traves de un gradiente altitudinal en las estribicaciones orientales de los Andes ecuatorianos. PhD diss., Departamento de Ciencias Biologicas, Pontifica Universidad Catolica del Ecuador.

Cisneros, L. M., K. R. Burgio, L. M.Dreiss, B. T. Kingbeil, B. D. Patterson, S. J. Presley, and M. R. Willig. 2014. Multiple dimensions of bat biodiversity along an extensive tropical elevation gradient. Journal of Animal Ecology, 83:1124–1136.

Cisneros, L. M., M. E. Fagan, and M. R. Willig. 2015. Effects of human-modified landscapes on taxonomic, functional and phylogenetic dimensions of bat biodiversity. Diversity and Distributions, 21:523–533.

Cisneros, L. M., M. E. Fagan, and M. R. Willig. 2016. Environ-

mental and spatial drivers of taxonomic, functional and phylogenetic characteristics of bat communities in human-modified landscapes. Peer J: DOI 10.7717/peerj.2551.

Colwell, R. K., and D. C. Lees. 2000. The mid-domain effect: geometric constraints on the geography of species richness. Trends in Ecology and Evolution, 15:70–76.

Currie, D. J., G. G. Mittelbach, H. V. Cornell, R. Field, J.-F. Guégan, B. A. Hawkins, D. M. Kaufman et al. 2004. Predictions and tests of climate-based hypotheses of broad-scale variation in taxonomic richness. Ecology Letters, 7:1121–1134.

Czaplewski, N. J., and G. S. Morgan. 2012. New basal noctilionoid bats (Mammalia: Chiroptera) from the Oligocene of subtropical North America. Pp. 162–209, in: Evolutionary History of Bats: Fossils, Molecules and Morphology (G. F. Gunnell and N. B. Simmons, eds.). Cambridge University Press, Cambridge.

Dávalos, L. M. 2006. The geography of diversification in the mormoopids (Chiroptera: Mormoopidae). Biological Journal of the Linnean Society, 88:101–118.

Dávalos, L. M. 2007. Short-faced bats (Phyllostomidae: Stenodermatina): a Caribbean radiation of strict frugivores. Journal of Biogeography, 34:364–375.

Dávalos, L. M., and A. L. Russell. 2012. Deglaciation explains bat extinction in the Caribbean. Ecology and Evolution, 12:1–7.

Davies, T. J., and L. B. Buckley. 2011. Phylogenetic diversity as a window into the evolutionary and biogeographic histories of present-day richness gradients for mammals. Philosophical Transactions of the Royal Society B, 366:2414–2425.

Davies, T. J., V. Savolainen, M. W. Chase, J. Moat, and T. G. Barraclough. 2004. Environmental energy and evolutionary rates in flowering plants. Proceedings of the Royal Society B–Biological Sciences, 271:2195–2200.

Dias, D., C. E. L. Esbérard, and R. Moratelli. 2013. A new species of *Lonchophylla* (Chiroptera, Phyllostomidae) from the Atlantic Forest of southeastern Brazil, with comments on *L. bokermanni*. Zootaxa, 3722:347–360.

Diniz-Filho, J. A. F., L. M. Bini, T. F. Rangel, R. D. Loyola, C. Hof, D. Nogués-Bravo, and M. B. Araújo. 2009. Partitioning and mapping uncertainties in ensembles of forecasts of species turnover under climate change. Ecography, 32:897–906.

Fine, P. V. A. 2015. Ecological and evolutionary drivers of geographic variation in species diversity. Annual Reviews of Ecology, Evolution, and Systematics, 46:369–392.

Fisher, R. A., A. S. Corbet, and C. B. Williams. 1943. The relation between the number of species and the number of individuals in a random sample of an animal population. Journal of Animal Ecology, 12:42–58.

FitzJohn, R. G., W. P. Maddison, and S. P. Otto. 2009. Estimating trait-dependent speciation and extinction rates from incompletely resolved phylogenies. Systematics Biology, 58:595–611.

Fleming, T. H. 1973. Numbers of mammal species in North and Central American forest communities. Ecology, 54:555–563.

Franklin, J. 2010. Mapping Species Distributions — Spatial Inference and Prediction. Cambridge University Press, Cambridge.

Fritz, S. A., O. R. P. Bininda-Emonds, and A. Purvis. 2009.

Geographical variation in predictors of mammalian extinction risk: big is bad, but only in the tropics. Ecology Letters, 12: 538–549.

Gaston, K. J., and R. A. Fuller. 2009. The sizes of species' geographic ranges. Journal of Applied Ecology, 46:1–9.

Gotelli, N. J., M. J. Anderson, H. T. Arita, A. Chao, R. K. Colwell, S. R. Connolly, D. J. Currie et al. 2009. Patterns and causes of species richness: a general simulation model for macroecology. Ecology Letters, 12:873–886.

Gotelli, N. J., and R. K. Colwell. 2001. Quantifying biodiversity: procedures and pitfalls in the measurement and comparison of species richness. Ecology Letters, 4:379–391.

Graham, G. L. 1983. Changes in bat species diversity along an elevational gradient up the Peruvian Andes. Journal of Mammalogy, 64:559–571.

Graham, G. L. 1990. Bats versus birds: comparisons among Peruvian volant vertebrate faunas along an elevational gradient. Journal of Biogeography, 17:657–668.

Grimm, V., E. Revilla, U. Berger, F. Jeltsch, W. M. Mooij, S. F. Railsback, H. H. Thulke, J. Weiner, T. Wiegand, and D. L. DeAngelis. 2005. Pattern-oriented modeling of agent-based complex systems: lessons from ecology. Science, 310: 987–991.

Handley, C. O., Jr. 1976. Mammals of the Smithsonian Venezuelan project. Brigham Young University Science Bulletin, Biological Series, 20:1–91.

Hannah, L. 2011. Climate Change Biology. Elsevier, Heidelberg.

Hawkins, B. A., C. M. McCain, T. J. Davies, L. B. Buckley, B. L. Anacker, H. V. Cornell, E. I. Damschen et al. 2012. Different evolutionary histories underlie congruent species richness gradients of birds and mammals. Journal of Biogeography, 39:825–841.

Hernández, C. E., E. Rodríguez-Serrano, J. Avaria-Llautureo, O. Inostroza-Michael, B. Morales-Pallero, D. Boric-Bargetto, C. B. Canales-Aguirre, P. A. Marquet, and A. Meade 2013. Using phylogenetic information and the comparative method to evaluate hypotheses in macroecology. Methods in Ecology and Evolution, 4:401–415. DOI: 10.1111/2041-210X.12033.

Hijmans, R. J., S. E. Cameron, J. L. Parra, P. G. Jones, and A. Jarvis. 2005. Very high resolution interpolated climate surfaces for global land areas. International Journal of Climatology, 25:1965–1978.

Hortal, J., F. de Bello, J. A. F. Diniz-Filho, T. M. Lewinsohn, J. M. Lobo, and R. J. Ladle. 2015. Seven shortfalls that beset large-scale knowledge of biodiversity. Annual Review of Ecology, Evolution and Systematics, 46:523–549.

Hubbell, S.P. 2001. The Unified Neutral Theory of Biodiversity and Biogeography. Princeton, NJ: Princeton University Press.

Iñiguez Dávalos, L. I. 1993. Patrones ecológicos en la comunidad de murcielagos de la Sierra de Manantlan. Pp. 355–370 in: Avances en el estudio de los mamíferos de México (R. A. Medellín and G. Ceballos, eds.). Publicaciones Especiales, Asociación Mexicana de Mastozoologia, vol. 1. Asociación Mexicana de Mastozoologia, Mexico D.F.

Kaufman, D. M. 1995. Diversity of New World mammals: universality of the latitudinal gradients of species and bauplans. Journal of Mammalogy, 76:322–334.

Klingbeil, B. T., and M. R. Willig. 2009. Guild-specific responses of bats to landscape composition and configuration in fragmented Amazonian rainforest. Journal of Applied Ecology, 46:203–213.

Lewinsohn, T., and P. I. Prado. 2002. Biodiversidade Brasileira—Síntese do Estado Atual do Conhecimento. Editora Contexto, São Paulo.

López-González, C., S. J. Presley, A. Lozano, R. D. Stevens, and C. L. Higgins. 2015. Ecological biogeography of Mexican bats: the relative contributions of habitat heterogeneity, beta diversity, and environmental gradients to species richness and composition patterns. Ecography, 38:261–272.

Losos, J. B. 2008. Phylogenetic niche conservatism, phylogenetic signal and the relationship between phylogenetic relatedness and ecological similarity among species. Ecology Letters, 11: 995–1003.

MacArthur, R. H. 1957. On the relative abundance of bird species. Proceedings of the National Academy of Sciences, USA, 43:293–295.

Magurran, A. E. 2004. Measuring Biological Diversity. Blackwell Publishing, Malden, MA.

Marmion, M., M. Parviainen, M. Luoto, R. K. Heikkinen, and W. Thuiller. 2009. Evaluation of consensus methods in predictive species distribution modelling. Diversity and Distributions, 15:59–69.

May, R. M. 1975. Patterns of species abundance and diversity. Pp. 81–120 in: Ecology and Evolution of Communities (M. L Cody and J. Diamond, eds.). Harvard University Press, Cambridge, MA.

McCain, C. M. 2007a. Area and mammalian elevational diversity. Ecology, 88:76–86.

McCain, C. M. 2007b. Could temperature and water availability drive elevational species richness? A global case study for bats. Global Ecology and Biogeography, 16:1–13.

McCain, C. M., and J.-A. Grytnes. 2010. Elevational gradients in species richness. In: Encyclopedia of Life Sciences. John Wiley and Sons, Chichester.

McNab, B. K. 1969. The economics of temperature regulation in Neotropical bats. Comparative Biochemistry and Physiology, 31:227–268.

Mena, J. L., L. F. Aguirre, J. P. Carrera, and S. Solari. 2011. Small mammal diversity in the tropical Andes: an overview. Pp. 260–275 in: Climate Change and Biodiversity in the Tropical Andes (S. K. Herzog, R. Martínez, P. M. Jorgensen, and H. Tiessen, eds.). Inter-American Institute for Global Change Research, [São José dos Campos, São Paulo, Brazil].

Meyer, C. F. J., and E. K. V. Kalko. 2008. Assemblage level responses of phyllostomid bats to tropical forest fragmentation: land-bridge islands as a model system. Journal of Biogeography, 35:1711–1726.

Mittelbach, G. G., D. W. Schemske, H. V. Cornell, A. P. Allen, J. M. Brown, M. B. Bush, S. P. Harrison et al. 2007. Evolution and the latitudinal diversity gradient: speciation, extinction and biogeography. Ecology Letters, 10:315–331.

Morand, S. 2000. Geographic distance and the role of island area and habitat diversity in the species-area relationships of four Lesser Antillean faunal groups: a complementary note to Ricklefs and Lovette. Journal of Animal Ecology, 69:1117–1119.

Munkemüller, T., F. C. Boucher, W. Thuiller, and S. Lavergne. 2015. Phylogenetic niche conservatism: common pitfalls and ways forward. Functional Ecology 29:627–639.

Muñoz Arango, J. 1990. Diversidad y habitos alimenticios de murcielagos en transectos altitudinales a traves del Cordillera Central do los Andes en Colombia. Studies in Neotropical Fauna and Environment, 25:1–17.

Muylaert, R. L., R. D. Stevens, and M. C. Ribeiro. 2016. Threshold effect of habitat loss on bat richness in Cerrado-forest landscapes. Ecological Applications, 26:1854–1867.

Muylaert, R. L., R. D. Stevens, C. E. L. Esbérard, M. A. R. Mello, G. S. T. Garbino, L. H. Varzinczak, D. Faria et al. 2017. Atlantic bats: a dataset of bat communities from the Atlantic Forests of South America. Ecology, 98:3227.

Parlos, J. A., R. M. Timm, Z. J. Swier, H. Zeballos, and R. J. Baker. 2014. Evaluation of the paraphyletic assemblage within Lonchophyllinae, with description of a new tribe and genus. Occasional Papers, Museum of Texas Tech University, 320:1–23.

Patrick, L. E., and R. D. Stevens. 2014. Investigating sensitivity of phylogenetic community structure metrics using North American desert bats. Journal of Mammalogy, 95:1240–1253.

Patrick, L. E., and R. D. Stevens. 2016. Phylogenetic community structure of North American desert bats: influence of environment at multiple spatial and taxonomic scales. Journal of Animal Ecology, 85:1118–1130.

Patten, M. A. 2004. Correlates of species richness in North American bat families. Journal of Biogeography, 31:975–985.

Patterson, B. D., V. Pacheco, and S. Solari. 1996. Distribution of bats along an elevational gradient in the Andes of southeastern Peru. Journal of Zoology, 240:637–658.

Patterson, B. D., D. G. Stotz, S. Solari, J. W. Fitzpatrick, and V. Pacheco. 1998. Contrasting patterns of elevational zonation for birds and mammals in the Andes of south-eastern Peru. Journal of Biogeography, 25:593–607.

Peixoto, F. P., P. H. P. Braga, M. V. Cianciaruso, J. A. F. Diniz-Filho, and D. Brito. 2014. Global patterns of phylogenetic beta diversity components in bats. Journal of Biogeography, 41:762–772.

Peixoto, F. P., F. Villalobos, and M. Cianciaruso. 2017. Phylogenetic conservatism of climatic niche in bats. Global Ecology and Biogeography 26, 1055–1065. DOI: 10.1111/geb.12618.

Pennell, M. W., and L. J. Harmon. 2013. An integrative view of phylogenetic comparative methods: connections to population genetics, community ecology, and paleobiology. Annals of the New York Academy of Sciences, 1289:90–105. DOI: 10.1111/nyas.12157.

Pereira, M. J. R., and J. M. Palmeirim. 2013. Latitudinal diversity gradients in New World bats: are they a consequence of niche conservatism? PLOS ONE, 8:e69245.

Pinto, C. M., M. R. Marchán-Rivadeneira, E. E. Tapia, J. P. Carrera, and R. J. Baker. 2013. Distribution, abundance and roosts of the fruit bat Artibeus fraterculus (Chiroptera: Phyllostomidae). Acta Chiropterologica, 15:85–94.

Preston, F. W. 1948. The commonness, and rarity, of species. Ecology, 29:254–283.

Pyron, R. A., G. C. Costa, M. A. Patten, and F. T. Burbrink. 2015. Phylogenetic niche conservatism and the evolutionary basis of ecological speciation. Biological Reviews, 90:1248–1262.

Pyron, R. A., and J. J. Wiens. 2013. Large-scale phylogenetic analyses reveal the causes of high tropical amphibian diversity. Proceedings of the Royal Society B, 280:20131622.

Ramoni-Perazzi, P., M. Muñoz-Romo, L. F. Chaves, and T. H. Kunz. 2012. Range prediction for the giant fruit-eating bat, Artibeus amplus (Phyllostomidae: Stenodermatinae) in South America. Studies on Neotropical Fauna and Environment, 47:87–103.

Ricklefs, R. E. 2004. A comprehensive framework for global patterns in biodiversity. Ecology Letters, 7:1–15.

Ricklefs, R. E., and I. J. Lovette. 1999. The roles of island area per se and habitat diversity in the species-area relationships of four Lesser Antillean faunal groups. Journal of Animal Ecology, 68:1142–1160.

Riedinger, V., J. Müller, J. Stadler, W. Ulrich, and R. Brandl. 2013. Assemblages of bats are phylogenetically clustered on a regional scale. Basic and Applied Ecology, 14:74–80.

Rodríguez, P., and H. T. Arita. 2004. Beta diversity and latitude in North American mammals: testing the hypothesis of covariation. Ecography, 27:547–556.

Rodriguez-Duran, A., and T. H. Kunz. 2001. Biogeography of West Indian bats: an ecological perspective. Pages 355–368 in: Biogeography of the West Indies: Patterns and Perspectives (C. A. Woods and F. E. Sergile, eds.). CDC Press, Boca Raton, FL.

Rohde, K. 1992. Latitudinal gradients in species diversity: the search for the primary cause. Oikos, 65:514–527.

Rojas D., Á. Vale, V. Ferrero and L. Navarro. 2012. The role of frugivory in the diversification of bats in the Neotropics. Journal of Biogeography, 39:1948–1960.

Rojas, D., O. M. Warsi, and L. M. Dávalos. 2016. Bats (Chiroptera: Noctilionoidea) challenge a recent origin of extant Neotropical diversity. Systematic Biology, 65:432–448.

Rolland, J., F. L. Condamine, F. Jiguet, and H. Morlon. 2014. Faster speciation and reduced extinction in the tropics contribute to the mammalian latitudinal diversity gradient. PLOS Biology, 12:e1001775.

Rosa, D. T. C. 2014. Redistribuição de morcegos endêmicos da Amazônia em cenários de mudanças climáticas. Master's thesis, Rio de Janeiro Federal University, Rio de Janeiro, Brazil.

Rosenzweig, M. L. 1995. Species Diversity in Space and Time. Cambridge University Press, Cambridge.

Rowe, R. J. 2005. Elevational gradient analyses and the use of historical museum specimens: a cautionary tale. Journal of Biogeography 32:1883–1897.

Sánchez-Cordero, V. 2001. Elevation gradients of diversity for rodents and bats in Oaxaca, Mexico. Global Ecology and Biogeography, 10:63–76.

Schulze, M. D., N. E. Seavy, and D. W. Whitacre. 2000. A comparison of the phyllostomid bat assemblages in undisturbed Neotropical forest and in forest fragments of a slash-and-burn farming mosaic in Peten, Guatemala. Biotropica, 32:174–184.

Shi, J. J., and D. L. Rabosky. 2015. Speciation dynamics during the global radiation of extant bats. Evolution, 69:1528–1545.

Simberloff, D. S. 1976. Experimental zoogeography of islands: effects of island size. Ecology, 57:629–648.

Simmons, N. B. 2005. Order Chiroptera. Pp. 312–529 *in*: Mammal Species of the World: A Taxonomic and Geographic Reference. 3rd ed. Vol. 1. (D. E. Wilson and D. M. Reeder, eds.). John Hopkins University Press, Baltimore.

Smith, B. T., R. W. Bryson, D. D. Houston, and J. Klicka. 2012. An asymmetry in niche conservatism contributes to the latitudinal species diversity gradient in New World vertebrates. Ecology Letters, 15:1318–1325.

Stephens, P. R., and J. J. Wiens. 2003. Explaining species richness from continents to communities: the time-for-speciation effect in emydid turtles. American Naturalist, 161:112–128.

Stevens, R. D. 2004. Untangling latitudinal richness gradients at higher taxonomic levels: familial perspectives on the diversity of New World bat communities. Journal of Biogeography, 31:665–674.

Stevens, R. D. 2006. Historical processes enhance patterns of diversity along latitudinal gradients. Proceedings of the Royal Society of London B, 273:2283–2289.

Stevens, R. D. 2011. Relative effects of time for speciation and tropical niche conservatism on the latitudinal diversity gradient of phyllostomid bats. Proceedings of the Royal Society B, 278:2528–2536.

Stevens, R. D. 2013. Gradients of bat diversity in Atlantic Forest of South America: environmental seasonality, sampling effort and spatial autocorrelation. Biotropica, 45:764–770.

Stevens, R. D., and H. N. Amarilla-Stevens. 2012. Seasonal environments, episodic density compensation and dynamics of structure of chiropteran frugivore guilds in Paraguayan Atlantic Forest. Biodiversity and Conservation, 21:267–279.

Stevens, R. D., S. B. Cox, R. E. Strauss, and M. R. Willig. 2003. Patterns of functional diversity across an extensive environmental gradient: vertebrate consumers, hidden treatments, and latitudinal trends. Ecology Letters, 6:1099–1108.

Stevens, R. D., and M. M. Gavilanez. 2015. Dimensionality of community structure: phylogenetic, morphological and functional perspectives along biodiversity and environmental gradients. Ecography, 38:861–875.

Stevens, R. D., and J. S. Tello. 2014. On the measurement of dimensionality of biodiversity. Global Ecology Biogeography, 23:1115–1125.

Stevens, R. D., J. S. Tello, and M. M. Gavilanez. 2013. Stronger tests of mechanisms underlying geographic gradients of biodiversity: insights from the dimensionality of biodiversity. PLOS ONE, 8:e56853.

Stevens, R. D., and M. R. Willig. 2000. Density compensation in New World bat communities. Oikos, 89:367–377.

Stevens, R. D., and M. R. Willig. 2002. Geographical ecology at the community level: perspectives on the diversity of New World bats. Ecology, 83:545–560.

Taddei, V. A., and B. K. Lim. 2010. A new species of *Chiroderma* (Chiroptera, Phyllostomidae) from northeastern Brazil. Brazilian Journal of Biology, 70:381–386.

Tavares, V. C., A. L. Gardner, H. E. Ramírez-Chaves, and P. M. Velazco. 2014. Systematics of *Vampyressa melissa* Tomas, 1926 (Chiroptera: Phyllostomidae), with descriptions of two new species of *Vampyressa*. American Museum Novitates, 3813:1–27.

Teixeira, T. S. M., D. T. C. Rosa, D.T.C., D. Dias, R. Cerqueira, and M. M. Vale. 2013. First record of *Lonchophylla peracchii* (Dias, Esbérard, Moratelli, 2013) (Chiroptera, Phyllostomidae) in São Paulo State, Southeastern Brazil. Oecologia Australis, 17:424–428.

Teixeira, T. S. M., M. M. Weber, D. Dias, M. L. Lorini, C. E. L. Esbérard, R. L. M. Novaes, R. Cerqueira, and M. M. Vale. 2014. Combining environmental suitability and habitat connectivity to map rare or data deficient species in the tropics. Journal for Nature Conservation, 22:384–390.

Tello, J. S., and R. D. Stevens. 2010. Multiple environmental determinants of regional species richness and effects of geographic range size. Ecography, 33:796–808.

Tello, J. S., and Stevens, R. D. 2012a. Murcielagos, características ambientales, y efectos de mitad de dominio. Pp. 91–104 *in*: Investigacion y Conservación Sobre Murciélagos en el Ecuador (D. G. Tirira and S. F. Burneo, eds.). Publicaciones Especiales Sobre Los Mamiferos del Ecuador, Quito.

Tello, J. S., and R. D. Stevens. 2012b. Can stochastic geographical evolution re-create macroecological richness-environment correlations? Global Ecology and Biogeography, 21:212–223.

Tokeshi, M. 1990. Niche apportionment or random assortment: species abundance patterns revisited. Journal of Animal Ecology, 59:1129–1146.

Tokeshi, M. 1993. Species abundance patterns and community structure. Advances in Ecological Research, 24:112–186.

Tuomisto, H. 2010. A diversity of beta diversities: straightening up a concept gone awry. Pt. 1. Defining beta diversity as a function of alpha and gamma diversity. Ecography, 33:2–22.

UNEP (United Nations Environment Programme). 2014. Nairobi, KE: UNEP. Available from: http://www.unep.org/.

Velazco, P. M. 2005. Morphological phylogeny of the bat genus *Platyrrhinus* Saussure, 1860 (Chiroptera: Phyllostomidae) with the description of four new species. Fieldiana Zoology, 105:1–53.

Velazco, P. M., and A. L. Gardner. 2009. A new species of *Platyrrhinus* (Chiroptera: Phyllostomidae) from western Colombia and Ecuador, with emended diagnoses of *P. aquilus*, *P. dorsalis*, and *P. umbratus*. Proceedings of the Biological Society of Washington, 122:249–281.

Velazco, P. M., A. L. Gardner, and B. D. Patterson. 2010. Systematics of the *Platyrrhinus helleri* species complex (Chiroptera: Phyllostomidae), with descriptions of two new species. Zoological Journal of the Linnean Society, 159:785–812.

Velazco, P. M., and B. D. Patterson. 2008. Phylogenetics and biogeography of the broad-nosed bats, genus *Platyrrhinus* (Chiroptera: Phyllostomidae). Molecular Phylogenetics and Evolution, 49:749–759.

Velazco, P. M., and B. D. Patterson. 2013. Diversification of the yellow-shouldered bats, genus *Sturnira* (Chiroptera, Phyllostomidae), in the New World tropics. Molecular Phylogenetics and Evolution, 68:683–698.

Velazco, P. M.,and B. D. Patterson. 2014. Two new species of yellow-shouldered bats, genus *Sturnira* Gray, 1842 (Chiroptera, Phyllostomidae) from Costa Rica, Panama and western Ecuador. ZooKeys, 402:43–66.

Villalobos, F., and H. T. Arita. 2010. The diversity field of New World leaf-nosed bats (Phyllostomidae). Global Ecology and Biogeography, 19:200–211.

Villalobos, F., A. Lira-Noriega, J. Soberón, and H. T. Arita. 2014. Co-diversity and co-distribution in phyllostomid bats: evaluating the relative roles of climate and niche conservatism. Basic and Applied Ecology, 15:85–91.

Villalobos, F., and T. F. Rangel. 2014. Geographic patterns of biodiversity: macroecological approaches for a complex phenomenon. Pp. 1–11 *in*: Frontiers in Ecology, Evolution and Complexity (M. Benitez, O. Miramontes, and A. Valiente-Banuet, eds.). Copit-arXives, Mexico City.

Villalobos, F., T. F. Rangel and J. A. F. Diniz-Filho. 2013. Phylogenetic fields of species: cross-species patterns of phylogenetic structure and geographical coexistence. Proceedings of the Royal Society B–Biological Sciences, 280:20122570.

Weber, M. M., and C. E. V. Grelle. 2012. Does environmental suitability explain the relative abundance of the tailed tailless bat, *Anoura caudifer*? Natureza e Conservação, 10:221–227.

Weber, M. M., R. D. Stevens, J. A. F. Diniz-Filho, and C. E. V. Grelle. 2017. Is there a correlation between abundance and environmental suitability derived from ecological niche modelling? A meta-analysis. Ecography, 40:817–828.

Weber, M. M., R. D. Stevens, M. L. Lorini, and C. E. V. Grelle. 2014. Have old species reached most environmentally suitable areas? A case study with South American phyllostomid bats. Global Ecology and Biogeography, 23:1177–1185.

Weber, M. M., L. C. Terribile, and N. C. Cáceres. 2010. Potential geographic distribution of *Myotis ruber* (Chiroptera, Vespertilionidae), a threatened Neotropical bat species. Mammalia, 74:333–338.

Wiens, J. J. 2008. Commentary on Losos (2008): niche conservatism déjà vu. Ecology Letters, 11:1004–1005.

Wiens, J. J., and M. J. Donoghue. 2004. Historical biogeography, ecology and species richness. Trends in Ecology and Evolution, 19 639–644.

Wiens, J. J., G. Parra-Olea, M. Garcia-Paris, D. B. Wake. 2007. Phylogenetic history underlies elevational biodiversity patterns in tropical salamanders. Proceedings of the Royal Society of London B, 274:919–928.

Willig, M. R., and C. P. Bloch. 2006. Latitudinal gradients in species richness: a test of the geographic area hypothesis at two ecological scales. Oikos, 112:163–173.

Willig, M. R., D. M. Kaufman, and R. D. Stevens. 2003. Latitudinal gradients in biodiversity: pattern, process, scale, and synthesis. Annual Review of Ecology, Evolution and Systematics, 34:273–309.

Willig, M. R., and S. K. Lyons. 1998. An analytical model of latitudinal gradients of species richness and an empirical test for marsupials and bats in the New World. Oikos, 81:93–98.

Willig, M. R., S. J. Presley, C. P. Bloch and H. H. Genoways. 2009. Macroecology of Caribbean bats: effects of area, elevation, latitude and hurricane disturbance. Pp. 216–264 *in*: Island Bats: Evolution, Ecology and Conservation (T. H. Fleming and P. A. Racey, eds.). University of Chicago Press, Chicago.

Willig, M. R., and E. A. Sandlin. 1991. Gradients of species density and species turnover in New World bats: a comparison of quadrat and band methodologies. Pp. 81–96 *in*: Latin American Mammalogy: History, Biodiversity and Conservation (M. A. Mares and D. J. Schmidly, eds.). University of Oklahoma Press, Norman.

Willig, M. R., and K. W. Selcer, K. W. 1989. Bat species density gradients in the New World: a statistical assessment. Journal of Biogeography, 16:189–195.

Conservation

23

Challenges and Opportunities for the Conservation of Brazilian Phyllostomids

Enrico Bernard,
Mariana Delgado-Jaramillo,
Ricardo B. Machado,
and Ludmilla M. S. Aguiar

Introduction

Brazil is the fifth largest country in the world, with an area of more than 8.5 million square km. This continental-sized, biome-rich country harbors one of the highest bat species richness in the world (Nogueira et al. 2018). Currently, more than 180 species of bats are known in Brazil (Nogueira et al. 2018) and at least 44 genera and 93 species are phyllostomids. From the large *Vampyrum spectrum* (>150 grams) and *Chrotopterus auritus* (60–100 g) to the smaller *Mesophylla macconnelli* and *Vampyressa thyone* (<10 grams both), the variety of phyllostomids in Brazil is quite impressive. Since 2005, at least 10 new bat species (including two new genera) were described for Brazil, and four are phyllostomids: *Xeronycteris vieirai, Dryadonycteris capixaba, Lonchophylla peracchii,* and *Lonchopylla inexpectata* (Moratelli and Dias 2015; Nogueira et al. 2018). Descriptions of other new species are on the way, and genetic studies are showing that several of the Brazilian "species" are in fact complexes of species; therefore, the bat richness in the country remains open for new additions.

But, like many other species, bats are also threatened in Brazil: of the 1,173 nationally threatened animal species, seven are bats and four are phyllostomids: *Lonchorhina aurita, Glyphonycteris behnni, Lonchophylla dekeyseri,* and *X. vieirai* (MMA 2014). Although large extensions of the original six terrestrial Brazilian biomes are still preserved (~72% of Amazonia, ~73% of Pantanal, e.g., http://plataforma.mapbiomas .org/map#coverage), the conservation scenario is problematic for others: only ~7% remains for the Atlantic Forest, for instance (Ribeiro et al. 2009). Moreover, the situation within some of the biomes is heterogeneous: the eastern portion of Amazonia is under severe human pressure and most of the southern portion of Cerrado is already lost (www.mapbiomas.org). While in some parts of the country immediate conservation interventions are still not necessary, in others, securing the last remaining natural fragments is urgently needed. As a consequence, different parts of Brazil experience different conservation demands, and, consequently, their bat populations are exposed to pressures and threats under different severity.

Facing the different remaining percentages and conservation pressures and threats experienced by all Brazilian terrestrial biomes, and given the importance bats play in the ecology of those areas, a better knowledge of the distribution of Brazilian phyllostomids is useful for identifying, for example, where hotspots of species richness are, what the conservation situation of these hotspots is, and which should be considered priority for conservation interventions. To contribute to such analysis, this chapter is divided in two parts. In the first, we present a panorama of the phyllostomids in Brazil by using species distribution modeling and geographic information system techniques, which allows us to identify areas with higher and lower potential species richness. We also estimate the impact of land conversion on conservation-sensitive species and conduct a gap analysis to select new sites for conservation and improve the current Brazilian protected-areas system with phyllostomids as a focal group. In the second part, we present and discuss major pressures and threats affecting the protection of the Brazilian phyllostomids, as well as the challenges and opportunities that must be pursued to guarantee their protection.

Part 1: Species Modeling

Data Processing

We initially considered modeling the potential distribution for all phyllostomid species for Brazil (Nogueira et al. 2018). However, nine species were eliminated from our analysis due to the scarce (<10) number of records or due to taxonomic uncertainty between records of congener species (e.g., *Lonchophylla* spp., for which new species were recently described, changing the current distribution knowledge for the entire genus in Brazil—see Moratelli and Dias [2015]). Therefore, 82 species were considered in our exercise: 74 had their potential distribution modeled, based on a larger number of points, and for eight we had to use the minimum convex polygon due to the scarce number of records (see table 23.1, and the detailed methodological explanation below).

We searched for species records in the databases Web of Science (http://www.webofknowledge.com), Google Scholar (https://scholar.google.com.br), the Instituto Chico Mendes Institute de Conservação da Biodiversidade (ICMBio) (http://www.icmbio.gov.br), Species-Link (http://www.splink.org.br), VertNet (http://www.vertnet.org), and Global Biodiversity Information Facility (GBIF) (http://www.gbif.org). The compiled points were checked and filtered for location consistency and taxonomy. Generally, records from localities based on museum specimens are likely to exhibit spatial autocorrelation and suffer from environmental biases (Araújo and Guisan 2006; Loiselle et al. 2008). To mitigate such problems, we filtered localities that were within 25 km of one another under the same environmental conditions, keeping the most localities possible (Boria et al. 2014).

We then generated different distribution models for each species based on a set of 19 bioclimatic variables from WorldClim 1.4 (http://www.worldclim.org/). We also considered the Normalized Difference Vegetation Index (NDVI—a proxy for measuring vegetation cover), plus elevation at a 5 km^2 resolution. In order to avoid colinearity among bioclimatic variables (i.e., when two variables are highly correlated), we calculated the Pearson correlation index among the 21 variables available and, for those with ≥80% of correlation, eliminated the variable with the lowest contribution. After this process, we only considered nine noncorrelated bioclimatic variables derived from temperature and rainfall (mean daily temperature range, isothermality, temperature seasonality, maximum temperature in the warmest month, mean temperature in the wettest quarter, mean temperature in the coldest quarter, annual rainfall, rainfall in the driest quarter, rainfall in the warmest quarter).

Using the software MaxEnt 3.2.1 (Phillips and Dudík 2008; Phillips et al. 2006), we generated different distribution models for each of the 82 species considered. Instead of employing default settings for the regularization multiplier, we made different tests to find the species-specific tuning for our modeling settings and evaluated the best models produced (Boria et al. 2014; Radosavljevic and Anderson 2014). We've set the program to use 80% of the data to calibrate the model and 20% for the test, using $n - 1$ replicates, where n is the number of records of occurrences as suggested by Pearson et al. (2007). To assess the predictive capacity of the models, we used the area under the curve (AUC), where models with the best performance in the predictions will have AUC values close to 1,

Table 23.1. Species of phyllostomids for Brazil (sensu Nogueira et al. 2018)

| Species | Distribution | Conservation Sensitive Species | | | | | Area | | | | |
		Priority	Endemic	Endangered	Monotypic Genera	Gleaning Animalivore	Potential Distribution (in km²)	Remaining Area (in km²)	Remaining Area (%)	Protected Area (%)	Goal used (%)
Ametrida centurio Gray 1847	MaxEnt	x	...	...	x	...	2,572,745.23	1,750,791.21	68.05	10.21	10.00
Anoura caudifer (É. Geoffroy 1818)	MaxEnt	...	...	...	...	...	...	...	...	...	...
Anoura geoffroyi Gray 1838	MaxEnt	...	...	...	...	...	...	...	...	...	...
Artibeus concolor Peters 1865	MaxEnt	...	...	...	...	...	...	...	...	...	...
Artibeus fimbriatus Gray 1838	MaxEnt	...	...	...	...	...	...	...	...	...	...
Artibeus lituratus (Olfers 1818)	MaxEnt	...	...	...	...	...	...	...	...	...	...
Artibeus obscurus (Schinz 1821)	MaxEnt	...	...	...	...	...	...	...	...	...	...
Artibeus planirostris (Spix 1823)	MaxEnt	...	...	...	...	...	...	...	...	...	...
Carollia benkeithi Solari and Baker 2006	MaxEnt	...	...	...	...	...	...	...	...	...	...
Carollia brevicauda (Schinz 1821)	MaxEnt	...	...	...	...	...	...	...	...	...	...
Carollia perspicillata (Linnaeus, 1758)	MaxEnt	...	...	...	...	...	...	...	...	...	...
Chiroderma doriae Thomas 1891	MaxEnt	...	...	...	...	...	...	...	...	...	...
Chiroderma trinitatum Goodwin, 1958	MaxEnt	...	...	...	...	...	...	...	...	...	...
Chiroderma villosum Peters 1860	MaxEnt	...	...	...	...	...	...	...	...	...	...
Chiroderma vizottoi Taddei and Lim 2010	MCP	x	x	...	...	...	105,894.95	80,395.74	75.92	4.23	26.67
Choeroniscus minor (Peters 1868)	MaxEnt	...	...	...	...	...	...	...	...	...	...
Chrotopterus auritus (Peters 1856)	MaxEnt	x	...	...	x	x	6,388,148.6	3,445,177.84	53.93	6.28	26.67

(continued)

Table 23.1. Continued

| Species | Distribution | Conservation Sensitive Species | | | | | Area | | | | |
		Priority	Endemic	Endangered	Monotypic Genera	Gleaning Animalivore	Potential Distribution (in km²)	Remaining Area (in km²)	Remaining Area (%)	Protected Area (%)	Goal used (%)
Dermanura anderseni (Osgood, 1916)	MaxEnt	...	...	...	...	...	...	...	...	...	...
Dermanura bogotensis (Andersen, 1906)	MaxEnt	...	...	...	...	...	...	...	...	...	...
Dermanura cinerea Gervais 1856	MaxEnt	...	...	...	...	...	...	...	...	...	...
Dermanura gnoma (Handley 1987)	MaxEnt	...	...	...	...	...	...	...	...	...	...
Desmodus rotundus (É. Geoffroy 1810)	MaxEnt	x	...	...	x	...	6,990,369.2	3,759,088.25	53.78	5.94	10.00
Diaemus youngii (Jentink 1893)	MaxEnt	x	...	...	x	...	6,535,481.57	35,525,59.8	54.36	6.31	10.00
Diphylla ecaudata Spix 1823	MaxEnt	x	...	...	x	...	7,663,165.02	4,215,019.92	55	5.83	10.00
Dryadonycteris capixaba Nogueira et al. 2012	MaxEnt	x	x	...	x	...	417,132.12	134,804.08	32.32	2.66	35.00
Glossophaga commissarisi Gardner 1962	MaxEnt	...	...	...	...	...	...	...	...	...	...
Glossophaga longirostris Miller 1898	—	...	...	...	...	...	...	...	...	...	...
Glossophaga soricina (Pallas, 1766)	MaxEnt	...	...	...	...	...	...	...	...	...	...
Glyphonycteris behnii (Peters 1865)	MCP	x	...	x	...	...	2,099,900.6	1,007,507.49	47.98	3.13	43.33
Glyphonycteris daviesi (Hill 1964)	MaxEnt	...	...	...	...	...	...	...	...	...	...
Glyphonycteris sylvestris Thomas 1896	MaxEnt	...	...	...	...	...	...	...	...	...	...
Hsunycteris thomasi (J. A. Allen 1904)	MaxEnt	x	...	...	x	...	3,535,785.73	2,456,029.61	69.46	9.46	10.00
Lampronycteris brachyotis (Dobson 1879)	MaxEnt	x	...	...	x	...	2,722,561.16	1,880,401.23	69.07	9.61	26.67

Species											
Lichonycteris degener Miller 1931	MaxEnt	...	...	...	...	...	...	...	...	...	...
Lionycteris spurrelli Thomas 1913	MaxEnt	×	...	...	×	×	4,775,576.55	2,763,259.93	57.86	7.54	10.00
Lonchophylla bokermanni Sazima et al. 1978	...	...	×	...	...	...	...	...	...	...	...
Lonchophylla dekeyseri Taddei et al. 1983	...	...	...	...	...	...	...	...	...	...	...
Lonchophylla mordax Thomas 1903	...	...	...	...	...	...	...	...	...	...	...
Lonchophylla peracchii Dias et al. 2013	...	...	×	...	...	...	...	...	...	...	...
Lonchorhina aurita Tomes 1863	MaxEnt	×	...	×	...	×	8,098,970.6	4,568,703.16	56.41	6.19	43.33
Lonchorhina inusitata Handley and Ochoa 1997	...	...	...	...	...	×	...	...	...	...	...
Lophostoma brasiliense Peters 1866	MaxEnt	×	...	...	...	×	6,7963,30.63	3,712,974.51	54.63	6.08	18.33
Lophostoma carrikeri (J. A. Allen 1910)	MaxEnt	×	...	...	...	×	4,116,216.55	2,551,612.29	61.99	19.41	18.33
Lophostoma schulzi (Genoways and Williams 1980)	MCP	×	...	...	...	×	254,487.62	218,392.47	85.82	8.14	18.33
Lophostoma silvicola d'Orbigny 1836	MaxEnt	×	...	...	...	×	7,183,254.59	4,232,058.98	58.92	6.75	18.33
Macrophyllum macrophyllum (Schinz 1821)	MaxEnt	×	...	...	×	×	3,573,519.67	1,478,943	41.39	3.24	10.00
Mesophylla macconnelli Thomas 1901	MaxEnt	×	...	...	×	×	4,592,399.07	3,233,417.57	70.41	9.67	10.00
Micronycteris brosseti Simmons and Voss 1998	MCP	×	...	...	...	×	547.99	325.04	59.31	0	35.00
Micronycteris hirsuta (Peters 1869)	MaxEnt	×	...	...	...	×	3,045,065.41	1,951,308.02	64.08	9.02	18.33
Micronycteris homezorum Pirlot 1967	MCP	×	...	...	...	×	11,584.46	4633.12	39.99	0.51	18.33

(continued)

Table 23.1. Continued

	Conservation Sensitive Species						Area				
Species	Distribution	Priority	Endemic	Endangered	Monotypic Genera	Gleaning Animalivore	Potential Distribution (in km²)	Remaining Area (in km²)	Remaining Area (%)	Protected Area (%)	Goal used (%)
Micronycteris megalotis (Gray 1842)	MaxEnt	x	...	...	...	x	6,845,438.34	3,753,604.95	54.83	6.12	18.33
Micronycteris microtis Miller 1898	MaxEnt	x	...	...	...	x	4,791,816.27	2,593,636.97	54.13	5.72	18.33
Micronycteris minuta (Gervais 1856)	MaxEnt	x	...	...	...	x	4,366,530.46	2,150,380.63	49.25	5.07	18.33
Micronycteris sanborni Simmons 1996	MaxEnt	x	x	...	...	x	2,125,806.22	1,028,846.97	48.4	2.42	43.33
Micronycteris schmidtorum Sanborn 1935	MaxEnt	x	...	...	...	x	4,693,426.13	2,843,242.72	60.58	7.99	18.33
Mimon bennettii (Gray 1838)	MaxEnt	x	...	...	...	x	2,152,187.87	665,837.13	30.94	2	18.33
Mimon crenulatum (É. Geoffroy 1803)	MaxEnt	x	...	...	...	x	5,380,830.19	3,142,355.38	58.4	6.66	18.33
Neonycteris pusilla (Sanborn 1949)	MPC	x	...	...	x	...	12.56	12.56	100	0	43.33
Phylloderma stenops (Peters 1865)	MaxEnt	x	...	...	x	...	5,947,886.54	3,582,904.52	60.24	7.21	10.00
Phyllostomus discolor (Wagner 1843)	MaxEnt	...	...	...	...	...	...	...	...	...	...
Phyllostomus elongatus (É. Geoffroy 1810)	MaxEnt	...	...	...	...	...	...	...	...	...	...
Phyllostomus hastatus (Pallas, 1767)	MaxEnt	...	...	...	...	...	...	...	...	...	...
Phyllostomus latifolius (Thomas 1901)	MaxEnt	...	...	...	...	...	...	...	...	...	...
Platyrrhinus angustirostris Velazco et al. 2010	...	...	...	...	...	...	...	...	...	...	...
Platyrrhinus aurarius Handley and Ferris 1972	...	...	...	...	...	...	...	...	...	...	...

Species	Method								
Platyrrhinus brachycephalus Rouk and Carter 1972	MaxEnt	...	...	...	...	...	...	...	...
Platyrrhinus fusciventris Velazco et al. 2010	...	...	...	...	...	...	...	...	...
Platyrrhinus incarum (Thomas 1912)	MaxEnt	...	...	...	...	...	...	...	...
Platyrrhinus infuscus (Peters 1880)	MaxEnt	...	...	...	...	...	...	...	...
Platyrrhinus lineatus (É. Geoffroy 1810)	MaxEnt	...	...	...	...	...	...	...	...
Platyrrhinus recifinus (Thomas 1901)	MaxEnt	×	×	...	2,354,936.21	708,091.91	30.07	1.88	26.67
Pygoderma bilabiatum (Wagner 1843)	MaxEnt	×	...	×	2,319,756.8	672,453.03	28.99	1.97	10.00
Rhinophylla fischerae Carter 1966	MaxEnt	...	...	...	...	...	...	...	...
Rhinophylla pumilio Peters 1865	MaxEnt	...	...	...	...	...	...	...	...
Scleronycteris ega Thomas 1912	MCP	×	×	×	540,428.14	407,499.88	75.4	13.56	10.00
Sphaeronycteris toxophyllum Peters 1882	MCP	×	×	×	2,025,436.86	1,618,672.81	79.92	10.14	10.00
Sturnira lilium (É. Geoffroy 1810)	MaxEnt	...	...	...	...	...	...	...	...
Sturnira magna de la Torre 1966	MaxEnt	...	...	...	...	...	...	...	...
Sturnira tildae de la Torre 1959	MaxEnt	...	...	...	...	...	...	...	...
Tonatia bidens (Spix 1823)	MaxEnt	×	...	×	4,112,300.34	2,081,885.63	50.63	4	18.33
Tonatia saurophila Koopman and Williams 1951	MaxEnt	×	...	×	3,963,820.51	2,461,976.5	62.11	7.57	18.33
Trachops cirrhosus (Spix 1823)	MaxEnt	×	...	×	7,740,233.25	4,378,030.17	56.56	6.4	26.67
Trinycteris nicefori (Sanborn 1949)	MaxEnt	×	...	×	3,515,915.96	2,354,189.9	66.96	8.46	26.67
Uroderma bilobatum Peters 1866	MaxEnt	...	...	...	...	...	...	...	...

(continued)

Table 23.1. Continued

Species	Distribution	Conservation Sensitive Species					Area				
		Priority	Endemic	Endangered	Monotypic Genera	Gleaning Animalivore	Potential Distribution (in km²)	Remaining Area (in km²)	Remaining Area (%)	Protected Area (%)	Goal used (%)
Uroderma magnirostrum Davis 1968	MaxEnt	...	...	...	...	...	...	...	...	...	...
Vampyressa pusilla (Wagner 1843)	MaxEnt	...	...	...	...	...	...	...	...	...	...
Vampyressa thyone Thomas 1909	...	...	...	...	...	...	...	...	...	...	...
Vampyriscus bidens (Dobson 1878)	MaxEnt	...	...	...	...	...	...	...	...	...	...
Vampyriscus brocki (Peterson 1968)	MaxEnt	...	...	...	...	...	...	...	...	...	...
Vampyrodes caraccioli (Thomas 1889)	MaxEnt	...	...	...	...	...	...	...	...	...	...
Vampyrum spectrum (Linnaeus, 1758)	MaxEnt	x	...	x	x	x	4,238,321.98	2,947,434.19	69.54	8.08	35.00
Xeronycteris vieirai Gregorin and Ditchfield 2005	MaxEnt	x	x	x	x	...	651,489.89	337,724.87	51.84	1.53	60.00

Note: Data for 82 species were used to assess their potential distribution and richness in the country. Their distribution was estimated either based on species distribution modeling (assigned here as "MaxEnt," since we used the software MaxEnt 3.2.1 for that purpose [Phillips and Dudik 2008; Phillips et al. 2006]) or using the minimum convex polygon technique (here assigned MCP) due to scarce number of records. A subset of 40 conservation-sensitive species (see p. 422 for the rationale on the species selection) was used for a refined analysis on their potential distribution, the remaining natural area within such distribution, and the percentage of area within protected areas. For those species, conservation goals between 10% and 60% of the potential distribution were used.

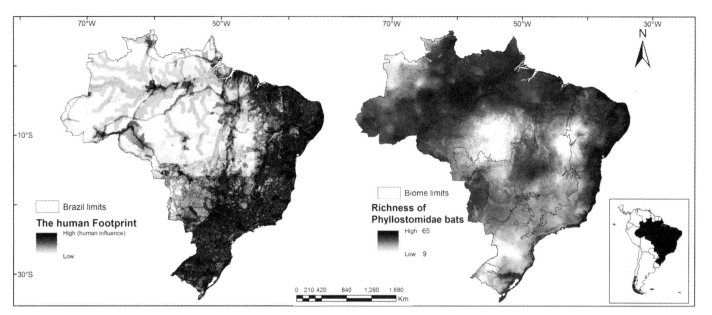

Figure 23.1. The human footprint (*left*) and Phyllostomidae richness patterns for Brazil (*right*) obtained by species distribution models based on data for 83 species. In an alarming way, some of the areas with highest potential species richness in Brazil are also under intense human pressure. This is clearly the case of the Atlantic Forest along the coastal part of Brazil, as well as some parts of Amazonia, Cerrado, and Pantanal. Data for the human footprint was obtained from WCS and CIESIN (2005).

whereas AUC values close to 0.5 indicate models equal to or worse than random (Phillips and Dudík 2008). Because AUC does not directly quantify overfitting, we quantified it by calculating the difference between the calibration and the evaluation AUCs (Warren and Seifert 2011): the smaller the difference between the two, the lesser the overfitting present in the model. We also evaluated models by qualitative visual examination of the resulting maps, based on expert knowledge of the distribution where the species are known to occur.

To mitigate the occurrence of false positives (commission errors) in the potential distribution generated, we excluded biomes where the species was not recorded. The remaining models were converted into binary presence-absence values based on two established threshold values. We used a minimum training presence threshold (Pearson et al. 2007) for species considered to be more generalist and having wider distributions and a tenth-percentile presence threshold (Anderson and Gonzalez 2011; Radosavljevic and Anderson 2014) for more restricted species. Finally, we overlapped the distribution of those individual binaries in order to generate a map of species richness for all species evaluated.

Species Richness among Regions

Based on a mean AUC of 0.88 ± 0.08 for tests and 0.90 ± 0.06 for training, the richness map for the 82 species modeled highlighted the central part of Amazonia and the coast of the Atlantic Forest as the species richest regions (fig. 23.1, *right*). In contrast, the southern part of Amazonia, the transition between Caatinga and Cerrado, and the southern part of the country in the Pampa biome were the less speciose. Phyllostomid species richness varied between nine and 65 species per 100 km^2 (modal 50 species/100 km^2—fig. 23.1).

Our analysis indicated that areas with the highest potential species-richness values (>60 spp. of phyllostomids) are in the Atlantic Forest coast, mainly in its northeastern region. This is worrisome, considering that this is one of the most threatened and fragmented parts of the Atlantic Forest in Brazil (Ribeiro et al. 2009). In fact, most of the nationally threatened species of bats in Brazil are present or restricted to the Atlantic Forest (MMA 2014). Due to their flight capacity and high mobility, there is a false perception that bats would be less affected by habitat loss and fragmentation. However, some species of bats have high foraging and roosting specificity (e.g., Kalko et al. 1999), and thus this argument does not hold since the responses of Neotropical bats to environmental disturbance can be highly variable (e.g., Cunto and Bernard 2012; Jung and Kalko 2011; Meyer and Kalko 2008; Meyer et al. 2008, 2016). Therefore, the conservation status of phyllostomids in the Atlantic Forest needs to be better assessed in order to identify whether some of those species are experiencing higher risks than supposed (see "Data Deficiency and Its Implications for Classifying the Species Conservation Status" below).

Interestingly, parts of the Caatinga dry forests of northeastern Brazil were highlighted in our analysis, with a potential to harbor >55 species of phyllostomids. If confirmed, such high species richness will remedy the historical neglect of this area's importance for bats in Brazil (Bernard et al. 2011). In fact, some of the new taxa of nectar-feeding bats recently described in the country have records in the Caatinga (Nogueira et al. 2012; Dias et al. 2013; Rocha et al. 2014), and at least one genus (*Xeronycteris*) and one species (*L. inexpectata*) seem to be restricted to that area.

Challenges When Modeling Bat Species Richness and Distribution in Brazil

Species distribution models are not perfect and they frequently require *in situ* validation and refinement (e.g., Elith and Leathwick 2009; Wisz et al. 2013). This is a challenging task in Brazil. Despite significant progress over the past 15 years—which has resulted in the description of new species and genera (e.g., Dias et al. 2013; Fazzolari-Correa 1994; Feijó et al. 2015; Gregorin and Dietchfield 2005; Gregorin et al. 2006; Moratelli and Dias 2015; Moratelli et al. 2011; Nogueira et al. 2012; Taddei and Lim 2010) and in the extension in hundreds or even thousands of kilometers in the known distribution of several species of bats in Brazil (e.g., Gregorin and Loureiro 2011; Nogueira et al. 2008; Pimenta et al. 2010; Rocha et al. 2014)—the information on the occurrence and distribution of bat species in the country is still heterogeneous and fragmented. Nearly 60% of the Brazilian territory does not have a single formal record of bat species (Bernard et al. 2011). The best sampled biome is the Atlantic Forest (~80% of sampling coverage), and the Amazon is the least sampled (~23%). None of the Brazilian biomes can be considered minimally sampled for bats: the average known number of species per 3,000 km² varies from 4.8 in the Pampa to 13.7 in the Atlantic Forest (Bernard et al. 2011). This is very low considering that, as presented above, our modeling for phyllostomids only projected a modal of 50 species/100 km². Alarmingly, if the current pace of inventories is maintained, it will take 33 years before the entire country has at least one formal record of bat species, but another 200 years until we

can consider Brazil minimally sampled for its bat fauna (Bernard et al. 2011). This time span is too long, especially considering that several of the knowledge gaps for bats in Brazil are located along deforestation frontiers. Brazilian biologists interested in bat conservation are facing a real risk of being late, without the possibility of knowing which species of bats are present in large portions of the country before those areas are significantly and definitively altered and affected by deforestation.

Focusing on Conservation-Sensitive Species

Different species often experience different pressures and threats, and the need for conservation actions can vary greatly between them. Identifying the most threatened species is often necessary for a better decision-making process (Margules and Pressey 2000). Identifying threatened species is also necessary owing to the fact that we are dealing with a very species-rich group in a continental-sized country: not all Brazilian phyllostomids are experiencing the same risks. Therefore, using the data generated for the 82 species for which we've assessed the distribution, we selected a subset of 40 species we considered as conservation priorities (table 23.1). These species were selected based on their conservation needs, using four criteria: (1) species listed either in the Brazilian official list of endangered species (MMA 2014) or in the 2016 International Union for Conservation of Nature (IUCN) Red List (www.redlist.org); (2) species belonging to monotypic genera, (3) species endemic to Brazil; and (4) species ecologically considered as more vulnerable, like gleaning animalivores (Farneda et al. 2015; Klingbeil and Willig 2009; Meyer and Kalko 2008; Peters et al. 2006).

We set different goals for each one of those 40 species. A conservation goal involves determining the size of a species distribution area that must be included within a reserve system so that the species can be considered sufficiently protected. This valuation is arbitrary, considering that species differ in their life histories, habitat requirements, current conservation status, and vulnerability. So there is no scientific consensus on a general value for an overall sufficient level of classification. In our analysis, the goals were calculated separately for each species, taking into account six criteria: (1) species listed as vulnerable in the official threatened species list of Brazil were assigned 1.5 points; (2) species listed as vulnerable in the IUCN Red List were assigned 1 point, while those listed as near threatened were assigned 0.5 points; (3) species endemic to Brazil were assigned 1.5; (4) species with a distribution <600 km² were assigned 1 point; (5) species belonging to a monotypic genus were assigned 0.5 points; and (6) ecologically vulnerable species were assigned 1 point. The total points accrued for each species was calculated, and those with the major priority (3.5 points) were assigned with the highest goals, while species with the lowest-priority level (0.5 points) had lower goals. We also tried different goals (between 10% and 60% of the species distributions) used in previous studies and proposed as ideal scenarios for conservation (Esselman and Allan 2011; Friendlander et al. 2003; Lessmann 2011; Pawar et al. 2007).

Based on the current system of fully protected areas, conservation goals of 10–60% could not be achieved for 33 of the 40 priority species. Twelve of those species have less than 25% of their conservation goal achieved, and 14 species have reached less than 75% of their conservation goal (table 23.1). The situation of some species is also worrisome. In an unpublished analysis we conducted, we identified that 10 out of the 40 priority species here selected have more than 50% of their distribution already deforested and 23 species have more than 25% (MD-J, unpublished data). Alarmingly, *Pygoderma bilabiatum*, *Platyrrhinus recifinus*, *Mimon bennettii*, and *Dryadonycteris capixaba* have lost more than 66% of the natural vegetation cover within their potential distribution area. The conservation status of *D. capixaba* was not yet assessed by IUCN, but in its description authors stated that this bat is locally rare and occurs in only two vegetation formations, both now highly fragmented (Nogueira et al. 2012). For *X. vieirai*, an endemic, nationally threatened, and monotypic species, 48% of its distribution is already disturbed by human activities. Even so, *X. vieirai* is still labeled as data deficient by the IUCN (Solari 2015). In fact, the truth is that we lack good data on the conservation status of several Brazilian species (see "Data Deficiency and Its Implications for Classifying the Species Conservation Status" below).

Gap Analysis, the Selection of New Sites for Conservation, and the Improvement of the Current Brazilian Protected-Areas System

Despite some controversy, protected areas are still the best strategy for the maintenance of biodiversity and ecological services (e.g., Chape et al. 2005; Gaston et al. 2008; Hannah et al. 2007; Watson et al. 2014). Brazil has one of the largest protected-areas systems in the world, with ~2.2 million km² (WDPA 2016). But how are those protected areas performing for the 40 conservation-sensitive bat species we selected? By using the boundaries of the fully protected areas in Brazil (data for 2011—http://mapas.mma.gov.br/i3geo /datadownload.htm) we performed a species-focused gap analysis to determine which and to what extent each of the priority species were represented in the Brazilian protected-areas system and what percentage of their distribution must be protected to achieve the goals set for each. We also overlapped the species distribution models generated with the map of deforestation in Brazil (based on 2009 data—http://siscom.ibama .gov.br/monitorabiomas) so we could quantify the habitat loss within each potential distribution and calculate the area of occupation of each species.

Thirteen out of the 40 priority species had less than 5% of their distribution within fully protected areas; 32 species had less than 10% (table 23.1). All endemic species evaluated had less than 4.5% of their distribution within fully protected areas. There were no records of *Neonycteris pusilla* and *Micronycteris brosseti* in any of the current Brazilian fully protected areas. Once again, the situation of *X. vieirai* is problematic, since less than 2% of its potential distribution lies within protected areas. So in order to improve the *in situ* conservation for some species, expanding the current national protected-areas system would be necessary in Brazil.

Once again, modeling tools can help to identify the best scenarios for such expansion (Margules and Pressey 2000). In an approach based on systematic conservation planning (Margules and Pressey 2000), we used Marxan software version 2.43 (Ball et al. 2009) for the selection of conservation priority areas for the 40 priority species. We used 10 km × 10 km grid cells and considered the human footprint shapefile (WCS and CIESIN 2005), as well as a shapefile of mining areas in Brazil (Departamento Nacional de Produção Mineral [National Department of Mineral Production]—http://www.dnpm.gov.br/conteudo .asp?IDSecao=62&IDPagina=46) to calculate the conservation "cost" of each cell: the assigned cost was inversely proportional to its environmental impact on the establishment of conservation areas. In our analysis, we also considered: (1) the shapefiles with the current

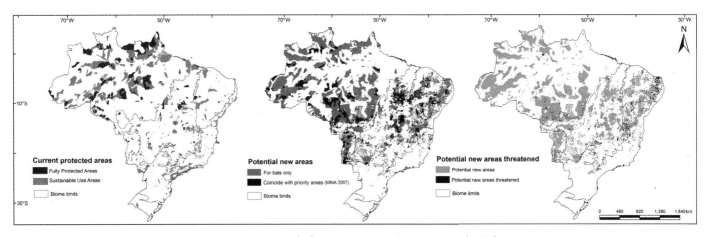

Figure 23.2. Current protected areas for bats in Brazil (*left*), new potential areas for conservation (*middle*), and potential new areas already threatened by mining concessions or in the buffer zones of wind parks (*right*). In order to improve the protection of bats in the country, the current Brazilian protected-areas system needs to be expanded, mostly with the creation of new parks and reserves in the southern part of Amazonia and along the contact zone between Cerrado and Caatinga.

sustainable-use protected areas (data for 2011—http://mapas.mma.gov.br/i3geo/datadownload.htm); (2) the shapefiles with known caves in the country (data for 2014—ICMBio/CECAV 2016), which we selected because at least 38 phyllostomid species were already recorded using caves in Brazil (Guimarães and Ferreira 2014); and (3) the shapefile containing the priority areas for the conservation, sustainable use, and sharing of benefits from the biodiversity in Brazil, which are candidate areas for protection identified by the Brazilian Ministry of Environment since 1990, considering biodiversity targets, sustainable-use targets, and persistence targets and processes (MMA 2016).

The program was parameterized to include all the grid cells inside currently fully protected areas and to avoid redundancy by spatially aggregating solutions to complement the current protected-areas system. Finally, we ran 100 replicates of 100 million interactions each. For visual comparison and analysis, we used the "best" protected-area solution map for each input data scenario (Ardron et al. 2008).

To consider scenarios of 10–60% of conservation of the species' potential distribution area, the strictly protected-areas system in Brazil would have to be expanded from the current 7% of the country's continental territory to 38.7%. Fifty-nine percent of that expansion would have to occur in the Amazon biome, followed by Cerrado (22%) and the Caatinga (11%). But a 5.5-fold increase in one of the largest protected-areas system in the world means that millions of hectares would have to be declared protected. Interestingly,

45% of the proposed areas for conservation of the species we analyzed match with the current established sustainable-use protected areas (fig. 23.2), and 26% match with areas already long identified as national priorities for biodiversity conservation by the Brazilian Ministry of Environment (fig. 23.2). So 26% of the proposed expansion lies in areas already identified as important ones and whose protection would benefit not just bats but other large parts of the Brazilian biota as well. This is a major opportunity: bat experts should join conservationists focused on other taxa in Brazil and, together, should advocate for the immediate protection of those priority areas. In fact, after a period of stagnation in the creation of new areas and even with the downgrading, downsizing, and degazettement of existing ones (Bernard, Penna et al. 2014; Pack et al. 2016), the Brazilian government needs to consider the expansion of the national protected-areas system to achieve some of the international agreements to which Brazil is a signatory, like the Aichi Biodiversity Targets, and the more recent Paris Agreement.

Where to Expand the Protected Areas?

Twenty-nine percent of the expansion of the protected-areas system identified in our analysis would be composed of completely unprotected areas, 62% of them in Amazonia, 22% in the Cerrado, 11% the Caatinga, 3% in the Pantanal, and 2% in the Atlantic Forest. A careful analysis of other attributes could help to improve the decision on which areas could contribute most for better conservation of bats. In this process, cave protection must be considered (see "Recent Changes in the Cave Protection Legislation in Brazil" below). Our analysis indicated that 1,237 currently unprotected caves should and could be added to the current protected-areas system in Brazil. Close to 700 of those caves coincide with the priority areas for conservation proposed by the Ministry of Environment but are unprotected so far. If considered, those 1,237 caves could join a set of another 1,500 caves located within current fully protected areas, plus an additional 2,290 caves within current sustainable-use protected areas, elevating to 5,724 the number of caves within Brazilian protected areas. The selection of which caves should be protected will demand specific analysis considering that, for example, among the 2,990 caves located within sustainable-use protected areas, approximately 1,100 (37%) coincide with mining blocks. At any rate, cave protection is both a necessity and an opportunity for the improvement of bat conservation in Brazil.

Part 2: Pressures, Threats and Challenges to Improve Bat Conservation in Brazil

As presented and discussed above, habitat loss and degradation definitively threats most of the Brazilian bats and some phyllostomids in particular. But what are other major pressures and threats affecting bats in Brazil? In 2012, a group of Brazilian bat biologists conducted a horizon-scanning exercise on the conservation of bats in the country (Bernard et al. 2012). Horizon scanning has been adopted to identify issues of relevance, potential threats, and opportunities for biodiversity conservation, as well as to identify options for public policies, to describe possible future scenarios of environmental and social change, and to consider how these changes may affect conservation objectives (see Sutherland et al. 2008 and included references). In the example mentioned above, it is important to raise awareness of issues never considered before and to help set priorities for bat conservation in Brazil.

Some of those topics deserve special attention, considering the impact they may have on the overall conservation of Brazilian bat species and on some phyllostomids in particular. Here we detail four of those issues: (1) recent changes in the cave protection legislation in Brazil; (2) the status of the interaction between

bats and wind farms; (3) the problem of data deficiency and its implications for classifying the species conservation status; and (4) the impact of pest exterminator companies on bats in Brazil.

Recent Changes in the Cave Protection Legislation in Brazil

Estimates suggest that Brazil may harbor over 300,000 caves (Piló and Auler 2011). However, only about 10% of them have been officially recorded (ICMBio/CECAV 2016). Until October 2008, the Brazilian legislation considered all caves part of the Brazilian speleological heritage and, therefore, subject to full protection. However, the Presidential Decree 6640 of November 2008 changed the law (Brasil 2008) and determined that in order to be protected any cave should undergo a classification process according to their relevance (Brasil 2008). Brazilian caves are now classified as having maximum, high, medium, or low relevance (MMA 2017). This classification must be determined by the analysis of ecological, biological, geological, hydrological, paleontological, scenic, historical-cultural, and socioeconomic attributes assessed under local and regional approaches, and only those caves categorized as maximum relevance would be fully protected a priori. Considering the country's cave potential, the prior categorization of all the Brazilian caves as a prerogative for their protection is unfeasible and, in practice, Decree 6640 reduced their protection, as caves in the categories of high, medium, and low relevance lost formal protection and may be legally exploited and destroyed after an environmental licensing process.

Due to the country's size and the existence of large areas that have never been sampled, the Brazilian caves and their associated fauna are mostly poorly known (e.g., Silva and Ferreira 2016). At least 72 species of bats were already recorded using caves in Brazil, including 39 phyllostomids (Guimarães and Ferreira 2014; Oliveira et al. 2018). The nationally threatened *X. vieirai* and *G. behnni* were already recorded using caves, while *L. dekeyseri* and *L. aurita* are considered cave-dependent species (Guimarães and Ferreira 2014). But the real number of cave-associated species is certainly higher.

Subpopulations of cave-dependent bat species can be locally extinct by the destruction of their shelters

(Aguiar et al. 2006). Therefore, the reduction of protection and increased exploitation of Brazilian caves allowed by Decree 6640 is a real threat not only to the conservation of bat species using those shelters but also to the rich, poorly known, and frequently endemic and species-rich ecosystems associated with the bat guano. The reversion of Decree 6640 is very unlikely, considering that Brazil's Deputy Chamber and Senate are deeply influenced by sectors contrary to the conservation of Brazil's biodiversity, like mining (e.g., Fearnside 2016). One way to alleviate the impact of Decree 6640 is to improve the classification process of the relevance of these caves, taking into account specific demands for the conservation of bats.

Currently, any cave recorded as having of one of the seven nationally threatened species will receive the maximum relevance status (MMA 2017). But other local and regional criteria also apply, as specified in the legislative act of 02/2017 (MMA 2017), published by the Ministry of the Environment nearly one year after Decree 6640. Concerning the presence of bats, caves will have their relevance elevated if they (1) roost populations of pollinators, seed dispersers, or insectivorous species; (2) roost populations for a period longer than 30 consecutive days; (3) contain large deposits of guano; (4) harbor new or undescribed species; (5) harbor high species richness; (6) harbor high relative species abundance; (7) harbor mandatory trogloxene species (i.e., species that definitively need the caves to complete part of their life cycle, like mandatory cave bats); (8) harbor exceptionally large populations; and (9) harbor rare species.

Overall, the entire process of cave protection in Brazil has become more complex and will require greater capacity for data generation and synthesis by those who were willing to follow the established standards. Although the intention was to have a more direct process, based on measurable variables, the legislative act of 02/2009 has some vague definitions that must be urgently refined and better described. The correction of that legislative act provides some opportunities to improve the conservation of caves in Brazil, and all those who are interested in bats in the country should be aware of what this regulation says. A new proposal has been under evaluation since 2013, but a final version of that document has not been reached yet. Considering that contrary sectors want to weaken the country's

environmental regulations to clear the way for rapid development of energy facilities, mines, and agriculture (e.g., Tollefson 2016), it is very important that this process of improving the current legislative act be resumed as soon as possible by the Ministry of Environment. A clear and effective legal framework is necessary as soon as possible, and if well conducted, with the broad participation of the Brazilian bat expert community, there is the possibility that the final document will evolve for a better and stronger format, which is essential for the effective protection of Brazilian bats.

Bats and Wind Farms

Nearly 71,000 km^2 of the Brazilian territory have wind speeds suitable for power generation (Amarante et al. 2001). There are currently 601 wind farms operating in Brazil, with an installed generation potential of 15 gigawatts (ABEEólica 2019). Although it represents ~9% of the current national energy matrix (ABEEólica 2019), wind generation is the fastest growing energy source in Brazil. Supported by public policies to promote the installation of new wind farms in the country, another 4.6 gigawatts are under construction (ABEEólica 2019). This means that thousands of wind turbines will go into operation in Brazil in the coming years.

Although necessary as an alternative energy source, wind turbines have been identified by studies in several parts of the world as collision hazards for bats, causing the death of animals with a frequency and magnitude that may pose significant threats to the conservation of bat populations (see Arnett et al. 2016 and references therein). This situation is unlikely to be different in Brazil, and not surprisingly, the study of the interaction between bats and wind parks has been identified as a priority for the conservation of bats in the country (Bernard et al. 2012).

Overall, 70% of the areas with the highest wind potential are data gaps, lacking elementary information on species presence (Bernard, Paese et al. 2014). Despite having specific and mandatory legislation dated from 2014, Brazil's federal and state legislative acts have a vague and relaxed approach regarding the possible impacts of wind farms on bats (Valença and Bernard 2015). Larger wind parks can be fractioned in smaller units and licensed based on simplified and less rigorous studies. The Brazilian legislation clearly specifies neither the procedures and the minimum necessary effort for pre- and post-installation of wind farms nor what measures should be adopted for the mitigation of their impacts. Few parks in the country had their effects on bats systematically monitored. The only study publicly available, a 4-year monitoring of the Osório wind farm in the southernmost part of the country, reported nine species of bats killed, including *Artibeus lituratus* (Barros et al. 2015). The presence of a large frugivore (*A. lituratus*) among the killed species in that park should be a cautionary message to monitor more closely the interactions between bats and wind farms in Brazil: the profile of the species killed in the country could be different than the already recorded vespertilionids and molossids. In northeastern Brazil, the region holding the greatest potential for wind generation and also the largest parks installed in the country, there are simply no published data on mortality of bats by wind turbines, raising concerns as to whether the monitoring there is being properly conducted. The few personal communications from consultants working in the region confirm that bats are being found dead in some parks, but no official data is yet publicly available.

There is an urgent need to change the current licensing process of wind farms in Brazil, including more rigorous data collection and making those data public. Brazilian environmental agencies should require de facto environmental impact assessments for the installation of wind parks, especially when they are located in data gap areas (Bernard, Paese et al. 2014; Valença and Bernard 2015). Alarmingly, recent bills proposed by the Brazilian legislative chambers could completely mutilate or eliminate entirely the environmental licensing process in Brazil (Fearnside 2016). This is unacceptable from the perspective of bat conservation; the technical rigor in the licensing of wind farms must be proportional to the investment being made in this technology: the wind energy industry is spending millions of dollars in Brazil, but none of that money is flowing to the research and conservation of bats. Alliances involving state and federal agencies, private industry, academic institutions, and nongovernmental organizations to assess and minimize bat mortality at wind farms are urgently needed in Brazil. The United States' Bats and Wind Energy Cooperative (www.batsandwind.org) is an interesting model that could be replicated in Brazil. This is an open opportunity and a goal that should be pursued, with a great potential to bring positive results for the conservation of Brazilian bats.

Data Deficiency and Its Implications for Classifying the Species Conservation Status

The Instituto Chico Mendes de Conservação da Biodiversidade (ICMBio), an independent arm of the Brazilian Ministry of Environment, is the office responsible for the elaboration of the Brazilian threatened species list. The institute uses the internationally accepted classification proposed by the IUCN (IUCN 2012), which is based on five criteria: A—population size reduction (past, present, and/or projected); B—geographic range size and fragmentation, decline, or fluctuations; C—small population size and fragmentation, decline, or fluctuations; D—very small or restricted population; and E—quantitative analysis of extinction risk (e.g., population viability analysis). For each of these criteria, there are additional subcriteria that allow a better definition of the level of threat and its causes.

The detailed examination of these criteria leads to a worrying finding: the Brazilian experts do not have national quantitative data on population decline for almost all the species of bats in the country (criterion A), have no inventory data on the number of sexually reproductive individuals (criterion C), do not have demographic studies that allow estimation of the number of mature individuals for the vast majority of the species (criterion D), and consequently, do not have population viability analyses either (criterion E) except for *L. dekeyseri* (Aguiar et al. 2006).

So, in practice, the main driver used to classify the level of threat for species in Brazil is related to the reduction in habitat quality by deforestation or degradation since these processes can result in reductions of populations and in the potential distribution of the species (see Bernard et al. 2013). In fact, five out of the seven nationally threatened species of bats in Brazil were classified solely based on criterion A and not on measured population declines per se but, rather, on projected, inferred, or suspected population reductions likely to be seen in the future: *Furipterus horrens*, vulnerable A3c; *Natalus macrourus*, vulnerable A3c; *G. behnii*, vulnerable A4c; *L. aurita*, vulnerable A3c; *X. vieirai*, vulnerable A4c. The other species are *Eptesicus taddeii*, classified as vulnerable, but based on two criterion

(A4c and B2ab[i,ii,iii]), and *L. dekeyseri* (endangered C2a[ii]), the only bat species in Brazil classified based on an observed, estimated, projected, or inferred continuing population decline (MMA 2014). In other words, we are classifying the conservation status of Brazilian bats based on *the possible implications* of habitat loss and degradation. In a utopic scenario, where there would be no habitat loss or degradation in the country, almost all species of Brazilian bats would be then classified as data deficient, due to the total lack of basic information about quantitative declines in national populations (criterion A), on the number of sexually reproductive individuals (criterion C), on estimations of national population sizes (criterion D), and on the total absence of population viability analyses for the Brazilian bats (criterion E). This is disturbing.

As for criterion B, the significant data gaps on species distribution in the country severely affect the quality of precise estimations (see Aguiar et al. 2015; Bernard et al. 2011). The various recent expansions of the known distribution for several species in the country, some of them covering thousands of kilometers, will directly influence how experts will calculate the species' extent of occurrence (EOO), defined as the area contained within the shortest continuous imaginary boundary that can be drawn to encompass all the known, inferred, or projected sites of present occurrence of a taxon, excluding cases of vagrancy for those species (IUCN 2012). But, perhaps, more important than the EOO, is the area of occupancy (AOO), defined as the area within its EOO that is occupied by a taxon, excluding cases of vagrancy (IUCN 2012). In other words, the AOO represents the area of suitable habitat currently occupied by the taxon, and its measure reflects the fact that a taxon will not usually occur throughout the entire area of its EOO, due to the existence of portions of obviously unsuitable habitat. Although several species of bats have very large EOO in Brazil, most of their AOO are already severely lost and disturbed.

The lack of necessary basic information for a better classification of the conservation status of the Brazilian bat species points to the need for the immediate establishment of bat monitoring programs in the country. Brazil does not yet have a national bat banding and monitoring program, like the existing one for birds (www.icmbio.gov.br/cemave/sna.html). The discussion for implementing such program extends now for more than 20 years, but there are no concrete moves

so far. There are several isolated bat-marking initiatives in progress in the country (Barros et al. 2012), but the lack of a systematic and standardized banding and monitoring program for bats in Brazil accentuates some elementary data gaps: it is not known which Brazilian species are able to move within the country and how much; it is still not clear if Brazilian bats perform migratory movements; and there are no indicators or assessments to determine whether national populations are experiencing size fluctuations.

More information on the existence of migratory movements among bat species in Brazil is necessary to address several existing knowledge gaps. Issues such as shelter fidelity, social behavior, food ecology, and longevity can be answered by tagging individuals (e.g., Fleming 1988; Handley et al. 1991; Martino et al. 2006). The evaluation of the effects of wind farms and mines on possible migratory routes of insectivorous, frugivorous, or nectarivorous species would also be improved if we have had a systematic banding and monitoring program in the country (e.g., Popa-Lisseanu and Voigt 2009). The existence of information on bat migration among municipalities in Brazil would improve and facilitate the implementation of joint intervention initiatives in, for example, the monitoring and control of rabies outbreaks (Schneider et al. 2009). Moreover, the lack of information on bat migration between Brazil and its neighboring countries precludes the establishment of joint bilateral conservation initiatives. Just recently, in 2015, Brazil fully joined the Convention on the Conservation of Migratory Species of Wild Animals (www.cms.int) and based on this convention, it was possible to establish, for example, the Agreement on the Conservation of Populations of European Bat (known as Eurobats—www.eurobats.org). Therefore, there is no doubt that the establishment of a national bat-banding and -monitoring program in Brazil is not just necessary but also urgent.

Bats and the Pest Exterminators Industry

As wild animals, bats enjoy formal protection according to the current legislation in Brazil (e.g., Law 5197/1967, or the Federal Constitution, Article 225, paragraph 1, item 8). However, there is a strong pressure to treat these animals as urban and rural pests, thereby making them potential targets for pest exterminator companies. Legislative act Norm 141/2006 issued by IBAMA

(Brazilian Environmental Agency) states that in observance of the law and other applicable regulations, bats species can be controlled by health, agriculture, and environmental government agencies without the need for special authorization only if they are in urban and peri-urban areas. Pest control of *Desmodus rotundus*, for example, can occur only in areas where rabies is considered endemic or areas experiencing a potential risk of rabies outbreaks, but those areas need to be officially declared as such by the Ministries of Agriculture and Health, following official standards. However, such a norm is frequently ignored. Extermination practices, frequently performed by untrained and unskilled workers, do not distinguish species or situations, and extermination operations can occur even when there is neither the authorization from the environmental agencies nor the express statement that bats are being considered "harmful." There are frequent media announcements on the internet, TV, and newspapers in which extermination companies offer "*desmorcegação*," that is, extreme bat removal from homes, industries, or other structures. The pest exterminators industry in Brazil remains in business based mainly on misinforming the public regarding bats and counts on lack of oversight by the authorities in controlling their activities and practices and the products used. Prohibited substances are widely advertised, along with banned practices, which are performed by ill-prepared and malicious companies. The cumulative long-term effects of such harmful practices on bats have not been properly quantified in Brazil (Bernard et al. 2012), but such practices can undoubtedly expose some bat populations to the real risk of local extinctions (Aguiar et al. 2010).

Control of vampire bats is especially problematic in Brazil. There is widespread use of "pasta vampiricida," a noxious mix of vaseline and anticoagulant, the employment of which is officially supported by the Brazilian Ministry of Agriculture (MAPA 2009). That "control" method includes the capture of vampire bats, coating them with pasta vampiricida, and then releasing them so they can be poisoned by self- or mutual grooming. Bats poisoned thus will die of internal bleeding. According to the Brazilian Ministry of Agriculture, for each vampire bat that is given pasta vampiricida, another 20 will die (MAPA 2009). Controversial and considered inefficient (Streicker et al. 2012; Johnson et al. 2014), pasta vampiricida should be applied under very specific conditions: during a confirmed outbreak of rabies, only

a veterinarian or the equivalent should apply it, and it should not be applied directly to any bat but, instead, to specific body parts of horses, cows, or other large animals being attacked by hematophagous bats, thereby sparing non-hematophagous species (MAPA 2009). Unfortunately, pasta vampiricida can be easily purchased anywhere in Brazil and applied directly to bats of several species by untrained people. As a result, bats are being killed regardless of species, feeding habits, or the presence of the rabies virus. Pasta vampiricida should be seen as an extreme measure, to be used solely after other management measures have failed.

The increasing size of the Brazilian cattle herds can also directly influence the biological and nonbiological factors related to the transmission of rabies virus by *D. rotundus* (Schneider et al. 2009). Among the biological factors are the increased availability of food sources (cattle), which directly affects the size of vampire bat populations. Among the nonbiological factors are the environmental changes associated with the process of meat production and the expansion of its boundaries through deforestation and fragmentation; the living and working conditions of the people involved in these activities (usually with inadequate housing, which increases their chances of being bitten by vampire bats); cattle workers' poor access to prophylactic measures against the transmission of the rabies virus; and increased attempts to control populations of hematophagous bats, with deleterious consequences for other, nontarget bat species (Schneider et al. 2009). Vampire bat control is identified as one of the major causes of bat mortality worldwide (O'Shea et al. 2016). Even so, research on new methods to control vampire bats seems to have stagnated (e.g., Almeida et al. 2008; Lee et al. 2012; Schneeberger and Voigt 2015).

Currently, there is evidence that other environmental factors are affecting vampire bats in Brazil. Recently, Galetti et al. (2016) pointed out that an alarming rise in the numbers and distribution of invasive feral pigs and wild boars (*Sus scrofa*) in the country may favor *D. rotundus*, since vampire bats are feeding on the blood of these large animals. They estimated that the probability of vampire bat attacks on feral pigs was as high as 10% in a given night. This is an issue, considering that rural areas of Brazil face an unprecedented invasion of feral pigs, with an increase of 500% in their populations since 2007 (Pedrosa et al. 2015). Since *S. scrofa* is a growing source of blood for vampire bats, the population of *D. rotundus* is also likely to increase. Furthermore, Galetti et al. point out, there is additional reason for concern based on the fact that *D. rotundus* is a reservoir for other viruses with epidemiological potential, including hantavirus and coronavirus (Galetti et al. 2016). In another recent study, using fecal DNA, Ito et al. (2016) identified human blood in the diet of *Diphylla ecaudata* in the Caatinga dry forests of northeastern Brazil—a novel prey for this species. Their study showed that *D. ecaudata*'s diet is more flexible than expected. However, the record of humans as prey and the absence of blood from native species in their samples may reflect a low availability of wild birds in the study site, reinforcing the impact of human activities on local ecological processes involving vampire bats. As indicated here, Brazil faces severe changes in its natural habitats. The impact of such environmental changes on vampire bats needs to be better assessed but, in any case, changing the traditional way vampire bats are being controlled in Brazil should be a priority for bat conservation in the country due to the potential benefits that banning those old practices have to several species.

Perspectives for Bat Conservation in Brazil: The Need for Better Communication

The conservation of bats in Brazil is a challenge: overall, bats are not seen as charismatic animals by the people. Bats are frequently associated with rabies transmission in the country, and changing the major threats and pressures affecting them will require confronting some of the most economically important sectors of the Brazilian economy, especially agribusiness and mining. But such a task cannot be neglected or ignored by Brazilian society. Due to their species richness and amplitude of ecological interactions, bats in general—and phyllostomids in particular—participate in both top-down and bottom-up ecological processes, interacting with a wide range of other species (Gardner 1977) and providing environmental services essential for humans, like pollination and seed dispersal (Lobova et al. 2009) and pest control (e.g., Boyles et al. 2011; Wanger et al. 2014).

In the Neotropics, 62 plant families have at least one species dispersed by bats (Smith et al. 2004), and at least 858 plant species depend on bats for pollination or dispersal (Geiselman et al. 2002; www.batplant.org). Several of those species are present in Brazil, and some

of them are key species, such as *Ficus* spp., on which a large number of other animal species depend. *Carollia*, *Sturnira*, and *Artibeus* play important roles in the seed rain of pioneer species, such as *Cecropia* spp., *Piper* spp., *Solanum* spp., and *Vismia* spp., which are among the most abundant plants during primary and secondary successions. In a scenario of hyperfragmentation and severe defaunation, like that in the Atlantic Forest, bats are directly involved in maintaining a genetic flow essential to several remaining plant species (Melo et al. 2006; Mendes et al. 2016; Santo-Silva et al. 2013, 2015; Tabarelli et al. 2010). This role has to be better publicized.

Moreover, it is very likely that, when feeding on a myriad of insects, bats are providing valuable services for Brazilian agribusinesses. Similarly, it is very likely that they may also be acting as predators of insects involved in the transmission of diseases to humans in the country. Brazilian bats need better propaganda, and the role they play for free for humans is a great opportunity to publicize them. Societies and associations, like Sociedade Brasileira para o Estudo de Quirópteros (the Brazilian Bat Society—www.sbeq.net) or the Red Latinoamericana para Conservación de los Murciélagos (www.relcomlatinoamerica.net) could play a leading role in the dissemination of better information on Brazilian bats. In this way, the task of conserving bats could be made less difficult by obtaining better estimates of, for example, the affects some species mays have on the economy of Brazil (e.g., Boyles et al. 2011; Ghanem and Voigt 2012; Maas et al. 2015; Wanger et al. 2014). This topic offers great potential for research in the country (Bernard et al. 2012), and the resulting data would be decidedly useful in conservation actions and environmental education. If the real value of services provided by bats is made clearer, people may better realize how important it is to protect and conserve them. The message should be obvious: ultimately, in a megadiverse country like Brazil, conserving bats is essential for preserving the ecological relationships in which these animals participate, as well as for the maintenance of vital environmental services they provide for free for humans.

Acknowledgments

EB would like to thank the Brazilian Conselho Nacional de Desenvolvimento Científico e Tecnológico (CNPq) for the fellowship grant, and MD-J would like to thank Coordenação de Aperfeicoamento de Pessoal de Nível Superior (CAPES) for the fellowship. We thank Departamento de Zoologia, Centro de Biociências, and Universidade Federal de Pernambuco for supporting our studies on bats in Brazil. We thank Maria João Ramos Pereira, Marco Mello, and one anonymous reviewer for their constructive comments on this chapter.

References

ABEEólica (Associação Brasileira de Energia Eólica). 2019. Números do setor. http://abeeolica.org.br/dados-abeeolica/.

Aguiar, L. M. S., D. C. Brito, and R. B. Machado. 2010. Do current vampire bat (*Desmodus rotundus*) population control practices pose a threat to Dekeyser's nectar bat's (*Lonchophylla dekeyseri*) long-term persistence in the Cerrado? Acta Chiropterologica 12 (2): 275–282.

Aguiar, L. M. S., R. B. Machado, and A. D. Ditchfield. 2006. Plano de Ação para a conservação do morceguinho do Cerrado *Lonchophylla dekeyseri*. Relatório para o Ministério do Meio Ambiente, PROBIO/MMA, Brasil.

Aguiar, L. M. S., R. O. Rosa, R. B. Machado, and G. Jones. 2015. Effect of chronological addition of records to species distribution maps: the case of *Tonatia saurophila maresi* (Chiroptera, Phyllostomidae) in South America. Austral Ecology, 40:836–844.

Almeida, M. F., L. F. A. Martorelli, C. C. Aires, R. F. Barros, and E. Massad. 2008. Vaccinating the vampire bat *Desmodus rotundus* against rabies. Virus Research, 137 (2): 275–277.

Amarante, O. A. C., M. Brower, J. Zack, and A. L. de Sá. 2001. Atlas do Potencial Eólico Brasileiro. Ministério das Minas e Energia, Brasília.

Anderson, R. P., and I. Gonzalez. 2011. Species-specific tuning increases robustness to sampling bias in models of species distributions: an implementation with Maxent. Ecological Modelling, 222:2796–2811.

Araújo, M. B., and A. Guisan. 2006. Five (or so) challenges for species distribution modelling. Journal of Biogeography, 33:1677–1688.

Ardron, J. A., H. P. Possingham, and C. J. Klein. 2008. Marxan Good Practices Handbook. Pacific Marine Analysis and Research Association, Vancouver.

Arnett, E. B., E. F. Baerwald, F. Mathews, L. Rodrigues, A. Rodríguez-Durán, J. Rydell, R. Villegas-Patraca, and C. C. Voigt. 2016. Impacts of wind energy development on bats: a global perspective. Pp. 295–323 in: Bats in the Anthropocene: Conservation of Bats in a Changing World (C. C. Voigt and T. Kingston, eds.). Springer International Publishing, New York.

Ball, I., H. Possingham, and M. Watts. 2009. Marxan and relatives: software for spatial conservation prioritisation. Pp. 185–195 in: Spatial Conservation Prioritisation: Quantitative Methods and Computational Tools (A. Moilanen, K. Wilson, and H. Possingham, eds.). Oxford University Press, Oxford.

Barros, M. A. S., J. L. Luz, and C. E. L. Esbérard. 2012. Situação atual da marcação de morcegos no Brasil e perspectivas para a criação de um programa nacional de anilhamento. Chiroptera Neotropical, 18:1074–1088.

Barros, M. A. S., R. G. Magalhães, and A. M. Rui. 2015. Species composition and mortality of bats at the Osório wind farm, southern Brazil. Studies on Neotropical Fauna and Environment, 50 (1): 31–39.

Bernard, E., L. M. S. Aguiar, D. Brito, A. P. Cruz-Neto, R. Gregorin, R. B. Machado, M. Oprea, A. P. Paglia, and V. C. Tavares. 2012. Uma análise de horizontes sobre a conservação de morcegos no Brasil. Pp. 19–35, in: Mamíferos do Brasil: Genética, Sistemática, Ecologia e Conservação. Vol. 2 (T. R.O. Freitas and E. M. Vieira, eds.). Sociedade Brasileira de Mastozoologia, Rio de Janeiro.

Bernard, E., L. M. S. Aguiar, and R. B. Machado. 2011. Discovering the Brazilian bat fauna: a task for two centuries? Mammal Review, 41:23–39.

Bernard, E., J. L. Nascimento, and L. M. S. Aguiar. 2013. Flagging a species as threatened: the case of *Eptesicus taddeii*, an endemic bat from the Brazilian Atlantic Forest. Biota Neotropica, 13 (2): 314–318. http://www.biotaneotropica .org.br/v13n2/en/abstract?short-communication+bn 01413022013.

Bernard, E., A. Paese, R. B. Machado, and L. M. S. Aguiar. 2014. Blown in the wind: bats and wind farms in Brazil. Natureza e Conservação, 12 (2): 106–111.

Bernard, E., L. A. Penna, and E. Araújo. 2014. Downgrading, downsizing, degazettement, and reclassification of protected areas in Brazil. Conservation Biology, 28 (4): 939–950.

Boria, R. A., L. E. Olson, S. M. Goodman, and R. P. Anderson. 2014. Spatial filtering to reduce sampling bias can improve the performance of ecological niche models. Ecological Modelling, 275:73–77.

Boyles, J. G., P. M. Cryan, G. F. McCracken, and T. H. Kunz. 2011. Economic importance of bats in agriculture. Science, 332:41–42.

Brasil, Presidência da República. 2008. Decreto nº 6.640, de 7 de novembro de 2008. Modifica o Decreto 99556/1990. DOU de 10/11/2008. http://www.planalto.gov.br/ccivil_03 /_Ato2007-2010/2008/Decreto/D6640.htm.

Chape, S., J. Harrison, M. Spalding, and I. Lysenko. 2005. Measuring the extent and effectiveness of protected areas as an indicator for meeting global biodiversity targets. Philosophical Transactions of the Royal Society of London B, 360:443–455.

Cunto, G. C., and E. Bernard. 2012. Neotropical bats as indicators of environmental disturbance: what is the emerging message? Acta Chiropterologica, 14:143–151.

Dias, D., C. E. L. Esbérard, and R. Moratelli. 2013. A new species of *Lonchophylla* (Chiroptera, Phyllostomidae) from the Atlantic Forest of southeastern Brazil, with comments on *L. bokermanni*. Zootaxa, 3722 (3): 347–360.

Elith, J., and J. R. Leathwick. 2009. Species distribution models: ecological explanation and prediction across space and time. Annual Review of Ecology, Evolution, and Systematics, 40:677–697.

Esselman, P. C., and J. D. Allan. 2011. Application of species distribution models and conservation planning software to the design of a reserve network for the riverine fishes of northeastern Mesoamerica. Freshwater Biology, 56:71–88.

Farneda, F. Z., R. Rocha, A. López-Baucells, M. Groenenberg, I. Silva, J. M. Palmeirim, P. E. Bobrowiec, and C. F. Meyer. 2015. Trait-related responses to habitat fragmentation in Amazonian bats. Journal of Applied Ecology, 52:1381–1391.

Fazzolari-Correa, S. 1994. *Lasiurus ebenus*, a new vespertilionid bat from Southeastern Brasil. Mammalia, 58:119–123.

Fearnside, P. M. 2016. Brazilian politics threaten environmental policies. Science, 353:746–748.

Feijó A., P. A. Rocha, and S. L. Althoff. 2015. New species of *Histiotus* (Chiroptera: Vespertilionidae) from northeastern Brazil. Zootaxa, 4048 (3): 412.

Fleming, T. H. 1988. The Short-Tailed Fruit Bat: A Study in Plant-Animal Interactions. University of Chicago Press, Chicago.

Friendlander, A., J. Salader, R. Appeldoorn, P. Usseglio, C. Mc-Cormick, S. Bejarano, and A. Mitchell-Chui. 2003. Designing effective marine protected areas in Seaflower Biosphere Reserve, Colombia, based on biological and sociological information. Conservation Biology, 17:1769–1784.

Galetti, M., F. Pedrosa, A. Keuroghlian, and I. Sazima. 2016. Liquid lunch—vampire bats feed on invasive feral pigs and other ungulates. Frontiers in Ecology and the Environment, 14 (9): 505. DOI: 10.1002/fee.1431.

Gardner, A. L. 1977. Feeding habits. Pp. 243–349 *in*: Biology of Bats of the New World Family Phyllostomidae. Pt. 2 (R. J. Baker, J. K. Jones Jr., and D. C. Carter, eds.). Texas Tech Press, Lubbock.

Gaston, K. J., S. F. Jackson, L. Cantu-Salazar, and G. Cruz-Pinon. 2008. The ecological performance of protected areas. Annual Review in Ecology, Evolution and Systematics, 39:93–113.

Geiselman C. K., S. A. Mori, and F. Blanchard. 2002 onwards. Database of Neotropical Bat/Plant Interactions. http://www .nybg.org/botany/tlobova/mori/batsplants/database /dbase_frameset.htm.

Ghanem, S. H., and C. C. Voigt. 2012. Increasing awareness of ecosystem services provided by bats. Advances in the Study of Behavior, 44:279–302.

Gregorin, R., and A. D. Dietchfield. 2005. New genus and species of nectar-feeding bat in the tribe Lonchophyllini (Phyllosto-mydae: Glossophaginae) from northeastern Brazil. Journal of Mammalogy, 86:403–414.

Gregorin, R., E. Gonçalves, B. K. Lim, and M. D. Engstrom. 2006. New species of disk-winged bat *Thyroptera* and range extension for *T. discifera*. Journal of Mammalogy, 87:238–246.

Gregorin, R., and L. O. Loureiro. 2011. New records of bats for the state of Minas Gerais, with range extension of *Eptesicus chiriquinus* Thomas (Chiroptera: Vespertilionidae) to southeastern Brazil. Mammalia, 75: 291–294.

Guimarães, M. M., and R. L. Ferreira. 2014. Morcegos caverní-colas do Brasil: novos registros e desafios para conservação. Revista Brasileira de Espeleologia, 2 (4): 1–33.

Handley, C. O., Jr., D. E. Wilson, and A. L. Gardner 1991. Demog-raphy and natural history of the common fruit bat, *Artibeus*

jamaicensis, on Barro Colorado Island, Panama. Smithsonian Contributions to Zoology, 511:1–173.

Hannah, L., G. Midgley, S. Andelman, M. Araujo, G. Hughes, E. Martínez-Meyer, R. Pearson, and P. Williams. 2007. Pro-tected area needs in a changing climate. Frontiers in Ecology and the Environment, 5:131–138.

ICMBio/CECAV (Instituto Chico mendes de Conservação da Biodiversidade/Centro Nacional de Pesquisa e Conservação de Cavernas). 2016. Cadastro Nacional de Informações Espe-leológicas. http://www.icmbio.gov.br/cecav/canie.html.

Ito, F., E. Bernard, and R. A. Torres. 2016. What is for dinner? First report of human blood in the diet of the hairy-legged vampire bat *Diphylla ecaudata*. Acta Chiropterologica 18 (2): 509–515.

IUCN (International Union for Conservation of Nature). 2012. IUCN Red List Categories and Criteria: Version 3.1. 2nd ed. IUCN—the World Conservation Union, Gland, Switzerland

Johnson, N., N. Aréchiga-Ceballos, and A. Aguilar-Setien. 2014. Vampire bat rabies: ecology, epidemiology and control. Viruses, 6 (5): 1911–1928.

Jung, K., and E. K. V. Kalko. 2011. Adaptability and vulnerability of high flying Neotropical aerial insectivorous bats to urban-ization. Diversity and Distributions, 17:262–274.

Kalko, E. K. V., D. Friemal, C. O. Handley Jr., and H. U. Schintz-ler. 1999. Roosting and foraging behavior of two Neotrop-ical gleaning bats, *Tonatia silvicola* and *Trachops cirrhosus* (Phyllostomidae). Biotropica, 31:344–353.

Klingbeil, B. T., and M. R. Willig. 2009. Guild-specific responses of bats to landscape composition and configuration in frag-mented Amazonian rainforest. Journal of Applied Ecology, 46 (1): 203–213.

Lee, D. N., M. Papes, and R. A. van Den Bussche. 2012. Present and potential future distribution of common vampire bats in the Americas and the associated risk to cattle. PLOS ONE, 7 (8): e42466. DOI: 10.1371/journal.pone.0042466.

Lessmann, J. 2011. How to maximize species conservation in continental Ecuador: recommendations for expanding the national network of protected areas. Biodiversidad en Areas Tropicales y su Conservación, Universidad Internacional Menendez Pelayo, Quito.

Lobova, T. A., C. K. Geiselman, and A. S. Mori. 2009. Seed dispersal by bats in the Neotropics. Memoirs of the New York Botanical Garden, vol. 101. New York Botanical Garden, Bronx.

Loiselle, B. A., P. M. Jorgensen, T. Consiglio, I. Jiménez, J. Blake, L. G. Logmann, and M. O. Montiel. 2008. Predicting species distributions from herbarium collections: does climate bias in collection sampling influence model outcomes? Journal of Biogeography, 35 (1): 105–116.

Maas, B., D. S. Karp, S. Bumrungsri, K. Darras, D. Gonthier, J. C. C. Huang, C. A. Lindell et al. 2015. Bird and bat preda-tion services in tropical forests and agroforestry landscapes. Biological Reviews of the Cambridge Philosophical Society, 91 (4): 1081–1101.

MAPA (Ministério da Agricultura, Pecuária e Abastecimento). 2009. Controle da raiva dos herbívoros: manual técnico.

Ministério da Agricultura, Pecuária e Abastecimento, Secretaria de Defesa Agropecuária, Brasília.

Margules, C. R., and R. L. Pressey. 2000. Systematic conservation planning. Nature, 405:243–253.

Martino, A. M. G., J. Aranguren, and A. Arends. 2006. New longevity records in South American microchiropterans. Mammalia, 70 (1–2): 166–167.

Melo, F. P. L., R. Dirzo, and M. Tabarelli. 2006. Biased seed rain in forest edge: evidence from the Brazilian Atlantic Forest. Biological Conservation, 132:50–60.

Mendes, G., V. Arroyo-Rodríguez, W. R. Almeida, S. R. R. Pinto, V. D. Pillar, and M. Tabarelli. 2016. Plant trait distribution and the spacial reorganization of tree assemblages in a fragmented tropical forest landscape. Plant Ecology, 217:31–42.

Meyer, C. F., J. Frund, W. P. Lizano, and E. K. Kalko. 2008. Ecological correlates of vulnerability to fragmentation in Neotropical bats. Journal of Applied Ecology, 45:381–391.

Meyer, C. F., and E. K. Kalko. 2008. Assemblage-level responses of phyllostomid bats to tropical forest fragmentation: land-bridge islands as a model system. Journal of Biogeography, 35:1711–1726.

Meyer, C. F., M. J. Struebig, and M. R. Willig. 2016. Responses of tropical bats to habitat fragmentation, logging, and deforesta-tion. Pp. 63–103, *in*: Bats in the Anthropocene: Conservation of Bats in a Changing World (C. C. Voigt and T. Kingston, eds.). Springer International Publishing, New York.

MMA (Ministério do Meio Ambiente). 2014. Portaria No. 444, de 17 de Dezembro de 2014. http://www.mma.gov .br/biodiversidade/especies-ameacadas-de-extincao /atualizacao-das-listas-de-especies-ameacadas.

MMA (Ministério do Meio Ambiente). 2016. Portaria No. 223, de 21 de junho de 2016. http://www.mma.gov.br /biodiversidade/biodiversidade-brasileira/%C3%A1reas -priorit%C3%A1rias/item/10724.

MMA (Ministério do Meio Ambiente). 2017. Instrução Norma-tiva No. 02, de 30 de Agosto de 2017. http://www.icmbio .gov.br/cecav/downloads/legislacao.html.

Moratelli, R., and D. Dias. 2015. A new species of nectar-feeding bat, genus *Lonchophylla*, from the Caatinga of Brazil (Chirop-tera, Phyllostomidae). ZooKeys, 514:73–91.

Moratelli, R., A. L. Peracchi, D. Dias, and J. A. Oliveira. 2011. Geographic variation in South American populations of *Myotis nigricans* (Schinz, 1821) (Chiroptera, Vespertilioni-dae), with the description of two new species. Mammalian Biology, 76:592–607.

Nogueira M. R., I. P. Lima, G. S. T. Garbino, R. Moratelli, V. C. Tavares, R. Gregorin, and A. L. Peracchi. 2018. Updated checklist of Brazilian bats: version 2018.1. Comitê da Lista de Morcegos do Brasil—CLMB. Sociedade Brasileira para o Estudo de Quirópteros (Sbeq). http://www.sbeq.net /updatelist.

Nogueira, M. R., I. P. Lima, A. L. Peracchi, and N. B. Simmons. 2012. New genus and species of nectar-feeding bat from the Atlantic Forest of Southeastern Brazil (Chiroptera: Phyl-lostomidae: Glossophaginae). American Museum Novitates, 3747:1–30.

Nogueira, M. R., A. Pol, L. R. Monteiro, and A. L. Peracchi.

2008. First record of Miller's mastiff bats, *Molossus pretiosus* (Mammalia: Chiroptera), from the Brazilian Caatinga. Chiroptera Neotropical, 14:346–353.

Oliveira, H. F. M., M. Oprea, and R. I. Dias. 2018. Distributional patterns and ecological determinants of bat occurrence inside caves: a broad scale meta-analysis. Diversity, 10:49.

O'Shea, T. J., P. M. Cryan, D. T. S. Hayman, R. K. Plowright, and D. G. Streicker. 2016. Multiple mortality events in bats: a global review. Mammal Review, 46 (3): 175–190.

Pack, S. M., M. N. Ferreira, R. Krithivasan, J. Murrow, E. Bernard, and M. B. Mascia. 2016. Protected area downgrading, downsizing, and degazettement (PADDD) in the Amazon. Biological Conservation, 197:32–39.

Pawar, S., M. S. Koo, M. F. Ahmed, S. Chaudhuri, and S. Sarkar. 2007. Conservation assessment and prioritization of areas in Northeast India: priorities for amphibians and reptiles. Biological Conservation, 136:346–361.

Pearson, R. G., C. J. Raxworthy, M. Nakamura, and A. T. Peterson. 2007. Predicting species distributions from small numbers of occurrence records: a test case using cryptic geckos in Madagascar. Journal of Biogeography, 34:102–117.

Pedrosa, F., R. Salerno, F. V. B. Padilha, and M. Galetti. 2015. Current distribution of invasive feral pigs in Brazil: economic impacts and ecological uncertainty. Natureza e Conservação 13 (1): 84–87.

Peters, S. L., J. R. Malcolm, and B. L. Zimmerman. 2006. Effects of selective logging on bat communities in the Southeastern Amazon. Conservation Biology, 5:1410–1421.

Phillips, S. J., R. P. Anderson, and R. E. Schapire. 2006. Maxi-mum entropy modeling of species geographic distributions. Ecological Modelling, 190:231–259.

Phillips, S. J., and M. Dudík. 2008. Modeling of species distri-butions with Maxent : new extensions and a comprehensive evaluation. Ecography, 31 (2): 161–175.

Piló, L. B., and A. Auler. 2011. Introdução à Espeleologia. Pp. 7–23, *in*: III Curso de Espeleologia e Licenciamento Ambiental. CECAV/Instituto Chico Mendes de Conservação da Biodiversidade, Brasília.

Pimenta, V. T., C. T. Machel, B. S. Fonseca, and A. D. Ditchfield. 2010. First occurrence of *Lonchophylla bokermanni* Sazima, Vizotto and Taddei, 1978 (Phyllostimidae) in Espírito Santo State, Southeastern Brazil. Chiroptera Neotropical, 16 (2): 840–842.

Popa-Lisseanu, A. G., and C. C. Voigt. 2009. Bats on the move. Journal of Mammalogy, 90 (6): 1283–1289.

Radosavljevic, A., and R. P. Anderson. 2014. Making better Max-ent models of species distributions: complexity, overfitting and evaluation. Journal of Biogeography, 41:629–643.

Ribeiro, M. C., J. P. Metzger, A. C. Martensen, F. J. Ponzoni, and M. M. Hirota. 2009. The Brazilian Atlantic Forest: how much is left, and how is the remaining forest distributed? Implications for conservation. Biological Conservation, 142 (6): 1141–1153.

Rocha, P. A., A. Feijó, D. Dias, J. Mikalauskas, J. Ruiz-Esparza, and S. F. Ferrari. 2014. Major extension of the known range of the capixaba nectar-feeding bat, *Dryadonycteris capixaba* (Chiroptera, Phyllostomidae): is this rare species widely distributed in eastern Brazil? Mastozoología Neotropical, 21 (2): 361–366

Santo-Silva, E. E., W. R. Almeida, F. P. Melo, and M. Tabarelli. 2013. The nature of seedling assemblages in a fragmented tropical landscape: implications for forest regeneration. Biotropica, 45:386–394.

Santo-Silva, E. E., K. D. Withey, W. R. Almeida, G. Mendes, A. V. Lopes, and M. Tabarelli. 2015. Seeding assemblages and the alternative successional pathways experienced by Atlantic Forest fragments. Plant Ecology and Diversity, 8:483–492.

Schneeberger, K., and C. C. Voigt. 2015. Zoonotic viruses and conservation of bats. Pp. 263–292, *in*: Bats in the Anthropocene: Conservation of Bats in a Changing World (C. C. Voigt and T. Kingston, eds.). Springer International Publishing, New York.

Schneider, M. C., P. C. Romijn, W. Uieda, H. Tamayo, D. F. Silva, A. Belotto, J. B. Silva, and L. F. Leanes. 2009. Rabies transmitted by vampire bats to humans: an emerging zoonotic disease in Latin America? Pan American Journal of Public Health, 25:260–269.

Silva, M. R., and R. L. Ferreira. 2016. The first two hotspots of subterranean biodiversity in South America. Subterranean Biology, 19:1–21.

Smith, N., S. A. Mori, A. Henderson, D. W. Stevenson, and S. V. Heald. 2004. Flowering Plants of the Neotropics. Princeton University Press, published in association with the New York Botanical Garden, Princeton, NJ.

Solari, S. 2015. *Xeronycteris vieirai*. The IUCN Red List of Threatened Species 2015: e.T136321A22021092. http://dx.doi.org /10.2305/IUCN.UK.2015-4.RLTS.T136321A22021092.en.

Streicker, D. G., S. Recuenco, W. Valderrama, J. G. Benavides, I. Vargas, V. Pacheco, R. E. Condori et al. 2012. Ecological and anthropogenic drivers of rabies exposure in vampire bats: implications for transmission and control. Proceedings of the Royal Society B–Biological Sciences, 279:3384–3392.

Sutherland, W. J., M. J. Bailey, I. P. Bainbridge, T. Brereton, J. T. A. Dick, J. Drewitt, N. K. Dulvy et al. 2008. Future novel threats and opportunities facing UK biodiversity identified by horizon scanning. Journal of Applied Ecology, 45:821–833.

Tabarelli, M., A. V. Aguiar, L. C. Girão, C. A. Peres, and A. V. Lopes. 2010. Effects of pioneer tree species hyperabundance on forest fragments in northeastern Brazil. Conservation Biology, 24:1654–1663.

Taddei, V. A., and B. K. Lim. 2010. A new species of *Chiroderma* (Chiroptera; Phyllostomidae) from northeastern Brazil. Brazilian Journal of Biology, 70 (2): 381–386.

Tollefson, J. 2016. Political upheaval threatens Brazil's environmental protections. Nature, 539:147–148.

Valença, R. B., and E. Bernard. 2015. Another blown in the wind: bats and the licensing of wind farms in Brazil. Natureza e Conservação 13:117–122.

Wanger, T. C., K. Darras, S. Bumrungsri, T. Tscharntke, and A. M. Klein. 2014. Bat pest control contributes to food security in Thailand. Biological Conservation, 171:220–223.

Warren, D. L., and S. N. Seifert. 2011. Ecological niche modelling in Maxent: the importance of model complexity and the performance of model selection criteria. Ecological Applications, 21:335–342.

Watson, J. E. M., N. Dudley, D. B. Segan, and M. Hockings. 2014. The performance and potencial of protected areas. Nature, 515:67–73.

WCS (Wildlife Conservation Society) and CIESIN (Center for International Earth Science Information Network)— Columbia University. 2005. Last of the Wild Project, Version 2, 2005 (LWP-2): Global Human Footprint Dataset (Geographic). NASA Socioeconomic Data and Applications Center (SEDAC), Palisades, NY. http://dx.doi.org/10.7927 /H4M61H5F.

WDPA (World Database on Protected Areas). 2016. Available at: https://www.protectedplanet.net/.

Wisz, M. S., J. Pottier, W. D. Kissling, L. Pellissier, J. Lenoir, C. F. Damgaard, C. F. Dormann et al. 2013. The role of biotic interactions in shaping distributions and realised assemblages of species: implications for species distribution modelling. Biological Reviews, 88 (1): 15–30.

24

Threats, Status, and Conservation Perspectives for Leaf-Nosed Bats

Jafet M. Nassar,
Luis F. Aguirre,
Bernal Rodríguez-Herrera
and Rodrigo A. Medellín

Introduction

The conservation status of bats worldwide is mainly influenced by anthropogenic factors to which these organisms are especially susceptible because of their slow population growth rates, longevity, and high metabolic rates (Voigt and Kingston 2016). Thus, under the current rate of human intervention in practically all terrestrial ecosystems, bat conservation represents a major challenge, and the level of difficulty to achieve positive results can be very high in many countries where bats continue to be stigmatized (Kingston 2016). This pessimistic scenario contrasts with growing evidence demonstrating the multiple ecological functions and ecosystem services provided by bats (Ghanem and Voigt 2012; Kunz et al. 2011). In this regard, leaf-nosed bats stand out as versatile providers of ecosystem services when compared to other bat families worldwide, because besides consumption of arthropods (Giannini and Kalko 2004; chap. 14, this vol.), many species are pollinators and seed dispersal agents of hundreds of tropical plants (Kunz et al. 2011; Muscarella and Fleming 2007). In addition, phyllostomid bats are considered to be valuable indicators of habitat disruption in a wide range of forest types in the Neotropics due to their diversity and differential dependence on forested environments (Jones, Jacobs et al. 2009).

Conservation of phyllostomid bats requires implementation of conservation strategies and actions in a vast region of the New World, including tropical and subtropical America, all of Mexico, part of southwestern United States, and the Caribbean. The most recent diagnosis of extrinsic factors affecting bats in a large proportion of this region is an evaluation conducted by the Latin American and Caribbean Bat Conservation Network (RELCOM) in 2010 (RELCOM 2016a). Founded in 2007, this network of scientists, students, and members of environmental organizations has become the strongest regional initiative for bat conservation in Latin America and the Caribbean. It involves the coordinated actions of 22 bat conservation programs from 24 countries (Aguirre et al. 2014). The diagnosis by RELCOM incorporates the major threats affecting bats according to the *Global Action Plan for Microchiropteran Bats* (Hutson et al. 2001), a review chapter on conservation ecology of bats (Racey and Entwistle 2003), and a recent volume covering major topics related to bat conservation worldwide (Voigt and Kingston 2016). As a result, we consider that the five major extrinsic threat categories influencing the conservation status of leaf-nosed bats are (1) habitat loss and fragmentation, (2) roost disturbance and destruction, (3) human-bat conflicts, (4) environmental contaminants, and (5) emerging threats.

Threats

Habitat Loss and Fragmentation

Habitat loss and fragmentation represent the major threat affecting biodiversity and ecosystem functioning worldwide (Sala et al. 2000; Wade et al. 2003). This process includes reduction and fragmentation of natural habitats and their conversion into human-modified environments, a pattern expanding at increasing rates in tropical regions (Achard et al. 2014). Several anthropogenic activities can be identified as main drivers of habitat loss and fragmentation in phyllostomid territory: agriculture and cattle ranching, logging, mining, urban and industrial expansion, and energy development megaprojects (e.g., wind parks and dams).

Over the last few decades, rate of habitat loss and fragmentation directly and indirectly linked to human activities has increased substantially (NRC 2001, 2003). In the case of forest cover change, a trend of increasing forest loss has been detected in the tropical domain, with some of the most affected areas located in South America (Hansen et al. 2013; Heywood and Watson 1995). Considering only the Amazon, more than 2 million ha are deforested every year (Fearnside et al. 2005). Degradation of tropical and subtropical forests can affect bats in complex ways at population and ensemble levels, changing species abundance and ensemble composition and structure in different ways depending on the taxonomic group, life-history and ecological traits, vegetation structure and composition of fragments, quality of the vegetation matrix containing these fragments, and season (Gorresen et al. 2005; Meyer and Kalko 2008; Meyer et al. 2016; Rex et al. 2011; Willig et al. 2007).

Responses of phyllostomid bats to habitat fragmen-

tation vary among guilds and species. While certain species and feeding guilds are highly susceptible to habitat disruption, others can adapt to it or even benefit from it. Species richness and abundance of phyllostomids in mixed habitats, containing mature forest fragments embedded in agricultural vegetation and secondary vegetation patches or in agroforestry crop systems, can be similar or higher than in well-preserved forests (García-Morales et al. 2013; Willig et al. 2007). Pioneer plants in early and intermediate successional patches and several crop species provide phyllostomid bats with food, increasing their abundance (Castro-Luna and Galindo-González 2011). In the case of omnivorous and fruit- and flower-feeding bats, species richness and abundance can be positively associated with crop systems under tree shade (e.g., cocoa and coffee) and with crops and early secondary forests close to mature forest (García-Morales et al. 2013; Williams-Guillén and Perfecto 2010; Willig et al. 2007); however, the effect can turn negative with intensification of crop management and conversion of forests into monocultures and pastures (García-Morales et al. 2013). Examples of species associated with fragmented sites include *Leptonycteris yerbabuenae*, *Glossophaga soricina*, *G. morenoi*, *G. commissarisi*, *Lonchophylla thomasi*, and *Choeroniscus godmani* among nectarivores; most common species of *Sturnira*, *Artibeus*, and *Carollia*, plus *Uroderma bilobatum* and *Vampyrodes major*, among frugivores; *Lonchorhina aurita* and *Micronycteris schmidtorum*, among insectivores; and *Desmodus rotundus* among sanguinivores (Avila-Gómez et al. 2015; García-García and Santos-Moreno 2014; Klingbeil and Willig 2009). Conversely, in the case of gleaning animalivores (e.g., Phyllostominae), species' relative abundances are often positively associated with well-preserved forests, and bats in this guild are among the first to disappear from them after disturbance (García-Morales et al. 2013; Klingbeil and Willig 2009; Medellín et al. 2000). Examples of species highly susceptible to forest fragmentation include *Lophostoma silvicolum*, *Mimon crenulatum*, *Trachops cirrhosus*, *Glyphonycteris sylvestris*, and *Vampyrum spectrum* among gleaning animalivores; *Hylonycteris underwoodi* among nectarivores; and *Artibeus aztecus* and *Sturnira ludovici* among frugivores.

Despite evidence linking a significant number of phyllostomid species to disturbed habitats, several cautions are raised by Willig et al. (2007) concerning their presence there. Presence of some species in altered habitats might reflect their use as corridors, while mature forests could be their main sources of roosts and food. Completeness of inventories of bat fauna in a mixed-habitat area based on mist netting likely decreases as vertical structure increases; thus, a higher percentage of captures of locally occurring bat species is expected in crop and early secondary forest patches compared to mature forests. Finally, increase in abundance of frugivorous and nectarivorous bats in disturbed habitats within the forest matrix could potentially alter the kinds of biotic interactions in which other species that are naturally less abundant participate.

Roost Disturbance and Destruction

A critical habitat component for bats includes spaces used for protection, resting, and mating and as maternity roosts. The broad variety of roosts used by phyllostomid bats includes tree parts, rock crevices, caves, and different human-made structures that can be inhabited temporarily or permanently (Furey and Racey 2016; chap. 18, this vol.). Among natural roosts, caves are critical for the survival of a large proportion of bat species worldwide (Furey and Racey 2016). In the Neotropical region, large caves are considered important targets for conservation action, because they can host large colonies of bats and species-rich assemblages (Brunet and Medellín 2001). But even small caves can house species-rich bat assemblages; however, cave use by bats has been poorly documented for a large part of the Neotropics (see Voss et al. 2016). Unfortunately, caves are exposed to high levels of disturbance because they are frequently used for mining, storage, commercial, recreational, religious, and cultural purposes (Furey and Racey 2016; Mickleburgh et al. 2002). Detrimental actions performed inside caves inhabited by bats include frequent disturbance with noise and artificial lights, intensive guano harvesting, burning of objects, poisoning, closing of exits, and forced bat exclusion (Ladle et al. 2012; Mancina et al. 2007).

The phyllostomids most vulnerable to roost disturbance are those mainly dependent on caves or human-made structures for roosting. Mancina et al. (2007) described the risks faced by several cave-dwelling species in Cuba (e.g., *Phyllonycteris poeyi*, *Erophylla sezekorni*, *Brachyphylla nana*), where roost caves, especially those identified as hot caves (see Rodríguez-Durán 1995; and chap. 18, this vol.), are exposed to factors that change

their microclimate. Maintaining the needed microclimatic conditions characteristic of hot caves is highly dependent on presence of large bat colonies, and any type of disturbance reducing colony size represents a risk factor.

Long-nosed bats (*Leptonycteris* spp.) are also among the phyllostomid species frequently affected by roost destruction and disturbance (IUCN 2016). These gregarious species can inhabit caves in colonies that contain thousands to hundreds of thousands of individuals (Fleming and Nassar 2002). Caves used by *Leptonycteris curasoae* as roosts on Margarita Island, Gran Chimana Island, and the Paraguaná Peninsula, Venezuela, show evidence of vandalism, and because caves are scarce in these places, bats invade abandoned houses where they are easily killed (fig. 24.1; JMN, pers. obs.).

Phyllostomid bats that roost in different parts of trees in tropical forests can also be affected by roost loss under high levels of deforestation and land conversion, as occurs in the Amazon rainforest (Voss et al. 2016).

Human-Bat Conflicts

As human populations grow, people's demand for space and resources increases and natural habitats become more fragmented, pushing bats into closer contact with humans. Under these circumstances, chances for human-bat conflicts increase in three forms: (1) transmission of infectious diseases to humans and domestic animals, (2) occupation of human-made structures, and (3) damage to fruit crops. Each of these conflicts may trigger bat control measures that have negative effects on their populations.

Bats are natural reservoirs of many pathogens, and they have recently been associated with a broad spectrum of emerging infectious zoonotic diseases of viral origin that affect humans, domestic animals, and wildlife (FAO 2011; Schneeberger and Voigt 2016). Most groups of zoonotic viruses identified in bats have been detected in Europe, Asia, Africa, and Australia (Shi 2013). A recent paper by Olival et al. (2017), however, proposed that Neotropical bats are potential hosts for a large number of poorly known viruses and zoonoses, triggering an alarm through the public media across Latin America. Contrasting with this prediction, available evidence indicates that only the rabies virus (genus *Lyssavirus*) represents an important health issue linked to bats in the New World (Banyard et al. 2014). The

Figure 24.1. Colony of lactating and juvenile individuals of the Curaçaoan long-nosed bat (*Leptonycteris curasoae*), temporarily roosting inside a rural small house at Juan Gil, Falcon State, northwestern Venezuela. Photo by J. M. Nassar.

rabies virus in vampire bats (*D. rotundus*), from Mexico to northern Argentina, has the most significant impact on humans and domestic animals (Kuzmin et al. 2011). This problem will escalate with time in certain regions of Central and South America, where high rates of land conversion are transforming forests into pastures, forcing vampire bats to switch from wildlife to livestock and domestic animals as food sources (Schneider et al. 2009). As a consequence of this conflict, many attempts have been made to control vampire populations in Latin America (Streicker et al. 2012); however, when improperly conducted, these control campaigns can have negative effects on nontargeted bat species, including many phyllostomid bats (Rupprecht et al. 2006).

Other zoonotic viruses, not yet associated with public health problems in the Americas, have also been found in leaf-nosed bats, including coronaviruses (Asano et al. 2016), hantaviruses (Sabino-Santos et al. 2015), paramyxoviruses (FAO 2011), togaviruses, bunyaviruses, arenaviruses, and herpesviruses (Calisher et al. 2006). Phyllostomid bats can also be vectors and reservoirs of haemoparasites (*Trypanosoma cruzi* and *T. evansi*) that are responsible for transmitting diseases to humans and domestic animals (Lisboa et al. 2008; Silva-Iturriza et al. 2013). Finally, in a few cases,

phyllostomid bats have been reported as infected with *Histoplasma capsulatum* (Taylor et al. 1999), a fungal pathogen that can be present in the bat's guano and is responsible for causing histoplasmosis in humans that have been in contact with it; however, there is no evidence specifically showing that this infection is transmitted from bats to humans (Voigt et al. 2016).

Overall, with the exception of uncommon rabies transmission from vampire bats, bat disease transmission to humans is exceedingly rare in Latin America; however, misleading disease speculation has the potential to cause significant negative impact on many species and populations of bats in the region (Tuttle 2017). Responsible and prudent treatment of disease-related findings involving bats by the media and by public health authorities is crucial to prevent the rise of new human-bat conflicts.

The second important cause of human-bat conflict resulting in significant death tolls among bat populations is when bats roost temporarily or permanently in human-made structures. Under certain circumstances, this phenomenon can result from low numbers of natural roosts near urbanized areas resulting from loss of roost sites or roost disturbance. Historically, this problem has resulted in the use of eradication control measures (Barclay 1980). Only in countries where advanced legislation protects synanthropic bats from these extreme measures do solutions to this problem not put bat colonies at risk (Voigt et al. 2016). However, this is not the case in many countries in Latin America and the Caribbean, where environmental and wildlife laws often do not favor bat conservation (see chap. 23, this vol.).

Environmental Contaminants

Many anthropogenic activities generate environmental contaminants that can affect bat health, occasionally leading to reduced fecundity, death, and population declines (O'Shea and Johnston 2009). Pesticides represent the main type of contaminants affecting bats worldwide. Bats ingest them when drinking polluted waters or by feeding on plants (fruit and nectar) and arthropods previously exposed to them (Bayat et al. 2014). The effects of these substances can be expressed in many ways, including neuronal disorders (Clark 1988), increased metabolic rates (Swanepoel et al. 1998), malfunction of reproductive organs (Stahlschmidt and Brühl 2012), and immunosuppression and endocrine disruption (Kannan et al. 2010).

Evidence of bat exposure to organochlorine and organophosphorus pesticides and their harmful consequences on different target organs exists since the 1960s (O'Shea and Johnston 2009). Despite of the banning of these pesticides in many countries, in several developing nations they are still frequently used (Konradsen et al. 2003). This practice represents a potential risk to phyllostomid bats with foraging habits linked to agricultural areas. For example, application of fenthion (organophosphorus) on cultivated fruits in Brazil can generate hepatic disorders in *Artibeus lituratus* (Amaral et al. 2012). In contrast, new generation pesticides (e.g., pyrethroids, neonicotinoids) are characterized by their low residual effects and toxicity to wild species of vertebrates; however, their potential negative impacts on bats have rarely been evaluated (Bayat et al. 2014).

Heavy metals also raise concerns because of their potential negative effects on bats (Mickleburgh et al. 2002; Zukal et al. 2015). Poisoning effects by heavy metals in bats and other vertebrates have been expressed in a broad variety of ways, including liver and renal pathologies, DNA damage, hemochromatosis, paralysis, and alterations in cholinergic functions (Zukal et al. 2015). Several harmful heavy metals (e.g., cadmium, chromium, copper, lead, and zinc) have been detected in fruit- and nectar-feeding bats, mostly from the Paleotropics (Zukal et al. 2015).

Of the various ways in which heavy metals can become incorporated into the environment, disposal of mercury and cyanide into local natural drainages and rivers can be potentially hazardous to bats in several countries of Central and South America (Veiga et al. 2014). These contaminants are released in large amounts in connection with gold-mining operations, resulting in negative environmental and health impacts (Veiga et al. 2014; Velásquez-López et al. 2011). However, we currently do not know the extent to which species-rich assemblages of phyllostomid bats are being affected by mercury and cyanide poisoning in gold-mining centers and their areas of influence.

Emerging Threats

Anthropogenic activities of more recent development, or not previously considered for lack of evidence, pose increasing concern for their potential impact on bats in the New World. For example, wind power has recently become one of the most important sources of renewable energy worldwide (Ackermann 2005). Although originally promoted as a "green" energy alternative, the impact of wind energy facilities on wildlife and habitats can be substantial (Arnett 2012; NRC 2007). In the case of bats, wind turbines have proven to be lethal at many wind farms in temperate North America and Europe (Arnett et al. 2016). Bats seem to be attracted to turbines because they confuse them with roosts or because insects consumed by insect-feeding bats concentrate near them (Cryan et al. 2014; Kunz et al. 2007). As a consequence of being attracted to the turbines, bats get killed by direct collision, barotraumas, and possibly ear damage (Grodsky et al. 2011; Rollins et al. 2012).

For other regions of the world, information on bat fatalities associated with wind farms is quite limited (Arnett et al. 2016); however, the rate at which wind power centers are being constructed in many countries can give us an idea of the potential risks to which bat populations will be exposed in the near future. From Mexico to Chile and in the Caribbean, the trend suggests a fast increase in the number of wind farm facilities that will be established in the coming years

(Kozulj 2010; Wind Power 2014). Up to now, published reports of bat fatalities in this region are restricted to few locations, all of them involving phyllostomid bats among the affected species. The Isthmus of Tehuantepec, Oaxaca, Mexico, is becoming a focal area of wind farms. In one of them, 203 bat fatalities, including eight species of leaf-nosed bats, were recorded during four years of monitoring (Bolívar-Cimé et al. 2016). In South America, 336 bat fatalities were recorded in a large wind power complex in Rio Grande do Sul State, southern Brazil, during three years of monitoring (Barros et al. 2015). In eastern Puerto Rico, 11 of 13 species of bats that live on the island have suffered fatalities linked to a wind farm, including five fruit- and nectar-feeding species of phyllostomids (Rodríguez-Durán and Feliciano-Robles 2015).

When considering the problem of wind turbines and bats in Latin America and the Caribbean, phyllostomid bats represent an important challenge. Preconstruction bat surveys to help locate bat-safe areas mostly rely on ultrasound detection to monitor bat activity; however, echolocation calls produced by most phyllostomids (whispering bats) are not easily detected using acoustic devices (Rodríguez-Durán and Feliciano-Robles 2015).

The second emerging threat that should be considered in relation to cave-dwelling phyllostomids is predation by exotic species. Invasion by exotics is the second most important cause of biodiversity loss worldwide (Pimm et al. 1995). The effect of exotic species on bats has been barely investigated; however, there is potential risk that exotic taxa could affect their population status. Any invasive species with detrimental effects on bats during their roosting time has the potential to become a sizable threat to them. In Cuba and Puerto Rico, frequent cat (*Felis catus*) predation on bats has been recorded at the entrance of caves inhabited by large colonies (Mancina 2011; Rodríguez-Durán et al. 2010). Rats (several species in the genus *Rattus*) can also be potential predators of bats in caves frequented by these rodents. The Pacific rat (*R. exulans*) affects colonies of two species of *Mystacina* in New Zealand (O'Donnell 2008). One of us (JMN) has observed rats jumping on bats trapped on mist nets and rapidly killing them at the entrance of caves on the islands of Curaçao and Bonaire.

The third emerging threat with proven incidence on the New World is global climate change. For more than 40 years, global climate change has been an issue of general concern for its potential consequences on biodiversity (Walther et al. 2002). However, available studies providing evidence on the effects of climate change on Neotropical bats are still scarce and of recent manufacture. Bats in general, and leaf-nosed bats in particular, are predicted to be affected by global warming through alteration of their geographic ranges of distribution, with concomitant effects on their reproductive success, behavior, and biotic interactions (Rodríguez-Rocha et al. 2017). By means of the expected increase in environmental temperature, lowland species are predicted to expand their altitudinal range or switch to higher elevations. Thus, intermediate and high altitude specialist taxa could be exposed to competitive exclusion by more resilient and abundant generalist species, undergoing population decline and extinction. La Val (2004) demonstrated changes in spatial distribution of bat species along an altitudinal gradient in Costa Rica, showing that 24 species with mostly lowland distribution in the 1970s colonized higher elevations over the subsequent decades. Among them, several species of phyllostomids typical of low elevations were included (e.g., *Vampyrodes caraccioli*, *Glyphonycteris sylvestris*, *Mimon bennettii*, *Carollia castanea*, *D. rotundus*, and *Platyrrhinus helleri*).

Climate change is also expected to modify the latitudinal range of phyllostomid species with important conservation implications. The best documented example is the predicted change in spatial distribution of the common vampire *D. rotundus*, reservoir of bovine paralytic rabies and restricted in latitude and elevation to mean minimum temperatures of 10°C (McNab 1973). Zarza et al. (2017) developed niche models to identify current and future distribution patterns of *D. rotundus* in Mexico, and their predictions for the 2050 and 2070 climate scenarios indicated a reduction in the distribution of the species in southern Mexico and an expansion in distribution in the central and northern portions of this country. Thus, rabies disease is expected to spread to new areas in central and northern Mexico with the arrival of *D. rotundus*, bringing with it the bat-human conflict and its harmful effects on the native bat fauna.

Global warming has an additional potentially negative impact on the bat fauna of the Caribbean region through its effect of increasing the potential intensity and destructive power of hurricanes (Emanuel 2005). Hurricanes and tropical storms that transit across the Caribbean every year represent a sizable threat to bats. Small islands are particularly susceptible to drastic population declines or local extinctions of bats if directly exposed to hurricanes, although inter-island recolonization may help recover their populations (Jones et al. 2001; Willig et al. 2009). Tree-roosting and plant-feeding phyllostomids can be vulnerable to hurricanes if they affect a considerable fraction of the plant communities present on an island. If the number of plants used as roosts and food drops drastically, nectar and fruit bats have to invest more energy searching for sufficient food to survive until plant communities recover (Gannon and Willig 2009).

Finally, concerning the white-nose syndrome, an emerging threat to temperate hibernating bats that has killed several millions in the United States and Canada (Turner et al. 2011), no study has shown evidence of effects on phyllostomids, as they do not live in areas with long, cold winters, and they do not hibernate.

Global and Regional Conservation Status

Global Conservation Status

Globally, bat conservation status has being assessed following the standards of the International Union for Conservation of Nature and Natural Resources (IUCN), available at their official World Wide Web page (www.iucnredlist.org). This system was designed to determine the relative risk of extinction of species, and its main purpose is to catalog and highlight those plants and animals that are facing a high risk of global extinction (i.e., those listed as "critically endangered," "endangered," and "vulnerable"). The IUCN Red List also includes information on species that are categorized as extinct or extinct in the wild, taxa that cannot be evaluated because of insufficient information ("data deficient"), and species that are either close to meeting the threatened thresholds or that would be threatened if there were not ongoing taxon-specific conservation programs ("near threatened").

The Bat Specialist Group completed the reassessment of the IUCN Red List status of bat species in 2016. Out of 216 extant species of phyllostomid bats recognized in the current volume (see chap. 4, this vol.), a total of 13 were assigned to threatened categories (table 24.1): one as critically endangered (*Phyllonycteris aphylla*), five as endangered (*Chiroderma improvisum*,

Leptonycteris nivalis, Lonchophylla dekeyseri, Lonchorhina fernandezi, and *Sturnira nana*), and seven as vulnerable (*Choeroniscus periosus, L. curasoae, Lonchorhina marinkellei, Lonchorhina orinocensis, Musonycteris harrisoni, Platyrrhinus chocoensis*, and *Vampyressa melissa*). Together, they represent 6% of all phyllostomids and 8% of all evaluated taxa within the family (fig. 24.2). These figures represent a proportionally low fraction of threatened species for the family at a global level.

Two species (*P. aphylla* and *C. improvisum*) are restricted to one or two Caribbean islands, where roost disturbance and destruction and the devastating effects of hurricanes and volcanic activity are mainly responsible for population declines. Two other species (*L. nivalis* and *M. harrisoni*) are distributed in dry ecosystems of southwestern United States and Mexico, where they are exposed to deterioration of habitat quality and reduced area of occupancy. The remaining nine species are mostly restricted to South America, specifically in the Andean Region and the Amazon Basin, where they are relatively rare and known from few and small populations. The main threat affecting these species is habitat loss and fragmentation through land conversion. The rest of the evaluated taxa falls within the near threatened ($N = 11$, 5%), data deficient ($N = 21$, 10%), and least concern ($N = 124$, 59%) categories. Forty-seven taxa (22%), in 15 genera (*Anoura*: 4 species, *Artibeus*: 4, *Carollia*: 2, *Chiroderma*: 1, *Dryadonycteris*: 1, *Hsunycteris*: 3, *Lichonycteris*: 1, *Lonchophylla*: 4, *Lonchorhina*: 1, *Lophostoma*: 2, *Micronycteris*: 3, *Platyrrhinus*: 6, *Sturnira*: 9, *Uroderma*: 3, *Vampyressa*: 2, and *Vampyrodes*: 1) still await status assessment.

Regional and Country-Level Conservation Status

At regional and country levels, evaluation of the conservation status of bats in the New World has been conducted using different classification systems, in most cases, as part of a major group analysis (e.g., Rodríguez et al. 2015), and in a few cases, separately or exclusively examining bats (e.g., Burneo et al. 2015).

Some countries (e.g., Argentina, Ecuador, and Venezuela) have used IUCN's international standards, while others have created their own methodologies to determine risk level. The United States has the U.S. Fish and Wildlife Service Listing (U.S. Fish and Wildlife Service 2016), implemented under the Endangered

Table 24.1. Global conservation status of phyllostomid bats according to IUCN Red List categories and criteria (IUCN 2016)

Subfamily	Tribe	Subtribe	Species	Categories	Criteria	Year	NA	CA	SA	CR
Glossophaginae	Brachyphyllini	Phyllonycterina	*Phyllonycteris aphylla*	CR	C2a(i)	2015				x
Stenodermatinae	Stenodermatini	Vampyressina	*Chiroderma improvisum*	EN	B1ab(i,ii,iii)	2016				x
Glossophaginae	Glossophagini		*Leptonycteris nivalis*	EN	A2c	2016	x			
Lonchophyllinae	Lonchophyllini		*Lonchophylla dekeyseri*	EN	C2a(i)	2016			x	
Lonchorhininae			*Lonchorhina fernandezi*	EN	D	2016			x	
Stenodermatinae	Sturnirini		*Sturnira nana*	EN	B2ab(iii)	2016			x	
Glossophaginae	Choeronycterini	Choeronycterina	*Choeroniscus periosus*	VU	A3c	2015			x	
Glossophaginae	Glossophagini		*Leptonycteris curasoae*	VU	A2c	2015			x	x
Lonchorhininae			*Lonchorhina marinkellei*	VU	D2	2016			x	
Lonchorhininae			*Lonchorhina orinocensis*	VU	A4c	2016			x	
Glossophaginae	Choeronycterini	Choeronycterina	*Musonycteris harrisoni*	VU	C1	2015	x			
Stenodermatinae	Stenodermatini	Vampyressina	*Platyrrhinus chocoensis*	VU	A2c	2015		x	x	
Stenodermatinae	Stenodermatini	Vampyressina	*Vampyressa melissa*	VU	A2c	2015			x	
Glossophaginae	Choeronycterini	Choeronycterina	*Choeronycteris mexicana*	NT		2008	x	x		
Stenodermatinae	Stenodermatini	Ectophyllina	*Ectophylla alba*	NT		2015		x		
Glossophaginae	Glossophagini		*Leptonycteris yerbabuenae*	NT		2016	x	x		
Lonchophyllinae	Lonchophyllini		*Lonchophylla hesperia*	NT		2015			x	
Lonchophyllinae	Lonchophyllini		*Platalina genovensium*	NT		2016			x	
Stenodermatinae	Stenodermatini	Vampyressina	*Platyrrhinus ismaeli*	NT		2016			x	
Stenodermatinae	Stenodermatini	Vampyressina	*Platyrrhinus matapalensis*	NT		2016			x	
Rhinophyllinae			*Rhinophylla alethina*	NT		2008			x	
Stenodermatinae	Stenodermatini	Stenodermatina	*Stenoderma rufum*	NT		2016				x
Stenodermatinae	Sturnirini		*Sturnira mordax*	NT		2008		x	x	
Phyllostominae	Vampyrini		*Vampyrum spectrum*	NT		2008	x	x	x	

Note: Categories considered are critically endangered (CR), endangered (EN), vulnerable (VU) and nearly threatened (NT). Geographic regions with presence of the species: North America (NA), Central America (CA), South America (SA), and Caribbean (CR). For complete interpretation of categories and criteria, see IUCN (2012). Phyllostomid classification follows Cirranello and Simmons (chap. 4, this vol.).

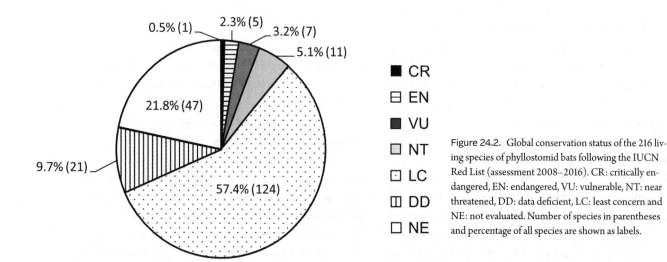

Figure 24.2. Global conservation status of the 216 living species of phyllostomid bats following the IUCN Red List (assessment 2008–2016). CR: critically endangered, EN: endangered, VU: vulnerable, NT: near threatened, DD: data deficient, LC: least concern and NE: not evaluated. Number of species in parentheses and percentage of all species are shown as labels.

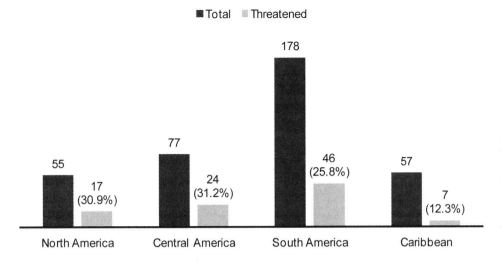

Figure 24.3. Number of total (*black bars*) and threatened (*gray bars*) species of phyllostomid bats at the region level, including percentage of threatened species in parentheses. Threatened species include critically endangered, endangered, and vulnerable according to the IUCN system or their equivalent categories according to Risk Assessment Method and Evaluation Method for Degree of Threat systems.

Species Program. Species selected as in need of federal protection are classified as endangered or threatened, depending on their status and degree of threat. Mexico and most countries in Central America use the Risk Assessment Method (Método de Evaluación de Riesgo), originally formulated in Mexico to evaluate the country's flora and fauna (Sánchez et al. 2007; SEMARNAT 2010). This classification method also considers two categories: endangered of extinction and threatened. Bolivia uses the Evaluation Method for Degree of Threat (Método de Evaluación del Grado de Amenaza), created in that country as the result of a series of adaptations from the IUCN Red Listing and other methods available region wide (Aguirre et al. 2009). Finally, other countries and territories, as is the case for many Caribbean islands, have not assessed bats' conservation status at a local level.

North America

In North America, out of 55 phyllostomids recognized for the region, 15 genera and 17 species (31%) are threatened (fig. 24.3; supplementary table S24.1). Long-nosed bats, *Leptonycteris nivalis* and *L. yerbabuenae*, have been considered endangered in the United States and threatened in Mexico, although the latter has been approved for delisting and is awaiting publication in Mexico's Diario Oficial de la Federación (Medellín and Torres Knoop 2013); *M. harrisoni* and *V. spectrum*, both in Mexico, are considered endangered of extinction; the remaining 13 species are classified as threatened in Mexico.

In general, threats to phyllostomids across their North American range are shared, with habitat degradation and roost disturbance being most important. However, because of the natural history of some species, they may be at greater risk than others. For this region, we focus on the conservation of four species linked to dry ecosystems and that are dependent on flowers and fruit.

Leptonycteris yerbabuenae and *L. nivalis*, long-distance migratory species from southern and central Mexico to northern Mexico and southern United States, were considered at risk in the late 1980s and early 1990s in the United States and Mexico, respectively, due to their low colony numbers or complete absence in a limited number of caves surveyed in Mexico and Arizona (Wilson 1985). Threats that were listed as affecting the two species included habitat disruption, roost disturbance and destruction, misguided efforts to kill vampire bats, or sheer vandalism (U.S. Fish and Wildlife Service 1994, 1995). Wilson (1985) reported that both species of *Leptonycteris* are particularly sensitive to disturbance, disappearing or drastically decreasing in numbers after humans visit roosts. Their endangered classification in the late 1980s prompted bat biologists to conduct research on the species that yielded contrasting results. On one hand, a significant number of caves known historically to be used by *L. yerbabuenae* were observed to be depleted of bats or with evidence of anthropogenic impact, suggesting demographic decline (Medellín and Walker 2003; Medellín et al. 2009). On the other hand, cave surveys conducted by Wilkinson and Fleming (1996) in Arizona and Mexico, in combination with estimates of its genetically effective population size at the time of surveys, indicated that this species had not experienced a population bottleneck in the 1970s and 1980s.

A third migratory species, the hog-nosed bat (*Choeronycteris mexicana*), was listed as threatened in Mexico in the 1990s (Diario Oficial de la Federación 1994). This little-known species, although widespread, faces the same threats as *Leptonycteris* species in North America and has the same susceptibility to disturbance at its roosts.

With the implementation of recovery plans for *Leptonycteris* species (U.S. Fish and Wildlife Service 1994, 1995) and creation in 1994 of the Bat Conservation Program of Mexico (Medellín 2003), conservation actions in favor of *L. yerbabuenae* and *L. nivalis* in the United States and Mexico began in earnest. Three-pronged strategies, based on strong environmental education, a diverse and long-term research agenda, and an effective, on-the-ground conservation action implementation, set the stage to enforce the conservation of both species. More than 20 years of data from the 13 largest known colonies of *L. yerbabuenae* indicate stability or growth in all of them. As a result, Mexico delisted *L. yerbabuenae* in 2013 (Medellín and Torres Knoop 2013). Delisting is also currently underway in the United States.

Leptonycteris nivalis remains a focus of conservation concern. There is a single known mating roost for this species. Torres Knoop (2014) performed ecological niche modeling and identified additional potential mating roosts that are in process of being verified. Ammerman et al. (2009) reported on a thermal imaging census of the species in the northernmost known roost, Emory Peak cave in Texas, and estimated fewer than 3,000 bats at the peak of the colony in July, a number similar to that reported by Easterla (1972) in the early 1970s. Much research and conservation actions are still needed to make progress in the recovery of *L. nivalis*.

Finally, *M. harrisoni*, endemic to the Mexican west coast, represents another taxon of special conservation concern among North American phyllostomids. It is a highly specialized bat morphologically but with a generalized diet (Tschapka et al. 2008); it is rare throughout its range; and it is closely associated with tropical dry forests in western Mexico, where it seems to depend on banana plantations and dry forests in good conservation state for food (González-Terrazas et al. 2012). Conservation programs for this bat are still pending.

Central America

In Central America, bat conservation status has been assessed with the Risk Assessment Method protocol in Costa Rica, El Salvador, Guatemala, Honduras, and Nicaragua. For Panama, risk assessment of bats has been conducted by a local expert (R. Samudio, pers. comm.). No information is available for Belize yet.

Out of 77 phyllostomids recognized for the region, 23 genera and 24 species (31%) are threatened at some level (fig. 24.3; supplementary table S24.1). Three species were assigned the dual status of critically endangered in one country and threatened in another country: *Artibeus inopinatus*, *Phylloderma stenops*, and *Tonatia saurophila*. All three were classified as critically endangered in Nicaragua, where up to 10 different threats were identified affecting bats in comparison with two to seven threats reported for the other countries (Rodríguez-Herrera and Sánchez 2015). *Artibeus incomitatus*, formerly considered critically endangered due to habitat loss and a very restricted distribution (Pino and Samudio 2008), is currently considered a synonym of *A. watsoni* (see chap. 4, this vol.). The remaining 21 species were categorized as threatened.

Central America has experienced high rates of deforestation and land conversion (Rodríguez-Herrera and Sánchez 2015). In addition, indiscriminate extermination of bat colonies and destruction of their roosts have had a negative impact on cave-dwelling species. For this region, three species associated with forests are of particular mention due to their susceptibility to habitat degradation.

Ectophylla alba is endemic to Central America and has a high degree of habitat specialization, using only tents as roosts from mainly two species of plants in very specific microhabitats and feeding on one species of fig (Rodríguez-Herrera et al. 2011). Roost destruction has a strong negative impact on this species, because this bat spends considerable time and energy building leaf tents. Educational material produced using *E. alba* as flagship species has helped explain the importance of roosts and environmental services provided by bats in Central America.

Artibeus inopinatus, also endemic to Central America, is associated with threatened dry forests on the Pacific coast. Its restricted distribution is a consequence of intensive deforestation. In El Salvador, only 3% of forest areas used by this species are inside protected lands (Rodríguez-Herrera and Sánchez 2015). Initiatives to protect this bat focus on two lines of action: (1) field studies in Honduras to expand knowledge of its natural history, population status, and distribution; and (2) environmental education activities and workshops in Nicaragua, where *A. inopinatus* is being used as flagship species to protect dry forests.

Despite of its broad distribution, *P. stenops* is another species of special concern in Central America, where it is relatively rare. It is restricted to well-preserved habitats, which makes it particularly susceptible to forest degradation. Extensive areas in Central America correspond to a mosaic of primary and secondary forest and agricultural areas, which are not suitable for this species. *Phylloderma stenops* will be proposed among the taxa with conservation priority for Costa Rica's new wildlife law.

All conservation initiatives described previously for phyllostomid bats in Central America are part of a regional master plan formulated by the Bat Conservation Programs of Costa Rica, Honduras, Guatemala, El Salvador, and Nicaragua (Rodríguez-Herrera and Sánchez 2015). This regional strategy seeks to develop coordinated multidisciplinary conservation actions in all countries in the region and rests on a key feature: local capacity building of its associates.

South America

In South America, bat conservation status has been assessed using different approaches, depending on the country. Argentina, Brazil, Chile, Colombia, Ecuador, Peru, and Venezuela have followed the IUCN system, Bolivia the Evaluation Method for Degree of Threat system, and Paraguay and Uruguay have followed classifications based on national wildlife regulations. No information is available for the bats of Suriname and the Guianas.

Of the 177 phyllostomids recognized for the region, 25 genera and 46 species (26%) are threatened in at least one of the countries where they have been assessed (fig. 24.3; supplementary table S24.1). Five species are considered to be critically endangered, three of them based on assessments in Ecuador (*Choeroniscus periosus*, *Lonchophylla chocoana*, and *L. orcesi*), one in Peru (*Gardenycteris koepckeae*), and one in Paraguay

(*Anoura caudifer*). Twelve species have been categorized as endangered (*Artibeus ravus*, *Chiroderma doriae*, *Lonchophylla dekeyseri*, *L. hesperia*, *Lonchorhina aurita*, *L. fernandezi*, *Platalina genovensium*, *Platyrrhinus chocoensis*, *P. lineatus*, *P. matapalensis*, *P. recifinus*, and *Sturnira nana*), most of which occur in Andean countries. The remaining 29 taxa have been categorized as vulnerable.

Across South America, bats are facing a broad variety of threats, including habitat conversion, persecution, roost disturbance, contamination, proliferation of wind farms, and introduction of exotic species (Aguirre et al. 2016; Jones, Jacobs et al. 2009). Some of the highest rates of deforestation and land conversion for agriculture and cattle breeding occur in this region (*sensu* FAO 2009), and many phyllostomid bats strongly dependent on woody vegetation for roosting and feeding are being affected (Aguirre et al. 2016). Several countries, including Chile and Bolivia, have identified bat-human conflicts as the second major threat affecting bats.

Because of the large and heterogeneous territory covered by South America, it is very difficult to identify species that could be labeled as the most threatened in this subcontinent (but see chap. 23, this vol., for Brazil). Many locally threatened species can be considered not at risk in other parts of their geographic distributions. In addition, bat species are sometimes rare or absent from some countries or eco-regions for many ecological reasons unrelated to anthropogenic factors (Voss et al. 2016). One of the most threatened species in South America is *Platalina genovensium*. This cactophilic nectar-feeding bat likely is migratory along the coast of Peru and northern Chile (Ossa et al. 2016). In Peru, mining activities might be affecting their populations mostly through loss of columnar cacti and roost availability (Pari et al. 2015). Also, use of bats in traditional medical practices might be affecting some populations of *P. genovensium* near major towns (Sahley and Baraybar 1996). In 2013, the Bat Conservation Program of Peru convened a workshop to devise a management plan for the conservation of *P. genovensium* (H. Zamora, pers. comm.). This plan identified four major working areas: (1) protective legislation at different spatial scales, (2) management and protection activities directed at roosting sites and feeding areas, (3) research addressing natural history and migratory routes, and (4) education and public outreach.

An endangered species that has been receiving

attention in South America is Tomes long-nosed bat (*L. aurita*). Even though this species is categorized as least concern (IUCN 2016), it is considered endangered in Bolivia (Vargas et al. 2009). In this country, there is a lack of suitable roosting sites for this cave specialist. The Action Plan for the Conservation of Threatened Bats of Bolivia includes several initiatives through research, conservation, and education directed at the recovery of this species (Aguirre et al. 2010).

Although still in its early stages, a South American bat conservation strategy is being created with participation of all bat conservation program members of RELCOM in South America. The strategy is organized in a series of conservation actions adapted to the main problems identified in four major eco-regions: Caribbean coast, Andes, Amazon, and temperate South America.

Caribbean

For the Caribbean region, Willig et al. (2009) listed the bat fauna of the Bahamas islands, Greater Antilles, and Lesser Antilles, comprising a total of 65 islands. Additional information on the bat fauna of five islands near the Venezuelan coast (Bekker 1996; Gomes and Reid 2015; Petit et al. 2006; Simal et al. 2015) is also available for a total of 70 Caribbean islands. But for only nine of them did we find published information on local assessments of bat conservation status. Of 57 phyllostomids reported for these islands, seven genera and seven species (12%) are threatened (fig. 24.3; supplementary table S24.1). Two species have been classified as critically endangered: *Phyllonycteris aphylla* from Jamaica and *L. curasoae* from Curaçao; three species have been classified as endangered: *C. improvisum* in Guadeloupe and Montserrat and *Brachyphylla cavernarum* and *Stenoderma rufum* on Saint John and Saint Thomas; and three species have been classified as vulnerable: *Erophylla bombifrons*, *Monophyllus redmani*, and *S. rufum* on Puerto Rico.

The Caribbean region shows a low proportion of threatened species of phyllostomids compared to the continental regions that have been examined. This observation contrasts with the well-accepted notion that islands are considered among the most fragile habitats on the planet and where most historical extinctions of bats have occurred (Jones, Mickleburgh et al. 2009; Willig et al. 2009). As previously mentioned, local assessments of bat conservation status are missing for the majority of the Caribbean islands, and this limits the information needed to have a more realistic figure of the number of threatened bats in them.

A high proportion (~50%) of Caribbean bats are endemic to the region (Koopman 1989), and many of these species have small geographic ranges and populations, which in some cases are confined to one or a few islands (Gannon and Willig 2009; Willig et al. 2009). This insular condition exposes bats to population decline and extinction due to a combination of factors, including habitat loss, roost disturbance, catastrophic stochastic events (cyclonic and volcanic activity), extended dry periods, random demographic effects, limited genetic diversity, and restricted space and population sizes for coping with accelerated rates of anthropogenic activity (Jones, Mickleburgh et al. 2009; Willig et al. 2009). A closer examination of the effects of these factors on the bat fauna on an island-by-island basis would likely show more species of phyllostomids under risk at a local level.

The most threatened species of phyllostomids are confined to few islands and are exposed to habitat loss and the effects of catastrophic events. *Sturnira angeli* and *C. improvisum* are of particular concern, because they both inhabit the islands of Guadeloupe and Montserrat (Simmons 2005), where they have become so rare that they have not been captured for many years (Pedersen et al. 2013). Their declines in population size have been attributed to the effect of volcanic activity and hurricanes. *Stenoderma rufum* is considered under risk in three of five islands where it is found (Puerto Rico, Saint John, and Saint Thomas). Main threats to this frugivorous bat are a combination of habitat loss and hurricanes. Its dependence on canopy trees for roosting and several wild fruit species makes it particularly vulnerable after a hurricane hits an island and destroys extensive areas of natural vegetation (Gannon and Willig 2009). On the islands of Aruba and Curaçao, *L. curasoae* is exposed to high levels of human disturbance at its diurnal and maternity roosts, and its foraging areas are under high pressure due to land conversion (Nassar and Simal 2014; Petit et al. 2006).

Another group of Caribbean phyllostomids of conservation concern are those restricted to a single island and confronting current or potential future threats. *Phyllonycteris aphylla*, a cave-dwelling species restricted to Jamaica, disappeared from several of its known roosts and is thought to be represented by fewer than 250 adults (Koenig and Dávalos 2015). *Micronycteris buriri*, which is not considered to be threatened currently but is geographically restricted to Saint Vincent (Larsen et al. 2011), has been collected primarily in forested areas and banana plantations that could be seriously damaged by hurricane activity.

Conservation initiatives to counteract threats affecting these and other bats in the Caribbean are scarce and have not been focused on particular species but on the total assemblage of bats on an island or bats associated with particular habitats or cave systems. A few islands have formulated action plans and conservation initiatives specifically directed to protect bats or wildlife in general (Cottam et al. 2009; Jones, Mickleburgh et al. 2009; PRDNER 2015; Speer et al. 2015). On eight of them (Aruba, Bonaire, Cuba, Curaçao, Dominican Republic, Puerto Rico, and Trinidad and Tobago), interested parties have created their bat conservation programs under the support and coordination of RELCOM.

Perspectives for Conservation

Conservation of phyllostomid bats represents a major challenge for several reasons: (1) the high number of potential target species involved, (2) a broad spectrum of conservation problems, (3) multiple geographic scales at which conservation problems are expressed and need to be solved, (4) diverse socioeconomic and sociopolitical circumstances in the region, which affect the development of any conservation initiative, (5) insufficient human resources adequately trained to conduct a wide number of duties, and (6) a lack of funding to support the high volume of programs and projects formulated every year.

Regional conservation problems, such as bat conservation, are more effectively handled through networks of partners with unified general goals and methods. Our experience working on behalf of bat conservation under the coordinated actions established by RELCOM indicates that long-term effective actions rest on three pillars, representing three major action lines that branch out in multiple ways: (1) research, (2) education and communication, and (3) conservation actions. It is by a strong investment in resources and capacity building in each area, and through the interplay of their products, that major achievements in bat conservation can be obtained.

Here we summarize our perspectives for the development of conservation initiatives for phyllostomid bats in the near future, taking into consideration their current conservation status, main threats, and human resources and organizations already established in the region.

Research

The recently published book *Bats in the Anthropocene: Conservation of Bats in a Changing World* (Voigt and Kingston 2016) offers a vast and detailed compilation of future research directions and questions that have been recommended in connection with the main conservation problems confronted by bats around the world, including the critical threats identified in this chapter that affect phyllostomid bats. Here, we focus on the attention these threats have received from a research perspective in terms of investment of funds and the main topics of investigation addressed by national bat conservation programs affiliated with RELCOM (known in Spanish as Programas para la Conservación de los Murciélagos).

For financial investment on research, we used a database provided by Bat Conservation International covering 10 years of funding in the region (2003–2013). Bat Conservation International is one of the main international agencies that have provided financial support to conservation-oriented bat research projects in Latin America and the Caribbean for several decades. Bat Conservation International's funding has been allocated primarily to support research questions linked to habitat loss (48%) and human-bat conflicts (15%)—figures that are in concordance with the relative importance of these threats to bats in the region (fig. 24.4a). An important fraction of funds (26%) has also helped develop bat research not directly connected with these threats, covering diverse autoecological questions about threatened taxa. Less funding has gone to support projects addressing conservation issues linked to roost disturbance and destruction (8%), environmental contaminants (1%), and emerging threats (1%). Research questions it has supported have been centered on ecosystem services, community ecology, and feeding ecology (58% of total investment), while comparatively less funding has been directed toward other critical threats, such as roosting ecology (9%) and bat health (1%). This disproportionate investment among priority research areas does not necessarily mean that this funding agency has favored some research topics over others of equal conservation relevance—it also reflects researchers' choices.

To explore research preferences of national bat conservation programs working throughout the region, we reviewed RELCOM's triennial technical report (2010–2013), available at the network's World Wide Web page (www.relcomlatinoamerica.net). This document covers research activities conducted by 15 programs (ABC islands [Aruba, Bonaire, and Curaçao], Argentina, Bolivia, Chile, Colombia, Costa Rica, Cuba, Ecuador, Guatemala, Honduras, Mexico, Peru, Puerto Rico, Uruguay, and Venezuela). Habitat loss received the greatest research effort (30%), followed by studies addressing questions related to roost disturbance (25%) and human-bat conflicts (25%) (fig. 24.4b). Studies addressing problems concerned with environmental contaminants (4%) and emerging threats (1%) received comparatively lower research interest. Regarding research topics selected, main research questions focused on community ecology (17%) and population ecology (17%), followed by roosting ecology (14%), feeding ecology (11%), and emerging threats (8%).

In line with the observed trend and understanding the complexity of impacts associated with habitat loss and fragmentation at different geographic scales across the region, conservation-oriented research should continue examining how these factors affects the stability of bat assemblages, the ecosystem processes and services in which these organisms participate, and the relationship between land conversion and spread of zoonotic viral diseases among humans. These studies should also be accompanied by research focused on landscape management, testing the beneficial effects of riparian corridors and agroforestry systems on the recovery of bat populations, assemblages, and metacommunities (Meyer et al. 2016).

Research lines receiving less attention should be the focus of more investment in the near future. Emerging threats on bats throughout Latin America and the Caribbean, such as the effect of wind turbines and hydropower facilities, need to be included among priority research goals. For instance, in Central America, there are 21 wind projects operating in five of seven countries in the region, which may increase to 61 in five years (Wind Power 2014). Wind turbines in operation in Brazil will triple in five years (Beltrão and Bernard 2015). In addi-

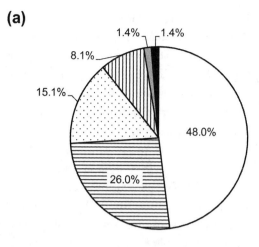

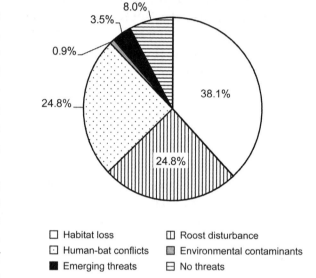

☐ Habitat loss ⊞ Roost disturbance
⊡ Human-bat conflicts ▨ Environmental contaminants
■ Emerging threats ⊟ No threats

Figure 24.4. Percentage of Bat Conservation International's financial support allocated to conservation-oriented research on major threats affecting bats in Latin America and the Caribbean from 2003 to 2013 (a), and percentage of research efforts addressing major threats affecting bats by 15 Latin American and Caribbean bat conservation programs between 2010 and 2013 (b).

tion, a sizable increase in the number of dams projected to be built on the basin's rivers of Amazonian countries will likely result in further forest loss and alteration of habitat components used by bats in the region (Lees et al. 2016). We need to learn how phyllostomid species respond to dams and their short- and long-term impacts on the environment.

Another research area that has been poorly addressed corresponds to emerging diseases associated with bats. Concern for this potential problem affecting both bat and human populations needs to be considered in research agendas. In order to counteract sensationalist news accounts spread through the global media and linking outbreaks of infectious diseases to bats, there is need to test bats as potentially important vectors of pathogens that could affect human populations throughout the New World.

It is important to increase attention and investment directed at evaluating the conservation status of species of phyllostomid bats that are not yet included in the IUCN Red List or are in need of further examination. Sixty-seven species still remain without evaluation or have been categorized as data deficient (fig. 24.2). Since several of these species are endemic, occur in restricted habitats, or possess ecological similarities with other taxa already classified as threatened, it is quite likely that some of them will fall within the threatened categories, too. In addition, evaluation of the conservation status of bats in Caribbean islands under high anthropogenic pressure and those exposed to the impact of hurricanes should also be considered a priority.

Finally, with regard to long-term research planning, emphasis should be given to addressing the effects of climate change on phyllostomids, especially those species dependent on constrained habitats along altitudinal gradients. We expect that the effects of climate change on the most susceptible habitats used by phyllostomid bats will also include interactions with other threats affecting many species.

Education and Communication

Across the Americas and the Caribbean, bats are the least charismatic mammals (Gareca et al. 2007). Several myths and traditions about bats, along with misconceptions about their true nature, exist in most parts of Latin America, and this has posed a threat to their populations. For example, several phyllostomid bats are randomly captured for medicinal purposes in the Andean region (Lizarro et al. 2010), including nectar-feeding bats (*Anoura*), fruit-eating bats (*Carollia*), and occasionally vampire bats (*D. rotundus*). Traditional local medicine men use the blood of bats to heal "head problems" (i.e., epilepsy), a practice that has turned into a wildlife trade of bats in popular markets, locally known as witch markets (Lizarro et al. 2010). In addition, the most common misconception about bats in Latin America is that they are all rabid or are carriers of diseases and should be killed (Gareca et al. 2007).

Environmental education can contribute signifi-

cantly to overcome conservation problems that phyllostomid bats are facing across the entire region. Raising public awareness through environmental education, public outreach, general communication, and diffusion strategies are among the most important actions conservationists can undertake to support protection, recovery, and management of bat populations (Aguirre et al. 2014). It has been suggested that environmental education that teaches people about bats and their habitat needs can sometimes be even more important than research in achieving conservation goals (Jacobson and McDuff 1998; Jacobson et al. 2006).

Environmental education has become a crucial strategy for the protection of phyllostomid bats in the Neotropics over the last three decades and has been gaining momentum with the development of local and regional programs for bat conservation throughout the continent. Initiatives such as those implemented by RELCOM and their local associates have proven that focusing their efforts on local communities and schools near key roosts or particular habitats for bats is important for a successful long-term conservation effects (Aguirre et al. 2014). Specific actions have been taken, especially in the case of bats that are threatened, to change people's misconceptions and to explain the importance of phyllostomids. The role of bats in maintaining ecological processes that benefit not only bats but also the ecosystem and local human economies are particularly emphasized in environmental education programs and public outreach activities. In addition, education initiatives are linked with detailed management plans to permanently protect these sites through formal legislation and local community stewardship (Navarro et al. 1996).

In Latin America, many conservation projects include an extensive distribution of educational materials, such as posters and fliers. Printed materials directed primarily toward schoolchildren include short storybooks featuring phyllostomid bats that are particularly charismatic, with narratives about their importance for ecosystem services. Featured bats include *L. yerbabuenae* in Mexico (Navarro 2001), *E. alba* in Central America (Cordero-Schmidt et al. 2010), and *L. aurita* in South America (Galarza and Aguirre 2011). Television and radio programs have proven to be important tools for reaching people and changing their misconceptions. The award-winning radio show *Aventuras al Vuelo*, created in Mexico, has been broadcast widely in several countries in Latin America with great success. As part of changing peoples' misconceptions, workshops are regularly organized at every level, from elemenatary schools to universities. In the period 2010–2013, for example, 55 projects related to environmental education and communication were organized by RELCOM's conservation programs.

Conservation Actions

Conservation actions comprise a wide variety of activities, directly or indirectly applied to bat populations and their habitats that contribute to help reverse and neutralize the negative effects of threats on them. Several conservation initiatives have been implemented in recent years across Latin America and the Caribbean, targeting some of the major factors affecting bats.

In general, habitat loss in the Neotropics is being counteracted in two main ways: first, through protection of natural areas and restoration of disturbed lands; and second, through landscape management of agricultural areas. In the first case, a regional program designed by RELCOM has been implemented to identify and create Important Areas for Bat Conservation (AICOM, for its acronym in Spanish; Aguirre and Barquez 2013). Conservation programs involved in the creation of an AICOM develop activities for its protection and management with the involvement of local environmental authorities. By the end of year 2016, 64 AICOMs had been identified, and several of them have become areas protected by national and local governments or by environmental nongovernment organizations (RELCOM 2016b). In addition, use of agroecosystems and creation of vegetation corridors are gaining approval as methods to help increase bat diversity in many agricultural areas in the Neotropics (de la Peña-Cuellar 2015; Meyer et al. 2016). Coffee and cacao plantations, traditionally cultivated under shade, generate complex and adequate vegetation matrices that maintain diverse phyllostomid assemblages.

Actions aimed at protecting roost sites are essential for bat conservation. Cave disturbance and destruction is being counteracted with legal protection of caves; however, despite the fact that many countries in the region possess numerous caves used by bats as roosts, most caves are not part of national systems of protected areas. To promote cave conservation in the region, the same program that is creating AICOMs is also designed to identify and certify protection of Important Sites for Bat Conservation (SICOM, for its acronym in Spanish). Until 2016, 16 SICOMs had been named, and local multidisciplinary groups are developing projects for their protection and management (RELCOM 2016b).

A more ambitious strategy in the same direction is the creation of systems of protected caves based on the environmental legislations of each country. These include: (1) fauna sanctuaries such as Santuario de Fauna Cuevas de Paraguaná, Paraguaná Peninsula, northwestern Venezuela (Delfín et al. 2011); (2) fauna reserves such as Bonaire Cave and Karst Nature Reserve, Bonaire Island, Dutch Caribbean (CARIBSS 2016); and (3) national parks such as Parque Nacional Barra Honda, Guanacaste, Costa Rica, (Artavia Durán 2016).

Concerning emerging threats in the Neotropics, exponential development of wind energy facilities in many countries necessitates planning and implementation of global solutions for bat conservation. With this in mind, RELCOM has created an international committee to evaluate the impact of wind energy facilities in the region and to propose a series of measures to mitigate and reduce their effects on bats (RELCOM 2016c). To achieve these goals, the committee has developed two tools: a master plan to manage bat-wind energy conflicts in Latin America and the Caribbean and guidelines to evaluate the impact of wind energy facilities on bats. The challenge for the coming years is to be able to use these tools in a coordinated way.

Conclusions

Increasing rates of habitat loss and a wide variety of anthropogenic activities continue to cause vast environmental impacts and ecosystem alterations across Latin America and the Caribbean. At the same time, research, conservation, and environmental education have grown exponentially in the region, supported by a very significant critical mass of local scientists, educators, decision makers, and students. These two factors are opposing factors in a serious problem that affects biodiversity. Phyllostomid bats represent an extremely significant portion of the biological diversity in most of the continent and can be used as indicators to measure the state of forest conservation (Medellín et al. 2000). Their ecological and behavioral attributes make them easily sampled and high-resolution environmental indicators because of their close connection to forests. Habitat degradation, roost disturbance, and vandalism remain as the main threats affecting phyllostomids across their entire range of distribution. The first of these threats may be manageable with alternative strategies, such as adopting an umbrella species approach. For example, jaguars may function as a useful umbrella species protecting many smaller species that coexist with them (Medellín et al. 2016), and by focusing conservation efforts on jaguars, many bat species and other plant and animal groups will benefit. Growing recognition of AICOMs and SICOMs across the continent is another tool that is very promising for the future of phyllostomids and bats in general. The increasing number of scientists, students, and professionals implementing conservation projects for phyllostomids or including them in their projects is also enhancing conservation success. Recovering threatened species is very possible, but once recovered, those species must continue to be monitored into the future. Extratropical phyllostomids are a bit different in that protecting them via umbrella species does not necessarily apply. Specifically, conservation of nectar bats has been approached in a number of ways, always including research, public education, and conservation actions. Recently, academics, land owners, producers, bartenders, and distributors joined forces for tequila and mezcal production, introducing the "bat-friendly" concept, benefiting pollinating phyllostomids (Trejo-Salazar et al. 2016) and agaves by permitting 5% of agaves to flower and feed bats and promoting agave genetic diversity.

The conservation toolkit continues to expand, with new research approaches such as metagenomics, in-depth phylogenies, microbiome, and other studies. Considering the enhanced workforce of bat researchers and environmentalists across the continent, the emerging threats, and the sense of community from Argentina and Chile to Canada, a surprising result may be the protection of many species of phyllostomid bats through multidisciplinary recovery plans. Partnership must be an essential part of the plans at all stages. Collaborative networks as the working framework is proving to be highly successful in the case of bat conservation in the New World.

Acknowledgments

We wish to thank Valéria da Cunha Tavares and Armando Rodríguez-Durán for their valuable comments and suggestions on an earlier version of this chapter. Onil Ballestas provided assistance preparing the figures.

References

Achard, F., R. Beuchle, P. Mayaux, H. J. Stibig, C. Bodart, A. Brink, S. Carboni et al. 2014. Determination of tropical deforestation rates and related carbon losses from 1990 to 2010. Global Change Biology, 20:2540–2554.

Ackermann, T. 2005. Historical development and current status of wind power. Pp. 7–24 *in*: Wind Power in Power Systems (T. Ackermann, ed.). John Wiley & Sons, West Sussex.

Aguirre, L. F., R. Aguayo, J. Balderrama, C. Cortéz, T. Tarifa, P. Van Damme, L. Arteaga, and D. Peñaranda. 2009. El método de evaluación del grado de amenaza para especies (MEGA). Pp. 7–17 *in*: Libro rojo de la fauna silvestre de vertebrados de Bolivia (Ministerio de Medio Ambiente y Agua, ed.). Ministerio de Medio Ambiente y Agua, La Paz, Bolivia.

Aguirre, L. F., and R. M. Barquez. 2013. Critical areas for bat conservation: Latin American conservationists build a grand strategy. Bats, 31:10–12.

Aguirre, L. F., R. A. Medellín, and B. Rodríguez-Herrera. 2016. From threat to opportunity—strategies for bat conservation in the Neotropics. Pp. 140–153 *in:* Tropical Conservation. Perspectives on Local and Global Priorities (A. A. Aguirre, and R. Sukumar, eds.). Oxford University Press, New York.

Aguirre, L. F., M. I. Moya, L. L. Arteaga B., M. I. Galarza M., A. Vargas E., K. Barboza Marquez, D. A. Peñaranda, J. C. Pérez-Zubieta, M. F. Terán V., and T. Tarifa. 2010. Plan de acción para la conservación de los murciélagos amenazados de Bolivia. BIOTA-PCMB, MMAA-VBCC-DGB, UICN-SSC-BSG, CBG-UMSS, Cochabamba, Bolivia.

Aguirre, L. F., J. M. Nassar, R. M. Barquez, R. A. Medellín, L. Navarro, A. Rodríguez-Durán, and B. Rodríguez-Herrera. 2014. De esfuerzos locales a una iniciativa regional: La Red Latinoamericana y del Caribe para la Conservación de los Murciélagos (RELCOM). Ecología en Bolivia, 49:45–50.

Amaral, T. S., T. F. Carvalho, M. C. Silva, M. S. Barros, M. C. Picanço, C. A. Neves, and M. B. Freitas. 2012. Short-term effects of a spinosyn's family insecticide on energy metabolism and liver morphology in frugivorous bats *Artibeus lituratus* (Olfers, 1818). Brazilian Journal of Biology, 72:299–304.

Ammerman, L. K., M. McDonough, N. I. Hristov, and T. H. Kunz. 2009. Census of the endangered Mexican long-nosed bat *Leptonycteris nivalis* in Texas, USA, using thermal imaging. Endangered Species Research, 8:87–92.

Arnett, E. B. 2012. Impacts of wind energy development on wildlife: challenges and opportunities for integrating science, management. Pp. 213–237 *in*: Wildlife Science: Connecting Research with Management (J. P. Sands, S. J. De Maso, L. A. Brennan, and M. J. Schnupp, eds.). CRC Press, Boca Raton, FL.

Arnett, E. B., E. F. Baerwald, F. Mathews, L. Rodrigues, A. Rodríguez-Durán, J. Rydell, R. Villegas-Patraca, and C. C. Voigt. 2016. Impacts of wind energy development on bats: a global perspective. Pp. 295–323 *in*: Bats in the Anthropocene: Conservation of Bats in a Changing World (C. C. Voigt, and

T. Kingston, eds.). Springer International Publishing, Cham, Switzerland.

Artavia Durán, E. J. 2016. Grupos de voluntarios de diversas nacionalidades reciben talleres de educación ambiental con respecto a murciélagos en el Parque Nacional Barra Honda, Nicoya, Guanacaste, Costa Rica. Boletín de la Red Latinoamericana y del Caribe para la Conservacíon de los Murcielagos, 70:20.

Asano, K. M., A. Santana Hora, K. Côrrea Scheffer, W. O. Fahl, K. Iamamoto, E. Mori, and P. E. Brandão. 2016. Alpha-coronavirus in urban Molossidae and Phyllostomidae bats, Brazil. Virology Journal, 13:110.

Avila-Gómez, E. S., C. E. Moreno, R. García-Morales, I. Zuria, G. Sánchez-Rojas, and M. Briones-Salas. 2015. Deforestation thresholds for phyllostomid bat populations in tropical landscapes in the Huasteca region, Mexico. Tropical Conservation Science, 8:646–661.

Banyard, A. C., J. S. Evans, T. R. Luo, and A. R. Fooks. 2014. Lyssaviruses and bats: emergence and zoonotic threat. Viruses, 6:2974–2990.

Barclay, R. M. R. 1980. Comparison of methods used for controlling bats in buildings. Journal of Wildlife Management, 44:502–506.

Barros, M. A. S., R. G. de Magalhaes, and A. M. Rui. 2015. Species composition and mortality of bats at the Osorio wind farm, southern Brazil. Studies on Neotropical Fauna and Environment, 50:31–39.

Bayat, S., F. Geiser, P. Kristiansen, and S. C. Wilson. 2014. Organic contaminants in bats: trends and new issues. Environment International, 63:40–52.

Bekker, J. P. 1996. Basisrapport zoogdierkundig onderzoek Aruba. Veere.

Beltrão Valença, R., and E. Bernard. 2015. Another blown in the wind: bats and the licensing of wind farms in Brazil. Natureza e Conservação, 13:117–122.

Bolívar-Cimé, B., A. Bolívar-Cimé, S. A. Cabrera-Cruz, O. Muñoz-Jiménez, and R. Villegas-Patraca. 2016. Bats in a tropical wind farm: species composition and importance of the spatial attributes of vegetation cover on bat fatalities. Journal of Mammalogy, 97:1197–1208.

Brunet, A. K., and R. A. Medellín. 2001. The species-area relationship in bat assemblages of tropical caves. Journal of Mammalogy, 82:1114–1122.

Burneo, S. F., M. D. Proaño, and D. G. Tirira, eds. 2015. Plan de acción para la conservación de los murciélagos del Ecuador. Programa para la Conservación de los Murciélagos del Ecuador y Ministerio del Ambiente del Ecuador, Quito.

Calisher, C. H., J. E. Childs, H. E. Field, K. V. Holmes, and T. Schountz. 2006. Bats: important reservoir hosts of emerging viruses. Clinical Microbiology Reviews, 19:531–545.

CARIBSS (Caribbean Speleological Society). 2016. Bonaire Cave and Karst Nature Reserve. http://www.caribss.org/en /bonaire/recreational-cave-park/.

Castro-Luna, A. A., and J. Galindo-González. 2011. Enriching agroecosystems with fruit-producing tree species favors the abundance and richness of frugivorous and nectarivorous bats in Veracruz, Mexico. Mammalian Biology, 77:32–40.

Clark, D. R., Jr. 1988. How sensitive are bats to insecticides? Wildlife Society Bulletin, 16:399–403.

Cordero-Schmidt, E., L. Navarro., and B. Rodríguez-Herrera. 2010. Clarita. Editorial Tirimbina, Sarapiquí, Costa Rica.

Cottam, M., J. Olynik, J. Blumenthal, K. D. Godbeer, J. Gibb, J. Bothwell, F. J. Burton et al. 2009. Cayman Islands National Biodiversity Action Plan 2009. Department of Environment, Cayman Islands Government.

Cryan, P. M., P. M. Gorresen, C. D. Hein, M. R. Schirmacher, R. H. Diehl, M. M. Huso, D. T. S. Hayman et al. 2014. Behavior of bats at wind turbines. Proceedings of the National Academy of Sciences, USA, 111:15126–15131.

de la Peña-Cuellar, E., J. Benítez-Malvido, L. D. Avila-Cabadilla, M. Martínez-Ramos, and A. Estrada. 2015. Structure and diversity of phyllostomid bat assemblages on riparian corridors in a human-dominated tropical landscape. Ecology and Evolution, 5:903–913.

Delfín, P., J. Ochoa-G., and A. Castillo. 2011. Santuario de Fauna Silvestre Cuevas de Paraguaná, Venezuela: lineamientos técnicos para su diseño. Terra, 2:13–45.

Diario Oficial de la Federación. 1994. NOM-059-ECOL-1994: Especies y subespecies de flora y fauna silvestres terrestres y acuáticas en peligro de extinción, amenazadas, raras y las sujetas a protección especial, y que establece especificaciones para su protección. Diario Oficial de la Federación, May 16, 1994.

Easterla, D. A. 1972. Status of Leptonycteris nivalis (Phyllostomatidae) in Big Bend National Park, Texas. Southwestern Naturalist, 17:287–292.

Emanuel, K. A. 2005. Increasing destructiveness of tropical cyclones over the past 30 years. Nature, 436:686–688.

FAO (Food and Agriculture Organization of the United Nations). 2009. State of the world's forests. Communication Division, FAO, Rome.

FAO. 2011. Investigating the role of bats in emerging zoonoses: balancing ecology, conservation and public health interests (S. H. Newman, H. E. Field, C. E. de Jong, and J. H. Epstein, eds.). FAO Animal Production and Health Manual No. 12. FAO, Rome.

Fearnside, P. M., W. F. Laurance, A. K. M. Albernaz, H. L. Vasconcelos, and L. V. Ferreira. 2005. Response [A delicate balance in Amazonia]. Science, 307:1045.

Fleming, T. H., and J. M. Nassar. 2002. Population biology of the lesser long-nosed bat, Leptonycteris curasoae, in Mexico and northern South America. Pp. 283–305 in: Columnar Cacti and Their Mutualists: Evolution, Ecology, and Conservation (T. H. Fleming and A. Valiente-Banuet, eds.) University of Arizona Press, Tucson.

Furey, N. M., and P. A. Racey. 2016. Conservation ecology of cave bats. Pp. 463–500 in: Bats in the Anthropocene: Conservation of Bats in a Changing World (C. C. Voigt and T. Kingston, eds.). Springer International Publishing, Cham, Switzerland.

Galarza, M. I., and L. F. Aguirre. 2011. Horacio y la leyenda de la espada Mágica. Biota-PCMB, Cochabamba.

Gannon, M. R., and M. R. Willig. 2009. Islands in the storm: disturbance ecology of plant-visiting bats on the hurricane-prone island of Puerto Rico. Pp. 281–301 in: Island Bats:

Evolution, Ecology, and Conservation (T. H. Fleming and P. A. Racey, eds.) University of Chicago Press, Chicago.

García-García, J. L., and A. Santos-Moreno. 2014. Efectos de la estructura del paisaje y de la vegetación en la diversidad de murciélagos filostómidos (Chiroptera: Phyllostomidae) de Oaxaca, México. Revista de Biología Tropical, 62:217–239.

García-Morales, R., L. Chapa-Vargas, E. I. Badano, J. Galindo-González, and K. Monzalvo-Santos. 2013. Evaluating phyllostomid bat conservation potential of three forest types in the northern Neotropics of Eastern Mexico. Community Ecology, 15:158–168.

Gareca, E., G. Rey-Ortiz, and L. F. Aguirre. 2007. Relación entre el conocimiento acerca de los murciélagos y las actitudes de cinco grupos sociales de Cochabamba. Pp. 99–103 in: Historia Natural, Distribución y Conservación de los Murciélagos de Bolivia (L. F. Aguirre, ed.). Fundación Simón I. Patiño, Santa Cruz, Bolivia.

Ghanem, S. H., and C. C. Voigt. 2012. Increasing awareness of ecosystem services provided by bats. Advances in the Study of Behavior, 44:279–302.

Giannini, N. P., and E. K. V. Kalko. 2004. Trophic structure in a large assemblage of phyllostomid bats in Panama. Oikos, 105:209–220.

Gomes, G. A., and F. A. Reid. 2015. The Bats of Trinidad and Tobago: A Field Guide and Natural History. Trinibats, [Trinidad and Tobago].

González-Terrazas, T. P., R. A. Medellín, M. Knoernschild, and M. Tschapka. 2012. Morphological specialization influences nectar extraction efficiency of sympatric nectar-feeding bats. Journal of Experimental Biology, 215:3989–3996.

Gorresen, P. M., M. R. Willig, and R. E. Strauss. 2005. Multivariate analysis of scale-dependent associations between bats and landscape structure. Ecological Applications, 15:2126–2136.

Grodsky, S. M., M. J. Behr, A. Gendler, D. Drake, B. D. Dieterle, R. J. Rudd, and N. L. Walrath. 2011. Investigating the causes of death for wind turbine-associated bat fatalities. Journal of Mammalogy, 92:917–925.

Hansen, M. C., P. V. Potapov, R. Moore, M. Hancher, S. A. Turubanova, A. Tyukavina, D. Thau et al. 2013. High-resolution global maps of 21st-century forest cover change. Science, 342:850–853.

Heywood, V. H., and R. T. Watson, eds. 1995. Global Biodiversity Assessment. Cambridge University Press, New York.

Hutson, A. M., S. P. Mickleburgh, and P. A Racey, eds. 2001. Global Action Plan for Microchiropteran Bats. IUCN, Gland, Switzerland.

IUCN. 2016. The IUCN Red List of Threatened Species. Version 2016–2. http://www.iucnredlist.org.

Jacobson, S. K., and M. D. McDuff. 1998. Conservation education. Pp. 237–255 in: Conservation Science and Action (W. J. Sutherlnad, ed.). Blackwell Science Ltd., Oxford.

Jacobson, S. K., D. McDuff, and M. Monroe. 2006. Conservation Education and Outreach Techniques. Oxford University Press, Oxford.

Jones, G., D. S. Jacobs, T. H. Kunz, M. R. Willig, and P. A. Racey. 2009. Carpe noctem: the importance of bats as bioindicators. Endangered Species Research, 8:93–115.

Jones, K. E., K. E. Barlow, N. Vaughan, A. Rodríguez-Durán, and M. R. Gannon. 2001. Short-term impacts of extreme environmental disturbance on the bats of Puerto Rico. Animal Conservation, 4:59–66.

Jones, K. E., S. P. Mickleburgh, W. Sechrest, and A. L. Walsh. 2009. Global overview of the conservation of island bats—importance, challenges, and opportunities. Pp. 496–530 in: Island bats: Evolution, Ecology and Conservation (T. H. Fleming and P. A. Racey, eds.). University of Chicago Press, Chicago.

Kannan, K., S. H. Yun, R. J. Rudd, and M. Behr. 2010. High concentrations of persistent organic pollutants including PCBs, DDT, PBDEs, and PFOS in little brown bats with white-nose syndrome in New York, USA. Chemosphere, 80:613–618.

Kingston, T. 2016. Cute, creepy, or crispy—how values, attitudes, and norms shape human behavior toward bats. Pp. 571–595 in: Bats in the Anthropocene: Conservation of Bats in a Changing World (C. C. Voigt and T. Kingston, eds.). Springer International Publishing, Cham, Switzerland.

Klingbeil, B. T., and M. R. Willig. 2009. Multi-scale responses of frugivorous and gleaning animalivorous bats to landscape composition and configuration in fragmented landscapes of lowland Amazonia. Journal of Applied Ecology, 46:203–213.

Koenig, S., and L. Dávalos. 2015. Phyllonycteris aphylla. The IUCN Red List of Threatened Species 2015: e.T17173A22133396. http://dx.doi.org/10.2305/IUCN.UK.2015-4.RLTS .T17173A22133396.en.

Konradsen, F., W. van der Hoekb, D. C. Cole, G. Hutchinson, H. Daisley, S. Singh, and M. Eddleston. 2003. Reducing acute poisoning in developing countries—options for restricting the availability of pesticides. Toxicology, 192:249–261.

Koopman, K. F. 1989. A review and analysis of the bats of the West Indies. Pp. 635–644 in: Biogeography of the West Indies: Past, Present, and Future (C. A. Woods, ed.). Sandhill Crane Press, Gainesville, FL.

Kozulj, R. 2010. La participación de las fuentes renovables en la generación de energía eléctrica: inversions y estrategias empresariales en América Latina y el Caribe. CEPAL, Naciones Unidas, Santiago de Chile.

Kunz, T. H., E. B. Arnett, W. P. Erickson, A. R. Hoar, G. D. Johnson, R. P. Larkin, M. D. Strickland, R. W. Thresher, and M. D. Tuttle. 2007. Ecological impacts of wind energy development on bats: questions, research needs, and hypotheses. Frontiers in Ecology and the Environment, 5:315–324.

Kunz, T. H., E. Braun de Torrez, D. M. Bauer, T. A. Lobova, and T. H. Fleming. 2011. Ecosystem services provided by bats. Pp. 1–38 in: The Year in Ecology and Conservation 2011 (R. A. Ostfeld, and W. H. Schlesinger, eds.). Annals of the New York Academy of Sciences, Wiley, New York.

Kuzmin, I. V., B. Bozick, S. A. Guagliardo, R. Kunkel, J. R. Shak, S. Tong, and C. E. Rupprecht. 2011. Bats, emerging infectious diseases, and the rabies paradigm revisited. Emerging Health Threats Journal, 4:7159.

Ladle, R. J., J. V. L. Firmino, A. C. M. Malhado, and A. Rodríguez-Durán. 2012. Unexplored diversity and conservation potential of Neotropical hot caves. Conservation Biology, 26:978–982.

Larsen, P. A., L. Siles, S. C. Pedersen, and G. G. Kwiecinskic. 2011.

A new species of *Micronycteris* (Chiroptera: Phyllostomidae) from Saint Vincent, Lesser Antilles. Mammalian Biology, 76:687–700.

La Val, R. K. 2004. Impact of global warming and locally changing climate on tropical cloud forest bats. Journal of Mammalogy, 85:237–244.

Lees, A. C., C. A. Peres, P. M. Fearnside, M. Schneider, and J. A. Zuanon. 2016. Hydropower and the future of Amazonian biodiversity. Biodiversity and Conservation, 25:451–466.

Lisboa, C. V., A. P. Pinho, H. M. Herrera, M. Gerhardt, E. Cupolillo, and A. M. Jansen. 2008. *Trypanosoma cruzi* (kinetoplastida, Trypanosomatidae) genotypes in Neotropical bats in Brazil. Veterinary Parasitology, 156:314–318.

Lizarro, D. M., I. Galarza, and L. F. Aguirre. 2010. Tráfico y comercio de murciélagos en Bolivia. Revista Boliviana de Ecología y Conservación Ambiental, 27:63–75.

Mancina, C. A. 2011. Introducción a los murciélagos. Pp. 123–133 *in*: Mamíferos de Cuba. (R. Borroto-Páez and C. A. Mancina, eds.) UPC Print, Vaasa, Finland.

Mancina, C. A., L. M. Echenique-Díaz, A. Tejedor, L. García, A. D. Álvarez, and M. A. Ortega-Huerta. 2007. Endemics under threat: an assessment of the conservation status of Cuban bats. Hystrix Italian Journal of Mammalogy, 18:3–15.

McNab, B. K. 1973. Energetics and the distribution of vampires. Journal of Mammalogy, 54:131–144.

Medellín, R. A. 2003. Diversity and conservation of bats in México: research priorities, strategies, and actions. Wildlife Society Bulletin, 31:87–97.

Medellín, R. A., A. Abreu-Grobois, M. C. Arizmendi, E. Mellink, E. Ruelas, E. Santana C., and J. Urbán. 2009. Conservacion de especies migratorias y poblaciones transfronterizas. Pp. 459–515 *in*: Capital Natural de México. Vol. 2: Estado de Conservación y Tendencias de Cambio (R. Dirzo, R. González, and I. J. March, comps.). Conabio, Mexico City.

Medellín, R. A., C. Chávez, A. de la Torre, H. Zarza, and G. Ceballos, eds. 2016. El Jaguar en el siglo XXI. Fondo de Cultura Económica e Instituto de Ecología, UNAM, Mexico City.

Medellín, R. A., M. Equihua, and M. A. Amin. 2000. Bat diversity and abundance as indicators of disturbance in Neotropical rainforests. Conservation Biology, 14:1666–1675.

Medellín, R. A., and L. Torres Knoop. 2013. Evaluación del Riesgo de extinción de *Leptonycteris yerbabuenae* de acuerdo al numeral 5.7 de la NOM-059-SEMARNAT-2010. Internal report to SEMARNAT, Instituto Nacional de Ecología, Mexico City.

Medellín, R. A., and S. Walker. 2003. Nightly wings, nectar sips. Endangered Species Bulletin, U. S. Fish and Wildlife Service, 28:16–17.

Meyer, C. F. J., and E. K. V. Kalko. 2008. Assemblage-level responses of phyllostomid bats to tropical forest fragmentation: land-bridge islands as a model system. Journal of Biogeography, 35:1711–1726.

Meyer, C. F. J., M. J. Struebig, and M. R. Willig. 2016. Responses of tropical bats to habitat fragmentation, logging, and deforestation. Pp. 63–103 *in*: Bats in the Anthropocene: Conservation of Bats in a Changing World (C.C. Voigt and T. Kingston, eds.). Springer International Publishing, Cham, Switzerland.

Mickleburgh, S. P., A. M. Hutson, and P. A. Racey. 2002. A review of the global conservation status of bats. Oryx, 36:18–34.

Muscarella, R., and T. H. Fleming. 2007. Seed dispersal by fruit bats: its importance in the conservation and regeneration of tropical forests. Biological Reviews, 82:573–590.

Nassar, J. M., and F. Simal. 2014. Studying bats and cultivating conservation on Aruba. Bats 32:10–12.

Navarro, L. 2001. Flores para Lucía, la Murciélaga. Bat Conservation International, University of Texas Press, Austin.

Navarro, L., J. Arroyo, and R. Medellín. 1996. Bat awareness in Mexico begins with children. Bats, 14 (3): 3–6.

NRC (National Research Council). 2001. Grand Challenges in Environmental Sciences. National Academy Press, Washington, DC.

NRC. 2003. Neon: Addressing the Nation's Environmental Challenges. National Academy Press, Washington, DC.

NRC. 2007. Environmental Impacts of Wind-Energy Projects. National Academy Press, Washington, DC.

O'Donnell, C. 2008. *Mystacina tuberculata*. The IUCN Red List of Threatened Species 2008: e.T14261A4427784. http://dx.doi.org/10.2305/IUCN.UK.2008.RLTS.T14261A4427784.en.

Olival, K. J., P. R. Hosseini, C. Zambrana-Torrelio, N. Ross, T. L. Bogich, and P. Daszak. 2017. Host and viral traits predict zoonotic spillover from mammals. Nature, 546:646–650. http://www.nature.com/doifinder/10.1038/nature22975.

O'Shea, T. J., and J. J. Johnston. 2009. Environmental contaminants and bats. Investigating exposure and effects. Pp. 500–528 *in*: Ecological and Behavioral Methods for the Study of Bats. 2nd ed. (T. H. Kunz and S. Parsons, eds.) Johns Hopkins University Press, Baltimore, Maryland.

Ossa, G., K. Vilchez, and P. Valladares. 2016. New record of the rare long-snouted bat, *Platalina genovensium* Tomas, 1928 (Chiroptera, Phyllostomidae), in the Azapa valley, northern Chile. Check List, 12:2 DOI: http://dx.doi.org/10.15560/12.2.1850.

Pari, A., K. Pino, C. E. Medina, E. López, and H. Zeballos. 2015. Murciélagos de Arequipa Historia Natural y Conservación. Natural History Museum of the National University of San Agustin, Arequipa, Peru.

Pedersen, S. C., H. H. Genoways, G. G. Kwiecinski, P. A. Larsen, and R. J. Larsen. 2013. Biodiversity, Biogeography, and Conservation of Bats in the Lesser Antilles. Mammalogy Papers: University of Nebraska State Museum. Paper 172. Museum, 172. DigitalCommons@University of Nebraska, Lincoln.

Petit, S., A. Rojer, and L. Pors. 2006. Surveying bats for conservation: the status of cave-dwelling bats on Curaçao from 1993 to 2003. Animal Conservation, 9:207–217.

Pimm, S. G., G. Russell, J. Gittleman, and T. Brooks. 1995. The future of biodiversity. Science, 269:347–350.

Pino, J., and R. Samudio. 2008. *Artibeus incomitatus*. The IUCN Red List of Threatened Species 2008: e.T2133A9271071.http://dx.doi.org/10.2305/IUCN.UK.2008.RLTS.T2133A9271071.en

PRDNER (Puerto Rico Department of Natural and Environmental Resources). 2015. Puerto Rico Wildlife Action Plan: Ten Year Review. Departamento de Recursos Naturales y Ambientales, San Juan, Puerto Rico.

Racey, P. A., and A. C. Entwistle. 2003. Conservation ecology of bats. Pp. 680–743 *in:* Bat Ecology (T. H. Kunz and M. B. Fenton, eds.). University of Chicago Press, Chicago.

RELCOM (Red Latinoamericana y del Caribe para la conservación de los Murciélagos). 2016a. Estrategia para la conservación de los murciélagos de Latinoamérica y el Caribe. RELCOM, Cochabamba, Bolivia. www.relcomlatinoamerica.net/images/PDFs/Estrategia.pdf.

RELCOM. 2016b. AICOMs y SICOMs. RELCOM, Cochabamba, Bolivia. www.relcomlatinoamerica.net/index.php/que-hacemos/conservacion/18-relcom/33-aicomsysicoms.

RELCOM. 2016c. Centrales eólicas. RELCOM, Cochabamba, Bolivia. www.relcomlatinoamerica.net/component/content/article/18-relcom/36-centrales-eolicas.html.

Rex, K., R. Michener, T. H. Kunz, and C. C. Voigt. 2011. Vertical stratification of Neotropical leaf-nosed bats (Chiroptera: Phyllostomidae) revealed by stable carbon isotopes. Journal of Tropical Ecology, 27:211–222.

Rodríguez, J. P., A. García-Rawlins, and F. Rojas-Suárez. 2015. Libro Rojo de la Fauna Venezolana. Provita y Fundación Empresas Polar, Caracas, Venezuela. http://animalesamenazados.provita.org.ve

Rodríguez-Durán, A. 1995. Metabolic rates and thermal conductance in four species of Neotropical bats roosting in hot caves. Comparative Biochemistry and Physiology, 110:347–355.

Rodríguez-Durán, A., and W. Feliciano-Robles. 2015. Impact of wind facilities on bats in the Neotropics. Acta Chiropterologica, 17:365–370.

Rodríguez-Durán, A., J. Pérez, M. Montalbán, and J. M. Sandoval. 2010. Predation by free-roaming cats on an insular population of bats. Acta Chiropterologica, 12:359–362.

Rodríguez-Herrera B., G. Ceballos, and R. A. Medellín. 2011. Ecological aspects of the tent building process by *Ectophylla alba* (Chiroptera: Phyllostomidae). Acta Chiropterologica, 13:365–372.

Rodríguez-Herrera, B., and R. Sánchez, eds. 2015. Estrategia centroamericana para la conservación de los murciélagos. Escuela de Biología, Universidad de Costa Rica, San José.

Rodríguez-Rocha, M., B. Rodríguez-Herrera, and S. Vilchez-Mendoza. 2017. Potential impacts of climate change on bats in the Neotropical rain forest. IUCN Bat Specialist Group Newsletter, 3:10–11.

Rollins, K. E., D. K. Meyerholz, G. D. Johnson, A. P. Capparella, and S. S. Loew. 2012. A forensic investigation into the etiology of bat mortality at a wind farm: barotrauma or traumatic injury? Veterinary Pathology, 49:362–371.

Rupprecht, C. E., C. A. Hanlon, and D. Slate. 2006. Control and prevention of rabies in animals: paradigm shifts. Developments in Biologicals, 125:103–111.

Sabino-Santos, G., Jr., F. Gonçalves Motta Maia, T. M. Vieira, R. de L. Muylaert, S. Miranda Lima, C. Barros Gonçalves, P. Doerl Barroso et al. 2015. Evidence of hantavirus infection among bats in Brazil. American Journal of Tropical Medicine and Hygiene, 93:404–406.

Sahley, C., and L. Baraybar. 1996. Natural history of the long-snouted bat, *Platalina genovensium* (Phyllostomatidae: Glossophaginae) in southwestern Peru. Vida Silvestre Neotropical 5:1–9.

Sala, O. E., F. S. Chapin III, J. J. Armesto, E. Berlow, J. Bloomfield, R. Dirzo, E. Huber-Sanwald et al. 2000. Global biodiversity scenarios for the year 2100. Science, 287:1770–1774.

Sánchez O., R. Medellín, A. Aldama, B. Goettsch, J. Soberón Mainero, and M. Tambutti. 2007. Método de Evaluación del Riesgo de extinción de las especies silvestres en México (MER). SEMARNAT, INE, Instituto de Ecología, UNAM, CONABIO, Mexico City.

Schneeberger, K., and C. C. Voigt. 2016. Zoonotic viruses and conservation of bats. Pp. 263–292 *in:* Bats in the Anthropocene: Conservation of Bats in a Changing World (C. C. Voigt and T. Kingston, eds.). Springer International Publishing, Cham, Switzerland.

Schneider, M. C., P. C. Romijn, W. Uieda, H. Tamayo, D. Fernandes da Silva, A. Belotto, J. Barbosa da Silva, and L. F. Leanes. 2009. Rabies transmitted by vampire bats to humans: an emerging zoonotic disease in Latin America? Revista Panamericana de Salud Pública, 25:260–269.

SEMARNAT (Secretaría de Medio Ambiente y Recursos Naturales). 2010. Norma Oficial Mexicana NOM-059-SEMARNAT-2010. Protección ambiental-Especies nativas de México de flora y fauna silvestres-Categorías de Riesgo y especificaciones para su inclusión, exclusión o cambio-Lista de especies en Riesgo. Diario Oficial de la Federación 30/12/2010, Mexico City.

Shi, Z. L. 2013. Emerging infectious diseases associated with bat viruses. Science China Life Science, 56:678–682.

Silva-Iturriza, A., J. M. Nassar, A. M. García-Rawlins, R. Rosales, and A. Mijares. 2013. *Trypanosoma evansi* kDNA minicircle found in the Venezuelan nectar-feeding bat, *Leptonycteris curasoae* (Glossophaginae), supports the hypothesis of multiple origins of that parasite in South America. Parasitology International, 62:95–99.

Simal, F., C. De Lannoy, L. García-Smith, O. Doest, J. A. De Freitas, F. Franken, I. Zaandam et al. 2015. Island-island and island-mainland movements of the Curaçaoan long-nosed bat, *Leptonycteris curasoae*. Journal of Mammalogy, 96:579–590.

Simmons, N. B. 2005. Order Chiroptera. Pp. 312–529 *in:* Mammal Species of the World (D. E. Wilson and D. M. Reeder, eds.). Johns Hopkins University Press, Baltimore.

Speer, K. A., J. A. Soto-Centeno, N. A. Albury, Z. Quicksall, M. G. Marte, and D. L. Reed. 2015. Bats of the Bahamas: natural history and conservation. Bulletin of the Florida Museum of Natural History, 53:45–95.

Stahlschmidt, P., and C. A. Brühl. 2012. Bats at risk? Bat activity and insecticide residue analysis of food items in an apple orchard. Environmental Toxicology and Chemistry, 31:1556–1563.

Streicker, D. G., S. Recuenco, W. Valderrama, J. Gomez Benavides, I. Vargas, V. Pacheco, R. E. Condori Condori et al. 2012. Ecological and anthropogenic drivers of rabies exposure in vampire bats: implications for transmission and control. Proceedings of the Royal Society B, 279:3384–3392.

Swanepoel, R. E., P. A. Racey, R. F. Shore, and J. R. Speakman. 1998. Energetic effects of sublethal exposure to lindane on pipistrelle bats (*Pipistrellus pipistrellus*). Environmental Pollution, 104:169–177.

Taylor, M. L., C. B. Chávez-Tapia, R. Vargas-Yañez, G. Rodríguez-Arellanes, G. R. Peña-Sandoval, C. Toriello, A. Pérez, and M. R. Reyes-Montes. 1999. Environmental conditions favoring bat infection with *Histoplasma capsulatum* in Mexican shelters. American Journal of Tropical Medicine and Hygiene, 61:914–919.

Torres Knoop, L. 2014. Refugio de apareamiento de *Leptonycteris nivalis*: modelación y búsqueda de un recurso limitante. Facultad de Ciencias, UNAM, Mexico City.

Trejo-Salazar, R. E., L. E. Eguiarte, D. Suro-Piñera, and R. A. Medellín. 2016. Save our bats, save our tequila: industry and science join forces to help bats and agaves. Natural Areas Journal, 36:523–530.

Tschapka, M., E. Sperr, L. A. Caballero, and R. A. Medellín. 2008. Diet and cranial morphology of *Musonycteris harrisoni*, a highly specialized nectar-feeding bat in western Mexico. Journal of Mammalogy, 89:924–932.

Turner, G. G., D. M. Reeder, and J. T. H. Coleman. 2011. A five-year assessment of mortality and geographic spread of white-nose syndrome in North American bats and a look to the future. Bat Research News, 52:13–27.

Tuttle, M. D. 2017. Fear of bats and its consequences. Journal of Bat Research and Conservation, 10.

U.S. Fish and Wildlife Service. 1994. Mexican Long-Nosed Bat (*Leptonycteris nivalis*) Recovery Plan. U.S. Fish and Wildlife Service, Albuquerque, NM.

U.S. Fish and Wildlife Service. 1995. Lesser Long-Nosed Bat Recovery Plan. U.S. Fish and Wildlife Service, Albuquerque, NM.

U.S. Fish and Wildlife Service. 2016. Endangered Species. http://www.fws.gov/endangered/.

Vargas, A., K. Barboza, and L. F. Aguirre. 2009. *Lonchorhina aurita* Tomes, 1963. Pp. 467–468 *in*: Libro Rojo de la Fauna Silvestre de Vertebrados de Bolivia (L. F. Aguirre, R. Aguayo, J. Balderrama, C. Cortez, and T. Tarifa, eds.). Ministerio de Medio Ambiente y Agua, La Paz, Bolivia.

Veiga, M. M., G. Angeloci, M. Hitch, and P. C. Velásquez-López. 2014. Processing centres in artisanal gold mining. Journal of Cleaner Production, 64:535–544.

Velásquez-López, P. C., M. M. Veiga, B. Klein, J. A. Shandro, and K. Hall. 2011. Cyanidation of mercury-rich tailings in artisanal and small-scale gold mining: identifying strategies to manage environmental risks in southern Ecuador. Journal of Cleaner Production, 19:1125–1133.

Voigt, C. C., and T. Kingston. 2016. Bats in the Anthropocene. Pp. 1–12. *in*: Bats in the Anthropocene: Conservation of Bats in a Changing World (C. C. Voigt and T. Kingston, eds.). Springer International Publishing, Cham, Switzerland.

Voigt, C. C., K. L. Phelps, L. F. Aguirre, M. C. Schoeman, J. Vanitharani, and A. Zubaid. 2016. Bats and buildings: the conservation of synanthropic bats. 427–462 *in*: Bats in the Anthropocene: Conservation of Bats in a Changing World (C. C. Voigt and T. Kingston, eds.). Springer International Publishing, Cham, Switzerland.

Voss, R. S., D. W. Fleck, R. E. Strauss, P. M. Velazco, and N. B. Simmons. 2016. Roosting ecology of Amazonian bats: evidence for guild structure in hyperdiverse mammalian communities. American Museum Novitates, 3870:1–43.

Wade, T. G., K. H. Riitters, J. D. Wickham, and K. B. Jones. 2003. Distribution and causes of global forest fragmentation. Conservation Ecology, 7:7.

Walther, G. R., E. Post, P. Convey, A. Menzel, C. Parmesan, T. J. C. Beebee, J. M. Fromentin, O. Hoegh-Guldberg, and F. Bairlein. 2002. Ecological responses to recent climate change. Nature, 416:389–395.

Wilkinson, G. S., and T. H. Fleming. 1996. Migration and evolution of lesser long-nosed bats *Leptonycteris curasoae*, inferred from mitochondrial DNA. Molecular Ecology, 5:329–339.

Williams-Guillén, K., and I. Perfecto. 2010. Effects of agricultural intensification on the assemblage of leaf-nosed bats (Phyllostomidae) in a coffee landscape in Chiapas, Mexico. Biotropica, 42:605–613.

Willig, M. R., S. J. Presley, C. P. Bloch, and H. H. Genoways. 2009. Macroecology of Caribbean bats: effects of area, elevation, latitude, and hurricane-induced disturbance. Pp. 216–264 *in*: Island Bats: Evolution, Ecology, and Conservation (T. H. Fleming and P. A. Racey, eds.). University of Chicago Press, Chicago.

Willig, M. R., S. J. Presley, C. P. Bloch, C. L. Hice, S. P. Yanoviak, M. M. Díaz, L. A. Chauca, V. Pacheco, S. C. Víctor, and S. C. Weaver. 2007. Phyllostomid bats of lowland Amazonia: effects of habitat alteration on abundance. Biotropica, 39:737–746.

Wilson, D. E. 1985. Status report: *Leptonycteris nivalis* (Saussure) Mexican long-nosed bat. U.S. Fish and Wildlife Service, Denver Wildlife Research Center, National Museum of Natural History, Washington, DC.

Wind Power. 2014. Wind turbines and wind farms database. http://www.thewindpower.net/.

Zarza, H., E. Martínez-Meyer, G. Suzán, and G. Ceballos. 2017. Geographic distribution of *Desmodus rotundus* in Mexico under current and future climate change scenarios: implications for bovine paralytic rabies infection. Veterinaria Mexico, 4 (3). DOI: http://dx.doi.org/10.21753/vmoa.4.2.390.

Zukal, J., J. Pikula, and H. Bandouchova. 2015. Bats as bioindicators of heavy metal pollution: history and prospect. Mammalian Biology, 80:220–227.

CONTRIBUTORS

Danielle M. Adams, Department of Biology, University of Maryland, College Park, dadams37@gmail.com

Ludmilla M. S. Aguiar, Laboratório de Biologia e Conservação de Morcegos, Departamento de Zoologia, Universidade de Brasília, Campus Darcy Ribeiro, ludmillaaguiar@unb.br

Luis F. Aguirre, Centro de Biodiversidad y Genética, Universidad Mayor de San Simón Casilla, Bolivia, laguirre@fcyt.umss.edu.bo

Justine J. Allen, Department of Ecology and Evolutionary Biology, Brown University, justine_allen@brown.edu

Lucila I. Amador, Unidad Ejecutora Lillo, CONICET-Fundación Miguel Lillo, Tucumán, Argentina, amadorlucila@gmail.com

Robert M. R. Barclay, Department of Biological Sciences, University of Calgary, barclay@ucalgary.ca

Enrico Bernard, Laboratório de Ciência Aplicada a Conservação dea Biodiversidade, Departament de Zoologia, Universidade Federal de Pernambuco, Brasil, enrico.bernard@ufpe.br

Signe Brinkløv, Department of Biology, University of Southern Denmark, signe@biology.sdu.dk

Gerald G. Carter, Smithsonian Tropical Research Institute, Balboa, Panama, ggc.bats@gmail.com

Andrea L. Cirranello, Department of Mammalogy, Division of Vertebrate Zoology, American Museum of Natural History, andreacirranello@gmail.com

Ariovaldo P. Cruz-Neto, Department of Zoology, São Paulo State University, Campus de Rio Claro, ariovaldopcruz@gmail.com

Nicolas J. Czaplewski, Oklahoma Museum of Natural History, University of Oklahoma, nczaplewski@ou.edu

Liliana M. Dávalos, Department of Ecology and Evolution, Stony Brook University, lmdavalos@gmail.com

Mariana Delgado-Jaramillo, Laboratório de Ciência Aplicada a Conservação dea Biodiversidade, Departament de Zoologia, Universidade Federal de Pernambuco, Brasil, marianadelgado13@yahoo.es

M. May Dixon, Department of Integrative Biology, University of Texas at Austin, marjorie.m.dixon@gmail.com

Elizabeth R. Dumont, School of Natural Sciences, University of California, Merced, edumont@ucmerced.edu

Sergio Estrada-Villegas, Department of Biological Sciences, Marquette University, estradavillegassergio@gmail.com

Theodore H. Fleming, Emeritus Professor of Biology, University of Miami, tedfleming@dakotacom.net

Inga Geipel, Smithsonian Tropical Research Institute, Balboa, Panama, inga.geipel@gmail.com

Norberto P. Giannini, Unidad Ejecutora Lillo, CONICET-Fundación Miguel Lillo, Tucumán, Argentina, and Facultad de Ciencias Naturales e Instituto Miguel Lillo, Universidad Nacional de Tucumán, norberto@amnh.org

Gregg F. Gunnell,* Division of Fossil Primates, Duke Lemur Center

Claire T. Hemingway, Department of Integrative Biology, University of Texas at Austin, cheming@utexas.edu

John W. Hermanson, Department of Biomedical Sciences, College of Veterinary Medicine, Cornell University, jwh6@cornell.edu

L. Gerardo Herrera M., Estación de Biología de Chamela, Instituto de Biología, Universidad Nacional Autónoma de México, gherrera@ib.unam.mx

Ricardo B. Machado, Laboratório de Planejamento Sistemático para a Conservação, Departamento de Zoologia, Universidade de Brasília, Campus Darcy Ribeiro, rbmac@unb.br

Angela M. G. Martino, Centro de Investigaciones en Ecología y Zonas Aridas (CIEZA), Universidad Nacional Experimental Francisco de Miranda, Venezuela, amg.martino@gmail.com

Rodrigo A. Medellín, Instituto de Ecología, Universidad Autónoma de México, medellin@iecologia.unam.mx

Marco A. R. Mello, Department of Ecology, Institute of Biosciences, University of São Paulo, marmello@usp.br

Reyna Leticia Moyers Arévalo, Unidad Ejecutora Lillo, CONICET-Fundación Miguel Lillo, Tucumán, Argentina, laettitia@gmail.com

Nathan Muchhala, Department of Biology, University of Missouri–Saint Louis, n_muchhala@yahoo.com

Renata L. Muylaert, Department of Ecology, Biosciences Institute, State University of São Paulo, renatamuy@gmail.com

Jafet M. Nassar, Centro de Ecologia, Instituto Venezolano de Investigaciones Científicas, Venezuela, jafet.nassar@gmail.com

Christopher Nicolay, Department of Biology, University of North Carolina, cnicolay@unca.edu

Rachel A. Page, Smithsonian Tropical Research Institute, Balboa, Panama, PageR@si.edu

John M. Ratcliffe, Department of Biology, University of Toronto Mississauga, j.ratcliffe@utoronto.ca

Armando Rodríguez-Durán, Departamento de Ciencias Naturales, Universidad Interamericana, Bayamón, Puerto Rico, arodriguez@bayamon.inter.edu

Bernal Rodríguez-Herrera, Escuela de Biología, Universidad de Costa Rica, bernal.rodriguez@ucr.ac.cr

Danny Rojas, Department of Natural Sciences and Mathematics, Pontifica Universidad Javeriana Cali, Colombia, rojasmartin.cu@gmail.com

Stephen J. Rossiter, School of Biological and Chemical Sciences, Queen Mary University of London, s.j.rossiter@qmul.ac.uk

Romeo A. Saldaña-Vazquez, Facultad de Ciencias Biológicas, Benemérita Universidad Autónoma de Puebla, México, romeo.saldana@gmail.com

Nancy B. Simmons, Department of Mammalogy, Division of Vertebrate Zoology, American Museum of Natural History, simmons@amnh.org

Richard D. Stevens, Department of Natural Resources Management, Texas Tech University, Richard.Stevens@ttu.edu

Sharon M. Swartz, Department of Ecology and Evolutionary Biology, Brown University, sharon_swartz@brown.edu

Jeneni Thiagavel, Department of Biology, University of Toronto Mississauga, jeneni.thiagavel@utoronto.ca

Marco Tschapka, Institute for Evolutionary Ecology and Conservation Genomics, University of Ulm, Germany, marco.tschapka@uni-ulm.edu

Paúl M. Velazco, Division of Paleontology, American Museum of Natural History, pvelazco@amnh.org

Fabricio Villalobos, Laboratório de Macroecologia Evolutiva y Red de Biología Evolutiva, Instituto de Ecología, México, and Laboratorio de Ecologia Teórica e Síntese, Instituto de Ciências Biológicas, Universidade Federal de Goiás, Campus Samambaia, Brasil, fabricio.villalobos@gmail.com

Marcelo M. Weber, Laboratório de Vertebrados, Departamento de Ecologia, Universidade Federal De Rio De Janeiro, Brasil, mweber.marcelo@gmail.com

Gerald S. Wilkinson, Department of Biology, University of Maryland, wilkinso@umd.edu

* Deceased.

SUBJECT INDEX

abductor pollicis brevis, 258

acoustic cues, 30, 124, 188–89, 197–98, 241, 246, 264, 278. *See also* object cues; olfaction: cues; spatial: cues; vision: cues

adaptations, 3–4, 13, 25, 36–37, 100, 106–7, 113, 115–16, 118, 152, 195, 239–40, 247–49, 257, 267, 273–78

adaptive radiation, 3–4, 7, 13, 18, 105–7, 109, 113, 114, 118, 169, 257

aerial hawking, 5, 12, 17, 29, 31, 34–35, 124, 134, 194, 241. *See also* hawking

aerial insectivores, 12, 198, 216, 241, 251. *See also* gleaning foraging mode

aerodynamic control, 152–56, 162–63, 199, 294; aerodynamic forces, 152, 158, 161–62

Africa, 9, 13, 18, 27, 36–37, 106, 248, 437; African, 3, 13, 36

African Rift Valley cichlids, 3

Agouti Cave, Grand Cayman, 72

AIC. *See* Akaike information criterion

Akaike information criterion (AIC), 107, 116, 224

allometric trends, 171, 173, 176, 248

allopatry, 30, 135. *See also* sympatry

Amazon, 8, 96, 98, 263, 267, 326, 355, 393, 403, 436–37, 441, 446

Amazon Basin, 8, 326, 393, 441

Amazon rainforest, 145

American Museum of Natural History, 47–48, 51, 87, 223

Americas, 3–4, 20, 52, 83, 354, 383, 438, 449

aminopeptidase, 113, 170–71

ammonia, 319

amphibians, 99, 135, 398

ANCOVA, 212

Anderson, Knud, 47

Andes, 8, 14, 18, 33, 97–100, 287, 350, 393, 446

angiosperms, 15–19

animal-eating, 189

animalivorous, 37, 66–67, 91, 111, 114, 129, 171, 174, 177, 179, 181, 211–12, 240–41, 246–51, 328–29, 341, 353

anisotropy, 152

annual reproductive output, 206, 212

Antarctica, 8–9, 36

anthesis, 284

anthropostructures, 318, 320

anticoagulant, 267–68, 429

Antigua, 68–72

Antilles, 9, 17, 29, 68, 71, 78–79, 96–97. *See also* Greater Antilles; Lesser Antilles

apomorphic characters and states, 34, 36, 126, 131–32

Arctic, 8

area under the curve (AUC), 414

Argentina, 7, 64, 66–68, 257, 264–65, 275, 301, 329, 348, 391, 438, 441, 451

Arizona, 69, 208, 337–38, 443

Arkenstone Cave, Arizona, 69

arm bands, 326

armwing skin, 155. *See also* plagiopatagium

Arredondo 2A, Florida, 69

Aruba, 446–48

Ashton Cave, Bahamas, 71–72

aspect ratio, 12, 123, 244–45, 249, 282–83, 334–35

Asunción, 74, 76

asynchronous reproduction, 231

Atlantic Forest, 4, 50, 96, 303, 335, 355, 383, 396, 401, 403, 413, 421–22, 425, 430

AUC. *See* area under the curve

auditory, 187–89, 191, 193, 195, 197–99, 247–48

Australasia, 13, 25–26, 28–29, 31, 36–37, 171, 250, 312, 437

Australian marsupials, 3

Azara, Felix, 45

Bahamas, 13, 68, 71–72, 79, 392

Bahia, Brazil, 74–76

Baja California, 207, 337

Baker, Robert J., 4, 373

Baker et al. 2000 paper, 88–89, 91–92; Baker et al. 2003 paper, 10, 12, 52, 89–90, 92, 94–95, 99, 113

banana, 190, 288, 444, 447

Banana Hole, Bahamas, 71–72

bandwidth, 124, 191, 195–96

Barbados, 68, 71

barbets, 18

Barro Colorado, Panama, 327, 335, 340, 354, 379

basal metabolic rate (BMR), 100, 144, 170–71, 180–85

bat fruits, 15, 18, 195–96, 298–99, 303
Bat Specialist Group, 44. *See also*
    International Union for Conservation
    of Nature and Natural Resources
bat-friendly, 451
bat-fruit interactions, 303–4. *See also*
    bat-plant interactions
bat-plant interactions, 18, 288, 381
bat-plant networks, 375–76, 381
Bayamon, Puerto Rico, 71
Bayesian analysis, 89, 92, 95, 98, 107, 112
Bay Point Cliff, California, 69
beetles, 13, 67, 240, 251, 336
behavioral flexibility, 199, 245, 247
Belize, 50, 68–70, 264–65
Big and Little Exuma Islands, 71
bimodal polyestry, 206, 208–9, 331,
    339–40
BioGeoBEARS, 96
biological catapults, 160. *See also* tendons
biomechanics, 31, 151, 163
biota, 7–8, 182, 405, 424
birds, 5, 10, 13–14, 17–18, 28, 36, 45, 47,
    66, 76, 98–99, 115, 151–52, 162, 175–
    76, 179, 197, 202, 212, 230, 239–40,
    242, 246–48, 250–51, 257, 262–63,
    276–78, 281, 283–84, 295–96, 317–
    18, 325, 336, 349–50, 359, 362, 381,
    398, 428, 430
births, 205–15, 222, 231, 265–66, 281,
    331, 336–40, 352, 359; birth-death
    dynamics, 110; birth mass, 211–12,
    215; birth synchrony, 210
bite force, 109, 111, 118, 196, 260, 276–77,
    296–98
blood glucose level, 116
BMR. *See* basal metabolic rate
Bodden Cave, Grand Cayman, 71–72
body mass, 37, 115–16, 123–33, 135–43,
    152, 154, 161, 170, 175, 182, 189, 196,
    199, 206, 208, 212–15, 222–29, 232–
    33, 260–61, 263, 283, 296, 298–99,
    326. *See also* body size
body size, 5, 11–12, 17, 66–67, 115–16,
    123–36, 143, 163, 191, 205–7, 210,
    214–15, 222–26, 230, 240–43, 248,
    250, 281–82, 331, 334, 348–49, 353.
    See also body mass
Bolivia, 35, 301, 320, 354, 357, 443,
    445–46, 448
Bonaire, 440, 447–48, 450
Botero-Castro et al. 2013 paper, 89, 91,
    94–95

Brazil, 4, 8, 33, 35, 44–48, 50, 68, 76, 96,
    209, 216, 257, 267, 301, 303, 327, 329,
    335, 339, 349, 354–57, 360, 362, 383,
    403, 413–15, 421–30, 439–40, 445,
    448; legislation, 425–26
British Museum, 44, 46–48
broadband FM (frequency modulated),
    33–34
Brownian motion model, 106, 171, 224
Brownian motion rate, 106, 112, 171
Burma Quarry, Antigua, 71–72
butterflies, 13, 99, 276

$^{13}C/^{12}C$ ratio, 54
Caatinga, 421–22, 424, 430
Cadena-G., Alberto, 48
California, 64, 70, 207, 257, 338
call wavelength, 195
Camagüey, Cuba, 71–73
Canada, 8, 441, 451
canine size, 229, 233
Canonical Phylogenetic Ordination,
    126–28
captivity, 4, 31, 182, 210, 212–13, 215,
    260, 263, 265–66, 278, 331
carbohydrate, 130, 171, 175–76, 273, 278
carbon dioxide, 8–9, 176
Caribbean islands, 4, 9, 68, 97, 338, 388,
    441, 449
carnivores, 13, 14, 66, 79, 94, 106, 114,
    123–24, 135–36, 196, 210–11, 222,
    240–41, 248–49, 254–55, 295
Carter, Dilford, 4
caulicarpy, 295
Cayman Brac, 71–72
Cebada Cave, Belize, 69–70
Cenozoic, 4, 7–11, 18, 64, 77
Centinela del Mar, 74
Central America, 8, 14, 33, 64–65, 68–
    69, 76–79, 96–99, 351, 383, 442–45,
    448–49
Central and South America, 7, 9, 32–33,
    44, 77, 97, 438–39
Central Andes, 14, 99
Chacaljas cave, Mexico, 70
Chichen-Itza Natural Water Well,
    Mexico, 70
Chile, 33, 439, 445, 448, 451
Chinandega, Realejo, Nicaragua, 70
chiropterophily, 283–87, 291
chitin, 173
chitinase, 173, 185–86
Chocó, 97–99

cicadas, 13
CIPRES, 51
cladogenesis, 35, 90, 93, 98–100, 133, 364
Clare et al. 2011 paper, 52
clearwing butterflies, 99
climate, 8, 18, 37, 43, 79, 95–97, 134,
    205–6, 210, 249, 303, 320, 350, 354–
    55, 362–63, 365, 392, 395, 402–3,
    440, 449
cluttered, 13, 193–94, 199, 244–45, 298
coevolution, 18, 171
cognitive ecology, 188, 199; cognitive
    strategies, 246
Coleby Bay Cave, Bahamas, 71–72
Colima, Mexico, 48
Colombia, 33, 48, 50, 64, 66, 68, 74, 94,
    327, 355, 393, 403, 445, 448
columnar cacti, 14, 208–9, 280–81, 283,
    287, 337, 445
communal roosts, 32. *See also* day roosts;
    ephemeral roosts; maternity roosts;
    mating roost
competition, 13, 18, 36–37, 78, 105,
    221–23, 227–33, 267, 281–83, 288,
    303, 348–50, 352, 363, 404
computed tomography (CT), 51–52
"concentrated" population, 329, 341
Conch Bar Cave, Bahamas, 71–72
Condoto, Colombia, 74
cones, 107, 189, 280. *See also* rods
controlling for phylogeny, 113, 182, 188,
    225
Convention on the Conservation of
    Migratory Species of Wild Animals,
    428
convergent, 2, 26, 29–31, 36, 50, 88, 92,
    95, 106–7, 110–13, 115, 117–18, 241,
    274, 284
coronaviruses, 5, 383–84, 430, 438
Costa Rica, 3, 17, 50, 209–10, 223,
    264–65, 282, 300–301, 303, 320, 327,
    329, 339–40, 354, 363, 440, 444–45,
    448, 450
Crab Cave, Grand Cayman, 71
Cretaceous, 8, 10, 18
critically endangered, 36, 441–46
crown (clade), 14, 16, 29, 33, 67, 79, 124,
    127, 133
crown (dental), 261
Cuajiniquil cave, Costa Rica, 339
Cuba, 35, 47, 68, 71–73, 78, 97, 135,
    208, 341, 437, 440, 447–48; Cuban
    National Museum, 47

Cueva Catedral, Puerto Rico, 71–73

Cueva de Centenario de Lenin, Cuba, 71–73

Cueva de Clara, Puerto Rico, 71–73

Cueva de El Abra, Mexico, 69–70

Cueva de la Brújula, Venezuela, 74

Cueva de La Presita, Mexico, 69–70

Cueva del Centenario de Lenin, Cuba, 71–73

Cueva del Guácharo, Venezuela, 74

Cueva del Jagüey, Cuba, 71–72

Cueva del Perro, Puerto Rico, 71–73

Cueva de los Indios, Cuba, 73

Cueva de los Masones, Cuba, 71–72

Cueva de Los Murciélagos, Venezuela, 74

Cueva de Paredones, Cuba, 71–73

Cueva de Quebrada Honda, Venezuela, 74–76

Cueva de Silva, Puerto Rico, 71

Cueva El Abron, Cuba, 73

Cueva GEDA, Cuba, 73

Cueva Grande de Judas, Cuba, 71–73

Cueva Lamas, Cuba, 71

Cueva La Presita, Mexico, 69

Cueva Monte Grande, Puerto Rico, 71–73

Cueva San Josecito, Mexico, 69. *See also* San Josecito Cave

Cueva Tenebrosa, Cuba, 72–73

Cuevas Blancas, Cuba, 71

Culebra Island, Puerto Rico, 73

culling, 267–68

Curaçao, 71, 207–8, 215, 438, 440, 446–47

cyclones, 446, 452. *See also* volcanic activity

cytb (cytochrome b), 89, 94, 98–99, 109, 113, 118, 333–34. *See also* mitochondrial genes

Daiquiri Cave, Cuba, 71–73

Dairy Cave, Jamaica, 71–73

Darwin's finches, 3, 106

Datzmann et al. 2010 paper, 91, 94–95

day roosts, 283, 299, 317, 325–26, 328, 337, 341. *See also* communal roosts; ephemeral roosts; maternity roosts; mating roost

deforestation, 5, 19, 31, 250, 262, 317, 335, 356, 383, 422–23, 427, 429, 437, 444–45

delayed embryonic development, 210–11, 231

dental characters, 65–68, 89, 95, 112–13, 259

dentition, 32, 35, 66–68, 248, 259–60, 273, 275–76, 280

deserts, 8–9, 18, 275, 280, 285, 287–88

Desmodus salivary plasminogen activator (DSPA), 261

desmokinase, 261

desmolaris, 261

dietary breadth, 115, 170, 180, 242, 250, 257, 275

dietary trends, 171, 251

"diffuse" populations, 326, 329, 341

digestive capacity, 109, 115, 118, 183, 298–99

digestive system, 261, 349

Digital Morphology, 52

dimorphism, 221–33, 336

Diquini Cave, Haiti, 73

disease emergence, 382

dispersal: extinction-cladogenesis analysis, 93, 98–99; vicariance analysis, 98–99

distance-sensing systems, 187. *See also* acoustic cues; olfaction: cues; vision: cues

diurnal, 16, 278, 318, 446

divergence, 3, 29, 94, 96–99, 103, 111, 125, 130–31, 134–36, 260–61, 358, 397

diversity gradients, 89, 396–98, 401–2, 405

Dobson, George Edward, 47

Dolphin Cave, Grand Cayman, 72

Dominican Republic, 4, 447

draculin, 261

drift, 350–53, 356, 358–60, 365–66, 380, 384

DRYAD database, 51

DSPA. See *Desmodus* salivary plasminogen activator

ear, ears, 29–30, 44, 188, 198, 242–43, 248–49, 252, 254, 439

Early Holocene, 69

Early Oligocene, 35, 63–64, 94

Earth, 3, 8–9, 18, 336, 347, 363, 405

earth history, 99–100. *See also* Early Holocene; Early Oligocene; Eocene; Holocene; Late Pleistocene; Miocene; Oligocene; Paleocene; Pleistocene; Pliocene

East Cave, Bahamas, 71–72

Ebola virus, 383–84

echolocation, 11, 13, 30–32, 34–35, 123–24, 133, 135, 188–99, 241–42, 246–47, 263–64, 278–79, 281, 285, 307, 440

ecological niche modeling (ENM), 402–5, 444

ecological theory of adaptive radiation, 105, 113. *See also* adaptive radiation

ecosystem services, 4–5, 18, 382, 384, 435, 447

ecotones, 14

ectoloph, 67

ectoparasites, 13, 232, 316, 376

Ecuador, 33, 48, 50, 354, 393, 403, 441, 445, 448

Edwards, George, 44–45

El Dudu, Hispaniola, 71–72

elevation gradients, 146, 349–50, 355, 368, 370, 388, 392–94

elongated jaws, 276

Eltonian shortfalls, 373, 383

enamel, 260

endangered species, 32, 36, 380, 403, 416, 420, 422, 428, 441–46

Eocene, 7–10, 26–30, 63–64, 66, 77, 79, 133–34, 359

ephemeral roosts, 33, 221, 230, 329–30. *See also* communal roosts; day roosts; maternity roosts; mating roost

epiphytes, 14, 32, 131, 282, 296

Espirito Santo, Brazil, 74

Eurobats (Conservation of Populations of European Bats), 428

European bats, 27, 44, 326

European introductions, 262, 320

European museums, 47–48

European scientists, 44–45

evolutionary radiation, 5. *See also* adaptive radiation

excretory physiology, 169–70, 181–82

excretory system, 170, 176, 178–79

extinction, 37, 78–79, 105, 134, 275, 317, 332–33, 352, 360–61, 375, 381, 427, 429, 440–41, 443, 446; rates, 96–97, 359, 398; risk, 249–50, 359, 427; threshold, 362–63. *See also* local extinction

false positives, 421

field metabolic rate (FMR), 180

Field Museum, 48

firmicutes, 118. *See also* microbiome

fishing, 34, 199

fission-fusion, 265, 380

flagellicarpy, 286, 295

flexor pollicis brevis, 258

flight speed, 124, 154, 161–62, 258, 264

floral morphology, 285

Florida, 25, 29, 35, 69, 94, 97, 257

FM (echolocation), 31–35, 191, 196, 279. *See also* frequency-modulated calls

FMR. *See* field metabolic rate

food abundance, 354

food sharing, 266–67

foraging behavior, 11, 29, 195, 199, 241, 243, 249, 264, 287–88, 300, 336, 341; mode, 5, 12–13, 18, 134, 278; selection-assimilation, 298; territory, 281

fossils, 19, 21, 25–29, 33, 35–41, 58, 63–68, 77–86, 94–95, 97, 99–103, 105, 119–21, 129, 144–46, 183, 199, 252, 257, 270, 289, 458

fragmentation, 5, 250, 288, 336, 342, 350, 356–58, 362–63, 379, 392, 421, 427, 429–30, 436, 441, 448

France, 27

French Guiana, 4, 17, 327

frequency-modulated calls, 13, 30, 32, 191. *See also* echolocation; FM (echolocation)

frog predation, 247–48

fructose, 115, 169, 173–74, 176, 284

frugivores, 4, 10, 13, 17–18, 50, 116–17, 127, 131–34, 136, 172, 175, 182, 190, 194–95, 197, 210, 212, 215–16, 239, 243–44, 246, 286, 295, 297–98, 302–3, 327–32, 334–35, 341–42, 354–57, 427, 436

frugivory, 17–18, 30, 68, 106, 114, 117, 170–71, 188, 205, 274–76, 303, 375, 378, 381–82

GABI. *See* Great American Biotic Interchange

Galapagos finches, 106. *See also* Darwin's finches

gammaproteobacteria, 118

Garivaldino Archaeological Site, Brazil, 75

garlic scent, 279, 285

Gatun Lake, Panama, 329, 335

GBIF. *See* Global Biodiversity Information Facility

GenBank, 52

generalist, 195, 199, 239–41, 247–51, 281, 298, 318, 360, 421, 440

Geoffroy, Étienne, 45

geographic dispersal, 9–10, 26, 28, 37, 64, 77, 90, 96, 99–100, 135, 250, 267, 325–26, 331–32, 334, 338, 342, 350–53, 356, 360–61, 363–66, 379–80, 393, 396, 398

Gervais, Paul, 47

gestation, 206, 208, 210–13, 215, 265

gleaning foraging mode, 12–13, 18, 33–34, 37, 88, 134, 195, 198–99, 241, 245, 249, 251, 354, 356–57

Global Biodiversity Information Facility (GBIF), 51, 414

glucose, 109, 115–17, 169, 173–76, 183, 266, 278, 284

GLUT2 (facilitated glucose transporter), 174

GLUT5 (facilitated fructose transporter), 174

Golden Age, 52

Goldfuss, Georg August, 46

Gondwana, 27–28, 36

Google Scholar, 333, 377

GPS telemetry, 384

Grand Cayman, 71–72

Gray, John Edward 46. *See also* British Museum

Great American Biotic Interchange (GABI), 64, 68, 76–77, 99

Great Britain, 27

Great Exuma Island, Bahamas, 71

Greater Antilles, 9, 12, 17, 35, 68, 79, 134, 208, 361, 392, 446. *See also* Cuba; Hispaniola; Jamaica; Puerto Rico

Grenada, 99

Griffiths 1982 paper, 48, 50

guilds, 11, 114, 135, 170–71, 240–42, 244–47, 250–51, 279–80, 282, 302, 318, 348, 355, 357–58, 362, 371, 393, 436

"gulpers," 298–300. *See also* "mashers"

Gundlach, Juan, 47

Habana, Cuba, 73

habitat destruction, 19, 342

habitat size, 359–60

Haiti, 68, 72–73

Handley 1976 paper, 9

hantavirus, 383, 430, 438

harems, 29, 36, 210, 224, 229–30, 232–33, 235, 299–300, 331, 336–41

Has Cave, Mexico, 70

Hawaiian honeycreepers, 3

hawking, 12, 29, 194, 276. *See also* aerial hawking

herbivores, 14, 22

Hernández-Camacho, Jorge Ignacio, 48

Hernández-Camacho and Cadena-G. 1978 paper, 48

heterozygosity, 333

hexose, 115, 174–76, 182, 284, 288

highlands, 133

hippocampus, 187, 197

Hispaniola, 68, 71–73

Historia Naturalis Brasiliae, 44

Holocene, 65, 68–79, 257

homeostasis, 114, 116–17

homeothermy, 317

home range, 124, 329, 363, 384

host-pathogen interactions, 377, 382, 384

hovering, 14, 16–17, 131, 151, 162, 176, 193, 196, 199, 231, 241, 277, 281–82

humerus, humeri, 65, 156

hummingbird, 14, 16–18, 176, 277–78, 285–87

Hunts Cave, Bahamas, 71–72

Hurricane Georges, 332

hurricanes, 332–33, 332, 345, 392, 440–41, 446–47, 449

icehouse Earth, 7–8

ICMBio. *See* Instituto Chico Mendes Institute de Conservação da Biodiversidade

iDigBio, 51

implantation, 208, 210, 231

Impossível-Ioio Cave, Brazil, 75

Inciarte Tar Pit, Venezuela, 75–76

incisors, 14, 269, 276

India, 27

infection dynamics, 380

inferior colliculi, 187, 191, 195

Inferno, Mexico, 70

infrared, 107, 198, 259

Inglis, Florida, 69

insect control, 4, 18

insect pollination, 14

insectivorous, 13–14, 18, 35, 66–67, 88, 94, 116–17, 124, 130, 165, 173, 184–86, 194, 197–200, 211, 215–16, 240, 251–52, 255, 270, 274, 277–78, 290, 294, 371, 380, 428. *See also* insect control

Instituto Chico Mendes Institute de

Conservação da Biodiversidade (ICMBio), 414, 425, 427–28
intake response, 109, 115, 172
International Union for Conservation of Nature and Natural Resources (IUCN), 31, 399–400, 402, 422–23, 427, 441–33, 445, 449
intramembranous muscle, 153–54, 158, 163
Iporanga, Brazil, 75
Isla de la Juventud, Cuba, 73. *See also* Isle of Pines
Isle of Pines, Cuba, 84. *See also* Isla de la Juventud
Isthmus of Panama, 64, 77, 99
Isthmus of Tehuantepec, Mexico, 440
IUCN. *See* International Union for Conservation of Nature and Natural Resources

Jacksons Bay Caves, Jamaica, 71–73
Jalisco, Mexico, 393
Jamaica, 9, 35, 44, 68, 71–73, 134, 207–8, 210, 321–22, 446
Jatun Uchco, Peru, 74
Jones, J. Knox, Jr., 4
Jones et al. 2002 paper, 91

katydids, 13, 203, 240, 245–46
key evolutionary innovation, 23, 105
kidney morphology, 170, 176
kinematics, 160–61
King Cave, Bahamas, 71–72

La Chepa, Dominican Republic, 4
La Selva, Costa Rica, 3, 282, 354
La Venta, Colombia, 32, 66, 68, 94–95. *See also* Laventan
Lacépède, Bernard-Germain-Etienne de, 45
lactation, 206, 209–13, 215, 266, 331
Laguna Santa, Venezuela, 45
Lamarao, Brazil, 74
Lara Cave, Mexico, 70
laryngeal echolocation, 187–88
Late Pleistocene, 69–76, 78, 257
Laurasia, 27
Laventan, 30, 66. *See also* Miocene
lek, lekking, 29, 32, 36, 229–31, 279, 312
Lesser Antilles, 9, 68, 71, 79, 99, 333, 361, 392, 446
life expectancy, 212, 331, 339
life histories, 3, 120, 144–45, 163, 205,

210–13, 215–19, 247, 252, 330, 332, 342
light wavelength, 189, 279–80
likelihood (statistics), 89, 94–95, 98
limb kinematics, 163
Linnaeus, Karl, 44
lipids, 172–73, 182, 266, 301
lithostructures, 318–19, 328
litter size, 205–6, 208, 211–12, 214–15, 222
livestock, 4, 257, 262–64, 267–68, 438
local extinction, 98–99, 352, 361–62, 365. *See also* extinction
locomotor behavior, 163
locomotor forces, 163
Loltún Cave, Mexico, 69–70
low-duty-cycle echolocation, 32
low intensity, 124, 279
lowlands, 9, 98, 338, 340, 350
Lyssavirus, 437

Madagascar, 26, 29–30, 36
major histocompatibility complex (MHC), 333
Malagasy, 25, 30, 334
maltase, 109, 113–14
Mammal Species of the World, 53
mandibular pits, 259–61
Manu National Park, 394
"mashers," 298–300. *See also* "gulpers"
maternal prey transfers, 196
maternity roosts, 265, 331, 337–38. *See also* communal roosts; day roosts; ephemeral roosts; mating roost
mating roost, 444. *See also* communal roosts; day roosts; ephemeral roosts; maternity roosts
Mato Grasso, Cuiaba, Brazil, 75
Max Planck Institute of Animal Behavior, 384
MDE. *See* mid-domain effect
mechanical advantage, 109, 111, 118, 244, 297–98
membrane stiffness, 154–55
metacommunities, 353, 355, 361, 363, 365, 377–79, 448
metacone, 67
metapopulations, 325
MERS (Middle East respiratory syndrome), 383
Mexico, 7, 12, 14, 18, 34–35, 48, 64, 68–70, 76, 86, 99, 207–9, 257, 263, 265, 275, 281–82, 287, 301, 303, 312, 327,

329, 337–38, 341, 350, 357, 359, 393, 402, 404, 435, 438–41, 449
MHC. *See* major histocompatibility complex
microbiome, 109, 117–18, 451
microbiota, 182
microclimate, 311, 317, 320, 328, 360, 437
microsatellite, 333, 341
mid-domain effect (MDE), 393–95, 401
Miller, Gerrit Smith, Jr., 26–27, 47
Miller's Cave, Grand Cayman, 71–72
Ministry of Environment of Brazil, 424–27
Miocene, 8–11, 14, 16–18, 29–33, 36–37, 65–68, 77–79, 94–95, 97, 99, 359
mitochondrial d-loop, 98, 333–34
mitochondrial genes, 89, 92, 94, 98–99, 109, 113, 333–35. See also *cytb* (cytochrome b)
mitochondrial ribosomal RNA genes, 89, 92, 117, 333
"mobile links," 380
molar, 33, 35, 66–67, 248–49, 276, 302
monestry, 206–9, 210, 212–16, 330–32, 337. *See also* polyestrous
Monkey Beds, 64. *See also* La Venta
monographs (1970s), 3, 5, 292, 373
monophyly, 49, 88, 91–92, 94, 99, 105, 112–13, 125, 240, 242
Monte Carlo permutations, 126–27
Monteverde, Costa Rica, 17, 303
MorphoBank, 51–52, 65
Morphosource, 52
moths, 13, 33, 251, 276, 285, 287
Moza, Cuba, 73
Mt. Orizaba, Veracruz, 69
musculoskeletal morphology, 163
Muséum National (Paris), 45
Museum of Natural History (Berlin), 384
mutualistic interactions, 377–78, 381
mutualistic networks, 374, 381

NALMA (North American Land Mammal Ages), 66, 77–78
nasal echolocation, 13, 195. *See also* laryngeal echolocation
natural selection, 3, 112–13, 125, 206, 351
Nature Serve, 399–400, 441. *See also* International Union for Conservation of Nature and Natural Resources
NDVI (Normalized Difference Vegetation Index), 414
Neartic, 16, 358

necklaces, 326

nectar density, 282–83

nectarivores, 10, 13, 17, 50, 67, 87, 94, 111–13, 115, 117–18, 130, 134, 172–73, 175, 179–80, 182, 194, 197–98, 215–16, 239, 244, 246, 250, 260, 273–83, 286–88, 298, 327, 341, 355–56, 403–4, 422, 436, 439–40, 445

nectarivory, 5, 17–18, 30, 67, 91, 106, 114, 136, 170–71, 188, 205, 273–77, 287, 375, 381

Neogene, 10, 64, 66, 78

Neotropical: bats, 249, 251, 279, 285, 347, 349, 358, 391, 399; biomes, 7, 98; birds, 10, 17; forests, 4, 244, 302; noctilionoids, 25, 27, 32, 37, 128; rainforests, 11–12, 135

Neotropics, 12, 15–18, 34, 37, 64, 135, 242, 287, 301, 303, 319–20, 336, 348, 350, 355, 357, 362, 430, 435, 437, 449–50

nephron, 109, 114, 178

nestedness, 98, 359, 375–76, 378–79, 384

neural representations, 187

New Mexico, 69

New Trout Cave, West Virginia, 69

New Zealand, 25–26, 28–29, 31, 36, 259, 275, 312, 440

Nicaragua, 68, 70, 444–45

niche conservatism, 96–97, 393, 396–98, 401–2

niche matrices, 348–49

nitrogen balance, 172–73

nocturnal, 11, 16, 284–85, 318, 380

North American Land Mammal Ages. See NALMA

Northeast Brazil, 349

noseleaf, 13, 30, 44, 46, 194–95, 198, 259

nostril movements, 191

null model, 349, 401

object cues, 246. See also spatial: cues

odoriferous, 279, 338

Old Fort Yuma, California, 70

old-growth forests, 357

Old World, 13, 18, 25, 28, 30, 35, 37, 110, 117–18, 246, 282, 284–85, 296

Old World fruit bats, 188, 242, 295

Oleg's Bat Cave, Hispaniola, 71–73

olfaction, 32, 187–91, 197–99, 229, 232, 264, 268, 279, 285, 287, 298–99; bulb, 187, 191, 198; cues, 30, 190, 193, 195, 197–98, 248, 264, 278, 286, 288; receptors, 107, 109, 118

Oligocene, 8, 14, 25, 30–31, 35, 77–78, 97

OLS (ordinary least square) regression, 171, 177–78, 180–81

omnivory, 79, 114, 117, 124, 127, 136, 188–89, 195, 197, 211, 242, 248–49, 251, 275, 281, 336

opsins, 279–80. See also cones; rods

optima, 106–8, 111–12, 298. See also Ornstein-Uhlenbeck (OU) models

Orange Valley, St. Ann Parish, Jamaica, 73

ordinary least square (OLS) regression, 171, 177–78, 180–81. See also phylogenetic generalized least squares (PGLS) regression

Orinoco, 8

Ornstein-Uhlenbeck (OU) models, 106–8, 111–12

OU models. See Ornstein-Uhlenbeck (OU) models

Owen, Robert D., 49

Pacific lowlands, 98. See also Chocó

Pacific Ocean, 8, 333

paleobiogeography, 68, 76

Paleocene, 7–9, 11, 20

Paleotropical, 17–18, 210

Pallas, Peter Simon, 45

Panama, 64, 66, 77, 79, 99, 201, 209–10, 250, 264–65, 301, 312, 327, 329, 331, 334, 340–41, 354, 360, 379, 444

Panamanian land bridge, 9, 14. See also Great American Biotic Interchange (GABI)

paracone, 67

Paraguay, 45, 68, 74, 76, 327, 335, 355, 361

paraphyly, 49, 88, 92, 94, 239

parasitism, 267, 287, 316, 375, 384; blood feeding, 257; flies, 382; plant, 32

parsimony, 89, 98

passive integrated transponders (PIT) tags, 326

passive listening, 32, 195–96, 263. See also acoustic cues

pasta vampiricida, 429

Patagonia, 66, 77, 317

pathogen, 152, 267, 361, 380, 382–83, 437–38, 449. See also host-pathogen interactions

Patton's Fissure, Cayman Brac, 71–72

PCA. See principal component analysis

PD. See phylogenetic distance

peak frequency, 124, 194

Peru, 33, 48, 50, 64, 66, 68, 74–75, 77, 267, 349, 355, 393–94, 445, 448

pest control, 429–30. See also insect control

Peter Cave, Cayman Brac, 72

Peters, Wilhelm, 46–47

PGLS. See phylogenetic generalized least squares regression

phyletic change, 126, 129–36

PHYLIP, 51

phylogenetic distance (PD), 98, 117, 243

phylogenetic generalized least squares regression (PGLS), 171, 174, 178–81, 223–26, 228

phylogeography, 89, 98

physiology, 4–5, 113, 115, 151, 169–70, 176, 181–83, 248, 257, 264, 267, 278, 288, 303, 311, 373

phytophagous, 127, 130, 188, 190, 198, 303, 360

phytostructures, 318–20

Pinacate Biosphere Reserve, Sonoran Desert, 337

Piso, Willem, 44–45

PIT tags. See passive integrated transponders (PIT) tags

plagiopatagialis, plagiopatagiales, 152–58, 163

plagiopatagium, 31, 152, 158–59, 163

plasticity, 182, 188, 244

plateau, 393

Pleistocene, 8, 31, 35, 64–66, 68–79, 95, 99, 129, 257; refugia, 31, 95–96

plesiomorphy, 105–6, 109

Pliocene, 8, 29, 64, 77, 79

Pollard Bay Cave, Cayman Brac, 71–72

pollen, 4, 13–14, 29, 32, 36, 124, 169, 171, 173, 180, 193, 208, 210, 239, 241, 249, 251, 275, 278, 284–87, 330–31, 341, 348

pollination, 4, 14, 16, 18, 245, 273, 284–88, 374–76, 381, 384, 430

polyestrous, 206, 208–10, 212–16, 331–32, 339–40. See also monestrous

population dynamics, 31, 329–31, 326

Port of Spain, Trinidad and Tobago, 73

Portland Cave, Jamaica, 71–72

postcopulatory selection, 222–23, 230

postmetacrista, 66–67

postparacrista, 67

postpartum estrus, 206, 209–10, 339
Potter Creek Cave, California, 69
poultry, 257
precipitation, 355, 391, 394–95, 404
precopulatory selection, 224, 227, 230
pregnancy, 123, 206, 209–12, 222, 231, 265, 331, 337–38
premetacrista, 67
premolar, 65, 67–68
preparacrista, 67
prey selection, 264. *See also* natural selection; sexual selection
primary forest, 356, 367
Principal Component Analysis (PCA), 170–71, 173–74, 177, 179, 181
proprioceptive, 154
protovampires, 14
Puerto Rico, 9, 47, 68, 71–73, 207–8, 332, 341, 440, 446–48

quadrupedal, 19, 31, 36–37
Quaternary, 10, 33, 64–65, 68–69, 71, 74, 79, 95–96, 100

rabies virus, 19, 267–71, 382, 428–30, 437–38, 453
radio-tracking, 264, 337, 339
RAG2 (recombination-activating gene 2), 88–89, 92, 94, 98
rainfall, 206, 209–10, 215, 302, 355, 414
rainforests, 11, 33, 100, 135, 287
raking phases, 34
RCV. *See* relative cortex volume
recombination-activating gene 2. See *RAG2*
reconciled area analysis, 98
Reddick 1, Florida, 69
Red Latinoamericana para Conservación de los Murciélagos (RELCOM), 430, 435, 446–50
relative cortex volume (RCV), 170–71, 177–78, 181, 183
relative medullary thickness (RMT), 113, 170–71, 177–78, 181, 183
relative medullary volume (RMV), 170–71, 177–78, 181, 183
RELCOM. *See* Red Latinoamericana para Conservación de los Murciélagos
reproductive phenology, 217
resource defense, 222, 265
richness, 96, 98, 117, 420–21. *See also* species richness
Rio de Janeiro, Brazil, 74

Rio Grande do Sul, Brazil, 440
RMT. *See* relative medullary thickness
RMV. *See* relative medullary volume
RNA, 117, 383
Roberts, Alphonse, 47
Rocky Mountains, 8
rods, 189, 279. *See also* cones
Rojas et al. 2016 paper, 90, 92–93, 95–97, 107, 130, 171, 224
roosts, 30–36, 109–10, 189, 193, 196–97, 216, 221, 223–24, 228–31, 233, 250, 264–65, 267, 283, 299–302, 311–13, 316–20, 325–26, 328–32, 336–42, 437–39, 441, 443–44, 446, 449–50; colonies, 325; disturbance, 5, 436–38, 441, 443, 445–48, 451; permanence, 223–28, 328; selection, 320; types, 110, 311–12, 316–19, 328–29, 342
rostrum, 222, 244, 275, 277, 287

Saint John, 446
Saint Thomas, 446
Saint Vincent, 71, 99, 447
SALMA (South American Land Mammal Ages), 64–66, 77
Sangão Archaeological Site, Brazil, 75–76
San Josecito Cave, Mexico, 69. *See also* Cueva San Josecito
San Miguel Island, California, 69
Santa Rosa, Costa Rica, 327, 339–40
São Francisco River, Brazil, 74–75
São Paulo, Brazil, 74–76
SARS-like virus (severe acute respiratory syndrome–like virus), 383
savannas, 8–9, 13, 18, 327, 353, 357, 384
Schluter, Dolph, 105, 113
Seba, Albertus, 44
secondary forests, 31, 300, 356–57, 360, 436–37, 445
seed dispersal, 4, 18, 245, 298–304, 374–75, 381–82, 384, 430
seed trap, 302
selection (evolutionary), 3, 89, 116, 133, 135, 144, 190, 222–25, 246, 266, 277, 280, 285–86, 298, 337, 350–53, 356, 361–62, 365–66, 380–81; pressures, 18, 191, 244–45, 286. *See also* natural selection; sexual selection
sensory: cues, 195, 264, 280; ecology, 18, 151, 154, 187–91, 195–99, 242, 247–48, 251, 278, 311, 373; modality, 110, 189–91, 197–98, 246, 279; systems, 187–89, 197, 281, 285–86

Serra da Mesa, Brazil, 74–76
1700s, 52
sex ratio, 336–39
sexual control, 221–23, 229, 231, 336, 340–41
sexual dimorphism, 221–28, 230–33
sexual maturity, 117, 211, 265, 331, 336
sexual selection, 222–23, 228–33
SGLT1 (sodium-dependent glucose-galactose transporter), 174
Shearwater Cave 1 and 2, Cayman Brac, 72
Sierra Diablo Cave, Texas, 69
Sierra Madre, 18
Sierra Mazateca, Mexico, 393
Simpson, George Gaylord, 48, 105, 113
Sir Harry Oakes Cave, Bahamas, 71–72
site selection, 162, 423, 425. *See also* natural selection; sexual selection
skin stiffness, 154, 156
snout, 44, 188, 244, 273–75, 277, 287. *See also* rostrum
social behavior, 265, 336, 428; bonds, 266; learning, 245–46
Sociedade Brasileira para o Estudo de Quirópteros, 430
Sonora, Arizona, 338. *See also* Sonoran Desert
Sonoran Desert, Arizona, 11, 209, 281, 288, 337
South America, 7–9, 14, 17–18, 28, 32–37, 44, 47, 64, 68, 74, 76–79, 96–99, 207–8, 281, 287, 334, 351, 358, 391, 393, 396, 398, 400–401, 436, 438–42, 445–46, 449
South American Land Mammal Ages. *See* SALMA
Southeast Asia, 13, 18, 248
spatial: bias, 404; cues, 188, 198, 246; memory, 187–89, 197–98, 246, 263, 280. *See also* acoustic cues; object cues; olfaction: cues; vision: cues
species richness, 4–5, 9, 11, 13, 79, 250, 273, 304, 327–28, 348, 350–51, 354–60, 362, 391–403, 413–14, 421–22, 426, 430, 436
sperm competition, 222–24, 230–31, 283
Spot Bay Cave, Cayman Brac, 72
Spukil Cave, Mexico, 70
stasis, 124–25, 128–30, 133, 135
stiffness of the skin, 154–56, 247
St. Michael Parish, Barbados, 71
St. Michel, Haiti, 71

St. Michel Cave, Mexico, 70–71
streblid flies, 316
stretchiness of the wing, 152
sucrose, 113, 115, 169, 173–74, 176, 284
sugar density, 172–73
superior colliculi, 187, 190
Surinam, 50, 68, 74–75, 445
survival rates, 215, 326, 331, 340
sweeps, 30–31, 35, 191, 247. *See also* frequency-modulated calls
sympatry, 33, 134–36. *See also* allopatry
synchrony, 154, 160, 210, 215, 231, 265. *See also* asynchronous reproduction

Talara Tar Seeps, Peru, 75
tanagers, 10, 17–18, 263
Tanzania, 30
taxonomic diversification, 95, 106–7, 112, 118
taxonomy, 26, 35, 43, 47, 49, 51, 90–92, 95, 373, 384, 400, 414
teeth, 32, 63–68, 79, 222–23, 244, 248–49, 259–62, 274–76, 302, 339. *See also* crown (dental); dental characters
telemetry, 363, 373, 384
temperature, 7–9, 18, 152, 198, 206, 278, 303, 317–20, 350, 355, 360, 363, 391, 393–97, 404–5, 414, 440
tendons, 44, 157, 160–61
Terlingua Cave, Texas, 69–70
termite nests, 224, 229, 233, 311, 317, 319–20, 328, 330, 336
terrestrial, 14, 17, 29, 31–32, 36–37, 64, 66, 123–24, 189, 199, 215, 242, 259, 295, 325, 332, 341, 359, 362, 413–14, 435
Tertiary, 63, 97
testes, 222–23, 227–33, 283, 312
Texas, 69–70, 405, 444
thermal characteristics, 317, 319–20
thermal imaging, 326, 444
Thomas, Michael Rogers Oldfield, 47
thorn forests, 9
ticks, 13

Tlapacoya, Mexico, 69, 79
TNC, 96, 396–99
torpor, 32, 278, 317, 355
toucans, 17–18
trawlers, 12, 191, 251. *See also* aerial insectivores
TreeBASE, 51
trehalase, 109, 113–14, 170–71, 173–74, 183
trends, 48, 129, 134, 175, 179, 181, 188, 199, 208, 214, 240, 242, 250–51, 275, 318, 351, 356, 359, 381, 404, 436, 439, 448. *See also* allometric trends; dietary trends
Trinidad, 210, 223, 263–64, 336–37
Trinidad and Tobago, 73, 447
tropical dry forests, 8–9, 18, 208, 329, 444
túngara frog, 195

U-Bar Cave, New Mexico, 69
ultrasound, 193, 195, 286, 311, 440
ultraviolet (UV), 107, 189–90, 279–80, 288
understory, 13, 16–17, 33, 132, 196, 241, 247, 249, 293, 299, 302, 323, 332, 339–40, 346, 354–55
United States, 7, 12, 51, 68, 86, 207, 281, 348, 391, 427, 435, 441, 443–44
U_{osm} (urine concentration capacity), 170–71, 179, 181
uplands, 10, 339. *See also* highlands
urine concentration capacity. *See* U_{osm}
Uruguay, 391, 445, 448
U.S. National Museum, 47–48, 223

vampire bats, 13–14, 18, 79, 90, 94, 111, 129, 171, 175–76, 182, 215, 257–68, 328, 429–30, 438, 443, 449; control, 19, 267, 428–30, 438. See also *Desmodus* salivary plasminogen activator; desmolaris; draculin; protovampires
Velazco et al. 2010 paper, 48–49
Venezuela, 9–10, 68, 74–76, 208, 215,

257, 275, 355, 437–38, 441, 445–46, 448, 450
Veracruz, Mexico, 69, 303, 327
"vernier" control of posture, 154
viral outbreaks, 380, 383
vision, 30, 187–91, 195, 197–98, 231–32, 263–64, 279–81, 285, 287–88, 326; cues, 195, 197–98, 264, 280. *See also* object cues; visual nuclei
visual nuclei, 191
VMS. *See* von Mises stress, 109, 298
volcanic activity, 31, 333, 441, 446. *See also* cyclones
volcanic islands, 9
von Mises stress (VMS), 109, 298
vortex, 146, 161–62, 165–67
vulnerable, 31, 332, 422, 427, 437, 441–43, 445–46

Wallace, Alfred Russel, 99
Wallingford Cave, Jamaica, 72
wavelengths, 189, 279–80. *See also* vision
Web of Science, 333, 414
Western Europe, 27
West Indies, 7, 9–10, 14, 16, 19, 35, 40–41, 64, 68, 71–72, 78, 81, 83, 85, 101, 223, 317, 320, 332, 340, 343, 367
West Virginia, 69, 257
Wetterer et al. 2000 paper, 50, 87–91, 112–13
"whispering" bats, 38, 124, 144, 200, 289, 291, 440
Willemstad, Curaçao, 71
William V (William of Orange), 45
Wilson, Edward O., 43
wind farms, 425–28, 439–40, 445
wing loading, 123, 244–45, 297–98, 334, 360
WorldClim 1.4, 414

Yerbabuena, Mexico, 69

Zealandia, 28, 36

TAXONOMIC INDEX

Ametrida centurio, 56, 415
Anoura, 10, 14, 48, 53, 130, 173, 188–89, 210, 217, 231, 239, 274–77, 289, 291, 293, 314, 404, 415, 441, 449
Anoura aequatoris, 53
Anoura cadenai, 53
Anoura carishina, 53
Anoura caudifer, 11, 53, 289, 404, 415
Anoura cultrata, 53, 130, 314
Anoura fistulata, 14, 53, 239, 275, 277
Anoura geoffroyi, 11, 53, 188–89, 210, 217, 275, 289, 415
Anoura latidens, 53
Anoura luismanueli, 53
Anoura peruana, 53
Anourina, 53, 274
Ardops nichollsi, 56, 316
Ariteus flavescens, 56, 145, 316
Artibeus, 4, 16, 38, 55–56, 85–87, 96, 101–2, 115–18, 127, 133, 143–44, 146–47, 153, 163, 165, 172, 184–85, 188, 200–202, 211, 216–18, 228–29, 234–35, 251, 254, 266, 268, 270, 273, 289, 291, 298–99, 306, 316, 321–22, 326, 329, 332, 340–45, 354, 388, 415, 427, 430, 436, 439, 441, 451, 454
Artibeus anderseni, 55
Artibeus cinereus, 55
Artibeus concolor, 55, 415
Artibeus fimbriatus, 55, 415
Artibeus glaucus, 55, 146
Artibeus gnomus, 56

Artibeus hirsutus, 56, 147
Artibeus inopinatus, 56, 147
Artibeus intermedius, 116
Artibeus jamaicensis, 38, 56, 102, 115, 117, 144, 165, 172, 184–85, 200–202, 217–18, 229, 234–35, 251, 254, 266, 289, 306, 316, 322, 332, 340–41, 344–45
Artibeus lituratus, 56, 96, 153, 163, 306, 415, 439, 451
Artibeus obscurus, 4, 56, 96, 101, 415
Artibeus phaeotis, 56
Artibeus planirostris, 56, 144, 216, 415
Artibeus toltecus, 56
Artibeus watsoni, 56, 143, 172, 216, 234, 266, 321, 329, 342–43

Brachyphylla, 47, 53, 81, 94, 97, 102, 127, 131, 146–47, 173, 217, 274–75, 278, 290–91, 293, 314
Brachyphylla cavernarum, 53, 71, 131, 146, 217
Brachyphylla nana, 53, 173, 278, 293, 314
Brachyphyllina, 53, 274
Brachyphyllini, 53, 97, 274

Carollia, 4, 16–17, 52–53, 57, 69, 94, 96–98, 101–2, 112, 115, 117–18, 131, 146, 158–59, 163, 167, 172, 184, 189, 200–203, 216, 218, 230–31, 234–36, 250, 252, 273, 289, 293, 298–302, 306–7, 315, 326, 338–40, 343, 345–46, 353, 415, 430, 436, 440–41, 449

Carollia benkeithi, 53, 415
Carollia brevicauda, 53, 57, 69, 96, 172, 218, 415
Carollia castanea, 53, 315, 346, 440
Carollia perspicillata, 4, 11, 44, 52–53, 117, 131, 135, 137, 158–59, 167, 175, 189, 190, 200, 202–3, 206, 207, 209, 216, 218, 225, 230–31, 235–36, 250, 252, 289, 298, 301, 307, 326, 338–39, 343, 415
Carollia subrufa, 53
Carolliinae, 4, 16–17, 53, 68–69, 88, 94, 152, 173, 315, 327–28, 330–31, 341
Centurio senex, 56, 231, 298, 315
Chiroderma, 56, 84–85, 132, 146, 172, 185–86, 301–2, 306–7, 315, 415, 434, 441
Chiroderma doriae, 56, 84, 146, 415
Chiroderma improvisum, 56, 441
Chiroderma salvini, 56, 315
Chiroderma trinitatum, 56
Chiroderma villosum, 56, 84–85, 172, 415
Chiroderma vizottoi, 56, 415
Choeroniscus, 53, 274, 314, 415, 436, 441
Choeroniscus godmani, 53
Choeroniscus minor, 53, 314, 415
Choeroniscus periosus, 53, 441
Choeronycterina, 53, 274
Choeronycterini, 53, 274
Choeronycteris, 53, 69, 85, 97, 131, 172, 274–75, 290, 314
Choeronycteris mexicana, 53, 69, 172, 275, 314
Chrotopterus, 13, 55, 66–67, 96, 127, 129, 211, 217, 227–28, 235, 253, 314, 348, 413, 415

Chrotopterus auritus, 13, 55, 66–67, 211, 217, 227–28, 235, 314, 348, 413, 415
Cubanycteris silvai, 78

Dermanura, 133, 147, 153, 321, 416
Dermanura cinerea, 416
Dermanura gnoma, 416
Desmodontinae, 53, 68–69, 78–79, 85, 94, 99, 124–26, 128, 327–28, 330–31
Desmodus, 19, 36, 41, 47, 53, 69, 78–79, 81–82, 84–86, 96, 102, 117–18, 129, 147, 165, 167, 184, 202, 210–11, 217, 230–31, 239, 253, 257, 263–72, 328, 344, 416, 429, 431, 436, 456
Desmodus archaeodaptes, 78
Desmodus draculae, 85, 129, 147
Desmodus rotundus, 19, 41, 45, 53, 69, 82, 102, 165, 202, 211, 217, 230, 239, 253, 257, 268–72, 416, 429, 431, 436, 456
Desmodus stocki, 79, 81
Diaemus youngii, 271, 416
Diphylla ecaudata, 53, 69, 217, 257, 268–71, 416
Dryadonycteris, 52–53, 97, 274, 413, 416, 433, 441
Dryadonycteris capixaba, 53, 413, 433

Ectophylla alba, 56, 132, 144, 146, 234, 315–16, 321–22
Ectophyllina, 56, 132
Emballonuridae, 27, 123, 236, 344
Emballonuroidea, 26–27
Enchisthenes hartii, 56
Erophylla, 9, 53, 94, 218, 229, 235, 274–75, 292–93, 314, 317, 322, 437
Erophylla bombifrons, 53, 314, 317
Erophylla sezekorni, 53, 218, 229, 235, 275, 292–93, 437

Furipteridae, 12, 25–26, 33–34, 39–41, 128–29, 142
Furipterus horrens, 33, 41, 142, 427

Gardnerycteris crenulatum, 55, 314
Glossophaga, 14, 46, 54, 69, 80, 94, 96–97, 101, 107, 115, 131, 161, 163, 165–67, 172, 183–84, 186, 189, 197–98, 200, 215, 218–19, 246, 252, 274–75, 277, 285, 288–89, 291–93, 314, 328, 369, 416, 436
Glossophaga commissarisi, 54, 172, 184
Glossophaga leachii, 54

Glossophaga longirostris, 54, 215, 218, 292–93
Glossophaga morenoi, 54
Glossophaga soricina, 45, 54, 69, 96, 107, 165–67, 186, 189, 197–98, 200, 246, 252, 275, 288–89, 291–92, 314, 328, 369, 416, 436
Glossophaginae, 9, 15, 53, 66–69, 78, 88, 94, 101, 124–27, 130–31, 152, 167, 218–19, 235, 273–74, 277, 287–90, 292–94, 314, 327, 330–31, 345, 433, 455
Glossophagini, 54, 97, 274
Glyphonycterinae, 54, 94, 125–27, 130–31, 315, 327–28
Glyphonycteris behnii, 54, 416
Glyphonycteris daviesi, 54, 130, 145, 251, 315
Glyphonycteris sylvestris, 54, 436, 440

Hsunycterini, 54, 274
Hsunycteris, 49, 52, 54, 131, 138, 274, 314, 328, 416
Hsunycteris cadenai, 54, 274
Hsunycteris pattoni, 54
Hsunycteris thomasi, 54, 138, 274, 281, 314, 328, 416
Hylonycteris, 48, 53, 67, 131, 274, 314, 328, 436
Hylonycteris underwoodi, 53, 138, 207, 274, 280, 282, 283, 314, 328, 436

Lampronycteris brachyotis, 13, 22, 55, 147, 249, 255
Leptonycteris, 54, 69, 94, 97, 115, 131, 161, 167, 172, 183, 197, 216–18, 231–32, 235, 266, 274–75, 278, 288, 290–93, 307, 314, 317, 330, 337, 342–46, 395, 437–38, 441, 451–52, 454–56
Leptonycteris curasoae, 54, 115, 216–18, 232, 235, 290, 292–93, 307, 314, 343–46, 438, 452, 455–56
Leptonycteris nivalis, 54, 69, 183, 278, 288, 291, 293, 342, 395, 441, 451–52, 456
Leptonycteris yerbabuenae, 54, 69, 161, 167, 172, 197, 231, 290, 292, 317, 330, 337, 343, 346, 454
Lichonycteris, 54, 131, 274, 417, 441
Lichonycteris degener, 54
Lichonycteris obscura, 54
Lionycteris, 54, 88, 131, 274, 315, 417
Lionycteris spurrelli, 54, 131, 315, 417

Lonchophylla, 54, 88, 94, 131, 139, 144, 147–48, 183, 269, 274, 315, 326, 403, 413–14, 417, 431, 433, 436, 441
Lonchophylla bokermanni, 54, 433
Lonchophylla chocoana, 54, 131, 139
Lonchophylla concava, 54
Lonchophylla dekeyseri, 54, 183, 413, 431, 441
Lonchophylla fornicata, 54
Lonchophylla handleyi, 54
Lonchophylla hesperia, 54
Lonchophylla inexpectata, 54
Lonchophylla mordax, 54, 315, 417
Lonchophylla orcesi, 54
Lonchophylla orienticollina, 54
Lonchophylla peracchii, 54, 413
Lonchophylla robusta, 54
Lonchophyllinae, 15–16, 54, 66, 68, 78–79, 94–95, 98, 113, 118, 124–27, 131, 148, 173, 273–74, 277, 287, 314–15, 327–28, 330–31
Lonchophyllini, 54, 102, 148, 274, 290
Lonchorhina aurita, 54, 139, 194, 413, 417, 436
Lonchorhina inusitata, 54
Lonchorhina marinkellei, 54
Lonchorhina orinocensis, 54, 130, 314, 441
Lonchorhininae, 54, 68, 95, 118, 124–27, 130, 240, 314, 328, 330
Lophostoma, 55, 58, 66–67, 85, 130, 139, 148, 216, 229, 234, 241, 246, 251, 314, 317, 321, 417, 436, 441
Lophostoma brasiliense, 55, 139
Lophostoma carrikeri, 55, 148, 417
Lophostoma schulzi, 55, 251, 417
Lophostoma silvicolum, 55, 85, 216, 229, 234, 241, 314, 317, 321, 417, 436

Macrophyllum macrophyllum, 13, 38, 42, 55, 139, 144, 147, 192, 193, 194, 198, 200, 241, 251, 255, 289, 314
Macrotinae, 54, 68, 94, 124–28, 328, 330–31
Macrotus californicus, 54, 139, 189, 200, 216, 231, 234, 266, 317, 321
Macrotus waterhousii, 4, 54, 117
Mesophylla macconnelli, 56, 139, 315, 413
Micronycterinae, 55, 58, 68, 94–95, 118, 124–27, 173, 196, 240, 327–28, 330–31
Micronycteris, 20, 55, 58, 82, 85, 88, 96, 102, 114, 129, 139, 143, 146–47, 195–

97, 201–2, 240–41, 249, 253–54, 322, 328–29, 342, 417–18, 436, 441, 454
Micronycteris brosseti, 55, 129, 139
Micronycteris hirsuta, 55, 417
Micronycteris matses, 55
Micronycteris megalotis, 55, 96, 249
Micronycteris microtis, 20, 55, 85, 143, 146, 195–97, 201–2, 240–41, 253–54, 329, 342
Micronycteris minuta, 55, 418
Micronycteris sanborni, 55
Micronycteris schmidtorum, 55, 436
Mimon bennettii, 55, 140, 146, 418, 440
Mimon crenulatum, 418
Molossidae, 36, 94, 123, 236, 344, 346, 452
Monophyllus, 4, 9, 54, 78, 83, 94, 130, 140, 210, 231, 274–76, 293, 314, 317, 332
Monophyllus plethodon, 54, 140
Monophyllus redmani, 4, 54, 130, 210, 231, 293, 314, 317, 332
Mormoopidae, 3, 7, 9, 25–26, 34–35, 40–41, 101, 142–43, 322, 327–28, 330–31
Mormoops megalophylla, 41, 85
Musonycteris, 54, 97, 140, 147, 274–75, 293, 314, 441, 456
Musonycteris harrisoni, 54, 140, 275, 293, 314, 441, 456
Mystacinidae, 25–27, 35, 39, 143, 275
Myzopodidae, 7, 25–26, 38, 41, 128, 143, 145

Natalidae, 26
Neonycteris pusilla, 54, 418
Noctilio albiventris, 34, 39–40, 143, 236, 317, 322
Noctilio leporinus, 38–39, 41, 84, 130, 229, 234
Noctilionidae, 7, 25–26, 34–35, 38, 40–41, 84, 127, 143, 234, 236, 327–28, 330–31, 370
Noctilionoidea, vii, 3, 7, 9, 21, 25–27, 39, 41, 84, 98, 102, 120, 123, 133, 146, 185, 254, 292, 306, 328, 391, 395, 401
Notonycteris magdalenensis, 78
Notonycteris sucharadeus, 78

Palynephyllum antimaster, 78
Phylloderma, 55, 78, 127, 130, 140, 217, 251, 418
Phylloderma stenops, 55, 140, 217, 251, 418
Phyllonycterina, 53, 274

Phyllonycteris, 53, 78, 131, 140, 145, 274–75, 291, 314, 437, 441, 453
Phyllonycteris aphylla, 53, 140, 145, 453
Phyllonycteris poeyi, 53, 291, 314, 437
Phyllops falcatus, 56, 140, 144, 316
Phyllops vetus, 85
Phyllostomidae, viii, 3–4, 7, 15–16, 19–20, 22, 25–26, 36, 38, 40, 52–53, 58, 66, 69, 77, 79, 81–85, 98, 100–103, 106, 117, 119–20, 123–27, 133, 139–48, 151–52, 164, 167, 169, 183, 186, 200–203, 205, 210, 215–19, 223, 234–36, 252–55, 269, 289–94, 306–7, 321–22, 343–48, 350–51, 353, 364, 367, 369–70, 388, 391, 394, 401–2, 404, 431, 433–34, 452–56
Phyllostomus, 13, 16, 36, 55, 67, 124, 127, 130, 140, 188–89, 195, 201, 203, 210, 217–18, 223, 229, 233–35, 248–49, 252, 254, 264, 266, 273, 302, 314, 328–29, 344–46, 418
Phyllostomus discolor, 13, 55, 140, 188, 195, 201, 217, 223, 234, 314, 328
Phyllostomus elongatus, 55
Phyllostomus hastatus, 36, 55, 130, 189, 203, 210, 218, 235, 248–49, 252, 254, 264, 329, 344–46
Phyllostomus latifolius, 55, 130
Platalina, 54, 87, 97, 140, 148, 274, 315, 454–55
Platalina genovensium, 54, 140, 148, 315, 454–55
Platyrrhinus, 56, 87, 96–99, 103, 127, 132, 140, 146–47, 212, 216, 315, 418–19, 440–41
Platyrrhinus albericoi, 56, 127, 140
Platyrrhinus angustirostris, 56
Platyrrhinus aurarius, 56, 418
Platyrrhinus brachycephalus, 56
Platyrrhinus chocoensis, 56, 441
Platyrrhinus dorsalis, 56
Platyrrhinus fusciventris, 56
Platyrrhinus helleri, 56, 315, 440
Platyrrhinus incarum, 56, 419
Platyrrhinus infuscus, 56, 419
Platyrrhinus ismaeli, 56, 146
Platyrrhinus lineatus, 56, 212, 216, 419
Platyrrhinus matapalensis, 56
Platyrrhinus recifinus, 56, 419
Platyrrhinus umbratus, 56
Platyrrhinus vittatus, 56
Pteronotus davyi, 39, 142

Pteronotus macleayii, 40
Pteronotus parnellii, 38–39, 153
Pteronotus quadridens, 317
Pygoderma bilabiatum, 56, 141, 236, 343, 367

Rhinophylla alethina, 55, 141
Rhinophylla fischerae, 419
Rhinophylla pumilio, 55, 315, 419
Rhinophyllinae, 55, 94, 118, 126, 131, 315, 327–28, 330–31

Scleronycteris, 54, 87, 274, 419
Scleronycteris ega, 54, 419
Speonycteridae, 25, 35
Sphaeronycteris toxophyllum, 56, 84, 232, 315
Stenoderma rufum, 56, 133, 141, 210, 299, 316, 332, 368
Stenodermatina, 56, 97–98, 132
Stenodermatinae, 9, 16–17, 55, 68, 78, 94–95, 97, 118, 124–27, 132, 145, 152, 173, 232, 302, 306, 315–16, 327–28, 330–31, 341, 343, 367
Stenodermatini, 55, 66, 68, 132–33
Sturnira, 16, 57, 96–99, 103, 115, 132, 141, 145, 147, 173, 185, 218, 232, 234, 273, 299, 302, 306, 315, 344, 348, 353–54, 363, 419, 430, 436, 441
Sturnira erythromos, 57, 145
Sturnira lilium, 57, 96, 115, 173, 218, 234, 306, 315, 344, 348, 363, 419
Sturnira ludovici, 57, 185, 363, 436
Sturnira magna, 57, 419
Sturnira mordax, 57
Sturnira nana, 57, 441
Sturnira tildae, 57, 419
Sturnirini, 131–32

Thyroptera tricolor, 33, 38, 41–42, 316–17, 321, 323
Thyropteridae, 25–27, 33, 38, 41, 128, 143
Tonatia, 20, 55, 58, 66–67, 80–81, 94, 141, 146, 188, 241, 249, 252, 255, 314, 328, 419, 431
Tonatia bidens, 55, 141, 249, 252, 314, 419
Tonatia saurophila, 55, 188, 241, 328, 419, 431
Trachops cirrhosus, 20, 55, 66, 142, 195, 200, 241, 246–47, 252, 254–55, 314, 341, 344, 419, 436
Trinycteris nicefori, 54, 130, 142, 315, 419

Uroderma, 56–57, 84, 96–101, 142, 146, 211, 228, 315, 317, 322, 419–20, 436, 441

Uroderma bilobatum, 56, 100–101, 142, 211, 228, 315, 317, 419

Uroderma magnirostrum, 57, 142

Vampyressa melissa, 57, 142, 441

Vampyressa pusilla, 57, 315, 420

Vampyressa thyone, 57, 413, 420

Vampyressina, 56, 132

Vampyrini, 55

Vampyriscus bidens, 57, 142, 420

Vampyriscus brocki, 57, 420

Vampyriscus nymphaea, 57, 229, 315

Vampyrodes caraccioli, 57, 142, 315, 440

Vampyrum, 13, 55, 66–67, 95, 123, 127, 129, 142, 205, 227, 236, 240, 254–55, 413, 420, 436

Vampyrum spectrum, 13, 55, 66, 123, 142, 205, 227, 236, 240, 254–55, 413, 420, 436

Vespertilionidae, 123, 205, 236, 270, 275

Vespertilionoidea, 26–27, 41

Xeronycteris, 52, 54, 97, 274–75, 413, 420, 434

Xeronycteris vieirai, 54, 275, 413, 420, 434